Handbook of
Neurochemistry

SECOND EDITION

Volume 1
CHEMICAL AND CELLULAR
ARCHITECTURE

Handbook of
Neurochemistry

SECOND EDITION

Edited by Abel Lajtha
Center for Neurochemistry, Ward's Island, New York

Handbook of
Neurochemistry

SECOND EDITION

Volume 1
CHEMICAL AND CELLULAR
ARCHITECTURE

Edited by
Abel Lajtha

Center for Neurochemistry
Ward's Isalnd, New York

SPRINGER SCIENCE+BUSINESS MEDIA, LLC

Library of Congress Cataloging in Publication Data

Main entry under title:

Handbook of neurochemistry.

Includes bibliographies and index.
Contents: v. 1. Chemical and cellular architecture.
1. Neurochemistry — Handbooks, manuals, etc. I. Lajtha, Abel. [DNLM: 1. Neurochemistry. WL 104 H434]

QP356.3.H36 1982	612'.814	82-493
ISBN 978-1-4757-0616-1	ISBN 978-1-4757-0614-7 (eBook)	AACR2
DOI 10.1007/978-1-4757-0614-7		

Contributors

A. M. Benjamin, Division of Neurological Sciences, University of British Columbia, Vancouver, British Columbia V6T 1W5, Canada

Kenneth A. Bonnet, Department of Psychiatry, New York University School of Medicine, New York, New York 10016

Alan A. Boulton, Psychiatric Research Division, University Hospital, Saskatoon, Saskatchewan S7N OXO, Canada

R. S. Bourke, Division of Neurosurgery, Albany Medical College, Albany, New York 12208

M. J. Brownstein, Laboratory of Clinical Science, National Institute of Mental Health, Bethesda, Maryland 20205

Arsélio P. Carvalho, Center for Cell Biology, Department of Zoology, University of Coimbra, 3049 Coimbra Codex, Portugal

Bert Csillik, Department of Anatomy, University Medical School, Szeged, Hungary

Csaba Fajszi, Institute of Biophysics, Biological Research Center, Hungarian Academy of Sciences, 6701 Szeged, Hungary

H. Gainer, Laboratory of Developmental Neurobiology, National Institute of Child Health and Human Development, Bethesda, Maryland 20205

E. Martin Gál, Neurochemical Research Laboratories, Department of Psychiatry, University of Iowa, Iowa City, Iowa 52242

Robert M. Gould, Laboratory of Membrane Biology, Institute for Basic Research in Mental Retardation, Staten Island, New York 10314

L. Hertz, Department of Pharmacology, University of Saskatchewan, Saskatoon, Saskatchewan S7N OWO, Canada

Augusto V. Juorio, Psychiatric Research Division, University Hospital, Saskatoon, Saskatchewan S7N OXO, Canada

H. K. Kimelberg, Division of Neurosurgery, and Department of Biochemistry and Anatomy, Albany Medical College, Albany, New York 12208

László Latzkovits, Institute of Experimental Surgery, Medical School of Szeged, 6701 Szeged, Hungary

Dan Matsumoto, Laboratory of Membrane Biology, Institute for Basic Research in Mental Retardation, Staten Island, New York 10314

Gary Mattingly, Laboratory of Membrane Biology, Institute for Basic Research in Mental Retardation, Staten Island, New York 10314

Hanna M. Pappius, Donner Laboratory of Experimental Neurochemistry, Montreal Neurological Institute, and Department of Neurology and Neurosurgery, McGill University, Montreal, Quebec H3A 2B4, Canada

Thomas L. Perry, Department of Pharmacology, University of British Columbia, Vancouver, British Columbia V6T 1W5, Canada

Leonid Pevzner, The Saul R. Korey Department of Neurology, Albert Einstein College of Medicine, Bronx, New York 10461

N. Seiler, Centre de Recherche Merrell International, 67100 Strasbourg, France

James H. Wood, Cerebral Blood Flow Laboratories, Division of Neurosurgery, Emory University Clinic, Atlanta, Georgia 30322

Preface

After the completion of the first edition of this series, this editor thought that a new edition would not be warranted in less than 15, perhaps 20, years, but it seems that we live in a time in which rapid changes are the norm and findings in a field such as neurochemistry develop exponentially. The task of a future editor attempting to get a comprehensive neurochemical handbook for the year 2000 would be even less enviable, but by then information processing may be very different.

The approach, the design, and the areas covered by each volume and each chapter are necessarily arbitrary, and it is likely that other editors or authors would have approached the coverage or the organization in a different manner. It is hoped, however, that readers will find the series helpful for beginning or for continuing work. There may be some overlap among the various chapters, but insisting on single coverage of an area would at times have restricted treatment to only one point of view and might have truncated and hurt the logical flow of some of the chapters.

Chapters in this series do not cover small areas in detail, but each covers a subject that could have been expanded to a book or could have been discussed in a week-long symposium. Still, one definition of a handbook is that it can be lifted and carried by hand. This series may be on the borderline of such a definition, but perhaps it fits "that by being brief it is of more help." Although they facilitate the finding of information and directions, such short presentations always gravely restrict details or background of the information presented. For these restrictions authors are not to be blamed. Clearly, full coverage is not possible, but the emphasis is on good chapters that are of support to those who turn to the book.

Most important, the editor wishes to thank the authors whose hard work is presented here; they are busy, have deadlines and unexpected emergencies, and, no doubt, this series has added just another difficult task to the long list. For their excellent contributions and cooperation we all are indeed grateful. The Handbook reflects not only the excitement of past findings but also, through the rapidly expanding present, the exciting possibilities of the future in our field.

Abel Lajtha

Contents

Chapter 4

Ammonia

A. M. Benjamin

Chapter 5

Water Spaces

Hanna M. Pappius

Chapter 6

Cerebral Amino Acid Pools

Thomas L. Perry

Chapter 7

Neuropeptides: An Overview

   M. L. Brownstein and H. Gainer

Chapter 8

Brain Trace Amines

   Alan A. Boulton and Augusto V. Juorio

Chapter 9

Polyamines

  N. Seiler

Chapter 10

Cyclic Nucleotides in the Central Nervous System
Kenneth A. Bonnet

Chapter 11

Biopterin

E. Martin Gál

Chapter 12

Neurons

Bert Csillik

Chapter 15

The Schwann Cell
Robert M. Gould, Dan Matsumoto, and Gary Mattingly

Chapter 16

Physiological Neurochemistry of Cerebrospinal Fluid
James H. Wood

Cation Transport

László Latzkovits and Csaba Fajszi

1. INTRODUCTION

1.1. Basic Concepts of Cation Transport

According to the traditional concept[1-3] of cation transport, there are "active" and "passive" fluxes: the former drives cations uphill (against an electrochemical gradient) at the expense of ATP consumption, whereas the latter moves cations downhill (in the direction of the electrochemical gradient) by simple diffusion across membrane "imperfections" or "pores." This traditional concept of active and passive cation fluxes has proved to be inadequate for two main reasons.[4-11] (1) It has been demonstrated that the "active" pump transporting both Na^+ and K^+, usually uphill, by direct consumption of ATP can also drive cation movements "on the level" (i.e., in the absence of any concentration gradient) or even downhill.[4,5] (2) Evidence has been collected that demonstrates that "passive" fluxes of cations are highly organized and are closely associated with important physiological functions[6]: many of them take place as part of counter- or cotransport mechanisms.[5-9] Thus, the energy of the electrochemical gradient of the cation actually moving downhill is not dissipated but is mostly consumed in promotion of the transport of different compounds (e.g., sugars, amino acids, other cations), in some cases even against a concentration gradient. In this way, "passive" fluxes of cations moving downhill can build up a concentration gradient for other cations without any waste of ATP.[5-9] Selectivity of the membrane for some "passive" cation fluxes enables it to convert the energy of primary ionic gradients into the energy needed for the maintenance of resting membrane potential as well as for cell excitation.[6,10,11]

The above facts gave the impetus for the construction of more appropriate new concepts and models of cation transport, but the various attempts have introduced quite divergent operational definitions of rather fictitious entities

László Latzkovits • Institute of Experimental Surgery, Medical School of Szeged, 6701 Szeged. Hungary. *Csaba Fajszi* • Institute of Biophysics, Biological Research Center, Hungarian Academy of Sciences, 6701 Szeged, Hungary.

such as mobile carriers, channels, mobile or fixed sites, carrier-mediated transport (either facilitated diffusion or exchange diffusion), and a great variety of ion pumps.[5-11] Recent advances in membrane biochemistry have made possible the correlation of some of these entities to definite macromolecular assemblies.[12] In addition, a detailed enzymological basis has been established for ion pumps.[4,5,13-15] As a result, authors have been able to define primary entities of cation transport that can be interpreted in terms of membrane biochemistry and summarize adequate features of the previous concepts[6] such as pumps, dissipators, ground permeability, and leaks. The last two are of little physiological importance as far as cation transport by intact cells is concerned, and their nature and role will be discussed briefly in connection with supramolecular structures existing within the membrane. Characteristics of pumps and dissipators are outlined in Table I.

The present chapter will be devoted to the physiological role and regulation of pumps and dissipators in the context of neurochemistry.

1.2. Do Cation Transport Phenomena That Can Be Exclusively Attributed to Any Function of the Nervous System Exist at All?

For decades, nerve excitation, the most authentic function of the nervous system, has been known to be closely correlated with changes of cation transport across the neuronal membrane.[16] Changes in cation transport associated with neuronal excitation possess some unique features.[17] Moreover, some non-

Table I
Essential Features of Primary Assemblies Organizing Cation Transport in the Intact Cell

Pumps	Dissipators
Energetics	
Directly consume ATP, the primary source of energy in the cells. Convert metabolic into electrochemical energy.	Utilize energy of the electrochemical gradient, the secondary source of energy in the cells. Convert electrochemical energy into another form while dissipating it to some extent.
Kinetics	
Show vectorial activation exerted by cations and saturation kinetics.	Exhibit primarily saturation kinetics.
Macromolecular assemblies correlated with them	
Na^+,K^+-activated ATPase and Ca^{2+},Mg^{2+}-activated ATPase.	Carriers and channels.
Level of knowledge about their structure	
There is essential information about the enzymes.	The actual molecular structure of either carriers or channels is poorly understood.
Primary functions in the cells	
Establishing uneven cation distribution between cells and environment.	Counter- or cotransport, establishing resting membrane potential and cell excitability.
Traditional concepts correlated with them	
"Active" cation transport	"Passive" cation movements

neuronal elements in the nervous system, such as glial cells, also exhibit quite peculiar characteristics with regard to cation transport.[18,19] These observations have finally led to the more or less tacit conception that cation transport is highly specialized in the nervous system.

This conception is undoubtedly true at the level of phenomenological description. Nevertheless, to the best of our knowledge today, the entire complexity of cellular cation transport and related phenomena can be traced back to definite primary entities of cation transport (see Table I and Section 1.1). Very likely, these primary entities are highly uniform at all levels of living cells. However, their function is controlled by certain specific submicroscopic (supramolecular) and morphological structures (described in Section 2.2) that lead to significant differences in the characteristics of cation transport in various cells.

Therefore, the question that introduced this section can be answered by a compromise: there are no primary entities of cation transport that can be exclusively attributed to any function of the nervous system; instead, cation transport is a complex process established by specific neuronal structures at different levels of organization and is closely related to the regulation of important functions in the nervous system.

2. GENERAL CORRELATES OF CATION TRANSPORT

2.1. Role of Cations and Significance of Cation Transport in the Control of General Cell Functions

At the level of molecular biochemistry, the cellular functions of cations and of cation movements can be summarized as follows:

1. Contribution to the supramolecular organization of both intracellular and extracellular compartments (e.g., cytoplasm, matrix of the extracellular microenvironment, etc.).[20,21]
2. Mutual regulatory interrelationships with pumps and dissipators, i.e., setup of conformational changes of definite membrane structures (e.g., subunits of an ATPase, carriers accomplishing co- and/or countertransport, cation channels).[4-11]
3. Cooperation with the signaling systems inside and outside of the cells (e.g., alteration of the function of hormonal receptors, that of membrane acceptors responsible for cell recognition, interactions with the metabolism of cyclic nucleotides, transmission of intercellular signals).[22-27]
4. Mutual control with cell metabolism.[23-28]

The above functions of cations and cation transport consequently lead to well-defined roles performed at the level of cell physiology, as follows:

1. Determination of osmotic conditions and cell volume regulation.[29-31]
2. Maintenance of resting membrane potential and excitation phenomena.[10,11,17,32-35]

3. Uptake of some metabolites (sugars, amino acids) and reuptake of neurotransmitters.[5-9,36]
4. Regulation of many of the cellular responses to various stimuli (e.g.,
   hormonal or immunologic activation, signals in cell-to-cell communication) and of cell multiplication.[37-40]
5. Changes in general metabolic activity (e.g., oxygen consumption, turnover of proteins, etc.) under certain conditions.[41]

2.2. Supramolecular and Morphological Structures Involved in Cation Movements

Both supramolecular and morphological correlates of cation transport at
different levels of organization have been reviewed recently.[12,20,42,43] Here, a
brief summary is given pointing out some peculiarities in the nervous system.

2.2.1. Supramolecular Structures

2.2.1a. Organization of the Membrane with Respect to Cation Transport.
It is generally postulated that all biological membranes possess an extremely
low ground permeability for cations which is nearly identical in different cell
types.[6,17] This permeability equals that of a pure lipid bilayer with homogeneous
distribution of constituents, i.e., without any mosaiclike organization. In the
lipid bilayer of the cell membrane, pumps and dissipators (see Table I), i.e.,
characteristic, perhaps genetically determined, macromolecular assemblies,
are embedded.[5,6,11,15] On the one hand, their occurrence in the membrane
brings about a unique supramolecular organization of other membrane components (proteins, lipids, lipoproteins) which by themselves carry no transport
function. On the other hand, supramolecular structures constructed this way
influence the functions of both pumps and dissipators, thus resulting in unique
transport characteristics.[11,12,15,28]

It is to be noted that in addition to these supramolecular structures of
highly specific transport functions, other supramolecular organizational principles also operate within the membrane. Recent evidence demonstrates that
artificial membranes consisting exclusively of lipid constituents show a high
level of dynamic "self-organization" under certain conditions.[44-46] When such
structures, which also exist in the living membrane, encage proteins, they can
produce "leaky faults" in the membrane; thus, nonselective leak transport of
cations determined solely by the electrochemical gradients takes place.[47,48]
However, it is very unlikely that leak transport of cations bears great significance under physiological conditions either in excitable or nonexcitable cells,
although it may disturb transport kinetics. The nonsaturable, linear components
of some of the cation movements may result from the presence of "leaky
faults."

*2.2.1b. Organization of the Cytoplasm and Intracellular Microenviron-
ment of the Membrane.* Macromolecular associations build up a highly organ-

ized supramolecular structure within the cytoplasm.[20,42,49] Some of the macromolecular assemblies are interconnected with the membrane, thus directly affecting transport functions.[28,48,49] This ultrastructure of cytoplasm establishes a microcompartmentation of both cell water and metabolites as well as ions. The ionic microcompartmentation is achieved by a heterogeneous distribution of fixed charges primarily for cation binding.[20,21,42]

There are extremely divergent views about the role of the foregoing cytoplasmic ultrastructure as far as cation transport is concerned. Some authors[17] claim that there is no relevant cation binding in the cytoplasm, whereas others[20] have developed models that are able to account for all of the features of cation transport only by invoking cation and ATP compartmentation in the cytoplasmic ultrastructure without taking into consideration any function of the membrane.

We will later point out evidence that in the cells of the nervous system the membrane provides the pacemaker of cation movements (see Section 4.4); nevertheless, supramolecular organization of the cytoplasm is also of paramount significance in relation to cation transport.

2.2.1c. Organization of the Extracellular Microenvironment of the Membrane. There is a general agreement that the intercellular space, i.e., the extracellular microenvironment of the cells, has definite supramolecular organization which is very important in the regulation of cation transport.[42,43,50–52] It seems that the microenvironment of the cells serves as a "communication channel"[51] in cell-to-cell interactions. Outstanding significance should be ascribed to this function in the nervous system where a wide variety of cell activities, including cation transport, are controlled by cell communications.

2.2.2. Morphological Structures

One genuine feature of the nervous system is its morphological heterogeneity at both microscopic and macroscopic level. In relation to cation transport, many functions of the nervous system have recently been considered to be interactions between specific compartments. The role of morphological entities (e.g., blood–brain barrier, cells, and subcellular organelles) in the control of cation transport will be discussed in the context of particular problems in subsequent paragraphs.

3. VALIDITY AND RELIABILITY OF THE FORMAL TREATMENTS USED FOR ANALYZING CATION TRANSPORT PHENOMENA

No study dealing with cation transport can avoid formal treatment of experimental data (e.g., computation of transport rates). Both theory and practice of such treatments have recently developed into a separate field of biophysics. A detailed discussion, therefore, would exceed the length of the present chapter. Nevertheless, at least two sorts of formal analyses have important relationships to neurochemistry; thus, we should allude to problems concerning their application.

Formal analysis is valid when the formulas applied for computation really fit the actual mechanism of cation transport taking place in the system. It is reliable when it provides sound numerical values of transport parameters in terms of statistics.

3.1. Problems Concerning the Application of Michaelis–Menten Kinetics

The aim and technique of the Michaelis–Menten treatment of general transport phenomena have recently been reviewed in detail.[8,9] Here, it seems reasonable to point out some relevant problems which are frequently neglected in studies dealing with the molecular mechanisms of various components of cation transport.

3.1.1. Validity of Michaelis–Menten Kinetics in Cation Transport Studies

The Michaelis–Menten equation is strictly valid only if a "mobile carrier" brings about the transport of a single cation and the empty carrier is immobile.[8,9] (The Michaelis–Menten treatment of transport processes finds its roots in the concept of "carrier-mediated" transport. In fact, a certain class of dissipators can be described "as if they were" mobile carriers.)

In the case of a carrier performing countertransport of two cations, fluxes can be also fitted to the Michaelis–Menten equation, but the values of both V_{max} and K_m for the inward transport should hyperbolically depend on the concentration of the countercation in the cell.[8,9] The reverse is true as well.[8,9] If the empty carrier is also mobile, it operates as an "inner leak"[8]; i.e., the flux becomes the sum of two saturable components.[9]

The fluxes of cotransported species depend on the concentrations of both permeants on both sides of the membrane, even in the simplest case when only the ternary complex (carrier plus two permeants) is mobile.[9]

When two transport processes are running simultaneously, the flux is the sum of two Michaelis–Menten components or of a Michaelis–Menten and a linear component.[53,54]

The presence of an unstirred layer in the vicinity of the membrane also distorts the kinetic curves.[54]

For all of these cases of deviations from the Michaelis–Menten kinetics, it is possible to separate the different transport components and to determine the parameters. Unfortunately, the simplicity of the relationship is then lost, and the parameters may be obtained only by computer fitting. For such a treatment, however, it is well understood that "the theoretical possibilities far exceed the experimental techniques available to extract data necessary for intelligent choices between them."[9]

The oversimplified application of Michaelis–Menten kinetics in some studies on characteristics of the Na^+,K^+-dependent ATPase is also questionable. It has been proved that this enzyme represents a highly organized supramolecular assembly in the membrane.[13–15] Therefore, differences in K_m and/or V_{max} among Na^+,K^+-dependent ATPases present in the membrane are inconclusive in deciding a question such as to what are the factual differences among these enzymes. We have no proper model of Na^+,K^+-ATPase at present that would be suitable for sufficient kinetic treatment.

3.1.2. Reliability of the Michaelis–Menten Treatment: Estimation of the Parameters in the Equation

Both V_{max} and K_m are routinely determined by the Lineweaver–Burk (double-reciprocal) plot. It has, however, been shown in simulation experiments that the double-reciprocal plot provides the poorest reliability among the possible kinds of linearization.[55,56] It is also inappropriate for testing for deviations from the Michaelis–Menten function (e.g., when there is a linear component or the sum of two saturable functions).[53]

The plot of v (initial rate) versus v/s (initial rate over concentration of transported compound) yields significantly more precise estimates. (This type of treatment is known as the Scatchard analysis.)

Recently, new methods have been proposed which yield particularly good results: combination of pairs of measurements[57] and the "direct linear plot"[58–61] may be of special interest in cation transport studies.

3.2. Tracer Kinetic Treatment[62–64]

Validity of the tracer kinetic analysis depends on the following two main features of the system under study (either *in vitro* or *in vivo*): whether the system is in a steady state, and whether a well-defined compartment structure can be correlated with the living system.

If steady state is established, the dynamics of tracer distribution among various compartments can be described by linear differential equations. The solution of these equations results in the sum of exponential terms, i.e., in a quite simple function of time.[62–66]

If the compartment structure has been given, and such a function is fitted to the experimental data, the values of the fluxes and the corresponding transport coefficients can be determined. The solution is very simple, since every process is of first order in the specific activity of the tracer irrespective of the actual mechanism.[67] Therefore, we can not gain any information about the order of the cation movement unless the analysis is repeated at several different steady states established by manipulating some physiological parameters (e.g., extracellular concentration of cations). The transport coefficients are apparent first-order reaction rates.[67]

If there is no well-defined compartment structure that can be correlated with the living system under study, the tracer kinetic treatment turns out to be a "model analysis" of the system.[63,68] The number of exponential terms provides the number of kinetically significant compartments. Occasionally, such an approach can provide important qualitative information about some aspects of the mechanism of cation transport.[69-71]

If the system is not in a steady state (e.g., after inhibition or activation of either pumps or dissipators or during cell activation), the change of tracer concentration or of the specific activity in various compartments of the living system cannot be described as the sum of simple exponential terms.[63,64,68] In fact, no general treatment exists for such systems. At most, one can try to fit the differential equations of a defined compartment structure directly to experimental data by a computer simulation of the tracer distribution. This way, important qualitative information can be obtained about the mechanism of cation transport actually taking place.[63,64,68,69] Nevertheless, quantitative parameters such as rate constants and fluxes have no validity at all.

Another possibility for treating non-steady-state systems can be achieved, of course, by using some well-defined simplifications. A brief outline of such an approach is given here for studying cation transport *in vitro* in a two-compartment system.

At the very early period of cation transport (e.g., 2–3 min), several measurable, determinable factors such as intracellular concentration of cation (c), that of the tracer (I), and the specific activities both in the cells (S_c) and in the medium (S_m) are linear functions of time. (Of course, these prerequisites of the treatment can be checked experimentally.) Consequently, within this period, the influx (V_i) and efflux (V_e) rates can be assumed to be constant in time. Hence, the initial values of these fluxes can be calculated, as follows:

$$V_i = (1/t)(\Delta I - \Delta c \cdot S_c^*)/(S_m^* - S_c^*)$$

$$V_e = (1/t)(\Delta I - \Delta c S_m^*)/(S_m^* - S_c^*)$$

where t is the time of observation (within the short initial period), ΔI is the change of the intracellular tracer concentration during this time, Δc is the change of the intracellular cation concentration, and S^* is the value of S at $t/2 = \frac{1}{2}(S^0 + S')$. By use of these equations, both influx and efflux rates can be determined from a single experiment.

Reliability of the tracer kinetic parameters obtained by graphic curve analysis ("curve peeling"[64]) is very poor, especially when there are more than two compartments. Reliable estimates may be obtained by direct computer fitting of the sum of exponential terms.

4. REGULATION OF CELLULAR CATION TRANSPORT

4.1. Movements of Na$^+$ and K$^+$ via Na$^+$,K$^+$-ATPase-Determined Ion Pump[4,5,15]

Transport of both Na$^+$ and K$^+$, which directly depends on the presence of ATP in the cells, is accomplished via a Na$^+$,K$^+$ pump determined by the

Na^+,K^+-ATPase (E.C. 3.6.1.3) in the membrane. The overall pump–ATPase system has five "modes of operation." In intact cells under physiological ionic conditions, the Na^+,K^+ pump primarily performs coupled movements of Na^+ and K^+ against a concentration gradient at the expense of ATP splitting ("normal" mode of operation). However, it has four further modes of operation, i.e., can perform four different kinds of transport phenomena, under "unusual conditions," as follows: ATP synthesis at the expense of coupled downhill transport of Na^+ and K^+ ("backward run of the pump"), pump-mediated Na^+–Na^+ exchange, pump-mediated K^+–K^+ exchange, and uncoupled extrusion of Na^+ from the cells against an electrochemical gradient with simultaneous ATP hydrolysis.[4,5]

All these modes of operation have certain common characteristics[4,5]: (1) they can be inhibited by both ouabain and "anti-pump" antibodies or antisera; (2) both Na^+ and K^+ exert a vectorial activation or inhibition of the rates of processes; i.e., similar changes in the concentration of the same cation produce inverse effects when occurring inside or outside of the cell; and (3) ATP plays a central role, although pump-mediated Na^+–Na^+ or K^+–K^+ exchanges do not alter the ATP concentration despite the fact that these "modes of operation" also require the presence of ATP.

4.1.1. Interrelationship between Na^+,K^+ Pump and Na^+,K^+-ATPase

The specific domain of the membrane, where the above Na^+ and K^+ movements take place, has a three-level organization (see Section 2.2.1a).

4.1.1a. Na^+,K^+-ATPase. Detailed descriptions of the enzyme structure have been published[5,13–15] (see also Chapter 3 in Vol. 4). However, it is important to point out that although the enzyme has thus far been purified from different organs of various species, no relevant variations in its composition could be revealed. Therefore, the Na^+,K^+-ATPase can be considered fairly uniform in various cells. This is further supported by the fact that even xenogenic antibodies exert an inhibitory effect on the enzyme.[4,15]

4.1.1b. Lipid Structure. A well-defined lipid structure closely associated with the Na^+,K^+-ATPase establishes a unit assembly that performs the functions of the Na^+,K^+ pump.

The purified Na^+,K^+-ATPase seems to require some "obligatory" lipid associates in order to exhibit enzyme activity. Recent reports provide controversial data about both cell specificity and the nature of the lipid components necessary for enzyme activation.[15] Nevertheless, there is no doubt that lipids play an outstanding role in the control of enzyme activity. Moreover, electron microscopic, freeze–fracture, and spin-label studies have proved that lipids closely associated with Na^+,K^+-ATPase provide a well-organized structure encaging the protein moiety.[15] This unit assembly is, to the best of our knowledge, the Na^+,K^+ pump.

4.1.1c. Supramolecular Organization of the Membrane in Mutual Interaction with the Na^+,K^+ *Pump.* There is at present primarily indirect evidence suggesting the existence and role of this "tertiary" structure which can involve dissipators and membrane-bound enzymes. The first part of this evidence is provided by the facts that, on the one hand, the Na^+,K^+ pump itself seems to show only poor organ or species specificity. Reconstitution experiments, when purified Na^+,K^+-ATPases were incorporated into artificial phospholipid vesicles, have shown that even such artificial Na^+,K^+ pumps can in many respects simulate the characteristics of the Na^+ and K^+ transport accomplished by cellular membranes.[15] On the other hand, cation movements by the Na^+,K^+ pump in the membrane of various living cells show striking divergent features as well.[17,19] Therefore, it is reasonable to assume a structure within the membrane that controls the pump activity by mutual interactions.

The second line of evidence is a more direct one. Several authors have demonstrated that there is an ATP-generating system associated with the inner surface of the membrane. The system involves glyceraldehyde-3-phosphate dehydrogenase (E.C. 1.2.1.9) and phosphoglycerate kinase (E.C. 2.7.2.3) and establishes a membrane-located pool of both ATP and ADP in the vicinity of Na^+,K^+-ATPase.[28,70,72] This membrane compartment of ATP and ADP links the ATP-generating enzymes in the membrane to the Na^+,K^+ pump as a functional unit.[28]

Very likely, dissipators (channels and carriers) are also involved in the above "tertiary" structure. Their activity can bring about separate regulation of cation concentrations in the vicinity of Na^+,K^+ pump in a highly dynamic way (see Section 4.4 and Figs. 1–3).

It has to be mentioned that very recently an elegant hypothesis was developed suggesting that the Na^+,K^+ pump itself might serve as a dissipator.[6] According to this "all pump model" of Na^+,K^+-ATPase-dependent cation transport, a certain number of Na^+,K^+-pumping sites in the membrane accomplish "on-the-level" or downhill movements of Na^+ and K^+ without splitting ATP as if they were carriers with co- or countertransport capabilities.

4.1.2. Binding Sites of the Na^+,K^+ Pump for Cations and the Ouabain Receptor

The Na^+,K^+ pump spans the membrane in all cells; i.e., it reaches the inner and outer surface.[13–15] Both surfaces hold specific cation-binding sites of highly different affinities for various cations: the inner has high-affinity sites for Na^+ and low-affinity sites for K^+, whereas the outer shows inverse affinities for both Na^+ and K^+.[4,13–15] Divergent affinities of cation-binding sites play a prominent role in the performance of vectorial regulation of the pump activity by cation concentration changes.[4] Affinity of cation-binding sites is influenced by the ATP concentration in the vicinity of Na^+,K^+ pump independently of the ATP-splitting activity of the pump,[4]

Recent results demonstrate both the high selectivity and the heterogeneity of cation-binding sites. For instance, both the K^+-binding site on the outer surface and the Na^+-binding site on the inner surface have very low affinity

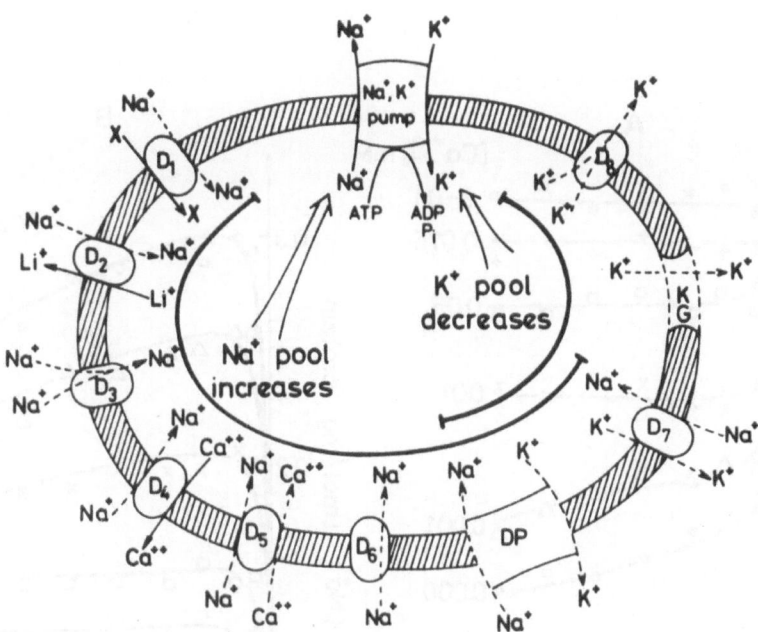

Fig. 1. Dissipators in the intact cell in the mode of operation leading to the activation of the Na$^+$,K$^+$ pump. Large arrows indicate direction and source of activation; solid arrows, uphill cation movements; broken arrows, downhill cation movements; X, organic compound (sugar, amino acid, etc.); DP, Na$^+$,K$^+$ pump assembly operating as a dissipator (e.g., under ouabain inhibition); KG, K$^+$ gating (and/or K$^+$ channels); D$_1$, Na$^+$-dependent uptake of organic substrates; D$_2$, Na$^+$:Li$^+$ countertransport; D$_3$, Na$^+$:Na$^+$ cotransport; D$_4$, Na$^+$:Ca^{2+} countertransport; D$_5$, Na$^+$:Ca^{2+} cotransport; D$_6$, Na$^+$ channels; D$_7$, Na$^+$:K$^+$ countertransport; D$_8$, K$^+$:K$^+$ cotransport. Any of the above dissipators can perform, under certain conditions, an inverse mode of operation, thus inhibiting the Na$^+$,K$^+$ pump.

for Li$^+$, as compared to that for K$^+$ and Na$^+$, respectively.[73] By Li$^+$-binding studies, different "catalytic" and "regulatory" K$^+$ sites have been demonstrated on the Na$^+$,K$^+$-ATPase of mammalian brain.[74] Further studies suggested that Li$^+$ interacts with only one of the K$^+$-binding sites of the Na$^+$,K$^+$-ATPase isolated from *E. electricus*.[75] It is very likely that conformational changes of the enzyme, determined by cooperative molecular transitions in the Na$^+$,K$^+$-transporting domain of the membrane, significantly alter cation-binding affinities.[13-15]

With respect to such conformational changes, the presence of membrane-bound ATP (and/or ADP) should be of outstanding importance.[4,13-15]

Particular physiological importance should be ascribed to one of the two outer K$^+$-binding sites which has been revealed as the primary ouabain receptor.[75] The number of ouabain receptors on the membranes of both mammalian red blood cells and nucleated somatic cells seems to be genetically controlled.[76-78] The great importance of this fact is emphasized by recent findings, as follows:

1. Digitalislike endogenous activity has been demonstrated in mammalian brain.[79]

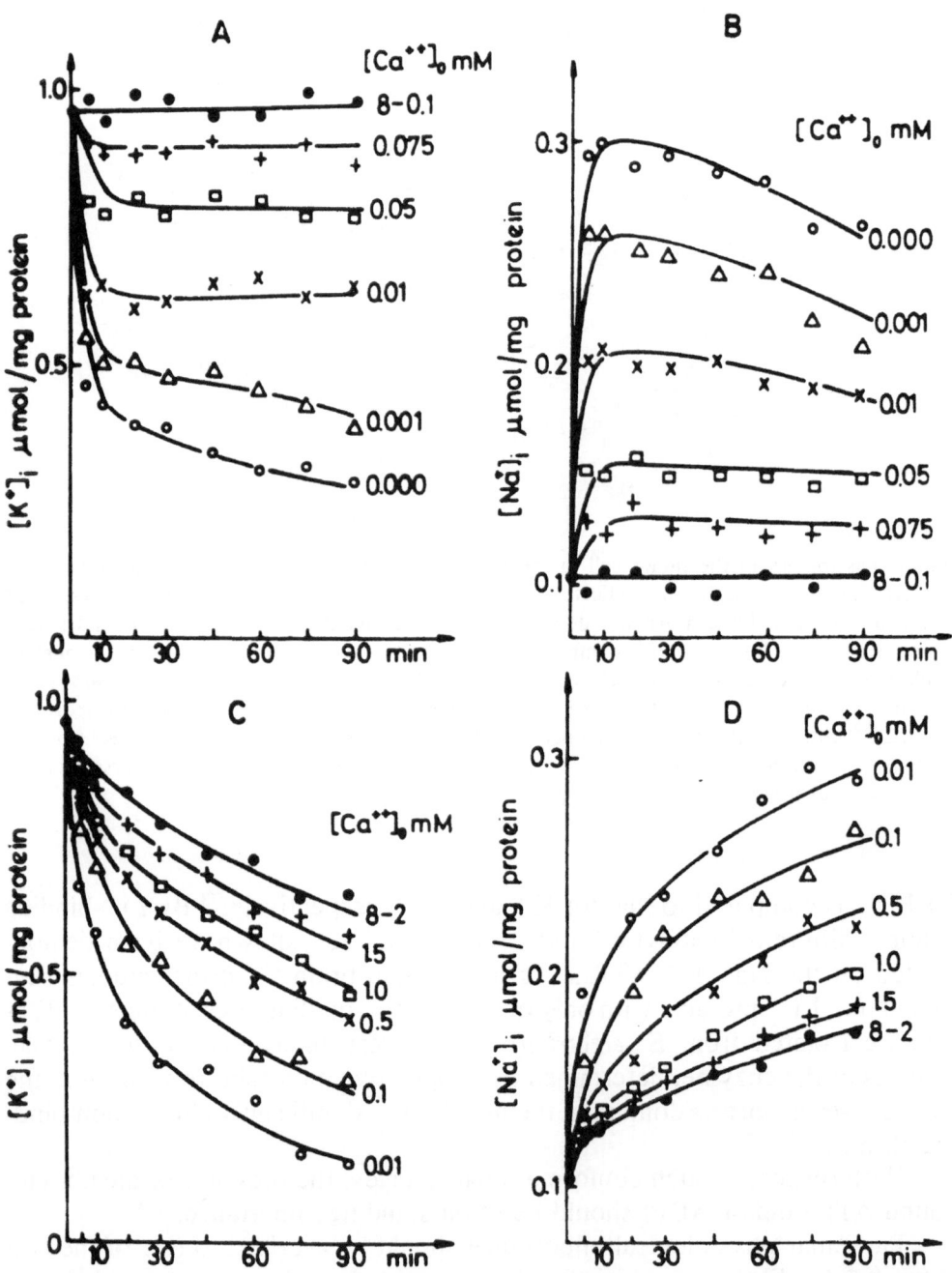

Fig. 2. Time course of K$^+$ and Na$^+$ concentrations in cultured glial cells as a function of external Ca^{2+} concentration. K$^+$ concentrations: (A) in the absence of ouabain; (C) in the presence of 10^{-4} M ouabain. Na$^+$ concentrations: (B) in the absence of ouabain; (D) in the presence of 10^{-4} M ouabain. Experimental procedures were as in ref. 71. Values are averages of 8–15 experiments; S.D. values are less than 2–3%.

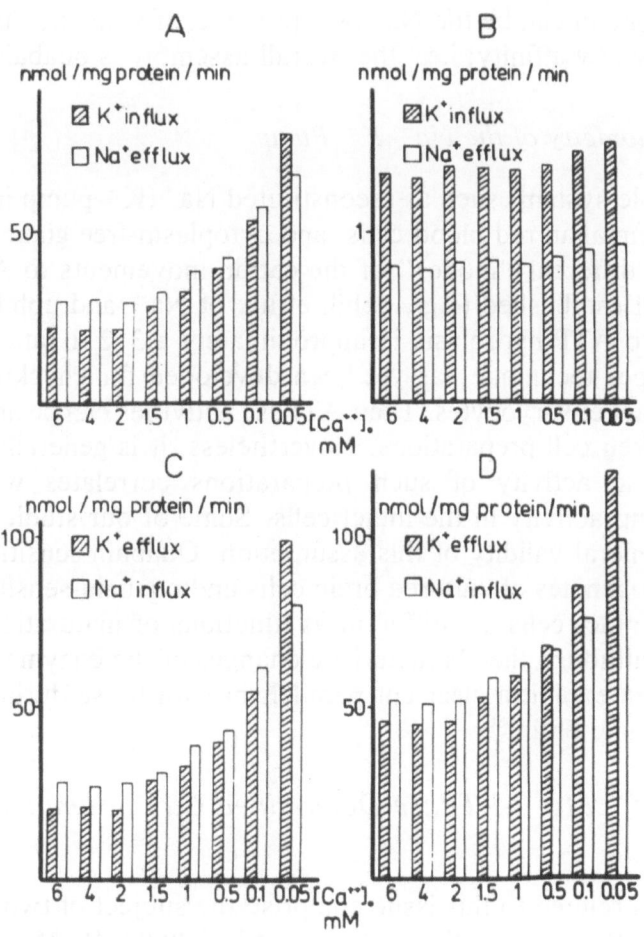

Fig. 3. Initial rate values of cation fluxes in cultured glial cells as a function of external Ca^{2+} concentration. Uphill fluxes: (A) in the absence of ouabain; (B) in the presence of 10^{-4} M ouabain. Downhill fluxes: (C) in the absence of ouabain; (D) in the presence of 10^{-4} M ouabain. Experimental procedures were as in ref. 71. Values are averages of 10–15 experiments; S.D. values are less than 2–3%.

2. Ouabain has been proved to represent an exogenous, highly effective regulator of cell proliferation and cell transformation.[27,37–39]
3. The ouabain sensitivity of the Na^+,K^+-ATPase assembly changes as a function of cell maturation.[80,81]
4. Oncogenic transformation of either brain cells or other somatic cells results in a decrease both in the ouabain sensitivity of the overall Na^+,K^+ pump activity and in [^{3}H]ouabain binding.[81,82]

The Na^+,K^+ pump under ouabain inhibition or under the effect of possible digitalislike endogenous substances,[79] operates as a dissipator, accomplishing, for instance, downhill Na^+ influx.[6,77] Cells that are genetically ouabain resistant possess a low level of Na^+,K^+-ATPase activity; however, they perform Na^+,K^+-dissipator activities.[76,77] Thus, one can assume that the genetic control of Na^+ and K^+ transport determines possible ratios between Na^+,K^+-pumping and Na^+,K^+-dissipating domains of the membrane. Such a Na^+,K^+-

dissipating domain can be the Na^+,K^+ pump itself when its ouabain receptor has extremely low affinity; i.e., the overall assembly is ouabain resistant.[6]

4.1.3. Stoichiometry of the Na⁺,K⁺ Pump

For simple systems such as reconstituted Na^+,K^+ pump in artificial lipid vesicles, mammalian red blood cells, and cytoplasm-free giant axon, the stoichiometry of different "modes" of the cation movements to ATP hydrolysis has been well established (e.g., uphill efflux of Na^+ and uphill influx of K^+ are coupled to ATP hydrolysis in approximately a $3:2:1$ ratio).[4,5,15,17] However, no proper technique has yet been developed for checking this stoichiometry in intact eukaryocytes. Their ATPase activities can be approached only by using broken cell preparations. Nevertheless, it is generally assumed that higher ATPase activity of such preparations correlates with augmented Na^+,K^+ pump activity in the intact cells. Some of our studies seem to contradict the general validity of this assumption. Ouabain-sensitive ATPase activity in homogenates of cultured brain cells and ouabain-sensitive K^+ uptake of the same intact cells are different as functions of maturation (i.e., time of cultivation), although the characteristic changes of the enzyme activity during cell cultivation exhibit a clear-cut parallelism with those during development of the embryonic brain.[80]

4.2. The Ca²⁺,Mg²⁺-ATPase-Determined Ca²⁺ Pump: Transport of Ca²⁺

Problems related to this issue comprise the subject of two other chapters in this series (Chapter 3, this volume, and Volume 4). Moreover, relevant reviews have recently been published that deal in detail with both general and particular correlates of Ca^{2+} transport.[5,6,17,23–26,83,84] Five of these reviews focus their interest particularly on Ca^{2+} transport phenomena,[5,6,17,83,84] whereas the others are primarily devoted to interactions between cyclic nucleotides and Ca^{2+}.[23–26] Therefore, only a brief summary of essential features of the Ca^{2+} pump is given here.

4.2.1. Similarities between the Na⁺,K⁺ Pump and the Ca²⁺ Pump

Beyond doubt, many essential similarities exist between the Na^+,K^+ pump and the Ca^{2+} pump in the plasma membrane. The Ca^{2+} pump is also determined by an enzyme (Ca^{2+},Mg^{2+}-ATPase) that spans the membrane and consumes energy from splitting ATP specifically for driving outward transport of Ca^{2+} against a concentration gradient. The transported Ca^{2+} asymmetrically activates the enzyme only at the inner surface of the membrane where ATP hydrolysis also takes place.[5,83,84] The rate of Ca^{2+} transport is a quite simple function of free Ca^{2+} concentration in the vicinity of highly specific Ca^{2+}-binding sites located on the inner side of the membrane. Affinity of the Ca^{2+}-binding sites may be high or low for Ca^{2+}, and it is influenced by ATP. The first step in the outward transport of Ca^{2+} is its specific binding to the enzyme.

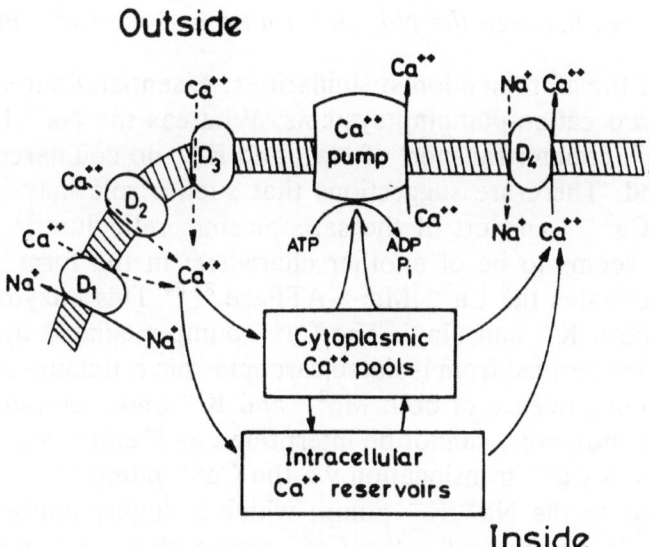

Fig. 4. Transmembrane and cytoplasmic pathways of Ca^{2+} movements. Large arrow indicates direction and source of Ca^{2+} pump activation; solid arrows, transmembrane uphill and cytoplasmic Ca^{2+} movements; broken arrows, transmembrane downhill Ca^{2+} movements; D_1, Ca^{2+}:Na^+ countertransport; D_2, "first messenger" (e.g., hormone)-sensitive Ca^{2+} channel; D_3, electric-potential-sensitive Ca^{2+} channels; D_4, Na^+:Ca^{2+} countertransport.

and Ca^{2+} translocation is accompanied by phosphorylation of the enzyme at the expense of ATP hydrolysis.[5,83,84] Recently, successful reconstitution of the Ca^{2+} pump assembly from synaptosomes into artificial lipid vesicles has been reported.[83]

In sarcoplasmic reticulum, the uphill Ca^{2+} movement performed by the Ca^{2+} pump proved to be reversible and permitted the synthesis of one molecule of ATP at the expense of downhill translocation of each two Ca^{2+} ions.[83,84]

4.2.2. The Ca^{2+} Pump and the Na^+:Ca^{2+} Exchange System

It is an intriguing analogy between the two cation pumping systems that both the Ca^{2+} pump and the Na^+,K^+ pump seem to be able to serve as dissipator systems. For example, in the sarcoplasmic reticulum, the Ca^{2+} pump can permit "passive" movement of Ca^{2+}.[83] In light of this finding, it is worthwhile to point out the close interrelationship that seems to exist between the Ca^{2+} pump assembly and the dissipator system performing Na^+:Ca^{2+} exchange.[17,84] The latter represents machinery that accomplishes uphill Ca^{2+} transport at the expense of the electrochemical energy of an inversely directed Na^+ gradient. Its functioning depends on the presence of ATP, although it does not require any splitting of ATP.[17,84] Most cells utilize both systems simultaneously to achieve uphill extrusion of free Ca^{2+} from the cell interior.[17,84] It is quite plausible that the two systems bring about a mutual regulatory unit, including the Na^+,K^+ pump as well, which provides for continuous reconstitution of the Na^+ gradient decreased by Na^+:Ca^{2+} exchange[84] (see Section 4.4 and Figs. 1 and 4).

4.2.3. Differences between the Na^+, K^+ Pump and the Ca^{2+} Pump

In spite of the aforementioned similarities, essential differences also exist between the two cation-pumping systems. Whereas the Na^+, K^+ pump performs the uphill countertransport of Na^+ and K^+, no countercation for Ca^{2+} has been found. There are suggestions that such a role may be ascribed to Mg^{2+} in the Ca^{2+} transport of the sarcoplasmic reticulum.[83] However, the role of Mg^{2+} seems to be of another character: in the form of a Mg–ATP complex, it activates the Ca^{2+}, Mg^{2+}-ATPase.[83,84] This enzyme can also be activated by both K^+ and Na^+. The Ca^{2+}-pump-mediated uphill uptake of Ca^{2+} by vesicles derived from isolated sarcoplasmic reticulum is accompanied by a simultaneous release of both Mg^{2+} and K^+ under certain conditions.[84] These findings, however, cannot be interpreted as if either Mg^{2+} or K^+ were countercations in Ca^{2+} translocation via the Ca^{2+} pump.[84]

In contrast to the Na^+, K^+ pump, which is highly uniform in different membranes from various cells, the Ca^{2+} pump shows essentially divergent characteristics even when it is isolated from different membranes of the same cell.[5,17,83,84]

Some of these differences may reflect incomplete isolation of the Ca^{2+} pump as a unit assembly. The significance of phospholipid ingredients of the Ca^{2+} pump has been studied on sarcoplasmic reticulum, and it has been revealed that they play an important role in the regulation of the overall characteristics of the pump.[83,84] Most of the essential differences among various membranes, however, as far as features of the Ca^{2+} pump are concerned, cannot be regarded as artifacts of the experimental technique.

4.3. Cation Movements via Dissipators, Passive Fluxes of Cations, and Transport of Mg^{2+} and Li^+

Either uphill or downhill (or "on-the-level") transport of any cation can take place via dissipators.[6] Nevertheless, it is very likely that only two of the dissipator-mediated uphill cation movements actually have great physiological importance in setting up the uneven cation distribution between cell and environment. One of these is the uphill extrusion of Ca^{2+} from the cell via the $Na^+ : Ca^{2+}$ exchange system described above (see Section 4.2.2). The other is a highly similar countertransport system that performs $Na^+ : Mg^{2+}$ exchange.[17] The latter can accomplish either uphill or downhill movement of Mg^{2+} depending on the direction of the Na^+ gradient. Although the two systems show similar characteristics, they are probably mediated by different "carriers."[17]

In all other cases, dissipator-mediated uphill fluxes of cations seem to have a "detoxifying" role: nonphysiological cations, e.g., Li^+, can be expelled from the cell interior by their function.[6,73] Dissipators are, however, primarily involved in the performance of downhill cation movements. According to recently collected evidence, these passive cation transports represent highly organized mechanisms that accomplish important functions in the regulation of cation movements in intact cells[6] (see Section 4.4 and Figs. 1 and 4).

The classification of dissipators is formalistic at present, since their actual molecular structure has been poorly revealed; however, pharmacological and electrophysiological studies provide a great help in this respect.[6–9,85–87] Thus, the following dissipator-mediated cation movements can be distinguished: carrier-mediated co- and countertransport of cations (where one of the cations can be substituted for nonionic compounds such as sugars, amino acids, neurotransmitters) and uncoupled cation fluxes across specific channels in the membrane.[6–9,85–87]

Characteristics of the passive cation transports strongly suggest that both carriers and channels represent cation-binding sites or "receptors" for cations.[6,17,87] The affinities of various cations for "dissipator binding sites" can be essentially different from those for pump binding sites. For instance Li^+ has a lower affinity for the inner Na^+ binding site of the Na^+,K^+-ATPase than the Na^+ has. However, it has much higher affinity for the inner binding site of the $Na^+ : Na^+$ countertransport system than the Na^+ itself has.[73] ATP in the membrane can significantly alter cation affinities of the "dissipator binding sites" even at the *trans*-side of the membrane.[5,6,17] Thus, membrane-bound ATP is an important regulator in passive cation transport, although no ATP splitting can be detected when pump activities are blocked and only dissipators operate.[5,6]

4.3.1 Co- and Countertransport of Cations

Dissipators performing co- and/or countertransport of cations are listed in Figs. 1 and 4 in the context of mutual regulatory interrelationship with pumps (see Section 4.4). The great number of them may be surprising at first glance. Nevertheless, as far as we know the mechanisms of the passive cations movements today, the possibility is quite plausible that each physiological cation (and many "foreign"cations as well) can be co- and countertransported by an individual "carrier mechanism." The "carriers" themselves might be quite similar, and specificity of cation binding sites, level of the membrane-bound ATP, and experimental conditions (i.e., cation concentrations both inside and outside of the cells) will give preference for one of the possible co- or countertransport mechanisms.[6–9,17]

The primary physiological role of co- and countertransport is the uptake of different substrates important for cell metabolism.[8,9,36] They certainly play an outstanding role, for instance, in the reuptake of some neurotransmitters. This issue will be dealt with in appropriate chapters of the present *Handbook*.

4.3.2. Cation Channels

Channels (also listed in Figs. 1 and 4) represent specialized supramolecular assemblies in the membrane that permit the highly selective uncoupled downhill movements of single cations. As a result of their cation selectivity, the energy of the electrochemical gradient of the cation moving downhill is not dissipated but is mostly consumed for the maintenance of the resting potential as well as for the development of excitation phenomena.[11] A channel can be "closed" or "open" depending on specific gating mechanisms.[85,86] The channels for various cations can be regarded as specific drug or hormonal receptors.[84,87] Some hormonal receptors (e.g., the acetylcholine receptor) have also been proved to include transmembrane cation channels.[88] Channels as regulators for passive cation movements also cooperate with pumps and other dissipators both in resting and activated cells (see Section 4.4).

Very likely, Na^+ and K^+ channels work on the same principles,[85] whereas the various Ca^{2+} channels function in a different way.[84] They certainly represent conformational changes of definite membrane constituents,[11,89] and the possibility can not be excluded that the Na^+,K^+-ATPase is somehow involved in the mechanism of the voltage-induced channel opening.[90]

4.3.3. Specific Ca^{2+}-Sensitive Gating for Passive K^+ and Na^+ Movements

Alterations of cation permeability in response to changes of Ca^{2+} concentration inside or outside of the cells show certain asymmetric ("vectorial") characteristics: an increase in internal Ca^{2+} concentration significantly increases the "passive" cation permeability of the membrane, and in some respects, a similar result is evoked by a decrease of external Ca^{2+} concentration.[5,6,91,92] Nevertheless, there are essential differences between the augmentation of the "passive" cation permeability initiated by $[Ca^{2+}]_i$ increase and that by $[Ca^{2+}]_o$ decrease.

4.3.3a. Selective Increase in K^+ Efflux by Internal Ca^{2+} ("Gárdos Effect"[6]). It was first observed in human erythrocytes, which do not contain any free Ca^{2+} under physiological conditions, that an increase of $[Ca^{2+}]_i$ (up to 1–10 μM) obtained by various loading procedures results in a highly selective, significant passive outflow of K^+ from the cells unaccompanied by any uptake of Na^+.[5,6] Recently, extensive studies have been devoted to this effect[5,6,92] since evidence has demonstrated its important role in nerve excitation phenomena.[92] (Indirect evidence suggests its involvement even in immunologic cell activation.[37,40] Although the mechanism of the "Gárdos effect"[6] is far from being clarified, it seems to be a specific and selective dissipator function in which Mg^{2+} cannot substitute for Ca^{2+}. The internal free Ca^{2+} either opens a selective K^+ channel (present both in erythrocytes and neurons) or triggers a specific K^+ carrier.[5,6,92] There are data suggesting that increase of $[Ca^{2+}]_i$ above a certain threshold permits an augmentation of the passive Na^+ influx as well.[92]

4.3.3b. Increase of Both K^+ Efflux and Na^+ Influx by External Ca^{2+}. It has been found by several authors that "passive" permeability of various eucaryotic cells for monovalent cations is controlled by external Ca^{2+}.[91] Concerning the mechanism of this role of Ca^{2+}, the general belief is that the decrease of $[Ca^{2+}]_o$ causes "leaky faults" in the membrane and thus "leak" transports of cations.[91]

Our recent findings, however, contradict such a simple explanation. We have observed that a decrease of $[Ca^{2+}]_o$ from 2 mM to 0.01 mM trigger cyclic AMP formation in cultured glial cells.[93] (A similar effect of low-Ca^{2+} medium on other cells has been also reported.[22] This is the event that induces an increase in both passive efflux of K^+ and passive influx of Na^+ by cultured glial cells exposed to low-Ca^{2+} medium[80,93] (see Section 4.4 and Figs. 2 and 3). Moreover, the decrease of $[Ca^{2+}]_o$ results in a significant decrease of Ca^{2+} uptake by the cultured glial cells, whereas Ca^{2+} release is almost independent of $[Ca^{2+}]_o$ changes (see Section 5.3). Consequently, a transient diminution in free $[Ca^{2+}]_i$ should also be expected in the cells when $[Ca^{2+}]_o$ decreases. Such changes of $[Ca^{2+}]_i$ can well explain the triggering of cyclic AMP formation.[22–26] Either the transiently elevated level of cyclic AMP itself or an excess free Ca^{2+} in the cytoplasm released by the cyclic AMP from intracellular reservoirs[23–26] can result in an increase in the "passive" cation permeability by specific gating mechanisms.[23–26]

4.3.4. "Leak" Transport of Cations

4.3.4a. Downhill Transport of Li^+ and Na^+ via the Anion Exchange Pathway in the Form of Complexes with CO_3^{2-} (Replacing the H^+ of Bicarbonate). This mechanism is responsible for a large percentage of the passive Li^+ uptake by various cells[73] and may play a role in the HCO_3^--stimulated Na^+ secretion performed by the choroid plexus.[94]

4.3.4b. Downhill Movements of Cations across Membrane "Imperfections" (See Sections 1 and 2.2.1a). According to our best knowledge of cation transport today, there is no reason to believe that "imperfections" of the intact cell membrane would enable significant uncontrolled passive diffusion, i.e., "leak" transport, of physiological cations.[6] In the absence of either a pump or a dissipator mechanism specialized for a certain cation flux, no such flux occurs at all. For instance, intact human erythrocytes do not possess any transport machinery for the performance of downhill Ca^{2+} influx. Thus, the level of Ca^{2+} uptake by intact human erythrocytes equals the very low amount of ground permeability when Ca^{2+} extrusion by the Ca^{2+} pump is blocked.[5,6] However, this "resistance" of the membrane to Ca^{2+} influx needs the presence of membrane-bound ATP at the inner side.[5,6] It should be emphasized that this membrane-bound ATP is not consumed in maintaining the low level of Ca^{2+} influx; i.e., membrane resistance against Ca^{2+} uptake in the intact cells is independent of the Ca^{2+} pump activity.[5,6]

In contrast to the Ca^{2+} influx, both the Na^+ influx and the K^+ efflux performed by intact erythrocytes exceed the ground permeability by about two orders of magnitude.[5,6] However, these "passive" cation movements (sometimes mentioned as "leak" transport) are driven by well-defined co- and/or countertransport mechanisms, i.e., by dissipator assemblies[5,6,76,77] (see Section 4.3.1).

The actually leak-transport-like component of passive Li^+ uptake by various cells[73] may be interpreted as an indication of "leaky faults" existing in the membrane for "foreign" cations. Nevertheless, the possibility has not been excluded that even "leak" uptake of Li^+ is mediated by one of the available dissipators.

4.4. Regulatory Interrelationships among Mechanisms Controlling Cation Transport

Changes brought about as results of either dissipator or pump activities can mutually activate and/or inhibit these activities. Figure 1 depicts a set of dissipators working in such a manner that changes in both K^+ and Na^+ concentrations evoked by their function increase the activity of the Na^+,K^+ pump. Of course, any of the dissipators can perform the opposite mode of operation as well. Thus, a highly flexible, dynamic regulatory system is established. Since the actual mode of operation of the single pump and dissipator assemblies as well as their activities depend on the concentration of both cations and ATP in their "microenvironment" (both inside and outside of the cell), compartmentation of cations and ATP contributes, particularly in eucaryocytes, to the flexibility of the cation transport regulation.

A similar scheme is given in Fig. 4 which lists dissipators affecting the Ca^{2+} pump.

Our findings shown in Figs. 2 and 3 may serve as an example of increased dissipator activity that is balanced by secondary augmentation of the Na^+,K^+ pump activity. In low-Ca^{2+} medium, cultured glial cells lose K^+ and take up Na^+ (see Section 4.3.3b). However, in the absence of ouabain, the cells soon achieve a new ionic equilibrium; i.e., altered levels of $[K^+]_i$ and $[Na^+]_i$ do not change further in time (Fig. 2), indicating a compensatory increase of both K^+ reuptake and Na^+ extrusion according to the scheme of Fig. 1. The compensatory mechanism is ouabain sensitive and can be exhausted by maintaining an extremely low level of Ca^{2+} in the medium (Fig. 2).

When the $[Ca^{2+}]_o$ decrease is relatively moderate (from 2 mM to 0.1 mM), in the absence of ouabain neither $[K^+]_i$ nor $[Na^+]_i$ is altered in the cultured glial cells (Fig. 2). Nevertheless, influx as well as efflux of both $^{42}K^+$ and $^{22}Na^+$ is increased as a function of $[Ca^{2+}]_o$ (Fig. 3A,C). The presence of ouabain does not influence the $[Ca^{2+}]_o$ sensitivity of downhill cation movements (K^+ efflux and Na^+ influx, Fig. 3C,D), whereas uphill cation fluxes (K^+ influx and Na^+ efflux) are not $[Ca^{2+}]_o$ dependent in the presence of ouabain (Fig. 3A,B). These findings also confirm the above interpretation: the increase of the uphill movements of both K^+ and Na^+ by cultured glial cells exposed to low-$[Ca^{2+}]_o$ media is a compensatory event reacting to the increase in passive K^+ outflow and Na^+ uptake.

5. PECULIARITIES OF CELLULAR CATION TRANSPORT IN THE NERVOUS SYSTEM

The heterogeneity of cellular elements and the presence of highly specialized morphological structures develop peculiar transport characteristics in the nervous system, although, as has been discussed in the foregoing paragraphs, regulation of cellular cation transport here depends on the same principles as that in other tissues. A coherent account of these peculiarities is somewhat impeded by the great diversity of experimental objects and techniques applied in studies of cation transport in the nervous system.

5.1. Characteristics of Neuronal Cation Transport

It is highly questionable whether the neuron represents a single functional unit as far as cation transport is concerned. Membranes of the cell body, axon, nerve terminal, and dendrites seem to have different characteristics with regard to transport phenomena.[17,84,92,95] Moreover, differential stimuli from the environment, such as changes of ionic milieu in various regions of the microenvironment by either simple diffusion of cations or glial–neuronal interaction,[50–52,96–98] can affect different areas of the neuronal membrane. A significant number of different model systems (e.g., giant axons,[17] *Aplysia* neurons,[92] brain cortex slices,[41] bulk-isolated nerve cells,[41] synaptosomes,[84] and cultured cells of neuronal character[99,100]) have provided a huge amount of information about particular problems of cation movements taking place through specialized areas of the neuronal membrane. However, regulatory interplay of these "local" transport events is far from being understood.

5.1.1. Pump Activities in Neuronal Cells

Neurons seem to possess a genetically determined commitment to be relatively ouabain resistant as compared to either glial cells or some other eucaryocytes.[78,80,101,102] (Data have been reported suggesting the possibility that this property of neuronal cells mights be somehow correlated with the genetic suppression of mitotic activity.[101]) The ATP-splitting activity of the Na^+,K^+ pump in the neuronal membrane is relatively insensitive to changes in the concentration of the external K^+.[19,81] These facts suggest a low number of K^+-binding sites present on the neuronal membrane (see Section 4.1.2). In agreement with such a conclusion, neuronal cells in culture accomplish lower K^+-transporting activity than do glial cells.[41,80,99]

Very likely, the Na^+,K^+ pump activity varies in different membrane areas as a function of the secondary structure of the pump determined by the actual lipid composition (see Section 4.1.1).

The Ca^{2+} pump operates in the neurons in close correlation with both the $Na^+:Ca^{2+}$ exchange system and the Na^+,K^+ pump[17,83,84] (see Sections 4.2 and 4.4 and Fig. 4). It plays a significant role in the control of synaptic functions.[50,83,84] Electrophysiological properties of the dendrites seem to depend particularly on the Ca^{2+}-transporting systems.[50,95]

5.1.2. Dissipators in Neuronal Cells

The high level of resting potential in neurons[41] and the neuronal excitability together indicate "that the additional dissipative fluxes during activity should be only a fraction of the dissipative fluxes at rest"[6]; i.e., dissipators should have a high activity even in resting neurons.[11] The passive cation fluxes representing these dissipator functions can be performed by channels[84-90] (see Section 4.3.2), Ca^{2+}-sensitive, specific gating mechanisms[92] (see Section 4.3.3), and co- or countertransport pathways (see Sections 4.2 and 4.3). Free Ca^{2+} in the cells and hormonal effects mediated via the second messenger system play a significant regulatory role in resting cation permeability[23-26,84-92,103,104] (see Section 4, and Chapter 3 in Volume 1 and Volume 6 in the present *Handbook*).

A schematic outline of the possible regulatory cirucits that can operate among pumps, dissipators, and the second messenger system (represented here by the cyclic AMP-generating and -splitting mechanisms) is given by Fig. 5. In the organization of these regulatory circuits, the compartmentalization of both cations and ATP plays a prominent role (see Section 2.2.1b). It is striking that mitochondria representing the ATP-generating system are mostly located in the vicinity of the membrane in the giant axon.[105] (Otherwise, the giant axon seems to represent an exception in this respect, since cation compartmentalization has been reported to be irrelevant in its cytoplasm.[17])

The high level of dissipator activity in the neuron presumes, of course, an equally high level of pump activity in the resting cells for maintaining ionic equilibrium.

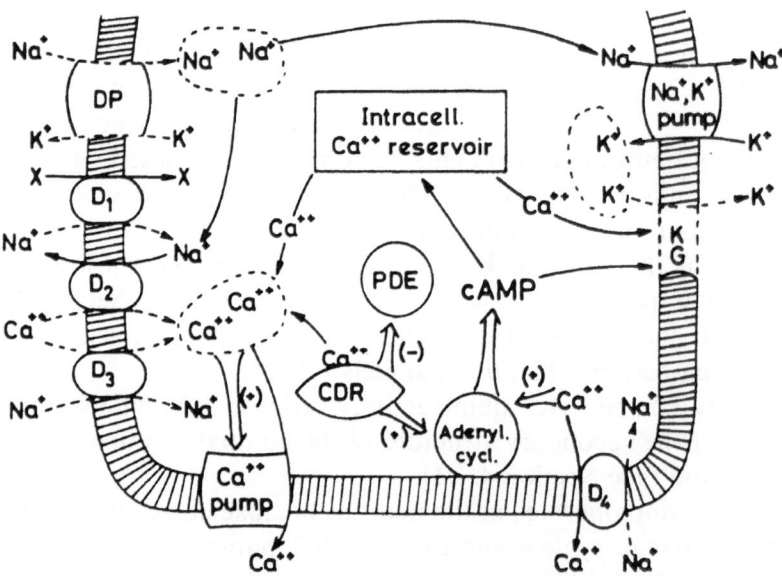

Fig. 5. Essential interactions among pumps, dissipators, and the second messenger system. PDE, phosphodiesterase; CDR, Ca^{2+}-dependent regulator protein; DP, Na^+,K^+ pump as a dissipator (e.g., under the influence of ouabain); KG, Ca^{2+}-sensitive and/or cyclic-AMP-induced K^+ gating; D_1, Na^+-dependent uptake of different substrates; D_2, $Ca^{2+}:Na^+$ countertransport; D_3, $Ca^{2+}:Na^+$ cotransport; D_4, $Na^+:Ca^{2+}$ countertransport; solid arrows, transmembrane uphill and intracytoplasmic cation movements; broken arrows, transmembrane downhill cation movements; large arrows, direction and source of activation (+) or inhibition (−).

5.1.3. Electronic Character of Cation Transport

Both the Na^+,K^+ pump and the Ca^{2+} pump, as well as many of the dissipators, e.g., the $Na^+:Ca^{2+}$ and $Na^+:Mg^{2+}$ countertransport systems and cotransport of K^+ and Na^+ (or K^+-K^+ and Na^+-Na^+ cotransports), are evidently electrogenic.[4,17] Beyond doubt, the electrogenic character of these cation movements contributes to the establishment of the resting potential. Nevertheless, the resting membrane potential is primarily $[K^+]_o$ dependent[41]; thus, the pumps and dissipators determining the uneven distribution of K^+ between the neuron and its environment are of paramount importance.

5.2. Characteristics of the Cation Transport by Glial Cells

Extensive studies have recently been devoted to the role of glia (primarily astroglia) in stabilizing the extracellular ionic composition in the brain.[41,50–52,97,106] This seems to be the specialized transport function of the glia.

5.2.1. Pump Activities in Glial Cells

5.2.1a. Characteristics of the Na^+,K^+ Pump. All glial cells irrespective of origin (bulk-isolated cells, cultured cell lines, primary cultures of glia from embryonic or newborn brain) have the Na^+,K^+ pump in their membrane for performing coupled uphill movements of Na^+ and K^+.[19,32,33,71,99,107,108] The Na^+,K^+ pump in the glia behaves "as if it were specialized" for transporting primarily K^+. Unfortunately, a precise estimation of the pump stoichiometry has inherent difficulties (see Section 4.1.3). For instance, in K^+-uptake experiments in which ^{42}K or ^{86}Rb is employed as tracer in cultured glial cells, no isotope equilibrium is reached within 2–3 h (i.e., the stationary specific activity inside the cells is much lower than in the medium), suggesting compartmentation of the K^+ pool.[71] The same conclusion was reached by simultaneously studying ^{42}K and ^{22}Na uptake.[32] When compartmentation of cation pools actually exists, the reliability of the uptake rate values is highly questionable (see Section 3.2), and judgments on pump stoichiometry have no validity.

In contrast with neuronal cells, ATP-splitting activity of the Na^+,K^+ pump in the membrane of glial cells exhibits a high sensitivity to changes in the external K^+ concentration.[19,81] Moreover, both ATP-hydrolyzing activity and K^+-transporting capability of the glial Na^+,K^+ pump proved to be exquisitely sensitive to ouabain.[19,41,80,81] These facts together indicate that several K^+-binding sites are present on the glial membrane (see 4.1.2). Since the studies providing the above findings were performed on "*in situ*" Na^+,K^+ pump, i.e., on either intact cells in culture or bulk-isolated cells, the conclusion that neuronal and glial Na^+,K^+-ATPases would represent two different enzymes is not justified (see Section 4.1.1). A detailed comparison of enzymologic, kinetic, and immunologic properties of the purified enzymes, combined with reconstitution experiments, were only able to decide the question whether neuronal and glial ATPases represent different enzymes. The above facts, however, may suggest that the number of Na^+,K^+ pumps is significantly different in the membrane of the two cell species.

The basis of the high sensitivity of the Na^+,K^+-pumping activity of the glial membrane to external K^+ has not been revealed; nevertheless, there is much evidence to demonstrate that it plays a prominent role in the performance of glial removal of excess K^+ from the glial microenvironment.[41,50–52,97,106]

5.2.1b. Ca^{2+}-Pump-Mediated Ca^{++} Transport in Glia. The presence of both the Ca^{2+} pump and the $Na^+:Ca^{2+}$ countertransport mechanism has been demonstrated in glial cells.[100,109] Calcium extrusion from glial cells may be significantly influenced by the internal compartmentalization of the Ca^{2+}.[109] In addition to subcellular organelles that store Ca^{2+}, some proteins such as the S-100 protein can bind significant amount of Ca^{2+}.[97] Increased external K^+ concentration results in Ca^{2+} release from glia.[109] Thus, the glia can serve as a Ca^{2+} reservoir for the intercellular cleft.

5.2.2. Dissipator Functions in Glial Cells

Recent findings have demonstrated that glial cells are "permeable" to Na^+ as well as to K^{+} [32,33,99,107] (see Section 4.4, Figs. 2 and 3); i.e., dissipator-mediated passive Na^+ movements should take place. Nevertheless, it is still a great challenge for neurochemists dealing with transport phenomena to account for the clear-cut evidence that demonstrates the primary importance of K^+ movements in the physiology of glia.[50–52,97,106] Although the high K^+ sensitivity of the glial Na^+,K^+ pump system should play an important role in the organization of K^+ transport in glial cells, the existence of specific K^+-gating mechanisms is inevitable, since Na^+ pool sizes are too small to remove excess potassium by pump-mediated $Na^+:K^+$ exchange alone.

Very likely, a $Na^+:Na^+$ countertransport system operates in glia, since in cultured glial cells as well as in glial–neuronal mixed populations in culture, $Li^+:Na^+$ countermovement has been observed accomplishing an uphill extrusion of Li^+ from the cells.[110,111] Other data can be interpreted as suggesting a $Na^+:Ca^{2+}$ countertransport mechanism performing Ca^{2+} influx.[109] Our own studies (Á. Rimanóczy and A. Juhász, unpublished results) demonstrate that cultured glial cells from chicken embryonic brain take up Ca^{2+} much more rapidly than glial–neuronal mixed populations. The Ca^{2+} uptake obeys Michaelis–Menten kinetics; thus, the influx is highly dependent on the external Ca^{2+} concentration[93] (see Section 4.3.3b). All these facts suggest that the glial membrane is capable of performing most transport functions found in other eucaryocytes.

5.3. Glial–Neuronal Interactions in Cation Transport

Recently, particular attention has been paid to this issue, and as a result relevant reviews have been published.[97,106] Therefore, we restrict ourselves to special topics concerning the role of glial–neuronal interaction in the organization of cellular cation transport.

In some respects the glial–neuronal interaction in cation transport resembles the metabolic communication that exists between ouabain-resistant and ouabain-sensitive cells interconnected by gap junctions[27]; although there is no direct contact between neurons and glia, they are separated by a narrow intercellular space. The common features are as follows: the primarily important event accomplished by the cell interaction is the cell-to-cell K^+ transmission from one cell to another; the K^+ carries a signal between the cells; and the interaction is mutual, with the two cells influencing each other.

It has been demonstrated by a mixed population of cultured glial and neuronal cells that under steady-state conditions, when spontaneous neuronal firing is very unlikely, K^+–K^+ exchange occurs between neuronal and glial cells.[71] Thus, transport characteristics of the mixed populations are different from those of either glial or neuronal cultured cells.[71,80] We do not know what the signal is between the cultured glia and neuron that brings about this K^+–K^+ exchange under ionic steady state. (A certain intercellular space exists between neuron and glia in the cultures as well; see Fig. 1 in ref. 111.)

In vivo, changes of cation concentrations (K^+, Ca^{2+}, and Na^+) in the intercellular cleft as a result of either neuronal or glial activity provide the signal(s) in the glial–neuronal interaction in cation transport.[50–52,80,97,106] The most important result of the interaction is the removal of "excess K^+" released by neuronal firing from the intercellular cleft via glial uptake.[97,106] This can be achieved by two distinct mechanisms: "spatial buffering" and active uptake via the glial Na^+,K^+ pump.[97,106] It seems that these mechanisms operate simultaneously, although, depending on species characteristics and brain area, one of them can prevail.[50–52,97,106] There are data suggesting that glial–neuronal interaction is not an obligatory prerequisite for the removal of the "excess K^+" from the vicinity of the neuron: it can be achieved by simple diffusion across the intercellular cleft as well.[96]

In the regulation of glial–neuronal interaction in cation transport, we should ascribe great significance to the role of "first messenger" effects (e.g., neurotransmitters, hormones).[22–26,97,106] Such effects can bring about "permeability changes" of the glial membrane[22–26] in cooperation with the effects of alterations of external cation concentrations which can permit current-carried passive movements of K^+ resulting in the phenomenon of "spatial buffering." Our findings (see Sections 4.3.3b and 4.4) suggest that changes in both external and internal Ca^{2+} concentrations can set up specific K^+ gate opening and, consequently, Na^+,K^+ pump activation in cultured glial cells.[80,93] Moreover, it has been demonstrated that neuronal activity results in a decrease in Ca^{2+} concentration in the intercellular cleft, i.e., in the extracellular microenvironment of the glia, particularly under pathological conditions.[18,50,51] Thus, the above mechanism can have physiological and pathological significance. A scheme of this possible mechanism playing a role in the regulation of glial–neuronal interaction is given in Fig. 6. (Concerning this scheme, it should be pointed out that the existence of the particular processes has been proved; see Sections 4.3.3 and 4.4 and Figs. 2 and 3. But the operation of such regulatory circuits has not been postulated.)

This regulatory assembly may provide a biochemical background that per-

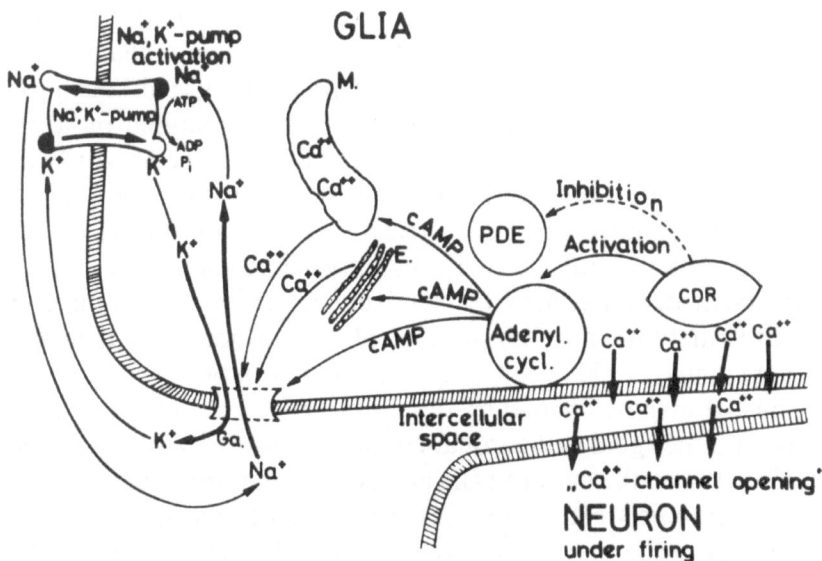

Fig. 6. Possible events induced by the decrease of external Ca^{2+} concentration in the vicinity of glial membrane caused by neuronal firing. PDE, phosphodiesterase; CDR, Ca^{2+}-dependent regulator protein; M., mitochondria; E., endoplasmic reticulum; Ga., Ca^{2+}-sensitive and/or cyclic-AMP-induced cation gating. High-affinity cation-binding sites on the Na^+,K^+ pump are shaded.

mits "current-carried K^+ movements via the glial cells. Moreover, it can contribute to the increase of Na^+,K^+ pump activity evoked by the direct effect of "excess" K^+.

6. CATIONIC ENVIRONMENT OF THE NEURON AND GLIA

6.1. Cation Transport by the Blood–Brain Barrier

This structure consisting of highly different elements provides the most important factor determining the ionic composition of the CSF which nearly equals that of the cellular microenvironment.[50–52,97] Functions of the blood–brain barrier will be dealt with in detail in Vol. 7 of this *Handbook*. Here we should point out that its elements have a distinctively high level of cation transport activity via both pumps and dissipators[94] for maintaining dissimilar ion composition of blood and CSF.

6.2. Role of External Cations in the Regulation of Cellular Cation Transport in the Nervous System

Most problems concerning this highly important issue have been dealt with in the foregoing paragraphs. As a summary, we should emphasize that at least two of the cations, K^+ and Ca^{2+}, in the extracellular space perform an "intercellular messenger" role in the brain.[50–52,97] Very likely, the physiological changes in their concentrations exert an effect by the activation or inhibition of pumps and dissipators; i.e., the basic structure of the membrane is not distorted. However, extreme changes of their external concentrations might result in formation of membrane "imperfections" which are not genuine ingredients of the membrane. Therefore, conclusions inferred from such experiments can be misleading.

7. CONCLUDING REMARKS: CATION TRANSPORT AND BRAIN FUNCTIONS

There is no space in the present chapter to deal with all of the important brain functions performed that are closely correlated with cation transport. However, it should be pointed out that many disorders of brain function have recently been revealed or suggested to be dependent on disturbances of cation transport mechanisms. These disorders, such as brain edema, epilepsy, and affective psychoses, will be the subject of other chapters in this *Handbook*.

REFERENCES

1. Rosenberg, T., 1948, *Acta Chem. Scand.* **2**:14–28.
2. Ussing, H. H., 1949, *Acta Physiol. Scand.* **19**:43–67.
3. Koefoed-Johnsen, V., and Ussing, H. H., 1960, *Mineral Metabolism*, Volume 1 (C. L. Comar and F. Bronner, eds.), Academic Press, New York, London, pp. 169–203.
4. Glynn, I. M., and Karlish, S. J. D., 1975, *Biochem. Soc. Spec. Publ.* **4**:145–158.
5. Sarkadi, B., and Tosteson, D. C., 1979, *Membrane Transport in Biology*, Volume 2 (D. C. Tosteson, ed.), Springer-Verlag, Berlin, Heidelberg, New York, pp. 117–160.
6. Lew, V. L., and Beaugé, L., 1979, *Membrane Transport in Biology*, Volume 2 (D. C. Tosteson, ed.), Springer-Verlag, Berlin, Heidelberg, New York, pp. 81–116.
7. Schultz, S. G., and Curran, P. F., 1970, *Physiol. Rev.* **50**:637–656.
8. Crane, R. K., 1977, *Rev. Physiol. Biochem. Pharmacol.* **78**:99–159.
9. Gunn, R. B., 1980, *Annu. Rev. Physiol.* **42**:249–259.
10. French, R. J., and Adelman. W. J., Jr., 1976, *Current Topics in Membranes and Transport*, Volume 8 (F. Bronner and A. Kleinzeller, eds.), Academic Press, New York, pp. 161–207.
11. Blumenthal, R., Changeux, J. P., and Lefever, R., 1970, *J. Membr. Biol.* **2**:351–374.
12. Nicolau, C., and Paraf, A. (eds.), 1977, *Structural and Kinetic Approach to Plasma Membrane Functions*, Springer-Verlag, Berlin, Heidelberg, New York.
13. Stein, W. D., Lieb, W. R., Karlish, S. J. D., and Eilam, Y., 1973, *Proc. Natl. Acad. Sci. U.S.A.* **70**:275–278.
14. Garrahan, P. J., and Garay. R. P., 1976, *Current Topics in Membranes and Transport*, Volume 8 (F. Bronner and A. Kleinzeller, eds.), Academic Press, New York, pp. 29–97.
15. Wallick, E. T., Lane, L. K., and Schwartz, A., 1979, *Annu. Rev. Phsyiol.* **41**:397–411.
16. Hodgkin, A. L., and Katz, B., 1949, *J. Physiol. (Lond.)* **108**:37–54.
17. Mullins, L. J., 1979, *Membrane Transport in Biology*, Volume 2 (D. C. Tosteson. ed.), Springer-Verlag, Berlin, Heidelberg, New York, pp. 161–210.
18. Prince, D. A., Pedley, T. A., and Ransom, B. R., 1978, *Dynamic Properties of Glia Cells*, (E. Schoffeniels, G. Franck, L. Hertz, and D. B. Tower, eds.), Pergamon Press, New York, pp. 281–303.
19. Franck, G., Grisar, T., Moonen, G., and Schoffeniels, E., 1978, *Dynamic Properties of Glia Cells* (E. Schoffeniels, G. Franck, L. Hertz, and D. B. Tower. eds.), Pergamon Press, New York, pp. 315–325.
20. Ling, G. N., 1977, *Mol. Cell Biochem.* **15**:159–171.

21. Walker, J. L., and Brown, H. M., 1977, *Physiol. Rev.* **57**:729–778.
22. Rasmussen, H., 1970, *Science* **170**:404–412.
23. Rasmussen, H., and Goodman, D. B. P., 1977, *Physiol. Rev.* **57**:421–509.
24. Rapp, P. E., and Berridge, M. J., 1977, *J. Theor. Biol.* **66**:497–525.
25. Berridge, M. J., 1975, *Advances in Cyclic Nucleotide Research*, Volume 6 (P. Greengard and G. A. Robison, eds.), Raven Press, New York, pp. 1–98.
26. Strewler, G. J., and Orloff, J., 1977, *Advances in Cyclic Nucleotide Research*, Volume 8 (P. Greengard and G. A. Robison, eds.), Raven Press, New York, pp. 311–361.
27. Ledbetter, M. L. S., and Lubin, M., 1979, *J. Cell Biol.* **80**:150–165.
28. Proverbio, F., and Hoffman, J. F., 1972, *Fed. Proc.* **31**:215–223.
29. Macknight, A. D. C., and Leaf, A., 1977, *Physiol. Rev.* **57**:510–573.
30. Kregenow, F. M., 1977, *Membrane Transport in Red Cells*, (J. C. Ellory and V. L. Lew, eds.), Academic Press, London, New York, pp. 383–426.
31. Bourke, R. S., and Nelson, K. M., 1972, *J. Neurochem.* **19**:633–685.
32. Kukes, G., Elul, R., and De Vellis, J., 1976, *Brain Res.* **104**:71–92.
33. Kimelberg, H. K., Bowman, C., Biddlecome, S., and Bourke, R. S., 1979, *Brain Res.* **177**:533–550.
34. Lux, H. D., 1980, *Antiepileptic Drugs: Mechanisms of Action* (G. H. Glaser, J. K. Penry, and D. M. Woodbury, eds.), Raven Press, New York, pp. 63–83.
35. Somjen, G. G., 1980, *Antiepileptic Drugs: Mechanisms of Action* (G. H. Glaer, J. K. Penry, and D. M. Woodbury, eds.), Raven Press, New York, pp. 155–167.
36. Wheeler, D. D., Callihan, C. S., and Wise, W. C., 1980, *J. Neurosci. Res.* **5**:201–216.
37. Kaplan, J. G., 1978, *Annu. Rev. Physiol.* **40**:19–41.
38. Leffert, H. L., and Koch, K. S., 1979, *Cell* **18**:153–163.
39. Whitfield, J. F., Boynton, A. L., Macmanus, J. P., Sikorska, M., and Tsang, B. K., 1979, *Mol. Cell. Biochem.* **27**:155–179.
40. Freedman, M. H., 1979, *Cell. Immunol.* **44**:290–313.
41. Hertz, L., and Schousboe, A., 1975, *Int. Rev. Neurobiol.* **18**:141–211.
42. Oschman, J. L., 1978, *Membrane Transport in Biology*, Volume 3 (G. Giebisch, ed.), Springer-Verlag, Berlin, Heidelberg, New York, pp. 55–93.
43. Ussing, H. H., and Leaf, A., 1978, *Membrane Transport in Biology*, Volume 3 (G. Giebisch, ed.), Springer-Verlag, Berlin, Heidelberg, New York, pp. 1–26.
44. Galla, H. J., and Sackmann, E., 1975, *Biochim. Biophys. Acta* **401**:509–529.
45. Jacobson, K., and Papahadjopoulos, D., 1975, *Biochemistry* **14**:152–161.
46. Van Dijck, P. W. M., De Kruijff, B., Verkleij, A. J., and Van Deenen, L. L. M., 1978, *Biochim. Biophys. Acta* **512**:84–96.
47. Kimelberg, H. K., and Papahadjopoulos, D., 1971, *J. Biol. Chem.* **246**:1142–1150.
48. Butler, K. W., Hanson, A. W., Smith, I. C. P., and Schneider, H., 1973, *Can. J. Biochem.* **51**:980–989.
49. Friedrich, P., Aprókovács, V. A., and Solti, M., 1977, *FEBS Lett.* **84**:183–186.
50. Nicholson, C., 1980, *Neurosci. Res. Prog. Bull.* **18**:177–322.
51. Nicholson, C., 1979, *The Neurosciences: Fourth Study Program* (F. O. Schmitt and F. G. Worden, eds.), MIT Press Journals, Cambridge, Massachusetts, London, pp. 457–476.
52. Somjen, G. G., 1979, *Annu. Rev. Physiol.* **41**:159–177.
53. Atkins, G. L., and Gardner, M. L. G., 1977, *Biochim. Biophys. Acta* **486**:127–145.
54. Winne, D., 1977, *Biochim. Biophys. Acta* **464**:118–126.
55. Dowd, J. E., and Riggs, D. S., 1965, *J. Biol. Chem.* **240**:863–869.
56. Endrényi, L., and Kwong, F. H. F., 1972, *Analysis and Simulation of Biochemical Systems* (H. C. Hemker and B. Hess, eds.), North-Holland, Amsterdam, pp. 219–237.
57. Fajszi, C., and Endrényi, L., 1974, *FEBS Lett.*, **44**:240–246.
58. De Miguel Merino, F., 1974, *Biochem. J.* **143**:93–95.
59. Eisenthal, R., and Cornish-Bowden, A., 1974, *Biochem. J.* **139**:715–720.
60. Cornish-Bowden, A., and Eisenthal, R., 1974, *Biochem. J.* **139**:721–730.
61. Cornish-Bowden, A., 1975, *Biochem. J.* **149**:305–312.
62. Solomon, A. K., 1960, *Mineral Metabolism*, Volume 1 (C. L. Comar and F. Bronner, eds.), Academic Press, New York, London, pp. 119–168.

63. Brownell, G. L., Berman, M., and Robertson, J. S., 1968, *Int. J. Appl. Radiat. Isot.* **19:**249–262.

64. Shipley, R. A., and Clark, R. E. (eds.), 1972, *Tracer Methods for In Vivo Kinetics*, Academic Press, New York, London.

65. Rubinow, S. I., and Winczer, A., 1971, *Math. Biosci.* **11:**203–247.

66. Lajtha, A., Latzkovits, L., and Toth, J., 1976, *Biochim. Biophys. Acta* **425:**511–520.

67. Reiner, J. M., 1953, *Arch. Biochem. Biophys.* **46:**53–79.

68. Fajszi, C., and Latzkovits, L., 1972, *Biophysik* **9:**64–69.

69. Latzkovits, L., Fajszi, C., and Szentistványi, I., 1972, Acta Biochim. Biophys. Acad. Sci. Hung. **7:**307–314.

70. Latzkovits, L., Szentistványi, I., and Fajszi, C., 1972, *Acta Biochim. Biophys. Acad. Sci. Hung.* **7:**55–66.

71. Latzkovits, L., Sensenbrenner, M., and Mandel, P., 1974, *J. Neurochem.* **23:**193–200.

72. Niehaus, W. G., Jr., and Hammerstedt, R. H., 1976, *Biochim. Biophys. Acta* **443:**515–524.

73. Ehrlich, B. E., and Diamond, J. M., 1980, *J. Membr. Biol.* **52:**187–200.

74. Schwann, A. C., and Albers, R. W., 1979, *J. Biol. Chem.* **254:**4540–4544.

75. Krishnan, N., and Albers, R. W., 1980, *J. Neurochem.* **35:**753–755.

76. Ellory, J. C., 1977, *Membrane Transport in Red Cells* (J. C. Ellory and W. L. Lew, eds.), Academic Press, London, New York, pp. 363–381.

77. Wiley, J. S., 1977, *Membrane Transport in Red Cells*, (J. C. Ellory and W. L. Lew, eds.), Academic Press, London, New York, pp. 337–361.

78. Siminovitch, L., 1976, *Cell* **7:**1–11.

79. Fishman, M. C., 1979, *Proc. Natl. Acad. Sci. U.S.A.* **76:**4661–4663.

80. Latzkovits, L., 1978, *Dynamic Properties of Glia Cells* (E. Schoffeniels, G. Franck, L. Hertz, and D. B. Tower, eds.), Pergamon Press, New York, pp. 327–336.

81. Grisar, T., Franck, G., and Schoffeniels, E., 1978, *Dynamic Properties of Glia Cells* (E. Schoffeniels, G. Franck, L. Hertz, and D. B. Tower, eds.), Pergamon Press, New York, pp. 359–369.

82. Banerjee, S. P., Bosman, H. B., and Morgan, H. R., 1977, *Exp. Cell Res.,* **104:**111–117.

83. Scarpa, A., and Carafoli, E. (eds.), 1978, *Annals of the New York Academy of Sciences,* Volume 307, *Calcium Transport and Cell Function,* New York Academy of Sciences, New York.

84. Sulakhe, P. V., and St. Louis, P. J., 1980, *Prog. Biophys. Mol. Biol.,* **35:**135–195.

85. Hill, B., 1976, *Annu. Rev. Physiol.,* **38:**139–152.

86. Colquhoun, D., 1978, *Cell Membrane Receptors for Drugs and Hormones: A Multidisciplinary Approach* (R. W. Straub and L. Bolis, eds.), Raven Press, New York, pp. 31–46.

87. Ritchie, J. M., 1978, *Cell Membrane Receptors for Drugs and Hormones: A Multidisciplinary Approach* (R. W. Straub and L. Bolis, eds.), Raven Press, New York, pp. 227–242.

88. Rash, J. E., Hudson, C. S., and Ellisman, M. H., 1978, *Cell Membrane Receptors for Drugs and Hormones: A Multidisciplinary Approach* (R. W. Straub and L. Bolis, eds.), Raven Press, New York, pp. 47–67.

89. Kallai-Sanfacon, M., and Reed, J. K., 1980, *J. Membr. Biol.* **54:**173–181.

90. Teissie, J., and Tsong, T. Y., 1980, *J. Membr. Biol.* **55:**133–140.

91. Wenner, C., and Hackney, J., 1976, *Arch. Biochem. Biophys.* **176:**37–42.

92. Meech, R. W., 1976, *Calcium in Biological Systems* (C. J. Duncan, ed.), Cambridge University Press, Cambridge, pp. 161–191.

93. Latzkovits, L., Rimanóczy, Á., Juhász, A., Torday, C., and Sensenbrenner, M., 1981, *Dev. Neurosci.* (in press).

94. Wright, E. M., 1978. *Membrane Transport in Biology*, Volume 3 (G. Giebisch, ed.), Springer-Verlag, Berlin, Heidelberg, New York, pp. 355–377.

95. Llinás, R., 1979, *The Neurosciences: Fourth Study Program*, (F. O. Schmitt and F. G. Worden, eds.), MIT Press, Cambridge, Massachusetts, pp. 555–571.

96. Lux, H. D., and Neher, E., 1973, *Exp. Brain Res.* **17:**190–205.

97. Varon, S. S., and Somjen, G. G., 1979, *Neurosci. Res. Prog. Bull.* **17:**1–239.

98. Nicholson, C., Phillips, J. M., and Gardner-Medwin, A. R., 1979, *Brain Res.* **169:**580–584.

99. Kimelberg, H. K., 1974, *J. Neurochem.* **22:**971–976.

100. Kürzinger, K., Stadtkus, C., and Hamprecht, B., 1980, *Eur. J. Biochem.* **103**:597–611.
101. Cone, C. D., and Cone, C. M., 1976, *Science* **192**:155–157.
102. Lodin, Z., Hartman, J., Kage, M. P., Korinková, P., and Booher, J., 1971, *Neurobiology* **1**:69–85.
103. Nathanson, J. A., 1977, *Physiol. Rev.* **57**:157–256.
104. Jenkinson, D. H., Haylett, D. G., and Koller, K., 1978, *Cell Membrane Receptors for Drugs and Hormones: A Multidisciplinary Approach* (R. W. Straub and L. Bolis, eds.), Raven Press, New York, pp. 89–105.
105. Darin de Lorenzo, A. J., Brzin, M., and Dettbarn, W. D., 1968, *J. Ultrastruct. Res.* **24**:367–384.
106. Schoffeniels, E., Franck, G., Hertz, L., and Tower, D. B. (eds.), 1978, *Dynamic Properties of Glia Cells*, Pergamon Press, Oxford, New York.
107. Kukes, G., De Vellis, J., and Elul, R., 1976, *Brain Res.* **104**:93–105.
108. Hertz, L., 1978, *Brain Res.* **145**:202–208.
109. Lazarewicz, J. W., Kanje, M., Sellström, A., and Hamberger, A., 1977, *J. Neurochem.* **29**:495–502.
110. Szentistványi, I., Janka, Z., Joó, F., Rimanóczy, Á., Juhász, A., and Latzkovits, L., 1979, *Neurosci. Lett.* **13**:157–161.
111. Szentistványi, I., Janka, Z., Rimanóczy, Á., Latzkovits, L., and Juhász, A., 1979, *Cell. Mol. Biol.* **25**:315–321.

2

Anion Transport in the Nervous System

H. K. Kimelberg and R. S. Bourke

1. INTRODUCTION

In the biosphere, chloride is the major extracellular anion, balancing the predominant cation sodium as well as other positive ions. Hence, Cl^- ions are most likely to accompany net movements of cations to preserve electrical neutrality and will therefore have a major role in regulation of cell volume. Chloride is also likely to be the major anion involved in carrying current during potential changes in excitable tissues. However, the electrical driving force and the permeability of an ion are as important as the amount of the ion present in determining the magnitude of its contribution to the total current flow. Also, such changes need not involve significant net movements of ions. Conversely, cell membrane potentials will determine how an ion is distributed between the cell and the extracellular fluid if the ion is both reasonably permeant and there is no active transport of the ion. Most cells in the CNS have large, inside-negative membrane potentials, and for an anion, this will result in low intracellular concentrations. Table I, taken from Tower,[1] shows the total content of the major anions and cations of mammalian brain. The aim of this chapter will be to review what is currently known about how the major inorganic anions, Cl^- and HCO_3^-, are distributed among the different compartments of the nervous system, how they are transported among these different cellular compartments, and how this distribution and transport affect cell function.

Historically, a rather extreme view of the compartmentation of Cl^- in animal tissues was taken: Cl^- was considered to be located only extracellularly in animal tissues.[2] This would require the presence of other small anions or anionic macromolecules inside cells to neutralize the positive charges of intracellular K^+ and other cations. Boyle and Conway's work in frog skeletal muscle[3] showed that Cl^- could be considered to be freely permeable and

H. K. Kimelberg • Division of Neurosurgery, and Department of Biochemistry and Anatomy, Albany Medical College, Albany, New York 12208.   *R. S. Bourke* • Division of Neurosurgery, Albany Medical College, Albany, New York 12208.

Table I
Major Inorganic Constituents of Cat Brain[a]

Type of sample	μmol/g fresh tissue					
	Na	K	Ca	Mg	Cl	HCO₃
Cerebral cortex	57	95	2.2	5.1	45	
Corpus callosum	49	73	2.2	5.1	36	12
CSF	141	3	1.2	1.2	130	21

[a] From ref. 1.

distributed according to a Donnan relationship with K^+; Na^+ was thought to be excluded from the cell interior by its being relatively impermeant. When Cl^- is in equilibrium with the membrane potential, it can initially modify rapid changes in transmembrane potentials caused by increased conductance to other ions, such as Na^+, that are out of equilibrium, as it seems to do in muscle.[4] However, these changes are transient, and only a small amount of Cl^- will move. It will then passively reequilibrate according to the Donnan potential established with K^+. This leaves unanswered the question of how Na^+, which will progressively accumulate inside the cell, ultimately gets back out again. According to current views, the membrane is not impermeable to Na^+ but is made effectively impermeable by the outward pumping of Na^+ by the $(Na^+ + K^+)$ pump to compensate for a small inward leak.[5]

The view of the transport of Cl^- and other anions in animal cells briefly summarized above has undergone considerable revision in the past decade. In nonnervous tissues such as erythrocytes and epithelia, mediated transport processes for Cl^- have been described.[6-8] These processes usually result in active transport of Cl^- in epithelial cells.[7-9] Active transport in this review will be considered to have occurred for any ion that is not in electrochemical equilibrium. Such a definition does not specify a particular mechanism and could be carried out by a pump directly using metabolic energy in the form of ATP primary active transport or a hetero-ion-exchange or co-transport process transfering the energy present in the gradient of one of the ions to the other transported ion (secondary active transport). In nervous tissue, investigators are only now beginning to apply concepts of anion transport from studies in other tissues. One focus of this chapter therefore, will be to relate the transport of anions into and within the brain, where appropriate, to current views on mediated and active transport of anions as derived from work in nonnervous tissues. The consequences of such transport processes in the CNS are likely to be significant since, for example, the Cl^- anion exchange process also transports HCO_3^- and OH^{-}[6-8] and thus could be involved in pH control in the CNS.

The results of earlier work on cation and anion compartmentation and transport in the CNS have been fully discussed in two major treatises by Van Harreveld[2] and Katzman and Pappius[10] and in chapters in the first edition of this series.[1,11,12] Van Harreveld,[2] in the preface to his 1966 monograph, noted that "no complete review of the literature is claimed." The usual geometric growth of knowledge makes a complete review, even for the relatively more limited field of anion transport, very difficult and inappropriate to the aim of this series. The focus of this chapter on inorganic anion transport in the nervous system, primarily involving Cl^- and HCO_3^-, will preclude discussion of other important topics involving anion transport such as the uptake and efflux of organic anions across the blood–brain barrier and the transport of organic anions, such as the putative transmitter glutamate, within the CNS. These topics will be discussed more appropriately in other chapters of this series.

2. ANION TRANSPORT IN NON-NERVOUS-SYSTEM TISSUES

To avoid discussing anion transport in the nervous system in a purely descriptive manner, it is necessary to provide some theoretical framework. The basic principles of anion transport have emerged primarily from studies in non-nervous-system tissues and cells. We shall, therefore, first briefly discuss diffusion and mediated mechanisms for anion transport in the major non-nervous tissues in which they have been studied.

2.1. Muscle

Much of the work on the electrical correlates of anion transport have focused on the predominant biological anion, Cl^-. Early work had assumed that, in a number of tissues, Cl^- was only extracellular, and its content in tissues was therefore used to estimate the extracellular space. The seminal study by Boyle and Conway[3] established that, at least in frog skeletal muscle, Cl^- could be viewed as being passively distributed according to the principles of the Donnan distribution,[2,5] in which the product of extracellular Cl^- and the permeant cation K^+ equal the product of their internal concentrations:

$$[K^+]_o \times [Cl^-]_o = [K^+]_i \times [Cl^-]_i \qquad [1]$$

In living cells, a large part of the intracellular anionic charge is considered to reside on impermeant macromolecules, the "stuff of life." Thus, the intracellular Cl^- concentration is low, whereas the intracellular K^+ concentration is high. The osmotic imbalance that results from this intracellular distribution of permeant and impermeant ions is offset by the high extracellular Na^+ concentration.[2,5] Sodium is rendered effectively impermeable by a small inward leak being offset by outward pumping of Na^+ on the ($Na^+ + K^+$) pump. This arrangement has been termed the "double Donnan" equilibrium.[5]

Equation 1 implies that both K^+ and Cl^- are in equilibrium with the membrane potential; i.e.,

$$E_m = (RT/nF) \ln([K^+]_o/[K^+]_i) = (RT/nF) \ln([Cl^-]_i/[Cl^-]_o) \qquad [2]$$

Since both $[K^+]_o$ and $[Cl^-]_i$ are low, the membrane potential inside the cell based on this relationship would be quite negative. Thus, if one has independently measured the membrane potential and the intracellular concentrations of K^+ or Cl^-, then it can be seen whether these ions are in electrochemical equilibrium. Under these conditions, the passive fluxes of each of these ions along its electrochemical gradient result in equal and opposite fluxes so that there is no net current or ion flux. The system is therefore in equilibrium state without any requirement for active transport processes. For frog skeletal muscle, Hodgkin and Horowicz[13] found that the distribution of K^+ and Cl^- fitted the above relationships reasonably well and that the permeabilities of K^+ and Cl^- were 1–2 and 4×10^{-6} cm/sec, respectively. Recent measurements in amphibian skeletal and cardiac ventricular muscle have given values of 1.4 to 6 mM for intracellular Cl^- concentrations,[14] consistent with passive transport of Cl^-. In other muscle tissue, Cl^- distribution is different. In mammalian cardiac muscle, $[Cl^-]_i$ is three to four times higher than expected from passive distribution.[15] Also, in this tissue, $[Cl^-]_i$ was lower in HCO_3^--free medium, and the increased $[Cl^-]_i$ found when HCO_3^- was added was inhibited by the anion exchange inhibitor 4-acetamido-4′-isothiocyanostilbene-2,2′-disulfonic acid (SITS).[15] These data (see Sections 2.2 and 2.3) suggest that the anion exchange system is capable of producing active accumulation of Cl^-. It should be noted that these processes are not limited to mammalian muscle; thus, in invertebrate (barnacle) muscle, Cl^- transport is also inhibited by SITS[16] as well as by furosemide.[17]

2.2. Erythrocytes

There have been extensive studies of anion transport in erythrocytes over the past decade. References to the original literature can be found in such recent reviews as those by Cabantchik *et al.*,[6] Gunn,[18] and articles in reference 8. Such studies have found that the rapid $Cl^- \leftrightarrow HCO_3^-$ exchange of red blood cells is catalyzed by a specific protein. This protein has been purified, and its structural arrangement in the membrane is becoming clarified.[6,8,18] The kinetics of this transport protein have also been studied in great detail for Cl^-, and to a lesser degree for HCO_3^-. This system shows saturation kinetics for both Cl^- [18] and HCO_3^-.[19] The carrier seems relatively nonspecific and carries anions other than Cl^- and HCO_3^-, the rapid transport of which is presumably its normal physiological function. Such anions include other halides, nitrate, and the divalent anions sulfate and phosphate, although their transport is much slower than Cl^- or HCO_3^-.[18] This system is inhibited by a variety of anionic compounds of which the most widely used are stilbene derivatives such as SITS (see Section 2.1) and a number of diuretics such as ethacrynic acid and furosemide. As pointed out by Cabantchik *et al.*,[6] the operation of such a hetero-ion-exchange system will tend to lead to equivalence of the intracellular/extracellular ratios of all the anions transported on this system.

$$r = [Cl^-]_i/[Cl^-]_o = [HCO_3^-]_i/[HCO_3^-]_o$$

$$= [SO_4^{2-}]_i/[SO_4^{2-}]_o = [OH^-]_i/[OH^-]_o \quad [3]$$

In red blood cells, these ratios are generally found to be equivalent. Also, the membrane potential is equivalent to the anion equilibrium potential. However, this is not because of the exchange system which carries no current and therefore will be unaffected by the membrane potential, but is because the free chloride permeability, although low, is still about 100 times larger than the cation permeabilities.[6]

In addition to the exchange system, a coupled KCl efflux, which is SITS insensitive and partly furosemide sensitive, has been described in red blood cells (e.g., see review, ref. 20). A further consequence of the anion exchange system transporting HCO_3^- and OH^- is that it can be involved in intracellular pH control.

2.3. Epithelia

In contrast to the apparent diffusional transport of Cl^- in frog skeletal muscle, and passive, mediated transport of Cl^- in erythrocytes, Cl^- transport in epithelia seems to be active in most cases. Intracellular Cl^- concentrations that are severalfold higher than those predicted from the membrane potentials have been found for a variety of epithelial cells.[7-9] This is in keeping with the functions of epithelia in absorbing or secreting NaCl to maintain ion and osmotic balance within the organism. Since control of brain osmolarity and secretion of NaCl into the CNS in CSF formation are major roles of the blood–brain barrier (see Section 3), parallels between epithelial function and the CNS in this regard should be expected. There is a vast literature on epithelial transport, and we shall simply summarize some of the main principles that seem to underlie epithelial chloride transport (see refs. 7 and 9 for recent brief reviews and a recent symposium proceedings[8] for further details and references).

In general, the role of Na^+-dependent uptake of Cl^- has been emphasized in which a linked neutral NaCl uptake mechanism has been proposed, especially for leaky epithelia.[7] It is envisaged that this system utilizes the free energy of the inward-directed Na^+ gradient to drive uphill transport of Cl^- (but see ref. 21 for an alternative viewpoint). In absorptive epithelia, the NaCl uptake site is visualized as being located on the luminal face. This process leads to active intracellular accumulation of Cl^- and is often furosemide sensitive. For absorbing epithelia, the Na^+ is pumped out of the basolateral membrane by the $(Na^+ + K^+)$ pump, and Cl^- can exit by moving down its electrochemical gradient. In secreting epithelia, NaCl uptake occurs at the basolateral or serosal face. Sodium is again pumped out at the basolateral face by the $(Na^+ + K^+)$ pump, and Cl^- passes down its electrochemical gradient at the luminal or mucosal surface.[7] If the Cl^- conductance of the mucosal membrane predominates the mucosal surface will be negative relative to the basolateral or serosal surface, and Na^+ could accompany Cl^- by paracellular movement in response to this potential difference across the epithelium.[22] Other work has identified an involvement for HCO_3^-/CO_2 buffer[23] and has shown inhibition by SITS[24] of transport of NaCl in epithelia. This also suggests a role for anion exchange transport in movements of NaCl across epithelia (see also papers in ref. 8).

3. TRANSPORT OF ANIONS INTO AND OUT OF THE CENTRAL NERVOUS SYSTEM

3.1. The Blood–Brain Barrier

An important feature of the transport of substances in and out of the CNS is that, unlike other tissues, there is a selective barrier to small molecules present in the blood. This has been termed the blood–brain barrier and has been a subject for intensive study since its first clear demonstration by Goldman in 1913 (see ref. 25, p. 38). There have been numerous books and reviews on this subject, and some of the most recent are those by Bradbury,[25] Davson,[26] and Chapter 4 in the text by Katzman and Pappius.[10] The reviews by Rapoport,[27] Oldendorf,[28] and Pardridge et al.[29] emphasize the role of the blood–brain barrier in the entry of drugs into the CNS.

A number of small-molecular-weight substances such as sucrose and EDTA are excluded from the brain, whereas water enters the CNS from the blood more slowly than from capillaries into other tissues. Thus, although water can cross the blood–brain barrier very fast, it is not flow-limited, and a finite permeability coefficient of $1-4 \times 10^{-4}$ cm/sec can be measured.[25,27] As Bradbury[25] has pointed out, this gives the blood–brain barrier the characteristics of a high-resistance epithelium. The polar substances that do enter the brain must therefore have specific transport systems. This selectivity allows the exclusion from the CNS of certain substances, permits others to pass in at varying and controllable rates (facilitated diffusion), and, when linked to an energy source, results in accumulation of substances (active transport). Ions are known to pass into the CNS relatively slowly compared to what might be expected from free diffusion in water, and thus, the blood–brain barrier forms what one might term an "ion curtain" around the brain.

Brightman and Reese[30] have shown the presence of tight junctions or zonula occludens between the endothelial cells which do not permit the passage into the CNS, after intravenous injection, of peroxidases with molecular weights as low as 2000 or colloidal lanthanum hydroxide. In contrast, after intraventricular injection of peroxidase, reaction product could be seen filling the extracellular space between the perivascular astrocytic endfeet and the perivascular space up to the region of the tight junctions between endothelial cells. It is inferred from these results that the tight junctions between endothelial cells constitute the structural basis of the blood–brain barrier, and that there is relatively free passage of even high-molecular-weight substances between the astroglial endfeet. Thus, the membranes of the endothelial cells facing the capillary lumen must be the sites of specific transport systems. In addition, tight junctions between the epithelial cells of the choroid plexus constitute the anatomical locus of the blood–CSF barrier.[25] Thus, brain extracellular and cerebrospinal fluids have separate barriers to the blood at the capillaries and choroid plexus, respectively. However, because of the ready exchange between these two fluids, there is clearly the possibility of processes at these two sites influencing each other.[31] In addition, although the view of the perivascular astroglial foot processes as the structural sites of the blood–brain barrier is no longer tenable, transport in these processes may act in series with transport in the endothelial cells of the capillaries to augment or modify passage of substances into and out of the brain.

3.2. Anion Transport and the Secretion of Cerebrospinal Fluid

Uptake of Cl^- from blood into brain and CSF has long been known to be very slow.[25] Recent studies on the kinetics of uptake of Cl^- from the vasculature in rats showed two components for both the cerebral cortex and the choroid plexus.[32,33] The slow components were thought to represent movement into the epithelial cell compartment for the choroid plexus and into the brain extracellular and cellular compartments for cerebral cortex. The $t_{1/2}$ values for this component ranged from 1 to 2 h. For CSF uptake, sampled from the cisterna magna, a fast component of $t_{1/2} = 0.18$ h and a slow component of $t_{1/2} = 1.2$ h were found. The former was thought to represent direct uptake across the blood/CSF barrier, and the latter isotope penetration from brain extracellular fluid into CSF.[33]

Cerebrospinal fluid production seems to involve the active transport of both Na^+ and Cl^-.[26,32,34–36] Treatment with both ouabain[34,35,37] and acetazolamide[35,38] reduces the influx of Na^+ and Cl^- from plasma into CSF, suggesting the involvement of both (Na + K) ATPase and carbonic anhydrase, although the effects of the two inhibitors are not additive.[34] It is of interest that carbonic anhydrase inhibitors reduce uptake of both Na^+ and Cl^-,[38] suggesting some link between the transport of these two ions. Although it is not possible to have active transport of a substance without a specific saturable transport mechanism, the converse is not true, since facilitated diffusion by definition does not actively accumulate a substance. It has, however, proved difficult to directly show saturation kinetics for the transfer of Cl^- from blood to CSF. One attempt to do this was in the study of Bourke *et al.*[39] who found a constant rate of uptake of ^{36}Cl from blood into CSF between plasma Cl^- concentrations of 60 to 120 mM suggesting a transport mechanism that saturated at 60mM plasma Cl^- or less. This study also provided evidence for active transport of Cl^- based on maintenance of CSF Cl^- levels when plasma Cl^- was reduced. One possibility for such active transport is that Cl^- follows active, electrogenic transport of Na^+ into the CNS. The magnitude of the Na^+-dependent potential would then become increasingly positive on the CSF side as plasma Cl^- decreased, resulting in a constant rate of Cl^- transport and maintenance of CSF Cl^- levels. That this was not so was shown by Abbott *et al.*[40] who found that reduction of plasma Cl^- had no effect on CSF Cl^- content and no consistent effect on the blood/CSF electrical potential. Indeed, in some cases, when the ratio of $[Cl^-]_{CSF}/[Cl^-]_{plasma}$ increased, the blood/CSF potential changed sign, and the CSF became negative. Based on Ussing's equation, the ratio of the passive fluxes of an ion across the blood–brain barrier with a potential difference of E (blood relative to CSF) is given by the following equation for a monovalent negative ion.[41]

$$J^{Cl^-}_{blood \rightarrow CSF} / J^{Cl^-}_{CSF \rightarrow blood} = ([Cl^-]_{blood}/[Cl^-]_{CSF})e^{-(EF/RT)} \qquad [4]$$

Since, in the steady state, the flux of Cl^- into CSF must equal the flux out,

$$[Cl^-]_{CSF}/[Cl^-]_{blood} = e^{-(EF/RT)} \qquad [5]$$

where R, T, and F have their usual meanings. Thus, for an E of -0.005 V, the ratio is 1.21. The fact that both at normal plasma Cl^- concentrations and pH values, ratios of 1.20 have been consistently found[36,37,39,40] is purely fortuitous in view of the fact that ratios of up to 2.13 were found when E was $+0.003$ V.[40] Thus, chloride does not appear to be transported passively between blood and CSF. Of course, at least one caveat to this treatment is that the origin of the blood/CSF potential measured in the ventricles or cisterna relative to blood and the relationship of this potential to the sites of CSF ion secretion, especially the extrachoroidal sites,[33,42] are unknown. Bradbury,[25] however, has argued that it must be mainly across the tight junction sites of the blood–brain barrier.

Since alterations in plasma Cl^- had no effect on plasma/CSF potential and vice versa, it is possible that transport of Cl^- into the CNS is by electrically neutral processes that could involve exchange for another anion or cotransport with Na^+. Such processes would be unaffected by changes in the plasma/CSF potential. There are now a number of examples of neutral active transport of Cl^- in epithelia involving either NaCl cotransport or anion exchange, usually for HCO_3, as discussed in Section 2.3. This leads to active accumulation of Cl^- inside the epithelia by one side of the cell. Chloride can then flow passively out of the other side down its electrochemical gradient or by electrically neutral exchange processes.[8] This model requires active transport but no free permeability to Cl^- on one side of the cell and free permeability to Cl^- or passive mediated exchange processes on the other side from which Cl^- is being secreted. It will result in intracellular Cl^- levels that are greater than expected from passive equilibration. Recent work has shown chloride levels in the choroid plexus of 67 mM which is 3.9 times greater than that expected from passive distribution, consistent with such a model being applicable to the choroid plexus epithelium.[32]

Ion transport processes that are presumably involved in CSF production have been investigated in an isolated sheet of bullfrog choroidal epithelium in an Ussing-type chamber by Wright.[43] On the basis of these studies, he proposed a model for ion transport across the epithelium which is shown in Fig. 1. The epithelium is viewed as being of a low-resistance type with major passive permeation pathways across the tight junctions. The driving force for net secretion is an electrogenic ($Na^+ + K^+$) pump located on the ventricular surface. The absence of any substantial potential under normal conditions is considered to be because of the low-resistance tight junctions and the relatively high permeability of the ventricular membrane to HCO_3^- which accompanies at least 50% of the Na^+ transported. In addition, the ventricular membrane is permeable to Cl^- which also accompanies part of the Na^+. The K^+ permeability is relatively small. Diffusion of CO_2 or H_2CO_3 from the serosal surface into the cell generates HCO_3^- and H^+ intracellularly, and H^+ exchanges for Na^+ at the serosal surface. Ouabain inhibits Na^+ secretion, but acetazolamide

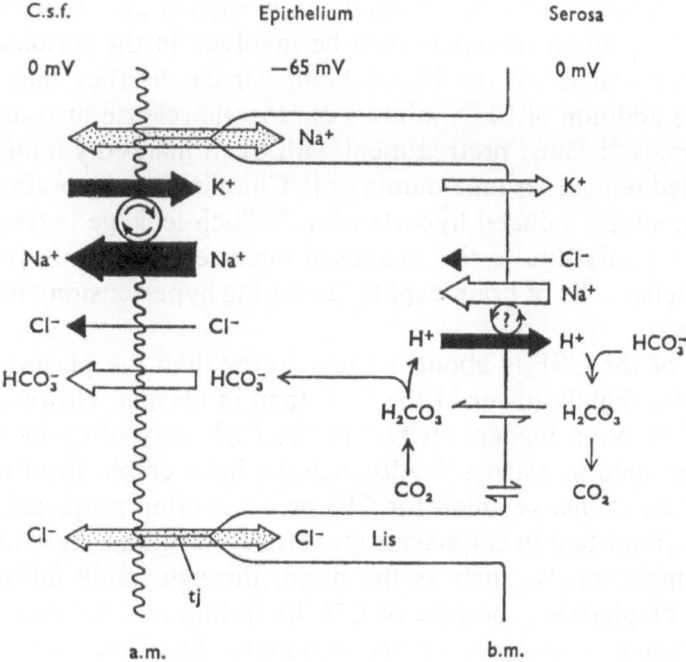

Fig. 1. Scheme for ion transport across bullfrog choroid plexus epithelium. Reprinted by permission of the publisher from E. M. Wright, Mechanisms of ion transport across the choroid plexus, *J. Physiol. (Lond.)* **226**:564. Copyright 1972 by Cambridge University Press.

is without effect. Therefore, it is considered that permeation of H_2CO_3 and/or noncatalyzed hydration of CO_2 is rapid enough to account for Na^+ secretion.

It is not clear how well the properties of this isolated frog choroidal epithelium correspond to its properties *in situ* or to the processes occurring in mammalian choroid plexus. The behavior of this system differs from that of the mammalian system in two ways. One is that acetazolamide inhibits CSF secretion in mammalian systems *in situ*,[35,38] and recent work has suggested that intracellular Cl^- in the choroidal epithelium is severalfold higher than expected from equilibration with the intracellular membrane potentials.[32] The addition of a $HCO_3^- \leftrightarrow Cl^-$ exchange system to the serosal surface in the model shown in Fig. 1 could result in active accumulation of Cl^- and also a partial exchange of HCO_3^- for Cl^-. An inhibitory effect of acetazolamide suggests that the rates of transport are sufficiently high that catalyzed hydration of CO_2 is required. Clearly, a number of schemes can be proposed, but any scheme has to accommodate such findings as the approximately +5 mV CSF/ blood potential having little effect on CSF Cl^- levels,[40] active accumulation of Cl^- within the choroidal epithelium,[32] and the probable location of the $(Na^+ + K^+)$ ATPase on the apical surface of the choroid plexus.[43,44] These processes relate only to CSF production at the choroid plexus. It is even less clear what transport processes occur at the other presumed major site of CSF production, the endothelial–astrocyte complex of the brain capillaries.[25,27] A speculative scheme for the transport processes occurring at this site incorporating some of the features discussed in relation to Fig. 1 is shown later in Fig. 6.

The results of a recent study raise the intriguing possibility that osmotic forces caused by anion transport may be involved in the formation of endo- and exocytotic vesicles at the blood–brain barrier, as they may be in other tissues where addition of SITS inhibits exocytotic release of transmitters and other substances.[45] Thus, pretreatment with SITS markedly inhibited leakage of [^{125}I]-labeled human serum albumin or [^{14}C]inulin into the brain parenchyma after experimentally induced hypertension.[46] Such leakage is transient and is thought to be partly due to the increased occurrence of pinocytotic vesicles in the endothelial cells of brain capillaries during hypertension and other pathological states.

The pH of the CSF is about 0.1 unit lower than the plasma pH, and so HCO_3^- may be slightly lower in the CSF than in plasma. However, since the P_{CO_2} of CSF is often higher, HCO_3^- in the CSF may often be the same or slightly higher than in plasma.[10] Although we have emphasized the transport of HCO_3^- as an exchange anion for Cl^- or, as is often proposed, as a co-ion for Na^+, it is important to emphasize that HCO_3 can gain access to membrane-delimited compartments, such as the brain, through freely diffusible CO_2.[47] Maren[38] has emphasized the role of CO_2 hydration in CSF formation, and it seems that about two-thirds of the estimated basal rate of formation of HCO_3^-, 0.53 mM/h in the cat, is catalyzed by carbonic anhydrase, since this amount is sensitive to carbonic anhydrase inhibitors. This amount of HCO_3^- corresponds to about 37% of the Na^+ entering into CSF, the remaining Na^+ probably being accompanied by Cl^- as suggested in the model proposed by Wright.[43] It is interesting that in the presence of a carbonic anhydrase inhibitor the uptake of all three ions was inhibited and that the relative proportions of the three ions entering the CSF remained the same.[38] Possibly, the hydration of CO_2 contributes more to CSF production than the net rate of HCO_3^- formation in the CSF suggests. Thus, as suggested above, part of the HCO_3^- produced could stimulate Cl^- entry to the choroidal epithelium by $Cl^- \leftrightarrow HCO_3^-$ exchange while the H^+ produced promotes $Na^+ \leftrightarrow H^+$ exchange, both occurring at the serosal surface. This would require hydration by the reaction $CO_2 + H_2O \leftrightarrow HCO_3^- + H^+$, which occurs in an environment of pH 6.5 to 7.5[48] rather than by the reaction $CO_2 + OH^- \rightarrow HCO_3^-$ which would predominate at pH > 10 and is favored by Maren[38] on kinetic grounds.

It should be noted that another important function of the choroid plexus epithelium is the energy-dependent transport of a variety of inorganic and organic anions from CSF to blood.[25,27] The *in vitro* bullfrog choroid plexus preparation shows some selectivity,[49] with the relative affinities for a number of anions being as follows:

$$ClO_4^- > ReO_4^- > BF_4^- > SCN^- > I^- > NO_3^- > Br^- > Cl^-$$

These relative affinities are not related either to their net transport or to the considerable accumulation in some cases of the anions within the plexus epithelium. Anion transport from CSF to blood is closely related to the ($Na^+ + K^+$) pump since such anion transport is inhibited by ouabain, the absence of K^+ from the ventricular surface, or the absence of Na^+ from either the ventricular or serosal surface. Active uptake of the anions could be by cotransport with Na^+ as has been shown to occur for a variety of substances in epithelia.[41,50] Since the direction of active anion transport is from the ventricular to the serosal surface, and the $(Na^+ + K^+)$ pump seems to be located at the ventricular surface,[43,44] then Na^+ will have to be pumped out across the same side as it came in. Outward pumping of Na^+ will also involve uptake of K^+, and therefore, exit of the anions at the serosal surface is likely to be with K^+, keeping K^+ in the CSF low as it is known to be.[37]

4. TRANSPORT OF ANIONS IN THE NERVOUS SYSTEM

4.1. Compartmentation of Anions

The major focus of studies on inorganic anion transport and compartmentation within the neuropil has been Cl^-. The values obtained for total Cl^- content of adult CNS tissue range from 27 to 47 μmol/g wet weight in a variety of species[1,2,51] (also see Table I). However, there has been considerable controversy regarding the distribution of Cl^- between the extracellular and intracellular compartments as well as the intracellular concentrations of Cl^- in different cell types. The history of this controversy is discussed in detail by Van Harreveld,[2] and the brief overview given here will essentially summarize this account.

Originally, the extracellular space in the CNS had been equated with the Cl^- space, which ranges from 25 to 35%, even when determined most precisely by being referred to the chloride concentration in the CSF.[52] Such a large extracellular space had been considered to be inconsistent with the apparent absence of space between the cells as seen in the first electron micrographs of the CNS.[2] The conflict between these two apparent experimental facts was compounded by the difficulty of interpreting electrophysiological data from either viewpoint. Such data indicated that, at least for the mammalian motor neuron, Cl^- was in equilibrium with a membrane potential of -70 mV,[53] which at an outside chloride concentration of 125 mM would mean that internal Cl^- should be 9 mM. The average total Cl^- content is around 40 μmol/g wet weight or approximately 40 mM, and it has been estimated that neurons constitute 35–50% of the total tissue volume.[54,55] An average intracellular concentration of 9 mM Cl^- in neurons would require that a considerable amount of Cl^- be located in a nonneuronal intracellular compartment in the absence of an appreciable extracellular space. It was suggested that this compartment could be the oligodendroglia,[56] since mammalian astroglia were known to have high negative potentials of up to -70 mV,[57] and thus a high $[Cl^-]_i$ in astrocytes would require active transport.

These conceptual problems were solved, in part, by the realization that the lack of an extracellular space as seen in electron micrographs of the brain was in fact caused by fixation artifacts whose dominant effect seemed to be swelling of the astroglia.[2] It is now clear that there is indeed a considerable extracellular space in mammalian brain which is highly species variable,[52] as shown in Table II for both experimentally determined and predicted values. In the rat, a value of 13.4% was recently determined for cerebral cortex[33] based on the uptake of ^{36}Cl by cerebral cortex as a function of the uptake of ^{36}Cl by the CSF after intraperitoneal injection. However, as will be discussed below, recent work has also suggested that Cl^- is actively accumulated in astroglial cells, in partial support of at least the Cl^- aspect of the original proposal that glia were high-NaCl cells.[56]

Presently, there seems to be general agreement concerning the relative distribution of Cl^- between the extracellular and intracellular space of the brain. However, the distribution of Cl^- among the different major cell types of the brain, neurons, astrocytes, and oligodendrocytes, remains very much a question for present and future research. There is even less information regarding Cl^- content and transport in brain endothelial cells, choroidal epithelium, and ependymal cells. As we look further into the complexities of CNS organization at the cellular level, we would certainly like to know whether there are differences in the distribution of Cl^- and other anions in different subclasses of these cells from different regions of the CNS, and even differences within different regions among the cells themselves. Although there has been some work on anion transport in different cell types based on studies of the whole tissue, further studies in this area will undoubtedly also emphasize the properties of isolated cell types.

A drawback to the cell biological approach is that one loses the ability to study the consequences of interactions within the nervous system on ion distribution and thus the perspective of the nervous system as an integrated tissue. This makes *in vivo* studies, although difficult to perform and interpret, of irreplaceable value. An early attempt to localize the sites of uptake of Cl^- in whole tissue was that of Van Harreveld[2] who used a AgCl precipitation method and light and electron microscopy to study uptake in normal brain tissue and brain tissue from animals after circulatory arrest. An increased density of

Table II
Calculated and Extrapolated Values for Chloride, Thiocyanate, Sucrose, and Inulin Spaces of Cerebral Cortex in Vivo[a]

Species	Species brain wt. (g)	Chloride content (µeq/g)	Chloride or SCN space (%)	Sucrose or inulin space (%)
Mouse	0.4	26.5	[17.0]	[8.5]
Rat	1.6	32.4	[23.0]	[14.5]
Guinea pig	4.75	36.9	27.5	19.4
Rabbit	9.3	39.7	30.1	22.5
Cat	31	44.7	34.9	28.0
Monkey	88.5	49.1	39.1	32.8
Sheep	109	[50.0]	[39.9]	33.8
Chimpanzee	380	[55.2]	[44.8]	39.5
Beef	380	55.2	[44.8]	[39.5]
Man	1320	[60.4]	[49.8]	[45.2]
Whale	6785	[67.3]	[56.3]	[52.7]

[a] Extrapolated values are given in brackets. From ref. 52.

staining of what was identified as apical dendrites in the cortex was observed, indicating increased uptake of Cl into these structures. In the cerebellum, increased staining of the Bergman glia which are glial fibrillary acidic protein positive and therefore astroglial,[58] in addition to staining of Purkinje cell processes, was also seen. Circulatory arrest is associated with large increases in cortical impedance which are considered to indicate a considerable decrease in extracellular space.[2] This supports the idea of increased Cl^- and cation uptake by cellular elements after circulatory arrest; these elements would then swell and thereby decrease the extracellular space. Van Harreveld's quantitative studies showed a 60% increase in the mean diameter of Bergmann glial fibers.[2] Recent studies using electron microscopy have shown that after ischemia there is an early, selective swelling of astroglia, especially the perivascular foot processes.[59,60] At later times there is also a swelling of neuronal elements.[60]

More recent studies have benefited from the use of ion-sensitive microelectrodes (ISMs) which can provide a continuous readout of extracellular ion levels in the exposed brain *in situ*. It has been shown that after anoxia[61] and spreading depression[62,63] in the rat and catfish there is a marked decrease in extracellular Cl^-, Na^+, and Ca^{2+} and an increase in K^+ (see Fig. 2). Decreased extracellular Na^+ and increased K^+ have also been reported by Hosmann *et al.*,[64] also using ISMs, after ischemia in the cat. These data suggest that under such conditions there is uptake of NaCl into a cell compartment with the subsequent swelling reducing the extracellular space by about 50%. Phillips

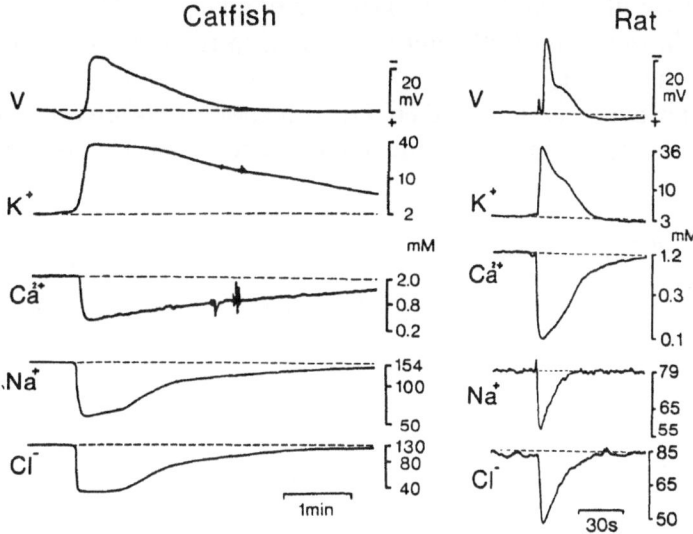

Fig. 2. Effects of spreading depression on extracellular Cl^-, Na^+, Ca^{2+}, and K^+ levels and slow potentials (*V*) of catfish and rat cerebellum. Ion levels determined with ion-specific electrodes. Spreading depression induced either by KCl microinjection or local bipolar electrode stimulation. Initial levels of NaCl were reduced in rat by superfusion with NaCl-depleted solutions. Note time course of spreading depression is much faster in the warm-blooded rat than in the cold-blooded catfish. Reprinted by permission of the publisher from C. Nicholson, Dynamics of the brain cell microenvironment, *Neurosci. Res. Program Bull.* **18**(2):287, MIT Press, Cambridge, Massachusetts. Copyright 1980 by The Massachusetts Institute of Technology.

and Nicolson[62] used trace anions in rat cerebellum to probe the size of the anion channel opened during induced spreading depression. They found that the concentrations in the extracellular space of anions up to a radius of 6.35 Å decreased in the same proportion as Cl^- (radius 3.6 Å), whereas anions of 7.30 Å diameter or greater increased in concentration by a similar amount. The dimensions of the anion pore revealed by these techniques was thus 6.35–7.30 Å.

Since we have observed (see Sections 4.2, 4.4) that K^+- or transmitter-stimulated uptake of NaCl in cat brain slices was inhibited by specific inhibitors of the anion exchange system of erythrocytes (see Section 2.2) and that astrocytes in culture also possess the same anion exchange system (see Section 4.5), it is intriguing to speculate that some of the uptake of Cl^- seen in spreading depression or anoxia is into astroglial cells via erythrocyte-type anion transport pores. In support of this viewpoint, another study[65] used CO_2, Cl^-, and pH electrodes applied to the surface of the exposed cerebral cortex of an anesthetized cat to measure directly extracellular Cl^- levels and to be able to calculate extracellular HCO_3^- from the pH and CO_2 levels. Increasing the inspired CO_2 resulted in a 1:1 increase in extracellular HCO_3^- and decrease in extracellular Cl^-, as shown in Fig. 3. The rate of decrease of extracellular pH was increased when the animals were treated with SITS. These data were interpreted as indicating that the hydration of CO_2 is preferentially intracellular, because of both intracellular buffers and carbonic anhydrase, and is likely to occur in glial cells. The HCO_3^- then formed leaves the glial cell via the anion exchange system in exchange for Cl^-. Work in primary astroglial cultures (see Section 4.5) supports this by showing the existence of an anion exchange system in these cells as well as the existence of a $Na^+ \leftrightarrow H^+$ exchange system that may aid intracellular buffering by leading to removal of intracellular H^+, perhaps into the blood.

As noted previously, Cl^- and HCO_3^- transport are often related. However, neither histochemical nor electrode measurements of HCO_3^- can be made.

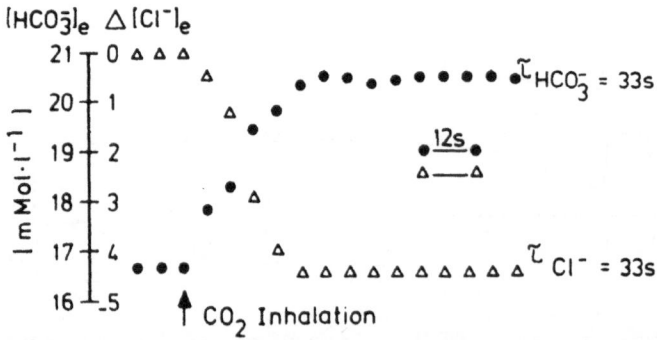

Fig. 3. Extracellular $[HCO_3^-]$ and $[Cl^-]$ recorded from exposed cerebral cortex of an anesthetized cat. Animal was artifically ventilated, and CO_2 was added to the inspired air at the time indicated; Cl^-, pH, and CO_2 were measured with surface electrodes, and HCO_3^- was calculated from the CO_2 and pH values. Reprinted by permission of the publisher from H. H. Loeschcke and H. R. Ahmad, Transients and steady state of chloride–bicarbonate relationships of brain extracellular fluid, in *Biophysics and Physiology of Carbon Dioxide* (C. Bauer, G. Gros, and H. Bartels, eds.), Springer-Verlag, Berlin, p. 443. Copyright 1980 by Springer-Verlag.

Bicarbonate concentrations are usually calculated from the Henderson–Hasselbach equation when the pH and P_{CO_2} are known,

$$pH = pK_a + \log([HCO_3^-]/P_{CO_2}S) \qquad [6]$$

where pK_a is the negative logarithm of the first dissociation constant of H_2CO_3 and S is the solubility coefficient of CO_2. Values of 6.13 and 0.0312 mmol/liter per mm Hg have been given[10] for pK_a and S, respectively, for CSF at pH 7.3 and 39°C, which should apply to the CNS as a whole. Values of lumbar CSF pH, P_{CO_2}, and HCO_3^- vary within a range of 7.28, 49.6 mm Hg, and 23.7 mM to 7.37, 42.1 mm Hg, and 23.3 mM, respectively, in man. Many mammalian species seem to fall within this range. However, the cat is anomalous in having both a low P_{CO_2} and HCO_3^- of 31.6 mm Hg and 16.4 mM, respectively.[10] However, since both values are low, this gives a normal CSF pH of 7.35. Average intracellular pH values in brain slices[66] of 6.95 at an external pH of 7.20 using the [^{14}C]dimethyloxazolidinedione method and 7.04 for brain *in vivo* using a fluorescent pH-sensitive dye[67] have been reported.

Since CO_2 is freely diffusible and thus can readily equilibrate across the cell membrane,

$$pH_o - pH_i = \log([HCO_3^-]_o/[HCO_3^-]_i) \qquad [7]$$

For a ΔpH of 0.30 and a $[HCO_3^-]_o$ of 23 mM, intracellular HCO_3^- will be 12 mM. It might be noted that the intracellular-to-extracellular ratio of 0.52 obtained is considerably larger than the value of 0.068 that would be in equilibrium with membrane potentials of -70 mV measured for motor neurons[53] and found for identified astroglia in cat brain *in situ*.[68] This requires either impermeability to HCO_3^- or a HCO_3^- pump. A H^+ pump would also result in active accumulation of HCO_3^-, since the pumping out of H^+ would lead to further hydration of the ubiquitous CO_2 and formation of a replacement H^+ and accumulation of HCO_3^-. Such a process would also require a low permeability to HCO_3^- to avoid excessive energy consumption.

The problem of a major discrepancy between the observation of a very small extracellular space based on the early electron micrographs of intact tissue and an abnormally large space based on Cl^- distribution was solved by the recognition that neither extreme position is correct and that Cl^- is an intracellular anion as well as the major extracellular anion. However, quantitative data on Cl^- distribution between intra- and extracellular space and its distribution between different cell types require further studies on defined *in vitro* or superfused *in vivo* preparations in which the ionic composition of the medium can be altered.

4.2. Anion Transport in Brain Slices

The earliest widely used *in vitro* brain tissue preparation was the brain slice which, it was found, could be maintained in a metabolically viable state and could be used for studying ion transport.[12,69,70] In addition, electrical activity is maintained,[12] and it has been recently found[71] that the membrane potentials of identified astrocytes in brain slices are around -70 mV and show a response to varying K_o^+ that is close to that predicted by the Nernst equation. Many of the earlier transport studies[72-77] focused on cation transport. Bourke and Tower[75,76] examined the effects of different ionic composition of the medium on the swelling and the associated uptake of Cl^-, K^+, and Na^+ in cat brain slices. Slices incubated in 27 mM K^+ medium buffered with 25 mM HCO_3^- showed a greater K^+ and lower Na^+ content and an unchanged Cl^- content as compared to slices in a 5 mM K^+ medium.[76] Increasing K^+ concentrations in the medium produced progressive tissue swelling which had a threshold at ~20 mM K^+ and which was prevented by replacement of Cl^- with the impermeant anion isethionate.[75] Later studies showed that the swelling found in high-K^+ solutions was associated with increased K^+, Na^+, and Cl^- contents.[78,79] These results were supported, as far as cation changes were concerned, by the work of Lund-Andersen and Hertz[80] who showed increased uptake of K^+ and Na^+ and tissue swelling with increasing K^+ concentrations in the media. These experiments were also performed in HCO_3^--buffered medium, but increased K^+ concentrations were obtained by adding KCl without reducing the Na^+ concentration, thus making the solutions hyperosmotic. If cells follow the Donnan relationship, this should result in their gaining K^+ and Cl^- equivalent to the increased amount in the extracellular solution without swelling.[3,13] Thus, it is significant that tissue swelling was again observed, a threshold effect at around 20 mM K^+ was found, and no swelling was observed when isethionate replaced chloride.[80] In addition, it is of interest that omission of Na^+ reduced the swelling at $[K^+]_o > 20$ mM[80] in agreement with the observation that at high $[K^+]_o$ increased uptake of Na^+ as well as K^+ and Cl^- is found.[78,79] Studies of these responses as a function of age,[81-83] as well as electron microscopy studies,[78,84] all suggested that a considerable proportion of this K^+-dependent swelling was located in astrocytes.

It has already been noted that the observation of swelling in the brain slice experiments in which KCl was increased by making the solution hyperosmotic[80] argued against such swelling being a result of Donnan forces. Other experiments showed that under conditions in which K^+ was increased with Na^+ reciprocally reduced, the uptake of Cl^- showed apparent saturation kinetics in terms of Cl^-.[85] The same study also found saturation in terms of K_o^+ when the rate of Cl^- uptake was measured at varying $[K^+]_o$ and fixed $[Cl^-]_o$, showing a dependence of Cl^- uptake on K^+ consistent with a coupled, electrically neutral KCl uptake system. However, it is also possible that the smaller increments in the initial rate of Cl^- uptake at higher $[K^+]_o$ result from the logarithmic dependence of the depolarization of the membrane potential on increasing $[K^+]_o$ and the effect of this on the electrochemical force driving inward movement of Cl^-. Indeed, if one applies the constant field equation[86] for calculating ion flux at a fixed $[Cl^-]_o$ of 130 and $[K^+]_i$ of 100 mM, and assuming that the membrane potential shows a Nernstian response to varying $[K^+]_o$, then apparent saturation kinetics are also obtained for Cl^- influx, with a $K_{1/2}$ of 30–40 mM K^+ (unpublished calculations). Other results, however, support the concept that some of the increased Cl^- uptake found when K_o^+ was increased and Na_o^+ reciprocally reduced was caused by mediated transport. Thus, uptake of NaCl is dependent on HCO_3^-, and this component is markedly temperature sensitive beginning at $30°C^{79}$ (see Fig. 4). It is also inhibited by SITS and ethacrynic acid derivatives[87] which are known to inhibit the anion exchange system in erythrocytes (see Section 2.2). We have also identified an anion exchange system in primary astroglial cultures, as will be discussed in Section 4.5.

4.3. Anion Transport in Superfused Cerebral Cortex

Cerebrocortical brain slices are metabolically active and can maintain levels of metabolites and ions that appear close to normal. There are, however, enough differences in ion content and inulin spaces,[12,70,76,77] as well as abnormal swelling of astroglia even in normal media[77,84] associated with increased NaCl and decreased K^+ levels,[70,76] to warrant duplication of studies in the intact brain not subject to a period of hypoxia required for the preparation of tissue slices. The work of Van Harreveld,[2] Nicolson,[63] and Hossman,[64] among others, already referred to in Section 4.1, has shown that during ischemia in intact animals there is a marked decrease in extracellular Na^+ and Cl^- and increased extracellular K^+. This is associated with a decreased extracellular space, implying cell swelling. Since astroglial swelling is pronounced after ischemia, as seen by electron microscopy,[59,60] it is tempting to suggest that the uptake of NaCl and the resultant swelling are predominantly into an astroglial compartment. In addition, increased swelling and uptake of Cl^- by exposed mammalian cortex after superfusion with solutions containing high K^+ concentrations has been found.[88,89] The swelling and uptake of Cl^- in intact cerebral cortex were reduced at lower temperatures, as they were in the *in vitro* cerebrocortical tissue studies. Furthermore, electron micrographs showed greatly expanded membrane-delimited compartments that were considered to represent astroglia, and the swelling and uptake of Cl^- were inhibited by the carbonic anhydrase inhibitor acetazolamide. These studies show that superfusion of the exposed, but otherwise intact cerebral cortex, with solutions containing high K^+ concentrations leads to a similar swelling and increased uptake of Cl^- as observed in brain slices *in vitro*.

4.4. Anion Transport and Putative Transmitters

The observation that part of the swelling of brain slices in response to increased K_o^+ was associated with increased uptake of NaCl suggested that part of the K_o^+ effect was indirect, stimulating some processes that involved coupled NaCl transport. The fact that addition of HCO_3^- under conditions of raised K_o^+ mainly stimulated increased uptake of NaCl, and that it was this increased uptake of NaCl that was inhibited by known inhibitors of the anion exchange system, raised the possibility that the K^+-stimulated process might be the anion exchange system.[87] A number of processes are stimulated by high K_o^+ in brain tissue including metabolism[90] and the release of transmitters.[91]

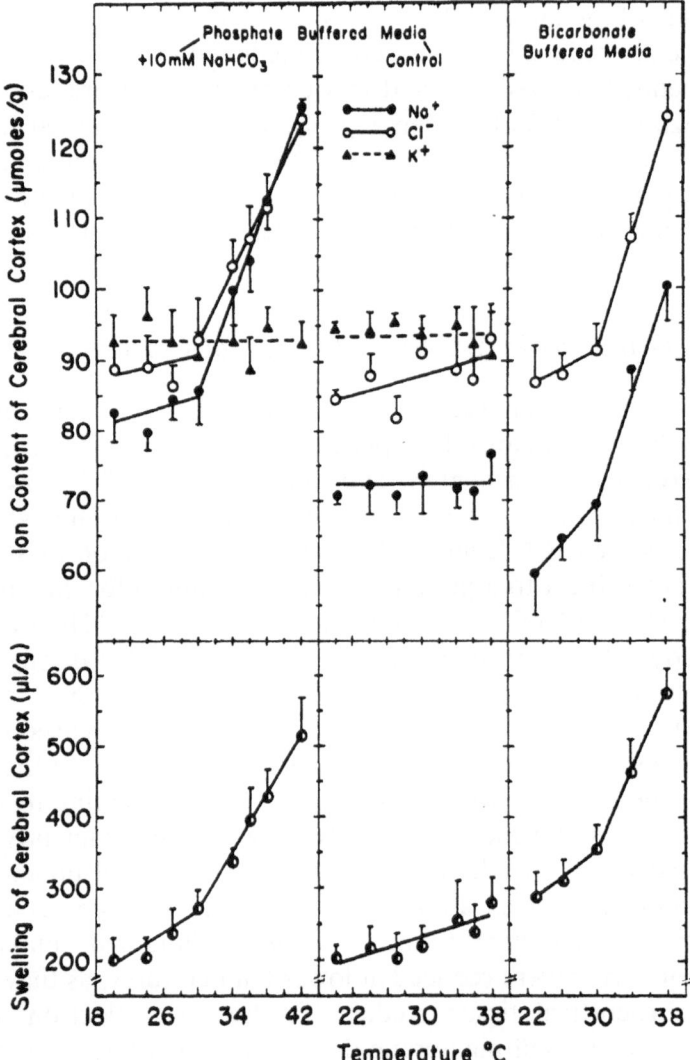

Fig. 4. Effect of temperature and HCO_3^- on swelling and ion content of monkey cerebral cortex slices. Slices were incubated for 1 h in either phosphate-buffered or HCO_3-buffered physiological balanced salt solutions as indicated; 10 mM HCO_3^- was added to the phosphate-buffered solution in the left-hand panel. Ion content is expressed in μmol/g initial wet weight. Swelling is the increase in wet weight after 1 h, expressed as μl water/g initial wet weight. Note that the slices show increased uptake of NaCl associated with increased water content only in media containing HCO_3^-. This effect is most marked at temperatures between 30 and 40°C. Note the absence of any effect on K^+ content. Reprinted by permission of the publisher from R. S. Bourke, H. K. Kimelberg, and L. R. Nelson, The effects of temperature and inhibitors on HCO_3^--stimulated swelling and ion uptake of monkey cerebral cortex, *Brain Res.* **105**:313. Copyright 1976 by Elsevier/North-Holland.

These effects, like K^+-induced swelling,[87] have a threshold effect at around 20 mm K^+, presumably related to a sufficient level of membrane depolarization. It has been shown that glioma cells and primary astroglial cultures respond to a number of putative neurotransmitters or neurohormones by increased cyclic AMP levels as well as other responses[92], similar to the effects of putative transmitters on brain slices.[93] Thus, the indirect effects of K^+ could result from release of transmitters from nerve endings because of K^+-induced depolarization which then act on some component of anion transport in astrocytes. There is precedence for effects of cyclic AMP[7,8] and norepinephrine[94] on Cl^- transport in epithelia, and we have observed that norepinephrine stimulates carbonic anhydrase activity in primary astroglial cells[95] associated with apparent increased phosphorylation of the enzyme.[96] Alternatively the effects of K^+ could be in whole or part a consequence of stimulated metabolism.

If the effect of high K_o^+ is indeed mediated by release of transmitters, then we should be able to duplicate the HCO_3^--dependent, SITS-inhibitable uptake of NaCl in brain slices by addition of transmitters at normal K_o^+. This indeed proved to be the case as we have shown for adenosine.[97] Also, the adenosine-stimulated uptake of NaCl in brain slices was dependent on the presence of HCO_3^- and was inhibited by SITS. There was no effect on K^+ content. Adenosine-stimulated swelling and ion uptake in brain slices in HCO_3^--containing medium are shown in Table III. The effect of norepinephrine is also shown, and it has similar effects to adenosine. In addition, we show that part of the effect of increased K_o^+ was inhibited by propranolol, suggesting that this concentration of K^+ was releasing endogenous norepinephrine which then interacts with a β receptor.[93] The effect of cyclic AMP on Cl^- transport in epithelia is usually to inhibit active NaCl accumulation in absorptive epithelia and to increase Cl^- secretion in secreting epithelia.[7] The effect of the addition of putative transmitters known to increase cyclic AMP levels in increasing NaCl uptake in brain slices is thus difficult to reconcile with these known effects of increased cyclic AMP in epithelia. Perhaps, if the scheme we show in Fig. 6 (Section 4.6) is correct, the putative transmitters are stimulating release from the endothelial cell of Cl^- which is then taken up by the astrocyte.

Recently, we have studied[98] the effect of addition of adenosine to the superfusion medium in the cat superfused cerebral cortex model (see Section

Table III
Effect of Putative Neurotransmitters and High K^+ on the Uptake of NaCl by Brain Slices[a]

Conditions	Percent swelling[b]	Cl^-[c]	Na^+[c]	K^+[c]
Control	10.7	86.2	99.4	73.9
+0.2 mM norepinephrine	19.8	95.5	111.5	75.1
+0.13 mM adenosine	20.3	90.7	112.4	63.9
+30 mM K^+	27.9	94.2	97.2	98.7
+30 mM K^+ +0.2 mM propranolol	18.5	80.6	83.6	99.3

[a] Uptake of ions by 0.5 mm thick cat cerebrocortical slices cut from the pial surface. The values shown were determined after the slices in the transmitter experiments had incubated for 1 h in a medium containing (mM): Na^+, 145; K^+, 4.5; Mg^{2+}, 0.4; Ca^{2+}, 1.3; Cl^-, 123; SO_4^{2-}, 1.2; HCO_3^-, 25; glucose, 10. For the 30 mM K^+ experiments, 20 mM HEPES replaced HCO_3^-, and Na^+ was reciprocally reduced. After 30 min, neurotransmitter to the concentration shown or $NaHCO_3$ to a concentration of 5 mM in the case of the 30 mM K^+ experiments was added. Temperature was 37°C, and initial pH 7.5 to 7.7. From refs 97 and 112, from which further experimental details can be obtained.
[b] Final minus initial wet weight/initial wet weight $\times$ 100.
[c] Concentrations expressed in μmol/g initial wet weight.

4.3). This resulted in swelling of the superficial cortical tissue which was associated with marked swelling of the perivascular astroglial processes as seen in electron micrographs (see Fig. 5A). Pretreatment with a nondiuretic ethacrynic acid derivative known to inhibit Cl^- uptake in brain slices prevented both the increase in wet weight and astroglial swelling.[98]

4.5. Anion Transport in Astroglial Cultures

Because the conditions that cause increased uptake of NaCl are also associated with marked astroglial swelling as determined by electron microscopy, we have inferred that the NaCl transport is predominantly into astrocytes. To provide further support for this possibility as well as a means of establishing the exact transport mechanisms, it is necessary to study pure preparations of these cells. Cell cultures of defined composition are most likely to be viable and maintain intact permeability barriers. There have been some transport studies on glial tumor lines,[99,100] but there is always uncertainty as to how closely the properties of such lines resemble those of normal astrocytes *in situ.* Thus, cells of the C_6 rat glioma line have average membrane potentials of -36 mV[100] which are quite different from the -70 to -90 mV potentials reported for astroglia *in situ.*[68,71,101] In contrast, the primary astroglial cultures[102] we have used show many of the characteristics of normal astroglia. They stain intensely for glial fibrillary acidic protein, and such GFAP-positive cells comprise at least 80% of the cells present in our cultures.[103] They have average membrane potentials of -70 mV and show a close to Nernstian response to varying K_o^+[104] similar to the properties of characterized mammalian astroglia *in situ*[68] and in brain slices *in vitro.*[71] Their morphology resembles that of astroglia *in situ* when induced to form processes by treatment with dibutyryl cyclic AMP or norepinephrine.[95] The GFAP-positive cells also contain carbonic anhydrase[105] which may[106] or may not[107] be present in astroglia *in situ.*

Such primary astrocyte cultures show SITS-sensitive Cl^- transport and saturation kinetics for Cl^- with an apparent $K_{1/2}$ for uptake of ^{36}Cl of 36–56 mM Cl_o^-.[108–110] The efflux of ^{36}Cl from preloaded cells is stimulated by the presence of Cl^- or HCO_3^- in the external medium.[109,110] In addition, we have determined that the intracellular concentration of Cl^- in these cells is around 40 mM, or about fourfold greater than expected for equilibration with their average membrane potentials of -70 mV.[109,110] In contrast, the Cl^- concentration of C_6 glioma cells has been estimated as 14 mM.[100] At 166 mM Cl_o^-, this gives a Cl^- equilibrium potential of -55 mV which is more negative than the average potential of -36 mV determined for these cells, consistent with active transport of Cl^- out of these cells. The value of 40 mM $[Cl^-]_i$ for the primary astrocytes is similar to the recent estimate of 46 mM Cl^- in glial cells *in situ* based on a measured value for total intracellular Cl^- content of 22 mmol/kg cell water in rat cerebral cortex, the assumption that Cl^- passively equilibrates in neurons ($[Cl^-]_i$ ~10 mM), and an estimate of 2:1 relative neuron–glial volume.

How such a distribution might vary between astroglia and oligodendroglia,

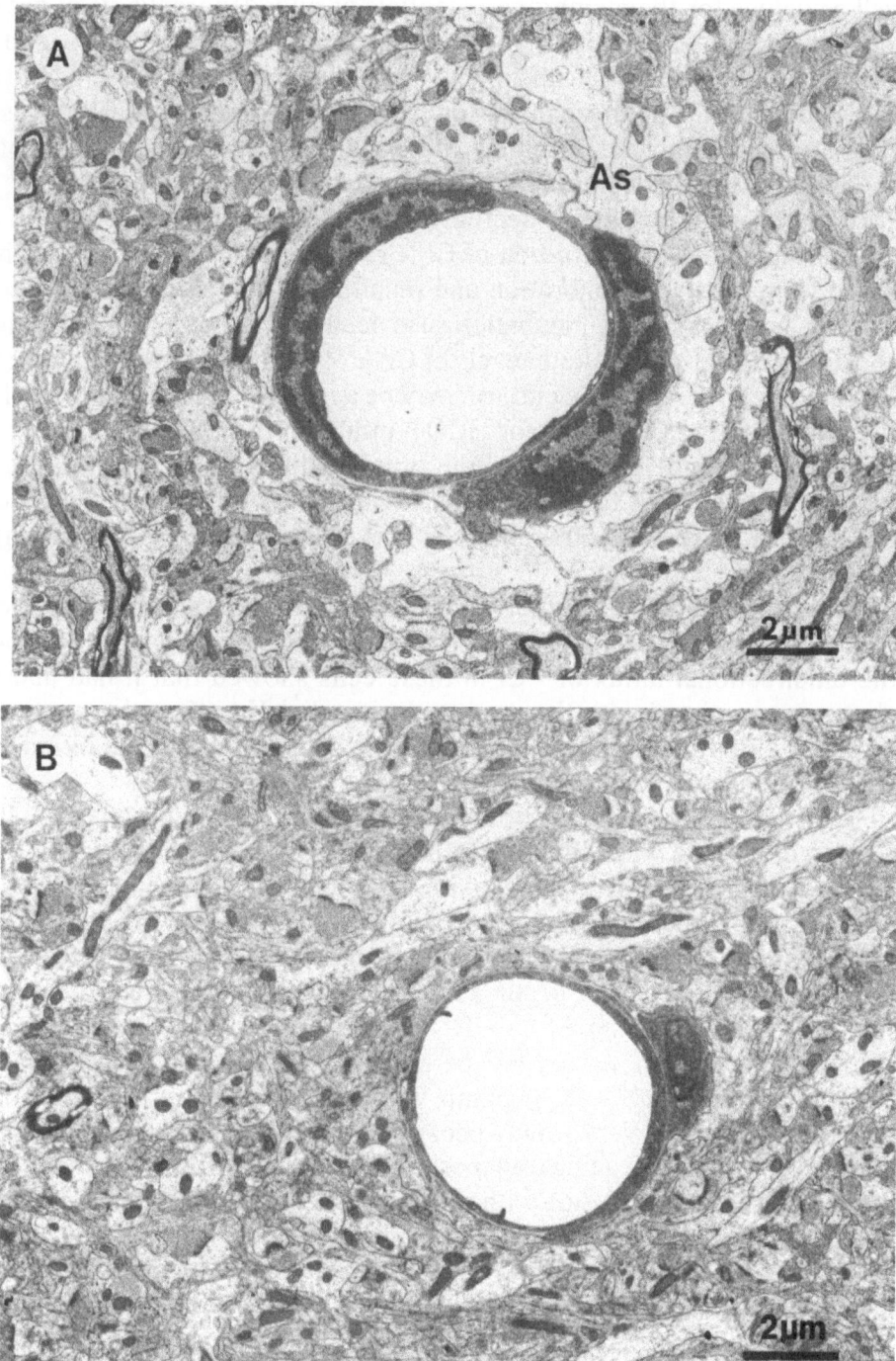

Fig. 5. A: Swelling of perivascular astrocyte processes in an electron micrograph from a 0.5 mm thick pial slice from the exposed cortex of an anesthetized cat that was perfused for 1 h with a solution containing 0.125 mM adenosine and then perfused–fixed. As, astrocytic processes. B: As in A except that adenosine was omitted. Reprinted by permission of the publisher from R. S. Bourke, J. B. Waldman, H. K. Kimelberg, K. D. Barron, B. D. SanFilippo, A. J. Popp, and L. R. Nelson, Adenosine-stimulated astroglial swelling in cat cerebral cortex *in vivo* with total inhibition by a non-diuretic acylaryloxyacid derivative, *J. Neurosurg.* **55:**364–370. Copyright 1981 by The American Association of Neurological Surgeons.

as well as values for the membrane potentials of oligodendroglia, is at present unknown. Thus, it would appear that astroglia may be high-Cl^- cells, and the data argue for a role of these cells in Cl^--secretory processes in the CNS such as CSF production. The mechanism for such active transport of Cl^- in astrocytes is at present unclear. Although SITS inhibited the initial rate of uptake of Cl^-, it only slightly decreased the steady-state concentration of Cl^-.[108] By contrast, furosemide markedly decreased both the initial rate of uptake and the final steady-state concentration of Cl^-, reducing it to levels close to those expected from passive equilibration and resulting in cell shrinkage.[110] Reduction of the temperature of incubation also leads to reduction in the rate of uptake and the final steady-state levels of Cl^-.[111] Such high intracellular levels of Cl^-, at least in astrocytes in culture, may be a means of reducing intracellular acidity by promoting Cl^- efflux for HCO_3^- influx on the exchange carrier, and addition of furosemide does result in internal acidification.[112] *In situ,* the exchange carrier may well function in the reverse direction under altered conditions such as increased extracellular acidity, with uptake of Cl^- and efflux of HCO_3^-.

There has been only one other detailed study of Cl^- transport in glial cells in culture, in a normal, established astrocyte line from Syrian hamsters. Studies on the unidirectional influx of ^{36}Cl in these cells showed that it had many of the features associated with the proposed astroglial K^+-dependent uptake of Cl^- and swelling *in situ* such as temperature sensitivity and inhibition by a variety of anions and acetazolamide.[113]

The measured membrane potentials of primary astrocyte cultures show little or no difference when the normal medium is changed to one lacking Cl^- or Na^+.[112] This is similar to the behavior described for astroglia in a *Necturus* optic nerve preparation[114] and suggests the Cl^- conductance to be small compared to those of other ions since a large change in E_{Cl} has taken place. Thus, accumulation of Cl^- has to occur as an electrically silent, hetero-anion-exchange system such as the $Cl^- \leftrightarrow HCO_3^-$ system or an active coupled cation–Cl^- uptake system. A very low permeability to Na^+ and Cl^- also implies that inhibiting the $(Na^+ + K^+)$ pump with ouabain will result in astroglial swelling occurring only very slowly because of accumulation of NaCl, and this may be offset by compensating transport processes. Electron microscopy studies of cat brain after superfusion with media containing [^3H]ouabain have shown that swelling of astroglia does not occur in regions penetrated by ouabain.[115] In agreement with this, we have found that addition of ouabain to primary astrocyte cultures does not affect cell volume as measured by [^{14}C]3-*O*-methylglucose.[112]

4.6. *Model of Anion Transport in Astrocytes*

Studies in cerebrocortical slices *in vitro* and in intact brain during ischemia and spreading depression have identified an important component of anion transport in the CNS, namely, uptake of NaCl into a cell compartment with resultant cell swelling and reduction in extracellular fluid volume.[63,64,78,87,112]

Based on studies with cerebrocortical slices *in vitro*, this uptake could be predominantly into astrocytes, involving HCO_3^- ions, and stimulated by several putative transmitters (see Sections 4.2–4.4). Investigations using primary astrocyte cultures confirmed the existence of both a $Cl^- \leftrightarrow HCO_3^-$ and a $Na^+ \leftrightarrow H^+$ exchange system in these cells.[108] Also, very large increases in cyclic AMP in response to addition of norepinephrine have been seen in these cultures.[92,95] Based on these characteristics, we have proposed a model[108] for increased uptake of NaCl into astrocytes in brain through simultaneous exchange of Na^+ and Cl^- for H^+ and HCO_3^-, respectively. Hydrogen ions and HCO_3^- can be produced by rapid intracellular hydration of CO_2 by astrocytes because of the presence of carbonic anhydrase[95,105] as was originally proposed by Tschirgi[116] and Giacobini[117] as a basis for NaCl secretion in CSF production. Relatively rapid exchanges of H^+ and HCO_3^- for the Na^+ and Cl^- may lead to accumulation of NaCl inside the cells at a faster rate than Na^+ can be removed by the $(Na^+ + K^+)$ pump, in part because of falling ATP levels in ischemia. Addition of HCO_3^- to the medium may further stimulate the process in brain slices since increased lactic acid production because of anoxia will cause more rapid production of CO_2 from the added HCO_3^-. Alternatively, since H_2CO_3 is the immediate ionic association product of $H^+ + HCO_3^-$, H_2CO_3 may immediately enter the cell, subsequently dissociating into $HCO_3^- + H^+$, as has been proposed for choroidal epithelium.[43] This mechanism should be independent of carbonic anhydrase, since only relatively instantaneous ionic association and dissociation reactions are involved if diffusion of H_2CO_3 is rapid enough to support the observed rates of NaCl transport. Also, the localization of carbonic anhydrase in astroglia is now controversial, since one group thinks it is only present in oligodendrocytes,[107,118] at least in rat cerebellum.

It can be appreciated that the existence of such exchange systems provides a means of controlling brain pH if the $Na^+ \leftrightarrow H^+$ exchange system is located preferentially at the perivascular foot processes leading to secretion of H^+ into and uptake of Na^+ from the blood. Chloride seems to be actively accumulated in astroglia[109,110] and *in situ* could be accumulated from the blood across the endothelial cell and then secreted into the brain by flowing down its electrochemical gradient into brain extracellular fluid at what would be the astroglial apical surface in the neuropil,[25] similar to proposals for secreting epithelia.[7,8]

A scheme summarizing these points is shown in a schematic of the astroglial–endothelial complex[25,119] in Fig. 6. The mechanism described above together with the orientation of the principal transport processes as shown will lead to secretion of NaCl across both the endothelial and astroglial cell. A further feature is that transport of HCO_3^- and H^+ into the region of the basement membrane (B.M.) will generate CO_2, some of which will diffuse into the endothelial cell and be hydrated back to form HCO_3^- and H^+. The endothelial membrane facing the capillary lumen is considered relatively impermeable to Na^+, H^+, Cl^-, and HCO_3^- as discussed in Section 3.2, and there is evidence that the $(Na^+ + K^+)$ ATPase is localized on the opposite, basal membrane.[120] The $(Na^+ + K^+)$ ATPase of both capillaries[121] and cultured astrocytes[122] show a similar $K_{1/2}$ for K^+ of 1–3 mM which corresponds to the external activation site of the $(Na^+ + K^+)$ pump. The $(Na^+ + K^+)$ ATPase also saturates at

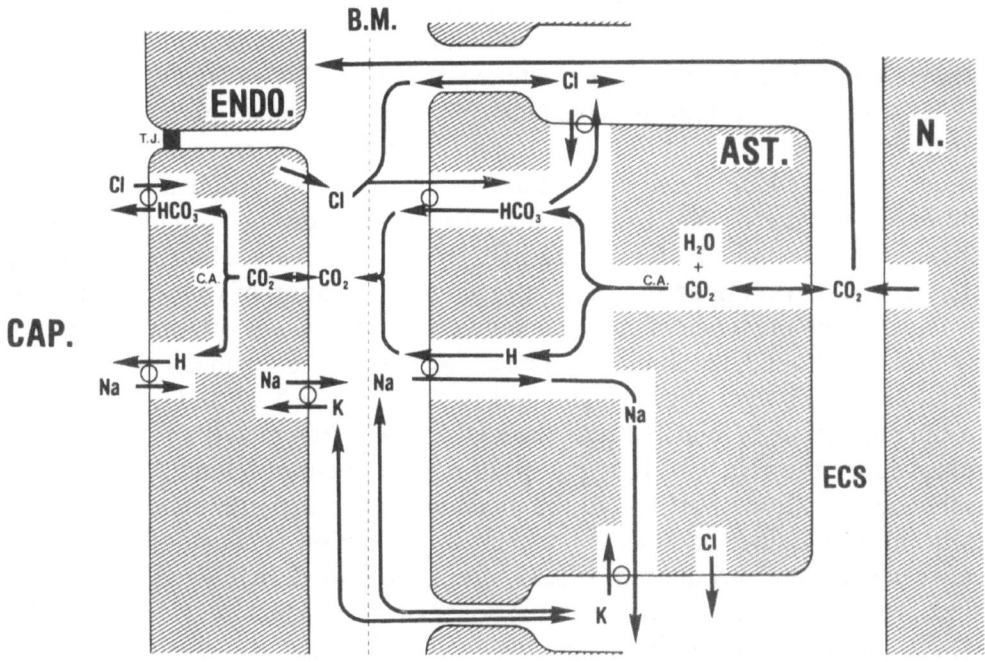

Fig. 6. Hypothetical scheme for the orientation of transport systems in endothelial cells and astrocytes leading to NaCl secretion and pH control in the CNS. CAP, capillary; T.J., tight junction; ENDO, endothelial cell; C.A., carbonic anhydrase; B.M., basement membrane; AST, astrocyte; ECS, extracellular space; N., neuron. See text for further details.

around 10 mM K,[121,122] suggesting that the pump in these cells, although responding to wide variation in internal Na^+, only responds within the normal range of $[K^+]_o$. We assume that the endothelial cell actively accumulates Cl^- at its luminal surface, that the basal membrane is freely permeable to Cl^-, and that the endothelial cell contains carbonic anhydrase.

Further assumptions are that $Cl^- \leftrightarrow HCO_3^-$ and $Na^+ \leftrightarrow H^+$ exchange systems are present in the endothelial cell and are localized as shown and that the $Cl^- \leftrightarrow HCO_3^-$ and $Na^+ \leftrightarrow H^+$ exchange systems of the astrocyte are also localized as shown. These are certainly a considerable number of assumptions, but this model does at least provide a reasonably plausible working hypothesis. Isolated brain capillary preparations have been described[121] in which it might be interesting to see if the $Cl^- \leftrightarrow HCO_3^-$ and $Na^+ \leftrightarrow H^+$ exchange systems and carbonic anhydrase exist. Recent work,[119] however, has reported the presence of astroglial endfeet on around 50% of isolated brain microvessels. It will also be very difficult to devise appropriate techniques and biological preparations in which the specific orientations we are suggesting can be tested. Cytochemical techniques may well be useful in this regard.

Since the model proposed above concerns active accumulation of anions, it may be appropriate to mention briefly a HCO_3^--stimulated ATPase activity found in a number of anion transport tissues (see ref. 123 for a recent review), and in neurons, glia, and choroid plexus from rat brain.[124] However, although such findings stimulated considerable interest in the possible role of such an enzyme in active anion transport in such tissues, its probable identity with mitochondrial ATPase[123,124] makes a role for such an enzyme in plasma membrane ion movements unlikely.

4.7. Chloride Content of Neurons

The problem of directly measuring the Cl^- content of neurons in the intact brain has already been discussed in Section 4.1 in relation to earlier concepts that the extracellular space of the CNS was very small and the attendant problem of assigning compartments for the 35 to 45 μmol Cl^-/g wet weight determined for brain tissue. One approach to this problem is the use of purely neuronal preparations such as squid axon. Membrane potentials, of course, have to be simultaneously measured to assess whether Cl^- is being actively transported under steady-state conditions of no net flux across the membrane (see equation 5). It has been found that $[Cl^-]_i$ in squid axon averages 108 mM based on analyses of extracted axoplasm.[125] The activity coefficient determined with Ag–AgCl electrodes was close to that for Cl^- in solution (0.7), and thus, the $[Cl^-]_i$ was about two to three times greater than expected from passive equilibration. This implies inward active transport of Cl^-, and Keynes[125] found that Cl^- influx was partially inhibited by 0.2 mM dinitrophenol, an uncoupler of oxidative phosphorylation, but was not affected by omission of Na_o^+. Recent work using a more sensitive internal dialysis method has shown that influx of Cl^- is directly proportional to $[Na^+]_o$, requires ATP, and is inhibited by furosemide.[126]

Intracellular Cl^- in centrally located neurons clearly has to be measured by other techniques, and estimates of $[Cl^-]_i$ in cat spinal motoneurons have been based on electrophysiological studies. Eccles[53] has suggested that at the resting membrane potential there is zero electrochemical gradient for Cl^-, based on the interpretation that the inhibitory postsynaptic potential (IPSP) in cat motoneurons is caused equally by contributions from Cl^- and K^+. Since the IPSP is -80 mV and E_K is -90 mV, the E_{Cl} was considered to have to be -70 mV, which is the measured resting membrane potential of the motoneurons. At external Cl^- of 125 mM, this gives a $[Cl^-]_i$ of 9 mM. Later work using injection of different combinations of ions into motoneurons also showed a contribution of Cl^- to the IPSP. Again it was concluded that the IPSP was caused by contributions from both K^+ and Cl^-.[127] There was also evidence for an inward $K^+ + Cl^-$ pump but no indication that E_{Cl} was normally more positive than E_m. In contrast to this viewpoint, Lux[128] postulated the existence of an outward Cl^- pump in cat spinal motoneurons resulting in E_{Cl} being more negative than E_m, and thus, the hyperpolarizing IPSP could be mainly caused by Cl^-. He found that intravenous infusion of ammonium salts depolarized the IPSP potentials and enhanced the depolarizing effects of intracellularly injected Cl^-.

As can be appreciated, such electrophysiological studies involve making a number of assumptions. A very promising recent approach to probing anion distribution in the CNS is the use of ion-specific microelectrodes (ISMs) already referred to in Section 4.1 in relation to measurements of extracellular anion levels. For intracellular recordings, much smaller tip diameter (<0.5 μm) electrodes have to be used for successful intracellular penetration. Furthermore, each cell must be impaled with both a conventional KCl electrode and an ISM, since the membrane potential has to be subtracted from the ISM reading to obtain the specific ion potential. The reader is referred to the book by Thomas[129] for further information on the uses and abuses of ISMs. Thus, he points out that when using a KCl microelectrode to measure membrane potential in conjunction with a Cl^- ISM, the KCl electrode leaks significant amounts of Cl^- into the cell.

Because of the requirement for dual impalement, ISMs have been used only in larger neuronal cells, particularly those found in some molluscs. In such preparations, Cl^- activities (a^i_{Cl}) of 37 have been found in the salt water *Aplysia*,[130] and calculated concentrations of 10–15 mM have been reported for the land snail, *Helix aspersa*.[131] Note that the value of 37 has to be divided by the activity coefficient (0.7) to give a value of 53 mM for the intracellular concentration of Cl^- in the *Aplysia* preparation. In *Aplysia*, the E_{Cl} was -56 mV,[130] and in *Helix*, E_{Cl} would be -50 to -60 mV based on $[Cl^-]_o$ of 108 mM.[131] These values are all more negative than the E_m values of -49 and -40 to -50 mV found for these two preparations, respectively. Similarly, Ascher *et al.*[132] found an E_{Cl} of -58 mV and E_m of -48 for *Aplysia* neurons. These studies provide, at least in molluscan neurons, direct evidence for outward pumping of Cl^- and have provided an opportunity to study its mechanisms. The possible nature of these mechanisms will be discussed in Section 4.9. They also support the idea that in mammalian motoneurons E_{Cl} may also be more negative than E_m and that the origin of the IPSP is an increase in Cl^- permeability. This has been discussed above and will also be discussed in the following section (4.8).

Electron probe X-ray microanalysis techniques can be used to measure the elemental chemical content of subcellular areas of tissue sections. For rat cerebral cortical cells, Cl^- contents of 38 to 45 and 39 to 58 meq/kg intracellular water were found for the cytoplasm and nucleus, respectively, of 1- and 24-month-old animals.[133] For 1-month-old animals, this compares to nuclear and cytoplasmic levels of 48 and 42 for Na^+ and 156 and 149 for K^+, all in meq/kg intracellular water. For frozen sections of frog spinal cord, mean values for motoneurons for Cl^-, Na^+, and K^+ were 43, 79, and 88 mmol/kg wet weight.[134] These values are considerably higher than those calculated for mammalian motoneurons based on electrophysiological studies. It could represent a species difference, but it was pointed out[134] that the values for Cl^- compared to a range of 8 to 34 mM determined from IPSP values and that a value for Na^+ of 17 mM was calculated from the peak of the action potential in frog motoneurons. A total ion content that is greater than that found by ion-specific microelectrodes, corrected to give concentrations using a reasonable estimate for the activity coefficient or based on electrical activity, is usually attributed to sequestration or binding of the ion. Clearly, further studies comparing total Cl^- content and activity in identified neurons in different nervous systems are required to resolve these questions.

4.8. *Effects of Chloride on Neuronal Potentials*

Since the available studies suggest that E_{Cl} in neurons is more negative than E_m, Cl^- should contribute to the resting potential of neurons and also affect the rate at which these potentials change during excitation if it has a significant resting conductance. The latter could occur even if E_{Cl} were identical to E_m since when E_m suddenly changes because of rapid alterations in the conductance of other ions, net Cl^- currents would occur until neutral transport of Cl^- resulted in E_{Cl} again being equal to E_m. However, the Cl^- current might be insignificant if Cl^- permeability is much smaller than those of the other ions contributing to the E_m. There appears to have been little work on the effects of Cl^- or other anions on the axonal action potential. However, in muscle, the increased excitability found in myotonia has been attributed to a greatly decreased Cl^- conductance which normally, with a mean of 6.6×10^{-4} mho/cm^2, is about threefold greater than K^+ conductance in goat intercostal muscles.[4] In squid axon, the relative K^+ and Cl^- permeabilities are in the opposite direction to muscle, with a ratio of 5.9 having been quoted.[125] This was based on a value for Cl^- influx of 9.2 pmol/cm^2 per sec, which could include electrically neutral NaCl cotransport. The Na-independent component may in fact represent the true conductive Cl^- permeability, and a value for this of around 1.4 pmol/cm^2 per sec has been found.[126] Reduction of $[Cl^-]_o$ has been shown to depolarize the membrane potential of molluscan neurons,[135] but in another case this occurred only if $[K^+]_o$ was also increased.[136]

It is well established that changes in Cl^- conductance have a marked effect on inhibitory postsynaptic potentials in both mammalian and invertebrate neurons.[127,128,135,137-139] These effects can be seen as a result of stimulated or spontaneous IPSPs[127,128,135] or by application of a variety of transmitters such as GABA and acetylcholine.[135,137-139] Eccles[53,127] has suggested that in cat motoneurons such IPSPs result from equal increased conductance to K^+ and Cl^-. However, others have obtained evidence in both mammalian and molluscan neurons that the hyperpolarization is predominantly caused by an increased Cl^- conductance, E_{Cl} being kept more negative than E_m by outward pumping of Cl^-. In cat spinal motoneurons,[128,140] intravenous injection of ammonium salts has been shown to inhibit such outward pumping of Cl^-. Ammonium salts will tend to alkalinize the inside of cells since NH_3 freely enters and forms NH_4^+ ions inside the cell.[141] This could inhibit a $Cl^- \leftrightarrow HCO_3^-$ exchange system acting in the direction of Cl^- out and HCO_3^- in as will be discussed in Section 4.9.

It has been shown in a large number of cases that the reversal potential, i.e., the potential at which the effect of stimulation of an IPSP or application of a transmitter changes from hyperpolarizing to depolarizing, can be altered by changes in external Cl^- concentrations or after intracellular injection of Cl^- (as examples see refs. 128 and 135). The GABA-induced IPSP can be abolished in Cl^--free solution.[138] Figure 7 shows the effects of changing $[K^+]_o$ and $[Cl^-]_o$ on the reversal potentials of spontaneous IPSPs or addition of acetylcholine in a neuron from *Helix aspersa* abdominal ganglion.[135] In this experiment, K^+ and Cl^- affected both the IPSP and ACh reversal potential,

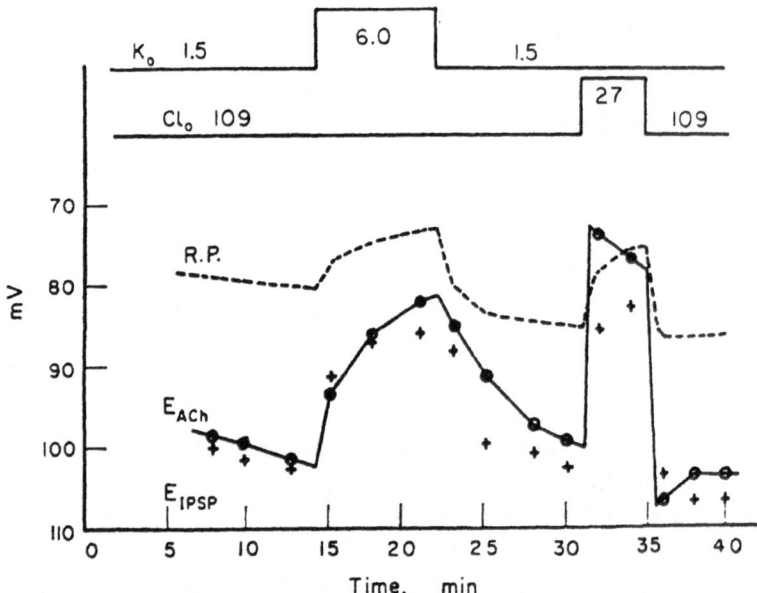

Fig. 7. Effect of variations in external Cl⁻ and K⁺ on resting potentials and reversal potentials of a snail neuron. Changes in external ionic concentrations are shown at the top of the figure where numbers represent concentrations in mM. Increasing K⁺ and decreasing Cl⁻ both brought about a decrease in the resting membrane potential (R.P.), as well as the reversal potentials to stimulation of the IPSP (+) or addition of acetylcholine (Ach) directly to the cell (O). The potentials were recorded with a conventional microelectrode bridge circuit which also enabled the cell to be hyperpolarized or depolarized by passing current through the recording electrode. This enabled the reversal potential for the IPSP or Ach response to be determined. Note that the decrease in the reversal potential with increasing K⁺ was much slower than that associated with decreasing Cl⁻. Reprinted by permission of the publisher from G. A. Kerkut and R. C. Thomas, The effect of anion injection and changes in the external potassium and chloride concentration on the reversal potentials of the IPSP and acetylcholine, *Comp. Biochem. Physiol.* **11**:205. Copyright 1964 by Pergamon Press.

but the effect of Cl⁻ was far more rapid. Other anions, as long as their diameters were no greater than 1.2 times that of the hydrated K⁺ ion, were also effective. Decreasing $[Cl^-]_o$ also reduced the resting membrane potential (R.P.), showing a contribution of this ion to the R.P. under nonstimulated conditions. Kerkut and Thomas[135] concluded from this study that approximately 90% of the current during the IPSP was carried by Cl⁻ and 10% by K⁺, and that ACh was the probable transmitter at the snail inhibitory synapse. Hyperpolarization of a presynaptic terminal by an apparent GABA-induced increased Cl⁻ conductance has also been described.[142]

4.9. Anion Transport Mechanisms in Neurons

Molluscan neurons are generally large and can be impaled with several microelectrodes. Using this technique, several workers have simultaneously studied membrane potentials, intracellular ion activities, and intracellular pH. Thomas[131] impaled the largest nerve cell of *Helix aspersa* with four microelectrodes and recorded intracellular pH and membrane potential and injected HCl by passing current between the remaining two electrodes, one of which contained HCl. The subsequent rate of intracellular alkalinization after injection of HCl was much slower in a HCO_3^-/CO_2-free Ringer as compared to Ringer containing HCO_3^-/CO_2. In further experiments, it was found that the intracellular alkalinization after HCl injection was inhibited when SITS was present. Also, the increase in pH_i after intracellular acidification caused by changing from HCO_3^-/CO_2-free to HCO_3^-/CO_2 Ringer was associated with a fall in Cl_i^-. However, after such treatment, pH_i did not recover in a Cl^--free medium (see Fig. 8).

An additional feature seen in the snail neuron is that when pH_i recovers subsequent to an intracellular acidification, there is an increase in $[Na^+]_i$, and also that recovery from intracellular acidification is inhibited when external Na^+ is removed. Thomas[131] favors a tightly coupled $Cl^- \leftrightarrow HCO_3^-$ and $Na^+ \leftrightarrow H^+$ exchange system to explain his data, with the energy for the entire pH-

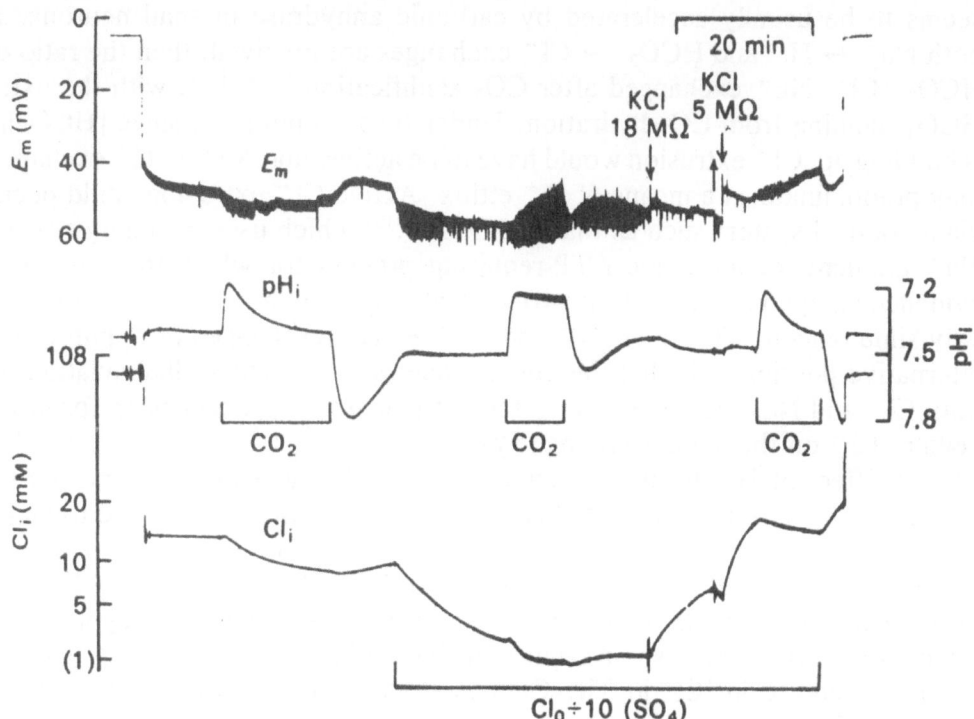

Fig. 8. Relationship between intracellular pH ($[pH]_i$) and intracellular Cl^- ($[Cl^-]_i$) in a snail neuron after intracellular acidification by replacement of HEPES-buffered Ringer with Ringer buffered with 21 mM $HCO_3^-/2.3\%$ CO_2. It can be seen that on replacement of HEPES Ringer with HCO_3^-/CO_2 Ringer, Cl^- left the cell, presumably in exchange for external HCO_3^-. In the low-Cl Ringer, 90% of the Cl^- was replaced with SO_4^{2-}. It can be seen that under these conditions the intracellular acidification on addition of HCO_3^-/CO_2 Ringer was not followed by intracellular realkalinization, presumably because there was insufficient internal Cl^- to exchange for external HCO_3^-. The E_m was recorded with a K_2SO_4-filled microelectrode. Internal Cl^- was then raised by insertion of an 18 MΩ KCl-filled electrode which was then removed and reinserted after the resistance had been reduced to 5 MΩ by breaking the tip; Cl_i^- rose, and the normal intracellular realkalinization after addition of HCO_3^-/CO_2 Ringer was again seen. Reprinted by permission of the publisher from R. C. Thomas, The role of bicarbonate, chloride and sodium ions in the regulation of intracellular pH in snail neurones. *J. Physiol. (Lond.)* **273**:325. Copyright 1977 by Cambridge University Press.

controlling system deriving from Na^+ entry down its electrochemical gradient. However, it is also possible to envisage separate $Cl^- \leftrightarrow HCO_3^-$ and $Na^+ \leftrightarrow H^+$ exchange systems similar to those shown in Fig. 6 for astroglia. In this model, after intracellular acidification, the following reaction proceeds to the right inside the cell:

$$HCO_3^- + H^+ \leftrightarrow CO_2 + H_2O \qquad [8]$$

This results in a decrease in intracellular HCO_3^- leading to exchange of extracellular HCO_3^- in for intracellular Cl^- out. The CO_2 formed from the dehydration reaction diffuses out of the cell, being rehydrated extracellularly by reaction [8] proceeding to the left. The overall effect is thus a 1:1:1 exchange of HCO_3^- in for Cl^- and H^+ out. Since 10 μM acetazolamide reduced both intracellular acidification and recovery after addition of CO_2,[143] CO_2 hydration seems to be usually accelerated by carbonic anhydrase in snail neurons. If both $Na^+ \leftrightarrow H^+$ and $HCO_3^- \leftrightarrow Cl^-$ exchanges are involved, then the ratio of $HCO_3^- : Cl^- : Na^+$ exchanged after CO_2 acidification is 2:1:1, with the extra HCO_3^- coming from CO_2 hydration. Under these conditions, since $[HCO_3]_i^-$ is building up, Cl^- extrusion would have to be active, and $Na^+ \leftrightarrow H^+$ exchange may predominate as a means of H^+ efflux. Active Cl^- extrusion could occur via a coupled system such as that of Thomas,[131] which uses the energy of the Na^+ gradient, or an active ATP-requiring process for which there is some evidence in squid axon.[144] With two separate systems, however, it is not clear why simultaneous $Cl^- \leftrightarrow HCO_3^-$ and $Na^+ \leftrightarrow H^+$ exchange are required. An alternative possibility is that the dependence of intracellular alkalinization on both Cl_i^- and Na_o^+ could be due to the ion pair $NaCO_3^-$ exchanging for intracellular Cl^- on the anion exchange system.[145]

An effect of HCO_3^- on intracellular Cl^- levels has also been reported by Russell for *Aplysia* neurons.[146,147] Here, the experiment was designed to make the interior of the neuron acidic by adding 50 mM NH_4Cl and then replacing it with NH_4-free medium in the absence of HCO_3^-/CO_2. As shown in Fig. 9A (top panel), NH_4^+-containing medium increased intracellular Cl^- slightly, and removal of NH_4^+, which should cause an intracellular acidification,[141] caused a slight decrease in $[Cl^-]_i$. The Cl_i^- was calculated from E_{Cl} by the Nernst equation. The lower panel (B) shows that the changes in $[Cl^-]_i$ were much greater when NH_4^+ medium was replaced with NH_4^+-free medium containing 10 mM HCO_3^-/CO_2. This effect was totally inhibited when 0.5 mM SITS was present. Figure 9 also shows very clearly that in these neurons E_{Cl} was more negative than E_m. Further experiments showed that the effect of HCO_3^- in reducing $[Cl^-]_i$ shown in Fig. 8B was just as marked when the hyperpolarization of the membrane was prevented by voltage clamping. Thus, intracellular acidification leads to extrusion of Cl^-, presumably by exchange with external HCO_3^- which promotes internal alkalinization. This system has also been shown to be present in squid axon where it is also inhibited by SITS and requires the presence of ATP in the internal dialysate.[144] The behavior of these systems in molluscan neurons thus resembles in its pH dependence that of the outward Cl^- pump in cat motoneurons, where addition of NH_4^+ leads to

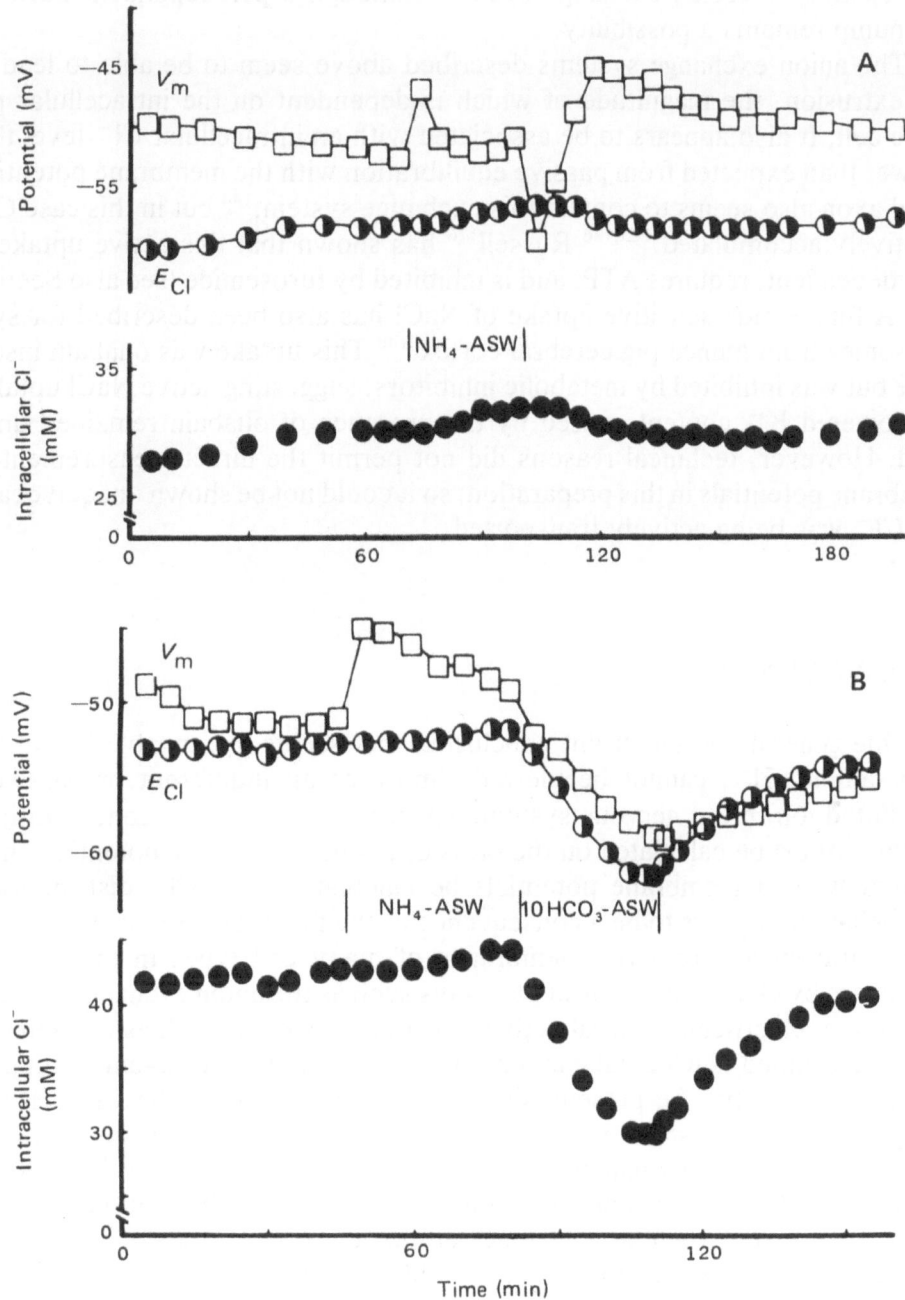

Fig. 9. Effects of HCO_3^- on Cl_i^- after changes in pH_i in an *Aplysia* neuron. In A (top panel), the neuron was originally bathed in Tris-ASW (artificial sea water) and then changed to ASW in which 60 mM of the NaCl was replaced with NH_4Cl. In B, the 10 HCO_3^--ASW was Tris-ASW in which 10 mM of the NaCl was replaced with 10 mM HCO_3^- and bubbled with 0.4% CO_2. See text for further details. Reprinted by permission of the publisher from J. M. Russell, Effects of ammonium and bicarbonate–CO_2 on intracellular chloride levels in *Aplysia* neurons, *Biophys. J.* **22**:133. Copyright 1978 by Rockefeller University Press.

increased $[Cl^-]_i$.[140] This effect may also reflect inhibition of Cl^- efflux on a coupled $Cl^- \leftrightarrow HCO_3^-$ exchange system, although a pH_i-dependent outward Cl^- pump remains a possibility.

The anion exchange systems described above seem to be able to lead to Cl^- extrusion, the magnitude of which is dependent on the intracellular pH of the cell. It also appears to be associated with an intracellular Cl^- level that is lower than expected from passive equilibration with the membrane potential. Squid axon also seems to contain this exchange system,[144] but in this case Cl^- is actively accumulated.[125,126] Russell[126] has shown that this active uptake is Na^+ dependent, requires ATP, and is inhibited by furosemide (see also Section 4.7). A furosemide-sensitive uptake of NaCl has also been described for synaptosomes from guinea pig cerebral cortex.[148] This uptake was ouabain insensitive but was inhibited by metabolic inhibitors, suggesting active NaCl uptake. The lowered K^+ content caused by the presence of ouabain remained unaltered. However, technical reasons did not permit the direct measurement of membrane potentials in this preparation, so it could not be shown unequivocally that Cl^- was being actively transported.

5. CONCLUSIONS

One general and important conclusion that seems inescapable is that the major anion, Cl^-, cannot be viewed simply as an indifferent or passively distributed ion in the nervous system. Thus, intracellular Cl^- concentrations cannot *a priori* be calculated on the basis of resting membrane potentials, nor, conversely, can membrane potentials be calculated from Cl^- distributions. This behavior appears to be a consequence of the fact that the major pathways of Cl^- transport across the membranes of many cell types in the nervous system are by electrically neutral pathways such as the anion exchange system. In addition, electrically neutral active transport processes such as furosemide-sensitive coupled NaCl uptake and perhaps other active cation–anion coupled uptake systems may be present. One advantage of most of the Cl^-, HCO_3^-, and possibly the transport of other anoins occurring by such processes rather than through large, permanently open conductive channels may be that net movements of permeant anions accompanied by permeable cations such as K^+ will not result when large changes in membrane potentials occur, as they often do in the CNS. Such transport would result in transient changes in cell volume. Rather much of the anion transport appears to occur by mediated processes that can be controlled by transmitters or hormonal influences. Furthermore, the metabolic state of the CNS can have an influence via changing levels of CO_2 and H^+. Alteration in levels of ATP or co-ions may also control the functioning of anion pumps.

Many of the mediated Cl^- and HCO_3^- transport processes occurring in nonneuronal cells of the nervous system, such as astrocytes and choroidal epithelium, seem to be related to the functions of the blood–brain barrier and to have a secretory function. In addition, such transport may have a role in both intra- and extracellular pH control within the neuropil. In this regard, the astrocyte, because of its location and the presence of the anion exchange system and possibly carbonic anhydrase, may function to preferentially hydrate CO_2 produced by neuronal metabolism, thereby producing HCO_3^- and H^+ to exchange with Cl^- and Na^+, principally from the blood. Such processes seem to be affected by transmitters and may relate to neuron–astroglia interactions (see Section 4.6). They also might be expected to occur in the endothelial cells of brain capillaries and also oligodendrocytes which contain considerable amounts of carbonic anhydrase.[107,118] However, no swelling of oligodendrocytes is seen under conditions where there is considerable swelling of astrocytes. If such swelling occurs by the mechanism we have outlined in Section 4.6, an absence of such swelling in oligodendrocytes might be related to the absence of an anion exchange system. Alternatively, the system in oligodendrocytes may run in the opposite direction with uptake of HCO_3^- and efflux of Cl^-. Thus, $Cl^- – HCO_3^-$ exchange can be involved in outward Cl^- pumping, as seen for neurons, resulting in cytoplasmic shrinkage associated with myelin formation or, in satellite oligodendrocytes, control of the ionic milieu.[118] However, further studies are clearly needed since, although there is now a growing body of work on anion transport in astrocytes, there remains no information on the anion transport or membrane potential properties of oligodendrocytes.

In neurons, as might be expected, active transport of Cl^- seems related to electrical activity. Chloride is actively pumped out of the cell, making E_{Cl} more negative than E_m in many neurons, allowing inhibitory hyperpolarizing potentials to occur by a transmitter-induced increase in Cl^- conductance. Resting Cl^- conductance in these neurons is generally low, but perhaps not as low as in astrocytes. In some molluscs at least, there is also good evidence for an anion exchange system in neurons, related on the one hand to intracellular pH control and on the other to the outward pumping of Cl^- required for hyperpolarizing inhibitory potentials (see Sections 4.8, 4.9).

Malfunction of anion transport may well be a concomitant of many pathological conditions of the nervous system. However, apart from the marked swelling of astrocytes after ischemia, hypoxia, or trauma which seems related to Cl^- transport,[112,149] there seem to have been no documented cases for such an involvement. However, problems of CSF production, pH regulation, and neuronal excitability could all potentially involve some aspects of Cl^- and HCO_3^- transport. Since swelling of astroglia involves uptake of NaCl, the dissipation of the inward Na^+ gradient could also impair Na^+-dependent accumulation of transmitters[150,151] and other substances by astrocytes. Thus, Cl^- and other anions may well play a more dynamic role in the nervous system that has been previously supposed. Future studies of the roles of anion transport will require more complex techniques, novel biological preparations, and more sophisticated analysis and interpretation. The rewards of such increased effort will undoubtedly be fresh insights into the roles of Cl^- and other anions in the nervous system under normal and abnormal conditions.

ACKNOWLEDGMENTS. We should like to thank Drs. D. O. Carpenter, C. Edwards, and D. L. Martin for reading the manuscript and for helpful suggestions and Mrs. Elvira Graham for typing the manuscript. The work of the authors of this review was supported by grant 13042 fron NINCDS.

We should also like to thank all the authors and publishers for permission to reproduce figures. Specific acknowledgments of permissions are given in the legends.

We thank Dr. Kevin Barron and Ms. Susan Easton for supplying the electron micrographs used in Fig. 5.

REFERENCES

1. Tower, D. B., 1969, *Handbook of Neurochemistry*, Volume 1 (A. Lajtha, ed.), Plenum Press, New York, pp. 1–24.
2. Van Harreveld, A., 1966, *Brain Tissue Electrolytes*, Butterworths, London.
3. Boyle, P. J., and Conway, E. J., 1941, *J. Physiol. (Lond.)* 100:1–63.
4. Bryant, S. H., and Morales-Aguilera, A. J., 1971, *J. Physiol. (Lond.)* 219:367–383.
5. MacKnight, A. D. C., and Leaf, A., 1977, *Physiol. Rev.* 57:510–562.
6. Cabantchik, Z. I., Knauf, P. A., and Rothstein, A., 1978, *Biochim. Biophys. Acta* 515:239–302.
7. Frizzell, R. A., Field, M., and Schultz, S. G., 1979, *Am. J. Physiol.* 236:F1–F8.
8. Brodsky, W. A. (ed.), 1980, *Anion and Proton Transport, Annals of the New York Academy of Sciences*, Volume 341, New York Academy of Sciences, New York.
9. Frizzell, R. A., and Duffey, M. E., 1980, *Fed. Proc.* 39:2860–2864.
10. Katzman, R., and Pappius, H. M., 1973, *Brain Electrolytes and Fluid Metabolism*, Williams & Wilkins, Baltimore.
11. Katzman, R., 1970, *Handbook of Neurochemistry*, Volume 4 (A. Lajtha, ed.), Plenum Press, New York, pp. 313–327.
12. Harvey, J. A., and McLlwain, H., 1969, *Handbook of Neurochemistry*, Volume 2, (A. Lajtha, ed.), Plenum Press, New York, pp. 115–136.
13. Hodgkin, A. L., and Horowicz, P., 1959, *J. Physiol. (Lond.)* 148:127–160.
14. Macchia, D. D., Polimeni, P. I., and Page, E., 1978, *Am. J. Physiol.* 235:C122–C127.
15. Vaughan-Jones, R. D., 1979, *J. Physiol. (Lond.)* 295:111–137.
16. Russell, J. M., and Brodwick, M. S., 1979, *J. Gen. Physiol.* 73:343–368.
17. Ashley, C. C., Ellory, J. C., Lea, T. J., and Ramos, M., 1978, *J. Physiol. (Lond.)* 285:52P–53P.
18. Gunn, R. B., 1979, *Mechanisms of Intestinal Secretion* (H. J. Binder, ed.), Alan R. Liss, New York, pp. 25–43.
19. Wieth, J. O., 1979, *J. Physiol. (Lond.)* 294:521–539.
20. Kregenow, F. M., 1981, *Ann. Review Physiol.* 43:493–505.
21. Kimmich, G. A., and Christin, C.-S., 1978, *Am. J. Physiol.* 235:C73–C81.
22. Silva, P., Stoff, J., Field, M., Fine, L., Forrest, J. N., and Epstein, F. H., 1977, *Am. J. Physiol.* 283:F298–F306.
23. Cremaschi, D., Henin, S., and Meyer, G., 1979, *J. Membr. Biol.* 47:145–170.
24. Brodsky, W. A., Durham, J., and Ehrenspeck, G., 1979, *J. Physiol. (Lond.)* 287:559–573.
25. Bradbury, M., 1979, *The Concept of a Blood–Brain Barrier*, John Wiley & Sons, Chichester, New York, Brisbane, Toronto.
26. Davson, H., 1976, *J. Physiol. (Lond.)* 255:1–28.
27. Rapoport, S. I., 1976, *Blood–Brain Barrier in Physiology and Medicine*, Raven Press, New York.
28. Oldendorf, W. H., 1974, *Annu. Rev. Pharmacol.* 14:239–248.
29. Pardridge, W. M., Connor, J. D., and Crawford, I. L., 1975, *CRC Crit. Rev. Toxicol.* 3:159–199.
30. Brightman, M. W., and Reese, T. S., 1969, *J. Cell Biol.* 40:648–677.
31. Davson, H., and Welch, K., 1971, *Ion Homeostasis of the Brain, Alfred Benzon Symposium III* (B. K. Siesjo and S. C. Sorensen, eds.), Academic Press, New York, pp. 9–21.
32. Smith, Q. R., Woodbury, D. M., and Johanson, C. E., 1981, *J. Neurochem.* 37:107–116.
33. Smith, Q. R., Johanson, C. E., and Woodbury, D. M., 1981, *J. Neurochem.* 37:117–124.
34. Davson, H., and Segal, M. B., 1970, *J. Physiol. (Lond.)* 209:131–153.

35. Csaky, T. Z., 1969, *Handbook of Neurochemistry*, Volume 2 (A. Lajtha, ed.), Plenum Press, New York, London, pp. 49–69.
36. Davson, H., 1969, *Handbook of Neurochemistry*, Volume 2 (A. Lajtha, ed.), Plenum Press, New York, London, pp. 23–48.
37. Held, D., Fencl, V., and Pappenheimer, J. R., 1964, *J. Neurophysiol.* **27**:942–959.
38. Maren, T. H., 1979, *J. Appl. Physiol.* **47**:471–477.
39. Bourke, R. S., Gabelnick, H. L., and Young, O., 1970, *Exp. Brain Res.* **10**:17–38.
40. Abbott, J., Davson, H., Glen, I., and Grant, N., 1971, *Brain Res.* **29**:185–193.
41. Stein, W. D., 1967, *The Movement of Molecules Across Cell Membranes*, Academic Press, New York, London, pp. 62–64.
42. Milhorat, T. H., Hammock, M. K., Fenstermacher, J. D., Rall, D. P., and Levin, V. A., 1971, *Science* **173**:330–332.
43. Wright, E. M., 1972, *J. Physiol. (Lond.)* **226**:544–571.
44. Smith, Q. R., and Johanson, C. E., 1980, *Am. J. Physiol.* F399–F406.
45. Pollard, H. B., Pazoles, C. J., Creutz, C. E., Ramu, A., Strott, C. A., Ray, P., Brown, E. M., Aurbach, G. D., Tack-Foldman, K. M., and Shulman, N. R., 1977, *J. Supramol. Struct.* **7**:277–285.
46. Hardebo, J. E., and Johansson, B. B., 1980, *Acta Neuropathol. (Berl.)* **51**:33–38.
47. Friis, M. L., Paulson, O. B., and Hertz, M. M., 1980, *Microvasc. Res.* **20**:71–80.
48. Kern, D. A., 1960, *J. Chem. Educ.* **37**:14–23.
49. Wright, E. M., 1974, *J. Physiol. (Lond.)* **240**:535–566.
50. Schultz, S. G., and Curran, P. F., 1970, *Physiol. Rev.* **50**:637–718.
51. Pappius, H. M., 1969, *Handbook of Neurochemistry*, Volume 2 (A. Lajtha, ed.), Plenum Press, New York, London, pp. 1–10.
52. Bourke, R. S., Greenberg, E., and Tower, D. B., 1965, *Am. J. Physiol.* **208**:682–692.
53. Eccles, J. C., 1955, *The Physiology of Nerve Cells*, The Johns Hopkins Press, Baltimore, pp. 22–29.
54. Hertz, L., and Schousboe, A., 1975, *Int. Rev. Neurobiol.* **18**:141–211.
55. Pope, A., 1978, *Dynamic Properties of Glia Cells* (E. Schoffeniels, G. Franck, L. Hertz, and D. B. Tower, eds.), Pergamon Press, Oxford, New York, pp. 13–20.
56. Katzman, R., 1961, *Neurology (Minneap.)* **11**:27–31.
57. Hild, W., Chang, J. J., and Tasaki, I., 1958, *Experientia* **14**:220–221.
58. Bignami, A., Dahl, D., and Rueger, D. C., 1980, *Advances in Cellular Neurobiology*, Volume 1 (S. Fedoroff and L. Hertz, eds.), Academic Press, New York, pp. 285–310.
59. Garcia, J. H., Kalimo, H., Kamijyo, Y., and Trump, B. F., 1977, *Virchows Archiv. [Cell Pathol.]* **25**:191–206.
60. Jenkins, L. W., Povlishock, J. T., Becker, D. P., Miller, J. D., and Sullivan, H. G., 1979, *Acta Neuropathol. (Berl.)* **48**:113–125.
61. Kraig, R. P., Nicholson, C., and Phillips, J. M., 1978, *Soc. Neurosci. Abstr.* **4**:66.
62. Phillips, J. M., and Nicholson, C., 1979, *Brain Res.* **173**:567–571.
63. Nicholson, C., 1980, *Neurosci. Res. Program Bull.* **18**(2):285–289.
64. Hossman, K.-A., Sakaki, S., and Zimmermann, V., 1977, *Stroke* **8**:71–81.
65. Loeschcke, H. H., and Ahmad, H. R., 1980, *Biophysics and Physiology of Carbon Dioxide* (C. Bauer, G. Gros, and H. Bartels, eds.), Springer-Verlag, Berlin, Heidelberg, New York, pp. 439–448.
66. Hertz, L., Schousboe, A., and Weiss, G. B., 1970, *Acta Physiol. Scand.* **79**:506–515.
67. Anderson, R. E., Michenfelder, J. D., and Sundt, T. M., Jr., 1980, *Anesthesiology* **52**:201–206.
68. Takato, M., and Goldring, S., 1979, *J. Comp. Neurol.* **186**:173–188.
69. Elliott, K. A. C., 1969, *Handbook of Neurochemistry*, Volume 2 (A. Lajtha, ed.), Plenum Press, New York, London, pp. 103–114.
70. Franck, G., 1972, *The Structure and Function of Nervous Tissue*, Volume VI (G. H. Bourne, ed.), Academic Press, New York, pp. 417–465.
71. Picker, S., Pieper, C. F., and Goldring, S., 1980, *Soc. Neurosci. Abstr.* **6**:326.
72. Elliott, K. A. C., 1955, *Can. J. Biochem. Physiol.* **33**:466–480.
73. Varon, S., and McIlwain, H., 1961, *J. Neurochem.* **8**:263–275.
74. Pappius, H. M., and Elliott, K. A. C., 1956, *Can. J. Biochem. Physiol.* **34**:1053–1067.

75. Bourke, R. S., and Tower, D. B., 1966, *J. Neurochem.* **13**:1071–1097.
76. Bourke, R. S., and Tower, D. B., 1966, *J. Neurochem.* **13**:1099–1117.
77. Ibata, Y., Piccoli, F., Pappas, G. D., and Lajtha, A., 1971, *Brain Res.* **30**:137–158.
78. Bourke, R. S., Kimelberg, H. K., West, C. R., and Bremer, A. M., 1975, *J. Neurochem.* **25**:323–328.
79. Bourke, R. S., Kimelberg, H. K., and Nelson, L. R., 1976, *Brain Res.* **105**:309–323.
80. Lund-Andersen, H., and Hertz, L., 1970, *Exp. Brain Res.* **11**:199–212.
81. Tower, D. B., and Bourke, R. S., 1966, *J. Neurochem.* **13**:1119–1137.
82. Franck, G., and Schoffeniels, E., 1972, *J. Neurochem.* **19**:395–402.
83. Schousboe, A., 1972, *Exp. Brain Res.* **15**:521–531.
84. Moller, M., Mollgard, K., Lund-Andersen, H., and Hertz, L., 1974, *Exp. Brain Res.* **22**:299–314.
85. Bourke, R. S., 1969, *Exp. Brain Res.* **8**:219–231.
86. Katz, B., 1966, *Nerve, Muscle and Synapse*, McGraw-Hill, New York, pp. 60–61.
87. Bourke, R. S., Daze, M. A., and Kimelberg, H. K., 1978, *Dynamic Properties of Glial Cells* (E. Schoffeniels, G. Franck, L. Hertz, and D. B. Tower, eds.), Pergamon Press, New York, pp. 337–346.
88. Bourke, R. S., Nelson, K. M., Naumann, R. A., and Young, O. M., 1970, *Exp. Brain Res.* **10**:427–446.
89. Bourke, R. S., and Nelson, K. M., 1972, *J. Neurochem.* **19**:663–685.
90. Takagaki, G., 1972, *J. Neurochem.* **19**:1737–1751.
91. Vargas, O., and Orrego, F., 1976, *J. Neurochem.* **26**:31–34.
92. Van Calker, D., and Hamprecht, B., 1980, *Advances in Cellular Neurobiology*, Volume 1 (S. Fedoroff and L. Hertz, eds.), Academic Press, New York, pp. 31–67.
93. Daly, J., 1977, *Cyclic Nucleotides in the Nervous System*, Plenum Press, New York, London.
94. Zadunaisky, J. A., 1979, *Mechanisms of Intestinal Secretion* (H. J. Binder, ed.), Alan R. Liss, New York, pp. 53–64.
95. Narumi, S., Kimelberg, H. K., and Bourke, R. S., 1978, *J. Neurochem.* **31**:1479–1490.
96. Church, G. A., Kimelberg, H. K., and Sapirstein, V. S., 1980, *J. Neurochem.* **34**:873–879.
97. Bourke, R. S., Kimelberg, H. K., and Daze, M. A., 1978, *Brain Res.* **154**:196–202.
98. Bourke, R. S., Waldman, J. B., Kimelberg, H. K., Barron, K. D., SanFilippo, B. D., Popp, A. J., and Nelson, L. R., 1981, *J. Neurosurg.* **55**:364–370.
99. Kimelberg, H. K., 1974, *J. Neurochem.* **22**:971–976.
100. Kukes, G., Elul, R., and De Vellis, J., 1976, *Brain Res.* **104**:71–92.
101. Orkand, R. K., 1977, *Handbook of Physiology—The Nervous System*, Volume I(2) (E. R. Kandel, ed.), American Physiological Society, Bethesda, pp. 855–875.
102. Sensenbrenner, M., 1977, *Cell, Tissue, and Organ Cultures in Neurobiology* (S. Fedoroff and L. Hertz, eds.), Academic Press, New York, pp. 191–213.
103. Stieg, P. E., Kimelberg, H. K., Mazurkiewicz, J. E., and Banker, G. A., 1980, *Brain Res.* **199**:493–500.
104. Kimelberg, H. K., Bowman, C., Biddlecome, S., and Bourke, R. S., 1979, *Brain Res.* **177**:533–550.
105. Kimelberg, H. K., Stieg, P. E., and Mazurkiewicz, J. E., 1981, *Trans. Am. Soc. Neurochem.* **12**:224.
106. Roussel, G., Delaunoy, J-P., Nussbaum, J-L, and Mandel, P., 1979, *Brain Res.* **160**:47–55.
107. Ghandour, M. S., Langley, O. K., Vincendon, G., and Gombos, G., 1979, *J. Histochem. Cytochem.* **27**:1634–1637.
108. Kimelberg, H. K., Biddlecome, S., and Bourke, R. S., 1979, *Brain Res.* **173**:111–124.
109. Kimelberg, H. K., 1981, *Biochim. Biophys. Acta*, **646**:179–184.
110. Kimelberg, H. K., and Biddlecome, S., 1980, *Soc. Neurosci. Abstr.* **6**:548.
111. Kimelberg, H. K., Biddlecome, S., Bourke, R. S., and Bowman, C., 1978, *Frontiers of Biological Energetics*, Volume I (P. L. Dutton, J. S. Leigh, and A. Scarpa, eds.), Academic Press, New York, pp. 563–572.
112. Kimelberg, H. K., Bourke, R. S., Stieg, P. E., Barron, K. D., Hirata, H., Pelton, E. W., and Nelson, L. R., 1981, *Head Injury: Basic and Clinical Aspects* (R. G. Grossman and P. L. Gildenberg, eds.), Raven Press (in press).

113. Gill, T. H., Young, O. M., and Tower, D. B., 1974, *J. Neurochem.* **23**:1011–1018.

114. Bracho, H., Orkand, P. M., and Orkand, R. K., 1975, *J. Neurobiol.* **6**:395–410.

115. Lowe, D. A., 1978, *Brain Res.* **148**:347–363.

116. Tschirgi, R. S., 1958, *Biology of Neuroglia* (W. F. Windle, ed.), Charles C Thomas, Spring-field, Illinois, pp. 130–138.

117. Giacobini, E., 1961, *Science* **134**:1524–1525.

118. Langley, O. K., Ghandour, M. S., Vincendon. G., and Gombos, G., 1980, *Histochem. J.* **12**:473–483.

119. White, F. P., Dutton, G. R., and Norenberg, M. D., 1980, *J. Neurochem.* **36**:328–332.

120. Firth, J. A., 1977, *Experientia* **33**:1093–1094.

121. Goldstein, G. W., 1979, *J. Physiol. (Lond.)* **286**:185–195.

122. Kimelberg, H. K., Narumi, S., Biddlecome, S., and Bourke, R. S. 1978, *Dynamic Properties of Glia Cells* G. Franck, L. Hertz, E. Schoffeniels, and D. B. Tower. eds.), Pergamon Press, Oxford, New York, pp. 347–357.

123. Bonting, S. L., de Pont, J. J. H. H. M., Van Amelsvoort, S. M. M., and Schrijen, J. J., 1980, *Ann. N.Y. Acad. Sci.* **341**:335–356.

124. Kimelberg, H. K., and Bourke, R. S., 1973, *J. Neurochem.* **20**:347–359.

125. Keynes, R. D., 1963, *J. Physiol. (Lond.)* **169**:690–705.

126. Russell, J. M., 1979, *J. Gen. Physiol.* **73**:801–818.

127. Eccles, J. C., Eccles, R. M., and Ito, M., 1964, *Proc. Soc. Lond. [Biol.]* **160**:197–210.

128. Lux, H. D., 1971, *Science* **173**:555–557.

129. Thomas, R. C., 1978, *Ion-Sensitive Intracellular Microelectrodes*, Academic Press, London, New York, San Francisco.

130. Russell, J. M., and Brown, A. M., 1972, *J. Gen. Physiol.* **60**:499–518.

131. Thomas, R. C., 1977, *J. Physiol. (Lond.)* **273**:317–338.

132. Ascher, P., Kunze, D., and Neild, T. O., 1976, *J. Physiol. (Lond.)* **266**:441–464.

133. Pieri, C., Z.-Nagy, I., Z.-Nagy, V., Giuli, C., and Bertoni-Fredda, C., 1977, *J. Ultrastruct. Res.* **59**:320–331.

134. Allakhverdov, B. L., Burovina, I. V., Chmykhova, N. M., and Shapovalov, A. I., 1980, *Neuroscience* **5**:2023–2031.

135. Kerkut, G. A., and Thomas, R. C., 1964, *Comp. Biochem. Physiol.* **11**:199–213.

136. Hagiwara, S., Kusano, K., and Saito, N., 1961, *J. Physiol. (Lond.)* **155**:470–489.

137. Chiarandini, D. J., and Gerschenfeld, H. M., 1967, *Science* **156**:1595–1596.

138. Ozawa, S., and Tsuda, K., 1973, *J. Neurophysiol.* **36**:805–816.

139. Carpenter, D. O., Swann, J. W., and Yarowsky, P. J., 1977, *J. Neurobiol.* **8**:119–132.

140. Llinás, R., Baker, R., and Precht, W., 1974, *J. Neurophysiol.* **37**:522–531.

141. Thomas, R. C., 1974, *J. Physiol. (Lond.)* **238**:159–180.

142. Kawai, N., and Niwa, A., 1978, *Brain Res.* **137**:365–368.

143. Thomas, R. C., 1976, *J. Physiol. (Lond.)* **255**:715–735.

144. Russell, J. M., and Boron, W. F., 1976, *Nature* **264**:73–74.

145. Becker, B. F., and Duhm, J., 1978, *J. Physiol. (London)* **282**:149–168.

146. Russell, J. M., 1978, *Biophys. J.* **22**:131–137.

147. Russell, J. M., 1980, *Ann. N.Y. Acad. Sci.* **341**:510–523.

148. Marchbanks, R. M., and Campbell, C. W. B., 1976, *J. Neurochem.* **26**:973–980.

149. Tower, D. B., 1979, *Advances in Neurology*, Volume 25 (M. Goldstein, L. Bolis, S. Gorini, C. Fieschi, and C. H. Millikan, eds.), Raven Press, New York. pp. 39–63.

150. Schousboe, A., 1977, *Cell, Tissue and Organ Cultures in Neurobiology* (S. Federoff and L. Hertz, eds.), Academic Press, New York, pp. 441–446.

151. Pelton, E. W., Kimelberg, H. K., Shipherd, S. V., and Bourke, R. S., 1981, *Life Sci.* **28**:1655–1663.

3

Calcium in the Nerve Cell

Arsélio P. Carvalho

1. INTRODUCTION

During the last decade, the study of Ca^{2+}-regulated functions has expanded explosively: Ca^{2+} is now regarded as the triggering agent of many biological reactions and has attained the status of a second messenger.[1,2] The functions of Ca^{2+} in the nervous system are many. At the cell surface, Ca^{2+} alters the membrane stability and permeability[3-5] and is involved in nerve impulse propagation.[4,6-8] On stimulation, there is an increased influx of Ca^{2+} into the nerve cell which couples excitation to neurotransmitter release.[4,9-11] The actions of the neurotransmitters on the postsynaptic membrane may also involve Ca^{2+} in the sequence of events that leads to the response of the postsynaptic cell.[12] It is now recognized that Ca^{2+} regulates the activity of enzymes such as adenylate cyclase,[13-15] phosphodiesterase,[14,15] protein kinases,[12,16-18] Ca^{2+}-ATPases,[19,20] and tyrosine hydroxylase,[21] among others. Furthermore, the action of narcotic drugs in the nervous system is mediated by alterations in the Ca^{2+} content of the nerve cell,[22-26] and the action of local anesthetics occurs through the displacement of Ca^{2+}.[5,27]

The major unifying concept of the mechanism of action of Ca^{2+} relies on the observation that most events in the cell triggered by Ca^{2+} are regulated through Ca^{2+}-binding proteins, of which calmodulin is the most ubiquitous.[28-31] It appears that the Ca^{2+} sensitivity of most Ca^{2+}-dependent enzymes and biological phenomena is in fact a sensitivity to the Ca-calmodulin complex[29,30,32-36] which in many instances controls the phosphorylation of specific proteins in the cell which trigger biological activity.[12,16,33,37-39]

Not all features of the Ca^{2+} action on the nerve cell are treated in this chapter because some nervous phenomena in which Ca^{2+} has clearly defined functions are interrelated with topics covered in other chapters in this series and can most appropriately be treated there. Space limitations also impose a selection of areas. The areas covered are those which, in terms of their accentuated development during the last 10 years, are clearly of importance.

Arsélio P. Carvalho • Center for Cell Biology, Department of Zoology, University of Coimbra, 3049 Coimbra Codex, Portugal.

Older areas, such as excitability and the action of local anesthetics, where Ca^{2+} was shown many years ago to play an important role, are not covered, but there are reviews available.[3,5,40] The reader is also referred to reviews on the effect of Ca^{2+} on membrane conductance.[4,7]

2. TECHNOLOGY FOR MEASURING CALCIUM

The intense interest in the physiological processes that are regulated by Ca^{2+} has instigated the development of many techniques for measuring this ion. Because of their widespread utilization, I review the principal techniques currently utilized to measure Ca^{2+} in neurochemical studies. These techniques permit measurement of total Ca^{2+}, free Ca^{2+}, and bound Ca^{2+}. In addition to the methods described in this section, the radiotracer method utilizing ^{45}Ca has been widely used with ample success in neurochemistry, but it will not be described here, since it does not deviate from other general radioisotope methods.

2.1. Atomic Absorption Spectrometry

The estimation of total Ca^{2+} content in brain tissue has been achieved by atomic absorption spectrometry.[41–43] In general, the assay methods that have been devised for the analysis of total Ca^{2+} in biological materials are free from interference, and their sensitivity approaches the micromolar range. The preparation of tissue for Ca^{2+} analysis requires either extraction with trichloroacetic acid[42] to liberate Ca^{2+} from its binding sites or, alternatively, ashing of the samples.[44] Interference by phosphate is avoided by using 1% lanthanum in all samples and standards.[45]

This method does not give information about the state (bound or free) or location of the Ca^{2+} and has limited application to kinetic studies of rapid processes. Atomic absorption spectrometry, however, is suitable for estimating calcium in whole brain homogenates or in synaptosomes and other fractions,[42,43] and the method is also adaptable for the analysis of other ions (Mg^{2+}, K^+, etc.) in the same sample.[43]

2.2. Calcium Electrodes

There have been several reports of the use of electrodes to follow ionized Ca^{2+} concentration,[46,47] and an ultrasensitive method was described by Madeira[48] which permits following changes in Ca^{2+} concentration of the order of 1–10 nmol even in the presence of a relatively large background of Ca^{2+} and other cations. The method also works at concentrations of Ca^{2+} as low as 10^{-6} M and was successfully applied to the study of Ca^{2+} transport by sarcoplasmic reticulum[48]; in principle, it is possible to apply it to studies of Ca^{2+} movements in other biological systems such as brain plasma membrane vesicles and synaptosomes.

There are now available Ca^{2+} electrodes whose selectivities for Ca^{2+} over H^+, Mg^{2+}, Na^+, and Zn^{2+} are 25,000, 26,000, 2800, and 7100, respectively.[49]

Calcium microelectrodes with tip diameters of 1 μm have now been prepared and used in a variety of applications[50-51] including measurements of the variations in the Ca^{2+} concentration in the extracellular space during brain activity.[51] Apparently, the extracellular Ca^{2+} concentration may decrease by 0.4 mM when a neural population is active.[51] Microelectrodes were recently utilized to measure intracellular free Ca^{2+} in the salivary gland of the insect *Phormia regina*, and continuous monitoring of the internal ionized Ca^{2+} concentration during serotonin-induced saliva release provided direct evidence for alterations in $[Ca^{2+}]_i$ during activity.[52] Some Ca^{2+} electrodes have been reported to measure Ca^{2+} activities in the range of 10^{-1} M to 10^{-6} M in unbuffered and in the range of 10^{-1} M to 10^{-8} M in Ca^{2+}-buffered systems, respectively.[49]

However, the most successful procedures currently employed to measure ionized Ca^{2+} in biological systems make use of the photoluminescent protein, aequorin, and of metallochromic indicators described in the next section.

2.3. Indicators for Measuring Ionized Calcium

2.3.1. Luminescent Protein (Aequorin)

Aequorin is a protein extracted from the jellyfish *Aequora aequorea*, and has the property of emitting light in the presence of Ca^{2+}. The properties and use of aequorin have been reviewed by Blinks *et al.*[53] This protein can be microinjected into a cell, and the internal Ca^{2+} concentration can be determined from its luminescence. Time resolution is in the millisecond range, and the sensitivity is at least down to 10^{-7} M Ca^{2+}. The mechanism of the light-emitting reaction is not known, but the method has been used to determine Ca^{2+} concentration in the range present inside cells.[53]

The relationship between a particular rate of light emission by aequorin–Ca complexes and the absolute concentration of ionized Ca^{2+} must be determined if quantitative results are to be obtained. The calibration of the aequorin method is performed by using Ca–EGTA buffers, and injection of such buffers into cells allows the fixation of the ionized Ca^{2+} at values determined by the ratio Ca–EGTA/free EGTA.[54] However, EGTA can also bind Mg^{2+} which is normally present in axons at concentrations 100,000 times higher than Ca^{2+}, and even a slight degree of binding can affect calibrations in Mg^{2+}-containing solutions.

Nevertheless, utilizing aequorin luminescence and Ca–EGTA buffers in squid axons, Baker *et al.*[6,55] estimated a value for intracellular ionized Ca^{2+} of about 0.3 μM and were able to monitor Ca^{2+} movements during stimulation. Similar experiments have been performed by Hallet and Carbone[56] who also detected luminescence transients associated with multiple action potentials. Llinás and Nicholson[10] also injected aequorin into presynaptic terminals of the squid giant synapse and showed transients in $[Ca^{2+}]_i$ associated with transmitter release. In a more refined experimental approach, DiPolo *et al.*[57] measured $[Ca^{2+}]_i$ in internally dialyzed squid axons.

The aequorin method has been used in a wide range of experiments to measure the concentration of ionized Ca^{2+},[53,58] but it has preferentially been employed in large cells such as squid axons[6,55] because only small quantities can be introduced into small cells, and the sensitivity of the method becomes considerably reduced.

The aequorin reaction is selective for Ca^{2+} and lanthanides, has a much lower sensitivity for Sr^{2+},[59] and, although Mg^{2+} does not initiate luminescence, it influences the luminescence induced by Ca^{2+}. The limit of time resolution of aequorin luminescence is of the order of 10 ms which imposes limitations on measuring the fastest physiological processes.[53,58]

2.3.2. Metallochromic Indicators

Water-soluble metallochromic indicators have been used to follow the concentration of ionized Ca^{2+} in cells or membrane suspensions[57,58,60,61] and also to estimate the Ca^{2+} uptake and binding by synaptosomes.[62,63] These dyes complex Ca^{2+} with characteristic shifts in their absorption spectra which can be measured by dual-wavelength differential spectroscopy.[61]

One of these indicators utilized in biological systems is murexide.[60–62] Murexide complexes very rapidly with Ca^{2+} (relaxation time <5 μs) and binds more strongly to Ca^{2+} than to Mg^{2+}, therefore being suitable to determine Ca^{2+} concentration in media containing Mg^{2+}. Murexide diffuses very rapidly and has a high molar absorption coefficient (1.25 mM^{-1} at pH 6.8), which allows the use of very low concentrations of murexide. Murexide has relatively low affinity for Ca^{2+} (apparent dissociation constant 3 × 10^{-5} M), which permits measuring Ca^{2+} concentrations without greatly perturbing the Ca^{2+} distribution. However, this property also restricts the sensitivity of murexide to levels of Ca^{2+} in the micromolar range.[61]

Another metallochromic indicator used to follow the ionized Ca^{2+} concentration is arsenazo III whose affinity for Ca^{2+} is one or two orders of magnitude greater than that of murexide.[57,61] The dye binds both Ca^{2+} and Mg^{2+} but shows different absorption changes when chelated by either cation. DiPolo et al.[57] employed arsenazo III to monitor the free Ca^{2+} in internally dialyzed squid axons and estimated the axoplasmic Ca^{2+} activity to be 50 nM. Therefore, the major advantage of arsenazo III is its high affinity for Ca^{2+} which results in large absorbance changes following relatively small Ca^{2+} concentration changes and thus makes possible measurements of Ca^{2+} in the nanomolar range. On the other hand, it exhibits poor selectivity and a slow rate of complexation with Ca^{2+}.[59] Furthermore, the high affinity for Ca^{2+} (Table I) is undesirable because arsenazo III becomes a significant Ca^{2+} buffer in the system and, at relatively high Ca^{2+} concentrations (>10 μM), the absorbance changes are no longer linear with ΔCa^{2+} concentrations.

Recently, the use of another metallochromic indicator in biological systems, antipyrylazo III, has been described.[64] Scarpa et al.[64] point out the following advantages of this dye over arsenazo III: (1) the dissociation constant is five times greater than that of arsenazo III (Table I); (2) neither the free nor the Mg^{2+}-bound forms of the dye have significant absorbance between 660 and 800 nm where a significant absorbance peak for the Ca^{2+}-bound form exists. Antipyrylazo III is, therefore, a "middle range" Ca^{2+} indicator which can be

Table I
Structural Formulas of Calcium Indicators and Dissociation Constants (K_D) or
Sensitivity Range for Calcium

Indicator	Structure	K_D	Suitable λ range (nm)
Murexide		1–3 mM	540–507
Arsenazo III		15–15 μM	675–685 or 650–685
Antipyrylazo III		95–380 μM	670–690 or 720–790
Chlorotetracycline		0.44 mM (in water) 9.0 μM (Ca^{2+}, 70% methanol) 25.0 μM (Mg^{2+}, 70% methanol)	$\lambda_{max} = 530$ nm
Chromophore of aequorin		Sensitivity range: 10^{-7}–10^{-4} M Ca^{2+}	

used for measuring, with adequate sensitivity, Ca^{2+} binding and transport in the Ca^{2+} concentration range where arsenazo III and murexide are less effective.

2.4. Chlorotetracycline, an Indicator for Bound Calcium

The principle of utilization of chlorotetracycline (CTC) to detect Ca^{2+} bound to biological membranes resides in the fact that CTC preferentially forms a complex with Ca^{2+} bound to the membrane phase,[42,58,65–68] and there is an increase in the fluorescence signal of CTC when it binds to the membrane. In apolar environments, the affinity of CTC for Ca^{2+} ($K_D = 9$ μM in 70% methanol) is greater than for Mg^{2+} ($K_D = 25$ μM in 70% methanol), suggesting that the method is relatively specific for Ca^{2+} over Mg^{2+}.[65]

The CTC fluorescence has been utilized to monitor Ca^{2+} movements in several biological systems including Ca^{2+} movements in brain membranes.[42,58,68,69] Most of the studies with CTC have been done with subcellular organelles that are able to accumulate Ca^{2+} by energy-linked transport.[66–70] In these cases, Ca^{2+} accumulates inside the membrane vesicles in high concentrations and binds to internal binding sites. Chlorotetracycline is able to migrate passively across the membrane and complexes with the cation on the membrane surface. This accumulation of Ca^{2+} is then reflected in an increase in CTC fluorescence.[58]

The employment of CTC to study Ca^{2+} movements in intact cells has not been extensively explored. However, Hallet *et al.*[69] reported alterations in CTC fluorescence during the generation of action potentials after administration of the probe externally or directly into the axoplasm of squid giant axons and lobster nerve fibers. These authors interpreted their results to mean that the fluorescence response to depolarization of the axon represents an alteration of the Ca^{2+} environment in the membrane, giving rise to a change in the fluorescence quantum yield.

Chlorotetracycline has also been utilized to follow the passive interaction of Ca^{2+} with brain membranes,[42] and, in optimal conditions, the changes in fluorescence have been successfully correlated with direct measurements of the changes in Ca^{2+} binding to the membranes. However, the parallelism between Ca^{2+} and CTC binding to the synaptosomal membranes, obtained in a predominantly sucrose medium, becomes less distinct when media simulating physiological conditions are utilized, which limits the usefulness of the method.[42]

Other limitations on the employment of CTC result from the Ca-binding properties of the molecule. Thus, at high Ca^{2+} and Mg^{2+} concentrations in the medium, the divalent cations may chelate most of the CTC added. This problem can be avoided by decreasing the concentration of divalent cations in the medium. Le Breton *et al.*[71] also tested the effect of CTC on platelet aggregation and concluded that CTC concentrations higher than 100 μM inhibit aggregation. Thus, when CTC is employed as an indicator, it should not be used above 100 μM, since higher levels may influence the response being monitored. In a recent paper, Gains[72] points out the difficulty in distinguishing between the fluorescence signal arising from CTC acting as a probe for divalent cations bound to specific sites on the membranes and that arising as a probe of neutral or negative regions of the membrane.

2.5. Methods for Localizing Calcium in the Cell

2.5.1. Fixation of Tissue

To determine the precise location of Ca^{2+} in the cell, it is necessary to carry out morphological studies at the ultrastructural level. However, before histological observation is possible, it is necessary to fixate the Ca^{2+} where it exists in the cell. This is achieved either by precipitating the Ca^{2+} by complexing molecules such as pyroantimonate and oxalate or by freeze-substitution or freeze-drying in the presence or absence of a fixing ion.[58,73,74]

The chemical fixation techniques in the presence of Ca^{2+}-precipitating agents are followed by dehydration, embedding, and sectioning. The potential flaw of this technique is that fixation can disrupt membrane permeability barriers and Ca^{2+}-binding proteins so that Ca^{2+} might migrate considerable distances before being precipitated. An alternative approach to fixate the Ca^{2+} where it exists in the cell is to freeze the tissue directly.[58,73] Freezing can be complete in milliseconds, so that Ca^{2+} does not migrate significantly before freezing occurs. Rapid freezing of the tissue in a liquid such as Freon® or isopentane is followed by direct microprobe analysis of the frozen hydrated tissue, or the water is withdrawn either by freeze-drying, in which ice is sublimated at low temperature, or by freeze-substitution, in which the tissue is treated with ethanol in order to substitute ethanol for ice.[58,73,74]

Pyroantimonate, in addition to precipitating Ca^{2+}, also precipitates Mg^{2+} and Na^+,[75] so that the appearance of a precipitate does not necessarily mean the presence of Ca^{2+}, although the pyroantimonate prevents the migration of Ca^{2+}, even if not specific for it, and also increases the electron density so that the precipitates may be observed easily with the electron microscope. But unlike oxalate, which does not easily cross biological membranes, pyroantimonate can fix Ca^{2+} at all sites in the cell because it crosses biological membranes readily. Oxalate is relatively specific for Ca^{2+}, so that an oxalate precipitate in tissue is most likely to be that of a Ca^{2+} salt. However, oxalate is known to increase Ca^{2+} accumulation in some cell organelles[76] which may introduce an error when it is used to fixate Ca^{2+} in the cell.

2.5.2. Identification of Calcium by Microanalytic Spectroscopy

The various fixation techniques attempt to prevent Ca^{2+} migration before the specimen is analyzed by electron microscopy. The identification of the electron-dense spots in the tissue is now accomplished by one of several microanalytical spectroscopic techniques, namely, electron probe, ion probe, proton probe, X-ray probe, and laser probe analysis.[58,77] Of these, the electron probe analysis has been the most widely employed because it requires only slight modifications of the normal electron microscope and has good sensitivity and resolution. The theory and practice of the technique have been described recently by Lechene and Warner.[77] The technique depends on the release of X-rays by the specimen being analyzed when a beam of high-energy electrons hits it. The X-rays released have energies and frequencies characteristic of the atom that emits them. Thus, it is possible to determine which elements are present on the basis of the X-ray spectrum obtained. The spatial limit of the electron probe technique is about 0.1 μm and depends on the minimum size of the probe and on the limits of detection of the system. With very small probes of diameter 0.1 μm and sections 0.1 μm thick, Ca^{2+} can be detected at a concentration of 24 μmol/g tissue.[58]

In practice, the electron probe has been utilized to detect intracellular

Ca^{2+} at sites where it is in high concentrations because of its limited ability to detect low Ca^{2+} concentrations. Intracellular Ca^{2+} granules lining the internal surface of squid axons and ganglia have been measured by this technique.[74,78,79] The technique has also been applied to localize Ca^{2+} in synaptosomes[80,300] and many other biological materials.[58]

2.5.3. Autoradiography

Autoradiography of ^{45}Ca may be used with the light and electron microscopes. The principle of the technique relies on introducing ^{45}Ca into the cell or biological material and, subsequently, fixing the cell to prevent redistribution of the isotope, followed by preparation of sections which are placed in direct contact with a photographic emulsion to develop spots caused by the radioactivity of the ^{45}Ca. Subsequent microscopic examination reveals both the morphology of the biological preparation and the appearance of dense silver grains from the exposed emulsion. The results permit evaluation of both the amount and distribution of the $^{45}Ca^{2+}$.[58,81,82]

This technique has been applied to many tissues, including nerve,[58,81,82] and is potentially a very sensitive technique. The technique is subject to all the fixation artifacts described previously.

2.6. Calcium–EGTA Buffers: A Practical Approach

Calcium–EGTA buffers containing stabilized concentrations of ionized Ca^{2+} have been utilized to control the intracellular free Ca^{2+}.[6,54,83] Very often these Ca^{2+} buffers are utilized to regulate the free Ca^{2+} in vitro to permit experiments to be performed at controlled $[Ca^{2+}]$. The advantage of EGTA is that it permits buffering of the medium with respect to free Ca^{2+} in the region of concentrations at which this ion activates many biological phenomena and enzymes such as Ca^{2+}-ATPase,[19,20] adenylate cyclase,[13–15] and phosphodiesterase.[14,15] It is fortunate that the dissociation constant for Ca–EGTA at physiological pH values is of the order of the concentrations of Ca^{2+} that activate these processes (i.e., about 10^{-7} M).[54,84] Furthermore, EGTA does not penetrate biological membranes easily, so that it can be utilized to buffer Ca^{2+} on one side of a biological membrane without disturbing the Ca^{2+} concentration across the membrane. Therefore, EGTA has been widely used as a Ca^{2+} buffer. In this section, I present some practical aspects related to the preparation and utilization of Ca–EGTA buffers.

The concentration of free Ca^{2+} can be stabilized to a desired value between pCa values of 5 and 8 by adjusting the ratio Ca–EGTA/free EGTA. At pH 7.0, the $-\log K$ for Ca–EGTA is about 6.7, K being the apparent dissociation constant of the following equation:

$$[Ca-EGTA] \overset{K}{\rightleftharpoons} [Ca] + [EGTA] \qquad [1]$$

This equation can be represented in a form analogous to that used for pH buffers:

$$pCa = pK + \log [EGTA]/[Ca–EGTA] \qquad\qquad [2]$$

in which $pCa = -\log [Ca]$.

In the range of pCa values above about 5, and in the presence of excess EGTA, the calculations can be done quite simply by approximation in which $[Ca–EGTA]$ is made equal to the total Ca^{2+} present, and the free $[EGTA]$ is equal to the total ligand concentration less the concentration of Ca^{2+} added.

The pH has a large influence on the Ca–EGTA dissociation constant: EGTA has four dissociable H^+, but normally only the binding of Ca^{2+} by the forms of EGTA with three and four negative charges is of importance. As a first approximation, the value of the $-\log K$ (pK) increases by two units for each pH unit increase in the pH range of 5–8, so that accurate pH values of the medium must be taken in calculating the true pCa values when using EGTA buffers. Thus, it is necessary to use the appropriate apparent pK value for the particular pH at which one works. Calculations of these apparent pK values have to be made, and these concepts are presented in several publications.[54,84]

Utilizing a true dissociation constant of approximately 10^{-11},[84] it can be calculated that for pH values of 6.75, 7.00, and 7.40, the apparent dissociation constants for the interaction between Ca^{2+} and EGTA are 6.20, 6.72, and 7.50, respectively.* Figure 1 represents the relationship existing between pCa and

---

* Calculation of the apparent dissociation constant, K, for CaEGTA at various pH values can be done as follows (Note: constants are only approximate, since values given in the literature do not agree). EGTA in the acid form has four dissociable H^+; i.e., H_4EGTA dissociates as follows:

$$H_4EGTA \underset{pK_1=2.00}{\rightleftharpoons} H^+ + H_3EGTA^- \underset{pK_2=2.68}{\rightleftharpoons} H^+ + H_2EGTA^{2-} \underset{pK_3=8.85}{\rightleftharpoons}$$

$$H^+ + HEGTA^{3-} \underset{pK_4=9.43}{\rightleftharpoons} H^+ + EGTA^{4-}$$

At physiological pH values, Ca^{2+} interacts predominantly with $EGTA^{4-}$, $EGTA^{3-}$, and $EGTA^{2-}$. Therefore, if we let EGTA be represented by A, the species that interact with Ca^{2+} are A^{4-}, A^{3-}, and A^{2-}. Therefore:

$$K = [A^{4-} + A^{3-} + A^{2-}][Ca^{2+}]/[ACa]$$

We also have the following dissociation constants according to the first equation for the total dissociation of H^+ from H_4EDTA:

$$[A^{4-}][Ca^{2+}]/[ACa] = 10^{-11} = K_{Ca}$$

$$[A^{4-}][H^+]/[HA^{3-}] = 3.72 \times 10^{-10} = K_4$$

$$[A^{3-}][H^+]/[HA^{2-}] = 1.41 \times 10^{-9} = K_3$$

Substituting in the equation for K, we have:

$$K = [A^{4-}][Ca^{2+}]/[ACa] + [Ca^{2+}][A^{4-}][H^+]/[ACa][K_4]$$

$$+ [Ca^{2+}][H^+][A^{4-}][H^+]/[ACa][K_3][K_4]$$

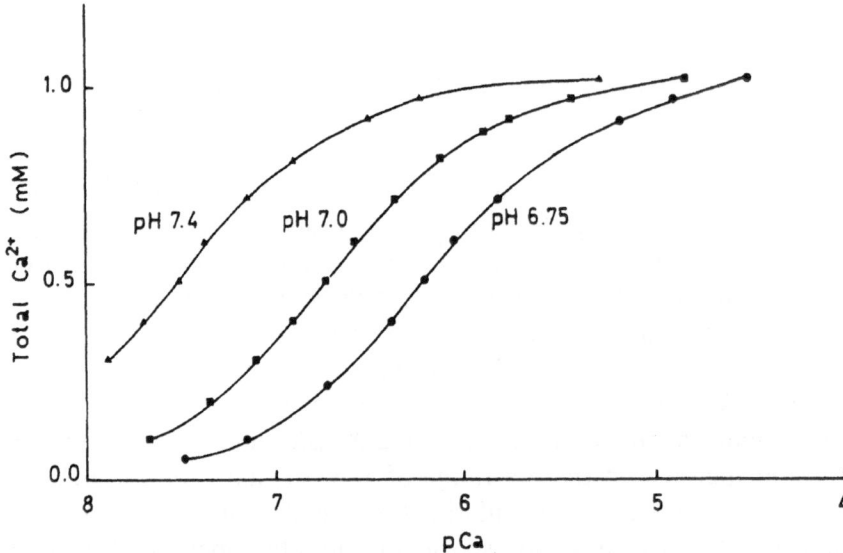

Fig. 1. The three curves relate the pCa to total Ca^{2+} added to a medium containing 1.0 mM EGTA at pH values of 6.75, 7.00, and 7.40. These are theoretical curves calculated with the aid of equation 2 of the text. The absolute dissociation constant for Ca–EGTA was taken as 10^{-11} M. The apparent dissociation constants were calculated for each pH value, as indicated in the footnote on page 77. The calculated apparent dissociation constants are 6.20, 6.72, and 7.50 for pH values of 6.75, 7.00, and 7.40, respectively (see text for details).

total Ca^{2+} (bound and free) in the medium when the concentration of total EGTA is 1.0 mM. The curves are constructed with the aid of equation 2, assuming that the concentration of Ca–EGTA is equal to the concentration of total Ca^{2+} added and that the concentration of free ligand (EGTA) is the difference between total EGTA and Ca–EGTA.

Since EGTA is about 100 times more specific for Ca^{2+} than for Mg^{2+}, the calculations are valid even in the presence of Mg^{2+}. However, one has to take into account the formation of Mg–EGTA complexes if a large excess of Mg^{2+} is used. One can neglect this interference if the $[Mg^{2+}]$ does not exceed the

---

Therefore:

$$K = K_{Ca} + K_{Ca}[H^+]/K_4 + K_{Ca}[H^+]^2/K_4 K_3$$

We now can calculate the value of K for any pH value by substituting the respective values in the last equation. The value for H^+ is obtained directly from the pH for which one wishes to calculate the value of K. An example of the method is the calculation of the value of K for pH 7.0:

$$K = 10^{-11} + 10^{-11}(10^{-7})/(3.72 \times 10^{-10}) + 10^{-11}(10^{-7})^2/$$

$$(3.72 \times 10^{-10})(1.41 \times 10^{-9}) = 1.93 \times 10^{-7}$$

Therefore, $pK = -\log 1.93 \times 10^{-7} = 6.72$ at pH 7.0.

[Ca^{2+}] by more than 100- to 200-fold. For higher ratios, the correction to be applied can be worked out.[84]

3. CALCIUM IONOPHORES

Changes in intracellular Ca^{2+} are of such fundamental physiological importance that substances that permit experimental variations of intracellular free [Ca^{2+}] constitute very valuable tools. In this respect, I have briefly described Ca–EGTA buffers and now briefly introduce two Ca^{2+} ionophores which facilitate the transfer of Ca^{2+} across biological membranes. The carboxylic antibiotics X-537A and A23187 (Fig. 2) have been widely used as Ca^{2+} ionophores in neurochemistry to alter the fluxes of Ca^{2+} across neural membranes, and relevant literature is referred to throughout this chapter.

The ionophore X-537A is not selective for Ca^{2+} since it also complexes other divalent and even monovalent cations which it transports across biological membranes.[85,86] In addition, X-537A also forms complexes with biogenic amines and facilitates their transfer across membranes.[85,87] In nonpolar media, such as the membrane phase, the monovalent or divalent cations are sandwiched between the polar faces of two X-537A molecules, whereas biogenic amines bind as a 1:1 complex to the polar face of the ionophore.[85,86] At low concentrations, X-537A catalyzes the electroneutral exchange of cations across the membrane and thus destroys existing cation gradients.

The lack of specificity of X-537A has discouraged its use in recent years as a Ca^{2+} ionophore. Instead, A23187 is now being preferentially utilized because it has a high specificity for divalent over monovalent cations and is relatively specific for Ca^{2+} over Mg^{2+}.[85,86] The Ca^{2+} complex of this ionophore is composed of one Ca^{2+} sandwiched between the polar faces of two A23187 molecules.[85,86,88] Although weak complexes are also formed with monovalent cations,[85,86] apparently they are not transferred across membranes with the exception of H$^+$ which acts as the counterion and is transferred across the membrane associated with A23187.[85,86,88] The transport of Ca^{2+} by A23187 is electroneutral in that two H$^+$ are exchanged for each Ca^{2+}.

A23187 is particularly useful to study the effect of Ca^{2+} on neurosecretion

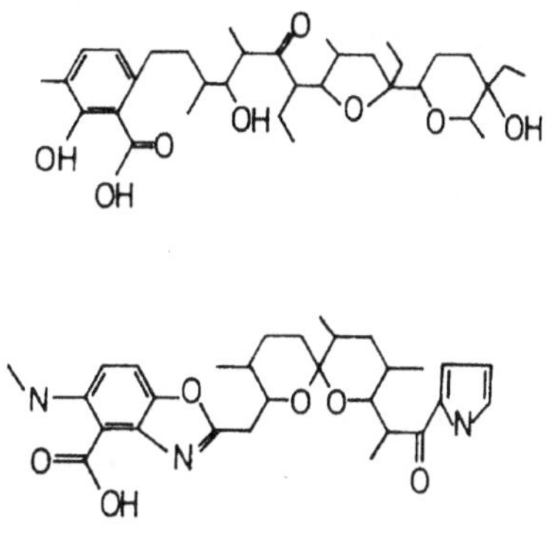

X-537A
(M.W. 589)

A23187
(M.W. 523)

Fig. 2. Structure and molecular weight of the calcium ionophores X-537A and A23187.

because it does not transport biogenic amines across membranes.[87] Thus, its effect on neurotransmitter release is considered to be mediated through its effect on Ca^{2+} influx.[89-91]

4. HOW NERVE CELL CALCIUM IS REGULATED

4.1. Ionized Calcium in Axons

Advantage has been taken of the large size of cephalopod axons to study Ca^{2+} buffering in axoplasm because it is possible to inject into single fibers Ca^{2+} indicators and Ca^{2+} buffers[92,93] so that the intracellular Ca^{2+} can be measured and controlled. Furthermore, a single axon yields 5–20 mg of axoplasm uncontaminated with extracellular Ca^{2+} which when centrifuged gives a supernatant and a sediment containing the intracellular organelles, so that the Ca^{2+} binding capacity of axoplasm can be separated into soluble and particulate components.[92,93]

The concentration of free Ca^{2+} has been determined by matching the intensity of luminescence of aequorin in a porous capillary inside the axon with the intensity obtained after dialysis of the axoplasm with Ca–EGTA buffer of known concentrations.[6,58] DiPolo *et al.*[57] also used a second method to measure free Ca^{2+} in axoplasm in which they measured the changes of absorbance of microinjected arsenazo III. The values for free Ca^{2+} obtained by the aequorin and arsenazo III techniques give values of 20 nM and 40–50 nM, respectively. Earlier studies by Hodgkin and Keynes[94] indicated that only a small portion of the total internal Ca^{2+} exists unbound and ionized. Thus, they observed that the mobility of ^{45}Ca injected into the axon in response to an imposed electric field yielded a value of 9×10^{-6} cm/s per V/cm, which is about 2% of that in free solution. In another type of study, Baker[95] found that the resting glow of aequorin injected into unstimulated squid giant axons in sea water was matched by the glow of a Ca–EGTA buffer corresponding to a free Ca^{2+} concentration of about 10^{-7} M.

It is of interest to know whether the neuron maintains the same level of ionized Ca^{2+} in all regions, including the cell body. Studies utilizing injections of arsenazo III[96] and Ca^{2+}-selective electrodes[97] in neurons of *Aplysia* give values of the order 10^{-7} M.

The small fraction of cytoplasmic Ca^{2+} that is free implies that the bulk of the Ca^{2+} is bound. Baker *et al.*[55] obtained evidence for this in experiments that showed that when aequorin was added to extruded squid axoplasm, addition of 10^{-4} M Ca^{2+} produced a transient increase in luminescence which disappeared within several seconds. Furthermore, electrical stimulation of axons into which aequorin had been injected led to an increase in $[Ca^{2+}]_i$ as detected by an increased luminescence whose intensity declined with a half-time of about 15 s. This decline was interpreted as reflecting a disappearance of free Ca^{2+} which became sequestered intracellularly.

4.2. Intracellular Calcium Buffers

The Ca^{2+} content in a fresh squid axon bathed in 3 mM Ca^{2+} sea water is about 50 μmol/kg axoplasm, whereas the free Ca^{2+} is about 20–40 nM,[92,93] corresponding to about 0.06% of the total. About 200 nM (about 0.5% of total) is bound to diffusible anions (ATP, aspartate, citrate, etc.), and about 70 nM is buffered by a Ca^{2+}-binding protein.[92,93] Baker and Schlaepfer[98] also describe a high-affinity nondialyzable axoplasm buffer that binds as much as 4 μm Ca^{2+}, corresponding to 8% of the total Ca^{2+}. The remaining Ca^{2+} is retained by axoplasm organelles, of which mitochondria account for a maximum of 3 μM, corresponding to about 6% of the total Ca^{2+}.[92,93] The remaining Ca^{2+}, which is 80–90% of the total (40 μmol/kg axoplasm) in fresh unloaded axons, is buffered by other organelles including endoplasmic reticulum.[92,93,99] These results are summarized in Table II.

The extent to which mitochondria serve to buffer the Ca^{2+} in the axoplasm depends on the Ca^{2+} load. Thus, at normal Ca^{2+} load, poisoning the mitochondrial Ca^{2+}-accumulating system produces only a two- to fourfold increase in free $[Ca^{2+}]_i$ in the axoplasm of fresh axons, whereas if the axons are first loaded with known amounts of exogenous Ca^{2+}, 99.5% of an exogenous load of 50 μmol/kg axoplasm is held in the mitochondria and can be released with the mitochondrial uncoupler FCCP.[92] With loads of 50–200 μM, the mean rise in ionized Ca^{2+} is about 0.6 nmol/μmol of load, or less than 1 part in 2000, which shows that mitochondria have a very high Ca^{2+}-buffering capacity when large loads of exogenous Ca^{2+} are imposed.[92] Under these conditions, inhibition of Ca^{2+} uptake by mitochondria causes a large increase in free Ca^{2+} in the axoplasm, in contrast to their effect in fresh unloaded axons.[92]

Although buffering of the imposed load of Ca^{2+} is much less when the Ca^{2+} uptake by mitochondria is poisoned by FCCP or cyanide, 90–95% of the imposed Ca^{2+} load is still buffered, and only 5–10% appears as free Ca^{2+}, which suggests that significant nonmitochondrial buffer exists in axoplasm, probably in other organelles.[92,93] It has been estimated that even after large Ca^{2+} loads, the mitochondria retain only about one-third of the total Ca^{2+} and

Table II
Distribution of Ca^{2+} in Squid Axon According to Data Reviewed by Brinley[92]

Region of axoplasm	Ca^{2+} concentration (μmol/kg axoplasm)	% of total
Total axoplasm	50	100
Free Ca^{2+}	0.03	0.06
Bound to nondialyzable axoplasm	4	8
Soluble Ca^{2+}-binding protein	0.07	0.14
Soluble anions	0.2	0.4
Mitochondria	3	6
Other organelles	≃40	≃80

that other organelles are the most significant buffering systems in axoplasm.[92] Brinley *et al.*[100] have shown that in near-physiological conditions the rate of Ca^{2+} uptake by mitochondria at physiological Ca^{2+} concentrations is rather low. In fact, the rate is very low at free Ca^{2+} concentrations below 200–300 nM, values much higher than it is reasonable to expect in the cell.[93,100] The results do not preclude rapid Ca^{2+} uptake by mitochondria located adjacent to the axolemma, where the concentration of Ca^{2+} may rise 10- to 20-fold during stimulation.[101] However, the available evidence is suggestive of a minor role for mitochondria in regulation of the Ca^{2+} concentration of the axoplasm at physiological intracellular Ca^{2+} concentrations.[92]

In vitro experiments using whole axoplasm have shown that if the free Ca^{2+} concentration is about 8 μM, the particulate matter of axoplasm takes up most of the Ca^{2+}, whereas at free Ca^{2+} concentrations of 0.18 μM, only about 40% of the Ca^{2+} is taken up.[93] Treatment of the axoplasm with FCCP shifts a significant part of the Ca^{2+} from the particulate to the soluble portion of the axoplasm, but about 70% of the Ca^{2+} remains in the particulate fraction, presumably associated with endoplasmic reticulum.[93] This retention of Ca^{2+} by endoplasmic reticulum apparently is ATP dependent as is the case in nerve terminals.[99,102]   Squid axons loaded with Ca^{2+} by stimulation in a medium of high Ca^{2+} and subsequently injected with oxalate contain electron-opaque deposits of Ca^{2+} in the endoplasmic reticulum as well as in mitochondria, as identified by electron probe X-ray microanalysis.[74,103] Thus, the endoplasmic reticulum of neurons probably is a Ca^{2+}-sequestering compartment. Furthermore, evidence is presented by Henkart[74] that Ca^{2+} may be released from intracellular stores in response to surface membrane stimuli.

4.3. Calcium Influx in the Axon

To maintain a low $[Ca^{2+}]_i$, the axolemma transports Ca^{2+} to the exterior.[83,104] In the absence of this transport system, one would expect that Ca^{2+} would accumulate inside the cell until it reached thermodynamic equilibrium which, for a membrane potential of 60 mV and $[Ca^{2+}]_o$ of 11 mM, would give a $[Ca^{2+}]_i$ of about 1 M. Influx across the squid axolemma was measured by Hodgkin and Keynes[94] to be 10^{-13} mol $Ca^{2+}/cm^2 \cdot s$ during rest. There are different channels for the entry of Ca^{2+} and Na^+ during the action potential, but the Na^+ channels may also be used by Ca^{2+} in internally perfused squid axon since tetradotoxin blocks the Ca^{2+} potential.[6,55]

Baker *et al.*,[55] using aequorin injected into squid axons, resolved the entry of Ca^{2+} into two phases during depolarization by voltage clamp: (1) an early rapid phase with a time course similar to that of Na^+ influx, sensitive to tetrodotoxin, and representing flow of Ca^{2+} through the Na^+ channels; (2) a second delayed phase of Ca^{2+} entry which increased steeply between 40 and 80 mV positive to the resting potential and then became smaller with further depolarization. This Ca^{2+} current declined during prolonged depolarization with a time constant of the order of seconds and was tetrodotoxin (TTX) insensitive.[6,55] The channel of this late phase of Ca^{2+} entry is considered to be the tetradotoxin-resistant presynaptic Ca^{2+} channel that functions in the mediation of transmitter release at the squid giant synapse.[10,105,106] The time constant for the late Ca^{2+} inactivation is similar to that of K^+ channel inactivation, but the recovery from inactivation is much slower.[6,107] The late phase entry of Ca^{2+} does not take place through the K^+ channels, and Baker[6] suggests that there is a separate Ca^{2+} channel which opens late in the action potential. Further characterization of these Ca^{2+} channels has been presented by several authors.[4,108] Entry of Ca^{2+} through these channels is unaffected by tetraethylamonium ions (TEA) which block the outward K^+ currents, but it is markedly reduced by externally applied Mg^{2+}, Mn^{2+}, and Co^{2+} which do not affect either the inward or outward K^+ currents. The organic Ca^{2+} antagonists verapamil (iproveratril) and D-600 also reduce the late Ca^{2+} entry.[6,108]

It is of interest that many of the properties of the late Ca^{2+} channel resemble those of the Ca^{2+}-dependent mechanism that initiates the release of neurotransmitters in that both are insensitive to TTX and TEA, are inhibited by Mg^{2+} and Mn^{2+}, and have similar voltage dependencies.[6,108]

The internal medium of the axon can be precisely controlled by internal dialysis while Ca^{2+} influx is measured continuously.[83] This has led to a more detailed study of the Ca^{2+} flux. In this method, a glass or plastic capillary tube with a central porous region is steered through the axon. The porous region is permeable to solutes of molecular weight up to 1000, thus allowing the exchange of solutes between the fluid passed through the capillary and the axoplasm. This technique has been used extensively to measure efflux and influx of different solutes in squid axon.[83,109]

At ionic concentrations near those found physiologically ($Na_i^+ = 40$ mM; $Ca_i^{2+} = 0.06$ μM; $Na_o^+ = 440$ mM; ATP = 2 mM), the Ca^{2+} influx is about 40 fmol/cm^2·s. This Ca^{2+} influx can be separated into two main components, one that requires ATP, depends on Na_i^+, and is activated by Ca_i^{2+} and another that persists in the absence of Na_i^+, Ca_i^{2+}, and ATP. Membrane depolarization increases this latter component which shows many similarities to the late Ca^{2+} channel described earlier.[55,83,109] This pathway accounts for most of the Ca^{2+} influx under physiological conditions.[83,109] Another characteristic of the Ca^{2+} influx is that the Na_i^+-dependent Ca^{2+} influx is modulated by $[Ca^{2+}]_i$. When $[Ca^{2+}]_i$ is elevated above physiological levels, the influx of Ca^{2+} increases provided that Na^+ is present inside. Thus, this extra influx of Ca^{2+} represents an exchange with Na_i^+ rather than with Ca_i^{2+}.[83,109]

4.4. Calcium Efflux in the Axon

The mechanisms to extrude Ca^{2+} from the nerve cell have been extensively studied with two principal preparations, synaptosomes[104,110,111] and the squid axon.[6,83,104,112] The prevalent conclusions are the following. (1) There is a Na^+-dependent Ca^{2+} efflux in the axon and in synaptosomes,[83,104,110–114] and a similar mechanism is also present in several other tissues.[83,104,110–113] This Ca^{2+} efflux in squid axons occurs in exchange for Na^+, does not require ATP although it is stimulated by it, and is inhibited by Na_i^+ and activated by Na_o^+. This system has a relatively low affinity for Ca_i^{2+}.[83] (2) There is another component for Ca^{2+} efflux from axons which has a high affinity for Ca_i^{2+}, depends on ATP, is not affected by Na_i^+, Na_o^+, Ca_o^{2+}, or Mg_o^{2+}, and is inhibited by internal vanadate which does not inhibit the efflux component with low affinity for Ca^{2+}.[83] These two systems insure that the axon maintains, over the long term, a constant low $[Ca^{2+}]_i$ in spite of short-term fluctuations because of neuronal activity and the constant tendency of Ca^{2+} to enter the axon to attain electrochemical equilibrium.

In the Na^+/Ca^{2+} exchange mechanism, the extrusion of Ca^{2+} depends on the presence of Na_o^+, and this effect is inhibited by Na_i^+, which suggests that it is the Na^+ gradient that drives the extrusion of Ca^{2+}.[83,104,112] Thus, the Na_o^+–Ca_i^{2+} countertransport occurs even in poisoned axons, since metabolic energy is not the direct source of energy, although it is necessary to maintain the Na^+ gradient through the Na^+–K^+ ATPase. Nevertheless, ATP has been shown to stimulate the Na^+-dependent Ca^{2+} efflux with a low apparent affinity ($K_{1/2} = 300$ μM ATP). It is assumed that the ATP interacts with the Na_i^+–Ca_i^{2+} countertransport mechanism but is not essential for operation of the mechanism, since the maximal rate of the Na_o^+-dependent Ca^{2+} efflux is little affected by ATP, and the phenomenon is observed even in ATP-depleted fibers.[83,115,116] Thus, it is assumed that ATP has only a regulatory role in the Na^+–Ca^{2+} countertransport mechanism.[115]

The component of Ca^{2+} efflux that is Na_o^+ independent and shows a high affinity for Ca^{2+} and ATP is inhibited by CN^-, which suggests its dependence on metabolism.[83] This Ca^{2+} efflux is "uncoupled" to the influx of other solutes and is strictly dependent on ATP (which activates with high affinity) and Mg_o^{2+} but is not affected by Na_i^+, Na_o^+, or Ca_o^{2+} and is inhibited by vanadate.[83,117] As already stated, this contrasts with the Na_o^+-dependent Ca^{2+} efflux which requires Na_o^+. It is of little importance at physiological $[Ca^{2+}]_i$ but becomes predominant at high $[Ca^{2+}]_i$; it is insensitive to vanadate but is strongly inhibited by Na_i^+. This second component can be further divided into ATP-independent and ATP-stimulated effluxes with low ATP affinity. A Ca_o-dependent Ca^{2+} efflux is also observed, but in unpoisoned axons and in axons dialyzed with ATP at near-physiological $[Ca^{2+}]_i$, very little Ca_o^{2+}-dependent Ca^{2+} efflux occurs.[83]

Is there a Ca^{2+} pump in the nerve cell membrane? This question has not been convincingly answered. DiPolo[118] demonstrated net Ca^{2+} extrusion from squid axons in the absence of ionic gradients by measuring both unidirectional fluxes in the same axon by a modification of the dialysis technique. In these experiments, the addition of ATP produced a net outward movement of Ca^{2+} in the absence of Na_o^+, Ca_o^{2+}, and Mg_o^{2+}.[83,118] Apparently, there is a Ca^{2+}-ATPase in the axon plasma membrane, since vanadate, an inhibitor of the Ca^{2+}-ATPase of red blood cells, inhibits the uncoupled Ca^{2+} efflux.[118] Other evidence in favor of the presence of a Ca^{2+}-ATPase in the plasma membrane of nerve cells came from studies with rat brain membranes which have a Ca^{2+}-dependent ATPase and transport Ca^{2+}.[19,20]

From an operational point of view, it is important that the affinity for Ca^{2+} of the ATP-dependent system is about 40 times higher than that of the Na^+/Ca^{2+} exchange mechanism.[83] This means that in the resting state, when $[Ca^{2+}]_i$

is low, the ATP-driven Ca^{2+} extrusion plays an important role in regulating the $[Ca^{2+}]_i$. At higher $[Ca^{2+}]_i$, the Na^+/Ca^{2+} exchange mechanism becomes of importance. It is estimated that at rest the amount of Ca^{2+} carried via the Na^+/Ca^{2+} exchange system does not exceed 25% of the total.[83]

The relationship between the Na_o^+-dependent Ca^{2+} efflux and the $[Na^+]_o$ best fits a cubic function of $[Na^+]_o$, which indicates that three Na^+ are required to activate the efflux of one Ca^{2+}.[104,112,115] With a stoichiometry of 3 Na^+ : 1 Ca^{2+}, the carrier could maintain a Ca^{2+} concentration gradient given by: $[Ca^{2+}]_o/[Ca^{2+}]_i = ([Na^+]_o/[Na^+]_i)^3 \exp(-V_m F/RT)$ where V_m is the membrane potential, and F, R, and T have their usual meaning. With a 60 mV resting potential and 10 : 1 Na^+ gradient, a Ca^{2+} gradient of 10^4 : 1 could be sustained; this is close to the value observed in squid axons.

4.5. Fluxes of Calcium in Synaptosomes and Synaptic Membrane Vesicles

Synaptosomes, the isolated components that represent the nerve terminals, have been widely used to study Ca^{2+} fluxes.[102,104,110,111,119,120] The overall results suggest that synaptosomes isolated from rat brain[102,104,110,111] and from the electric organ of *Torpedo marmorata*[299] maintain a low $[Ca]_i$ as does the intact cell and that the mechanism for Ca^{2+} extrusion probably is very similar to that described for the squid axon. This Ca^{2+} efflux utilizes the energy of the Na^+ gradient across the plasma membrane, and, in addition, there is a Ca^{2+}-ATPase which probably pumps Ca^{2+} to the exterior.[19,20,121]

The influence of Na^+ in the Ca^{2+} fluxes was initially studied in brain slices.[122,123] The results of these studies indicated that removal of external Na^+ increased the $[Ca^{2+}]_i$ and that Na_o^+ apparently increased the extrusion of Ca^{2+}, as was subsequently also shown with isolated synaptosomes.[104,110,111,124] A series of reports by Blaustein and collaborators[104,111,112] on fluxes of Ca^{2+} across the synaptosomal membrane indicate clearly that there is a Na_o^+-dependent Ca^{2+} efflux from synaptosomes which is independent of metabolism. Furthermore, synaptosomes exhibit a distinct voltage-sensitive Ca^{2+} influx. Thus, a high $[K^+]_o$ depolarizes the synaptosomal membrane, and this depolarization increases the Ca^{2+} influx about sixfold to about 1.8×10^{-13} mol/$cm^2 \cdot$s, and similar results are obtained if the depolarization is induced by veratridine.[125] This Ca^{2+} influx into synaptosomes was inhibited by about two-thirds by the Ca^{2+} antagonists verapamil and D-600 which block the Ca^{2+} channels.[126] There is present in synaptosomes a Ca^{2+}–Ca_o^{2+} exchange, also present in the squid axon.[110] Generally, the data for Ca^{2+} efflux in synaptosomes is consistent with a Ca^{2+} carrier mechanism that extrudes Ca^{2+} in exchange for Na^+ or Ca^{2+}, the latter being activated by Li^+, as is also the case in the squid axon.[110,112,127] The relationship between Ca^{2+} efflux and $[Na]_o$ is sigmoid and gives a value of 3 Na^+ : 1 Ca^{2+} as in the case of the axon.[104,110,111] Synaptosomes isolated from the electric organ of *Torpedo marmorata* display characteristics similar to those of synaptosomes isolated from rat brain.[299]

More recently, it has been shown that synaptic plasma membrane vesicles derived from synaptosomes after osmotic shock also display a Na^+/Ca^{2+} exchange mechanism.[128]

5. CALCIUM STORAGE AND BINDING BY SUBCELLULAR FRACTIONS

5.1. Mitochondrial and Nonmitochondrial Calcium Stores in Synaptosomes

Intracellular organelles buffer the cytoplasmic free Ca^{2+} in presynaptic nerve terminals. The Ca^{2+} is sequestered in mitochondria and nonmitochondrial organelles.[90,99,102,129,130] The nonmitochondrial organelles have a high affinity for Ca^{2+}, with uptake half-saturated at a Ca^{2+} concentration of about 0.35 μM, but they store only small amounts of Ca^{2+} (2–3 μmol/mg nerve terminal protein). On the other hand, mitochondria have a lower affinity (K_{Ca} = 10–100 μM) and a larger capacity of about 30 μmol/g protein.[99,102,129,130] The intrasynaptosomal organelles can buffer Ca^{2+} to about 0.4 μM when the total Ca^{2+} is about 30 μmol/g of protein or less.

The nonmitochondrial sequestration of Ca^{2+} requires ATP, and the structure that accumulates the Ca^{2+} has been identified as the smooth endoplasmic reticulum (SER),[99,102,129] although there are claims that synaptic vesicles isolated from the *Torpedo* electric organ accumulate Ca^{2+} in the presence of ATP.[131,132] Similar claims exist for coated vesicles,[133] but recent careful studies utilizing electron probe microanalysis to localize the Ca^{2+} accumulated within rat brain synaptosomes whose outer membranes had been made leaky by saponin produced evidence that, when mitochondrial poisons were present, electron-dense deposits accumulate only within structures identified as SER cisterns and vesicles, and none were found in the synaptic vesicles.[80]

The distribution of the Ca^{2+} in isolated nerve terminals has been estimated.[99,102] Of the total 2.2 μmol/g protein present in freshly prepared terminals, about 0.4 μmol/g protein is stored in the mitochondria since it is released by FCCP.[99,129] Nearly half of the Ca^{2+} (≈ 1 μmol/g protein) is released by the ionophore A23187, which suggests that this Ca^{2+} fraction is sequestered in the nonmitochondrial organelles such as SER.[99,102] The distribution of the Ca^{2+} that enters the nerve terminals after K^+ depolarization has also been studied, and it appears that in spite of the fact that mitochondria have a much higher Ca^{2+} capacity than the nonmitochondrial stores, about 50% of the Ca^{2+} accumulated in 10 s is stored in the nonmitochondrial stores, and only 20% is stored in mitochondria. However, as the Ca^{2+} load increases, a proportionately larger fraction of Ca^{2+} is sequestered in the mitochondria.[99,102,129,130] This suggests that at physiological Ca^{2+} concentrations the nonmitochondrial organelles, identified as SER, are especially important in buffering the internal free Ca^{2+} in the nerve terminals.[99,102]

Michaelson *et al.*[131] showed recently that the cholinergic synaptic vesicles isolated from the *Torpedo* electric organ transport Ca^{2+} by a Ca^{2+}-dependent ATPase. The uptake is ATP and Mg^{2+} dependent, follows saturation kinetics with a K_m of 50 μM, and is not inhibited by inhibitors of Ca^{2+} uptake by mitochondria. Maximal uptake is about 15 nmol Ca^{2+}/mg protein, and the Ca^{2+} is released by A23187, which suggests that Ca^{2+} is accumulated against a concentration gradient. To show that the uptake of Ca^{2+} is in fact by the synaptic vesicles, Michaelson *et al.*[131] followed the distribution of vesicular ACh and ATP-dependent Ca^{2+} uptake on a continuous sucrose gradient and showed that the two coincided. The significance of this Ca^{2+} uptake by synaptic vesicles to the regulation of the nerve terminal Ca^{2+} is not clear, and it is not known whether significant amounts of Ca^{2+} are accumulated in the synaptic vesicles at physiological Ca^{2+} concentrations. The fact that half-saturation of this transport process occurs at a Ca^{2+} concentration of 50 μM suggests that, as is the case in mitochondria, the synaptic vesicles do not normally play an active role in controlling the free Ca^{2+} concentration in the nerve terminal. Other studies[134] show that Ca^{2+} uptake by synaptic vesicles isolated from rat brain largely represents microsomal contamination.

In the neurohypophysis, the microvesicles accumulate Ca^{2+}, as can be visualized at the ultrastructural level.[135] Furthermore, isolated microvesicles from bovine neurohypophyses accumulate Ca^{2+} in an ATP-dependent manner.[136,137] It is suggested that the microvesicles in neurosecretory nerve terminals function as Ca^{2+}-accumulating organelles.[136,137]

Studies with mitochondria isolated from brain show that Ca^{2+} is accumulated by an energy-dependent process.[124,138–142] The capacity and affinity of the mitochondrial system for Ca^{2+} are somewhat controversial, but recent work of Nicholls and Scott[143] shows that under certain conditions mitochondria isolated from guinea pig cerebral cortex buffer the extramitochondrial free Ca^{2+} precisely when up to 200 nmol of Ca^{2+}/mg protein are accumulated in the matrix. The steady-state extramitochondrial free Ca^{2+} is maintained as low as 0.3 μM. Furthermore, these mitochondria respond rapidly to a slight perturbation in the external Ca^{2+} concentration, so that they appear well adapted for a putative role in the regulation of cytosolic free Ca^{2+}.[143] The rates of Ca^{2+} uptake by mitochondria are about 100-fold higher than the rates of Ca^{2+} uptake by vesicular preparations isolated from nerve endings.[143]

Another interesting aspect is that Na^+ in concentrations as low as 5 mM, in the presence of 120 mM K^+, causes the release of the Ca^{2+} accumulated by brain mitochondria.[142,144–146] There is an exchange between Na^+ and Ca^{2+} across the inner membrane, which also displays a Na^+–H^+ exchange activity, and the initial event in the accumulation of Ca^{2+} is an efflux of H^+, thereby generating an electrical potential difference which drives Ca^{2+} in.[143,144,146] The Na^+–Ca^{2+} exchange may be of physiological importance in releasing accumulated Ca^{2+} from mitochondria during activity. The role of mitochondria in regulating cytoplasmic Ca^{2+} in nerve cells remains controversial, since other work suggests that the mitochondrion is a low-affinity, high-capacity Ca^{2+} uptake system of importance only in the long-term regulation of intracellular Ca^{2+}.[92,93,99,102] This is not in accord with recent results showing that brain mitochondria *in vitro* can maintain a steady-state extramitochondrial free Ca^{2+} concentration as low as 0.3 μM.[143] Figure 3 summarizes the regulation of $[Ca^{2+}]_i$ by nerve cells.

5.2. ATP-Dependent Calcium Uptake by Synaptic Plasma Membranes

Transport of Ca^{2+} by isolated synaptic plasma membrane (SPM) vesicles might constitute a model for studying active Ca^{2+} efflux from the nerve terminal. Accumulation of Ca^{2+} in SPM vesicles implies that the vesicles are inside out, and this is supported by the observation that 80% of the nerve endings became inverted when synaptosomes were isolated from rat cerebral cortex.[147]

Rahamimoff and Abramovitz[148,149] recently showed that SPM transport Ca^{2+} and that this transport is ATP dependent. Similar results were reported by Saito *et al.*[150] who showed that the Ca^{2+} "binding" requires Mg^{2+}, is optimal at a Ca^{2+} concentration of 9×10^{-5} M in the presence of 10^{-4} M EGTA, and that 8–10 nmol Ca^{2+}/mg protein is accumulated. Ichida *et al.*[151] also showed that Ca^{2+} uptake by rat cerebral SPM is ATP and Mg^{2+} dependent. Oxalate (60 mM) increased this Ca^{2+} uptake after long incubations from values of about 8 nmol/mg protein to 52 nmol/mg protein, but this oxalate effect has not been reported by others who have reported ATP-dependent Ca^{2+} binding.[113,152]

There are contradictory reports on the role of a Ca^{2+}-ATPase in the transport of Ca^{2+} by SPM, but in some preparations, it is evident that Ca^{2+}-ATPase and Ca^{2+} uptake activity coexist,[152] which supports the proposal of a Ca^{2+} transport system mediated by a Ca^{2+}-ATPase. However, this relationship is not established, and critical studies on the Ca^{2+} transport mechanism of SPM are required. For example, it is not clear whether the ATP-dependent Ca^{2+} uptake represents intravesicular Ca^{2+} or Ca^{2+} bound to the membranes. The results of Rahamimoff and Abramovitz[149] suggest that in their preparation of

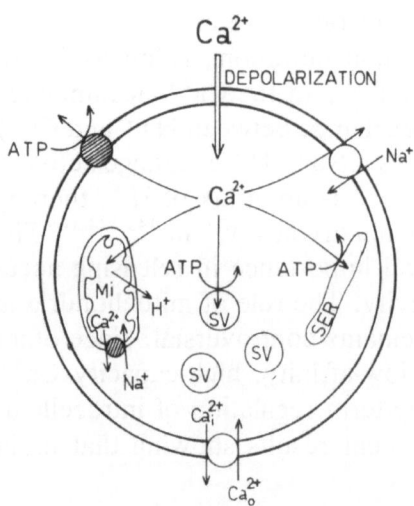

Fig. 3. Diagram illustrating how Ca^{2+} that enters the nerve terminal is handled by the various intracellular organelles and, ultimately, by the plasma membrane. Smooth endoplasmic reticulum (SER) and synaptic vesicles (SV) accumulate Ca^{2+} by a process that is ATP dependent, whereas mitochondria (M_i) utilize energy from oxidative phosphorylation to accumulate Ca^{2+} in exchange for H^+. The Na^+ gradient across the plasma membrane drives Ca^{2+} out. The Ca^{2+}-ATPase of the plasma membrane is probably also involved in the net extrusion of Ca^{2+}. In addition, there is a Ca_o/Ca_i exchange system present in the plasma membrane. The Na_i^+ may also serve as a triggering agent to release Ca^{2+} from mitochondria (see text for details).

SPM the Ca^{2+} is accumulated intravesicularly, since the Ca^{2+} ionophore, A23187, causes a 40% decrease in Ca^{2+} content previously accumulated. However, there are some doubts about the origin of their synaptosomal vesicles (i.e., neurotransmitter storage vesicles, endoplasmic reticulum, SPM, or a mixture of all). Both the Ca^{2+}-ATPase and Ca^{2+} uptake require Mg^{2+} in addition to ATP, and the systems, apparently, are linked together, since dicyclohexyl carbodiimide (DCCD), which inhibits ATPases, also inhibits Ca^{2+} transport.[149] The authors suggest, however, that the synaptosomal vesicles utilized might actually be of intrasynaptosomal origin and may function within the nerve terminal to regulate the intracellular concentration of Ca^{2+}.

Papazian *et al.*[153] reported the reconstitution of an ATP-dependent Ca^{2+} uptake system of synaptosomes in artificial lipid vesicles by a cholate dialysis procedure using excess exogenous phospholipid. The vesicles were then loaded with Ca^{2+} and oxalate in the presence of ATP and separated from the bulk of the preparation on density gradient. The purified Ca^{2+}-transporting vesicles contained two major proteins of 94,000 and 140,000 daltons. It is not clear whether these two proteins are part of the same transport system or represent two distinct Ca^{2+} transport systems from different regions of the nerve terminal.[153]

The SPM Ca^{2+}-ATPase[19,20] and the Ca^{2+} transport[20] system have been reported to require calmodulin. Kuo *et al.*[19] showed that after treatment of SPM with EGTA, addition of calmodulin is necessary for maximal activity of the Ca^{2+}-ATPase and of the Ca^{2+} transport system, which suggests that calmodulin regulates both processes. Calmodulin is also necessary to stimulate the Ca^{2+}-ATPase and Ca^{2+} uptake activities of other neuronal membrane fractions,[154] but the mechanisms by which calmodulin acts have not been investigated. It is suggested that it may mediate the phosphorylation of the Ca^{2+} "pump."[113]

5.3. ATP-Dependent Calcium Uptake by Microsomes

ATP-dependent Ca^{2+} binding by brain microsomes was reported in early studies by various groups of investigators.[152,155–161] This process, however, appeared to be different from the calcium uptake observed in sarcoplasmic reticulum in that activation by Ca^{2+}-precipitating agents, such as oxalate or phosphate, could not be obtained. Therefore, a different mechanism was proposed for brain microsomes, namely, that the presence of ATP activated Ca^{2-} binding to specific membrane sites.[156,160]

Later, Trotta and De Meis[162] reported an ATP-dependent Ca^{2+} uptake by rabbit brain microsomes that was stimulated by high concentrations of phosphate or oxalate. At 20–80 mM phosphate and 100 μM $CaCl_2$, phosphate promoted a fivefold increase in the amount of calcium stored at steady state to a maximum of 200–250 nmol Ca^{2+}/mg protein. Based on these and other data, these authors suggested that the ATP-dependent Ca^{2+} uptake in brain microsomes, as in muscle microsomes, is an active transport process, Ca^{2-} being accumulated as a free ion inside the vesicles.[162] However, relatively high phosphate or oxalate concentrations are required for Ca^{2+} uptake in brain microsomes, which raises some doubts regarding the interpretation of these results.

More recently, Trotta and De Meis[163] reported that the Ca^{2+}-ATPase of rabbit brain microsomes can catalyze an exchange reaction between the γ-phosphate of ATP and orthophosphate (ATP $\rightleftharpoons$ P_i exchange). This exchange reaction was studied both in brain microsomes that accumulate calcium and in an ATPase preparation solubilized with Triton X-100. The authors suggested the presence of two calcium-binding sites, of high and low affinities, respectively, located on the outer and inner surfaces of the membrane, which would regulate the relative rates of ATP hydrolysis and synthesis. In contrast to the Ca^{2+}-ATPase of sarcoplasmic reticulum, the Ca^{2+} affinity of the inner site was determined to be only 40 times higher than that of the outer site,[163] which might explain the lower efficiency of the cerebral microsomes in Ca^{2+} accumulation as compared to the skeletal muscle sarcoplasmic reticulum. The cerebral Ca^{2+}-ATPase could thus be reversed by a smaller gradient of Ca^{2+} than is the case for the muscle Ca^{2+}-ATPase system.[163]

In summary, there is general agreement that brain microsomal membranes, which predominantly represent fragments from endoplasmic reticulum, exhibit ATP-dependent Ca^{2+}-binding or uptake activity. However, it is still not clear whether this activity is part of a system coupled to a Ca^{2+}-ATPase similar to the Ca^{2+} pump of sarcoplasmic reticulum. The significance of the microsomal system as a potential mechanism for maintaining low Ca^{2+} concentration inside the nerve cell has recently been studied in osmotically shocked synaptosomal preparations as discussed in Section 5.1.

In Fig. 3 are illustrated the various processes regulating Ca^{2+} in the nerve terminal which involve the transfer of Ca^{2+} across cell organelle membranes or, ultimately, across the plasma membrane.

5.4. Passive Calcium Binding to Nerve Membranes

Several laboratories have studied the passive interaction of Ca^{2+} and other cations with synaptic plasma membranes isolated from brain.[62,164-169]

Fluorimetric studies[164-166] show that the fluorescence of 1-anilino-8-naphthalene sulfonate added to a synaptic plasma membrane suspension increases with ion concentration in the medium, divalent cations being more effective than monovalent ones in this respect.[166] The increase in fluorescence has been interpreted as representing either an increase of hydrophobicity of the membrane[165-166] or an increase in the number of probe binding sites.[166] However, conformation transitions which possibly accompany binding of cations to the membranes are not easily distinguished from nonspecific charge neutralization effects by anionic fluorescent probes.[166]

Madeira and Antunes-Madeira[165] showed that synaptic membranes bind Ca^{2+} (and other cations) at two binding sites with affinity constants of 0.87 and 74 μM, and maximum binding capacities of 115 and 90 nmol Ca^{2+}/mg protein, respectively.

A different approach utilized to follow Ca^{2+} interaction with brain membranes was developed by Kamino and co-workers[62,167-169] for rat brain synaptosomes. These authors utilized the murexide technique to measure free Ca^{2+} in synaptosomal suspensions in various ionic media. They showed that synaptosomes bind Ca^{2+} with a lower dissociation constant in NaCl than KCl media.[167]

Kamino and co-workers[168,169] further studied the effects of Mg^{2+} and some other cations on Ca^{2+} binding by synaptosomes and found that divalent cations competitively inhibit Ca^{2+} binding in the following order of decreasing affinity: $Mn^{2+} \gg Sr^{2+} > Cd^{2+}, Ba^{2+} > Mg^{2+}$. The effects of alkali metal ions on Ca^{2+} binding are noncompetitive and of cooperative nature, and the potency of inhibition is $K^+ > Rb^+ > Cs^+ > Li^+, Na^+ = 0$. Trivalent cations such as La^{3+} markedly reduce the Ca^{2+}-binding capacity of synaptosomes.[169]

Ruthenium red, which inhibits the binding of Ca^{2+} by a glycoprotein isolated from mitochondria,[170] combines firmly with Ca^{2+}-binding sites on synaptosomal membranes[62] and also inhibits Ca^{2+} binding by synaptic plasma membranes.[165] These data indicate that proteins, probably glycoproteins, are involved in the binding of Ca^{2+} by synaptosomal membranes.[165,166]

It is known that synaptosomes can accumulate Ca^{2+} passively, in a time-dependent manner,[110,119] and the effects of monovalent cations on this Ca^{2+} retention by synaptosomes are very different from the effects observed by Kamino et al.[62,167-169] Baumgold[171] has also shown that in addition to Ca^{2+} binding to external binding sites, Ca^{2+} is accumulated into vesicles formed by membranes isolated from lobster nerve.

Abood and Straw[172] studied the adsorption of $^{45}Ca^{2+}$ to a surface film of a hydrophobic protein derived from synaptic membranes and concluded that the behavior of various cations in this system significantly resembles their behavior in excitatory membranes. Abood et al.[173] further purified these synaptic membrane proteins and found that, although a number of them can bind Ca^{2+}, the greatest binding is associated with a component of a molecular weight of 16,000 which binds four Ca^{2+}/molecule and has a K_m of 1.5×10^{-5} M. An acidic tryptic peptide derived from this protein is responsible for the Ca^{2+} binding. However, the functional significance of this Ca^{2+}-binding protein is not known.

The protein-sensitized fluorescence of Tb^{3+} has been utilized to characterize Ca^{2+}-binding sites that have affinity for Ca^{2+} in the millimolar range in biological membranes[174] including brain membranes.[175,176] Hoss et al.[175] have shown that of the subcellular fractions isolated from brain, synaptic vesicles have the greatest number of sites monitored by Tb^{3+} fluorescence. These authors also suggested that morphine, which is known to deplete Ca^{2+} from nerve terminals *in vivo*, does not compete *in vitro* for these Ca^{2+}-binding sites in the synaptic vesicle fraction. Competition studies have shown that the order of affinity for the cation binding sites in synaptic vesicles is $Cu^{2+} > Mn^{2+} > Ca^{2+} > Mg^{2+}$, and Zn^{2+} is inactive.[176] Furthermore, Tb^{3+} inhibits Ca^{2+}-stimulated ATPase but not Mg^{2+}-stimulated ATPase, and it is suggested that the site monitored by Tb^{3+} fluorescence may be a component of the Ca^{2+}-ATPase of the synaptic vesicle fraction.[176]

Studies with microsomes isolated from mouse brain cortex show total cation-binding capacities of about 530 neq/mg protein at pH 7.5, and there are two types of cation-binding sites.[177] It was concluded that most of the binding sites of brain microsomes bind Ca^{2+}, Mg^{2+}, Na^+, and K^+ nonspecifically.[177]

6. REGULATION OF ENZYME ACTIVITY BY CALCIUM

6.1. Interaction of Calcium with Calmodulin

Calcium functions as a second messenger much like cyclic AMP and cyclic GMP.[1,2] Recently, many studies have been carried out on the biochemical mechanisms by which cyclic nucleotides and Ca^{2+} act at the molecular level, and the results are instructive because they help to define the nature of the modulating actions of these substances.

The Ca^{2+} actions as a second messenger result from its binding to specific proteins which may be considered Ca^{2+} receptors.[28,31] Calmodulin is the most widely distributed of these proteins and controls the effects of Ca^{2+} on many nerve cell enzymes. Among the enzymes whose stimulation by Ca^{2+} is mediated by calmodulin are brain adenylate and guanylate cyclases,[13-15] cyclic nucleotide phosphodiesterases,[14,15] the Ca^{2+}-ATPase of synaptic membranes,[19,20] and the protein kinases.[12,16-18] In this section, I concentrate on these enzymes because some coherent information has become available in this area during the last 5 years. However, much fragmentary information exists about the role of Ca^{2+} on the activation of many enzymes in the nervous system (Table III).

Calmodulin reacts with Ca^{2+} to form a complex which either acts directly on the enzyme system such as the Ca^{2+}-ATPase involved in Ca^{2+} transport or indirectly on a regulatory system such as the protein kinases.[30] In addition, the calmodulin–Ca^{2+} complex regulates the metabolism of the other second messenger, cyclic AMP, thereby coupling the two messengers, Ca^{2+} and cyclic AMP (in addition to cyclic GMP).[1,2,30] Calmodulin binds four Ca^{2+} per molecule with high affinity, and most studies suggest either multiple classes of sites for Ca^{2+} or negative cooperativity,[14,30,199,200] although a recent study suggests positive cooperativity.[199] The K_D values reported are between 10^{-7} and 10^{-5} M.[14,30,201]

Calmodulin is a monomeric protein with molecular weight of 15,000–19,000; it is heat stable and highly acidic.[14,30] The amino acid composition is characterized by absence of tryptophan and cysteine, a high threonine:serine ratio, 27 glutamate and 23 aspartate residues, and a single residue each of histidine and trimethyllysine.[14,30,202] The protein has a very characteristic UV absorption spectrum because of the absence of tryptophan and a high phenylalanine:tyrosine ratio. No major differences in amino acid composition have been detected among brain calmodulins isolated from various species.[202] Bovine, porcine, rat, rabbit, and chicken brain calmodulins also give identical profiles for the Ca^{2+}-dependent activation of phosphodiesterase, suggesting that calmodulin structure and function have been conserved throughout vertebrate evolution.[202]

When Ca^{2+} binds to calmodulin, the protein undergoes a large conformational change accompanied by 5–10% increase in α-helix content.[30,203] Lo-

Table III
Partial List of Enzymes or Processes Regulated by Calcium

Enzyme or process ·	Intervention of calmodulin	Reference
Regulation of enzyme activity		
Adenylate cyclase	+	13–15,178
Guanylate cyclase	+	179
Phosphodiesterase	+	14,29,35
Phospholipase A_2	+	180
Ca^{2+}-ATPase	+	19,20,181,182
Myosin light chain kinase	+	183
Protein kinases	+	184
Phosphorylase kinase	+	185
Tryptophan hydroxylase		186
Tyrosine hydroxylase		21
Nicotinamide adenine dinucleotide (NAD) kinase	+	187
Pyruvate dehydrogenase		188,189
NAD–isocitrate dehydrogenase		188,189
Oxoglutarate dehydrogenase		188,189
Other cell processes		
Excitation–contraction coupling		190
Depolarization–secretion coupling		10
Neurotransmitter release	+	11,89
Neurotransmitter synthesis		21,186
Exocytosis		191
Microtubule assembly–disassembly	+	30,192
Protein phosphorylation	+	12,16,18,32,34,37,39
K^+ channels		7
K^+ efflux		193
Permeability of gap junction		194
Glucose transport		195
Mitosis		196
Membrane stability and permeability		3–5,40
Acetylcholine system		
Acetylcholine receptor		197
Acetylcholinesterase		198
Narcotic analgesia		22–26

calized conformational changes have also been monitored by following the spectral changes associated with the tyrosyl[30,203] and phenylalanyl[204] residues as a function of Ca^{2+} concentration. Ultraviolet absorption changes of the tyrosyl residues associated with tyrosine 138 are complete when two of the four Ca^{2+}-binding sites are occupied.[203,205] Most spectral changes are complete when an average of 2 mol of Ca^{2+} is bound per mol of calmodulin.[30,201,203] Other transitions occur at an average site occupancy of three to four Ca^{2+} per molecule of calmodulin.[201,205]

The interaction between a given enzyme and the Ca^{2+} and calmodulin (CaM) occurs in at least two steps[14,29,30]:

$$CaM + nCa^{2-} \rightleftharpoons CaM \cdot Ca_n^{2+} \rightleftharpoons CaM^* \cdot Ca_n^{2+} \qquad [3]$$

The interaction between Ca^{2+} and calmodulin causes a conformational change in the calmodulin molecule, and the modified $CaM^* \cdot Ca_n^{2+}$ complex then interacts with the enzyme (E) to be activated:

$$CaM^* \cdot Ca_n^{2+} + E \rightleftharpoons (CaM^* \cdot Ca_n^{2+}) \cdot E^* \qquad [4]$$

The enzyme, which initially is inactive, itself undergoes a conformational change on interacting with the $CaM^* \cdot Ca_n^{2+}$ complex and becomes active.[14,29,30] Recent studies[206] show that the binding of Ca^{2+} to CaM exposes a domain with considerable hydrophobic character and that binding of hydrophobic ligands to this domain antagonizes CaM–protein interactions. This hydrophobic domain may serve as the interface for the Ca^{2+}-dependent binding of CaM to the enzyme being activated.[206]

The domain of calmodulin that is hydrophobic probably accounts for its binding of the antipsychotic tranquilizers, phenothiazenes, in the presence of Ca^{2+}.[206] Two moles of phenothiazine per mole of protein are bound with high affinity ($K_D = 1$ μM for trifluoperazine), and there is also some trifluoperazine binding of lower affinity.[30,206]

Brain has a high concentration of calmodulin (10^{-5} M) which is correlated with the important roles of Ca^{2+} in neural functions. There is in brain a soluble calmodulin-binding protein which prevents calmodulin-dependent activation of adenylate cyclase,[207,208] phosphodiesterase,[208,209] and other calmodulin-dependent enzymes.[30,208] This calmodulin-binding protein, in addition to binding calmodulin, also binds four Ca^{2+} per mole of protein with high affinity ($K_D \leq 10^{-6}$ M). This protein, designated calcineurin, is composed of two subunits: calcineurin A (mol. wt. 61,000), which interacts with calmodulin, and calcineuronin B (mol. wt. 15,000), which binds Ca^{2+}.[208] Other calmodulin-binding proteins have been described recently.[210]

6.2. Calcium and Regulation of Cyclic Nucleotide Level

Calcium in the presence of calmodulin regulates the activity of phosphodiesterase and adenylate cyclase[13–15,29,30] and thus influences the synthesis and degradation of cyclic AMP. Although there is no information on the role of calmodulin in the synthesis of cyclic GMP, guanylate cyclase is stimulated by Ca^{2+}.[179] It seems paradoxical that calmodulin stimulates both phosphodiesterase and adenylate cyclase, since these two enzymes catalyze opposing reactions. However, it now appears that guinea pig adenylate cyclase, in the presence of calmodulin, exhibits a biphasic response to Ca^{2+}. Thus, first there is a twofold stimulation of basal cyclase activity with half-maximal stimulation at 0.08 μM Ca^{2+}, and at higher Ca^{2+} concentrations, the cyclase activity is inhibited to 80–90% of basal activity, with half-maximal inhibition occurring at 0.3 μM Ca^{2+}.[15] Furthermore, the Ca^{2+} dependence of the phosphodiesterase which is calmodulin-dependent exhibits a half-maximal activation at 0.3 μM Ca^{2+}, a value identical to that of the half-maximal inhibition of the cyclase.[15] The Ca^{2+} concentrations at which these regulatory effects occur are within the range of the physiologically significant intracellular free Ca^{2+} levels.

Piascik *et al.*[15] recently proposed that immediately after stimulation of the cell, the intracellular free Ca^{2+} concentration rises above the resting levels, which are lower than 0.1 μM, and during this first phase after stimulation, Ca^{2+} forms a complex with calmodulin which stimulates adenylate cyclase activity bound to the plasma membrane, but as the intracellular free Ca^{2+} rises above 0.1 μM, the adenylate cyclase is inhibited. Concurrently, cyclic nucleotide (cyclic AMP and cyclic GMP) phosphodiesterase activity is stimulated at the higher Ca^{2+} concentrations. It is proposed that this cycle accounts for a coordinated regulation of adenylate cyclase and phosphodiesterase activities by Ca^{2+}.[15] It should be pointed out that stimulation of adenylate cyclase by Ca^{2+} may not be present in other tissues and thus may represent a specialized feature of neurosecretory tissue.[15,30]

Only one of several forms of existing cyclic nucleotide phosphodiesterase is activated by calmodulin.[30] The activation of the enzyme by calmodulin is Ca^{2+} dependent and occurs in two steps, as described generically in equations 3 and 4. The apparent K_m value of phosphodiesterase for calmodulin is of the order of 10^{-9} M, and the Ca^{2+} dependence of the enzyme activation is shifted to lower concentrations of Ca^{2+} when the calmodulin concentrations are raised.[14,30,199] Activation of phosphodiesterase appears to depend on the $CaM \cdot Ca_3^{2+}$ or $CaM \cdot Ca_4^{2+}$ species of the calmodulin–Ca^{2+} complex.[30,201] The Ca^{2+} concentration dependence of the adenylate cyclase stimulation is also dependent on the concentration of calmodulin,[14,30,211] which suggests that the mechanism of activation is as described in equations 3 and 4.

6.3. Calcium and Membrane Protein Phosphorylation

Calcium regulates the phosphorylation of a number of endogenous proteins in the nervous system, apparently through the activation of Ca^{2+}-sensitive kinases.[12,16,17,212,213] Thus, it appears probable that the increased influx of Ca^{2+} which triggers cellular activity mediates its effect through activation of these protein kinases which regulate phosphorylation of diverse proteins.[12,17,212] Thus, protein phosphorylation is a final common pathway for the action of many types of biological regulatory agents that affect the level of intracellular Ca^{2+}. This is in addition to other regulatory substances that affect phosphorylation through cyclic AMP, cyclic GMP, and probably other means.[12,16] In this section, I shall consider only the regulation of phosphorylation by Ca^{2+}.

Depolarization of nerve terminals isolated from rat cerebral cortex increases the influx of Ca^{2+} and the phosphorylation of specific endogenous proteins.[12,16,214] In addition to the depolarizing agents K^+ and veratridine, the Ca^{2+} ionophore A23187 also markedly increased the incorporation of ^{32}P into two specific proteins of 80,000 and 86,000 daltons which are present only in the nervous system.[12,214] This phosphorylation has an absolute Ca^{2+} dependence, can be reversed by the addition of EGTA, and is blocked by tetrodotoxin, which also blocks the increased Ca^{2+} influx.[214] The phosphorylation of these synaptosomal proteins occurs within intact synaptosomes because ^{32}P in osmotically shocked synaptosomes or $[\gamma\text{-}^{32}P]ATP$ added to the exterior of intact synaptosomes cannot serve as substrates.[214]

The mechanism by which Ca^{2+} stimulates the phosphorylation of synaptosomal proteins requires calmodulin. Thus, Ca^{2+} stimulates incorporation of ^{32}P into many protein bands in lysed synaptosome suspensions prepared by hypoosmotic shock provided the cytoplasm is also present, but the Ca^{2+}-dependent phosphorylation is lost if only the cytoplasm-free membranes are utilized.[184] Addition of boiled (or unboiled) synaptosomal cytoplasm to the membrane fraction permits recovery of the original activation of protein phosphorylation by Ca^{2+}.[184] Under these conditions, i.e., lysed synaptosomes, the highest levels of ^{32}P incorporation are observed in two proteins of about 51,000 and 62,000 daltons.[33,184] The stable factor required for Ca^{2+}-dependent protein phosphorylation is calmodulin which has been purified from various mammalian and invertebrate tissues.[202] Calcium, in the presence of calmodulin, stimulates protein phosphorylation through the activation of specific protein kinases,[16,17] and this mechanism may mediate some of the presynaptic actions of Ca^{2+}. However, Ca^{2+} and calmodulin also regulate protein phosphorylation indirectly by influencing the cyclic nucleotide levels by means of their action on the cyclic nucleotide phosphodiesterase and adenylate cyclase activities.[13–15]

It is of interest that the two proteins that are phosphorylated by depolarization of synaptosomes in the presence of Ca^{2+} apparently can also be phosphorylated in response to cyclic AMP.[12,215] Thus, these proteins, designated protein Ia and Ib,[12,215] are phosphorylated under the action of both Ca^{2+}-dependent and cyclic-AMP-dependent protein kinases. In brain slices, proteins Ia and Ib are also phosphorylated in a Ca^{2+}-dependent manner,[216] and it is likely that this phenomenon represents a real physiological event.

DeLorenzo *et al.*[33,217] have also shown that synaptic vesicles obtained from disrupted brain synaptosomes have a Ca^{2+}-dependent protein kinase system and that the Ca^{2+}-dependent protein phosphorylation and release of norepinephrine are both stimulated by calmodulin.[33] Furthermore, the Ca^{2+}- and calmodulin-dependent phosphorylation of specific vesicle proteins observed in the isolated vesicles was also shown to take place in synaptosomal preparations under conditions that stimulate depolarization-dependent Ca^{2+} uptake and neurotransmitter release.[33] On the basis of these studies it is argued that phosphorylation of two specific presynaptic nerve terminal proteins of about 51,000 and 62,000 daltons associated with synaptic vesicles may underlie the biochemical mechanism mediating the Ca^{2+} actions at the nerve terminal.[33] It is of interest that the anticonvulsant phenytoin inhibits this phosphorylation,[218] thus suggesting that the inhibiting effect of this agent on transmitter release from brain slices may result from the inhibition of protein phosphorylation mediated by Ca^{2+} and calmodulin.

The action of opiates may also involve effects on the Ca^{2+}-mediated phosphorylation of synaptic plasma membrane proteins. Thus, acute systemic administration into rats of morphine, methadone, or etorphine and the intracisternal administration of β-endorphin initially produced a large increase in the phosphorylation of several membrane proteins, especially those of 45,000, 82,000, and 86,000 daltons.[219] Subsequently, the opposite effect is observed in that the phosphorylation of these proteins is depressed below control levels.[219] Naloxone administered prior to the opioids prevents these changes.[219]

It is argued that the opioid effects on protein phosphorylation are mediated through Ca^{2+} changes induced by the opioids. Thus, synaptic Ca^{2+} levels fall after acute treatment with opioids,[23-25,219] and it is possible that the increased phosphorylation found in acute opioid administration results from the Ca^{2+} levels having fallen to values that optimally stimulate phosphorylation. The subsequent decrease in phosphorylation would be caused by a fall in Ca^{2+} to suboptimal concentrations.[219] It is of interest that the patterns in the phosphorylation induced by opioids on specific membrane proteins closely resemble those produced by Ca^{2+}.[219]

More recently, it was shown that incubation of purified cholinergic *Torpedo* synaptosomes with ^{32}P also results in specific incorporation of ^{32}P into a protein of about 100,000 daltons when the synaptosomes are depolarized by high K^+ or treated with the ionophore A23187 in the presence of Ca^{2+}.[213] Either of these treatments also causes increased influx of ^{45}Ca and increased efflux of acetylcholine. The kinetics of acetylcholine release reach maximal values about 45 s after the synaptosomes have been treated. These studies are particularly relevant because of the purity of the synaptosomal fraction utilized as compared with earlier synaptosomal fractions isolated from brain which are very heterogeneous and contain a mixture of synaptosomes with a variety of neurotransmitters. However, these results should be interpreted with care, because there is present in synaptic vesicles a Ca^{2+}-dependent ATPase of about 100,000 daltons whose phosphorylation is stimulated by Ca^{2+}.[220]

It is reported that one of the effects of phosphorylation of the synaptosomal membrane proteins is to decrease the permeability of the membranes to Ca^{2+}.[221,222] The results of other studies on the phosphorylation of high-molecular-weight proteins isolated from walking nerves of the shore crab show that Ca^{2+} at low concentration is necessary for optimal phosphorylation but that at high concentration Ca^{2+} is inhibitory. This phosphorylation is also sensitive to tetrodotoxin and veratridine, and the authors conclude that the phosphorylated proteins are involved in regulating the Na^+ conductance of the membrane.[223]

6.3.1. Calcium and Cyclic AMP as Second Messengers

The action of Ca^{2+} as a second messenger depends on its interaction with a Ca^{2+}-binding protein of which there exist several.[28,31] However, calmodulin is the most universal Ca^{2+}-receptor brain protein that, when complexed with Ca^{2+}, activates the various cellular enzymes, i.e., protein kinases,[12,16-18] phosphodiesterase,[14,15,30] adenylate cyclase,[13-15,30] and Ca^{2+}-ATPase.[20,21,29,30] This raises the interesting possibility that cellular enzymes that are activated through the Ca^{2+}–calmodulin complex all possess similar binding domains through which they recognize and respond to the complex. In the case of cyclic AMP as a second messenger, this nucleotide interacts with a regulatory subunit of the protein kinases present in brain.[221,224] Thus, the variety of cellular responses evoked by cyclic AMP are achieved as a result of the cyclic-AMP-stimulated protein kinases phosphorylating a variety of different intracellular proteins whose activity is modulated by phosphorylation.[16,29] Since the level of cyclic AMP in the cell is itself under the control of enzymes (phosphodiesterase and adenylate cyclase) that are modulated by Ca^{2+}–calmodulin complex, the regulation of the activity of the cell is rather complex. The interrelationships between changes in intracellular Ca^{2+} and cellular cyclic nucleotide levels are not yet clearly defined. Phosphodiesterases are stimulated by Ca^{2+} in the presence of calmodulin, and hormones and neurotransmitters that increase $[Ca^{2+}]_i$ are expected to decrease the level of cyclic AMP, but adenylate cyclase also responds to Ca^{2+}, although differently from phosphodiesterase.[15]

Furthermore, stimuli that bring about a rise in cytoplasmic Ca^{2+} also cause a transient increase in the level of cyclic GMP, probably through activation of guanylate cyclase by Ca^{2+},[2,179] even though the increased Ca^{2+} also causes activation of the cyclic nucleotide phosphodiesterase which has a higher affinity for cyclic GMP than for cyclic AMP.[14]

Discussions of some of these aspects, in which the function of Ca^{2+} and cyclic nucleotides are tentatively integrated, are found in recent reviews.[1,2,16,29] It is of particular relevance that calmodulin serves both as a mediator of Ca^{2+} functions and as a regulator of Ca^{2+}-dependent adenylate cyclase and phosphodiesterase.[29,30] The response time of the action of Ca^{2+} is very short, of the order of milliseconds, whereas that of the cyclic nucleotides it is of the order of seconds to minutes,[29] so that these intracellular messengers serve different functions in the cell.

6.4. Calcium-Stimulated ATPases

Two principal membrane fractions isolated from nerve tissue have been shown to contain ATPase stimulated by Ca^{2+}. These are membrane fractions rich in plasma membrane[113,121,151,225] and microsomal fractions.[113,220,226,227] Since these are distinct membrane fractions, their Ca^{2+}-ATPases will be reviewed separately, but it is not clear that we are dealing with different enzymes. In both cases, it is postulated that the Ca^{2+}-ATPases function to regulate the intracellular Ca^{2+} as part of a Ca^{2+} pump,[113] but whereas in synaptic plasma membranes the Ca^{2+} would be pumped to the exterior of the cell, in microsomes, the Ca^{2+} pumping translocates Ca^{2+} within the cell.

Recent reports have also described Ca^{2+}-ATPases in the synaptic vesicles isolated from *Torpedo* electric organ,[213] and it is suggested that they are involved in Ca^{2+} transport by these vesicles. Coated vesicles isolated from brain contain a Ca^{2+}-ATPase whose stimulation by Ca^{2+} is maximal at 8×10^{-7} M Ca^{2+}.[133] Apparently, these vesicles also take up Ca^{2+}, and the Ca^{2+} pump contains a protein of 100,000 daltons similar to that of sarcoplasmic reticulum.[213] Other ATPases that can be stimulated by either Ca^{2+} or Mg^{2+} have also been reported for the microsomal fraction of peripheral nerves.[228]

6.4.1. Calcium-Dependent ATPase of Synaptic Plasma Membranes

The Ca^{2+}-dependent ATPase of SPM also requires Mg^{2+}, and its low Ca^{2+} requirement suggest that the enzyme may be involved in Ca^{2+} transport. Thus, Duncan[121] reports a K_m of 4×10^{-7} M Ca^{2+} and an activation range of one unit of pCa. The activation of the enzyme in the presence of 3 mM Mg^{2+} is maximal at 2×10^{-6} M Ca^{2+}.[121] Older reports of Ca^{2+}-ATPase in a membrane fraction composed of plasma membranes exist in the literature,[225,226,229] but it is not always clear whether the compositions of the different preparations are equivalent; they are all rich in plasma membrane markers. Ohashi *et al.*[225] report an optimal concentration of Ca^{2+} (in the presence of 10^{-4} M EGTA) of about 8×10^{-5} M for a Ca^{2+}-ATPase present in synaptic membranes of guinea pig brain, and, in general, the specific activities reported by the various workers lie in the range of 6–18 μmol/mg protein per h.[113]

The coupling between this Ca^{2+}-ATPase and Ca^{2+} transport by SPM is not always evident. Thus, La^{3+}, Mn^{2+}, and ruthenium red inhibit the Ca^{2+}-ATPase but not the Ca^{2+}-uptake activity of SPM.[230] However, several recent reports now exist implicating a coupling between the Ca^{2+}-ATPase of SPM and the Ca^{2+}-accumulating activity, as is reviewed in Section 5.2 on Ca^{2+} transport by synaptic plasma membranes. Furthermore, it appears that both the Ca^{2+}-ATPase and the Ca^{2+} accumulation are regulated by calmodulin[19,20,113]

6.4.2. Calcium-Dependent ATPase of Microsomes

Brain microsomal preparations have been shown to exhibit Ca^{2+}-ATPase activity.[152,157,159–161,226,227] The properties of the Ca^{2+}-ATPase activity of brain microsomal fractions are somewhat variable, depending on the origin of the brain tissue from which the fractions are obtained. Robinson[226] pointed out some difficulties in the characterization of the Ca^{2+}-ATPase activity of the rat brain microsomal fraction. He observed, for instance, that in freshly prepared microsomes the maximal stimulation by Ca^{2+} of the "basal" Mg^{2+}-ATPase activity, measured in the presence of 4 mM $MgCl_2$ and 0.1 mM EGTA, was only about 20%. However, treatment of brain microsomes with deoxycholate (DOC) in a high-ionic-strength medium produced a stable preparation in which both the specific activity and the activity relative to the "basal" Mg^{2+}-ATPase were increased 2.5-fold. Moreover, in DOC-treated microsomes, $SrCl_2$ was more effective than $CaCl_2$ as an activator. Furthermore, KCl and NaCl stimulated the Ca^{2+}-ATPase activity in this preparation, in accord with previous findings.[159] The ATPase activity of DOC-treated membranes increased with temperature over the range 20–37°C, and the activation energies for the "basal" and Ca^{2+}-stimulated ATPase are 14 and 21 kcal/mol, respectively.[226]

The microsomal Ca^{2+}-ATPase activity is distinguished from the mitochondrial ATPase activity by its insensitivity to azide and oligomycin and from Na^+,K^+-ATPase activity by its insensitivity to ouabain.[226] The relative sensitivity to ruthenium red of the Ca^{2+}-ATPase suggested a link to Ca^{2+} transport, although the microsomal Ca^{2+} accumulation activity, also tested by Robinson,[226] was much more sensitive to the inhibitor than was the ATPase activity.

More recently, Robinson[220] reported that the Ca^{2+}-ATPase preparation from rat brain exhibits Ca^{2+}-dependent phosphorylation and that Ca^{2+} stimulates incorporation of ^{32}P into a single major protein of 100,000 daltons. Sermark and Vilhardt[231] solubilized the Ca^{2+}-ATPase of the microsomal fraction of cortical bovine brains with Triton X-100 and incorporated it in liposomes, and this preparation accumulates Ca^{2+} in the presence of ATP. The molar ratio between Ca^{2+} transport and ATP hydrolyzed is 1 : 3. Apparently, there are two Ca^{2+}-ATPase present, as can be detected by isoelectric focusing after solubilization with Triton X-100. One of the enzymes, isolated at pH 4.8, displays the highest affinity for Ca^{2+} ($K_m = 4.1 \times 10^{-8}$ M) but cannot be activated in the presence of excess Mg^{2+}, whereas the second enzyme, isolated at pH 6.3, can be activated by low concentrations of Ca^{2+} ($K_m = 1.2 \times 10^{-6}$ M) even in the presence of 1 mM Mg^{2+}.[231]

6.5. Other Enzymes Regulated by Calcium

There is much diverse information about the many enzymes that are regulated by Ca^{2+}. Table III contains a partial list of enzymes whose activity is influenced by Ca^{2+}. It is becoming increasingly evident that the action of Ca^{2+} on many of these enzymes is mediated through calmodulin or other Ca^{2+}-binding proteins.[14,28–30]

7. SOME CALCIUM FUNCTIONS: OVERVIEW

Some of the principal Ca^{2+} functions are summarized in Table III. In this section, I review in brief some of these functions; other functions will be referred to in other chapters of this series.

7.1. Excitation–Secretion Coupling

7.1.1. Calcium and Neurotransmitter Release in Vivo

The electrophysiological studies of Katz and Miledi[9,232,233] showed conclusively that depolarization increases the Ca^{2+} permeability at the presynaptic terminal with a consequent increase of the entry of Ca^{2+} which presumably triggers release of neurotransmitter.[4,9,11,232] Evidence that neurotransmitter release is brought about by elevated intracellular Ca^{2+} was obtained directly by microinjecting Ca^{2+} into nerve terminals [9,233] and also in studies in which an increased $[Ca^{2+}]_i$ was detected by light emission by aequorin injected into nerve terminals of the squid giant synapse.[10] Other studies at the motor nerve endings of the frog show that the rate of release of acetylcholine when an impulse arrives is highly dependent on $[Ca^{2+}]_o$ and that there is a sigmoid relationship between transmitter release and $[Ca^{2+}]_o$.[235] Neurotransmitter release highly dependent on $[Ca^{2+}]_o$ is also observed at the mammalian neuromuscular junction[236] and at the squid giant synapse.[237] The simplest interpretation for the sigmoid relationship is that several Ca^{2+} ions have to cooperate in the release of a quantum of transmitter, and analysis of some of the data indicate that three or four Ca^{2+} ions may be necessary to release one quantum.[11,235] It is not known how Ca^{2+} induces release, but it is postulated that it binds at some site in the presynaptic terminal.[238]

The "calcium hypothesis" as applied to presynaptic terminals has been tested by another approach to alter the $[Ca^{2+}]_i$ in the nerve terminals. Addition of X-537A to the frog nerve–muscle junction in the presence of 2.5 mM Ca^{2+} causes an increase in the amplitude of the end-plate potentials, presumably through the resultant increase in $[Ca^{2+}]_i$,[234] although this ionophore is not specific for Ca^{2+}.[85] On the other hand, reversal of the Ca^{2+} gradient between the inside and outside of the nerve terminal by EGTA added to the outside causes a decrease in transmitter release at the frog neuromuscular junction because of an efflux of Ca^{2+} from the terminal during depolarization of the membrane.[11,239,240]

At the neuromuscular junction and at the giant synapse of the squid, the morphology allows a clear-cut interpretation of the physiological results which show that Ca^{2+} plays a triggering role in the release of neurotransmitters. Similar results are now available for the neuron–neuron synapses of vertebrates. Thus, Dambach and Erulkar[241] showed that transmitter release at the spinal neuron of the frog is dependent on the presence of Ca^{2+} in the external medium. There remain some doubts about the role of $[Ca^{2+}]_o$ in the spontaneous liberation of transmitter. The miniature end-plate potential frequency in the frog neuromuscular junction is increased by inhibition of Ca^{2+} uptake by mitochondria.[11,242] Thus, mitochondria play a role in regulating transmitter release through their ability to regulate cytoplasmic Ca^{2+},[11,242] but it seems that under normal conditions the role of mitochondria is expected to be less than that of other nonmitochondrial Ca^{2+} stores.[92,93,102]

It is not known how Ca^{2+} causes the release of transmitter once it has entered the cell, but one hypothesis is that Ca^{2+} neutralizes negative fixed charges on the protoplasmic surfaces of the vesicles and of the axolemma, thus promoting fusion of the vesicles with the axolemma.[238,243] This mechanism requires that Ca^{2+} act very specifically, because there is a high concentration of Mg^{2+} in the nerve cell, and very large changes in transmitter release are brought about by micromolar concentrations of Ca^{2+}.

7.1.2. Calcium and Neurotransmitter Release by Synaptosomes

The most explored *in vitro* system in studies of neurotransmitter release has been the synaptosomal preparation. Synaptosomes accumulate neurotransmitters and can release them if depolarized by elevated K^+, veratridine, or electrical stimulation, in the presence of Ca^{2+}.[90,238,244,245] An extensive literature has accumulated, particularly during the last 5 years, regarding the release from synaptosomes of neurotransmitters and other substances. Thus, norepinephrine,[90,245] acetylcholine,[213,246] dopamine,[90,244,247] serotonin,[90] amino acids,[245,248,249] and ATP,[250,251] among others, are released from synaptosomes, and in nearly all cases, depolarization in the presence of Ca^{2+} is essential for this phenomenon to take place.[90]

The principal arguments in favor of the proposal that the release of transmitters from synaptosomes resembles the release from nerve terminals *in vivo* is that depolarization of synaptosomes by increased $[K^+]_o$ or by veratridine causes an increased uptake of Ca^{2+} by synaptosomes and an increased rate of neurotransmitter release that is Ca^{2+} dependent, and that this release is inhibited by Mg^{2+}.[90]

The utilization of veratridine as a tool to study the Ca^{2+}-dependent release of neurotransmitter has recently been questioned for the case of GABA and certain other amino acids.[252] Thus, the release of GABA from rat brain synaptosomes by veratridine is decreased, rather than increased, by Ca^{2+}.[252] The inhibitory effect of Ca^{2+} on the veratridine-induced release of GABA occurs at 37°C but not at 22–23°C, whereas the release of norepinephrine and dopamine by veratridine at 37°C is Ca^{2+} dependent. The authors[252] conclude that the evoked release of GABA (and glutamate) results more from veratridine-induced depolarization, which causes Na^+ influx, than from the accompanying influx of Ca^{2+} and suggest that the overall release of amino acids is reduced by the antagonism exerted by the divalent cation on the veratridine action at the Na^+ channel. In contrast, the influx of Ca^{2+} is of primary importance in triggering the exocytotic release of catecholamines, whereas depolarization itself has much less importance.[252]

There is now evidence that the neurotransmitter release occurs by exocytosis rather than through a carrier, since blockade of the amine carriers by specific inhibitors (desipramine or nomifensine) does not affect the release of norepinephrine or dopamine.[90] Further supporting evidence for the exocytotic release comes from the experiments of Fried and Blaustein[253] who showed that K^+ or veratridine depolarization of synaptosomes in the presence of Ca^{2+} causes a decrease in the number of synaptic vesicles contained in the synaptosomes. Some of these vesicles could be seen detaching from the membrane to form larger coated vesicles containing an extracellular marker (horseradish peroxidase or colloidal thorium dioxide) initially present in the exterior, which suggests that the interior of the vesicles had been in contact with the synaptosomal exterior, presumably during the fusion of its membrane with the axolemma during exocytosis.[243]

With the aid of the Ca^{2+} ioniphores X-537A and A23187,[85,90] influx of Ca^{2+} can be produced without depolarization, so that the release phenomenon observed under these conditions is easier to interpret. However, only A23187 displays a great selectivity for Ca^{2+}, so that its effect on neurosecretion is specifically attributed to its ability to take Ca^{2+} down its concentration gradient.[90] In contrast, X-537A also complexes monovalent cations and even organic amines which it can carry across biological membranes.[85,88] Thus, the Ca^{2+} ionophore A23187 is a much preferred ionophore to specifically trigger the event leading to physiological secretion, i.e., the entry of Ca^{2+} into the nerve terminal.

The release of catecholamines triggered by A23187 probably occurs mostly by an exocytotic process because it is Ca^{2+} dependent,[90] the amines released are unmetabolized,[91,254] and the release has been observed to occur accompanied by the release of dopamine-β-hydroxylase.[255] Furthermore, A23187-induced release is not affected by desipramine, an inhibitor of amine efflux mediated by the membrane carrier.[90]

Synaptosomes remain only a model that permits the study of brain excitation–secretion coupling *in vitro*. For the model to be valid, criteria established for well-studied stimulus–secretion systems should be fulfilled, and it is thus necessary to verify that Ca^{2+} is needed and sufficient for release,[256] that Mg^{2+} inhibits Ca^{2+}-dependent release,[257] and that the release is very rapid. Although not all these criteria are always met in synaptosomes, the role of Ca^{2+} in neurotransmitter release has often been shown.[90,111,245,247,258,259] Brain slices are another system that has been utilized to study *in vitro* release of neurotransmitters.[260]

The Ca^{2+} source for transmitter release may not always be external, and it has recently been shown that it is possible to evoke the release of [³H]GABA and [³H]glutamic acid from synaptosomes by increasing the release of Ca^{2+} by intrasynaptosomal mitochondria.[261] The increased efflux of mitochondrial Ca^{2+} could be induced by uncouplers of oxidative phosphorylation[261] and by increasing the $[Na^+]_i$.[262]

7.2. Calcium and the Action of Narcotics

7.2.1. Acute Narcotic Analgesia

Evidence suggests that Ca^{2+} plays an important role in mediating the action of narcotic analgesics.[22–26] Thus, Ca^{2+} antagonizes the inhibitory effects of morphine on the respiration of brain slices[263] and antagonizes the antinociceptive effects of narcotic drugs injected into mouse brain.[264,265]

Acute doses of morphine sulfate cause a loss of whole brain calcium.[21–25,301,266,267] The subcellular loss of Ca^{2+} after opiate administration occurs in fractions rich in nerve endings[21–25,266,267] and closely parallels the subcellular distribution of opiate receptor binding.[268] These studies are relevant to the observation that intracisternal injection of Ca^{2+} antagonizes the analgesic effect of morphine in mice, whereas injection of EDTA produces a weak analgesic effect which is diminished by equimolar concentrations of Ca^{2+}.[21,22,264,265] Furthermore, EGTA is an effective potentiator of analgesia, which suggests that it is the alteration in $[Ca^{2+}]$ that is of importance.[21,264] Furthermore, La^{3+}, which inhibits Ca^{2+} binding and movements across biological membranes, also produces analgesia by itself or potentiates the effect of morphine when injected intracerebrally.[265] Thus, it appears that Ca^{2+} is involved in narcotic analgesia, acute doses of narcotics having the effect of decreasing brain Ca^{2+} in specific brain regions.[22–26]

7.2.2. Chronic Narcotic Treatment

Chronic morphine treatment of rats and mice alters the response of the brain cell Ca^{2+} to morphine. Thus, under conditions of chronic treatment, Ca^{2+} increases significantly in the synaptosomal fraction,[21–25,266,267] and this increase can be prevented by simultaneous chronic administration of naloxone.[22–26] Furthermore, naloxone-precipitated withdrawal reverses the elevated levels of Ca^{2+} in the synaptosomal fraction.[26]

Binding of ^{45}Ca at low concentrations (10^{-7}–10^{-5} M) to synaptic plasma membranes *in vitro* was increased by acute morphine treatment and decreased by chronic treatment, and Yamamoto *et al.*[26] interpret the increased binding of ^{45}Ca at low Ca^{2+} $(10^{-7}$ M) as being caused by the release of endogenous Ca^{2+} from high-affinity sites by the acute morphine treatment which would increase the availability of these sites for ^{45}Ca. Madeira and Antunes-Madeira[165] showed that the high-affinity sites constitute only a small fraction of the total sites, so that changes in the occupation of these sites would not necessarily reflect a measurable change in the total Ca^{2+} content of SPM. The effects of $^{45}Ca^{2+}$ binding by SPM induced by chronic treatment are the opposite of those seen after acute treatment. Thus, a high-affinity group of binding sites (K_D = 47 μM) nearly disappears from the SPM, and the time course of this change parallels the development of tolerance and dependence after starting morphine administration. These $^{45}Ca^{2+}$-binding alterations at low medium Ca^{2+} $(10^{-7}$ M) observed for synaptic membrane preparations were not observed for intact synaptosomes, which suggested to the authors that the high-affinity sites affected by morphine are localized on the inner surface of the membrane, where they would function as part of the pump mechanism that maintains the intracellular Ca^{2+} low.[26]

Furthermore, it now appears that in mouse brain the Ca^{2+} pump of the intrasynaptosomal fraction that accumulates Ca^{2+} in the presence of ATP is also inhibited by 10^{-6} M morphine *in vitro*.[269] A similar effect is induced by the acute administration of morphine (10–20 mg/kg) to mice.[269] In contrast, after morphine pellet implantation (72 h) to induce tolerance and physical dependence, an enhancement of lysed synaptosomal $^{45}Ca^{2+}$ uptake occurred in the presence, but not in the absence, of ATP.[269] This enhancement of Ca^{2+} uptake is rectilinearly related to the degree of tolerance and dependence development. Both the acute and chronic effects of morphine on the ATP-dependent Ca^{2+} uptake by synaptosomes can be prevented by naloxone.[269] The uptake of $^{45}Ca^{2+}$ by intact synaptosomal fractions isolated from brains of mice is also reduced by morphine[270] and β-endorphin,[271] and naloxone prevents this effect in both cases.[270,271] After the animal was rendered tolerant and dependent on morphine, an enhancement of Ca^{2+} uptake is observed.[270]

These overall findings are taken as evidence that alterations in Ca^{2+} are somehow involved in regulating acute and chronic actions of morphine.[22,23] Acute narcotic treatment reduces synaptosomal Ca^{2+}, but continued administration of the narcotic replenishes the Ca^{2+} and eventually increases it above the control levels. This Ca^{2+} elevation represents one of the basic mechanisms underlying narcotic tolerance.[22-26] Thus, the changes in Ca^{2+} content and distribution in synaptosomes induced by acute and chronic opium treatment are considered to mediate the neurochemical events induced in the brain by opium.[22-26] Alterations in cell Ca^{2+} will influence neurotransmitter release and neurotransmitter-related activities under the control of adenylate cyclase, phosphodiesterase, and protein phosphorylation.[14-20] A proposed hypothesis[272] suggests that during development of morphine tolerance and dependence more adenylate cyclase enzyme molecules become available through enzyme induction in the postsynaptic region because of a reduced influx of Ca^{2+} caused by morphine. This would cause less activation of the calmodulin, which in turn would result in less cyclic AMP because of reduced activity of adenylate cyclase. On withdrawal, Ca^{2+} influx increases, and Ca^{2+} reacts with calmodulin and activates the abnormally high number of adenylate cyclase molecular present, which will cause a rise in cyclic AMP associated with the withdrawal syndrome.

At the level of the presynaptic terminal, acute narcotic administration causes a decrease in Ca^{2+} influx and binding, which may cause a decrease in neurotransmitter release, thus producing analgesia and other drug effects. A compensatory mechanism will then set in during prolonged administration of the drug so that the Ca^{2+} level rises above normal, and under these conditions, more opiate is necessary to produce a response. This adaptation would be equivalent to tolerance development, and the subsequent higher doses needed to produce an effect cause a further rise in Ca^{2+} in the nerve terminals. These high levels of Ca^{2+} will facilitate the release of neurotransmitters on drug discontinuance, which results in the hyperactivity characteristic of withdrawal.[22,23]

Sodium ion enhances opiate antagonist binding but reduces agonist binding.[273,274] It has been postulated that the Na^+ effect results from Na^+-induced displacement of Ca^{2+} from the synaptic membranes. Ross[23,24] proposes that agonists bind more readily to the divalent cation form of the receptor, and antagonists to the monovalent form. Calcium ions and other divalent cations, Mg^{2+}, Mn^{2+}, and Ni^{2+}, enhance opiate agonist binding, with Mn^{2+} having the most marked effect.[273] It is argued that Mn^{2+} produces a Ca^{2+}-associated type of receptor conformation.

7.3. Calcium and the Action of Ethanol and Barbiturates

Acute intraperitoneal administration of 1.5–2.5 g of ethanol per kg was initially reported to decrease brain Ca^{2+} content to about 50% of normal.[267] This Ca^{2+} depletion was observed in all brain areas studied, but in subfractionation studies it was shown to be restricted to the nerve ending, or synaptosomal, fraction.[267] Naloxone, the narcotic antagonist, reduced the Ca^{2+}-depleting effects of ethanol.[267]

The potencies of several alcohols in lowering brain Ca^{2+} do not parallel their pharmacological potencies, and the administration of Ca^{2+} increases rather than decreases (as would be expected) the effects of ethanol.[275] Furthermore, other workers have failed to observe any alterations in the subcellular localization of Ca^{2+} in brain after either acute or chronic ethanol treatment.[276] For example, Hood and Harris[277] did not find any alterations in subcellular localization of Ca^{2+} in mouse brain after acute or chronic injection of 4 g of ethanol per kg and report similar results for cortical tissue or synaptosomes of rats. More recently, Harris and Hood[278] showed that *in vitro* addition of ethanol (or pentobarbital) to synaptosomes isolated from rat or mouse brain inhibits the depolarization-dependent uptake of Ca^{2+} without affecting uptake under nondepolarizing conditions. Synaptosomes isolated from mice chronically ingesting ethanol are tolerant to the inhibitory effects of ethanol and pentobarbital, and synaptosomes from ethanol-tolerant or -dependent mice accumulate less Ca^{2+} in the absence of ethanol than do synaptosomes from control mice.[278] It is suggested that the known inhibitory effect of ethanol on the stimulated release of neurotransmitters and the development of tolerance to this effect may be mediated by the inhibition by ethanol of the depolarization-dependent influx of Ca^{2+}.

Pentobarbital also inhibits depolarization-induced $^{45}Ca^{2+}$ influx across synaptosomal membranes,[278–280] and it is suggested that central neurons system sedation produced by barbiturates may occur as a result of the decrease in the influx of Ca^{2+} into the nerve terminal.[279] However, chronic barbiturate administration resulted in an adaptation such that Ca^{2+} influx was less inhibited.[279] Haycock *et al.*[281] showed that pentobarbital markedly decreases the release of GABA and norepinephrine from isolated synaptosomes subsequent to KCl depolarization but does not inhibit the release caused by A23187, which is in accord with an effect of pentobarbital only on depolarization-induced Ca^{2+} influx. The depolarization-induced $^{45}Ca^{2+}$ influx into nerve endings isolated from brainstem, cerebellum, and cerebral cortex was 77%, 73%, and 39%, respectively, but no significant effect was observed on the $^{45}Ca^{2+}$ influx in synaptosomes isolated from striatum, hypothalamus, or midbrain.[279] This is in agreement with observations that the brainstem is one of the most sensitive areas to barbiturate inhibition.[282] Recently, Waller and Richter[283] reported that pentobarbital also inhibits the K^+-stimulated release of GABA, glutamate, aspartate, and acetylcholine from rat midbrain slices.

Depressant drugs also inhibit the active accumulation of $^{45}Ca^{2+}$ by the intrasynaptosomal structures that concentrate Ca^{2+} in the presence of ATP.[284] Ethanol (800 mM), pentobarbital (0.5 mM), and acetaldehyde (100 mM) significantly inhibit the ATP-dependent $^{45}Ca^{2+}$ uptake.[284] The inhibition of this $^{45}Ca^{2+}$ uptake may account for the increase in spontaneous release of neurotransmitters that is induced by high concentrations of depressants and anesthetics.[285,286] A similar effect is induced by sulfhydryl reagents which also inhibit ATP-dependent Ca^{2+} transport by isolated synaptosomes.[284] The drug concentrations that are required to increase the spontaneous release of neurotransmitters are similar to those that inhibit the ATP-dependent Ca^{2+} uptake by synaptosomes.[284] Inhibition of this Ca^{2+} uptake in the unstimulated cell would increase spontaneous release of neurotransmitters such as acetylcholine, whereas the increased Ca^{2+} influx in response to depolarization would be responsible for the burst of neurotransmitter release that follows depolarization-stimulated uptake of Ca^{2+} by intact synaptosomes is more sensitive than is the ATP-dependent uptake of Ca^{2+} by the intrasynaptosomal Ca^{2+} transport system.[284]

Chronic ethanol intake by rats causes a decrease in the ^{45}Ca-binding capacity of synaptosomal membranes.[287] The maximum number of binding sites decreases from about 11 nmol/mg of protein in controls to about 6 nmol/mg of protein in ethanol-treated animals, but there is no effect on the dissociation constant.[287] Furthermore, binding of Ca^{2+} to membranes isolated from ethanol-treated animals is relatively resistant to increasing concentrations of ethanol added *in vitro,* as compared to controls, which suggests an adaptive change.[287]

Other studies have shown that Ca^{2+} alters the biological effects of alcohol in the whole organism. Thus, intracerebroventricular administration of Ca^{2+} increased the sleeping time produced by ethanol and *t*-butanol as well as by chloral hydrate or by pentobarbital.[288] Lanthanum, a Ca^{2+} antagonist, antagonizes the hypothermic effect of ethanol.[288] Mice made dependent on ethanol go into convulsions on withdrawal, and Ca^{2+} suppresses this condition.[288] Thus, Ca^{2+} modulates the actions of ethanol in the whole organism, probably through alterations of Ca^{2+} in the brain.

7.4. Calcium and Fast Axoplasmic Transport

The anterograde transport of proteins in cytoplasm occurs at two distinct rates: fast and slow.[289-291] In general, most of the macromolecular substances show the fast transport rate, whereas soluble substances move at the slow rate.[289] The fast axoplasmic transport is independent of the size of the nerve fibers and is an energy-dependent process, requiring ATP.[291,292] Calcium has been shown to be an essential requirement for the process.[289-291,293] It was shown in desheathed cat nerve preparations that Ca^{2+} is essential for maintaining axoplasmic transport.

Recently, Hammerschlag[289,294] presented evidence that Ca^{2+} plays a role in axoplasmic transport of proteins at the initiation phase of axonal transport. This is the phase at which the proteins destined for transport must be selected from the population of newly synthetized proteins; it precedes the translocation phase.[289] Thus, incubation of neuronal cell bodies in a Ca^{2+}-free medium depresses the amount, but not the rate, of fast axonal transport of [^{3}H]protein.[289,294] These conditions do not affect protein synthesis or the general energy metabolism, and the effect can be reversed by transferring the preparations from the Ca^{2+}-free medium to normal medium. Furthermore, selective exposure of nerve trunks to a Ca^{2+}-free medium has no effect on [^{3}H]protein already undergoing translocation in the axon.[289,295] Assignment of the Ca^{2+} requirement to a somal site has been questioned, because fast axonal transport is inhibited in isolated sciatic nerve desheathed and incubated in a Ca^{2+}-free medium.[289,293] But the Ca^{2+}-requiring site in the desheathed nerve trunk has a divalent cation specificity distinct from that of the ganglionic site.[289,296] Differences in the divalent cation specificities of the axonal and ganglionic calcium requirements suggest that Ca^{2+} supports transport in nerves in a manner distinct from its role in maintaining transport in spinal ganglia.[289] Furthermore, initiation of fast transport as a Ca^{2+}-requiring step is common to the mobilization of all proteins destined for fast transport.[289]

Hammerschlag *et al.*[294] showed that $^{45}Ca^{2+}$ taken up by the frog dorsal root ganglion was transported at a rate characteristic of fast axoplasmic transport. The rapidly transported $^{45}Ca^{2+}$ is bound to a protein of 15,000 daltons which is probably calmodulin.[290,297] The mechanism of fast axoplasmic transport has been proposed to involve a "ratchet mechanism" somewhat similar to that of muscle contraction in which the microtubules and/or neurofilaments provide the rails,[293] and calmodulin may regulate this process.[290]

8. SUMMARY AND CONCLUSIONS

Because Ca^{2+} acts at so many levels in the cell, it is not yet possible to present a unified concept of the integrated role of Ca^{2+} in the nerve cell. The major concepts emerging from the information referred to in this chapter, relative to the regulation and functions of Ca^{2+} in the cell, are summarized in this section.

At the level of the membrane, Ca^{2+} is a stabilizing agent, and its removal from the external medium of nerve fibers makes the membrane more sensitive to excitation.[3–5,94] Thus, in the squid axon, the Na^+ and K^+ currents produced by a given depolarization are increased when the $[Ca^{2+}]_o$ is decreased: a fivefold reduction in the $[Ca^{2+}]_o$ induces a shift of 15 mV in the current/voltage curves.[94] The effect of Ca^{2+} probably results from its ability to diminish the electrostatic field of fixed negative charges on the surface of the nerve membrane near the ion channels.[298]

A depolarization pulse increases the influx of Ca^{2+}, and the resulting increase in $[Ca^{2+}]_i$ activates the K^+ channels, so that the K^+ conductance of the membrane increases.[7] This Ca^{2+} effect has also been shown by injecting Ca^{2+} into nerve cells of *Aplysia*,[4,7] *Helix*,[4,7] and vertebrates.[7] The inner surfaces of the membranes of neurons of the subesophageal ganglion of *Helix aspersa* are sensitive to free Ca^{2+} concentrations of 9×10^{-7} M, which produce a fall in resistance of 25%.[7] Actually, there are two components of outward K^+ current, one activated by Ca^{2+} and a second which is voltage dependent.[7] Furthermore, it appears that in snail neurons prolonged depolarization pulses or repetitive depolarizing steps cause depression of the outward K^+ current.[4,7]

Nerve cell depolarization increases the Ca^{2+} influx which can be resolved into two phases: an early phase with a time course similar to that of Na^+ influx which is sensitive to TTX, representing a flow of Ca^{2+} through the Na^+ channels[55]; and a second delayed phase of Ca^{2+} entry which increases steeply between 40 and 80 mV positive to the resting potential. This Ca^{2+} current is insensitive to TTX[6,55] and probably takes place through a separate Ca^{2+} channel.[6,108] The properties of the late Ca^{2+} channel resemble those of the Ca^{2+}-dependent mechanism that initiates the release of neurotransmitter. Both are insensitive to TTX and TEA, are inhibited by Mg^{2+} and Mn^{2+}, and have similar voltage dependencies.[6,108]

The Ca^{2+} that enters the nerve cell on depolarization is eventually extruded principally by two mechanisms: a Na^+/Ca^{2+} exchange mechanism in which Ca^{2+} is driven out by the electrochemical gradient of Na^+[83,104,110–113] and a less-well-defined Ca^{2+}-ATPase mechanism which is presumed to transport Ca^{2+} at the expense of ATP.[18,19,113] ATP also has a regulatory effect on the Na^+/Ca^{2+} exchange mechanism.[83,115] Thus, ATP activates the Na_o^+-dependent Ca^{2+} efflux and increases the affinity of the carrier for internal Ca^{2+} and for external Na^+.[83,115] The relationship between the Na_o^+-dependent Ca^{2+} efflux and the $[Na^+]_o$ fits best to a cubic function of Na^+, which indicates that three Na^+ are required to activate the efflux of one Ca^{2+}.[104,112,115] A Ca_o^{2+}/Ca_i^{2+} exchange also takes place, probably through the same carrier mechanism employed for the Na^+/Ca^{2+} exchange.[110,115] With a stoichiometry of 3 Na^+ : 1 Ca^{2+}, the electrochemical gradient of Na^+ can provide sufficient energy to maintain the free $[Ca^{2+}]_i$ in the physiological range (about 10^{-7} M).[115]

The total Ca^{2+} in squid axoplasm is about 50 μmol/kg axoplasm,[92,93] but the free $[Ca^{2+}]_i$ is only about 0.03 μM.[92,93] The remaining Ca^{2+} is held in mitochondria (3 μM), and other organelles ($\simeq$40 μM).[92,93] A small amount of Ca^{2+} is bound to nondialyzable axoplasm (4 μM), to soluble Ca^{2+}-binding proteins (0.07 μM), and to soluble anions (0.2 μM).[92,93] At near physiological conditions, the rate of Ca^{2+} uptake by mitochondria is low, and at $[Ca^{2+}]_i$ lower than 200–300 nm, little Ca^{2+} is taken up by mitochondria.[101] Furthermore, after poisoning of the mitochondria, about 70% of the axoplasmic Ca^{2+} still remains in the particulate fraction. Therefore, nonmitochondrial Ca^{2+} stores associated with cell organelles such as SER are the principal Ca^{2+}-buffering systems in the axon.[92,93,99,100,102] A similar conclusion is derived from the results obtained with synaptosomes isolated from rat brain.[63,99,102,129,130] The nonmitochondrial organelles in synaptosomes have a high affinity (uptake is half-saturated at 0.35 μM Ca^{2+}) but can store only about 3 μmol Ca^{2+}/g protein, whereas mitochondria require more than 10 μM Ca^{2+} for half-saturation but have a larger capacity ($\simeq$30 μmol Ca^{2+}/g protein).[99,102,129,130] The nonmitochondrial organelles can buffer $[Ca^{2+}]$ to about 0.4 μM in the presence of ATP when the Ca^{2+} load does not exceed 30 μmol Ca^{2+}/g protein.[63,99] Thus, SER is probably especially important in buffering cytoplasmic Ca^{2+} in the nerve terminal.[63,99,102,129,130] However, recent studies show that synaptic vesicles[143] and mitochondria[131] should not be disregarded in this respect.

The Ca^{2+} that enters the nerve cell triggers neurotransmitter release which has been studied both *in vivo*[4,9,11,232,234] and *in vitro*.[90,238,244,245] In addition, Ca^{2+} activates adenylate cyclase,[14,15] phosphodiesterase,[14,15] Ca^{2+}-ATPase,[19,20] protein kinases,[12,16–18] among other cell enzymes (Table III). Calmodulin modulates the action of Ca^{2+} on these enzymes.[14,30] The concerted action of Ca^{2+} on adenylate cyclase and phosphodiesterase in the presence of calmodulin is possible because the two enzymes have different sensitivities to Ca^{2+}.[15]

The Ca^{2+}-dependent phosphorylation of specific membrane proteins in synaptosomes has been implicated in neurotransmitter release.[33,217] This phosphorylation is also mediated by calmodulin. The results of other studies suggest that phosphorylation of membrane proteins decreases the permeability of the membranes to Ca^{2+}[221,222] and to Na^+.[223] A major unifying concept is that Ca^{2+} receptors exist in the cell in the form of specific Ca^{2+}-binding proteins.[28–31] It is the complex formed between Ca^{2+} and these proteins, of which calmodulin is of particular importance, that mediates the role of Ca^{2+} as a second messenger.[14,28–30]

Calcium antagonizes the analgesic effect of morphine.[22–26] Acute doses of morphine cause loss of brain Ca^{2+} in rat, predominantly in cell fractions rich in nerve endings.[22–26,266,267] Chronic morphine treatment of rats and mice increases Ca^{2+} in synaptosomal fractions.[22–26,266,267] Fluxes of Ca^{2+} across synaptosomal membranes[269,270] and Ca^{2+} binding to synaptic membranes[26] are also influenced by narcotic analgesics. These results are taken as evidence that Ca^{2+} is involved in regulating acute and chronic actions of morphine.[22–26] Alcohol and barbiturates are also reported to decrease the Ca^{2+} content at nerve endings when administered to rats[267] and, *in vitro*, to decrease the depolarization-dependent Ca^{2+} uptake by synaptosomes.[278] However, synaptosomes isolated from animals chronically ingesting ethanol are tolerant to the inhibitory effect of ethanol,[278] and less ^{45}Ca binding to the synaptosomal membranes occurs.[287] Recently, it was shown that ethanol and pentobarbital also decrease the ATP-dependent ^{45}Ca uptake by intrasynaptosomal structures.[284] It is concluded that Ca^{2+} contributes to the modulation of the actions of ethanol in the organism.[288]

Calcium regulates the fast axoplasmic transport of proteins by acting at the initiation phase, which is the phase when proteins destined for transport are selected from the population of newly synthetized proteins, and precedes the translocation phase.[289–291] Calmodulin may be involved in fast axonal transport.[290] It has been shown that microtubule assembly and disassembly is also regulated by Ca^{2+} and calmodulin.[192]

Some important functions of Ca^{2+} are not covered in this chapter. The reader is referred to Table III for a summary of those functions.

ACKNOWLEDGMENTS. The author is grateful to Prof. Vitor Madeira for reading the manuscript. The invaluable contribution of Ms A. R. Volpe in carrying out the literature search is greatly appreciated. This work was done while the author was the recipient of grants from INIC (Portuguese Ministry of Education and Science) and NATO (Grant No. 1513).

REFERENCES

1. Berridge, M. J., 1975, *Adv. Cyclic Nucleotide Res.* **6**:1–98.
2. Rasmussen, H., and Goodman, D. B. P., 1977, *Physiol. Rev.* **57**:421–509.
3. Shanes, A. M., 1958, *Pharmacol. Rev.* **10**:59–164.
4. Erulkar, S. D., and Fine, A., 1979. *Rev. Neurosci.* **4**:179–232.
5. Seeman, P., 1972, *Pharmacol. Rev.* **24**:583–655.
6. Baker, P. P., 1972, *Prog. Biophys. Mol. Biol.* **24**:179–223.
7. Meech, R. W., 1978, *Annu. Rev. Biophys. Bioeng.* **7**:1–18.
8. Ekert, R., and Tillotson, D., 1978, *Science* **200**:437–439.
9. Katz, B., and Miledi, R., 1967, *J. Physiol. (Lond.)* **189**:535–544.
10. Llinás, R., and Nicholson, C., 1975, *Proc. Natl. Acad. Sci. U.S.A.* **72**:187–190.
11. Rahamimoff, R., Erulkar, S. D., Alnaes, E., Meiri, H., Rotshenker, S., and Rahamimoff, H., 1975, *Cold Spring Harbor Symp. Quant. Biol.* **XL**:107–116.
12. Greengard, P., 1979, *Fed. Proc.* **38**:2208–2217.
13. Brostrom, M. A., Brostrom, C. O., Breckenridge, B. McL., and Wolff, D. J., 1978, *Adv. Cyclic Nucleotide Res.* **9**:85–99.
14. Wolff, D. J., and Brostrom, C. O., 1979, *Adv. Cyclic Nucleotide Res.* **11**:27–88.
15. Piascik, M. T., Wisler, P. L., Johnson, C. L., and Potter, J. D., 1980, *J. Biol. Chem.* **255**:4176–4181.
16. Greengard, P., 1978, *Science* **199**:146–152.
17. Yamauchi, T., and Fujisawa, H., 1980, *FEBS Lett.* **116**:141–144.
18. Petrali, E. H., Thiessen, B. J., and Sulakhe, P. V., 1980, *Int. J. Biochem.* **11**:21–36.
19. Kuo, C. H., Ichida, S., Matsuda, T., Kakiuchi, S., and Yoshida, H., 1979, *Life Sci.* **25**:235–240.
20. Sobue, K., Ichida, S., Yoshida, H., Yamazoki, R., and Kakiuchi, S., 1979, *FEBS Lett.* **99**:199–202.
21. Morgenroth, V. H., Boadle-Biber, M. C., and Roth, R. H., 1975, *Mol. Pharmacol.* **11**:427–435.
22. Chapman, D. B., and Way, E. L., 1980, *Annu. Rev. Pharmacol. Toxicol.* **20**:553–579.
23. Ross, D. H., and Cardenas, H. L., 1979, *Adv. Biochem. Psychopharmacol.* **20**:301–336.
24. Ross, D. H., 1978, *Calcium in Drug Action* (G. B. Weiss, ed.), Plenum Press, New York, pp. 241–259.

25. Ross, D. H., 1977, *Neurochem. Res.* **2**:581–593.
26. Yamamoto, H., Harris, R. A., Loh, H. H., and Way, E. L., 1978, *J. Pharmacol. Exp. Ther.* **205**:255–264.
27. Low, P. S. Lloyd, D. H., Stein, T. M., and Rogers, J. A., 1979, *J. Biol. Chem.* **254**:4119–4125.
28. Kretsinger, R. H., 1979, *Adv. Cyclic Nucleotide Res.* **11**:1–26.
29. Cheung, W. Y., 1980, *Science* **207**:19–27.
30. Klee, C. B., Crouch, T. H., and Richman, P. G., 1980, *Annu. Rev. Biochem.* **49**:489–515.
31. Kretsinger, R. H., 1976, *Annu. Rev. Biochem.* **45**:239–265.
32. Yamauchi, T., and Fujisawa, H., 1979, *Biochem. Biophys. Res. Commun.* **90**:1172–1178.
33. DeLorenzo, R. J., Freedman, S. D., Yohe, W. B., and Maurer, S. C., 1979, *Proc. Natl. Acad. Sci. U.S.A.* **76**:1838–1842.
34. Nishikawa, M., Tanaka, T., and Hidaka, H., 1980, *Nature* **287**:863–865.
35. Hanbauer, I., Gimble, J., Sankaran, K., and Sherard, R., 1979, *Neuropharmacology* **18**:859–864.
36. Vincenzi, F. F., 1979, *Proc. West. Pharmacol. Soc.* **22**:289–294.
37. Williams, M., 1979, *Trends Biochem. Sci.* **4**:25–28.
38. Bartfai, T., 1978, *Trends Biochem. Sci.* **3**:121–124.
39. Greengard, P., 1979, *Trends Pharmacol. Sci.* **1**:27–29.
40. Koketsu, K., 1969, *Neuroscience Research*, Volume 2 (S. Ehrenpreis and O. C. Solnitzky, eds.), Academic Press, New York, pp. 1–39.
41. Christian, G. D., and Feldman, F. J., 1970, *Atomic Absorption Spectroscopy*, Wiley-Interscience, New York.
42. Carvalho, C. A. M., 1978, *J. Neurochem.* **30**:1149–1155.
43. Schmidt, W. K., and Way, E. L., 1979, *J. Neurochem.* **32**:1095–1098.
44. Hanig, R. C., Tachiki, K. H., and Aprison, M. H., 1972, *J. Neurochem.* **19**:1501–1507.
45. Sanui, H., and Pace, N., 1966, *Appl. Spectrosc.* **20**:135–141.
46. Schwartz, H. D., 1975, *Clin. Chim. Acta* **64**:227–239.
47. Simon, W., Ammann, D., Dehme, M., and Morf, W. E., 1978, *Ann. N.Y. Acad. Sci.* **307**:52–70.
48. Madeira, V. M. C., 1975, *Biochem. Biophys. Res. Commun.* **64**:870–876.
49. Amman, D., Güggi, M., Pretsch, E., and Simon, W., 1975, *Anal. Lett.* **8**:709–720.
50. Nicholson, C., 1977, *Proc. Natl. Acad. Sci. U.S.A.* **74**:1287–1290.
51. Nicholson, C., 1980, *Fed. Proc.* **39**:1519–1523.
52. O'Doherty, J., Youmans, S. J., and Armstrong, W. McD., 1980, *Science* **209**:510–513.
53. Blinks, J. R., Prendergast, F. G., and Allen, D. G., 1976, *Pharmacol. Rev.* **28**:1–93.
54. Portzehl, H., Caldwell, P. C., and Rüegg, J. C., 1964, *Biochim. Biophys. Acta* **79**:581–591.
55. Baker, P. F., Hodgkin, A. L., and Ridgway, E. B., 1971, *J. Physiol. (Lond.)* **218**:709–755.
56. Hallet, M., and Carbone, E., 1972, *J. Cell. Physiol.* **80**:219–226.
57. DiPolo, R., Requena, J., Brinley, F. J., Jr., Mullins, L. J., Scarpa, A., and Tiffert, T., 1976, *J. Gen. Physiol.* **67**:433–467.
58. Caswell, A. H., 1977, *Int. Rev. Cytol.* **56**:145–181.
59. Shinomura, O., and Johnson, F. H., 1973, *Biochem. Biophys. Res. Commun.* **53**:490–494.
60. Ohnishi, T., and Ebashi, S., 1963, *J. Biochem.* **54**:506–511.
61. Scarpa, A., Brinley, F., Jr., Tiffert, T., and Dubyac, G. R., 1978, *Ann. N.Y. Acad. Sci.* **307**:86–112.
62. Kamino, K., Ogawa, M., Uyesaka, N., and Inouye, A., 1976, *J. Membr. Biol.* **26**:345–356.
63. Schweitzer, E. S., and Blaustein, M. P., 1980, *Biochim. Biophys. Acta* **600**:912–921.
64. Scarpa, A., Brinley, F. J., Jr., and Dubyak, G., 1978, *Biochemistry* **17**:1378–1386.
65. Caswell, A. H., and Hutchison, J. D., 1971, *Biochem. Biophys. Res. Commun.* **43**:625–630.
66. Caswell, A. H., 1972, *J. Membr. Biol.* **7**:345–364.
67. Carvalho, C. A. M., and Carvalho, A. P., 1977, *Biochim. Biophys. Acta* **468**:21–30.
68. Schaffer, W. T., and Olson, M. S., 1976, *J. Neurochem.* **27**:1319–1325.
69. Hallet, M., Schneider, A. S., and Carbone, E., 1972, *J. Membr. Biol.* **10**:31–44.
70. Caswell, A. H., and Hutchison, J. D., 1971, *Biochem. Biophys. Res. Commun.* **42**:43–49.
71. Le Breton, G. C., Sandler, W. C., and Feinberg, H., 1976, *Biochem. Biophys. Res. Commun.* **71**:362–370.

72. Gains, N., 1980, *Eur. J. Biochem.* **111**:199–202.

73. Ornberg, R. L., and Reese, T. S., 1980, *Fed. Proc.* **39**:2802–2808.

74. Henkart, M., 1980, *Fed. Proc.* **39**:2783–2789.

75. Klein, R. L., Yen, S. S., and Thureson-Klein, A., 1972, *J. Histochem. Cytochem.* **20**:65–78.

76. Podolsky, R. J., Hall, T., and Hatchett, S. L., 1970, *J. Cell Biol.* **44**:699–702.

77. Lechene, C. P., and Warner, R. R., 1977, *Annu. Rev. Biophys. Bioeng.* **6**:57–85.

78. Oschman, J. L., Hall, T. A., Peters, P. D., and Wall, B. J., 1974, *J. Cell Biol.* **61**:156–165.

79. Hillman, D. E., and Llinás, R., 1974, *J. Cell Biol.* **61**:146–155.

80. McGraw, C. F., Somlyo, A. V., and Blaustein, M. P., 1980, *Fed. Proc.* **39**:2796–2801.

81. Winegrad, S., 1970, *J. Gen. Physiol.* **55**:77–88.

82. Babel-Guerin, E., Bayenval, J., Droz, B., Durant, Y., and Hassing, R., 1977, *Brain Res.* **121**:348–352.

83. DiPolo, R., and Beaugé, L., 1980, *Cell Calcium* **1**:147–169.

84. Bartfai, T., 1979, *Adv. Cyclic Nucleotide Res.* **10**:219–242.

85. Pressman, B. C., 1976, *Annu. Rev. Biochem.* **45**:501–530.

86. Pfeiffer, D. R., Taylor, R. W., and Lardy, H. A., 1978, *Ann. N.Y. Acad. Sci.* **307**:402–423.

87. Kafka, M. S., and Holz, R. W., 1976, *Biochim. Biophys. Acta* **426**:31–37.

88. Baker, E. P., 1979, *Antibiotics*, Volume 1 (F. E. Hann, ed.), Springer-Verlag, New York, pp. 67–97.

89. Levi, G., Roberts, P. J., and Raiteri, M., 1976, *Neurochem. Res.* **1**:409–416.

90. Raiteri, M., and Levi, G., 1978, *Rev. Neurosci.* **3**:77–130.

91. Holz, R. W., 1975, *Biochim. Biophys. Acta* **375**:138–152.

92. Brinley, F. J., Jr., 1978, *Annu. Rev. Biophys. Bioeng.* **7**:363–392.

93. Brinley, F. J., Jr., 1980, *Fed. Proc.* **39**:2778–2782.

94. Hodgkin, A. L., and Keynes, R. D., 1957, *J. Physiol. (Lond.)* **138**:253–281.

95. Baker, P. F., 1976, *Fed. Proc.* **35**:2589–2595.

96. Thomas, M. V., and Gorman, A. L. F., 1977, *Science* **196**:531–533.

97. Owen, J. D., Brown, H. M., and Pemberton, J. P., 1976, *Biophys. J.* **16**:34a.

98. Baker, P. F., and Schlaepfer, W., 1975, *J. Physiol. (Lond.)* **249**:37P–38P.

99. Blaustein, M. P., Ratzlaff, R. W., and Schweitzer, E. S., 1980, *Fed. Proc.* **39**:2790–2795.

100. Brinley, F. J., Jr., Tiffert, T., and Scarpa, A., 1978, *J. Gen. Physiol.* **72**:101–127.

101. Mullins, L. J., and Requena, J., 1979, *J. Gen. Physiol.* **74**:393–413.

102. Blaustein, M. P., Ratzlaff, R. W., and Kendrick, N. K., 1978, *Ann. N.Y. Acad. Sci.* **307**:195–212.

103. Henkart, M. P., Reese, T. S., and Brinley, F. J., Jr., 1978, *Science* **202**:1300–1303.

104. Blaustein, M. P., 1974, *Rev. Physiol. Biochem. Pharmacol.* **70**:33–82.

105. Katz, B., and Miledi, R., 1969, *J. Physiol. (Lond.)* **203**:459–487.

106. Llinás, R., Blinks, J. R., and Nicholson, C., 1972, *Science* **176**:1127–1129.

107. Baker, P. F., Meves, H., and Ridgway, E. B., 1971, *J. Physiol. (Lond.)* **231**:527–548.

108. Kostyuk, P. G., 1980, *Neuroscience* **5**:945–959.

109. DiPolo, R., 1979, *J. Gen. Physiol.* **73**:91–113.

110. Blaustein, M. P., and Ector, A. C., 1976, *Biochim. Biophys. Acta* **419**:295–308.

111. Blaustein, M. P., and Oborn, C. J., 1975, *J. Physiol. (Lond.)* **247**:657–686.

112. Blaustein, M. P., 1976, *Fed. Proc.* **35**:2574–2578.

113. Sulakhe, P. V., and St. Louis, P. J., 1980, *Prog. Biophys. Mol. Biol.* **35**:135–195.

114. Ichida, S., Hata, F., Matsuda, T., and Yoshida, H., 1976, *Jpn. J. Pharmacol.* **26**:31–37.

115. Blaustein, M. P., 1977, *Biophys. J.* **20**:79–111.

116. DiPolo, R., 1977, *J. Gen. Physiol.* **69**:795–813.

117. DiPolo, R., Rojas, H., and Beaugé, L. A., 1979, *Nature* **281**:228–229.

118. DiPolo, R., 1978, *Nature* **274**:390–392.

119. Carvalho, C. A. M., *Life Sci.* **25**:73–82.

120. Carvalho, C. A. M., and Carvalho, A. P., 1979, *J. Neurochem.* **33**:309–317.

121. Duncan, C. J., 1976, *J. Neurochem.* **27**:1277–1279.

122. Bull, R. J., and Trevor, A. J., 1972, *J. Neurochem.* **19**:1011–1022.

123. Stahl, W. L., and Swanson, P. D., 1972, *J. Neurochem.* **19**:2395–2407.

124. Goddard, G. A., and Robinson, J., 1976, *Brain Res.* **110**:331–350.

125. Blaustein, M. P., 1975, *J. Physiol. (Lond.)* **247**:617–655.
126. Nachshen, D. A., and Blaustein, M. P., 1979, *Mol. Pharmacol.* **16**:579–586.
127. Blaustein, M. P., and Russel, J. M., 1975, *J. Membr. Biol.* **22**:285–312.
128. Rahamimoff, H., and Spanier, R., 1979, *FEBS Lett.* **104**:111–114.
129. Blaustein, M. P., Ratzlaff, R. W., Kendric, N. C., and Schweitzer, E. S., 1978, *J. Gen. Physiol.* **72**:15–41.
130. Blaustein, M. P., Ratzlaff, R. W., and Schweitzer, E. S., 1978, *J. Gen. Physiol.* **72**:43–66.
131. Michaelson, D. M., Ophir, I., and Angel, I., 1980, *J. Neurochem.* **35**:116–124.
132. Israel, M., Manaranche, R., Marsal, J., Meunier, F. M., Morel, N., Frachon, P., and Lesbats, B., 1980, *J. Membr. Biol.* **54**:115–126.
133. Blitz, A. L., Fine, R. E., and Toselli, P., 1977, *J. Cell Biol.* **75**:135–147.
134. Tsudzuki, T., 1979, *J. Biochem.* **86**:777–782.
135. Shaw, F. D., and Morris, J. F., 1980, *Nature* **287**:56–58.
136. Torp-Pedersen, C., Sarmark, T., Bundgaard, M., and Thorn, N. A., 1980, *J. Neurochem.* **35**:552–557.
137. Nordman, J. J., and Chevallier, J., 1980, *Nature* **287**:54–56.
138. Carafoli, E., and Lehninger, A. L., 1971, *Biochem. J.* **122**:681–690.
139. Lazarewicz, J. W., Haljamae, H., and Hamberger, A., 1974, *J. Neurochem.* **22**:33–45.
140. Tjioe, S., Bianchi, C. P., and Hanguard, N., 1970, *Biochim. Biophys. Acta* **216**:270–273.
141. Vickers, G. R., and Dowdall, M. J., 1976, *Exp. Brain Res.* **25**:429–445.
142. Carafoli, E., and Crompton, M., 1978, *Ann. N.Y. Acad. Sci.* **307**:269–284.
143. Nicholls, D. G., and Scott, I. D., 1980, *Biochem. J.* **186**:833–839.
144. Crompton, M., Moser, R., Lüdi, H., and Carafoli, E., 1978, *Eur. J. Biochem.* **82**:25–31.
145. Carafoli, E., 1979, *FEBS Lett.* **104**:1–5.
146. Nicholls, D. G., and Crompton, M., 1980, *FEBS Lett.* **111**:261–268.
147. Logan, J. G., and Waters, G. A., 1976, *J. Physiol. (Lond.)* **256**:30P–31P.
148. Rahamimoff, H., and Abramovitz, E., 1978, *FEBS Lett.* **89**:223–226.
149. Rahamimoff, H., and Abramovitz, E., 1978, *FEBS Lett.* **92**:163–167.
150. Saito, K., Uchida, S., and Yoshida, H., 1972, *FEBS Lett.* **22**:787–798.
151. Ichida, S., Kuo, C., Uchida, S., Nagai, K., and Yoshida, H., 1976, *Jpn. J. Pharmacol.* **26**:551–558.
152. Robinson, J. D., and Lust, W. D., 1968, *Arch. Biochem. Biophys.* **125**:286–294.
153. Papazian, D., Rahamimoff, H., and Goldin, S. M., 1979, *Proc. Natl. Acad. Sci. U.S.A.* **76**:3708–3712.
154. Iqbal, Z., Garg, B. P., and Ochs, S., 1979, *Neurosci. Abstr.* **5**:60.
155. Ohtsuki, I., 1969, *J. Biochem.* **66**:645–650.
156. Satomi, D., 1974, *J. Biochem.* **76**:391–396.
157. Nakamaru, Y., and Schwartz, A., 1971, *Arch. Biochem. Biophys.* **144**:16–29.
158. Diamond, I., and Goldberg, A. L., 1971, *J. Neurochem.* **18**:1419–1431.
159. Roufogalis, B. D., 1973, *Biochim. Biophys. Acta* **318**:360–370.
160. Nakamura, K., and Konishi, K., 1974, *J. Biochem.* **75**:1129–1133.
161. De Meis, L., Rubin-Altschul, B., and Machado, R. D., 1970, *J. Biol. Chem.* **245**:1883–1889.
162. Trotta, E. E., and De Meis, L., 1975, *Biochim. Biophys. Acta* **394**:239–247.
163. Trotta, E. E., and De Meis, L., 1978, *J. Biol. Chem.* **253**:7821–7825.
164. Gomperts, B., Lantelme, F., and Stock, R., 1970, *J. Membr. Biol.* **3**:241–266.
165. Madeira, V. M. C., and Antunes-Madeira, M. C., 1973, *Biochim. Biophys. Acta* **323**:396–407.
166. Krishnan, K. S., and Balaram, P., 1976, *Arch. Biochem. Biophys.* **174**:420–430.
167. Kamino, K., Uyesaka, N., and Inouye, A., 1974, *J. Membr. Biol.* **17**:13–26.
168. Kamino, K., Uyesaka, N., Ogawa, M., and Inouye, A., 1975, *J. Membr. Biol.* **23**:21–31.
169. Kamino, K., Uyesaka, N., Ogawa, M., and Inouye, A., 1975, *J. Membr. Biol.* **21**:113–124.
170. Sottocasa, G., Sandry, G., Panfili, E., Bernard, B., Gazzoti, P., Vasington, F. D., and Carafoli, E., 1972, *Biochim. Biophys. Res. Commun.* **47**:808–813.
171. Baumgold, J., 1979, *Biochim. Biophys. Acta* **558**:149–165.
172. Abood, L. G., and Straw, J., 1974, *Biochim. Biophys. Acta* **332**:85–96.
173. Abood, L. G., Hong, J. S., Takeda, F., and Tometsko, A. M., 1976, *Biochim. Biophys. Acta* **443**:414–427.

174. Mikkelsen, R. B., 1976, *Biological Membranes*, Volume 3 (D. Chapman and D. F. H. Wallach, eds.), Academic Press, New York, pp. 153–190.

175. Hoss, W., Okumura, K., Formaniak, M., and Tanaka, R., 1979, *Life Sci.* **24**:1003–1010.

176. Hoss, W., Okumura, K., Formaniak, M., and Tanaka, R., 1980, *Arch. Biochem. Biophys.* **203**:647–653.

177. Satomi, D., and Ito, K., 1971, *Sci. Papers Coll. Gen. Educ. Univ. Tokyo* **21**:175–187.

178. Cheung, W. Y., Bradham, L. S., Lynch, T. J., Lin, Y. M., and Tallant, E. A., 1975, *Biochem. Biophys. Res. Commun.* **66**:1055–1062.

179. Goldberg, N. D., and Haddox, M. K., 1977, *Annu. Rev. Biochem.* **46**:823–896.

180. Wong. P. Y. K., and Cheung, W. Y., 1979, *Biochem. Biophys. Res. Commun.* **90**:473–477.

181. Hinds, T. R., Larsen, F. L., and Vincenzi, F. F., 1978, *Biochem. Biophys. Res. Commun.* **81**:455–461.

182. Katz, S., and Remtulla, M. A., 1978, *Biochem. Biophys. Res. Commun.* **83**:1373–1379.

183. Dabrowska, R., Sherry, J. M. F., Aromatorio, D. K., and Hartshorne, D. J., 1978, *Biochemistry* **17**:253–257.

184. Schulman, H., and Greengard, P., 1978, *Proc. Natl. Acad. Sci. U.S.A.* **75**:5432–5436.

185. Cohen, P., Burchell, A., Foulkes, J. G., Cohen, P. T. W., Vanaman, T. C., and Nairn, A. C., 1978, *FEBS Lett.* **92**:287–293.

186. Knapp, S., Mandell, A. J., and Bullard, W. P., 1975, *Life Sci.* **16**:1583–1593.

187. Anderson, J. M., and Cormier, M. J., 1978, *Biochem. Biophys. Res. Commun.* **90**:596–605.

188. McCormack, J. G., and Denton, R. M., 1980, *Biochem. J.* **190**:95–105.

189. Denton, R. M., McCormack, J. G., and Edgell, N. J., 1980, *Biochem. J.* **190**:107–117.

190. Ebashi, S., 1976, *Annu. Rev. Physiol.* **38**:293–313.

191. Douglas, W. W., 1975, *Calcium Transport in Contraction and Secretion* (E. Carafoli, F. Clementi, W. Drabikowksi, and A. Margreth, eds.), North-Holland, Amsterdam, pp. 167–174.

192. Marcum, J. M., Dedman, J. R., Binkley, B. R., and Means, A. R., 1978, *Proc. Natl. Acad. Sci. U.S.A.* **75**:3771–3775.

193. Gardos, G., 1958, *Biochim. Biophys. Acta* **30**:653–654.

194. Deleze, J., and Loewenstein, W. R., 1976, *J. Membr. Biol.* **28**:71–86.

195. Clausen, T., Elbrink, J., and Dahl-Hansen, A. B., 1975, *Biochim. Biophys. Acta* **375**:292–308.

196. Berridge, M. J., 1976, *Symp. Soc. Exp. Biol.* **30**:219–231.

197. Eldefrawi, A. T., and Eldefrawi, M. E., 1977, *Calcium-Binding Proteins and Calcium Function* (R. H. Wasserman, R. A. Corradino, E. Carafoli, R. H. Kretsinger, D. H. MacLennan, and F. L. Siegel, eds.), North-Holland, New York, pp. 117–126.

198. Dudai, Y., and Silman, I., 1973, *FEBS Lett.* **30**:49–52.

199. Lin, Y. M., Lin, Y. P., and Cheung, W. Y., 1974, *J. Biol. Chem.* **249**:4943–4954.

200. Watterson, D. M., Harrelson, W. G., Keller, P. M., Strarief, F., and Vanaman, T. C., 1976, *J. Biol. Chem.* **251**:4501–4513.

201. Crouch, T. H., and Klee, C. B., 1980, *Biochemistry* **19**:3692–3698.

202. Watterson, D. M., Mendel, P. A., and Vanaman, T. C., 1980, *Biochemistry* **19**:2672–2676.

203. Klee, C. B., 1977, *Biochemistry* **16**:1017–1024.

204. Seamon, K., 1980, *Biochemistry* **19**:207–215.

205. Walsh, M., Stevens, F. C., Oikawa, K., and Kay, C. M., 1979, *Can. J. Biochem.* **57**:267–278.

206. LaPorte, D. C., Wierman, B. M., and Storm, D. R., 1980, *Biochemistry* **19**:3814–3819.

207. Wallace, R. W., Lynch, T. J., Tallant, E. A., and Cheung, W. Y., 1979, *J. Biol. Chem.* **254**:377–382.

208. Klee, C. B., Crouch, T. H., and Krinks, M. H., 1979, *Proc. Natl. Acad. Sci. U.S.A.* **76**:6270–6273.

209. Wang, J. H., and Desai, R., 1977, *J. Biol. Chem.* **252**:4175–4184.

210. Grand, R. J. A., and Perry, S. V., 1979, *Biochem. J.* **183**:285–295.

211. Lynch, T. J., Tallant, E. A., and Cheung, W. Y., 1977, *Arch. Biochem. Biophys.* **182**:124–133.

212. Greengard, P., 1976, *Nature* **260**:101–108.

213. Michaelson, D. M., and Avissar, S., 1979, *J. Biol. Chem.* **254**:12542–12546.

214. Krueger, B. K., Forn, J., and Greengard, P., 1977, *J. Biol. Chem.* **252**:2764–2773.

215. Sieghart, W., Forn, J., and Greengard, P., 1979, *Proc. Natl. Acad. Sci. U.S.A.* **76**:2475–2479.

216. Forn, J., and Greengard, P., 1978, *Proc. Natl. Acad. Sci. U.S.A.* **75**:5195–5199.

217. DeLorenzo, R. J., and Freedman, S. D., 1978, *Biochem. Biophys. Res. Commun.* **80**:183–192.
218. DeLorenzo, R. J., Emple, G. P., and Glaser, G. H., 1977, *J. Neurochem.* **28**:21–30.
219. Clouet, D. H., and Williams, N., 1980, *Endogenous and Exogenous Opiate Agonists and Antagonists*, (E. L. Way, ed.), Pergamon Press, New York, pp. 239–242.
220. Robinson, J. D., 1978, *FEBS Lett.* **87**:261–264.
221. Weller, M., 1979, *Protein Phosphorylation*, Pion Limited, London.
222. Weller, M., and Morgan, I. G., 1977, *Biochim. Biophys. Acta* **465**:527–534.
223. Schoffeniels, E., and Dandrifosse, G., 1980, *Proc. Natl. Acad. Sci. U.S.A.* **77**:812–816.
224. Nathanson, J., 1977, *Physiol. Rev.* **57**:157–256.
225. Ohashi, T., Uchida, S., Nagai, K., and Yoshida, H., 1970, *J. Biochem.* **67**:635–641.
226. Robinson, J. D., 1976, *Arch. Biochem. Biophys.* **176**:366–374.
227. Nakamaru, Y., 1968, *J. Biochem.* **63**:626–631.
228. Edstrom, A., Hanson, M., Prus, K., and Wallin, M., 1980, *J. Neurochem.* **35**:297–303.
229. Kadota, K., Mori, S., and Imaizumi, R., 1967, *J. Biochem.* **61**:424–432.
230. Ichida, S., Kuo, C. H., Matsuda, T., and Yoshida, H., 1976, *Jpn. J. Pharmacol.* **26**:39–43.
231. Sarmark, T., and Vilhardt, H., 1979, *Biochem. J.* **181**:321–330.
232. Katz, B., and Miledi, R., 1967, *J. Physiol. (Lond.)* **192**:407–436.
233. Miledi, R., 1973, *Proc. R. Soc. Lond. [Biol.]* **183**:421–425.
234. Kita, H., and van der Kloot, W., 1976, *J. Physiol. (Lond.)* **259**:177–198.
235. Dodge, F. A., and Rahamimoff, R., 1967, *J. Physiol. (Lond.)* **193**:419–432.
236. Hubbard, J. I., Jones, S. F., and Landau, E. M., 1968, *J. Physiol. (Lond.)* **196**:75–86.
237. Katz, B., and Miledi, R., 1970, *J. Physiol. (Lond.)* **207**:789–801.
238. Kelly, R. B., Deutsch, J. W., Carlson, S. S., and Wagner, J. A., 1979, *Annu. Rev. Neurosci.* **2**:399–446.
239. Rotshenker, S., Erulkar, S. D., and Rahamimoff, R., 1976, *Brain Res.* **101**:362–365.
240. Erulkar, S. D., and Rahamimoff, R., 1978, *J. Physiol. (Lond.)* **278**:501–511.
241. Dambach, G. E., and Erulkar, S. D., 1973, *J. Physiol. (Lond.)* **228**:799–817.
242. Alnaes, E., and Rahamimoff, R., 1975, *J. Physiol. (Lond.)* **248**:285–306.
243. Ceccarelli, B., and Hurlbut, W. P., 1980, *Physiol. Rev.* **60**:396–441.
244. Haycock, J. W., Levy, W. B., Denner, L. A., and Cotman, C. W., 1979, *Neuroscience* **4**:1341–1346.
245. Cotman, C. W., Haycock, J. W., and White, W. F., 1976, *J. Physiol. (Lond.)* **254**:475–505.
246. Michaelson, D. M., and Sokolovsky, M., 1978, *J. Neurochem.* **30**:217–231.
247. Raiteri, F., Cenito, A. M., Cervoni, R. del Camine, Ribera, M. T., and Levi, G., 1978, *Adv. Biochem. Psychopharmacol.* **19**:35–56.
248. Haycock, J., and Levy, W. B., 1978, *J. Neurochem.* **30**:1113–1125.
249. Fagg, G. E., and Lane, J. D., 1979, *Neuroscience* **4**:1015–1036.
250. White, T. D., 1978, *J. Neurochem.* **30**:329–336.
251. Pollard, H. B, and Pappas, G. D., 1979, *Biochim. Biophys. Res. Commun.* **88**:1315–1321.
252. Levi, G., Gallo, V., and Raiteri, M., 1980, *Neurochem. Res.* **5**:281–295.
253. Fried, R. C., and Blaustein, M., 1978, *J. Cell Biol.* **78**:685–700.
254. Colburn, R. W., Thoa, N. B., and Kopin, I. J., 1976, *Life Sci.* **17**:1395–1400.
255. Thoa, N. B., Costa, J. L., Moss, J., and Kopin, I. J., 1974, *Life Sci.* **14**:1705–1719.
256. Rubin, R. P., 1974, *Calcium and Secretory Process*, Plenum Press, New York.
257. Krnjevic, K., 1974, *Physiol. Rev.* **54**:418–540.
258. Elks, M. L., Youngblood, W. M., and Kizer, J., 1979, *Brain Res.* **172**:461–469.
259. Delangen, C. D. J., and Mulder, A. H., 1980, *Brain Res.* **185**:399–408.
260. Szerb, J. C., 1979, *J. Neurochem.* **32**:1565–1573.
261. Sandoval, M. E., 1980, *Brain Res.* **181**:357–367.
262. Sandoval, M. E., 1980, *J. Neurochem.* **35**:915–921.
263. Takemori, A. E., 1962, *J. Pharmacol. Exp. Ther.* **135**:89–93.
264. Harris, R. A., Loh, H. H., and Way, E. L., 1976, *J. Pharmacol. Exp. Ther.* **195**:488–498.
265. Iwamoto, E. T., Harris, R. A., Loh, H. H., and Way, E. L., 1978, *J. Pharmacol. Exp. Ther.* **206**:46–55.
266. Harris, R. A., Yamamoto, H., Loh, H. H., and Way, E. L., 1977, *Life Sci.* **20**:501–506.
267. Ross, D. H., Medina, M. A., and Cardenas, H. L., 1974, *Science* **186**:63–64.

268. Ross, D. H., Sherwood, C. L., and Cardenas, H. L., 1977, *Alcohol and Opiates* (K. Blum, ed.), Academic Press, New York, pp. 265–281.
269. Guerrero-Munoz, F., Guerrero, M. L., and Way, E. L., 1979, *J. Pharmacol. Exp. Ther.* **211**:370–374.
270. Guerrero-Munoz, F., Cerreta, F. K. V., and Guerrero, E., 1979, *J. Pharmacol. Exp. Ther.* **209**:132–136.
271. Guerrero-Munoz, F., Guerrero, M. L., and Way, E. L., 1979, *Science* **206**:89–91.
272. Sanghvi, I. S., and Gershon, S., 1977, *Biochem. Pharmacol.* **26**:1183–1185.
273. Pasternak, G. W., Snowman, A. M., and Snyder, S. H., 1975, *Mol. Pharmacol.* **11**:735–744.
274. Pert, C. B., and Snyder, S. H., 1973, *Proc. Natl. Acad. Sci. U.S.A.* **70**:2243–2247.
275. Erickson, C. K., Tyler, T. D., and Harris, R. A., 1978, *Science* **199**:1219–1221.
276. Harris, R. A., 1979, *Biochemistry and Pharmacology of Ethanol*, Volume 2 (E. Majchrowicz and E. P. Noble, eds.), Plenum Press, New York, pp. 27–41.
277. Hood, W. F., and Harris, R. A., 1979, *Biochem. Pharmacol.* **28**:3075–3080.
278. Harris, R. A., and Hood, W. F., 1980, *J. Pharmacol. Exp. Ther.* **213**:562–568.
279. Elrod, S. V., and Leslie, S. W., 1980, *J. Pharmacol. Exp. Ther.* **212**:131–135.
280. Blaustein, M. P., and Ector, A. C., 1975, *Mol. Pharmacol.* **11**:369–378.
281. Haycock, J. W., Levy, W. B., and Cotman, C. W., 1977, *Biochem. Pharmacol.* **26**:159–161.
282. Killman, E. K., 1962, *Pharmacol. Rev.* **14**:175–223.
283. Waller, M. B., and Richter, J. A., 1980, *Biochem. Pharmacol.* **29**:2189–2198.
284. Hood, W. F., and Harris, R. A., 1980, *Biochem. Pharmacol.* **29**:957–959.
285. Curran, M., and Seeman, P., 1977, *Science* **197**:910–911.
286. Baba, A., Fisherman, J. S., and Cooper, J. R., 1979, *Biochem. Pharmacol.* **28**:1879–1883.
287. Michaelis, E. K., and Myers, S. L., 1979, *Biochem. Pharmacol.* **28**:2081–2087.
288. Harris, R. A., 1979, *Pharmacol. Biochem. Behav.* **10**:527–534.
289. Hammerschlag, R., 1980, *Fed. Proc.* **39**:2809–2814.
290. Iqbal, Z., 1979, *Trends Neurosci.* **2**:311–312.
291. Wilson, D. L., and Stone, G. C., 1979, *Annu. Rev. Biophys. Bioeng.* **8**:27–45.
292. Ochs, S., 1976, *Basic Neurochemistry* (E. Siegel, R. W. Albers, R. Katzman, and B. W. Agranoff, eds.), Little, Brown & Co., Boston, pp. 429–444.
293. Ochs, S., Worth, R. M., and Chan, S. Y., 1977, *Nature* **270**:748–750.
294. Hammerschlag, R., Bakhit, C., Chiu, A. Y., and Dravid, A. R., 1977, *J. Neurobiol.* **8**:439–451.
295. Edstrom, A., 1974, *J. Cell Biol.* **61**:812–818.
296. Lavoie, P. A., Bolen, F., and Hammerschlag, R., 1979, *J. Neurochem.* **32**:1745–1751.
297. Iqbal, Z., and Ochs, S., 1978, *J. Neurochem.* **31**:409–418.
298. McLanghlins, S., Szabo, G., and Eisenman, G. J., 1971, *J. Gen. Physiol.* **58**:667–687.
299. Marsal, J., Esquerda, J. E., Fiol, C., Solsona, C., and Tomas, J., 1980, *J. Physiol. (Paris)* **76**:190–226.
300. Silbergeld, E., and Costa, J. L., 1979, *Exp. Neurol.* **63**:277–292.
301. Cardenas, H. L., and Ross, D. L., 1975, *J. Neurochem.* **24**:487–493.
302. Chan, S. Y., Ochs, S., and Worth, R. M., 1980, *J. Physiol.* **301**:477–504.

Ammonia

A. M. Benjamin

1. INTRODUCTION

Since the early demonstrations of the harmful effects of ammonia* on man and animals[1-3] and of the liberation of ammonia by stimulated nerve tissue,[4] a large literature concerned with the processes (operating both *in vivo* and *in vitro*) that are involved in the neural formation, transport, utilization, function, and toxicity of ammonia has accumulated. It indicates the dynamic nature of ammonia in the brain, its direct relationship to metabolism of glucose, amino acids, and proteins, its effects on cell energetics, its significant response to physiological change, its covert involvement in transmitter function, its influence on membrane properties and on cell morphology, and its implication at hyperconcentrations in convulsions and coma. Some of these aspects have been reviewed in the last two decades.[3,5-14] In this chapter, an attempt will be made at collation and synthesis of some of the early and present knowledge of the field.

2. METHODS OF ESTIMATION OF AMMONIA

Despite the existence of a number of sensitive and convenient methods for estimation of ammonia,[15-30] there are problems associated with the accurate determination of the free ammonia concentrations of biological materials including blood and brain. The problems encountered are of two kinds. One, that of contamination, arises because of the ubiquitous nature of ammonia and its liberation, during estimation, from labile tissue constituents such as proteins and glutamine. Such contamination can be minimized by use of scrupulously

---

* The term *ammonia* in this review expresses the sum of the ammonium ion (NH_4^+) and the undissociated ammonia (NH_3) in solution. The distinction between these two forms will be made in the text whenever necessary.

A. M. Benjamin • Division of Neurological Sciences, University of British Columbia, Vancouver, British Columbia V6T 1W5, Canada.

clean glassware, relatively pure reagents, and ammonia-free surroundings and by deproteinizing samples prior to assay. The other problem results from the rapid liberation or "burst" of ammonia that occurs in neural tissue post-mortem and, in the case of blood, also immediately after shedding. The first stage of liberated ammonia from blood which occurs within 5 min after shedding is presumed to arise enzymatically, chiefly from adenosine,[3,16] although proteins and glutamine may also be contributory.[31] The burst of brain ammonia post-mortem apparently occurs because of the presence of proteins and as a result of a suppressed glutamine synthesis caused by anoxia. Rapid freezing and deproteinization will help to diminish this problem.

Representative methods of ammonia estimation employ techniques of distillation,[15] microdiffusion,[16,17] ion exchange,[18,19] or enzymatic assay[20–22] in conjunction with those of titrimetry,[23] colorimetry,[16–19,23–26] spectrophotometry,[21,27] or fluorimetry.[20,22] More recent techniques involve analyses by automated systems,[28] by adaptation of conductometric methods,[29] and by use of ammonia-specific electrodes.[30] Colorimetry utilizes a wide variety of reagents that produce color on reaction with ammonia, such as ninhydrin,[24] phenol–hypochlorite,[19,25] hypobromide–phenosafranin,[26] or Nessler's reagent.[18,32] Enzymatic determination of ammonia usually employs the glutamate dehydrogenase reaction.[22] The relative merits and disadvantages associated with each technique have been discussed in the numerous articles that have appeared on ammonia estimation.[15–30]

3. AMMONIA CONCENTRATIONS IN VIVO AND IN VITRO

3.1. Ammonia Contents of Normal Brain, Blood, and Cerebrospinal Fluid

Values of ammonia content of the brain that appear in the literature prior to 1948 are mostly unreliable. With the advent of rapid freezing techniques using liquid air or liquid nitrogen which suppress changes post-mortem, values of brain ammonia between 0.15 and 0.35 μmol/g fresh wt. tissue have been reported for a variety of animal species[32–41] (Table I). There is a twofold increase of ammonia in mouse[22] or rat[32] brain if freezing of the brain occurs a few seconds after death, and a 3.5-fold increase in rat brain at the end of 30 s.[32] Values of 1.6 and 2 μmol/g fresh wt. tissue have been found in rat[42] and cat[43] brain, respectively, 2 to 3 min after death, and a value of 5 μmol/g guinea pig brain 20 min after death.[7,44]

Ammonia content of rat cerebellum (0.11 μmol/g) is lower than that of the frontal lobe (0.27 μmol/g) or of the brainstem (0.22 μmol/g).[45] The concentration of ammonia in immature rate brain is twice that of the adult and reaches the adult level at the age of 3 weeks.[46]

Wide variations (10–130 μM) in the normal blood ammonia contents of man and animals have been reported (Table I). The plasma concn. of ammonia is usually lower than that of whole blood because of the higher ammonia content of the erythrocytes; the erythrocyte/plasma concn. ratio for ammonia is about

Table I
Some Normal Ammonia Contents of Brain and Blood

Species	Brain (μmol/g initial wet wt.)	Blood[a] (mM)	Brain[b]/blood	Reference
Rat	0.16	0.094	1.70	32
	0.19	0.12	1.98	33
	0.20	0.13	1.95	34
	0.18	<0.01	>20	35
	0.21	0.067	4.00	36
Mouse	0.33	0.12	3.42	37
	0.24			22
Garden dormouse	0.35			38
Rabbit	0.18			39
Dog	0.26			40
	0.23			41
Human		0.02–0.11		See 5

[a] Blood ammonia content has been converted when necessary to mM.

[b] Brain content (μmol/g) converted to μmol/ml assuming tissue water is 80% of wet weight.

1.3–1.4.[47–49] There is little or no significant difference between CSF and plasma content of ammonia.[33,47] However, there apparently exists a brain-to-blood (or CSF) concn. ratio of ammonia greater than unity[12] (Table I).

3.2. Conditions Affecting Ammonia Levels in Vivo

The ammonia content of the brain analyzed after rapid freezing depends on the state of activity of the brain at the time of freezing. A reduction in functional activity seems to be associated with a reduced concn. of ammonia. There are significant suppressions of the brain ammonia content of rats under prolonged nembutal anesthesia (80%)[32] and of ammonia liberation by anesthetized nerve fibers.[4,6] Ammonia contents of rat brain during sleep[9] and of garden dormouse brain during hibernation[38] are diminished by 50%.

The brain,[8,9,32,50] like peripheral nerves,[4,6] forms ammonia following electrical stimulation. Certain agents capable of producing convulsions in animals enhance brain ammonia contents. These include methionine sulfoximine,[22,40] fluoroacetate,[40] thiosemicarbazide,[40] camphor,[6] picrotoxin,[32,51] bicuculline,[52] telodrin,[53] pentamethylene tetrazole,[40,54] bemigride,[54] lindane,[55] dieldrin,[51,55] heptachlor,[55] DDT,[55,56] and the *Lathyrus sativus* neurotoxin.[57] There may be concomitant increases in brain glutamine level. The rise of brain ammonia in many instances occurs in the preconvulsive state.[32,51]

Despite this fact and the fact that administration of ammonium salts to animals is well known to precipitate convulsions,[58,59] there is still uncertainty as to whether the increase in the level of ammonia from endogenous sources is the cause or a result of convulsions. This uncertainty exists in part because certain substances that afford protection to animals against toxic agents do so without diminishing the enhanced brain ammonia levels. It has been shown, for example, that protection against DDT poisoning by aminooxyacetic acid (presumably because of increased GABA levels)[56] and suppression of ammonium chloride toxicity by methionine sulfoximine occur with no concomitant diminution of brain ammonia levels. In addition, both methionine sulfoximine,[22,40] a convulsant, and aminooxyacetate,[40] an anticonvulsant, enhance cerebral ammonia concn. It appears, therefore, that whereas convulsions are by and large associated with increased brain ammonia levels, increased ammonia levels per se may not necessarily give rise to seizures if other extenuating circumstances come into play.

Anoxia,[32,39] high oxygen pressure,[60] hypercapnia,[61] hypoglycemia,[62] ischemia,[39] audiogenic seizures,[63] painful shocks to the extremities,[8,9,50] or deprivation of sleep[64] all enhance brain ammonia content. The increase in brain ammonia of mice subjected to convulsive electrical stimulation is suppressed in those mice that have received phenobarbital prior to the convulsive stimulation.[65]

Ammonia levels increase in the brain, CSF, and blood in hepatic encephelopathy[10,11] and after portocaval anastomosis.[45,66] It is enhanced in the CSF after epileptic seizures (see ref. 32). Hyperammonemia occurs in certain inborn errors of branched-chain amino acid catabolism, in inborn deficiencies of urea cycle enzymes, and in some disorders of lysine and orthinine metabolism.[10,12,13] Ammonia intoxication has been suggested as one causative factor in the encephalopathy of Reye's syndrome.[67]

3.3. Control of Ammonia Formation by Brain Tissue in Vitro

There is a large liberation of ammonia (14–17 μmol/g initial wet wt.) into substrate-free physiological saline medium at the end of 1 h incubation of brain cortex slices of the adult rat[42] (Table II) or guinea pig.[44] Addition of a substrate that is capable of supporting brain slice respiration, such as glucose, fructose, or pyruvate, strongly suppresses (>70%) ammonia formation (Table II). Whereas acetate, citrate, α-ketoglutarate, acetoacetate, or succinate have little effects on ammonia formation,[14] oxaloacetate (Table II), which supports brain slice respiration, causes a partial (45%) suppression.

Iodoacetate enhances ammonia output in the presence of glucose but not in its absence or in the presence of pyruvate.[14,44] The suppressive effect of glucose is less marked in presence of methionine sulfoximine, fluoroacetate, ouabain, malonate, or aminooxyacetate (Table II). Electron transport inhibitors such as azide, arsenite, cyanide, or amobarbital and uncouplers of oxidative phosphorylation such as sodium salicylate or 2,4-dinitrophenol inhibit the formation of ammonia in the absence of glucose but enhance ammonia liberation in its presence.[14,44] Ammonia formation in brain slices is suppressed under anoxia both in the presence and absence of glucose.[14,42,44]

It has been demonstrated that the increased aerobic formation of ammonia in brain slices in the absence of glucose over that occurring anaerobically results in large measure from endogenous glutamate oxidation[42,68,69] and that as the level of tissue glutamate diminishes, liberation of ammonia from glutamine takes place.[42] Partial suppression of ammonia formation by brain cortex

Table II
Ammonia, Amino Acids, and Carbohydrate Interrelations[a]

Additions to the medium	Glutamate	Glutamine	GABA	Aspartate	Alanine	Ammonia	Total N
Preincubation values[b]	11.8	4.4	2.0	3.4	0.6	1.6	28.2
Glucose-free[b]	3.6	1.8	1.7	9.8	0.7	16.8	36.2
Glucose (10 mM)[b]	9.9	6.7	2.8	3.6	1.5	5.2	36.4
Fructose (10 mM)	9.3	7.5	1.6	3.8	0.8	5.7	36.2
Pyruvate (5 mM)	9.0	6.8	2.4	3.5	3.2	5.8	37.5
Oxaloacetate (10 mM)	6.9	4.7	1.2	9.0	1.0	10.1	37.6
Amino oxyacetate (5 mM)	4.5	3.5	1.9	4.9	1.3	16.4	36.0
Ouabain (1 mM)	6.5	1.9	2.7	10.5	1.3	11.7	36.5
Aminooxyacetate (5 mM) + ouabain (1 mM)	7.8	4.2	1.9	4.9	1.2	11.9	36.1
With glucose (10mM)[b]							
2,4-Dinitrophenol (0.1 mM)	9.7	3.4	4.5	2.6	2.5	11.4	37.5
Ouabain (0.1 mM)	11.3	2.4	4.3	3.9	1.9	10.9	37.1
Amobarbital (1 mM)	10.7	3.0	6.3	2.5	1.5	8.0	35.0
Amino oxyacetate (5 mM)	3.9	6.8	1.7	5.0	1.0	11.6	36.8
Fluoroacetate (3 mM)	10.8	3.9	3.9	2.3	1.7	9.4	35.9
Methionine sulfoximine (5 mM)	10.9	2.5	2.8	4.0	1.4	11.2	35.3
Malonate (2 mM)	8.8	6.1	4.0	1.2	1.2	7.6	35.0

[a] Rat brain cortex slices incubated in oxygenated Ringer phosphate medium for 1 h at 37°C. Total (tissue + medium) values are expressed as μmol amino acid/g initial wet weight.

[b] Data from references 42, 72.

slices incubated in the absence of glucose is observed when regeneration of NAD(P) from NAD(P)H in mitochondria—which is required for glutamate oxidation—is suppressed (e.g., with amobarbital[14] or anoxia[42,44]), when the tissue content of endogenous glutamate is diminished by efflux into the incubation medium (e.g., with ouabain,[70] Table II), when the activity of glutamic dehydrogenase (e.g., with 5-bromofuroic acid[68]) or of glutaminase (e.g., with D-glutamate[44,71]) is inhibited, or when the synthesis of glutamine is stimulated (e.g., with L-glutamate[14,72]). Enhanced liberation of ammonia by brain cortex slices incubated in the presence of glucose (Table II) occurs when glucose oxidation is suppressed (e.g., with amino oxyacetate[72]), when (glial) reuptake of endogenous glutamate (released largely from neurons) is blocked[70] (e.g., with ouabain), or when synthesis of glutamine is inhibited either directly by competition (e.g., with methionine sulfoximine[42]) or indirectly by effects on glial energetics[73] (e.g., with fluoroacetate).

Results in Table II show that the sum of the values of amino acid N of glutamate, glutamine, GABA, and alanine and of ammonia N of incubating rat brain cortex slices is approximately constant under a variety of metabolic conditions. Fluctuations in the value of one constituent are compensated for by fluctuations in the values of other constituents. Thus, it is evident that an intimate relationship exists among ammonia, amino acids, and carbohydrate metabolism in the brain. The preincubation value of amino acid N plus ammonia N, however, increases during incubation (Table II), probably largely because of proteolysis (see Section 4.1.1).

Ammonia formation by infant (2-day-old) rat brain cortex slices incubated for 1 h in the absence of glucose is about 40% of that formed by the adult brain cortex slices; it is also less glucose sensitive.[42] It has been concluded that with maturity, the development of the glutamate–ammonia system is linked with the development of the citric acid cycle of events.[42]

4. AMMONIA METABOLISM IN BRAIN IN VIVO AND IN VITRO

4.1. Ammonia-Forming Processes: Sources of Ammonia in Brain

4.1.1. Proteins

Endogenous proteins have long been suggested as sources of cerebral ammonia both before and after proteolysis.[6,44] That proteolysis does indeed occur during incubation of brain tissue is evident from increases in the levels of essential amino acids.[42,68,69] Amino acids and ammonia arising from proteolysis amount to at least 10 μmol/g rat brain cortex slices in 1 h incubation, approximately 5.2 μmol/g of which is protein-derived ammonia.[42] It has been estimated that about 25% of the ammonia formed in substrate-free medium (i.e., 3.8 μmol/g guinea pig brain slices in 1 h and 5.6 μmol/g in 2 h) arises from protein amide bonds.[6] It has been shown that 16% of the protein-bound glutaminyl amide bonds, equal to 5.4 μmol ammonia/g guinea pig brain cortex slices, is deamidated *in situ*[74]; asparaginyl amide moieties of cerebral proteins are not deamidated. Of a variety of experimental conditions tested, only methionine sulfoximine intoxication prevented deamidation.[74]

Increased cerebral activity produced in rats fatigued by prolonged swimming results in enhanced glutamine and diminished protein amide nitrogen contents of the brain.[6] These effects are reversed following periods of rest.[6] Amidation of free carboxylic groups of cerebral proteins seems to occur *in vivo* on administration of NH_4Cl.[75] Attempts at reamidation of proteins *in vitro* by increasing the free ammonia and/or glutamine pools of guinea pig brain slices have not been successful.[74] One study[76] has claimed amidation of proteins in rat brain slices incubated aerobically in the presence of glucose. As there was little concomitant change in the level of ammonia or of glutamine, the source of the extra amide N, amounting to as much as 8 μmol/g above the unincubated control value, is not known.

4.1.2. Amino Acids

4.1.2a. Glutamic Acid. An initial study has suggested that the fall in the glutamate content of guinea pig brain slices incubated for 1 h in a glucose-free medium accounts for as much as 50% of the ammonia formed.[77] In this study, however, the accompanying increase of aspartate[68,69] which partly arises from endogenous glutamate had not been considered. Studies done using aminooxyacetate (to block endogenous glutamate transamination to aspartate) together with ouabain (added to diminish the glutamate content of the tissue) show that the fall of glutamate accounts for about 30% of the ammonia formed in rat brain cortex slices (Table II). This value will be somewhat higher if protein- and glutamine-derived glutamate are taken into account.

As the endogenous glutamate content of brain is largely neuronal,[70] it seems likely that ammonia liberation from endogenous glutamate *in vitro* has a neuronal origin. Brain slices liberate little or no free ammonia from exogenous L-glutamate taken up largely by glia[78] where it is partly transaminated to aspartate and α-ketoglutarate before oxidation occurs.[72] In the presence of 2.5 mM L-glutamate, it has been found that, of the total exogenous glutamate utilized by rat brain cortex slices in the absence of glucose, 49% is converted to aspartate, 37% is converted to glutamine, largely in glia, and the rest (about 14%) is fully oxidized by glutamic dehydrogenase, largely in neurons.[72]

4.1.2b. Glutamine. The fall of the endogenous glutamine content[42,72] of rat brain cortex slices incubated in glucose-free medium (Table II) accounts for 30% of the ammonia formed, if one molecule of glutamine is taken to give rise to two molecules of ammonia. A lower value has been observed with guinea pig brain cortex slices.[68,69] However, this is presumably because of loss of tissue glutamine (and other amino acids) into the medium during unnecessary storage of slices prior to incubation in fresh medium.[68] Liberation of ammonia from exogenous L-glutamine by brain slices occurs at about a tenth of the rate of that brought about by brain homogenates, presumably because of suppression of glutaminase by the high slice content of glutamate.[71] Phosphate-activated ammonia formation from L-glutamine (K_m, 1.6 mM) by rat brain homogenates is inhibited noncompetitively by L-glutamate (K_i, 0.45 mM) and competitively by NH_4^+ (K_i, 0.5 mM).[71] Calcium ions are stimulatory with max. stimulation occurring at 1 mM. The pH optimum is 8.2, with max. stimulation by Ca^{2+} occurring at pH 7.8.[71] Nerve terminals are rich in glutaminase activity which is confined to mitochondria.[71,79-81]

4.1.2c. Miscellaneous. Endogenous GABA has been suggested to account for some of the ammonia formed by guinea pig slices as its level falls during incubation in a glucose-free medium.[68] The fall is smaller with rat brain cortex slices (Table II) which give rise to little or no ammonia from exogenous GABA.[14] It has been suggested that all ammonia liberated by brain *in vitro* occurs from amino acids after conversion first to aspartate; the α-amino group of aspartate is then utilized in cyclic amination and deamination processes involving NAD or AMP.[82] Addition of L-aspartate to respiring rat brain cortex slices, however, has no effect on the rate of ammonia liberation.[14] It could be argued that exogenous L-aspartate does not enter the compartment of ammonia formation. However, raising the endogenous tissue content of aspartate by incubating rat brain cortex slices in glucose-free medium containing 105 mM KCl[70] (where there is 50% increase of aspartate content over the glucose-free control value) or lowering the aspartate level by use of metabolic inhibitors such as malonate[14] or aminooxyacetate (Table II) has little corresponding effect on the rate of ammonia formation by brain slices.

4.1.3. Purine Derivatives

Brain ammonia has been shown to increase (by 0.3 μmol/g) during the first minute after application of electric shock to rats. The increase occurs concomitant to increases of hypoxanthine compounds and diminutions of adenosine and total adenylic nucleotides; these effects are reversible.[83] These results together with others have led to the conclusion that ammonia formation takes place partly or possibly wholly (if increased glutamine content and diffusion of ammonia out of the brain are not taken into account) by operation of the purine nucleotide cycle.[83] Despite the presence of adenosine and adenylic deaminase activities in brain homogenates,[44,84] these observations are not compatible with those that show little significant change in the brain level of ATP (and also of ADP and AMP) during convulsions produced by hypoxia[85] or by administration of fluoroacetate[86] or methionine sulfoximine[22] in which there is enhanced liberation of ammonia. A compensatory reversible decrease of nucleic acid nitrogen which has been shown to occur in electrically stimulated brain cortex of cats *in vivo*[87] may partly account for the difference.

Ammonia formation by brain slices far exceeds the amount that can be formed by complete deamination of acid-soluble purine derivatives[44] some of which may arise from nucleic acid degradation during incubation.[88] It has been shown[9] that the contents of adenylic compounds of guinea pig brain slices are not diminished during incubation, nor are they increased on addition of ITP.

However, the apparent absence of an amination process may only be a consequence of a low rate of ITP uptake by brain slices.

4.1.4. Miscellaneous

Other deaminating enzymes present in brain (for references, see 7,9,44) are guanine and guanosine deaminases, NAD-deaminase, glucosamine 6-phosphate deaminase, and amine oxidase. These enzymes have not been considered to be major contributors to the ammonia pool of brain both *in vivo* or in incubating brain slices.

4.2. Ammonia-Utilizing Processes

The enzymes carbamyl phosphate synthetase, glutamic dehydrogenase, and glutamine synthetase are largely responsible for direct utilization of ammonia in mammalian tissues. Unlike liver, brain is devoid of carbamyl phosphate synthetase I and ornithine transcarbamylase activities.[89,90] Thus, the urea cycle per se appears to be absent in brain, although urea synthesis from citrulline and arginine can take place.[14,89,90]

4.2.1. Glutamic Dehydrogenase

It has been shown that intravenous or carotid infusion of [^{15}N]ammonium solutions results in incorporation of ^{15}N into cerebral glutamate and glutamine, presumably by actions of glutamic dehydrogenase and glutamine synthetase,[91] and that the specific radioactivity of tissue glutamine is greater than that of its precursor glutamate. This and other observations have led to the concept of metabolic compartmentation in the brain.[91] According to this concept, the bulk of glutamine synthesized in brain occurs from a relatively small pool or compartment of glutamate. Radioactive ketogenic precursors such as acetate, butyrate, and propionate, show metabolic compartmentation (both *in vivo* and *in vitro*), indicating their conversion to glutamate in large part by glutamic dehydrogenase present in the small glutamate compartment, presumably glia.[42,70,91-93] Radioactive glucogenic precursors such as glucose, pyruvate, and lactate, on the other hand, show no metabolic compartmentation, the specific radioactivity of glutamine relative to that of glutamate being less than one.[91] Incorporation of label from these precursors into brain glutamate may occur partly by the action of glutamic dehydrogenases, particularly the enzyme(s) present in the large glutamate compartment consisting of neuronal components.[42,70,92,93]

The direction of neural glutamic dehydrogenase activities appears to be regulated in part by the tissue NAD(P)/NAD(P)H concentration ratio.[42,68,72] When the ratio is high, e.g., in the absence of glucose, oxidative deamination of glutamate occurs. In the presence of glucose, when the ratio falls, and α-ketoglutarate is not rate limiting, reductive amination of α-ketoglutarate is favored. The K_m or NH_4^+ for the enzyme that is localized in mitochondria is 8 mM.[94]

4.2.2. Glutamine Synthetase

ATP-dependent formation of glutamine is the major process of ammonia utilization in brain. Enhanced brain glutamine levels occur *in vivo* during increased brain activity[6–9,32,35,83] brought about by physical exercise or convulsive agents including ammonium salts. It also occurs in hepatic coma,[11] after portocaval anastomosis,[45,66] and in brain slices on incubation in the presence of ammonium salts.[71] Part of the increased glutamine in the presence of excess ammonia results from increased glutamine synthetase activity and part from diminished glutaminase activity.[71] (An enhanced concn. of glutamine in brain and CSF[95] may account for the increased CSF concn. of α-ketoglutaramate in patients with hepatic coma.[96])

Glutamine synthetase is confined largely to glia (presumably the astrocytes) as revealed by metabolic,[42,70,93] ultrastructural,[95,97] and immunocytochemical studies[98] and by studies using glia grown in culture[99] or formed in response to kainate-induced neuronal degeneration.[92] Unlike brain slices, nerve terminal preparations are unable to synthesize glutamine in the presence of glucose[42,100] or glutamate,[100] and early subcellular fractionation studies have shown a paucity of the enzyme in such preparations.[80,100–102] Recent work has indicated significant glutamine synthetase activity in a synaptosomal fraction,[81,103] but immunocytochemical studies suggest extrasynaptic contamination.[104] The presence of the bulk of the enzyme in synaptic mitochondria[103] is not compatible with its earlier localization in microsomal and soluble fractions[80,94,101,102] and with the large body of evidence establishing predominance of the enzyme in glia.[42,70,92,93,95,97–99]

The K_m of ammonia for glutamine synthetase is 0.18 mM,[105] 0.39[94] for brain, and 0.042 for the C_6 (glioma) cells.[106]

5. TRANSPORT OF AMMONIA ACROSS CELL MEMBRANES IN VIVO AND IN VITRO

Any study of the processes involved in the transport of ammonia across cell membranes has to take into consideration the undissociated NH_3 and its protonated counterpart (NH_4^+), as both species coexist at the physiological range of pH.

In early work on ammonia transport, it had been concluded that the rate of transport of NH_4^+ ions into erythrocytes could not be measured directly, since NH_3 crosses cell membranes rapidly.[107] Subsequent studies which are summarized below have dealt with the influence of pH on NH_3 transport, the apparent nonaccumulative uptake of ammonia in brain, activation of the sodium pump by NH_4^+ ions, and the permeability of the NH_4^+ ion across neural membranes.

5.1. pH and Ammonia Transport

A number of studies[31,108–111] have suggested that the transport of ammonia follows the pH gradient drug distribution hypothesis whereby cell membranes are relatively impermeable to the ionized form (NH_4^+) but allow passage to the nonionized form (NH_3) with relative ease.

The NH_3 molecule in solution is hydrated and then ionized. The net effect of these two processes is given by the reaction $NH_3 + H^+ \overset{K_a}{\rightleftharpoons} NH_4^+$ which may be expressed by the Hendersen–Hasselbalch equation as follows:

$$pH = pK_a + \log([NH_3]/[NH_4^+]) \tag{1}$$

where K_a is the dissociation constant of NH_4^+ and pK_a the pH at which 50% of the ammonia is in the ionic form. The pK_a value for human blood is reported to be 9.15.[48] At physiological pH (7.4), 98.3% of the ammonia (i.e., NH_3 + NH_4^+) is present as NH_4^+. From equation 1, the following relationship for the distribution of ammonia between intra- and extracellular fluids has been derived.[108]

$$\frac{[\text{Ammonia}]_{\text{intra}}}{[\text{Ammonia}]_{\text{extra}}} = \frac{1 + 10^{(pK_a - pH)_{\text{intra}}}}{1 + 10^{(pK_a - pH)_{\text{extra}}}} \tag{2}$$

A similar relationship has been given for the distribution of ammonia between CSF and blood.[31]

It is readily apparent from equation 2 that pH is the only variable that determines the steady-state distribution of ammonia, as pK_a may be assumed to be the same in both compartments. It is also apparent from the equation that for equal intra- and extracellular pH values, i.e., in the absence of a proton gradient, the intracellular-to-extracellular ratio of ammonia concns. will become unity.

Since the brain pH is lower than that of blood or CSF,[12,33,47] NH_3 entering the cell quickly forms the NH_4^+ ion, resulting in brain-to-blood concn. ratios of ammonia greater than one (Table I). This ratio will presumably be sustained provided the pH gradient is maintained, which may easily be achieved at low (i.e., normal) external ammonia concn. However, as the external concn. of ammonia increases, the amount of NH_3 crossing cell membranes will also increase. This would tend to diminish the existing pH gradient across the membrane. Homeostatic mechanisms, in turn, would operate to resist the change. Ammonium-induced lactic acid formation both *in vivo*[10] and *in vitro*[112] may be one such hemeostatic mechanism.

With diminution of the pH gradient across the membrane, the brain-to-blood (or incubation medium) concn. ratio will tend to unity. It is not known whether actual abolition of the pH gradient accounts for the apparent absence of ammonia uptake against a concn. gradient by incubating brain slices[112] or by brain *in vivo* after acute administration of ammonium salts to rats.[32,35,113,114]

A direct correlation has been reported between the alteration of blood pH and tissue ammonia concns.; during metabolic and respiratory alkalosis, brain and muscle ammonia concns. increase two- to threefold.[110] Intravenous injections of LD_{50} doses of ammonium salts (chloride, acetate, bicarbonate, carbonate) have indicated that the different rates of passage of ammonium salts across the blood–brain barrier are related to their different effects on blood pH.[109]

5.2. Apparent Nonaccumulative Uptake of Ammonia in Brain

It has been reported that the rise in brain ammonia in methionine sulfoximine-treated rats far exceeds what should be expected from extracellular ammonia and pH considerations.[33] Therefore, it appears likely that the pH gradient at the brain cell membrane may not be the only factor influencing the ammonia concn. of the brain.[47] It has been suggested that some ammonia (up to a max. of about 2 mM) arising from endogenous sources in brain may occur segregated in a relatively nondiffusable form localized possibly in a specific brain compartment(s)[14,112]; it could be bound by hydrogen bonding[115] to proteins.[12,115] (See reference 116 for somewhat similar conclusions with regard to hepatocytes.) This suggestion has been made to account for the fact that the concn. of ammonia in incubated brain slices is considerably higher than that of the medium after incubation (tissue-to-medium concn. ratio >4) but is not significantly affected by the absence of glucose, by anoxia, by ouabain, or by metabolic inhibitors including methionine sulfoximine.[14,112] Addition of increasing concns. (1–30 mM) of NH_4Cl to the incubation medium results in a rapid uptake of ammonia only equal to that in the incubation medium together with the difference when incubation occurs in the absence of added NH_4Cl (i.e., about 2 mM).[112] This uptake is not affected by the presence of ouabain or of metabolic inhibitors, by absence of glucose, or by anoxia.[14] Acute administration of ammonium salts to rats often results in a concn. of ammonia in the brain equal to or less than that in the blood.[32,35,113,114] Thus, an apparent nonaccumulative (= passive?) uptake of ammonia by brain can occur both *in vitro*[112] and *in vivo*.[32,35,113,114] The question arises as to whether such negative evidence is indicative of the lack of an active cerebral uptake process in the special case of ammonia. The rapid diffusion of NH_3 across brain cell membranes will prevent the accumulation of free NH_4^+ against a concn. gradient so that, in effect, an active uptake process for ammonia may exist but cannot be observed. It has been estimated that the diffusion of NH_3 across the membrane occurs at a rate faster than either that for CO_2 (1500-fold) of that for O_2 (30,000-fold).[117]

5.3. Ammonium Ions and the Sodium Pump

The following observations indicate that NH_4^+ ions may substitute directly for K^+ ions in driving the sodium pump of skeletal muscle, erythrocytes, and brain.

1. Both NH_4^+ and Rb^+ can substitute for K^+ in activating Mg-dependent $(Na^+ + K^+)$-ATPase,[14,118] although NH_4^+,[112] unlike Rb^+,[119] is not accumulated against its concn. gradient in brain slices. Acute administration of ammonium salts also appears to enhance cerebral $(Na^+ + K^+)$-ATPase activity.[35,120]

2. Ammonium can replace K^+ in causing sodium extrusion against a concentration gradient in toad skeletal muscle, and this process is ouabain sensitive.[121] Similarly, NH_4^+ can cause the extrusion of Na^+ from so-

dium-filled red cells.[122] However, NH_4^+ requires a concentration three to seven times that of K^+ to produce comparable effects.[121,122]

3. Intraperitoneal injection of NH_4^+ results in a decreased red cell content of K^+ that is larger than the amount of NH_4^+ taken up and an increased plasma $[K^+]$.[123] There is no accompanying change in the $[Na^+]$ of either the red cell or plasma.[123] Similarly, incubation of rat brain cortex slices for 30 min in the presence of 10 mM NH_4Cl brings about a diminution of cerebral $[K^+]$ far in excess of the uptake of NH_4^+ with little effect on cerebral $[Na^+]$.[112] These effects of NH_4^+ may be caused by competition between NH_4^+ and K^+ for the inward carrier presumably $(Na^+ + K^+)$-ATPase.[112] In infant brain, with only low $(Na^+ + K^+)$-ATPase activity, an equal exchange of K^+ ions for NH_4^+ ions occurs.[112] Such observations indirectly support the possibility of an "active" or carrier-mediated process of NH_4^+ transport in mature brain and in erythrocytes.

5.4. Membrane Permeabilities to Ammonium Ions

Studies of the electrical properties of a variety of membranes have assumed a finite permeability of the membrane to NH_4^+ ions. Calculations of the contribution of the NH_4^+ ion to the resting transmembrane potential (V) in brain at 37°C have been made using the Nernst equation as follows:

$$V = -61.5 \log \left[([K^+]_i + \alpha[NH_4^+]_i)/([K^+]_o + \alpha[NH_4^+]_o) \right] \text{ mV} \qquad [3]$$

where α is the permeability of NH_4^+ relative to K^+, and i and o are subscripts denoting the intra- and extracellular concns., respectively. The value of α obtained from membrane potential changes is 0.2[124] or 0.25[125] for the squid giant axon and 0.13 for the membrane of Ranvier's node.[126] With an α value of 0.2, apparent falls of V of 15 mV[35] and 4.5–6.5 mV[41] have been calculated, respectively, for brains of rats and dogs subjected to acute hyperammonemia.

Experiments done with squid giant axons have shown that NH_4^+ can substitute partially for Na^+ in maintaining axon excitability[124,127] and for K^+ in causing membrane depolarization.[124,128] Ammonium can also affect postsynaptic inhibition in spinal[129] and trochlear motoneurons,[130] presumably by suppressing a chloride extrusion mechanism.[129,130]

6. COVERT INVOLVEMENT OF AMMONIA IN BRAIN FUNCTION

It has been suggested that a cycle of events, the glutamate–glutamine cycle, exists in brain[42,70] whereby glutamate (GABA and aspartate) released from neurons during excitation is partly withdrawn from the extraneuronal space by uptake into glia, where glutamine synthesis is largely confined, and eventually returned in the form of electrophysiologically inactive glutamine. As neuronal formation and glial utilization of ammonia occur during operation of the glutamate–glutamine cycle, it is apparent that under normal conditions, the ammonium ion plays an important role in maintaining the steady-state levels of glutamate and its derived neurotransmitters in the neuron.[42] It should be pointed out, however, that as the level of ammonia rises above the normal, and as saturation of glutamine synthetase is approached, inhibition of neuronal glutaminase by excess ammonia would impair the operation of the glutamate–glutamine cycle.[71]

7. EFFECTS OF AMMONIA IN VIVO AND IN VITRO

7.1. Ammonia Toxicity

Studies of ammonia toxicity in a number of species have been reviewed in great detail[5] and will only be dealt with briefly. Values of LD_{50} and $LD_{99.9}$ of ammonium salts for the rat,[36,59] mouse,[50,131,133] chick,[132] dog,[133] and fish[134] have been reported. These values are influenced by a number of factors such as the route of administration[5] and the effects of the ammonium salt on blood pH.[131] Hypoxia,[37] alkalosis,[110] and hyperthermia[135] enhance ammonia toxicity; hypothermia[37,59,135] and acidosis[5] afford some protection.

The toxic syndrome appears to be quite similar in the various species studied.[5] Soon after administration of a lethal dose of an ammonium salt, the animal undergoes hyperventilation and clonic convulsions. This is followed by a fatal tonic extensor convulsion which may or may not occur after a time period during which the animal remains comatose. Survivors of an $LD_{99.9}$ dose recover completely, usually in 50–60 min.[5]

Urea cycle intermediates[5,36,59] such as L-arginine, L-citrulline, or ornithine afford protection against ammonia-induced convulsions. L-Glutamate,[137] L-aspartate,[5,138] and their N-acetyl and N-carbamyl derivatives[139] and also methionine sulfoximine[37] and α-methylglutamate[140] suppress ammonia intoxication. Mixtures of L-arginine and L-glutamate, ornithine and L-aspartate, and L-arginine and pyrrolidone carboxylate are, however, more effective suppressors.[5,138]

The toxic effects of ammonia in hepatic failure are exacerbated in the presence of fatty acids and mercaptans.[11] Lactulose and antibiotics such as neomycin have been used to diminish systemic ammonia concns. during acute and chronic hepatic encephalopathy (see refs. 5,10).

Three possible mechanisms in the CNS, each operating separately or in concert, appear to be the basis for brain dysfunction in ammonia intoxication. They are (1) direct action of ammonium on electrical activities of neurons, (2) suppression of energy-generating processes leading to functional and morphological changes of brain cells, particularly the astrocytes, and (3) diminution of neurotransmitter synthesis and release from nerve terminals.

7.2. Effects of Ammonium on Electrical Activities and Ion Content of Brain Cells

Studies done with guinea pig cerebellar slices[112] have shown that the frequencies of spontaneous action potentials are suppressed partially at $[NH_4^+]$

less than 2 mM and wholly at $[NH_4^+]$ exceeding 2 mM. At high $[NH_4^+]$ (>5 mM), an increased frequency of action potentials (excitation) precedes the inhibition. Deficiency of chloride in the superfusion medium causes inversion of inhibition by $[NH_4^+]$ to excitation. Reduction of $[K^+]$ or of $[Na^+]$ augments inhibition by NH_4^+ in a normal-chloride medium and diminishes excitation in a low-chloride medium. These effects of NH_4^+ occur after a lag period of at least 15 s and appear in part to be consequences of an ammonium-induced metabolic change. It has been suggested that NH_4^+-enhanced lactate formation, presumably in the neuron,[73] may affect a transient decrease of pH in the responsible neuron which causes a changed permeability of the membrane to Cl^- and possibly to Na^+ and K^+.[112]

Studies of the effects of NH_4^+ on the ion contents of incubated slices[112] have shown that NH_4^+ promotes an inward flow of Na^+ and an outward flow of K^+ in guinea pig cerebellar slices and in adult rat cerebral cortex slices. There is also inward movement of Cl^- in rat brain cortex slices which is biphasic. The first phase, presumably occurring in neurons, is seen with low $[NH_4^+]$ (1–2 mM) and consists of an increased tissue content of Cl^- with little or no water uptake. A second phase seen with high $[NH_4^+]$ (>5 mM), in which the uptake of chloride is directly proportional to the Na^+ and water uptakes, appears to take place largely in glia.

From the above studies, it has been concluded[112] that the changed ionic permeabilities of the neuron, particularly that of chloride, may be responsible for some of the acute effects of ammonium administration.

Other studies have suggested that hyperammonemia (which is usually accompanied by an abnormal EEG[9,10]) brings the resting transmembrane potential to the threshold of firing[35,41] and that a general increase in neuronal excitability could well account for the convulsions in rats seen during acute administration of ammonia.[35]

7.3. Effects of Ammonium on Brain Cell Energetics and Morphology

A number of investigators[141–143] have suggested that high ammonium concns. suppress energy-generating processes in brain mitochondria, and from their studies, the suppression seems to occur by one or more of the following mechanisms: (1) interference with oxidative decarboxylation of pyruvate and α-ketoglutarate[141]; (2) suppression of the malate–aspartate shuttle which transfers reducing equivalents from cytoplasm to mitochondria[142]; (3) depletion of α-ketoglutarate and, therefore, suppression of the citric acid cycle[143]; (4) depletion of NADH and, therefore, suppression of the electron transport chain.[5] Presumably, diminutions of α-ketoglutarate and NADH concentrations are consequences of enhanced formation of glutamate which is required for ATP-dependent synthesis of glutamine. Enhanced synthesis of glutamine would itself be partly responsible for ATP depletion.

However, despite its attractiveness, the notion of energy depletion during ammonia intoxication is not borne out by certain observations. For example, the observation of an ammonium-induced depletion of α-ketoglutarate[5,144] is not unequivocal.[35,113,142,145] (Part of the diminished contents of citric acid cycle intermediates may be replenished by ammonium stimulation of CO_2 fixation.[146])

There is little effect of NH_4Cl (<10 mM) on unstimulated respiration of rat brain cortex slices,[112] which is difficult to understand if ammonium ions were to impair the operations of both the citric acid cycle and the electron transport chain. The suggestion of ammonium suppression of the malate–aspartate shuttle made from observations *in vivo*[142] seems not to be supported by experiments carried out *in vitro*.[147] In contrast to mitochondria,[141] brain tissue incubated with ammonium salts (2–18 mM) have shown little or no impairment of pyruvate decarboxylation.[148] Finally, the adenylate energy charge of most brain areas (excluding the brainstem[113]) is unaffected in hyperammonemia induced by portocaval anastomosis[45,142] or by administration of ammonium salts[35,113]; levels of creatine phosphate, however, fall.[113,142] At external $[NH_4^+] < 2$ mM, tissue [ATP] is little affected both *in vivo*[35] and *in vitro*.[112] Nevertheless, despite these negative observations, the energy depletion hypothesis may still partly explain brain malfunction in ammonia intoxication, since it is possible to deduce a localized effect of high ammonium concns. on brain energetics as follows.

It has been suggested from experiments *in vitro* that the neuron is the major compartment of stimulated aerobic glycolysis and that the ammonium-enhanced rate of lactate formation takes place largely in the neuron,[73] presumably through direct stimulation of phosphofructokinase. Increased neuronal glycolysis generates ATP in the cytoplasm which may, at least in some measure, compensate for loss of neuronal ATP formation from impaired citric acid cycle and electron transport chain activities. The impairments cannot be large initially, because little glutamine synthesis occurs in the neuron and because in hyperammonemia there is no apparent increase in the tissue level of glutamate[35,40,66,113] which should occur with any large reductive amination of neuronal α-ketoglutarate. Moreover, any fall of NAD(P)H in neuronal mitochondria should favor the reverse reaction, as the bulk of the tissue glutamate is present in the neuron,[70] which is the anatomical location of the hypothetical energy cycle[91] and the major pool of tissue ATP.[73] These observations may explain why the tissue ATP content is unaffected in hyperammonemia induced by portocaval shunting or after acute administration of ammonium salts to animals.

Glycolysis in glia, on the other hand, appears to be less sensitive to metabolic control than that occurring in the neuron,[73] so that little ammonium-enhanced glycolytic formation of ATP takes place in glial cytoplasm. Glia also contain the minor pool of tissue ATP.[73] Thus, a significant fall in the glial [ATP] may not be reflected in the tissue ATP content but will be accompanied by profound changes in glial structure and function. The fall of glial [ATP] could occur due to suppression of the citric acid cycle[143] as reductive amination of α-ketoglutarate proceeds to replenish the (small[91]) pool of glial glutamate[70] that is continually being drained off by ATP-dependent glutamine synthethase. This view supports the demonstrations of increased glutamic dehydrogenase activity of astrocytes in hyperammonemia.[120,149]

It has been noted earlier that ammonia enhances water uptake by incubated brain cortex slices.[112] This occurs largely in glia, as neurons undergo little swelling.[150] The increased water uptake accompanies enhanced inflows of Na^+ and Cl^- and a loss of tissue K^+ which is larger than the gain of tissue

NH_4^+.[112] These changes of ionic contents must partly be consequences of an impaired sodium pump as a result of diminished glial ATP.

Similarly, in hyperammonemia produced by portocaval shunts, there is swelling of astrocytes[95,97] which resembles that resulting from the injection of ouabain[151] or from brain anoxia[152,153] or that seen in hepatic coma.[10,154] Hypertrophy and hyperplasia of astrocytes may occur together.[97] Electron microscopic studies have revealed[95,97] that the expanded cytoplasmic compartment of astrocytes is associated with proliferation of mitochondria and endoplasmic reticulum (organelles probably responsible for increased glutamate and glutamine synthesis, respectively) and with vacuolar and other changes. The severity of these morphological modifications of astrocytes parallels the serum ammonium levels[95,97] and presumably also the energy levels of the astrocyte. Neurons seem to undergo little morphological change in the early stages of chronic hyperammonemia.[95,97]

It appears, therefore, that changes in the structure of glia in chronic hyperammonemia are consequences of altered metabolic, energy-generating, and transport events in these cells.

7.4. Effects of Ammonium on Neurotransmitter Synthesis and Release

Convulsive dosages of ammonium salts have been shown to diminish the cerebral concn. of acetylcholine *in vivo*[155] and the rate of acetylcholine synthesis by rat brain cortex slices.[156] The high concn. of NH_4Cl (15 mM) used in the experiments *in vitro*[156] has been shown also to suppress mitochondrial oxidative decarboxylation of pyruvate,[141] and this effect presumably accounts for the diminished formation of the acetylcholine precursor, acetyl-CoA, in the cholinergic neuron. In contrast to these observations, however, other studies have shown little change in the level of acetylcholine[157] or of pyruvate decarboxylation[148] in the brain during development of acute ammonia-induced coma in which the levels of norepinephrine and serotonin are also unaffected.[157] Other studies have shown enhanced levels of serotonin and diminished levels of norepinephrine in brains of animals with severely damaged or absent livers (for references, see 11).

Ammonium chloride (2 mM) diminishes the levels of glutamate and aspartate in incubated rat brain cortex slices.[71] The level of GABA is unaffected. Infusion of NH_4Cl solutions for 45 min into dogs brings about similar changes in cerebral amino acid contents, although the fall of glutamate level is not as marked.[40] Acute administration of NH_4^+ to rats causes a fall of cerebral aspartate.[35] Whereas little change of glutamate content appears in 10 min after administration of ammonium salts, a significant fall is observed at the end of 30 min.[113] It is possible, however, that there could be falls in the neurotransmitter pools of glutamate in short periods of time. Such falls would not be apparent in glutamate estimations made on brain tissue as a whole, especially if the falls are compensated by rises at other locations (e.g., by increased reductive amination of α-ketoglutarate in glia). Small but significant decreases of brain glutamate and aspartate contents have been reported in acute[142] and chronic[45,158] hyperammonemia in rats with portocaval shunts. It has been shown also that NH_4^+ ions suppress the K^+-evoked release of endogenous glutamate from hippocampal slices.[159]

It is therefore conceivable that ammonium at concns. above the normal may cause a fall in the content and release of glutamate and related transmitters from excited neurons and that these effects occur in part by a block of glutamate synthesis from glutamine in the nerve terminal.

REFERENCES

1. Jacobson, C., 1910, *Am. J. Physiol.* **26**:407–412.
2. Mathews, S. A., 1922, *Am. J. Physiol.* **59**:459–460.
3. McDermott, W. V., 1957, *N. Engl. J. Med.* **257**:1076–1081.
4. Tashiro, S., 1922, *Am. J. Physiol.* **60**:519–543.
5. Kamin. H. (Chairman, Subcommittee on Ammonia), 1979, *Ammonia*, Committee on Medical and Biological Effects of Environmental Pollutants, National Research Council, University Park Press, Baltimore.
6. Vrba, R., 1957, *Rev. Czech. Med.* **3**:1–26.
7. Weil-Malherbe, H., 1962, *Neurochemistry*, 2nd ed. (K. A. C. Elliot, I. H. Page, and J. H. Quastel, eds.), Charles C. Thomas, Springfield, Illinois, pp. 321–330.
8. Tsukada, Y., 1966, *Prog. Brain Res.* **21A**:268–291.
9. Tsukada, Y., 1971, *Handbook of Neurochemistry*, Volume 5A (A. Lajtha, ed.), Plenum Press, New York, pp. 215–233.
10. Schenker, S., Breen, K. J., and Hoyumpa, A. M., Jr., 1974, *Gastroenterology* **66**:121–151.
11. Zieve, L., and Nicoloff, D. M., 1975, *Annu. Rev. Med.* **26**:143–157.
12. Hsia, Y. E., 1974, *Gastroenterology* **67**:347–374.
13. Colombo, J. P., 1971, *Congenital Disorders of the Urea Cycle and Ammonia Detoxication*, Monographs in Paediatrics, Volume 1, S. Krager, New York.
14. Benjamin, A. M., 1972, *Ammonia and Amino Acid Metabolism and Transport in the Brain in Vitro*, Ph.D. Thesis, University of British Columbia, Vancouver.
15. Burg, P. D. V., and Mook, H. W., 1963, *Clin. Chim. Acta* **8**:162–163.
16. Conway, E. J., 1957, *Microdiffusion Analysis and Volumetric Error*, C. Lockwood, London.
17. Seligson, E., and Hirahara, K., 1957, *J. Lab. Clin. Med.* **49**:962–974.
18. Hutchinson, J. H., and Labby, D. H., 1962, *J. Lab. Clin. Med.* **60**:170–178.
19. Fenton, J. C. B., and Williams, A. H., 1968, *J. Clin. Pathol.* **21**:14–18.
20. Rubin, M., and Knott, L., 1967, *Clin. Chim. Acta* **18**:409–415.
21. Kaltwasser, H., and Schlegel, H. G., 1966, *Anal. Biochem.* **16**:132–138.
22. Folbergrova, J., Passoneau, J. V., and Lowry, O. H., 1969, *J. Neurochem.* **16**:193–203.
23. Christian, G. D., and Feldman, F. J., 1967, *Clin. Chim. Acta* **17**:87–93.
24. Nathan, D. G., and Rodkey, F. L., 1957, *J. Lab. Clin. Med.* **49**:779–785.
25. McCullough, H. A., 1967, *Clin. Chim. Acta* **17**:297–304.
26. Stone, W. E., 1956, *Proc. Soc. Exp. Biol. Med.* **93**:589–591.
27. Mondzac, A., Ehrlich, G. F., and Seegmiller, J. E., 1965, *J. Lab. Clin. Med.* **66**:526–531.
28. Vaidyanath, N., Birkhahu, R., Border, J. R., McMenamy, R., Oswald, G., Trietley, G., and Yuan, T. F., 1976, *Anal. Biochem.* **70**:479–488.
29. Friedl, F. E., 1972, *Anal. Biochem.* **48**:300–306.
30. Proelss, H. F., and Wright, B. W., 1973, *Clin. Chem.* **19**:1162–1169.
31. Moore, E. W., Strohmeyer, G. M., and Chalmers, T. C., 1963, *Am. J. Med.* **35**:350–362.
32. Richter, D., and Dawson, R. M. C., 1948, *J. Biol. Chem.* **176**:1199–1210.
33. Hindfelt, B., 1975, *J. Neurol. Sci.* **25**:499–506.
34. Ehrlich, M., Plum, F., and Duffy, T. E., 1980, *J. Neurochem.* **34**:1538–1542.

35. Hawkins, R. A., Miller, A. L., Nielsen, R. C., and Veech, R. L., 1973, *Biochem. J.* **134**:1001–1008.
36. Roberge, A., and Charbonneau, R., 1968, *Rev. Can. Biol.* **27**:321–331.
37. Warren, K. S., and Schenker, S., 1964, *J. Lab. Clin. Med.* **62**:442–449.
38. Godin, Y., Mark, J., Kayser, C., and Mandel, P., 1967, *J. Neurochem.* **14**:142–144.
39. Thorn, W., and Heimann, J., 1958, *J. Neurochem.* **2**:166–177.
40. Tews, J. K., and Stone, W. E., 1965, *Prog. Brain Res.* **16**:135–163.
41. Benzi, G., Arrigoni, E., Strada, P., Pastoris, O., Villa, R. F., and Agnoli, A., 1977, *Biochem. Pharmacol.* **26**:2397–2404.
42. Benjamin, A. M., and Quastel, J. H., 1975, *J. Neurochem.* **25**:197–206.
43. Krebs, H. A., Eggleston, L. V., and Hems, R., 1949, *Biochem. J.* **44**:159–163.
44. Weil-Malherbe, H., and Green, R. H., 1955, *Biochem. J.* **61**:210–218.
45. Holmin, T., and Seisjo, B. K., 1974, *J. Neurochem.* **22**:403–412.
46. Oja, S. S., von Bonsdorff, A. R., and Lindroos, O. F. C., 1966, *Nature* **212**:937–938.
47. Hindfelt, B., 1975, *Clin. Sci. Mol. Med.* **48**:33–37.
48. Bromberg, R. A., Robin, E. D., and Forkner, C. E., 1960, *J. Clin. Invest.* **39**:332–341.
49. Caesar, J., 1962, *Clin. Sci.* **22**:33–41.
50. Tsukada, Y., Takagaki, G., Sugimoto, S., and Hirano, S., 1958, *J. Neurochem.* **2**:295–303.
51. Hathway, D. E., Mallinson, A., and Akintonwa, D. A. A., 1965, *Biochem. J.* **94**:676–686.
52. Chapman, A. G., Meldrum, B. S., and Seisjö, B. K., 1977, *J. Neurochem.* **28**:1025–1035.
53. Hathway, D. E., and Mallinson, A., 1964, *Biochem. J.* **90**:51–60.
54. Whistler, K. E., Tews, J. K., and Stone, W. E., 1968, *J. Neurochem.* **15**:215–220.
55. Omer, V. St., 1971, *J. Neurochem.* **18**:365–374.
56. Martin, M. A., Kar, P. P., and Anand, M., 1976, *J. Neurochem.* **27**:979–981.
57. Cheema, P. S., Padmanaban, G., and Sarma, P. S., 1971, *J. Neurochem.* **18**:2137–2144.
58. Torda, C., 1953, *J. Pharmacol. Exp. Ther.* **107**:197–203.
59. Greenstein, J. P., Winitz, M., Gullino, P., Birnbaum, S. M., and Otey, M. C., 1956, *Arch. Biochem. Biophys.* **64**:342–354.
60. Faiman, M. D., Nolan, R. J., Baxter, C. F., and Dodd, D. E., 1977, *J. Neurochem.* **18**:2137–2144.
61. Folbergrova, J., Macmillan, V., and Seisjö, B. K., 1972, *J. Neurochem.* **19**:2507–2517.
62. Agardh, C -D., Folbergrova, J., and Seisjö, B. K., 1978, *J. Neurochem.* **31**:1135–1142.
63. Leornard, B. E., 1965, *Biochem. Pharmacol.* **14**:1293–1298.
64. Haulica, I., Ababei, L., Teodorescu, C., Rosca, V., Moisiu, M., and Haller, C., 1970, *J. Neurochem.* **17**:823–826.
65. King, L. J., Carl, J. L., and Lao, L., 1974, *J. Neurochem.* **22**:307–309.
66. Williams, A. H., Kyu, M. H., Fenton, J. C. B., and Cavanagh, J. B., 1972, *J. Neurochem.* **19**:1073–1077.
67. Huttenloker, P. R., Schwartz, A. D., and Klatski, G., 1969, *Pediatrics* **43**:443–453.
68. Weil-Malherbe, H., and Gordon, J., 1971, *J. Neurochem.* **18**:1659–1672.
69. Weil-Malherbe, H., and Gordon, J. W., 1973, *Neuropharmacology*, **12**:367–381.
70. Benjamin, A. M., and Quastel, J. H., 1972, *Biochem. J.* **128**:631–646.
71. Benjamin, A. M., 1981, *Brain Res.* **208**:363–377.
72. Benjamin, A. M., and Quastel, J. H., 1974, *J. Neurochem.* **23**:457–464.
73. Benjamin, A. M., and Verjee, Z. H., 1980, *Neurochem. Res.* **5**:921–934.
74. Wherrett, J. R., and Tower, D. B., 1971, *J. Neurochem.* **18**:1027–1042.
75. Martinsen, E. E., and Tyakhepyl'd, L. Y., 1961, *Biokhimiya* **26**:984–992.
76. Kometiani, P. A., Klein, Y. E., Gvalia, N. V., and Gotsiridze, Y. G., 1970, *J. Neurochem.* **17**:1331–1337.
77. Takagaki, G., Hirano, S., and Tsukada, Y., 1957, *Arch. Biochem. Biophys.* **68**:196–205.
78. Benjamin, A. M., and Quastel, J. H., 1976, *J. Neurochem.* **26**:433–443.
79. Bradford, H. F., and Ward, H. K., 1976, *Brain Res.* **110**:115–125.
80. Salganicoff, L., and De Robertis, E., 1965, *J. Neurochem.* **12**:287–309.
81. Weiler, C. T., Nystrom, B., and Hamberger, A., 1979, *Brain Res.* **160**:539–543.
82. Buniatian, H. C., 1970, *Handbook of Neurochemistry*, Volume 3 (A. Lajtha, ed.), Plenum Press, New York, pp. 399–413.

83. Schultz, V., and Lowenstein, J. M., 1978, *J. Biol. Chem.* **253**:1938–1943.

84. Weil-Malherbe, H., and Green, R. H., 1955, *Biochem. J.* **61**:218–224.

85. Nelson, S. R., and Mantz, M. L., 1971, *Life Sci.* **10**:901–907.

86. Benitez, D., Pscheidt, G. R., and Stone, W. E., 1954, *Am. J. Physiol.* **176**:483–487.

87. Geiger, A., Yamasaki, S., and Lyons, R., 1956, *Am. J. Physiol.* **184**:239–243.

88. Vrba, R., and Folbergrova, J., 1959, *J. Neurochem.* **4**:338–349.

89. Sadasivudu, B., and Indira Rao, T., 1974, *J. Neurochem.* **23**:267–269.

90. Sadasivudu, B., and Indira Rao, T., 1976, *J. Neurochem.* **27**:785–794.

91. Berl, S., and Clarke, D. D., 1969, *Handbook of Neurochemistry*, Volume 2 (A. Lajtha, ed.), Plenum Press, New York, pp. 447–472.

92. Nicklas, W., Nunez, R., Berl, S., and Duvoisin, R., 1979, *J. Neurochem.* **33**:839–844.

93. Balazs, R., Patel, A. J., and Richter, D., 1973, *Metabolic Compartmentation in the Brain* (R. Balazs and J. E. Cremer, eds.), Macmillan, London, pp. 167–184.

94. Van den Berg, C., 1970, *Handbook of Neurochemistry*, Volume 3 (A. Lajtha, ed.), Plenum Press, New York, pp. 355–379.

95. Zamora, A. J., Cavanagh, J. B., and Kyu, M. H., 1973, *J. Neurol. Sci.* **18**:25–41.

96. Vergara, F., Plum, F., and Duffy, T. E., 1974, *Science* **183**:81–83.

97. Norenberg, M. D., and Lapham, L. W., 1974, *J. Neuropathol. Exp. Neurol.* **33**:422–435.

98. Martinez-Hernandez, A., and Norenberg, M. D., 1977, *Science* **195**:1356–1358.

99. Schousboe, A., Svenneby, G., and Hertz, L., 1977, *J. Neurochem.* **20**:999–1005.

100. Bradford, H. F., and Thomas, A. J., 1969, *J. Neurochem.* **16**:1495–1504.

101. Sellinger, O. Z., and De Balbain Vester, F., 1962, *J. Biol. Chem.* **237**:2836–2844.

102. Ward, H. K., and Bradford, H. F., 1979, *J. Neurochem.* **33**:339–342.

103. Dennis, S. C., Lai, J. C. K., and Clark, J. B., 1980, *Brain Res.* **197**:469–475.

104. Norenberg, M. D., and Martinez-Hernandez, A., 1979, *Brain Res.* **161**:303–310. .

105. Pamiljans, V., Krishnaswamy, P. R., Dumville, G., and Meister, A., 1962, *Biochemistry* **1**:153–158.

106. Pishak, M. R., and Phillips, A. T., 1980, *J. Neurochem.* **34**:866–872.

107. Jacobs, M. H., and Parpart, A. K., 1938, *J. Cell. Comp. Physiol.* **11**:175–192.

108. Milne, M. D., Scribner, B. H., and Crawford, M. A., 1958, *Am. J. Med.* **24**:709–729.

109. Warren, K. S., and Nathan, D. G., 1958, *J. Clin. Invest.* **37**:1724–1728.

110. Stabenau, J. R., Warren, K. S., and Rall, D. P., 1959, *J. Clin. Invest.* **38**:373–383.

111. Warren, K. S., 1962, *Nature* **195**:47–49.

112. Benjamin, A. M., Okamoto, K., and Quastel, J. H., 1978, *J. Neurochem.* **30**:131–143.

113. Hindfelt, B., and Seisjö, B. K., 1971, *Scand. J. Clin. Lab. Invest.* **28**:365–374.

114. Rosado, A., Flores, G., Mora, J., and Soberon, G., 1962, *Am. J. Physiol.* **203**:37–42.

115. Barker, A. V., 1968, *Biochim. Biophys. Acta* **168**:447–455.

116. Sainsbury, G. M., 1980, *Biochim. Biophys. Acta* **631**:305–316.

117. Robin, E. D., Travis, D. M., Bromberg, P. A., Forkner, C. E., Jr., and Tyler, J. M., 1959, *Science* **129**:270–271.

118. Skou, J. C., 1960, *Biochim. Biophys. Acta* **42**:6–23.

119. Bourke, R. S., and Tower, D. B., 1966, *J. Neurochem.* **13**:1099–1117.

120. Sadasivudu, B., Indira Rao, T., and Murthy, C. R., 1977, *Neurochem. Res.* **2**:639–655.

121. Beaugé, L. A., and Oritz, O., 1970, *J. Exp. Zool.* **174**:309–316.

122. Post, R. L., and Jolly, P. C., 1957, *Biochim. Biophys. Acta* **25**:118–128.

123. Albano, O., and Francavilla, A., 1971, *Gastroenterology* **61**:893–897.

124. Binstock, L., and Lecar, H., 1969, *J. Gen. Physiol.* **53**:342–361.

125. Hagiwara, S., Eaton, D. C., Stuart, A. E., and Rosenthal, W. P., 1972, *J. Membr. Biol.* **9**:373–384.

126. Hille, B., 1973, *J. Gen. Physiol.* **61**:669–686.

127. Tasaki, I., Singer, I., and Watanabe, A., 1965, *Proc. Natl. Acad. Sci. U.S.A.* **54**:763–769.

128. Ishida, Y., 1967, *Jpn. J. Pharmacol.* **17**:6–18.

129. Lux, H. D., 1971, *Science* **173**:555–557.

130. Llinás, R., Baker, R., and Precht, W., 1974, *J. Neurophysiol.* **37**:522–533.

131. Warren, K. S., 1958, *J. Clin. Invest.* **37**:497–501.

132. Wilson, R. P., Muhrer, M. E., and Bloomfield, R. A., 1968, *Comp. Biochem. Physiol.* **25:**295–301.

133. Wilson, R. P., Davies, L. E., Muhrer, M. E., and Bloomfield, R. A., 1968, *Am. J. Vet. Res.* **29:**897–906.

134. Wilson, R. P., Andersen, R. O., and Bloomfield, R. A., 1969, *Comp. Biochem. Physiol.* **28:**107–118.

135. Schenker, S., and Warren, K. S., 1962, *J. Lab. Clin. Med.* **60:**291–301.

136. Zuiedema, G. D., Gainsford, W. D., Kowalczyk, R. S., and Wolfman, E. F., Jr., 1963, *Arch. Surg.* **87:**578–582.

137. Sapirstein, M. R., 1943, *Proc. Soc. Exp. Biol. Med.* **52:**334–335.

138. Salvatore, F., and Bocchini, V., 1961, *Nature* **191:**705–706.

139. Niculescu, V., Bonciocoate, C., and Stancu, C., 1965, *Biochem. Pharmacol.* **14:**1635–1643.

140. Lamar, C., Jr., 1970, *Toxicol. Appl. Pharmacol.* **17:**795–803.

141. McKhann, G. M., and Tower, D. B., 1961, *Am. J. Physiol.* **200:**420–424.

142. Hindfelt, B., Plum, F., and Duffy, T. E., 1977, *J. Clin. Invest.* **59:**386–396.

143. Bessman, S. P., and Bessman, A. N., 1958, *J. Clin. Invest.* **34:**622–628.

144. Clark, G. M., and Eiseman, B., 1958, *N. Engl. J. Med.* **159:**178–180.

145. Shorey, J., McCandless, D. W., and Schenker, S., 1967, *Gastroenterology* **54:**706–711.

146. Waelsch, H., Berl, S., Rossi, C. A., Clarke, D. D., and Purpura, D. P., 1964, *J. Neurochem.* **11:**711–728.

147. Verjee, Z. H., and Benjamin, A. M., 1979, *XIth International Congress of Biochemistry*, National Research Council of Canada, Ottawa pp. 553.

148. Walker, C. O., and Schenker, S., 1970, *Am. J. Clin. Nutr.* **23:**619–632.

149. Norenberg, M. D., 1976, *Arch. Neurol.* **33:**265–269.

150. Gerschenfeld, H. M., Wald, F., Zadunaisky, J. A., and De Roberts, E., 1959, *Neurology (Minneap.),* **9:**412–425.

151. Cornog, J. L., Jr., Gonatas, N. K., Feierman, J. R., 1967, *Am. J. Pathol.* **51:**573–590.

152. Bakay, L., 1968, *Prog. Brain Res.* **29:**315–341.

153. Maxwell, O. S., and Kruger, L., 1965, *J. Cell Biol.* **25:**141–157.

154. Ware, A. J., D'Agostino, A. N., and Combes, B., 1971, *Gastroenterology* **61:**877–884.

155. Ulshafer, T. R., 1958, *J. Lab. Clin. Med.* **52:**718–723.

156. Braganca, B. M., Faulkner, P., and Quastel, J. H., 1953, *Biochim. Biophys. Acta* **10:**83–88.

157. Walker, C. O., Speeg, K. V., Levinson, J. D., and Schenker, S., 1971, *Proc. Soc. Exp. Biol. Med.* **136:**668–671.

158. Cremer, J. E., Heath, J. F., Teal, H. M., Woods, M. S., and Cavanagh, J. B., 1975, *Neuropathol. Appl. Neurobiol.* **3:**291–311.

159. Hamberger, A., Hedquist, B., and Nystrom, B., 1979, *J. Neurochem.* **33:**1295–1302.

Water Spaces

Hanna M. Pappius

1. INTRODUCTION

Precise information about the relative sizes of the extracellular and intracellular compartments of tissues and about dynamic relationships between these compartments is often required for the interpretation of physiological and pathological processes. In cerebral tissues, heterogeneity of the intracellular spaces provides an additional challenge in determining how fluid is distributed among the various compartments.

In the brain, the measurement of the size of the fluid compartments has been particularily difficult because of inherent technical and theoretical problems. The early estimates of considerable extracellular space (ECS) were based on the assumption that sodium and chloride were mostly extracellular.[1,2] Subsequently, the first electron micrographs of cerebral cortex showed little, if any, space between the cellular elements,[3] leading electron microscopists to doubt the very existence of ECS. The controversy sparked by these diametrically opposite views stimulated renewed interest in the whole question of fluid distribution in brain. The resulting careful reassessment of assumptions made in earlier investigations, as well as development of new methodologies, led to a consensus regarding the presence of a functional ECS in cerebral tissues amounting to about $20 \pm 5\%$ under normal physiological conditions. At the same time, it became clear that in brain rapid shifts of fluid between intra- and extracellular compartments can occur, particularily under unfavorable conditions of oxygenation as exemplified by conventional methods of fixation and dehydration, making morphological estimates of cerebral "spaces" quite unreliable unless special precautions are taken.

2. DEFINITION OF "SPACES"

A chemically determined fluid compartment of a tissue is defined as the volume of the tissue water that is available under steady-state conditions for

---

Hanna M. Pappius • Donner Laboratory of Experimental Neurochemistry, Montreal Neurological Institute, and Department of Neurology and Neurosurgery, McGill University, Montreal, Quebec H3A 2B4, Canada.

the distribution of the particular marker used. The space can be calculated by the following equation:

$$\text{Marker space } (\%) = (\text{Concentration of the marker in tissue/Concentration of the marker in reference fluid}) \times 100$$

For this equation to be meaningful, equilibration between the reference fluid and the compartment in question must have really occurred, the concentration of the marker in the reference fluid must be accurately known, and it must have remained unchanged during the equilibration period. The marker must be quantitatively excluded from the intracellular compartment if the "space" is to be an estimate of the volume of the extracellular fluid.

When distribution of the various conventional extracellular markers such as sodium, chloride, thiocyanate, iodide, sulfate, sucrose, or inulin is studied in brain, difficulties are encountered in fulfilling each of the postulates mentioned above.

3. METHODOLOGY

3.1. Inherent Problems

The various aspects of the physiology of the blood–brain barrier and of the cerebrospinal fluid (CSF) that are pertinent to the discussion of fluid compartmentation in brain have been exhaustively treated by several authors.[4–7] They will be only briefly reviewed here.

In studies of cerebral tissue spaces, the most serious problems are encountered in establishing real equilibrium. The inherent difficulties vary with the marker used.

The penetration of all ECS markers from blood to brain is slow, and steady state in most cases is approached only after hours of equilibration, necessitating constant infusion of the marker to maintain its unchanging level in the blood. With rapidly excreted substances such as sulfate, sucrose, and inulin, functional nephrectomy is also usually required. Thiocyanate is bound by plasma proteins, and the concentration of the free ion available for equilibration with the tissue must be determined.

In other tissues, extracellular fluid is thought to be an ultrafiltrate of plasma. In brain, the interstitial fluid is more closely related to the CSF, and thus, the latter is a more appropriate reference fluid for the estimation of cerebral ECS. Many factors contribute to considerable differences, even under steady-state conditions, in the concentration of the various extracellular markers in CSF as compared to plasma ultrafiltrate. Thus, sodium and chloride content of the CSF is maintained at levels higher than those in plasma ultrafiltrate by active secretory processes involved in the formation of CSF. Markers injected intravenously do not reach concentration levels in the CSF as high as those in the plasma if they are removed by bulk flow of the CSF faster than they can penetrate from blood into the CSF space. Some markers, specifically thiocyanate, iodate, and sulfate, are also actively removed from the CSF by transport systems in the choroid plexus. Further, transport processes at the cerebral capillary level which are now being elucidated[8] may also contribute to the concentration gradients of certain extracellular markers between the blood and the CSF. These concentration gradients maintained at steady state prevent real equilibration of markers from occurring between blood, on one hand, and the CSF and brain fluid compartments on the other. Concentration gradients in the reverse direction may develop when an extracellular marker introduced only into the CSF is lost into the blood.

As a result, to obtain a reasonably precise estimate of the distribution volume in brain of any specific marker, its levels in CSF must be known.

Intracellular penetration of sodium, chloride, and thiocyanate and the fact that sulfate is not completely inert metabolically are further complications that must be taken into account when these markers are used to delineate the extracellular space in brain.

3.2. Experiments in Vivo

To approximate the true equilibrium conditions necessary for estimation of the volume of distribution of extracellular markers in brain *in vivo* and to circumvent the question of which fluid to use as reference, the same effective concentration of several markers has been maintained in blood and in the CSF by a combination of intravenous infusion and ventriculocisternal perfusion.[9-17]

An alternate approach to the problem of measurement of extracellular space in brain has been the use of the diffusion profile of the marker.[18-21] With this technique, the marker space can be calculated by extrapolation to boundary conditions when the marker is perfused through the CSF space without the requirement for total equilibration between the fluid compartments under study. However, for this method to give a reasonable estimate of the size of the ECS, the marker must not be taken up by cellular elements, and it must diffuse freely through the extracellular space. The latter condition may not, in fact, exist because of restriction of movement of markers, especially those of large molecular size, resulting from the tortuosity of cerebral extracellular spaces and because of the bulk flow of brain extracellular fluid.[20,22,23]

Changes in extracellular space volume of brain have been evaluated by irrigating the cortical surface with appropriate markers and monitoring their concentration by ion-selective microelectrodes.[24] Extracellular space has also been determined by measurements of electrical impedance of the tissue.[5,25-28]

3.3. Experiments in Vitro

To avoid the complications caused by the blood–brain barrier, *in vitro* preparations have been used by numerous investigators.[6] The use of tissue slices, however, introduces artifacts resulting from damage of the tissue in slicing. To minimize the effects of damage, isolated intact frog brain,[29] rabbit retina,[30] and bowfin brain[31] have been used *in vitro* in studies on cerebral tissue spaces.

4. SPACES IN VIVO

4.1. Marker Spaces

4.1.1. Sodium and Chloride Spaces

The chloride space calculated from chloride distribution using CSF as reference fluid and corrected for intracellular chloride penetration was found to be between 21 and 24% in the rabbit[1,5,6,30] and in approximately the same range in rat[2] and cat.[32] Similar calculations based on sodium distribution gave an ECS of 24%.[1,5,6]

When entrance of [24]Na from blood to brain in rat was subjected to compartmental analysis, two [24]Na compartments could be distinguished.[33] The first, about 22%, was considered to be equivalent to the extracellular space, whereas the second, 6%, was thought to represent intracellular penetration of the ion.

4.1.2. Thiocyanate Space

Thiocyanate penetrates relatively rapidly from blood to brain, but the estimation of its volume of distribution is complicated not only by its penetration into the intracellular compartment[34] but also by active removal of this marker from the CSF.[9] When at any given plasma thiocyanate level steady-state conditions are reached, a concentration gradient for thiocyanate is established between the plasma and the CSF. This gradient decreases with increasing concentration of thiocyanate in the plasma as the transport system for the ion in the choroid plexus becomes saturated.[35,36] This explains the increase in thiocyanate space from 4 to 17% with increasing plasma concentration when plasma is used as a reference fluid.[37] Even with simultaneous intravenous infusion and ventriculocisternal perfusion at a low concentration of thiocyanate (1 mM), the volume of distribution of thiocyanate in cerebral tissues is significantly lower than at a higher concentration (10 mM), suggesting that some active transport mechanism out of the brain is operative.[6,15]

When thiocyanate space was measured in the cat after 6 h of intravenous infusion combined with ventriculocisternal perfusion at high concentration of thiocyanate (10 mM) and corrected for intracellular penetration of the marker, an ECS of 24% for cerebral cortex and 17% for subcortical white matter was found.[6,15]

Thiocyanate space of 20% (not corrected for intracellular penetration) was reported for rabbit brain when the plasma-to-CSF concentration gradient for thiocyanate was diminished, but not completely abolished, by inhibition of the choroid plexus transport with iodide or dinitrophenol.[38]

4.1.3. Iodide Space

Iodide is also actively removed from the CSF, and as a result, the steady-state concentrations of iodide in the CSF and in the extracellular fluid of the brain are much lower than that in the blood.[39,40] The suggestion that the iodide content of the CSF is a better approximation of brain extracellular fluid iodide than is its level in the plasma was, in fact, made in one of the earliest publications on the subject.[39] With simultaneous intravenous infusion and ventriculocisternal perfusion, either at high concentration of iodide or in the presence of perchlorate to inhibit choroid plexus transport of iodide, iodide space of 17 to 22.5% was obtained in rabbits and dogs.[13,40]

4.1.4. Sulfate Space

In contrast to sodium, chloride, thiocyanate, and iodide, sulfate penetrates into the brain only to a very limited extent. Levels of isotopically labeled sulfate in the CSF even hours after intravenous injection into nephrectomized animals are only one-sixth to one-fourth of the concentration of the marker in the blood. When earlier data from several authors were recalculated using the CSF as reference fluid, a sulfate space of 16 to 24% was obtained.[41]

In combined intravenous infusion and ventriculocisternal perfusion experiments of 6 h duration, a sulfate space of 12% was obtained in the cat.[12] However, in these studies, the cisternal effluent contained only 81% of the sulfate concentration of the perfusion solution, indicating that transport mechanisms removing sulfate from the CSF were not completely inhibited, and true equilibration between the compartments involved was not achieved. On the other hand, sulfate space of 21% was found by extrapolation from sulfate concentration profiles in similar experiments involving intravenous infusion and subarachnoid space perfusion in the rabbit.[16] This latter method does not require complete equilibration to have occurred for the estimation of a distribution space. In the rat with combined infusion–perfusion at high concentrations of thiosulfate in order to saturate potential outward transport mechanisms, a space of 16% was obtained.[42]

4.1.5. Inulin Space

Inulin injected intravenously does not pass the blood–brain barrier to any extent,[2,43] but when perfused through the CSF spaces, inulin will diffuse slowly into the brain.[18] In fact, the concentration profiles obtained in combined infusion–perfusion experiments indicated that ^{14}C-labeled inulin moved into two separate compartments,[19,20] a small one delineated during infusion-only experiments, apparently equilibrating with the blood, and a larger one equilibrating with the CSF. Thus, in spite of the fact that inulin is not actively removed by the so-called "sink" action of the CSF as are thiocyanate and iodide, combined intravenous infusion and ventriculocisternal perfusion are required for the determination of the volume of distribution of inulin in brain.

The values for inulin space obtained with experimental procedures that allowed equilibration to be approached, if not always reached, were 13% and 14% in the rat,[17,11] 16.5% and 18% in the rabbit,[16,20] 18% and 21% in the cat,[20,28,19] 14% and 19% in the dog,[18,20] and 18% in the monkey.[20]

4.1.6. Sucrose Space

Sucrose, like inulin, does not penetrate readily from blood to brain, and it equilibrates slowly between CSF and brain.[6] Thus, ventriculocisternal perfusion combined with intravenous infusion is required for true equilibration between the fluid compartments.

With appropriate methodology, sucrose space of 17% has been reported for the rat[17] and 20% for the rabbit, cat, dog, and monkey brain.[20,28]

4.1.7. Space Size in Relation to Brain Size

In experiments in which chloride, thiocyanate, inulin, and sucrose spaces were measured in the nine mammalian species, the size of the space was found to vary as a function of the logarithm of the average brain weight of the species.[44] However, these data are not reliable, since the markers were injected intracisternally, and subarachnoid fluid and subadjacent cerebral cortex were sampled simultaneously after an interval of time during which CSF levels of the markers decreased steadily. It is unlikely that true equilibrium between the tissue and the CSF was attained under these conditions.

On the other hand, "spaces" reported by several authors using a variety of animals do not indicate any obvious species differences, and in one study nearly identical inulin and sucrose spaces were found in four mammalian species.[20]

4.1.8. Spaces in Immature Brain Tissue

Changes in the space available for distribution of chloride, thiocyanate, sulfate, and inulin during development have been observed in chicks, rats, and cats.[45–49] The volume of distribution of all markers using plasma as a reference fluid decreased with maturation of the central nervous system.[45–48] Since the discrepancy between the concentration of the markers in the CSF and in the plasma increases progressively with age,[48] changes in apparent marker spaces determined in this way reflect changes in factors contributing to the low apparent permeability of the adult brain such as a progressive maturation of transport mechanisms at the choroid plexus and an increase in the rate of flow of CSF.[6]

In two studies marker spaces were determined using CSF as reference fluid.[48,49] When inulin and sucrose distribution volumes were calculated from brain and CSF levels of the ^{14}C-labeled markers 24 h after intravenous injection into nephrectomized rats, a decrease was noted from the fetal level of 51% and 59%, respectively, until 16 days of age, followed by a secondary increase to obviously excessive values of 55% and 75% in the adult.[48] It is clear from the data of this particular study that in postnatal rats, even after 24 h, CSF inulin and sucrose were not in equilibrium with plasma. The changes in the apparent distribution of the markers at different ages, especially the secondary increase in the later stage of development, may be explained by relative differences in the rate of maturation of processes involved in the restriction to the penetration of the marker into the brain, on one hand, and those responsible for the removal of the marker from the CSF, on the other.[6]

Similar interpretation can be offered for the findings in kittens in which ventriculocisternal perfusion of [^{14}C]inulin was carried out without intravenous injection of the marker.[49] In the newborn, inulin space of 26% was determined from tissue content after 3–4 h of perfusion and the concentration of inulin in the perfusate. At 1 to 2 months of age, the value was 6%, and in the adult, 14%. The authors suggested that the increase in inulin space in the adult cat as compared to 1- to 2-month-old kitten was caused by increased intracellular uptake of inulin. A more likely reason, in the absence of intravenous injection of inulin, is a somewhat more leaky blood–brain barrier for inulin in the younger animal in the direction from brain to blood.

4.2. Extracellular Space Determined by Electrical Impedance Measurements

The volume of the extracellular space in the central nervous system has also been deduced from measurements of specific impedance in brain tissue. This method is based on the assumption that the participation of the intracellular ions in current transport is hampered by cell membranes and, therefore, at low frequencies, the alternate currents used for impedance measurements in tissue are carried mainly by extracellular electrolytes. In this formulation, the relative volumes of the intracellular and the extracellular compartments are among the determining parameters of the specific impedance of the tissue. This would not be true if many cells in the tissue constituted a low-resistance pathway.[6,25] Theoretical and practical considerations of this approach to estimating the volume of the extracellular fluid have been discussed in detail by Van Harreveld.[5]

In the rabbit, rat, and cat, estimates of central nervous tissue extracellular space of 15 to 25% have been derived from measurements of cortical impedance.[5,25–28]

4.3. Lability of Fluid Compartments in Cerebral Tissue

Considerable increases in cortical impedance have been observed after a few minutes of asphyxia, during spreading depression, and during perfusion fixation *in vivo* for electron microscopy.[25,27,50,51] These changes have been interpreted to arise from a shift of extracellular water and electrolytes into the intracellular compartments, both glia and dendrites appearing swollen under these conditions.[5] Rapid shrinkage of the extracellular space during spreading depression and ischemia was also deduced from studies in which extracellular choline concentration was measured with ion-selective microelectrodes following cortical surface irrigation with mock CSF containing the marker.[24] Combined intravenous infusion and ventriculocisternal perfusion with isotopically labeled thiosulfate showed a reduction of the volume of distribution of the marker following intracarotid injection of the metabolic inhibitor 2,4-dinitrophenol.[42] All these results indicate that the distribution of water and electrolytes in cerebral tissues *in vivo* is labile and can change drastically under unfavorable physiological conditions.

This is of particular importance when considering electron micrographic evidence regarding ECS in brain. Whereas conventionally prepared electron micrographs show no spaces between the cellular elements, in electron micrographs of rapidly frozen tissue subjected to substitution fixation with osmium tetroxide at low temperatures, large extracellular spaces can be seen.[52] These spaces were reduced if the animal was submitted to 8 min of asphyxia prior to this type of fixation,[52] showing that the demonstration of ECS in brain depends on proper oxygenation of the tissue prior to and during fixation. Further, different fixatives affect the extracellular spaces differently,[53,54] and major changes occur in fluid distribution during perfusion,[55] precluding any direct relationship between the spaces in the normal and in the fixed tissue. Thus, unless special stringent conditions of fixation are employed and tested for effects on volume changes, electron micrographs of cerebral tissues cannot be accepted as quantitative representations of brain fluid compartments *in vivo*, and the absence of ECS in early electron micrographs must be considered a fixation artifact.

4.4. Abnormal Spaces

Edema is, by definition, an increase in tissue water content. A practical application of extracellular markers has been their use in elucidating which fluid compartment is affected in various types of cerebral edema. Thus, in so-called vasogenic cerebral edema, increased sulfate[56] and thiocyanate[57] spaces were demonstrated. Similarily, with [^{14}C]-inulin and [^{14}C]-sucrose, it has been shown that increased water content in experimental chronic and acute hydrocephalus is associated with increased extracellular space.[58,59] In contrast, in cerebral edema induced by triethyltin sulfate space was unchanged,[60] a finding compatible with electron microscopic evidence of massive intramyelinic vacuoles in this condition.[6]

5. SPACES IN VITRO

A detailed consideration of the numerous studies in which the distribution of extracellular markers in incubated cerebral tissue slices was measured shows that the delineation of fluid compartments in brain tissue *in vitro* is no easier than *in vivo*, although the problems encountered are different.[6]

Briefly, when appropriate techniques of preparation and incubation of cerebral slices were employed, three distinct fluid compartments could be delineated: a "space" that did not equilibrate with any of the markers used, another that equilibrated with all extracellular markers, and an intermediate one that equilibrated with some markers (e.g., sucrose and thiocyanate) but not with others (e.g., inulin and protein). The first of these presumably corresponds to the undamaged intracellular compartment; the second is equivalent to the true extracellular space and the grossly damaged areas of the tissue which have been shown to swell considerably on incubation. The so-called "third space," which does not have a counterpart *in vivo*, is thought to represent a fraction of non-inulin space that can equilibrate with smaller molecules and some ions and which must be considered as an intracellular compartment that has sustained some damage during slicing so that its permeability characteristics have been altered.

Which of the markers gives the best estimate of the intracellular space for any particular study remains problematical and must depend on the nature of the processes under investigation.

6. MORPHOLOGICAL CORRELATES OF MARKER DISTRIBUTION SPACES AND DELINEATION OF CELLULAR SPACES

To be meaningful, chemically delineated fluid spaces should be correlated with morphological compartments of the tissue. As already pointed out, because of the lability of brain fluid distribution, electron micrographs of cerebral tissues do not provide, at this time, a quantitative representation of the extracellular space of cerebral tissues *in vivo*. Nevertheless, there is now agreement between biochemical and morphological evidence that a definite functional extracellular space exists in the brain.

Several markers used for the estimation of extracellular space are known to penetrate to a limited extent into an intracellular compartment or compartments.[6] However, there is no indication that any of the markers equilibrate with an intracellular compartment. Thus, their limited intracellular penetration implies an overestimation of the extracellular space when their distribution volume is determined rather than a delineation of a particular intracellular compartment. Furthermore, the finding that the movement of several markers into cerebral tissue involves multiple rate components does not necessarily reflect their penetration into fixed, anatomically separated compartments. Suggestions that glial space may be estimated in this way can be questioned on theoretical grounds.[6]

Attempts at correlation of biochemically determined fluid spaces with morphological compartments have been somewhat more successful *in vitro*.[6,61] The enlarged inulin space in swollen incubated slices, also equilibrating with fluorescently labeled protein, was located at the margin of the slice,[62] whereas marked extracellular swelling in the superficial zone of tissue slices was consistently observed in electron micrographs.[61,63] Good agreement was also found between biochemically and morphologically observed increases in intracellular spaces. However, comparable increases in the non-inulin space were noted whether, depending on experimental conditions, morphological changes were primarily noted in neurons, in glia, or in both.[61]

7. CONCLUSION

On the basis of both biochemical and morphological evidence, measurable and functional extracellular space exists in cerebral tissues. As will be seen

Table I
Extracellular Space in Brain in Vivo

Measured by	Space (%)	References
Distribution of marker		
Sodium	22–24	5,33
Chloride	21–24	2,5,32
Thiocyanate	17–24	6,15
Iodide	17–22.5	13,40
Sulfate	16–24	5,16
Inulin	13–21	11,16–20,28
Sucrose	17–20	17,20,28
Mannitol	19.5–26	17,28
Electrical impedance	15–25	25–28

in Table I, when measured by a variety of chemical markers *in vivo*, this space represents between 15 and 25% of total tissue volume. Apparent differences among the distribution volumes in brain of the various conventional extracellular markers probably reflect differences in mechanisms affecting their equilibration in the tissue rather than actual differences in the size of the compartments they delineate. A similar range of results was obtained when extracellular space was determined from electrical impedance measurements.

No comparable chemical methods have been developed to estimate neuronal and glial volumes.

REFERENCES

1. Manery, J. F., and Haege, L. F., 1941, *Am. J. Physiol.* **134**:83–93.
2. Woodbury, D. M., Timiras, P. S., Koch, A., and Ballard, A., 1956, *Fed. Proc.* **15**:501–502.
3. Schultz, R. L., Maynard, E. A., and Pease, D. C., 1957, *Am. J. Anat.* **100**:369–407.
4. Davson, H., 1967, *Physiology of Cerebrospinal Fluid*, J. and A. Churchill, London.
5. Van Harreveld, A., 1966, *Brain Tissue Electrolytes*, Butterworth, Washington.
6. Katzman, R., and Pappius, H. M., 1973, *Brain Electrolytes and Fluid Metabolism*, Williams & Wilkins, Baltimore.
7. Bradbury, M., 1979, *The Concept of a Blood-Brain Barrier*, John Wiley & Sons, Chichester.
8. Betz, A. L., Firth, J. A., and Goldstein, G. W., 1980, *Brain Res.* **192**:17–28.
9. Pollay, M., and Davison, H., 1963, *Brain* **86**:137–150.
10. Oldendorf, W. H., and Davson, H., 1967, *Arch. Neurol.* **17**:196–205.
11. Woodward, D. L., Reed, D. J., and Woodbury, D. M., 1967, *Am. J. Physiol.* **212**:367–370.
12. Cutler, R. W. P., Lorenzo, A. V., and Barlow, C. F., 1968, *Arch. Neurol.* **18**:316–323.
13. Ahmed, N., and Van Harreveld, A., 1969, *J. Physiol. (Lond.)* **204**:31–50.
14. Davson, H., and Segal, M. B., 1969, *Brain* **92**:131–136.
15. Pappius, H. M., 1970, *Proceedings of the Wates Symposium on the Blood-Brain Barrier* (R. V. Coxon, ed.), Truex Press, Oxford (privately circulated), pp. 112–116.
16. Levin, E., Arieff, A., and Kleeman, C. R., 1971, *Am. J. Physiol.* **221**:1319–1326.
17. Amtorp, O., 1979, *J. Physiol. (Lond.)* **294**:81–90.
18. Rall, D. P., Oppelt, W. W., and Patlak, C. S., 1962, *Life Sci.* **2**:43–48.
19. Katzman, R., Schimmel, H., and Wilson, C. E., 1968, *Proc. Rudolph Virchow Med. Soc. City N.Y.* **26**:254–280.
20. Levin, V. A., Fenstermacher, J. D., and Patlak, C. S., 1970, *Am. J. Physiol.* **219**:1528–1533.

21. Pollay, M., and Kaplan, R. J., 1970, *Am. J. Physiol.* **219**:802–808.
22. Cserr, H. F., and Ostrach, L. H., 1974, *Exp. Neurol.* **45**:50–60.
23. Dunker, R. O., Harris, A. B., and Jenkins, D. P., 1976, *Brain Res.* **118**:199–217.
24. Hansen, A. J., and Olsen, C. E., 1980, *Acta Physiol. Scand.* **108**:355–365.
25. Ranck, J. B., 1963, *Exp. Neurol.* **7**:153–174.
26. Van Harreveld, A., Murphy, T., and Nobel, K. W., 1963, *Am. J. Physiol.* **205**:203–207.
27. Nevis, A. H., and Collins, G. H., 1967, *Brain Res.* **5**:57–85.
28. Fenstermacher, J. D., Li, C.-L., and Levin, V. A., 1970, *Exp. Neurol.* **27**:101–114.
29. Zadunaisky, J. A., and Curran, P. F., 1963, *Am. J. Physiol.* **205**:949–956.
30. Ames, A. III, and Nesbett, F. B., 1966, *J. Physiol. (Lond.)* **184**:216–238.
31. Friede, R. L., and Hu, K. H., 1971, *J. Physiol. (Lond.)* **218**:477–493.
32. Kibler, R. F., O'Neill, R. P., and Robin, E. D., 1964, *J. Clin. Invest.* **43**:431–443.
33. Levin, V., and Patlak, C. S., 1972, *J. Physiol. (Lond.)* **224**:559–581.
34. Coombs, J. S., Eccles, J. C., and Fatt, P., 1955, *J. Physiol. (Lond.)* **130**:326–373.
35. Streicher, E., Rall, D. P., and Gaskins, J. R., 1961, *Am. J. Physiol.* **206**:251–254.
36. Pappius, H. M., 1968, *Prog. Brain Res.* **29**:455–460.
37. Streicher, E., 1961, *Am. J. Physiol.* **201**:334–336.
38. Pollay, M., 1966, *Am. J. Physiol.* **210**:275–279.
39. Wallace, G. B., and Brodie, B. B., 1939, *J. Pharmacol. Exp. Ther.* **65**:220–226.
40. Bito, L. Z., Bradbury, M. W. B., and Davson, H., 1966, *J. Physiol. (Lond.)* **185**:323–354.
41. Van Harreveld, A., Ahmed, N., and Tanner, D. J., 1966, *Am. J. Physiol.* **210**:777–780.
42. Baethman, A., and Sohler, K., 1975, *J. Neurobiol.* **6**:73–84.
43. Morrison, A. B., 1959, *J. Clin. Invest.* **38**:1769–1777.
44. Bourke, R. S., Greenberg, E. S., and Tower, D. B., 1965, *Am. J. Physiol.* **208**:682–692.
45. Lajtha, A., 1957, *J. Neurochem.* **1**:216–227.
46. Barlow, C. F., Domek, N. S., Goldberg, M. A., and Roth, L. J., 1961, *Arch. Neurol.* **5**:102–110.
47. Vernadakis, A., and Woodbury, D. M., 1965, *Arch. Neurol.* **12**:284–293.
48. Ferguson, R. K., and Woodbury, D. M., 1969, *Exp. Brain Res.* **7**:181–194.
49. Shaywitz, B. A., and Escriva, A., 1972, *Neurology (Minneap.)*, **22**:238–245.
50. Van Harreveld, A., and Ochs, S., 1956, *Am. J. Physiol.* **187**:180–192.
51. Van Harreveld, A., and Biersteker, P. A., 1964, *Am. J. Physiol.* **206**:8–16.
52. Van Harreveld, A., Crowell, J., and Malhotra, S. K., 1965, *J. Cell. Biol.* **25**:117–137.
53. Torack, R. M., 1965, *Z. Zellforsch.* **66**:352–364.
54. Torack, R. M., Duffy, M. L., and Haynes, J. M., 1965, *Z. Zellforsch.* **66**:690–700.
55. Van Harreveld, A., and Khattab, R. I., 1969, *J. Cell Sci.* **4**:437–453.
56. Katzman, R., Gonatas, N., and Levine, S., 1964, *Arch. Neurol.* **10**:58–65.
57. Streicher, E., Ferris, P. J., Prokop, J. D., and Klatzo, J., 1964, *Arch. Neurol.* **11**:445–449.
58. Lux, W. E., Jr., Hochwald, G. M., Sahar, A., and Ransohoff, J., 1970, *Arch. Neurol.* **23**:475–479.
59. Levin, V. A., Milhorat, T. H., Fenstermacher, J. D., Hammock, M. K., and Rall, D. P., 1971, *Neurology (Minneap.)* **21**:238–246.
60. Katzman, R., Aleu, F., and Wilson, C., 1963, *Arch. Neurol.* **9**:178–187.
61. Moller, M., Mollgard, K., Lund-Andersen, H., and Hertz, L., 1974, *Exp. Brain Res.* **22**:229–314.
62. Pappius, H. M., Klatzo, I., and Elliott, K. A. C., 1962, *Can. J. Biochem. Physiol.* **40**:885–898.
63. Hokfelt, T., 1968, *Z. Zellforsch.* **91**:1–74.

Cerebral Amino Acid Pools

Thomas L. Perry

1. INTRODUCTION

This chapter is intended to present a survey of the free amino acids and related compounds that can be identified and quantitated in human brain and in the brains of commonly used species of laboratory mammals. Even though the word "free" may often be omitted, I refer only to those compounds that are detectable without hydrolysis of brain extracts. N-Acetylated amino acids, for instance, are not considered. For convenience, I have included with the amino acids several amines such as glycerophosphoethanolamine, phosphoethanolamine, and ethanolamine and several small peptides such as GSH, GSSG, and homocarnosine, since these amino compounds are readily measured together with true amino acids by many amino acid analyzer techniques.

The review describes interspecies similarities and differences in brain amino compounds, postmortem changes, regional distribution in the brain, and alterations associated with aging or produced by drugs. The chapter also describes some of the changes in brain amino acids that are encountered in human hereditary disorders. Omitted from this chapter is discussion of amino acid transport in brain, cerebral amino acid metabolism, and the neurotransmitter or neuromodulator roles of amino acids. These important subjects are covered in other chapters of the *Handbook of Neurochemistry*.

2. METHODS OF SEPARATING AND QUANTITATING AMINO COMPOUNDS IN BRAIN

A large variety of methods for quantitating amino acids in brain have been used by neuroscientists during the last two decades. Many methods have been designed for measurement of a single amino acid or of a few major compounds of interest. Wherever a wide variety of amino compounds have required measurement, methods employing an automatic amino acid analyzer have proved

---

Thomas L. Perry • Department of Pharmacology, University of British Columbia, Vancouver, British Columbia V6T 1W5, Canada.

useful. Such techniques, however, are more time consuming and expensive than some of the methods designed for only one or a few compounds.

2.1. Deproteinization of Brain Homogenates

For most analyses of free amino acids in brain, homogenates of brain must first be deproteinized. This has been accomplished with 75% or 80% ethanol,[1–3] picric acid,[4–6] sulfosalicylic acid,[7] trichloroacetic acid,[8–12] or perchloric acid.[13,14] With some of the methods, ethanol is later removed from the deproteinized extracts by evaporation, trichloroacetic acid by extraction with diethyl ether, and picric acid by preliminary passage of the extract through an anion-exchange resin. An advantage of deproteinizing brain homogenates with perchloric acid is that the excess perchlorate can readily be removed from the protein-free extract by adding potassium hydroxide until a pH of 2.8 to 3.0 is reached and then removing the insoluble potassium perchlorate by centrifugation.

2.2. Measurement of One or a Few Amino Acids

A number of methods have employed thin-layer chromatography (TLC) for measurement of the major free amino acids of brain. These include the determination of glutamic acid, aspartic acid, γ-aminobutyric acid (GABA), glycine, alanine, glutamine, and serine by two-dimensional TLC after conversion to their dinitrophenyl (DNP) derivatives.[8] Other methods have utilized two-dimensional TLC of the tritiated dansyl derivatives of brain amino acids.[15–17] Dinitrophenyl amino acids have also been converted to their methyl esters for subsequent separation and quantitation by gas–liquid chromatography.[10]

Many investigators have determined GABA content in brain by enzymatic–fluorometric techniques.[18–21] γ-Aminobutyric acid has also been quantitated in brain homogenates by a radioreceptor assay that uses [^{3}H]-GABA as a ligand.[22–24] Brain GABA[25–27] and taurine[11,12] have also been determined in many studies by brief liquid chromatographic techniques that employ the amino acid analyzer specifically to measure one or the other of these two compounds.

2.3. Measurement of a Large Number of Amino Acids

Wherever it is important to separate and quantitate a large number of amino acids and related compounds in human brain or in the brain of experimental animals, use of an automatic amino acid analyzer is preferable. In my laboratory, we have for many years used a technique in which brain homogenates are deproteinized with 0.4 M perchloric acid and then applied to a single column of a Technicon® automatic amino acid analyzer. The method as originally described[13] has only been altered in that homogenization of brain is now carried out with use of motor-driven Teflon® pestles in glass tissue grinders.

Excess perchlorate is removed from the deproteinized brain extract by adjusting the pH to 2.8 with KOH and then separating the clear supernatant from the precipitated potassium perchlorate by centrifugation. Amino acid analysis is carried out using a single column (140 × 0.6 cm) of Technicon Chromobeads®, Type B, a lithium citrate buffer elution system, and reaction of the separated amino acids with ninhydrin.[28]

The resolution of amino compounds obtainable with this method is illustrated in Fig. 1, in which the amino acid chromatogram of a biopsy specimen of human temporal cortex is traced. Some important advantages of the method are that the three brain compounds, glycerophosphoethanolamine, taurine, and phosphoethanolamine, are separated; that aspartic acid and GSH are separated; that asparagine, glutamic acid, and glutamine are separated; and that GSSG and glycine are separated. In addition, the column is operated at 35°C until glutamine has been eluted and thereafter at 70°C.[28] Use of a low operating temperature during the first part of the amino acid analysis is essential in order to avoid cyclization of glutamine to pyrrolidone carboxylate with resulting substantial underestimation of brain glutamine content.

2.4. Newer Techniques

Many amino acid analyzer systems could undoubtedly be designed to give as good or even better resolution of brain amino compounds. Several investigators[29-31] have developed single-column methods employing lithium citrate buffers which give excellent separations of the amino acids in physiological fluids. There is some urgency in adapting these methods to separate the very complex mixture of free amino compounds present in brain, because the Technicon Chromobeads®, Type B, resin which we have used is no longer commercially available. The more recently manufactured Chromobeads® give a much poorer resolution of amino compounds than those that were available in the 1960s. The new improved cation-exchange resins now available for amino acid analysis of physiological fluids[29-31] will not resolve the large variety of amino compounds in brain without further modification of the recommended eluting buffers.

In the development of improved new techniques for automated amino acid analysis of brain, it is crucial that GSH be separated from aspartic acid and that GSSG be separated from glycine. Large amounts of GSH and GSSG are present in living brain, and it is likely that many published reports give artifactually high aspartate and glycine values because of failure to separate these peptides from them. Satisfactory amino acid analyzer methods must also separate glycerophosphoethanolamine and phosphoethanolamine clearly from taurine, since all three compounds are always present in mammalian brain. Even though studies utilizing amino acid analyzer techniques had repeatedly shown, more than a decade ago, that both glycerophospoethanolamine and taurine occur in brain and that they have similar elution times from cation-exchange columns,[5,13,14,32] the "wheel" of glycerophosphoethanolamine's occurrence in brain was only recently rediscovered![7] Many published reports of taurine content in brain are unduly high because of failure of the method used

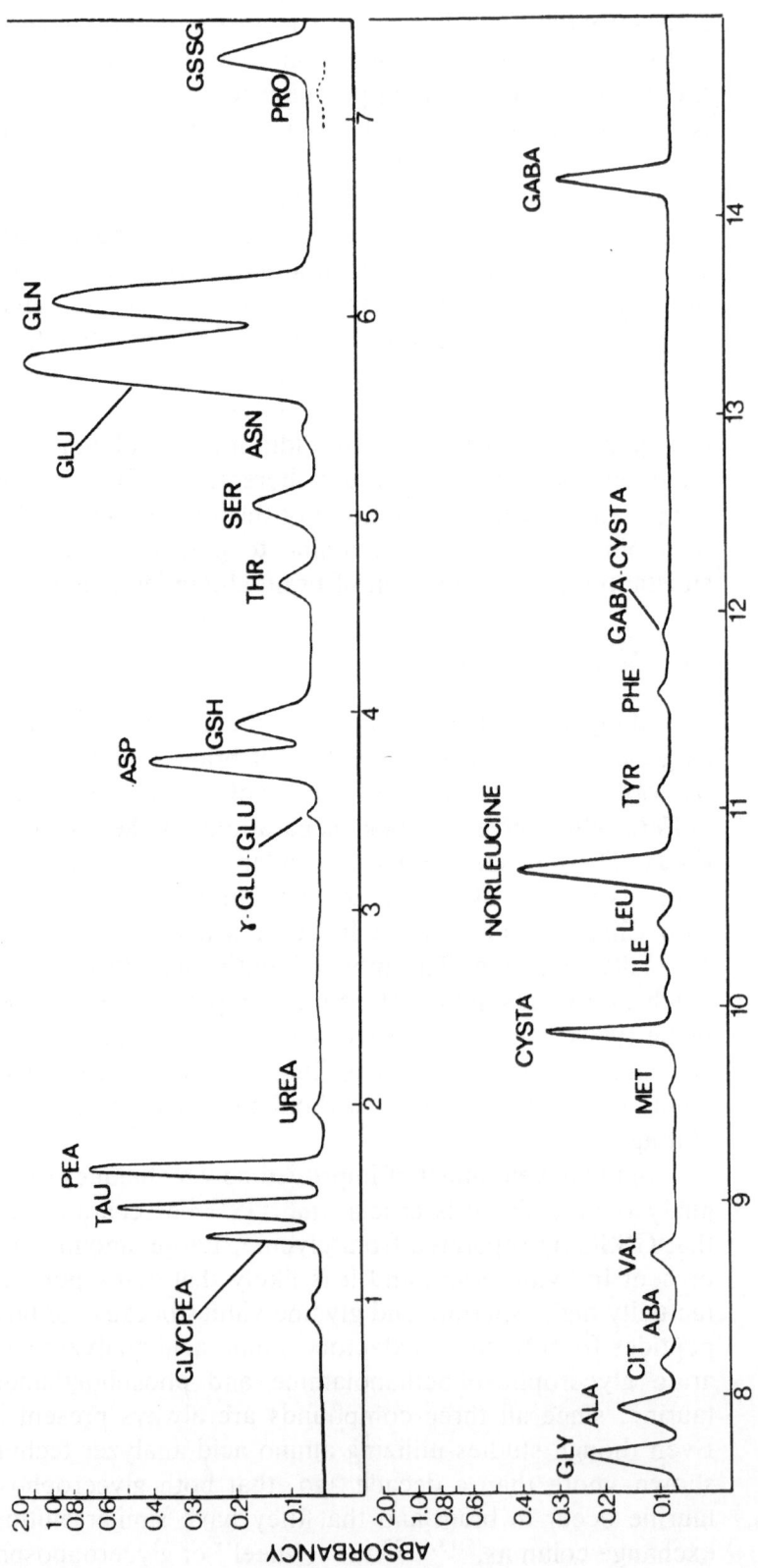

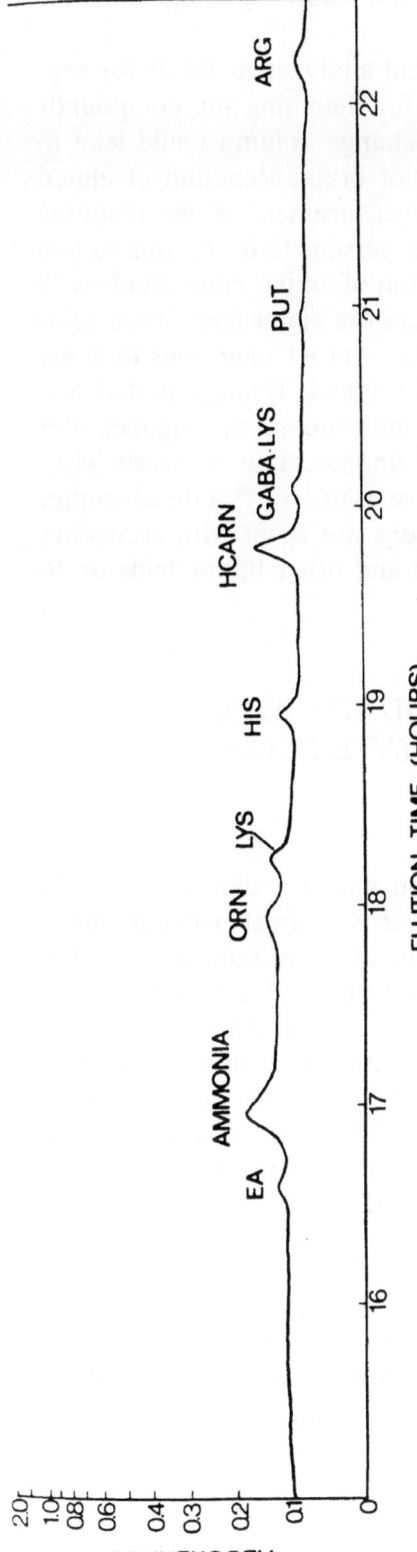

Fig. 1. Tracing of a chromatogram of the free amino compounds in a deproteinized extract of a biopsy specimen of human temporal cortex equivalent to 0.1 g wet weight of brain. Abbreviations used are: GLYCPEA, glycerophosphoethanolamine; TAU, taurine; PEA, phosphoethanolamine; γ-GLU-GLU, γ-glutamyl-glutamic acid; ASP, aspartic acid; GSH, reduced glutathione; THR, threonine; SER, serine; ASN, asparagine; GLU, glutamic acid; GLN, glutamine; PRO, proline; GSSG, oxidized glutathione; GLY, glycine; ALA, alanine; CIT, citrulline; ABA, α-amino-*n*-butyric acid; VAL, valine; MET, methionine; CYSTA, cystathionine; ILE, isoleucine; LEU, leucine; TYR, tyrosine; PHE, phenylalanine; GABA-CYSTA, γ-aminobutyrylcystathionine; GABA, γ-aminobutyric acid; EA, ethanolamine; ORN, ornithine; LYS, lysine; HIS, histidine; HCARN, homocarnosine; GABA-LYS, γ-aminobutyryllysine; PUT, putreanine; and ARG, arginine. (Taken from Perry,[116] by permission of Raven Press, New York.)

to separate taurine from glycerophosphoethanolamine or phosphoethanolamine or both.

Besides development of improved amino acid analyzer methods for separating brain amino compounds, new methods for detecting the compounds once they have been eluted from the cation-exchange column could lead to much greater versatility in biochemical studies of brain. Reaction of eluted amino compounds with *o*-phthalaldehyde and measurement of the resulting fluorescence produced by the adducts[7,31,33] increase sensitivity by one to two orders of magnitude over that provided by reaction of amino compounds with ninhydrin. This should make it possible to measure the major free amino acids of brain in very small amounts of tissue (1 or 2 mg wet wt.) and thus to study amino acid neurotransmitters in very small brain nuclei. Using *o*-phthalaldehyde detection, it should also be practical to measure many oligopeptides routinely present in living brain that are either undetectable or barely visualized on chromatograms with the methods we have used.[13,14,28] A disadvantage of using *o*-phthalaldehyde is that this reagent does not react with secondary amines, and thus, one cannot quantitate proline and other imino acids or N-methylated compounds that may occur in brain.

3. CHANGES IN BRAIN AMINO COMPOUNDS WITH INCREASING DEATH-TO-FREEZING INTERVALS

3.1. Experimental Animals

A number of investigators have studied the changes that occur in the contents of free amino acids in the brains of laboratory animals when the brain is allowed to stand for brief periods (10–45 min) at room temperature. The GABA content has been shown to rise appreciably in rat brain within 2 min of death unless the brain is either instantly frozen,[34] or fixed by microwave irradiation.[2] This rapid rise in GABA content presumably results from continued activity of the synthesizing enzyme for GABA, glutamic acid decarboxylase (L-glutamate 1-carboxylyase, E.C. 4.1.1.15) (GAD). The postmortem rise in brain GABA content can also be prevented in the mouse or rat by intravenous injection of 3-mercaptopropionic acid, a potent GAD inhibitor, shortly before killing of the animal.[3]

Studies in the dog,[5] rat,[35] and in rats, mice, and guinea pigs[36] all showed rises not only in GABA content but also in alanine when brain was allowed to stand at room temperature for brief periods. Little change was observed in the contents of other amino acids in these short periods after brain death. In the rat, we have observed 20 to 25% increases in alanine and GABA content in brain allowed to stand at room temperature for 2 min after sacrifice of the animal. After 30 min at room temperature, GABA and alanine content is almost doubled over values present during life; contents of valine, isoleucine, leucine, methionine, and ethanolamine are markedly increased, and the content of GSH is decreased.[37]

For meaningful studies of amino acids and related compounds in experimental animals, brain should be removed from the skull and frozen immediately. This can be accomplished within 20 to 30 s of sacrifice with small laboratory animals, and with larger animals, superficial brain areas can be removed surgically under anesthesia and frozen instantly. Microwave irradiation of the animal's head is an alternative method of killing which should terminate postmortem enzymatic activity,[2,3] but its suitability for studies of free amino acids other than GABA has not been reported.

3.2. Human Brain

Measurement of the free amino acid contents that probably occur during life can only be carried out for superficial regions of human brain. Biopsies of cerebral or cerebellar cortex can be immersed in liquid nitrogen immediately after neurosurgical removal. Such material is often available to neurochemists under circumstances that are ethically proper. Superficial brain area may have to be sacrificed by the neurosurgeon who removes a deep-seated tumor or brain abscess, and there is increasing use of diagnostic brain biopsies and of removal of epileptogenic foci from human brain. However, even if the specimen is instantly frozen when removed, biochemical changes may have already occurred during gradual compromise of blood supply as the biopsy is carefully dissected out and bleeders are cauterized by the neurosurgeon. It is, of course, rarely possible in human patients to obtain promptly frozen deep nuclei from brain.

Investigation in human brain of amino acid changes that are related to disease or to drug use requires knowledge of the rates at which contents of various compounds change between a patient's death and the time at which the autopsied brain is frozen. The gradual cooling of human brain in cadavers stored under typical mortuary temperatures has been explored by inserting thermocouples through the skull into cerebral cortex or more deeply into the thalamus and recording temperatures.[38] It requires 24 to 30 h before human brain reaches a stable temperature of 4°C under these conditions. Obviously, the more rapidly human brain is removed from the skull after death, and the sooner it is frozen, the greater the accuracy and variety of the biochemical studies that can be done. Skilled neuropathologists can accurately dissect frozen human brain (allowed to thaw partially to a waxy consistency) into a large number of precise anatomical regions.

In my laboratory, we have measured the rates at which various free amino compounds change in human brain after death under simulated mortuary conditions.[37] This has been done for biopsied brain by comparing the contents of amino compounds in instantly frozen biopsies of cerebral cortex with the contents in other portions of the same biopsies that were incubated at 35°C for periods ranging from 10 to 240 min. With autopsied human brain, comparable portions of the same cortical gyrus have been allowed to stand at either 0° or 4°C for periods up to 120 h and amino acid contents compared with those of an immediately frozen portion.

These experiments have shown that the contents of a number of free amino compounds are not altered in human brain either during incubation at 35°C for up 4 h or during storage at 0 to 4°C for up to 120 h after death. These unaltered compounds include taurine, phosphoethanolamine, glutamic acid, glutamine, α-amino-*n*-butyric acid, cystathionine, γ-aminobutyrylcystathionine, β-alanine, homocarnosine, γ-aminobutyryllysine, and putreanine. Two compounds whose contents are stable in human brain for the first 4 h after death undergo changes on prolonged cold storage of autopsied brain. Aspartate content rises progressively after 4 h, whereas glycerophosphoethanolamine content drops slowly after 4 h.

The GABA content increases more slowly after death in human brain than it does in the rat, the rise becoming apparent only at 30 min and reaching a maximum level by 1 to 3 h after death. Thereafter, GABA content remains stable in autopsied brain for up to 120 h under mortuary conditions.

The GSH content drops rapidly after brain death in man, being only one-half its original value after 4 h at 35°C and only about 10% of the value present during life after 120 h at 4°C. The steady drop in GSH is accompanied by rises in the brain contents of its hydrolysis products, glycine, cystine, and glutathione–cysteine mixed disulfide.

Contents of most of the amino acids that are components of proteins and peptides (with the exception of aspartic acid, glutamic acid, and glutamine) rise steadily in human brain during the first 4 h after brain death. Significant increases in the contents of alanine, lysine, and arginine, as well as of ethanolamine, are apparent as early as 30 min after brain death.

Figure 2 illustrates the changes that can be expected in human brain during the first 4 h after death for five compounds that are known or putative neurotransmitters. Taurine, aspartate, and glutamate contents remain unchanged. Glycine and GABA contents rise markedly. Taurine and glutamate contents remain the same or approximately the same as during life for long periods in autopsied brain. Aspartate contents in autopsied brain approximate those during life for up to 4 h but not thereafter. Brain GABA content can safely be compared among different patients when the death-to-freezing interval exceeds about 2 hr, since GABA content neither increases nor decreases on long storage under mortuary conditions.

The failure of such metabolically active substances as glutamic acid and glutamine to exhibit major alterations in content in human brain with increasing death-to-freezing intervals is surprising. However, it can prove useful both in studies of glutamate as a neurotransmitter and in assessing changes secondary to ammonia accumulation in patients dying with hepatic encephalopathy or with genetically determined disorders of the urea cycle.

4. FREE AMINO COMPOUNDS IN BRAINS OF LABORATORY ANIMALS

Table I presents the mean values of 35 free amino acids and related compounds in the brains of eight species of adult mammals that are commonly used in neurochemical research. The contents shown are those found when brain is very rapidly frozen after sacrifice of the animal, and they are likely to

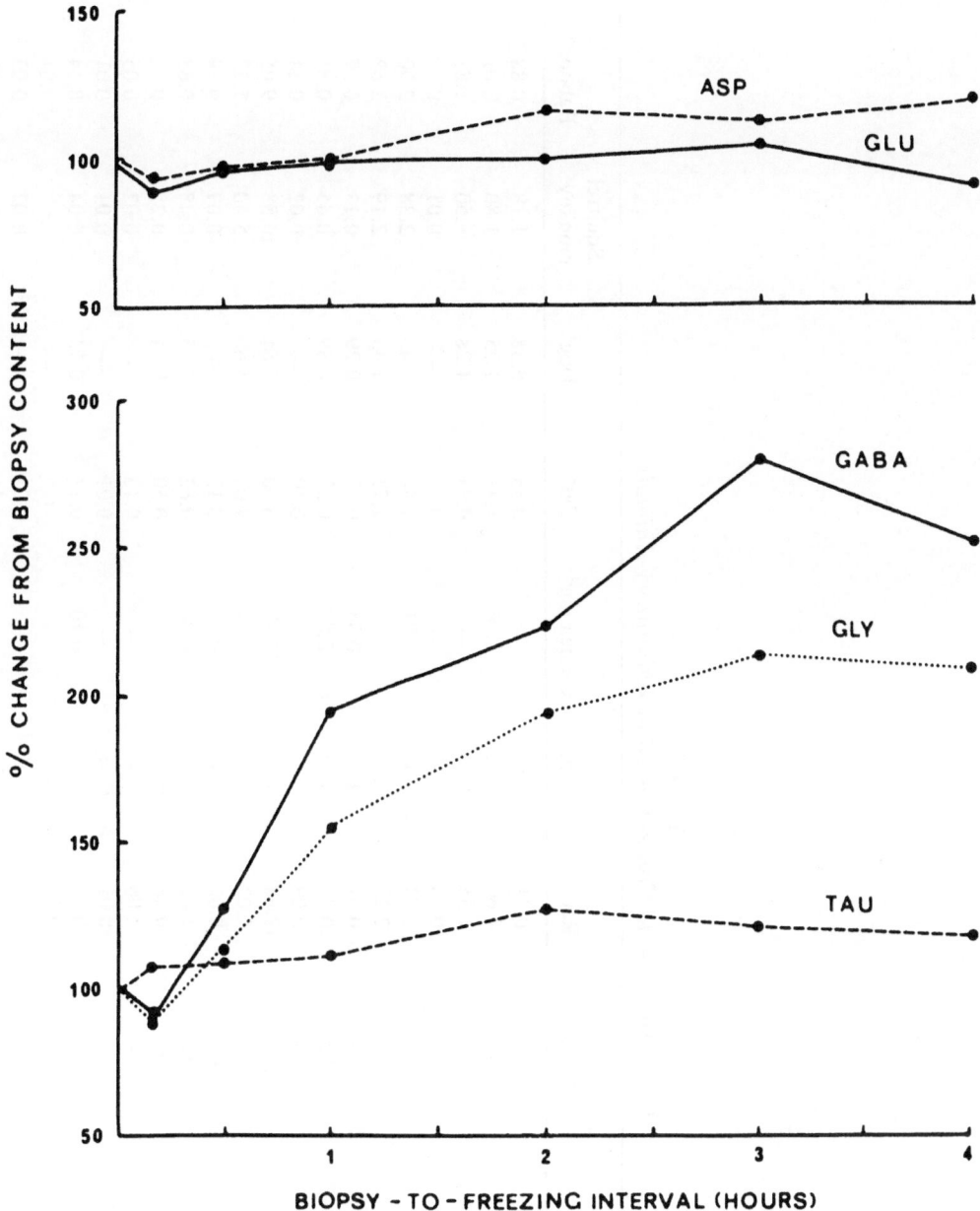

Fig. 2. Changes in contents of five known or putative amino acid neurotransmitters in human brain biopsies incubated at 35°C. Values plotted are the means after various incubation times and are expressed as percentages of the content of each amino acid in the instantly frozen portion of each biopsy. Excitatory amino acids are shown above, and inhibitory amino acids below. Abbreviations are those used in Fig. 1. (Taken from Perry *et al.*,[37] by permission of Raven Press, New York.)

approximate closely the values present during life. For the smaller species (mouse, rat, rabbit, and guinea pig), the contents are those found in homogenized whole brain, whereas contents shown for cat, dog, squirrel monkey, and baboon are for cerebral cortex.

Three free amino compounds that were listed as present in the brains of several laboratory mammals in the first edition of the *Handbook of Neuro-*

Table I
Brain Amino Acid Content in Adult Laboratory Animals

Compound	Mouse[a]	Rat[b]	Rabbit[c]	Guinea pig[d]	Cat[e]	Dog[f]	Squirrel monkey[g]	Baboon[h]
Glycerophosphoethanolamine	—	0.43	0.57	—	0.56	0.58	1.18	0.88
Taurine	8.32	4.60	1.04	1.50	1.31	1.25	1.80	2.01
Phosphoethanolamine	—	1.36	1.54	—	0.93	1.28	1.50	1.81
Hypotaurine	—	0.06	—	—	0	—	0.01	0
Aspartic acid	4.17	2.78	2.05	3.91	1.54	2.45	2.38	2.00
GSH[i]	—	2.23	0.58	—	1.73	1.45	2.19	2.62
Threonine	0.47	0.61	0.14	0.35	0.17	0.29	0.13	0.27
Serine	0.96	0.86	0.76	0.80	0.52	0.39	0.45	0.81
Asparagine	—	0.09	—	—	0.10	—	0.02	0.11
Glutamic acid	10.10	12.46	8.53	10.40	6.00	7.81	10.88	9.05
Glutamine	4.52	5.02	3.08	4.59	4.01	5.60	5.40	5.27
Proline	0.13	0.06	0.04	0.17	0.13	—	0.07	0.19
Glycine	1.31	1.02	0.97	0.70	0.62	0.55	0.48	0.63
Alanine	0.52	0.48	0.42	0.45	0.50	0.14	0.25	0.72
Citrulline	—	0.06	—	—	0.10	—	0.02	0.05
α-Amino-n-butyric acid	—	0.01	0.01	—	0.01	—	0.01	0.01
Valine	0.11	0.06	0.08	0.10	0.13	0.13	0.04	0.14
Cystine	—	tr	—	—	tr	—	tr	0.01
Methionine	—	0.02	0.01	—	0.01	—	0.02	0.03
Cystathionine	—	0.06	0.23	—	0.21	—	0.04	1.43

	a	b	c	d	e	f	g/h
Isoleucine	0.03	0.02	0.03	0.03	0.05	0.02	0.07
Leucine	0.05	0.05	0.06	0.08	0.08	0.04	0.11
Tyrosine	0.05	0.05	0.05	0.07	0.05	0.04	0.06
Phenlalanine	0.06	0.03	0.03	0.06	0.04	0.04	0.06
β-Alanine	—	0.01	0.03	tr	tr	0.01	0.04
γ-Aminobutyric acid	1.79	1.90	1.46	2.00	0.88	0.94	1.35
Tryptophan	—	0.01	—	tr	0.01	0.01	0.01
Ethanolamine	—	0.11	0.12	0.13	0.13	0.12	0.37
Ornithine	0.03	0.01	0.01	0.04	0.04	0.03	0.03
Lysine	0.26	0.22	0.04	0.05	0.10	0.10	0.27
Histidine	0.08	0.05	—	0.07	0.13	0.08	0.11
Homocarnosine	0.17	0.08	—	0.12	0	—	0.23
γ-Aminobutyryllysine	—	0	—	—	0	—	tr
Putreanine	—	0.01	—	—	0.02	—	0.03
Arginine	0.12	0.11	—	0.17	0.09	0.07	0.17

[a] Mouse, whole brain, μmol/g wet wt.[36]
[b] Rat, whole brain, μmol/g wet wt.[39]
[c] Rabbit, whole brain, μmol/g wet wt.[32]
[d] Guinea pig, whole brain, μmol/g wet wt.[36]
[e] Cat, frontal cortex, μmol/g wet wt.[40]
[f] Dog, cerebral cortex, μmol/g wet wt.[55]
[g] Saimiri sciureus, frontal cortex, μmol/g wet wt. (unpublished data).
[h] Papio papio, frontal cortex, μmol/g wet wt.[41]
[i] Total GSH and GSSG expressed as GSH.

chemistry[42] are not included in Table I. These are hydroxyproline, sarcosine, and glucosamine. In my laboratory, we have been unable to detect these three compounds with use of the amino acid analyzer in brain from rats, cats, squirrel monkeys, baboon, or man. They may be present in minute amounts, but they are not seen on chromatograms of brain where ninhydrin is used as the detecting agent, and both hydroxyproline and sarcosine as secondary amines would fail to react with *o*-phthalaldehyde in the more sensitive amino acid analyzer techniques using the latter reagent. Two amino acids not listed earlier[42] are readily detectable in the brains of some laboratory animals. These are hypotaurine[43,44] and putreanine.[45]

Inspection of Table I shows a striking similarity in the contents of many amino acids among species. This is so especially in the case of most of the amino acids that are components of proteins. However, there are some important interspecies differences. These may reflect real metabolic differences or may simply be artifacts introduced by sampling different brain regions or by differences in methods of measurement.

Taurine content is considerably higher in whole brain of mice and rats than it is in whole brain of the other two species of rodents for which data are presented or in cortex of the carnivores and primates. This probably reflects the presence of a much more active synthetic pathway from cysteine to taurine in rats and mice. Hypotaurine, the immediate precursor of taurine, is present in readily measureable amounts in rat brain. Taurine is synthesized *in situ* in rat brain,[46] whereas in the cat there is either very limited or no synthesis of taurine, the amino acid being obtained from dietary sources.[47,48] Taurine content becomes markedly reduced in the brains of kittens maintained on a taurine-deficient diet.[49] Cystathionine content is much higher in cerebral cortex of the baboon than in brains of the other species included in Table I.

The content of glutamic acid is lower in the cortex of the two carnivores listed in Table I than in the species of rodents and primates. The GABA content is higher in the four rodents' brains. However, this may be because data are shown for GABA content in whole brain. Since GABA content is considerably higher in some deep brain nuclei, the apparently low GABA content in the cat, dog, squirrel monkey, and baboon (where cortex only has been sampled) may be misleading. Homocarnosine (γ-aminobutyrylhistidine) is not detectable in the brains of cats[40] or of dogs (T. L. Perry, unpublished data), whereas it is easily quantitated in the brains of mice, rats, guinea pigs, and primates. The reason for the absence of homocarnosine from the brains of these two carnivores is unexplained. Carnosine (β-alanylhistidine) and homocarnosine are very difficult to separate with most amino acid analyzer techniques, including the one we have used.[28] We have found that carnosine is the predominant imidazole dipeptide present in the olfactory bulb of the mouse, rat, cat, and dog when the homocarnosine and/or carnosine is/are isolated, purified, and hydrolyzed, with subsequent determination of β-alanine or GABA freed by hydrolysis. Little or no carnosine is found in other brain areas outside the olfactory tract in these species. There is considerable evidence that carnosine may serve as an olfactory neurotransmitter or neuromodulator in the olfactory bulb and tract.[50–52]

The relatively high content of aspartic acid shown in Table I for whole brain of the mouse and the guinea pig and the relatively low GSH content shown for rabbit brain may be artifactual. If the amino acid analysis techniques used in these studies[32,36] had failed to resolve aspartic acid from GSH, the apparent values for these two compounds could have been distorted. Otherwise, there appears to be little interspecies variation in brain aspartate or GSH content.

5. FREE AMINO COMPOUNDS IN BIOPSIED HUMAN BRAIN

Table II presents the mean free amino acid contents of two different regions of human brain. These values are from biopsies examined in my laboratory during the last decade. The brain specimens were all frozen in liquid nitrogen within 10 to 30 s of the neurosurgeon's severing the blood supply. The vast majority of human brain biopsies are removed from frontal or temporal cortex. Since there are few differences chemically between these two regions, values for the two areas have been pooled. Table II shows separately the mean values for 14 single biopsies obtained from 14 nonepileptic patients, most of whom underwent surgery for removal of deep-seated tumors, and none of whom had any known inherited metabolic disorder. Also shown are the mean values for 51 epileptogenic foci removed from 34 patients with intractable focal epilepsy. Most amino compounds are present in similar amounts in the biopsies from both groups of patients. Finally, Table II shows mean values for the same compounds in single biopsies of cerebellar cortex from nine patients who did not have any generalized or hereditary cerebellar disease. The values shown in Table II are probably as close as one can practically and ethically come to measuring the contents of amino acids in the living brain of normal human beings.

Comparison of the values listed for human frontal–temporal cortex in Table II with those listed for frontal cortex in the squirrel monkey and baboon (Table I) shows that the contents of most compounds are very similar. However, phosphoethanolamine content is higher in human brain than in these two lower primates, and cystathionine content in the nonepileptic human brain is considerably higher than in the squirrel monkey. Aspartic acid and GABA content are each considerably lower in human brain than in frontal cortex of the two lower primates. The data shown in Table II for epileptogenic foci are probably representative of what is present in normal human frontal–temporal cortex except for five compounds. Glutamic acid, glutathione, glycine, and GABA contents are elevated, and cystathionine content is reduced. Changes in these five compounds may be related either to the biochemical abnormalities producing the epileptic discharges or to the anticonvulsant drugs employed in treating the patients.[53]

Hypotaurine is sometimes measureable in biopsied human brain, especially in the cerebellum (Table II). Putreanine has also been identified in human brain[54] and again is regularly detectable in cerebellar cortex. Several small peptides are also detectable in biopsied as well as in autopsied human brain.

Table II
Amino Acid Content[a] in Biopsied Human Brain

Compound	Frontal or temporal cortex, nonepileptic patients (14)	Frontal or temporal cortex, epileptogenic foci (51)	Cerebellar cortex (9)
Glycerophosphoethanolamine	0.89 ± 0.39	0.87 ± 0.33	0.63 ± 0.37
Taurine	1.34 ± 0.45	1.42 ± 0.43	3.31 ± 1.60
Phosphoethanolamine	1.76 ± 0.54	2.23 ± 0.77	1.44 ± 0.48
Hypotaurine	tr	tr	0.06 ± 0.06
Aspartic acid	1.18 ± 0.28	1.17 ± 0.29	0.89 ± 0.23
GSH[b]	2.23 ± 0.62	2.66 ± 0.36	2.98 ± 0.38
Threonine	0.28 ± 0.14	0.24 ± 0.06	0.44 ± 0.13
Serine	0.43 ± 0.08	0.46 ± 0.09	0.48 ± 0.19
Asparagine	0.06 ± 0.02	0.07 ± 0.03	0.07 ± 0.03
Glutamic acid	7.58 ± 1.86	9.44 ± 1.71	7.58 ± 1.42
Glutamine	5.42 ± 1.84	5.37 ± 1.07	7.26 ± 1.15
Proline	0.07 ± 0.04	0.07 ± 0.04	0.12 ± 0.07
Glycine	0.51 ± 0.19	0.72 ± 0.36	0.86 ± 0.69
Alanine	0.32 ± 0.13	0.40 ± 0.13	0.61 ± 0.25
Citrulline	0.06 ± 0.02	0.07 ± 0.03	0.11 ± 0.06
α-Amino-n-butyric acid	0.04 ± 0.04	0.03 ± 0.02	0.04 ± 0.04
Valine	0.14 ± 0.04	0.14 ± 0.03	0.15 ± 0.08
Cystine	0.01 ± 0.01	0.01 ± 0.01	0.01 ± 0.01
Methionine	0.02 ± 0.01	0.01 ± 0.01	0.02 ± 0.01
Cystathionine	1.06 ± 1.11	0.59 ± 0.57	0.99 ± 1.03
Isoleucine	0.03 ± 0.01	0.02 ± 0.01	0.03 ± 0.02
Leucine	0.07 ± 0.02	0.08 ± 0.02	0.08 ± 0.04
Tyrosine	0.05 ± 0.02	0.04 ± 0.02	0.05 ± 0.03
Phenylalanine	0.05 ± 0.02	0.04 ± 0.02	0.05 ± 0.03
β-Alanine	0.01 ± 0.01	0.01 ± 0.01	tr
γ-Aminobutyric acid	0.63 ± 0.21	0.82 ± 0.21	0.74 ± 0.13
Tryptophan	0.01 ± 0.01	0.01 ± 0.01	0.01 ± 0.01
Ethanolamine	0.14 ± 0.10	0.14 ± 0.09	0.20 ± 0.31
Ornithine	0.04 ± 0.01	0.02 ± 0.02	0.03 ± 0.02
Lysine	0.12 ± 0.04	0.10 ± 0.03	0.20 ± 0.06
Histidine	0.09 ± 0.03	0.11 ± 0.03	0.10 ± 0.04
Homocarnosine	0.28 ± 0.14	0.32 ± 0.14	0.47 ± 0.31
γ-Aminobutyryllysine	0.03 ± 0.03	0.02 ± 0.02	0.03 ± 0.04
Putreanine	tr	tr	0.01 ± 0.01
Arginine	0.08 ± 0.01	0.06 ± 0.02	0.09 ± 0.02

[a] Values (mean ± S.D.) expressed in μmol/g wet wt. Number of biopsies examined indicated in parentheses. tr = trace.
[b] Total GSH and GSSG expressed as GSH.

γ-Glutamylglutamic acid and γ-glutamylglutamine are routinely seen on amino acid analyzer chromatograms as a small common peak eluted just before aspartic acid (Fig. 1). γ-Aminobutyrylcystathionine[55] is eluted from the cation-exchange column just after phenylalanine (Fig. 1). In addition, there are present in low concentration in human brain several additional oligopeptides whose N-terminal amino acid is GABA but whose exact structure has not yet been worked out. These peptides ought to be easily quantifiable with amino acid analyzer techniques employing *o*-phthalaldehyde as detection reagent. What the physiological roles of γ-aminobutyrylcystathionine and similar peptides are, and whether or not their measurement is important, have yet to be determined.

6. REGIONAL DISTRIBUTION OF AMINO COMPOUNDS IN BRAIN

6.1. Human Brain

For many free amino compounds, it is not possible to provide meaningful information about variations in content among different regions of human brain. Only the superficial areas are available for biopsy. All of the amino acids whose content progressively rises with increasing death-to-freezing intervals (see Section 3.2) and GSH, whose content steadily drops, cannot be reliably compared between deeper regions in autopsied human brain.

Table II shows that there are some important differences between frontal–temporal cerebral cortex and cerebellar cortex. Taurine content is considerably greater in cerebellar cortex, and contents of homocarnosine, putreanine, and hypotaurine are also high in this region. Phosphoethanolamine, on the other hand, is lower in cerebellar cortex.

Table III summarizes data from my laboratory on the regional differences in free amino compounds in autopsied brain from a large number of adults who died without neurological or psychiatric disorders. Figures are shown for ten compounds. Seven of these (taurine, phosphoethanolamine, glutamic acid, glutamine, cystathionine, homocarnosine, and γ-aminobutyryllysine) have been shown not to change in content for long periods after brain death (see Section 3.2); GABA content reaches a stable maximum by about 2 h after death, and the brain specimens included in Table III had death-to-freezing intervals longer than 2 h. Aspartic acid content in human brain is unchanged for about 4 h after death, and the values shown for aspartate in Table III are limited to those specimens with a death-to-freezing interval of 4 h or less. Glycerophosphoethanolamine content falls slowly in autopsied human brain with prolonged death-to-freezing intervals; however, the regional differences shown for this compound in Table III are probably representative of those occurring during life.

Inspection of Table III shows that glycerophosphoethanolamine content is relatively high in the globus pallidus, substantia nigra, inferior olivary nucleus, and dentate nucleus. As dissected out, these nuclei contain considerable white matter. White matter has a much higher glycerophosphoethanolamine content than does gray matter, the highest glycerophosphoethanolamine content in human brain being found in the corpus callosum, which is purely white matter.[13] Phosphoethanolamine content is highest in the frontal cortex (much higher than in occipital cortex) and is relatively high in cerebellar cortex and in the caudate nucleus. Cystathionine content varies markedly among individual patients as shown by the large S.D. about each mean value. Contents of

Table III
Regional Distribution of Amino Compounds in Autopsied Human Brain[a]

Compound	Frontal cortex (25)	Occipital cortex (24)	Cerebellar cortex (25)	Caudate nucleus (30)	Putamen (16)
Glycerophosphoethanolamine	0.84 ± 0.33	0.76 ± 0.36	0.99 ± 0.41	0.99 ± 0.40	0.77 ± 0.32
Taurine	1.04 ± 0.38	0.95 ± 0.41	2.66 ± 1.03	1.22 ± 0.45	1.19 ± 0.23
Phosphoethanolamine	1.83 ± 0.51	1.08 ± 0.43	1.69 ± 0.45	1.45 ± 0.51	0.96 ± 0.44
Aspartic acid[b]	1.38 ± 0.54	1.92 ± 1.68	1.36 ± 0.99	0.81 ± 0.46	0.95 ± 0.61
Glutamic acid	8.36 ± 1.60	8.45 ± 1.23	9.03 ± 1.66	10.34 ± 1.82	11.45 ± 2.02
Glutamine	4.50 ± 1.36	4.90 ± 2.83	5.71 ± 1.54	4.35 ± 2.20	3.71 ± 1.13
Cystathionine	0.55 ± 0.53	1.65 ± 1.25	0.51 ± 0.61	0.73 ± 0.44	1.19 ± 0.68
γ-Aminobutyric acid[c]	1.69 ± 0.45	1.87 ± 0.53	1.66 ± 0.43	2.88 ± 0.83	2.80 ± 0.80
Homocarnosine	0.28 ± 0.18	0.42 ± 0.33	0.59 ± 0.33	0.22 ± 0.19	0.46 ± 0.20
γ-Aminobutyryllysine	0.03 ± 0.03	0.05 ± 0.05	0.03 ± 0.03	0.02 ± 0.02	0.04 ± 0.04

[a] Values (mean ± S.D.) expressed in μmol/g wet wt. Number of autopsied brains examined indicated in parentheses. tr = trace.

[b] Aspartic acid contents are shown only for brains with a death-to-freezing interval of 4 h or less. Number of brains examined is less than indicated for other compounds.

[c] γ-Aminobutyric acid contents shown are characteristic for autopsied human brain and are substantially higher than those probably present during life.

cystathionine are highest in the occipital cortex, globus pallidus, and dentate nucleus.

Of the amino acids that are known or possible neurotransmitters or neuromodulators and whose contents can meaningfully be compared in autopsied human brain, taurine content is strikingly higher in cerebellar cortex than in other brain regions. Aspartic acid content is highest in the occipital cortex and thalamus and relatively low in the caudate nucleus, mesolimbic system, inferior olivary nucleus, and dentate nucleus. Glutamic acid content is highest in the putamen and caudate nucleus and relatively high in the cerebellar cortex and the thalamus.

The most striking regional differences in human brain are those in GABA content. The highest GABA content is found in the globus pallidus, with high but progressively lower levels in the substantia nigra, dentate nucleus, nucleus accumbens, olfactory tubercle, and striatum. The high GABA content of the globus pallidus and the substantia nigra is to be expected because of the presence in these nuclei of a large population of GABAergic striatonigral neurons and interneurons. The GABA content is also understandably high in the dentate nucleus where the GABAergic Purkinje neurons from the cerebellar cortex terminate. Homocarnosine and γ-aminobutyryllysine, two of the N-terminal GABA dipeptides of brain, are present in large amounts in three areas characterized by high GABA contents: the dentate nucleus, substantia nigra, and globus pallidus. It is surprising, however, that contents of these two dipeptides are equally high in the inferior olivary nucleus, where GABA content is very low. By contrast, the contents of the dipeptides are relatively low in the caudate nucleus, nucleus accumbens, and olfactory tubercle, where GABA content

Table III. (Continued)

Globus pallidus (15)	Nucleus accumbens (15)	Olfactory tubercle (12)	Thalamus, medial–dorsal (20)	Substantia nigra (27)	Inferior olivary nucleus (10)	Dentate nucleus (17)
1.11 ± 0.52	0.85 ± 0.52	0.64 ± 0.45	0.93 ± 0.43	1.32 ± 0.45	2.07 ± 0.87	1.53 ± 0.76
1.22 ± 0.22	0.80 ± 0.31	0.70 ± 0.22	0.85 ± 0.28	1.05 ± 0.36	1.62 ± 0.55	1.13 ± 0.42
0.79 ± 0.40	0.89 ± 0.34	0.76 ± 0.29	0.58 ± 0.21	0.79 ± 0.23	0.45 ± 0.18	0.36 ± 0.17
0.93 ± 0.43	0.87 ± 0.33	0.72 ± 0.18	1.95 ± 1.68	1.62 ± 1.23	0.45 ± 0.21	0.56 ± 0.31
5.88 ± 1.94	6.78 ± 1.44	6.51 ± 1.38	9.01 ± 1.84	5.59 ± 1.41	5.16 ± 1.27	4.71 ± 1.49
5.61 ± 1.76	5.01 ± 1.88	4.13 ± 2.08	4.13 ± 1.30	4.12 ± 1.93	4.65 ± 1.80	4.10 ± 1.09
1.54 ± 0.74	0.64 ± 0.39	0.55 ± 0.33	1.23 ± 0.70	1.16 ± 0.74	1.11 ± 1.08	1.41 ± 0.99
7.19 ± 1.57	4.43 ± 1.47	3.11 ± 0.83	1.98 ± 0.70	6.09 ± 1.37	1.40 ± 0.32	4.80 ± 0.98
0.73 ± 0.38	0.26 ± 0.15	0.34 ± 0.21	0.64 ± 0.32	0.80 ± 0.33	1.05 ± 0.31	1.37 ± 0.47
0.05 ± 0.06	tr	0.01 ± 0.01	0.05 ± 0.05	0.11 ± 0.09	0.13 ± 0.15	0.11 ± 0.12

is relatively high. These discrepancies are not explained for homocarnosine by regional differences in activities of its synthesizing enzyme, homo-carnosine–carnosine synthetase [L-Histidine : β-alanine ligase (AMP), EC 6.3.2.11], or of its degrading enzyme, homocarnosinase.[56]

Fine mapping of the distribution of GABA in the human substantia nigra has been carried out by Kanazawa and his co-workers[57,58] who found GABA content highest in the zona reticulata at the sites of striatonigral nerve terminals. The distribution of GABA between right and left sides of autopsied human brain has been explored in a number of regions, and no lateral asymmetry in GABA contents was found.[59]

6.2. Regional Distribution of Brain Amino Acids in Laboratory Animals

A number of investigators have measured differences in the regional distribution of amino acids in rat brain.[2,3,23,60–62] Studies of the differences in contents of known and putative neurotransmitter compounds have been especially detailed for the different layers of the cerebellum, where efforts have been made to identify neurotransmitter candidates for the different types of cerebellar cortical neurons.[63–65] Table IV presents the mean contents of five known or possible neurotransmitter amino acids in various regions of rat brain. The distribution of the same compounds is shown for five areas of the cat's brain in Table V. We froze these biopsy specimens immediately after surgical removal, and values are probably similar to those prevailing during life. Other investigators have carefully explored the regional distribution of neurotransmitter amino acids in the spinal cord of the cat.[6,66]

GABA content is very high in the substantia nigra of the rat and relatively high in the nucleus accumbens and olfactory tubercle, thus resembling the regional distribution of GABA in human brain. However, in contrast to human

Table IV

Regional Distribution of Some Amino Acids in Rat Brain[a]

Compound	Cerebral cortex	Cerebellum[b]	Striatum[c]	Mesolimbic[d] area	Substantia nigra	Thalamus	Hypothalamus	Pons–medulla
Taurine	7.20[e]	5.88	8.05	6.29	—	—	—	2.40[e]
Aspartic acid	—	2.14	2.13	2.63	—	—	—	—
Glutamic acid	—	11.53	11.86	11.73	—	—	—	—
Glycine	—	0.67	—	0.63	—	—	—	—
γ-Aminobutyric acid	1.25[f]	1.32	1.73	2.48	4.26[f]	2.30[f]	3.76[f]	1.54[f]

[a] Mean contents are expressed in μmol/g wet wt.
[b] From ref. 61.
[c] Unpublished data.
[d] Includes nucleus accumbens and olfactory tubercle.[62]
[e] From ref. 60.
[f] From ref. 2.

Table V
Regional Distribution of Some Amino Acids in Cat Brain[a]

Compound	Association cortex[b]	Motor cortex[b]	Visual cortex[b]	Cerebellar cortex[c]	Caudate nucleus[b]
Taurine	1.31	1.33	1.65	3.00	2.03
Aspartic acid	1.54	1.55	1.75	2.23	1.19
Glutamic acid	6.00	5.50	5.91	9.61	8.55
Glycine	0.62	0.54	0.63	0.53	0.59
γ-Aminobutyric acid	0.88	0.83	1.01	0.86	1.29

[a] Mean contents are expressed in μmol/g wet wt.
[b] Ref. 40.
[c] Unpublished data.

brain, taurine content is not exceptionally high in the cerebellum of the rat. The regional distribution of known or putative neurotransmitter amino acids in cat brain is like that of human brain in that taurine content is high in the cerebellar cortex, aspartate content is low in the caudate nucleus, and glutamate content is relatively high in the cerebellar cortex and caudate nucleus.

7. CHANGES IN BRAIN AMINO ACID CONTENT WITH INCREASING AGE

7.1. Changes with Age in Laboratory Mammals

Comparison of the free amino acid contents of whole brain in newborn and adult laboratory mammals shows a very similar pattern of changes in all species studied. Quantitation of a wide variety of amino acids in newborn and adult brain has been reported for the mouse,[67] rabbit,[32] guinea pig,[68] cat,[42] and dog.[42] In these species, the contents of taurine and phosphoethanolamine both drop as the animal matures, whereas the contents of glycerophosphoethanolamine, aspartic acid, glutamic acid, and GABA all rise in brain as the animal matures. Many of the amino acids that are components of protein decrease in content in adult brain. Glutamine and cystathionine contents tend to be unchanged in the brains of newborn and adult animals. More recently conducted studies have confirmed these differences which were recorded in the first edition of the *Handbook of Neurochemistry*.[42] The GABA content in adult mouse brain is more than double that found in newborn mouse brain.[25] The mean taurine content of rat brain drops from 16.55 μmol/g wet wt. in newborn pups to 4.24 μmol/g wet wt. in adults.[46] Equally marked decreases in taurine content between neonates and adults have been reported for many brain regions in the rhesus monkey.[12] Sturman and his colleagues[69] have found in all species studied that the taurine content of brain is greater in the newborn animal than it is in the mature animal and that taurine content decreases gradually to attain adult values at about the time of weaning. They speculate that high taurine levels are retained in brain until synaptogenesis is complete. Putreanine content has been shown to increase in the brain of rhesus monkeys in parallel with increases in the contents of putrescine and spermidine.[70]

7.2. Changes with Age in Human Brain

Table VI shows the mean contents of ten amino compounds in four regions of autopsied brain from infants aged less than 14 months and from adults examined in our laboratory. Some of the infants suffered inherited metabolic disorders, but not ones likely to have altered the compounds listed. The adult patients died without neurological or psychiatric disease. As described in Section 3.2, the amino compounds listed are either not altered in content for substantial periods after death or, in the case of GABA, rapidly reach stable maximum levels in postmortem brain. It is therefore reasonable to compare amino acid contents between corresponding regions of infant and adult brain collected at autopsy.

Although the number of infants examined is small, and their ages at death (2 days to 14 months) too varied to obtain highly reliable mean values, it is nevertheless possible to see (Table VI) variations with age in human brain comparable to those seen in laboratory mammals. Taurine and phosphoethanolamine contents are much higher in the brain of infants than in adults. Hypotaurine, either undetectable or present only in trace amounts in adults, is frequently present in measureable quantities in the brain of infants. This is probably a reflection of a relatively rapid biosynthesis of taurine within the brain early in life. Glycerophosphoethanolamine content is relatively low in infant brain. This may well reflect the incomplete myelination of brain early in life. Contents of aspartic acid, glutamic acid, and homocarnosine all increase in human brain after infancy. On the other hand, we have found GABA content as high in infant brain as in adults. Glutamine content is similar in infant and adult human brain, whereas cystathionine content is greater in the cerebral cortex of adults than in infants.

7.3. Changes in Brain GABA Content in Older Patients

McGeer and McGeer[71] have found in many regions of autopsied human brain a decrease in activity of glutamic acid decarboxylase (GAD), the enzyme that synthesizes GABA, which correlates with advancing age. In general, thalamic and cortical areas showed the greatest declines in GAD activity, whereas the basal ganglia showed relatively less decline with age. These workers suggest that there are losses in GABAergic neuronal activity in normal aging.[71]

Since quantitation of GABA content in autopsied brain is potentially important in studies of several human neurological or psychiatric disorders, it is important to know whether or not there is a decrease in brain GABA content with aging that matches the reported drop in GAD activity. When we plotted GABA content against age at death in control autopsied brains, linear regression analysis indicated a significant decline with advancing age for the thalamus ($P < 0.01$) and for the frontal cortex ($P < 0.05$) but not for the occipital cortex, cerebellar cortex, caudate nucleus, substantia nigra, or nucleus accumbens.

Table VI

Amino Compounds in Infant and Adult Postmortem Human Brain[a]

Compound	Frontal cortex		Occipital cortex		Cerebellar cortex		Caudate nucleus	
	Infant[b] (16)	Adult (25)	Infant[b] (12)	Adult (24)	Infant[b] (12)	Adult (25)	Infant[b] (9)	Adult (30)
Glycerophosphoethanolamine	0.33 ± 0.27	0.84 ± 0.33	0.24 ± 0.06	0.76 ± 0.36	0.19 ± 0.12	0.99 ± 0.41	0.35 ± 0.32	0.99 ± 0.40
Taurine	2.79 ± 1.21	1.04 ± 0.38	2.94 ± 1.51	0.95 ± 0.41	5.15 ± 1.38	2.66 ± 1.03	2.59 ± 0.42	1.22 ± 0.45
Phosphoethanolamine	3.92 ± 2.13	1.83 ± 0.51	3.91 ± 2.16	1.08 ± 0.43	2.93 ± 1.08	1.69 ± 0.45	3.22 ± 1.59	1.45 ± 0.51
Hypotaurine	0.10 ± 0.12	tr	0.12 ± 0.13	tr	0.12 ± 0.15	tr	0.07 ± 0.08	tr
Aspartic acid[c]	0.65 ± 0.57	1.38 ± 0.54	0.61 ± 0.23	1.92 ± 1.68	0.72 ± 0.63	1.36 ± 0.99	0.47 ± 0.37	0.81 ± 0.46
Glutamic acid	5.39 ± 3.91	8.36 ± 1.60	5.82 ± 3.91	8.45 ± 1.23	6.61 ± 4.33	9.03 ± 1.66	5.37 ± 3.69	10.34 ± 1.82
Glutamine	4.19 ± 2.67	4.50 ± 1.36	5.04 ± 1.12	4.90 ± 2.83	6.69 ± 2.24	5.71 ± 1.54	4.16 ± 2.16	4.35 ± 2.20
Cystathionine	0.34 ± 0.24	0.55 ± 0.53	0.47 ± 0.44	1.65 ± 1.25	0.84 ± 0.68	0.51 ± 0.61	0.80 ± 0.43	0.73 ± 0.44
γ-Aminobutyric acid	1.89 ± 0.92	1.69 ± 0.45	2.12 ± 0.97	1.87 ± 0.53	1.65 ± 0.71	1.66 ± 0.43	1.68 ± 1.25	2.88 ± 0.83
Homocarnosine	0.04 ± 0.06	0.28 ± 0.18	0.06 ± 0.07	0.42 ± 0.07	0.24 ± 0.24	0.59 ± 0.33	0.08 ± 0.07	0.22 ± 0.19

[a] Values (mean ± S.D.) are expressed in μmol/g wet wt. Number of brains examined is shown in parentheses. tr = trace.
[b] Infants were all aged under 14 months. Several had inherited metabolic disorders, but not ones involving the amino compounds listed.
[c] Aspartic acid contents are shown only for brains with a death-to-freezing interval of 4 h or less. Number of brains examined is less than indicated for other compounds.

8. EFFECTS OF DRUGS ON BRAIN AMINO ACIDS

8.1. Drug Effects on the GABA System

A large variety of chemical agents and drugs have been found to elevate brain GABA content in experimental animals as a result of their acting as inhibitors of the enzyme γ-aminobutyrate aminotransferase (4-aminobutyrate : 2-oxoglutarate aminotransferase, E.C. 2.6.1.19) (GABA-T). Many of these drugs also act as enzyme inhibitors for GAD, but their effect is weaker on the synthesizing enzyme than on the degrading enzyme, GABA-T, and their net effect is to produce an increase in brain GABA content. Compounds that have been used to elevate brain GABA content include hydroxylamine,[19] hydrazine,[72] pargyline,[73] phenelzine,[74,75] L-cycloserine,[76] ethanolamine O-sulfate,[77,78] isoniazid,[75,79] aminooxyacetic acid,[1,17,79,80] γ-acetylenic GABA,[26,81] γ-vinyl GABA,[33,39] gabaculine, and isogabaculine.[82] Many of these drugs also increase β-alanine content in brain, since β-alanine is degraded by GABA-T, and they increase the content of homocarnosine in brain, presumably as a result of increasing GABA, one of the substrates of homocarnosine–carnosine synthetase.[56]

Sodium dipropylacetate (sodium valproate) increases brain GABA content briefly in experimental animals when injected intraperitoneally in large doses.[17,83,84] However, this widely used anticonvulsant drug is an inhibitor of succinic semialdehyde dehydrogenase (succinate semialdehyde : NAD(P)$^+$ oxidoreductase, E.C. 1.2.1.16) (SSADH), the second enzyme in the degradative pathway for GABA, rather than being a GABA-T inhibitor.[17,85] When sodium valproate is given to experimental animals orally in doses much higher than those customarily used in treating human epilepsy, brain GABA content is not elevated.[80] Morphine has been reported to increase GAD enzyme activity and to elevate GABA content in the thalamus and spinal cord of rats.[86]

Several chemical substances exert their neurotoxicity by causing a marked inhibition of GAD activity, with a resultant lowering of brain GABA content. These include 3-mercaptopropionic acid,[3] and allylglycine and its metabolite 2-keto-4-pentenoic acid.[87–90] Although many studies describe elevations or decreases in brain GABA content in animals produced by drugs, I am not aware of any observations of such drug effects having been observed yet in biopsy or autopsy specimens of human brain.

8.2. Drug Effects on Other Brain Amino Acids

Whenever measurement of a wide variety of amino compounds is carried out using amino acid analyzer techniques, one is apt to find drug effects on amino compounds that were not anticipated. Chronic administration of the GABA-T inhibitor, γ-vinyl GABA, not only produces the expected rise in contents of GABA, β-alanine, and homocarnosine in rat brain but also markedly increases hypotaurine content and significantly reduces contents of threonine and glutamine.[39] Presumably, γ-vinyl GABA inhibits enzymes other than GABA-T, among them hypotaurine dehydrogenase (hypotaurine : NAD$^+$ oxidoreductase, E.C. 1.8.1.3). Hydrazine administered chronically to rats not only increases GABA and β-alanine content in brain but also significantly raises the contents of hypotaurine, α-aminoadipic acid, alanine, cystathionine, tyrosine, and ornithine, and it lowers brain glutamine content.[72] Besides inhibiting GABA-T, hydrazine probably also inhibits hypotaurine dehydrogenase, ornithine-oxo-acid aminotransferase (E.C. 2.6.1.13), tyrosine aminotransferase (E.C. 2.6.1.5), alanine-oxo-acid aminotransferase (E.C. 2.6.1.12), and 2-aminoadipate aminotransferase (E.C. 2.6.1.39).

Administration of the neurotoxin 3-acetylpyridine to rats causes a significant decrease in taurine content in the cerebellum, probably as a result of decreased oxidation of cysteine to cysteine sulfinic acid in the synthetic pathway from cysteine to taurine. NAD^+ is a cofactor for cysteine dioxygenase (E.C. 1.13.11.20), and 3-acetylpyridine-NAD (formed after 3-acetylpyridine treatment) is ineffective as a cofactor.[61] When sodium valproate is administered in large doses intraperitoneally to rats, there is, besides a rise in brain GABA content, an unexpected marked decrease in brain aspartic acid content.[80]

Finally, some preliminary information is available as to the effects of several drugs commonly used in medicine on brain amino acids. Lloyd and Hornykiewicz[23] found that chronic administration of haloperidol or clozapine to rats did not alter GABA content in the substantia nigra, although acute administration of these drugs lowered GABA content. Chronic administration of L-DOPA did not affect GABA content in substantia nigra, although its acute administration raised GABA content in this area. We found that no changes were produced in the contents of GABA or 11 other amino compounds in the mesolimbic system of rat brain after chronic administration of the antipsychotic drugs chlorpromazine and haloperidol.[62] Some commonly used anticonvulsant drugs may interfere with the urea cycle and may sometimes produce ammonia intoxication and a rise in glutamine content in brain. There is preliminary evidence that phenobarbital and primidone,[91] as well as sodium valproate,[92] may have such an effect.

9. ABNORMALITIES OF BRAIN AMINO COMPOUNDS IN HUMAN DISEASE

9.1. Genetically Determined Enzyme Deficiencies

Abnormal contents of free amino acids and related compounds have been observed in a number of genetically determined disorders that are inherited in autosomal recessive or X-linked fashion. Each of these diseases involves an enzyme deficiency that in many cases has been clearly defined, although in a few disorders, such as hereditary tyrosinemia, the basic enzymatic error is poorly understood. As a result of the enzyme deficiency, which may occur in liver and other extraneural tissues, in brain alone, or in both, certain amino compounds either accumulate or are deficient in brain. These biochemical abnormalities in brain may in turn produce the neurological symptoms, mental deficiency, or psychiatric disturbances common in these disorders.

Phenylketonuria has long been a model of these hereditary disorders. What is remarkable is how little information is available about the derangement of amino acids in brain in untreated phenylketonuria despite the intense research interest in this disease during the last three decades and despite the fact that older retarded patients with untreated phenylketonuria die regularly in institutions. The only study I have been able to find is one by McKean and Peterson[93] in which only six amino compounds were measured in the autopsied brains of three patients. As would be expected, phenylalanine contents were elevated, and tyrosine contents were relatively low.

Other well-studied hereditary metabolic disorders for which surprising little information is available as to amino acid changes in brain are maple syrup urine disease and homocystinuria. Prensky and Moser[94] found valine, isoleucine, and leucine contents elevated in the brain of a single patient with maple syrup urine disease. In the most common form of homocystinuria, that is caused by a deficiency of the enzyme cystathionine β-synthase (E.C. 4.2.1.22), methionine content has been found modestly elevated and cystathionine content greatly reduced in autopsied brain, whereas homocystine was not detectable.[95,96] In hereditary tyrosinemia, a disorder in which both tyrosine and methionine concentrations are elevated in physiological fluids but in which the primary enzymatic failure may not lie in the metabolic pathway of either amino acid, methionine content is elevated in autopsied brain,[97] and we have recently found (T. L. Perry, unpublished observation) that both methionine and tyrosine contents are considerably elevated in brain. β-Alanine content was markedly elevated in the autopsied brain of the only reported patient with hyper-β-alaninemia,[98] and sarcosine was readily measured in the brain of a patient who died with hypersarcosinemia,[99] although this amino acid is normally undetectable in human brain.

Amino acid contents have been measured in autopsied brain in two of the five genetically determined disorders of the urea cycle.[100] In both ornithine transcarbamylase (E.C. 2.1.3.3) deficiency[100] and in argininosuccinase (E.C. 4.3.2.1) deficiency (argininosuccinic aciduria),[100,101] glutamine content is greatly elevated in autopsied brain. This is a result of the failure to detoxify ammonia through the urea cycle both in liver and in brain. Glutamate and glutamine contents rise in brain as a result of compensatory reactions between ammonia and α-ketoglutarate and then between ammonia and glutamate. Elevations in the contents of alanine and of α-amino-*n*-butyric acid in brain also occur, presumably as a result of similar reactions between ammonia and pyruvate or α-ketobutyrate.[101] Citrulline content in brain is decreased in ornithine transcarbamylase deficiency[100] and is increased in argininosuccinic aciduria.[101] Finally, argininosuccinic acid is found in very large amounts in autopsied brain of patients dying with argininosuccinic aciduria, although it is undetectable in control human brain.[101] It is likely that the appropriate biochemical abnormalities will eventually be found in brain from patients with the three remaining inherited disorders of the urea cycle. In addition, high glutamine content is routinely found in brain of patients dying with hepatic encephalopathy secondary to nongenetic forms of liver disease.

Two disorders inherited as autosomal recessives are characterized by abnormalities in the brain contents of oligopeptides. In pyroglutamic acidemia (5-oxoprolinemia), there is a fluctuating and often very marked deficiency in the activity of glutathione synthetase (E.C. 6.3.2.3), which results in a greatly reduced GSH content in many tissues.[102] We recently found, in a brain biopsy from a patient with pyroglutamic acidemia, that the GSH content of frontal cortex was only about one-third the normal level.[103] In homocarnosinosis,[104] a rare disorder in which homocarnosine concentrations are greatly elevated in CSF, we have also shown by examination of a brain biopsy that homocarnosine content of the frontal cortex was increased fourfold and that activity of the degrading enzyme homocarnosinase was absent.[105]

In glycine encephalopathy (nonketotic hyperglycinemia), an often fatal neurological disorder of young children that is inherited as an autosomal recessive, there is a marked accumulation of glycine in the brain.[106] In several other disorders, the ketotic hyperglycinemias, there may be marked increases in plasma glycine concentrations, although glycine is not elevated in the CSF or in brain. The accumulation of glycine in brain in glycine encephalopathy results from a complete absence of activity of the glycine cleavage enzyme complex in brain,[107] and the excessive glycine content itself may be responsible for the disrupted neuronal function in this disorder. Table VII shows the glycine contents of several brain regions from seven infants who died with glycine encephalopathy.

9.2. Brain Amino Compounds in Autosomal Dominant Disorders

Huntington's chorea was the first dominantly inherited neurological disease in which abnormalities were discovered in brain amino compounds at

Table VII
Glycine Content of Autopsied Brain in Seven Infants with Glycine Encephalopathy[a]

Subjects	Frontal cortex	Occipital cortex	Cerebellar cortex	Caudate nucleus	Putamen–globus pallidus	Cervical cord
Control infants	1.56 ± 0.45	1.85 ± 0.46	2.15 ± 0.96	1.67 ± 0.62	1.74 ± 0.66	2.49 ± 0.60
Patient 1, 10 months	4.51	3.55	10.05	—	7.43	8.90
Patient 2, 8 months	4.84	4.24	8.13	—	8.40	8.33
Patient 3, 13 days	11.29	14.24	17.66	13.07	12.30	14.66
Patient 4, 1 month	4.84	—	—	—	—	—
Patient 5, 18 days	2.92	—	—	—	—	—
Patient 6, 4 months	2.95	3.25	1.80	6.13	2.97	8.28
Patient 7, 14 days	5.60	6.63	15.05	9.27	10.27	11.43

[a] Glycine content is expressed in μmol/g wet wt., with mean ± S.D. given for control infants (aged 2 days to 14 months). Age at death is indicated for each glycine encephalopathy patient.

death. The GABA content has been shown to be significantly reduced in the basal ganglia of Huntington's chorea patients,[108,110] and this is accompanied by a decrease in homocarnosine content and an increase in glycerophosphoethanolamine content.[108-110] Table VIII shows the mean GABA contents of several regions of autopsied brain from a large number of adult control subjects and Huntington's chorea patients examined in my laboratory. There is a significant decrease in GABA content in the caudate nucleus, putamen, globus pallidus, substantia nigra, and occipital cortex in Huntington's chorea. It is believed that the low GABA content of these brain areas and a corresponding decrease in GAD enzyme activity[111] both reflect the loss of a population of GABAergic neurons. Why these neurons, as well as certain groups of neurons that utilize acetylcholine[109] or substance P[112] as their neurotransmitters, die prematurely in Huntington's chorea is still unknown.

We have recently discovered similar complex deficiencies of aspartic acid, glutamic acid, or GABA content in the cerebellar cortex, the dentate nucleus, and other areas of the brain in patients with dominantly inherited olivopontocerebellar atrophies.[113] These changes may well represent losses of particular populations of cerebellar neurons whose neurotransmitters are aspartate, glutamate, or GABA. At least three distinct disorders, with different biochemical characteristics, are represented among the dominantly inherited cerebellar diseases.[113]

9.3. Amino Acid Changes in Other Neuropsychiatric Diseases

Temporal lobe epilepsy presents an excellent opportunity for study of possible amino acid abnormalities in brain, since drug-resistant patients often require neurosurgical removal of active epileptogenic foci that have been carefully mapped by electrocorticography. Glutamic acid content is markedly elevated in some of these excised foci that have been instantly frozen in liquid nitrogen.[53] The mean glutamate content of epileptogenic foci removed from temporal and frontal cortex (Table II) is significantly greater ($P < 0.001$) than the mean glutamate content of biopsies from the same cortical regions in nonepileptic patients. Since glutamic acid is an excitatory neurotransmitter, the

Table VIII
GABA Content of Autopsied Brain in Huntington's Chorea[a]

	Caudate nucleus	Putamen	Globus pallidus	Substantia nigra	Occipital cortex
Control subjects	2.88 ± 0.15 (30)	2.80 ± 0.20 (16)	7.19 ± 0.41 (15)	6.09 ± 0.26 (27)	1.87 ± 0.11 (24)
Huntington's chorea patients	1.40 ± 0.17[b] (22)	1.38 ± 0.14[b] (7)	2.54 ± 0.31[b] (8)	2.51 ± 0.17[b] (22)	1.40 ± 0.09[c] (21)

[a] Values (mean ± S.E.M.) are expressed in μmol/g wet wt. Numbers of subjects examined for each region are shown in parentheses.
[b] $P < 0.001$.
[c] $P < 0.005$.

possibility exists that some cases of focal epilepsy may result from excessive glutamatergic neuronal activity.

When we examined autopsied brain from patients who died with a diagnosis of schizophrenia, we found that some of these patients had an unusually low GABA content in the nucleus accumbens and in the medial dorsal nucleus of the thalamus.[114] The deficiency in GABA content in these two brain regions could not be readily explained by the ages of the schizophrenic patients at death, by their immediate causes of death, or by the antipsychotic drug treatment they had received. The finding of a significantly reduced GABA content in schizophrenia, especially in the nucleus accumbens, raises the interesting possibility that some forms of schizophrenia involve an imbalance in the function of certain GABAergic neurons in a complex interacting system involving dopaminergic, cholinergic, GABAergic, and other neurons, as originally suggested by Roberts.[115]

The examples cited above of abnormalities in contents of amino compounds found in biopsy or autopsy specimens of brain from patients suffering from neurological or mental diseases should make it clear that much can be learned from careful biochemical studies of human brain. Expansion of such studies can not only help the neurochemist to understand the roles of a great variety of chemical substances in normal brain function, but it should in time lead to improved methods of preventing or treating important diseases of the brain that cause much human suffering.

ACKNOWLEDGMENTS. The original investigations reported in this review were supported by grants from the Medical Research Council of Canada. I am very much indebted to Mrs. Shirley Hansen, Mrs. Maureen Murphy, and Mrs. Janet MacLean for expert technical assistance over many years and for the preparation of the tables and figures for this chapter.

REFERENCES

1. Kuriyama, K., Roberts, E., and Rubinstein, M. K., 1966, *Biochem. Pharmacol.* **15**:221–236.
2. Balcom, G. J., Lenox, R. H., and Meyerhoff, J. L., 1975, *J. Neurochem.* **24**:609–613.
3. van der Heyden, J. A. M., and Korf, J., 1978, *J. Neurochem.* **31**:197–203.
4. Tallan, H. H., Moore, S., and Stein, W. H., 1956, *J. Biol. Chem.* **219**:257–264.
5. Tews, J. K., Carter, S. H. Roa, P. D., and Stone, W. E., 1963, *J. Neurochem.* **10**:641–653.
6. Duggan, A. W., and Johnston, G. A. R., 1970, *J. Neurochem.* **17**:1205–1208.
7. Tachiki, K. H., and Baxter, C. F., 1979, *J. Neurochem.* **33**:1125–1129.
8. Shank, R. P., and Aprison, M. H., 1970, *Anal. Biochem.* **35**:136–145.
9. Aprison, M. H., McBride, W. J., and Freeman, A. R., 1973, *J. Neurochem.* **21**:87–95.
10. Smith, J. E., Lane, J. D., Shea, P. A., McBride, W. J., and Aprison, M. H., 1975, *Anal. Biochem.* **64**:149–169.
11. Sturman, J. A., and Gaull, G. E., 1975, *J. Neurochem.* **25**:831–835.
12. Sturman, J. A., Rassin, D. K., Gaull, G. E., and Cote, L. J., 1980, *J. Neurochem.* **35**:304–310.
13. Perry, T. L., Berry, K., Hansen, S., Diamond, S., and Mok, C., 1971, *J. Neurochem.* **18**:513–519.
14. Perry, T. L., Hansen, S., Berry, K., Mok, C., and Lesk, D., 1971, *J. Neurochem.* **18**:521–528.
15. Starr, M. S., 1975, *J. Neurochem.* **24**:1229–1236.
16. Joseph, M. H., and Halliday, J. A., 1975, *Anal. Biochem.* **64**:389–402.

17. Emson, P. C., 1976, *J. Neurochem.* **27**:1489–1494.
18. Hirsch, H. E., and Robins, E., 1962, *J. Neurochem.* **9**:63–70.
19. Baxter, C. F., and Roberts, E., 1959, *Proc. Soc. Exp. Biol. Med.* **101**:811–815.
20. Kravitz, E. A., and Potter, D. D., 1965, *J. Neurochem.* **12**:323–328.
21. Graham, L. T., Jr., and Aprison, M. H., 1966, *Anal. Biochem.* **15**:487–497.
22. Enna, S. J., and Snyder, S. H., 1976, *J. Neurochem.* **26**:221–224.
23. Lloyd, K. G., and Hornykiewicz, O., 1977, *Life Sci.* **21**:1489–1496.
24. Ferkany, J. W., Butler, I. J., and Enna, S. J., 1979, *J. Neurochem.* **33**:29–33.
25. Levi, G., Amaldi, P., and Morisi, G., 1972, *Brain Res.* **41**:435–451.
26. Jung, M. J., Lippert, B., Metcalf, B. W., Schechter, P. J., Böhlen, P., and Sjoerdsma, A., 1977, *J. Neurochem.* **28**:717–723.
27. Schechter, P. J., Tranier, Y., Jung, M. J., and Böhlen, P., 1977, *Eur. J. Pharmacol.* **45**:319–328.
28. Perry, T. L., Stedman, D., and Hansen, S., 1968, *J. Chromatogr.* **38**:460–466.
29. Benson, J. V., Jr., Gordon, M. J., and Patterson, J. A., 1967, *Anal. Biochem.* **18**:228–240.
30. Benson, J. V., Jr., 1972, *Anal. Biochem.* **50**:477–493.
31. Benson, J. R., and Hare, P. E., 1975, *Proc. Natl. Acad. Sci. U.S.A.* **72**:619–622.
32. Agrawal, H. C., Davis, J. M., and Himwich, W. A., 1966–1967, *Brain Res.* **3**:374–380.
33. Jung, M. J., Lippert, B., Metcalf, B. W., Böhlen, P., and Schechter, P. J., 1977, *J. Neurochem.* **29**:797–802.
34. Minard, F. N., and Mushahwar, I. K., 1966, *Life Sci.* **5**:1409–1413.
35. Shank, R. P., and Aprison, M. H., 1971, *J. Neurobiol.* **2**:145–151.
36. Lajtha, A., and Toth, J., 1974, *Brain Res.* **76**:546–551.
37. Perry, T. L., Hansen, S., and Gandham, S. S., 1981, *J. Neurochem.* **36**:406–412.
38. Spokes, E. G. S., and Koch, D. J., 1978, *J. Neurochem.* **31**:381–383.
39. Perry, T. L., Kish, S. J., and Hansen, S., 1979, *J. Neurochem.* **32**:1641–1645.
40. Perry, T. L., Sanders, H. D., Hansen, S., Lesk, D., Kloster, M., and Gravlin, L., 1972, *J. Neurochem.* **19**:2651–2656.
41. Hansen, S., Perry, T. L., Wada, J. A., and Sokol, M., 1973, *Brain Res.* **50**:480–483.
42. Himwich, W. A., and Agrawal, H. C., 1969, *Handbook of Neurochemistry, Volume 1: Chemical Architecture of the Nervous System,* 1st ed. (A. Lajtha, ed.), Plenum Press, New York, pp. 33–52.
43. Peck, E. J., Jr., and Awapara, J., 1967, *Biochim. Biophys. Acta* **141**:499–506.
44. Perry, T. L., and Hansen, S., 1973, *J. Neurochem.* **21**:1009–1011.
45. Kakimoto, Y., Nakajima, T., Kumon, A., Matsuoka, Y., Imaoka, N., Sano, I., and Kanazawa, A., 1969, *J. Biol. Chem.* **244**:6003–6007.
46. Sturman, J. A., Rassin, D. K., and Gaull, G. E., 1977, *J. Neurochem.* **28**:31–39.
47. Hayes, K. C.; Carey, R. E., and Schmidt, S. Y., 1975, *Science* **188**:949–951.
48. Schmidt, S. Y., Berson, E. L., and Hayes, K. C., 1976, *Invest. Ophthalmol.* **15**:52–58.
49. Sturman, J. A., Rassin, D. K., and Gaull, G. E., 1978, *Taurine and Neurological Disorders* (A. Barbeau and R. J. Huxtable, eds.), Raven Press, New York, pp. 49–71.
50. Margolis, F. L., 1974, *Science* **184**:909–911.
51. Neidle, A., and Kandera, J., 1974, *Brain Res.* **80**:359–364.
52. Harding, J., and Margolis, F. L., 1976, *Brain Res.* **110**:351–360.
53. Perry, T. L., and Hansen, S., 1981, *Neurology (Minneap.)* **31**:872–876.
54. Perry, T. L., Hansen, S., and Kloster, M., 1972, *J. Neurochem.* **19**:1395–1396.
55. Perry, T. L., Hansen, S., Schier, G. M., and Halpern, B., 1977, *J. Neurochem.* **29**:791–795.
56. Kish, S. J., Perry, T. L., and Hansen, S., 1979, *J. Neurochem.* **32**:1629–1636.
57. Kanazawa, I., Miyata, Y., Toyokura, Y., and Otsuka, M., 1973, *Brain Res.* **51**:363–365.
58. Kanazawa, I., and Toyokura, Y., 1975, *Brain Res.* **100**:371–381.
59. Rossor, M., Garrett, N., and Iversen, L., 1980, *J. Neurochem.* **35**:743–745.
60. Lombardini, J. B., 1978, *Taurine and Neurological Disorders* (A. Barbeau and R. J. Huxtable, eds.), Raven Press, New York, pp. 119–135.
61. Perry, T. L., MacLean, J., Perry, T. L., Jr., and Hansen, S., 1976, *Brain Res.* **109**:632–635.
62. Perry, T. L., Hansen, S., and Kish, S. J., 1979, *Life Sci.* **24**:283–288.
63. McBride, W. J., Nadi, N. S., Altman, J., and Aprison, M. H., 1976, *Neurochem. Res.* **1**:141–152.

64. Nadi, N. S., McBride, W. J., and Aprison, M. H., 1977, *J. Neurochem.* **28**:453–455.
65. McBride, W. J., Rea, M. A., Fetten, D. L., and Rohde, B. H., 1980, *Neurochem. Res.* **5**:337–344.
66. Rizzoli, A. A., 1968, *Brain Res.* **11**:11–18.
67. Agrawal, H. C., Davis, J. M., and Himwich, W. A., 1968, *J. Neurochem.* **15**:917–923.
68. Agrawal, H. C., Davis, J. M., and Himwich, W. A., 1968, *J. Neurochem.* **15**:529–531.
69. Sturman, J. A., Rassin, D. K., and Gaull, G. E., 1977, *Life Sci.* **21**:1–22.
70. Kremzner, L. T., and Sturman, J. A., 1979, *J. Neurochem.* **33**:1115–1117.
71. McGeer, E. G., and McGeer, P. L., 1979, *GABA-Neurotransmitters: Pharmacochemical, Biochemical and Pharmacological Aspects* (P. Krogsgaard-Larsen, J. Scheel-Krüger, and H. Kofod, eds.), Munksgaard, Copenhagen, pp. 340–356.
72. Perry, T. L., Kish, S. J., Hansen, S., Wright, J. M., Wall, R. A., Dunn, W. L., and Bellward, G. D., 1981, *J. Neurochem.* **37**:32–39.
73. Schatz, R. A., and Lal, H., 1971, *J. Neurochem.* **18**:2553–2555.
74. Popov, N., and Matthies, H., 1969, *J. Neurochem.* **16**:899–907.
75. Perry, T. L., and Hansen, S., 1973, *J. Neurochem.* **21**:1167–1175.
76. Wood, J. D., Peesker, S. J., Gorecki, D. K. J., and Tsui, D., 1978, *Can. J. Physiol. Pharmacol.* **56**:62–68.
77. Fowler, L. J., 1973, *J. Neurochem.* **21**:437–440.
78. Fletcher, A., and Fowler, L. J., 1980, *Biochem. Pharmacol.* **29**:1451–1454.
79. Perry, T. L., Urquhart, N., Hansen, S., and Kennedy, J., 1974, *J. Neurochem.* **23**:443–445.
80. Perry, T. L., and Hansen, S., 1978, *J. Neurochem.* **30**:679–684.
81. Schechter, P. J., Tranier, Y., Jung, M. J., and Sjoerdsma, A., 1977, *J. Pharmacol. Exp. Ther.* **201**:606–612.
82. Schechter, P. J., Tranier, Y., and Grove, J., 1979, *Life Sci.* **24**:1173–1182.
83. Godin, Y., Heiner, L., Mark, J., and Mandel, P., 1969, *J. Neurochem.* **16**:869–873.
84. Simler, S., Ciesielski, L., Maitre, M., Randrianarisoa, H., and Mandel, P., 1973, *Biochem. Pharmacol.* **22**:1701–1708.
85. Harvey, P. K. P., Bradford, H. F., and Davison, A. N., 1975, *FEBS Lett.* **52**:251–254.
86. Kuriyama, K., and Yoneda, Y., 1978, *Brain Res.* **148**:163–179.
87. Horton, R. W., 1978, *Biochem. Pharmacol.* **27**:1471–1477.
88. Horton, R. W., Chapman, A. G., and Meldrum, B. S., 1978, *J. Neurochem.* **30**:1501–1504.
89. Reingold, D. F., and Orlowski, M., 1979, *J. Neurochem.* **32**:907–913.
90. Pajunen, A. E. I., Hietala, O. A., Baruch-Virransalo, E.-L., and Piha, R. S., 1979, *J. Neurochem.* **32**:1401–1408.
91. Perry, T. L., Hansen, S., and MacLean, J., 1976, *Clin. Chim. Acta* **69**:441–445.
92. Coutter, D. L., and Allen, R. J., 1980, *Lancet* **1**:1310–1311.
93. McKean, C. M., and Peterson, N. A., 1970, *N. Engl. J. Med.* **283**:1364–1367.
94. Prensky, A. L., and Moser, H. W., 1966, *J. Neurochem.* **13**:863–874.
95. Gerritsen, T., and Waisman, H. A., 1964, *Science* **145**:588.
96. Brenton, D. P., Cusworth, D. C., and Gaull, G. E., 1965, *Pediatrics* **35**:50–56.
97. Perry, T. L., Hardwick, D. F., Dixon, G. H., Dolman, C. L., and Hansen, S., 1965, *Pediatrics* **36**:236–250.
98. Scriver, C. R., Pueschel, S., and Davies, E., 1966, *N. Engl. J. Med.* **274**:636–643.
99. Gerritsen, T., and Waisman, H. A., 1978, *The Metabolic Basis of Inherited Disease*, 4th ed. (J. B. Stanbury, J. B. Wyngaarden, and D. S. Fredrickson, eds.), McGraw-Hill, New York, pp. 514–517.
100. Shih, V. E., 1978, *The Metabolic Basis of Inherited Disease*, 4th ed. (J. B. Stanbury, J. B. Wyngaarden, and D. S. Fredrickson, eds.), McGraw-Hill, New York, pp. 362–386.
101. Perry, T. L., Wirtz, M. L. K., Kennaway, N. G., Hsia, Y. E., Atienza, F. C., and Uemura, H. S., 1980, *Clin. Chim. Acta* **105**:257–267.
102. Marstein, S., and Perry, T. L., 1981, *Clin. Chim. Acta* **109**:13–20.
103. Marstein, S., Jellum, E., Nesbakken, R., and Perry, T. L., 1981, *Clin. Chim. Acta* **111**:219–228.
104. Gjessing, L. R., and Sjaastad, O., 1974, *Lancet* **2**:1028.
105. Perry, T. L., Kish, S. J., Sjaastad, O., Gjessing, L. R., Nesbakken, R., Schrader, H., and Løken, A. C., 1979, *J. Neurochem.* **32**:1637–1640.

106. Perry, T. L., Urquhart, N., MacLean, J., Evans, M. E., Hansen, S., Davidson, A. G. F., Applegarth, D. A., MacLeod, P. J., and Lock, J. E., 1975, *N. Engl. J. Med.* **292:**1269–1273.

107. Perry, T. L., Urquhart, N., Hansen, S., and Mamer, O. A., 1977, *Pediatr. Res.* **12:**1192–1197.

108. Perry, T. L., Hansen, S., and Kloster, M., 1973, *N. Engl. J. Med.* **288:**337–342.

109. Bird, E. D., and Iversen, L. L., 1974, *Brain* **97:**457–472.

110. Urquhart, N., Perry, T. L., Hansen, S., and Kennedy, J., 1975, *J. Neurochem.* **24:**1071–1075.

111. Bird, E. D., MacKay, A. V. P., Rayner, C. N., and Iversen, L. L., 1973, *Lancet* **1:**1090–1092.

112. Gale, J. S., Bird, E. D., Spokes, E. G., Iversen, L. L., and Jessell, T., 1978, *J. Neurochem.* **30:**633–634.

113. Perry, T. L., Kish, S. J., Hansen, S., and Currier, R. D., 1981, *Neurology (Minneap.)* **31:**237–242.

114. Perry, T. L., Kish, S. J., Buchanan, J., and Hansen, S., 1979, *Lancet* **1:**237–239.

115. Roberts, E., 1972, *Neurosci. Res. Program Bull.* **10:**468–481.

116. Perry, T. L., 1978, *Taurine and Neurological Disorders* (A. Barbeau and R. J. Huxtable, eds.), Raven Press, New York, pp. 441–451.

Neuropeptides
An Overview

M. J. Brownstein and H. Gainer

1. INTRODUCTION

In the four decades that have passed since Ernst and Berta Scharrer suggested that certain populations of central neurons secrete peptide hormones,[1] a multitude of biologically active peptides have been discovered and/or characterized in brain extracts (Table I). Some of these were first found in the brain, others in nonnervous tissue such as gut, pancreas, skin, or pituitary. When a peptide is discovered in the CNS that is present in other organs, it must be characterized. The form of the peptide in brain may be different from the form in the periphery. Furthermore, it is unsafe to assume that all peptides found in brain are made there; they may be made in the periphery and transported to and concentrated in the brain.

Many of the peptides listed in Table I have been chromatographically or chemically identified in brain extracts, but there are peptides in this table that have only been detected immunocytochemically or by means of assays that are not perfectly specific. Ideally, a biologically or immunologically reactive species should be subjected to a number of analytical procedures before it is attributed to the brain (Table II). Special care should be taken to identify modified forms of the peptide being studied, since a number of posttranslational alterations of peptides and proteins have been demonstrated: carboxymethylation, acetylation, sulfatation, phosphorylation, amidation, ADP ribosylation, and glycosylation.

2. ANATOMY OF PEPTIDERGIC NEURONS

The first step in understanding the biochemistry, physiology, and pharmacology of peptides is to investigate the anatomy of peptidergic systems. The

M. J. Brownstein • Laboratory of Clinical Science, National Institute of Mental Health, Bethesda, Maryland 20205.   *H. Gainer* • Laboratory of Developmental Neurobiology, National Institute of Child Health and Human Development, Bethesda, Maryland 20205.

Table I
Peptides Reported to Be Present in the Brain

Adrenocorticotropic hormone[2–7]	Luteinizing hormone-releasing
Angiotensin II[a,8,9]	hormone[28–30]
Bombesin[10]	α-Melanocyte-stimulating hormone[31–33]
Bradykinin[11]	Methionine enkephalin[22–25]
Carnosine[12]	Motilin[34]
Cholecystokinin[13,14]	Neurotensin[35–37]
Corticotropin-releasing hormone[b]	Oxytocin[38–41]
Dynorphin[15]	Prolactin[42,43]
β-Endorphin[16,17]	Secretin[44]
Gastrin[a,18]	Sleep peptide[45]
Glucagon[19]	Somatostatin[46–49]
Growth hormone[a,20]	Substance P[50–52]
Insulin[21]	Thyrotropin[a,20]
Leucine enkephalin[22–25]	Thyrotropin releasing hormone[53–57]
β-Lipotropin[26,27]	Vasopressin[38–41,58]
Luteinizing hormone[a,20]	Vasoactive intestinal polypeptide[59,60]

[a] Presence in brain questionable at present.
[b] W. Vale, personal communication.

development of immunohistochemical methods has allowed anatomic studies to be done rapidly at the light and electron microscopic levels. These techniques, like any others, have their shortcomings[61]; both false positives (staining of non-peptide-containing structures) and false negatives (failure to stain peptide-containing structures) crop up from time to time. False positives can be avoided by careful validation of one's methods (Table III). False negatives are more difficult to eliminate but can be reduced in number by varying one's fixation, antibody, and method of visualization. Peptidergic cell bodies, which are difficult to stain, can sometimes be visualized if animals are treated with colchicine to dam up peptides in the perikarya. It is impossible in a short chapter to describe the anatomy of all of the brain's peptidergic systems. Therefore, I shall confine my remarks to generalizations that have emerged from the anatomic studies.

With few exceptions (e.g., magnocellular hypothalamic neurons in the supraoptic and paraventricular nuclei), most peptidergic neuronal perikarya

Table II
Characterization of Peptides in Tissue Extracts

Extraction. An extraction procedure should be chosen that removes as much of the peptide from the tissue as possible and that protects the peptide against alteration.

Assays. Ideally, a variety of assays should be used: bioassays, radioimmunoassays. radioreceptor assays.

Chromatography. Several chromatographic analyses should be attempted. The size. charge, and hydrophobicity of the molecule should be established. The native peptide as well as chemical derivatives should be studied.

Chemical characterization. Amino acid analysis, N-terminal amino acid determination, "fingerprint" analysis after enzymatic cleavage, and amino acid sequencing are the ultimate tests of a peptide's identity.

Table III
Validation of Immunohistological Findings

Whenever possible affinity-purified or monoclonal antibodies should be used.

No matter what kind of antibody is employed, its structural specificity should carefully be determined.

More than one antibody should be used, and the antibodies chosen should react with different parts of the peptide under investigation.

Controls should be performed to check the specificity of the staining. The effects of omitting the first antibody, of replacing the first antibody with preimmune serum, and of adding an excess of the peptide that is being visualized as well as other related peptides should be established.

Immunohistological findings should be corroborated independently. For example, the effect of lesions on peptide levels should be studied.

---

look like their non-peptide-containing neighbors. Indeed, in some neurons, peptides and nonpeptide transmitters coexist and may be released simultaneously. Within neurons, the highest levels of peptides seem to be found in nerve terminals to which peptide-rich granules travel from the cell body (see Section 3).

Each of the peptides studied to date has a unique distribution in the nervous system. These distributions reflect the locations of peptidergic perikarya and their axons and terminal fields. These vary from species to species. For example, cholecystokinin is found principally in cells of the hypothalamus and preoptic area of frogs and in cerebral cortical neurons of mammals (J. Trubach and M. Beinfeld, unpublished data).

Based on anatomic observations, workers in a number of laboratories have suggested that peptides can play several roles. They can be used by neurons as neurotransmitters when they are released at synapses or as neurohormones when they are released into the bloodstream. They can also be used by glandular secretory cells as "paracrine" mediators that are released into the extracellular space to influence neighboring cells or as hormones that are released into the blood. Somatostatin, for instance, is released by hypothalamic neurons into the pituitary portal vessels to act as a growth hormone release-inhibiting hormone. It is also secreted by D cells in the pancreatic islets, and it acts locally to inhibit insulin and glucagon release. Finally, it serves as the neurotransmitter for neurons in the brain and gut. In the former, it may not invariably be released at well-differentiated synapses, but, like other peptides, it may sometimes be released into the extracellular space at "open synapses" to affect numerous target cells in a hormonal manner.

It should be evident that a peptide must satisfy similar criteria in order to be called a neurotransmitter, neurohormone, hormone, or paracrine mediator. These are listed in Table IV.

There is nothing about the distribution of peptides in the brain or about their mode of action that suggests that peptidergic cells form a discrete neuroendocrine system (i.e., that they act in concert to orchestrate specific endocrine or vegetative functions). They should be thought of as any other class of chemical messenger. Furthermore, it is not true that all cells that contain

Table IV
Criteria for Establishing Peptides as Neurotransmitters, Neurohormones, Hormones,
or Paracrine Mediators

The peptide must be present in the secretory tissue.
It must be synthesized there.
It must be released from cells.
It must occupy receptors on target cells and elicit a response.
There must be a mechanism to terminate the action of the peptide.
Drugs that affect the action of exogenously applied peptide must also affect the action of cells which are presumed to secrete the peptide.

peptides are of neural crest origin as Pearse originally suggested.[62] The pineal gland, anterior pituitary, and hypothalamus arise from the neuroectoderm or specialized ectodermal placodes, not from the neural crest.

3. BIOSYNTHESIS OF PEPTIDES

In theory, a small peptide could be synthesized in either of two ways. It could be made by an enzyme complex each member of which would add an amino acid to the growing chain. Glutathione and carnosine are built this way, but most neuropeptides seem to be made by ribosomes as parts of larger precursor molecules. The best characterized of these precursors is "proopiocortin," the 31,000-dalton glycoprotein that gives rise to ACTH, α-MSH, β-lipotropin, and β-endorphin.[63–66] It is unlikely that all peptide precursors are as rich in biologically active fragments as proopiocortin is. On the other hand, it may not be uncommon to find more than one copy of an active peptide per precursor molecule as in the case of the adrenal medullary enkephalin precursor.[67,68]

Three strategies have evolved for identifying and studying peptide precursor molecules. The simplest of these involves looking for "big" immunologically reactive forms of the peptide in question. Frequently, antibodies raised against small peptide species do not recognize N- or C-terminally extended forms of the peptide. When, as is often the case, the active peptide sequence is neighbored on either side by one or more basic amino acid residues, trypsin and carboxypeptidase B can be used to liberate the desired immunologically reactive molecule. This molecule can subsequently be identified by means of reverse-phase high-performance liquid chromatography.

A second method for detecting peptide precursors involves pulse labeling them with radioactive amino acids *in vitro* or *in vivo*. The advantage of this method is that the pulse-labeling paradigm allows one to follow the conversion of the putative precursor into smaller intermediates and fully processed molecules. Thus, this method provides definitive evidence of a precursor–product relationship.

Finally, precursors can be characterized by cell-free translation of their respective mRNAs or, more elegantly, by characterization of these RNA species themselves. Recent advances in the field of molecular genetics should allow rapid progress to be made in understanding the structure of peptide precursors and the genes that code for them. To date, however, rather little is known about the proneuropeptides. Those precursors about which there are some structural data include the precursors for vasopressin neurophysin,[69,70] oxytocin neurophysin,[69,70] and somatostatin.[71–74]

After the precursors are formed, they have to be broken down or processed to yield their active components. The enzymes that are responsible for processing have not been well studied. It is clear that trypsinlike enzymes must be involved in the processing of many precursors. These enzymes cleave proteins on the carboxy-terminal side of basic amino acid residues (arginine, lysine) and free those bits of the precursor surrounded by basic residues from the rest. Next a carboxypeptidase-B-like trimming enzyme should come into play. This enzyme removes basic amino acids from the carboxy-terminal end of the peptide.

$$X — B — B — P E P T I D E — B — B — X$$
$$\uparrow \quad \uparrow \qquad\qquad\qquad \uparrow \quad \uparrow \quad \uparrow$$
$$T \quad T \qquad\qquad\qquad CB \quad T \quad T$$

These two enzymes are sufficient to process peptides with unmodified termini, but other enzymes are needed in order to modify the peptides further, e.g., to amidate or acetylate them.

4. PEPTIDE RECEPTORS

Progress in studying receptors for neuropeptides has been hampered by the lack of high-affinity, high-specific-activity ligands, simple and specific bioassays, and well-characterized agonists and antagonists. The opiate receptors are better understood than other classes of peptide receptors because opiate pharmacology was in such an advanced state before the various opioid peptides were isolated. It is clear that more than one opiate receptor exists.[75] The μ receptor is found in the guinea pig ileum, the δ receptor in the mouse vas deferens. Morphine binds preferentially to μ receptors, which are thought to mediate analgesic response. The distributions of the opiate receptors (and others) in the brain have been studied by autoradiography at the light microscopic level.[76]

The peptides for which brain binding sites have been demonstrated include vasoactive intestinal polypeptide,[77] cholecystokinin,[78,79] insulin,[80] neurotensin,[81] and substance P.[82]

5. PEPTIDE DEGRADATION

After peptides are secreted onto target cells, their actions must be terminated. Unlike catecholamines and serotonin, peptides do not seem to be recaptured by the peptidergic cells that release them. Instead, they are probably degraded by peptidases. Since many of the biologically active peptides have derivatized N and C termini, endopeptidases must be involved in their metabolism. Endopeptidases have, in fact, been discovered that cleave several of the neuropeptides internally,[83] but little attention has been paid to date to showing that these enzymes are physiologically important. If specific inhibitors of the endopeptidases can be developed that are active *in vivo*, it may be possible to demonstrate the role of the enzymes convincingly. Furthermore, these inhibitors may prove useful clinically just as monoamine oxidase inhibitors have.

6. CONCLUDING REMARKS

Studies of neuropeptides are still in their infancy, but they have excited a great deal of interest in the community of neurobiologists (see refs. 84–89). The addition of the peptides to the list of putative neurotransmitters has quadrupled its size, and the work that has been done on peptides to date has contributed to new concepts of brain function and to revisions of old ones. Many questions about peptide neurochemistry remain to be answered, and these will surely be addressed in succeeding volumes of this *Handbook*.

REFERENCES

1. Scharrer, E., and Scharrer, B., 1940, *Proc. Assoc. Res. Nerv. Ment. Dis.* **20**:170–194.
2. Krieger, D. T., Liotta, A., and Brownstein, M. J., 1977, *Proc. Natl. Acad. Sci. U.S.A.* **74**:648–652.
3. Larsson, L. I., 1977, *Lancet* **8052/3**:1321–1323.
4. Nilaver, G., Zimmerman, E. A., Defendini, R., Liotta, A., Krieger, D. T., and Brownstein, M. J., 1979, *J. Cell Biol.* **81**:50–58.
5. Orwoll, E., Kendall, J. W., Lamorena, L., and McGilvra, R., 1979, *Endocrinology* **104**:1845–1852.
6. Pelletier, G., and Leclerc, R., 1979, *Endocrinology* **104**:1426–1433.
7. Watson, S. J., Richard, C. W. III, and Barchas, J. D., 1978, *Science,* **200**:1180–1182.
8. Changaris, D. G., Severs, W. B., and Keil, L. C., 1978, *J. Histochem. Cytochem.* **26**:593–607.
9. Fuxe, K., Ganten, D., Hökfelt, T., and Bolme, P., 1976, *Neurosci. Lett.* **2**:229–234.
10. Brown, M., Rivier, J., Kobayashi, R., and Vale, W., 1978, *Gut Hormones* (S. R. Bloom, ed.), Churchill Livingston, London, pp. 515–558.
11. Hori, S., 1968, *Jpn. J. Physiol.* **18**:772–787.
12. Margolis, F. L., 1974, *Science* **184**:909–911.
13. Innis, R. B., Correa, F. A., Uhl, G. R., Schneider, B., and Snyder, S. H., 1979, *Proc. Natl. Acad. Sci. U.S.A.* **76**:521–525.
14. Straus, E., and Yalow, R. S., 1978, *Proc. Natl. Acad. Sci. U.S.A.* **75**:486–489.
15. Goldstein, A., Tachibana, S., Lowney, L. I., Hunkapiller, M., Hood, L., 1979, *Proc. Natl. Acad. Sci. U.S.A.* **76**:6666–6670.
16. Krieger, D. T., Liotta, A., Suda, T., Palkovits, M., and Brownstein, M. J., 1977, *Biochem. Biophys. Res. Commun.* **76**:930–936.
17. Matsukuma, S., Yoshimi, H., Sucoka, S., Kataoka, K., Ono, T., and Ohgus, N., 1978, *Brain Res.* **159**:228–233.
18. Rehfeld, J. F., 1978, *Nature* **271**:771–773.
19. Sasaki, H., Ebitani, I., Tominaga, M., Yamatoni, K., Yawata, Y., and Hara, M., 1980, *Endocrinol. Jpn.* **1**:135–140.
20. Pacold, S. T., Kersteins, L., Hojvat, S., and Lawrence, A. M., 1978, *Science* **199**:804–806.
21. Havrankova, J., Schmechel, D., Roth, J., and Brownstein, M., 1978, *Proc. Natl. Acad. Sci. U.S.A.* **75**:5737–5741.

22. Hughes, J., 1976, *Brain Res.* **106**:189–197.
23. Sar, M., Stumpf, W. E., Miller, R., Chang, K. J., and Cuatrescasas, P. J., 1978, *J. Comp. Neurol.* **182**:17–38.
24. Simantov, R., Kuhar, M. J., Uhl, G. R., and Snyder, S. H., 1977, *Proc. Natl. Acad. Sci. U.S.A.* **74**:2167–2171.
25. Yang, H. Y., Hang, J. S., and Costa, E., 1977, *Neuropharmacology* **16**:303–307.
26. Watson, S. J., Barchas, J. D., and Li, C. H., 1977, *Proc. Natl. Acad. Sci. U.S.A.* **74**:5155–5158.
27. Zimmerman, E. A., Liotta, A., and Krieger, D. T., 1978, *Cell Tissue Res.* **186**:393–398.
28. Barry, J., Dubois, M. P., and Carette, B., 1974, *Endocrinology* **95**:1416–1423.
29. McCann, S. M., 1962, *Am. J. Physiol.* **202**:395–400.
30. Palkovits, M., Arimura, A., Brownstein, M. J., Schally, A. V., and Saavedra, J. M., 1974, *Endocrinology* **96**:554–558.
31. Dube, D., Lissitzky, J. C., Leclerc, R., and Pelletier, G., 1978, *Endocrinology* **102**:1283–1291.
32. Eskay, R. L., Giraud, P., Oliver, C., and Brownstein, M. J., 1979, *Brain Res.* **178**:55–67.
33. Jacobowitz, D. M., and O'Donohue, T. L., 1978, *Proc. Natl. Acad. Sci. U.S.A.* **75**:6300–6304.
34. Chey, W. Y., Escoffery, R., Roth, F., Chang, T. M., Yon, C. H., and Yajima, H., 1980, *Regul. Peptides [Suppl.]* **1**:519.
35. Carraway, R., and Leeman, S. E., 1976, *J. Biol. Chem.* **251**:1045–1052.
36. Kobayashi, R., Borwn, M., and Vale, W., 1977, *Brain Res.* **126**:584–588.
37. Uhl, G. R., Kuhan, M. S., and Snyder, S. H., 1977, *Proc. Natl. Acad. Sci. U.S.A.* **74**:4059–4063.
38. Buijs, R. M., Swaab, D. F., Dogterom, J., and VanLeeuwen, F. W., 1978, *Cell Tissue Res.* **186**:423–433.
39. Dogterom, J., Snijdewint, F. G. M., and Buijs, R. M., 1978, *Neurosci. Lett.* **9**:341–346.
40. Swanson, L. W., 1977, *Brain Res.* **128**:346–353.
41. Swanson, L. W., Sawchenko, P. E., 1980, *Neuroendocrinology* **31**:410–417.
42. Fuxe, K., Hökfelt, T., Eneroth, P., Gustafsson, J. A., and Skett, P., 1977, *Science* **196**:889–900.
43. Toubeau, G., Deselin, J., Parmentier, M., and Pasteels, J. L., 1979, *Neuroendocrinology* **29**:374–384.
44. O'Donohue, T. L., Charlton, C. G., Miller, R. L., Boden, G., and Jacobowitz, D. M., 1981, *Proc. Natl. Acad. Sci. U.S.A.* **78**:5221–5224.
45. Kastin, A. J., Olson, G. A., Schally, A. V., and Coy, D. H., 1980, *Trends Neurosci.* **3**:163–165.
46. Schoenenberger, G. A., and Monnier, M., 1977, *Proc. Natl. Acad. Sci. U.S.A.* **74**:1282–1286.
47. Brownstein, M., Arimura, A., Sato, H., Schally, A. V., and Kizer, J. S., 1975, *Endocrinology* **96**:1456–1461.
48. Elde, R. P., and Parsons, J. A., 1975, *Am. J. Anat.* **144**:541–548.
49. Vale, W., Rivier, C., Palkovits, M., Saavedra, J. M., and Brownstein, M. J., 1974, *Endocrinology* **94**:A128.
50. Brownstein, M. J., Mroz, E. A., Kizer, J. S., Palkovits, M., and Leeman, S. E., 1976, *Brain Res.* **116**:299–305.
51. Cuello, A. C., Emson, P. C., DelViacco, M., Gale, L., Iversen, L. L., Kanazawa, I., Paxinos, G., 1978, *Centrally Acting Peptides* (J. Hughes, ed.), Macmillan Press, New York, pp. 135–156.
52. Hökfelt, T., Johansson, O., Kellerth, J. O., Ljungdahl, A., Nilsson, G., Nygards, A., and Pernow, B., 1977, *Substance P* (U. S. von Euler and B. Pernow, ed.), Raven Press, New York, pp. 117–145.
53. Brownstein, M. J., Palkovits, M., Saavedra, J., Bassiri, R., and Utiger, R. D., 1974, *Science* **185**:267–269.
54. Hökfelt, T., Fuxe, K., Johansson, O., Jeffcoate, S., and White, N., 1975, *Eur. J. Pharmacol.* **34**:389–392.
55. Hökfelt, T., Fuxe, K., Johansson, O., Jeffcoate, S., and White, N., 1975, *Neurosci. Lett.* **1**:133–139.
56. Leppäluotto, J., Koivusalu, F., and Fraama, R., 1978, *Acta Physiol. Scand.*, **104**:175–179.
57. Oliver, C., Eskay, R. L., Ben-Jonathan, N., and Porter, J. C., 1974, *Endocrinology* **95**:540–553.
58. Zimmerman, E. A., Defendini, R., Sokol, H. W., and Robinson, A. G., 1975, *Ann. N.Y. Acad. Sci. U.S.A.* **248**:92–111.
59. Fuxe, K., Hökfelt, T., Said, S. I., and Mutt, V., 1977, *Neurosci. Lett.* **5**:241–246.

60. Larsson, L. I., Fahrenkrug, J., Schaffalitsky de Muckadell, O., Sundler, F., Hakanson, R., and Rehfeld, J. F., 1976, *Proc. Natl. Acad. Sci. U.S.A.* **73**:3197–3200.
61. Swaab, D. F., Pool, C. W., and VanLeeuwen, F. W., 1977, *J. Histochem. Cytochem.* **25**:388–389.
62. Pearse, A. G. E., 1969, *J. Histochem. Cytochem.* **17**:303–313.
63. Eipper, B. A., and Mains, R. E., 1978, *J. Biol. Chem.* **253**:5732–5744.
64. Nakanishi, S., Inoue, A., Kita, T., Nakamura, M., Chang, A. C. Y., Cohen, S. N., and Numa, S., 1979, *Nature* **278**:423–427.
65. Roberts, J. L., and Herbert, E., 1977, *Proc. Natl. Acad. Sci. U.S.A.* **74**:5300–5304.
66. Roberts, J. L., Phillips, M., Rosa, P. A., and Herbert, E., 1978, *Biochemistry* **17**:3609–3618.
67. Lewis, R. V., Stern, A. S., Kimura, S., Rossier, J., Stein, S., and Udenfriend, S., 1980, *Science* **208**:1459–1461.
68. Kimura, S., Lewis, R. V., Stern, A. S., Rossier, J., Stein, S., and Udenfriend, S., 1980, *Proc. Natl. Acad. Sci. U.S.A.* **77**:1681–1688.
69. Brownstein, M. J., Russell, J. T., and Gainer, H., 1980, *Science* **207**:373–378.
70. Russell, J. T., Brownstein, M. J., and Gainer, H., 1979, *Proc. Natl. Acad. Sci. U.S.A.* **76**:6086–6090.
71. Esch, F., Böhlen, P., Ling, N., Benoit, R., Brazeau, P., Guillemin, R., 1980, *Proc. Natl. Acad. Sci. U.S.A.* **77**:6827–6831.
72. Joseph-Brovo, P., Charli, J. L., Sherman, T., Boyer, H., Bolivar, F., and McKelvy, J. F., 1980, *Biochem. Biophys. Res. Commun.* **94**:1004–1012.
73. Noe, B. D., Fletcher, D. J., Spiess, J., 1979, *Diabetes* **28**:724–730.
74. Pradayrol, L., Jörnvall, H., Mutt, V., Ribet, A., 1980, *FEBS Lett.* **109**:55–58.
75. Lord, J. A. H., Waterfield, A. A., Hughes, J., Kosterlitz, H. W., 1977, *Nature* **267**:495–499.
76. Goodman, R. R., Snyder, S. H., Kuhar, M. J., Young, W. S. III, 1980, *Proc. Natl. Acad. Sci. U.S.A.* **77**:6239–6243.
77. Taylor, D. P., Pert, C. B., 1979, *Proc. Natl. Acad. Sci. U.S.A.* **76**:660–664.
78. Saito, A., Sankaran, H., Goldfine, I. D., Williams, J. A., 1980, *Science,* **208**:1155–1156.
79. Innis, R. B., Snyder, S. H., 1980, *Proc. Natl. Acad. Sci. U.S.A.* **77**:6917–6921.
80. Havrankova, J., Roth, J., Brownstein, M., 1978, *Nature* **272**:827–829.
81. Uhl, G. R., Bennett, J. P., Jr., Snyder, S. H., 1977, *Brain Res.* **130**:299–313.
82. Nakata, Y., Kusaka, Y., Segawa, T., Yajima, H., Kitagawa, K., 1978, *Life Sci.* **22**:259–268.
83. Marks, N., 1977, *Peptides In Neurobiology,* (H. Gainer, ed.), Plenum Press, New York, pp. 221–258.
84. Barker, J. L., and Smith, T. B. (eds.), 1980, *The Role of Peptides in Neuronal Function,* Marcel Dekker, New York.
85. Costa, E., and Trabucchi, M. (eds.), 1978, *The Endorphins,* Volume 18, Raven Press, New York.
86. Gainer, H., 1977, *Peptides in Neurobiology,* Plenum Press, New York.
87. Hughes, J., 1978, *Centrally Acting Peptides,* Macmillan, New York.
88. Martin, J. B., Reichlin, S., and Bick, K. L. (eds.), 1981, *Neurosection and Brain Peptides: Implications for Brain Functions and Neurological Disease,* Raven Press, New York.
89. Martin, J. B., Reichlin, S., and Brown, G. M., 1977, *Clinical Neuroendocrinology,* F. A. Davis, Philadelphia.

Brain Trace Amines

Alan A. Boulton and Augusto V. Juorio

1. INTRODUCTION

The neologism "trace amines" was first adopted by a study group held under the auspices of the American College of Neuropsychopharmacology at its Puerto Rico meeting in 1975. The first book[1] on this topic recorded the papers presented by the participants of this study group. The definition of a trace amine[2] appears to include those monoamines whose tissue concentrations lie between 0.1 and 100 ng/g. That it is an inappropriate term follows from the fact that some of the well-known trace amines can exist in some tissues in amounts considerably in excess of 100 ng/g and indeed, in the presence of appropriate monoamine oxidase inhibitors, many of them exceed this level; it is also a fact that some of the well-known putative monoamine neurotransmitters may be present in trace quantities in some locations.

That some of the trace amines may be important is not now disputed: they have been shown to possess inordinately fast turnover rates, and indeed, the accumulation rates of some of them[3] compare favorably with those of dopamine (DA), norepinephrine (NA), and 5-hydroxytryptamine (5-HT); they may be synthesized by mechanisms that interrelate the phenylalkyl with the phenolic and the catecholic amines; they are heterogeneously distributed both within the cell and across the brain; they are behaviorally active, and, in the presence of CNS stimulants, antipsychotics, and other drugs, their concentrations are differentially affected as are their uptake and release. Recently, they have been shown to exert potent potentiation and even reversing effects on the catecholamines and 5-HT when the latter are iontophoresed onto neurons. Because of this, they have been proposed to act as neuromodulators (neuroregulators, synaptic activators, etc.)[4,5] and to be implicated in functional psychoses and affective disorders as well as in certain other neurological and psychiatric conditions.

Alan A. Boulton and Augusto V. Juorio • Psychiatric Research Division, University Hospital, Saskatoon, Saskatchewan S7N 0X0, Canada.

In this review, the neurochemical, neurophysiological, neuropharmacological, and clinical properties of several of the more significant trace amines are discussed.

2. METHODS

Even a cursory glance at Tables I–III will reveal that different research groups obtain grossly dissimilar values for some of the trace amines in certain tissues. It must be said at the outset, therefore, that only those values that are obtained by using relatively (or absolutely) unambiguous and sensitive procedures can be considered reliable. In the main, these are mass spectrometry in conjunction with a separative procedure such as thin-layer chromatography [i.e., the so-called direct-probe high-resolution mass spectrometric integrated ion-current procedure (TLC-MS-IIC)][6–9] or gas chromatography (i.e., the so-called mass fragmentographic procedure)[10–12] after derivatization of the amine (and on occasion other functional groups) or radioenzymatic procedures that are coupled with derivatization of the labeled product followed by its chromatographic separation.[13–17]

In certain specific instances and only when the results obtained have been checked with a mass spectrometric procedure may less rigorous procedures be utilized. To date, these include the fluorimetric procedure for β-phenylethylamine (PE) by Suzuki and Yagi[18] and the gas chromatographic coupled with electron capture detection for para-tyramine (p-TA) and PE by Baker and co-workers[19,20] All other procedures, which include spectrophotometric detection of crude tissue extracts, immunologic, simple chromatographic coupled with ultraviolet or fluorescent detection, unmodified radioenzymatic, ion exchange, electrophoretic, and relatively unsophisticated gas chromatographic procedures, produce trace amine values that are too high. Additional comments concerning these procedures will be found later in the text as each amine is considered separately.

Because Volume 2 in this series will include detailed descriptions of many analytical techniques including mass spectrometry, radioenzymology, and gas chromatography, and since these latter sections will specifically address the question of trace amine analysis, further details have not been included here.

3. β-PHENYLETHYLAMINE

3.1. Presence and Regional Distribution

It has been known for most of this century that PE possesses a sympathomimetic effect,[21] is present in blood[22] and urine,[23] and exhibits a stimulating effect on behavioral activity (very similar to the so-called amphetamine stereotypy)[24] that is much enhanced in animals pretreated with a monoamine oxidase inhibitor.[25,26]

Phenylethylamine has now been shown to be present in the brains of human, rat, rabbit, and domestic fowl as well as in the ganglia of octopus, locust, and starfish (see Table I). Most workers find its concentration in whole brain to range from 0.4 to 8 ng/g (Table I). In contrast to these values, Mosnaim and co-workers and Fischer *et al.*[28,38,41,43] have claimed levels as high as 500 ng/g (see Table I) in an untreated and 2–3 μg/g in most treated brains (Table II). Even the very high rate of turnover observed for PE after monoamine oxidase inhibition[3] cannot explain these large differences found within the same species (Table I). As mentioned in Section 2, the estimation of small amounts of amines in brain and other tissues by methods that are not sufficiently specific or sensitive can yield enormously high values; this is clearly the case with respect to the values reported by the Mosnaim and Fischer groups.

The highest concentrations of PE in rat brain regions were found in the hypothalamus and caudate nucleus, with the values obtained ranging from 2 to 31 ng/g (Table III). This range of concentration is somewhat disquieting, but it is apparent that some investigators[10,12] obtain values that are consistently higher than those of others. The reasons for these differences are not clear, but it may be that in certain of the methods a reproducible inclusion of some unspecific material contributes to the PE level. Alternatively, since PE is metabolically so very active,[3,54] it is quite possible that its levels are subject to wide fluctuations depending on the environmental or behavioral conditions preceding death. It is a fact, however, that animals killed by microwave irradiation exhibit PE levels that are not different from those usually found (S. R. Philips, D. A. Durden, and A. A. Boulton, unpublished observations); neither were postmortem changes particularly marked.[27]

Table I

The Concentration of β-Phenylethylamine (PE), p-Tyramine (p-TA), m-Tyramine (m-TA), p- and m-Octopamine combined (OA), and Tryptamine (T) in Neural Tissues of Some Vertebrates and Invertebrates[a]

	PE (ng/g)	p-TA (ng/g)	m-TA (ng/g)	OA (ng/g)	T (ng/g)
Human (caudate nucleus)	1.5[27]	1.2[27]	3.0[27]	—	0.2[27]
Human (whole brain)	180[28]	—	—	—	—
Rat (whole brain)	1.8,[29] 1.5,[30] 1.1,[19] 1.7,[10] 8.1,[12] 5.0,[18] 492[38]	1.1,[31] 1.8,[20] 12.9,[36] 4.2[12]	0.32[32]	2.9[33]	0.5,[34] 22,[35] <0.5,[11] 20.9,[37] 69[39]
Rabbit (whole brain)	0.44,[40] 341,[41] 420[43]	0.36,[40] 186,[42] <10[44]	0.16[40]	—	0.05[40]
Domestic fowl (whole brain)	0.71[45]	0.50[45]	0.30[45]	—	<0.3[45]
Octopus (optic lobes)	3.0[46]	79.0[46]	0.6[46]	540[46]	<0.6[46]
Locust (ganglia)	<0.20[47]	72.0[47]	0.20[47]	2430[47]	<0.20[47]
Starfish (arm nerve)	4.4[48]	2.2[48]	<0.8[48]	260[48]	1251[48]

[a] The concentrations are in ng/g of fresh tissue. The superscripts indicate the source of the reference.

Table II
The Effect of Some Drugs on the Content of β-Phenylethylamine (PE) in the Rat or Rabbit Brain[a]

Drug	Dose (mg/kg)	Time (h)	Rat brain PE (ng/g)	Rabbit brain PE (ng/g)
Controls	—	—	1.5[30]	0.4,[40] 341,[41] 400[43]
Pargyline	75	1.5	75.5*[30]	—
Pargyline	100	3	—	11.9*[40]
Pargyline	150/day for 3 days		—	1210*[43]
d-Amphetamine	10	4	—	3494*[41]
Phenylalanine	1000	1	4.6*[30]	—
Pargyline + phenylalanine	75	1.5	630*[30]	
	1000	1		
Δ⁹-THC	3	1	—	2300*[43]

[a] The concentrations are given in ng/g of fresh tissue, and the asterisk (*) indicates that the results are statistically significantly different from their respective controls. The superscript indicates the source of reference.

3.2. Neurophysiological Studies

It has been claimed by Giardina et al.[55] that the administration of PE to rabbit cortical neurons produces either excitation, inhibition, or no effect (34, 38, and 28%, respectively) and that these effects were opposite to those of, and not mediated by, NA. These results are, however, controversial mainly because the currents used for the drug application were too high and considered likely to yield nonspecific responses and also because of the lack of definition between excitatory and inhibitory responses. In other investigations, it has been shown that PE iontophoresed onto striatal or cortical neurons of the rat brain reduced the rate of firing of the neurons in a fashion similar to the catecholamines.[56] In a more detailed study, it has been found that the administration of PE at subthreshold currents does not produce any effect by itself but markedly potentiates the decrease in the rate of firing produced by DA.[57]

3.3. Synthesis and Catabolism

The administration of high doses of phenylalanine produced a threefold increase in the concentration of PE in the rat brain. This became even more marked (420-fold increase) in the presence of pargyline (Table II). After an intraperitoneal injection of tritiated phenylalanine, the amount of radioactivity associated with PE accounted for less than 5% of the total brain tritium and declined rapidly with time[58]; as might be expected, the proportion increased markedly in the presence of a monoamine oxidase inhibitor.[58] The administration of deprenyl, a type B monoamine oxidase inhibitor, produced a large and specific [i.e., no increases were observed in the concentrations of p-TA, meta-tyramine (m-TA) or tryptamine (T)] increase in the concentration of PE in the rat striatum (Table IV). In contrast, in the presence of clorgyline, a type A monoamine oxidase inhibitor, PE levels did not increase (Table IV). These findings agree well with the enzymological study of Yang and Neff.[60]

Table III

The Regional Distribution of β-Phenylethylamine (PE), Phenylethanolamine (PE-OH), p-Tyramine (p-TA), m-Tyramine (m-TA), p-Octopamine (p-OA), m-Octopamine (m-OA), and Tryptamine (T) in the Rat Brain[a]

	PE (ng/g)	PE-OH (ng/g)	p-TA (ng/g)	m-TA (ng/g)	p-OA (ng/g)	m-OA (ng/g)	T (ng/g)
Caudate nucleus	8.0,[29] 2.1,[49] 4.2,[51] 18,[10] 30.7[12]	6.2,[35] 10,[10] <0.5[b]	19.2,[31] 8.0,[50] 11.6,[49] 13.6[12]	2.3,[50] 2.0[49]	—	<0.4[b]	2.93[34]
Hypothalamus	25.3,[29] 2.5,[40] 1.7,[49] 2.1,[51] 13,[10] 54.5[12]	25,[35] 8[10] <0.5[b]	11.3,[31] 13.9,[12] 1.9[49]	0.6[49]	3.4,[13] 2.8,[52] 4.9, 12.2[15]	<0.4,[b] 0.6, 1.3[15]	0.94[34]
Olfactory tubercles	1.74[49]	—	4.5[49]	0.93[49]	—	—	—
Hipppocampus	1.14,[49] 7.7[12]	7.7[35]	1.1,[49] 4[12]	0.3[49]	—	—	—
Brainstem	2.2,[29] 1.6,[51] 10[10]	6.7,[35] 3,[10] <0.4[b]	2.2,[31] 1.9,[49] 3100[53]	—	1.1[13]	<0.5[13]	0.24[34]
Cerebellum	3.4,[29] 2.1,[51] 8,[10] 6.8[13]	2.6,[35] 3[10]	2.3,[31] 1230,[53] 1.0[12]	—	—	—	0.27[34]
Rest	1.1,[29] 1.4[51]	—	1.6[31]	—	—	—	0.32[34]

[a] The concentrations are given in ng/g of fresh tissue. The superscripts indicate the source of the reference.
[b] Calculated from ref. 12.

Table IV

The Effects of Some Type A, Type B, or Mixed-Activity Monoamine Oxidase Inhibitors on the Concentration of β-Phenylethylamine, p-Tyramine, m-Tyramine, and Tryptamine in the Rat Striatum. The Rats Were Killed 4 h after the Drug Administration[a]

	Dose (mg/kg)	PE (ng/g)	p-TA (ng/g)	m-TA (ng/g)	T (ng/g)
Control	—	2.2	14.1	3.8	2.0
Clorgyline	10	2.0	63.2*	8.1*	5.9*
Deprenyl	1	7.7*	13.1	5.0	1.7
Tranylcypromine	1	20.7*	88.5*	11.9*	7.0*
Phenelzine	10	3.3	90.9*	13.1*	7.4*
Pargyline	75	213*	123*	15.4*	40.4*
Iproniazid	100	227*	49.2*	12.5*	9.5*
Catron	15	—	57.9*	14.4*	24.5*

[a] The concentrations are given in ng/g of fresh tissue, and the asterisk (*) indicate that the results are statistically significantly different from controls. The results are from Philips and Boulton.[59]

Other drugs that inhibit both types of MAO (tranylcypromine, pargyline, iproniazid, and catron) substantially increased PE levels in the rat striatum (Table IV). Similar effects have also been observed in the rabbit[40] (Table II), domestic fowl,[45] and optic lobes and other regions of the circumesophageal ganglia of the octopus.[46]

All of these results indicate that PE possesses a high turnover rate and relatively short half-life (1–5 min)[3,54]; its rate of accumulation in the presence of pargyline in the rat brain (1.5 nmol/g per h)[3] confirms this. What is perhaps unexpected but worthy of note is that the accumulation of PE is quite analogous to that of the catecholamines and 5-HT (see Table XIII). The current situation regarding the synthesis and catabolism of PE is summarized in Fig. 1.

3.4. Storage

Phenylethylamine is a soluble compound that readily passes through the blood–brain barrier[62]; it is present in both particulate and supernatant fractions isolated from rat brain.[63] Reserpine administration reduces the ability of intraneuronal granules to store catecholamines and 5-HT[64–66]; it does not, however, change the concentration of PE in the rat striatum (Table VIII). This finding suggests that PE is distributed equally throughout the various subcellular compartments and is not stored in specific granules.

3.5. Release

Pretreatment of rodents with PE decreases the concentration of brain catecholamines, especially in the synaptosomal fractions[67–70]; this suggests that PE induces the release of brain catecholamines.

3.6. Drug Effects

It has been claimed by Sabelli's group that the administration of *d*-amphetamine to rabbits produces an increase in their brain PE levels from 341 ng/g to 3494 ng/g[41] (see Table II). This group[41] has further claimed that rabbit brain PE, initially 400 ng/g, was increased to 2300 ng/g 3 h after the administration of Δ^9-tetrahydrocannabinol (see Table II). Because of the very high endogenous levels of PE, however, these claims should be interpreted with considerable caution. It is of interest that Danielson *et al.*[49] investigating the effect of acute and chronic administration of *d*-amphetamine to rats, could find no changes in PE levels in the striatum, olfactory tubercles, hypothalamus, or hippocampus, whereas there were clearly established and differential changes in *m*-TA and *p*-TA. It is also of interest that PE seems not to be affected by many of the drugs (antipsychotics, dopamine agonists and antagonists, etc.) that profoundly affect the tyramines, the catecholamines, and 5-HT.

4. TYRAMINES

4.1. Presence and Regional Distribution

Interest in tyramine (that is, the *p*-isomer) is not new or even recent; its synthesis was first described in the last century.[71,72] Somewhat later, it was

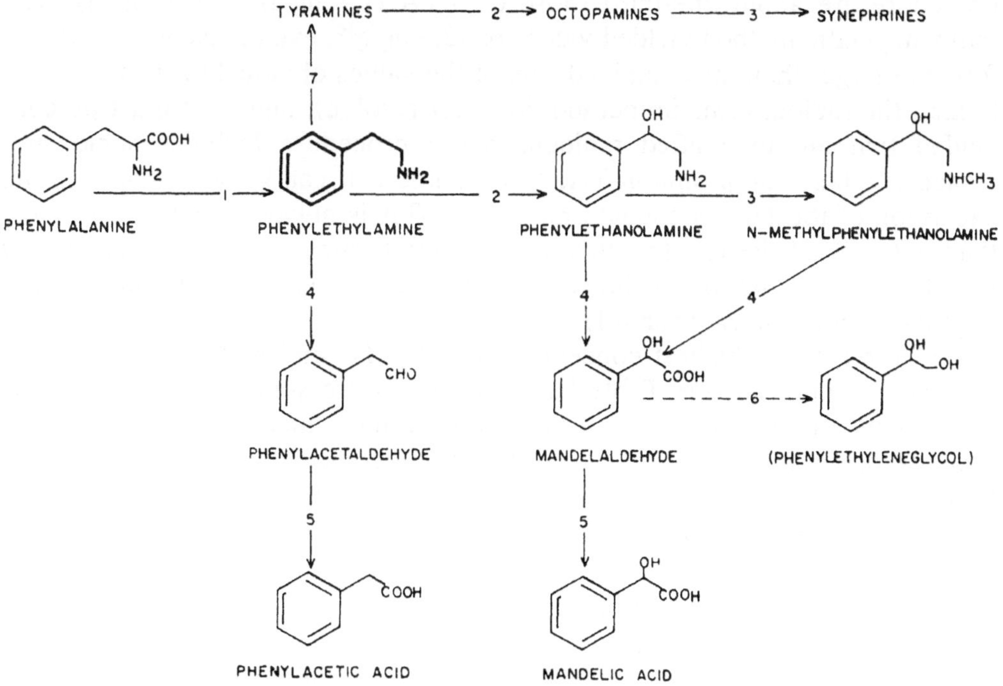

Fig. 1. Metabolism of phenylethylamine. 1, decarboxylase; 2, β-hydroxylase; 3, *N*-methyltransferase; 4, monoamine oxidase; 5, aldehyde oxidase; 6, alcohol dehydrogenase; 7, ring (de)hydroxylase. Taken from Boulton[61] and reprinted by the permission of University Park Press.

isolated from the salivary glands of the octopus.[73] Tyramines can exist in three isomeric forms: 2-, 3-, or 4-hydroxyphenylethylamine, usually referred to, respectively, as *o*-, *m*-, and *p*-TA. In this chapter, only *p*- and *m*-TA will be discussed. *o*-Tyramine has been identified as a constituent of mammalian urine[23,74–77] and is known to be a potent pharmacological agent mimicking the effects of amphetamine[78]; its concentration in the mammalian brain, however, is well below the limits of detection (A. V. Juorio, unpublished data).

With the exception of the work of Gunne and Jonsson,[44] early attempts to estimate the concentration of *p*-TA in mammalian tissues were inaccurate because the methods used lacked sensitivity and specificity. This applied to the fluorimetric method of Spector *et al.*[53] as well as to early work from this laboratory in which dansyl *p*-TA was assessed fluorimetrically and by mass spectrometry of nonderivatized rat brain extracts. The values obtained, 189 and 194 ng/g, respectively,[79–81] although agreeing with each other, were some 100 times too high. Similar high values have been obtained for the rabbit and dog brain (341 and 141 ng/g, respectively) using a radioimmunological assay.[42,82] Since these initial estimates have not been confirmed by more recent, more specific, and more sensitive (range 1–13 ng/g whole rat brain) procedures (i.e., thin-layer chromatographic–mass spectrometric,[31] gas chromatographic–mass spectrometric,[12] gas chromatographic,[20] or radioenzymatic[83]), they, along with all other high values, should be considered erroneous and therefore be disregarded.

The levels of *m*-TA (0.32 ng/g) in rat brain as obtained using the MS-IIC procedure are about one-third those of *p*-TA (1.1 ng/g) (see Table I). The radioenzymatic method yielded values of 12.9 ng/g[36]; even these values, therefore, are larger than the combined sum of the values obtained for *p*- and *m*-TA (which the radioenzymatic method does not resolve), and so it must be concluded that an unmodified radioenzymatic procedure includes some substance(s) other than *p*- and *m*-TA. From Table I, it can be seen that with the exception of the human caudate nucleus, *m*-TA is always significantly lower than *p*-TA in all tissues. In some invertebrate tissues, such as the ganglia of the octopus and the locust, however, *p*-TA can be up to 20–60 times higher than its level in the rat (Table I).

In the rat, the highest concentrations of *p*-TA and *m*-TA were found in the caudate nucleus (see Table III), and both were somewhat higher in the mouse caudate nucleus.[84–87] Following the caudate nucleus, the next highest areas were the hypothalamus and olfactory tubercles; the lowest levels were found in the hippocampus, brainstem, cerebellum, and "rest" of the brain, respectively (see Table III).

4.2. Neurophysiological Studies

The first iontophoretic application of TA (presumably the *p*-isomer) was by Bevan *et al.*[88] They described an excitation or depression of rat cortical neurons with latencies of onset and recovery times that were longer for TA than for NA. Each TA-sensitive cell was also shown to be sensitive to NA, and there was a high correlation between the directions of responses to TA and to NA; most cells excited or depressed by TA were also either excited or depressed by NA. Henwood *et al.*,[56] in 1979, showed that the iontophoretic application of somewhat larger amounts of *p*-TA to cortical or caudate neurons mostly produced a depression of the firing rate similar to that observed following application of DA and NA.

In more recent work, Jones and Boulton[57] have shown that the application of *p*-TA to rat cortical or caudate neurons using weak iontophoretic currents did not cause any change in the base-line firing rate but that when DA was applied at the usual concentration in the presence of these weak applications of *p*-TA, there was a marked potentiation in the DA response (see Fig. 2a). The responses of cortical neurons to NA were equally potentiated by *p*-TA (see Fig. 2a), but the responses of both cortical or caudate nucleus neurons to γ-aminobutyric acid were not changed by *p*-TA.[57] The responses of cortical neurons to dopamine were also potentiated by the application of subthreshold amounts of *m*-TA (see Fig. 2b).

4.3. Synthesis and Catabolism

It has been generally accepted that *p*-TA is formed to some extent by decarboxylation of *p*-tyrosine, and indeed, this has been shown to occur in the 0.25- to 2-h time period following administration of labeled *l*-tyrosine to the rat.[89] At 4 h after the administration of a large dose (500 mg/kg) of *p*-tyrosine, however, even in the presence of monoamine oxidase inhibition,[36,90] no further increase in *p*-TA was observed. In addition, *p*-TA has been found to be formed following the administration of phenylalanine, *l*-DOPA, DA, and PE.[76,77,89,91-95] Such findings indicate that the synthesis of *p*-TA is by dehydroxylation of catecholic precursors and by hydroxylation of PE as well as to some extent by decarboxylation of *p*-tyrosine. In a similar manner, the *m*-isomer of TA has been shown to arise by hydroxylation of PE and dehydroxylation of DA.[76,77] Although decarboxylation of *m*-tyrosine occurs very readily, this substance is not a normal constituent of the diet. Current views concerning the synthesis of the TAs and some other phenolic amines are shown in Fig. 3.

The catabolism of *p*-TA to *p*-hydroxyphenylacetic acid has been known since it was administered to dogs or perfused through isolated organs such as the liver or uterus.[96] It has also been used as a substrate to measure monoamine oxidase activity in the brain and other tissues.[97] Recent studies have shown that *p*-TA and *m*-TA are almost equally good substrates for monoamine oxidase in the rat caudate nucleus (see Table V), whereas the rate of oxidation for *o*-tyramine was markedly lower (see Table V). Following cerebral intraventricular introduction of radioactive *p*-TA to the rat, only a very small amount of amine could subsequently be detected in the brain; in the presence of a monoamine oxidase inhibitor, however, the amount was markedly increased.[99]

The administration of pargyline (100 mg/kg) produced a substantial increase (to nine or ten times the control values) in the concentrations of both *p*-TA and *m*-TA in the rabbit brain,[40] an increase similar to that observed in the rat brain.[36] More recently, it has been shown that the administration of clorgyline, tranylcypromine, phenelzine, pargyline, iproniazid, or catron pro-

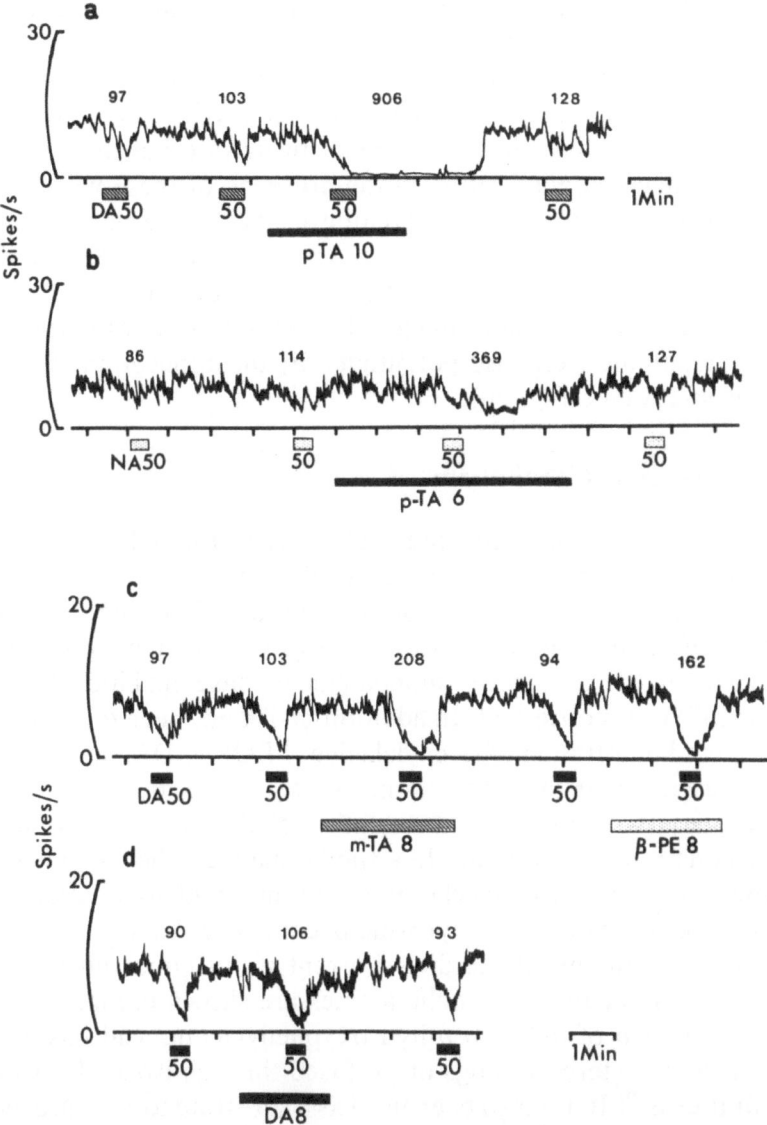

Fig. 2. Ratemeter records of the firing rate of two neurons. The bars beneath the traces indicate iontophoretic drug applications, and the figures refer to the intensity of the ejecting current in nanoamperes. The figures above the trace refer to the size of the responses as a percentage of the mean control response before application of *p*-TA. (a) A study completed on a neuron in the caudate nucleus. Applications of DA resulted in depression of the cell firing rate. During the application of *p*-TA (black bar), the response was greatly enhanced. Recovery occurred within 3 min of termination of *p*-TA application. (b) Firing rate record of a cortical neuron weakly depressed by NA. The depression was potentiated more than threefold during the application of *p*-TA. (c) Ratemeter record of a single cortical neuron. The cell was depressed by regular application of DA. The concurrent application of both *m*-TA and β-PE potentiated the DA response. (d) The concurrent application of DA, however, had little effect. These results are from Jones and Boulton[57] and are published with permission of the *Canadian Journal of Physiology and Pharmacology.*

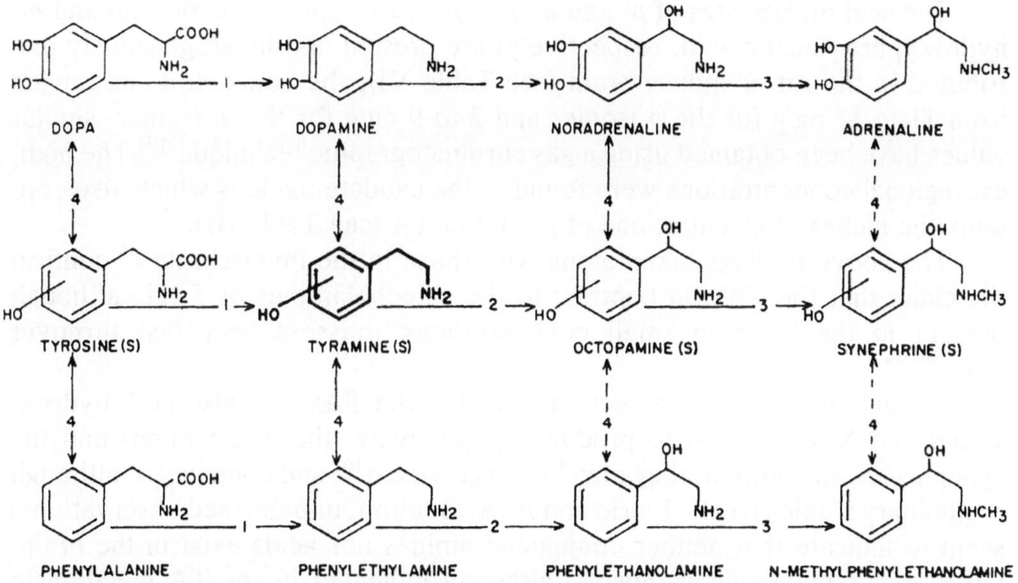

Fig. 3. Synthesis of the phenolic amines. 1, decarboxylase; 2, β-hydroxylase; 3, *N*-methyltransferase; 4, ring (de)hydroxylase. Taken from Boulton[61] and reprinted by the permission of University Park Press.

duces a substantial increase in the concentration of both the *p*- and *m*-isomers of TA in the rat caudate nucleus[59,100] (see Table IV), whereas deprenyl, a more specific type A monoamine oxidase inhibitor,[60] did not significantly change the concentration of either TA isomer (see Table IV) and inhibited the caudate nucleus enzyme to a lesser extent than did pargyline or clorgyline in the case of *p*-TA (see Table V). The mouse striatal concentrations of both TAs are also markedly increased (three to six times their control levels) following the administration of phenelzine or its *p*-chloro or *p*-fluoro derivatives.[101] Pargyline produced an effect that was more marked than that produced by any of the phenelzine derivatives tested; the administration of pargyline and phenelzine together led to increases that were not higher than those produced by pargyline alone.[101] Moderate doses of tranylcypromine and clorgyline also increased (1.6–4 times their control levels) mouse striatal tyramines.[86]

Table V

The Effect of Some Inhibitors on the Specific Monoamine Oxidase Activity towards the p-, m-, and o- Isomers of Tyramine of a Rat Striatal Mitochondrial Fraction[a]

	Dose (mg/kg)	Time (h)	Specific MAO activity (nmol/mg protein per min)		
			p-Tyramine	*m*-Tyramine	*o*-Tyramine
Control	—	—	2.4	1.9	0.28
Pargyline	20	4	<0.03*	<0.03*	<0.03*
Clorgyline	2	4	0.75*	1.4*	0.2
Deprenyl	2	4	1.6*	0.9*	0.1

[a] The asterisk indicates that the results are statistically significantly different from controls. The results are from Yu and Boulton.[98]

The acid metabolites of p- and m-TA (p-hydroxyphenylacetic acid and m-hydroxyphenylacetic acid, respectively) are present and heterogeneously distributed in the rat or mouse brain (see Table VI); the concentrations ranged from 11 to 32 ng/g for the p isomer and 2 to 9 ng/g for the m isomer; similar values have been obtained using a gas chromatographic technique.[104] The highest regional concentrations were found in the caudate nucleus which also contains the highest concentrations of p- and m-TA (see Table III).

The above findings taken along with those in the invertebrates[46] confirm the claim that the TAs, in contrast to the catecholamines or 5-HT, although present in the brain in small concentrations, possess very fast turnover rates.[40,99]

As will be discussed in Sections 6 and 7, the TAs may also be β-hydroxylated and N-methylated to produce, respectively, the octopamines and the synephrines. In addition, they may be N-acetylated[107] and conjugated, although preliminary studies (A. V. Juorio and A. A. Boulton, unpublished observations) seem to indicate that neither conjugated amines nor acids exist in the brain. Other minor metabolic pathways, although indicated in the TA metabolism pathways (see Fig. 4), are not discussed further here.

4.4. Storage

Tyramines are associated with synaptosomes obtained after subcellular fractionation of the rat brain[63]; in the presence of pargyline, the amounts in all fractions were increased but were more pronounced in the supernatant.

Both p-TA and m-TA in the rat caudate nucleus are reduced to less than 20% of their control values within 24 h of an injection of 1 or 10 mg/kg[50] of reserpine. A smaller dose (0.4 mg/kg) significantly decreases the content of both TAs in the caudate nucleus, with the effects becoming apparent within 45 min and persisting for at least 6 h in the case of p-TA and 19 days in the case of m-TA (see Table VII). Similar reductions in mouse striatal p- or m-TA

Table VI
The Regional Distribution of p-Hydroxyphenylacetic Acid (p-HPAA) and m-Hydroxyphenylacetic Acid (m-HPAA) in the Brain[a]

	Rat		Mouse	
Tissue	p-HPAA (ng/g)	m-HPAA (ng/g)	p-HPAA (ng/g)	m-HPAA (ng/g)
Whole brain	10.6,[102] 17.1,[103] 32.2,[105] 24.9[106]	2.3,[102] 9.4[103]	12.5[104]	4.1[104]
Hypothalamus	4.5[102]	1.2[102]	11.2[104]	3.8[104]
Caudate nucleus	28.3[102]	5.5[102]	27.9[104]	8.7[104]
Olfactory tubercles	—	—	20.2[104]	5.3[104]
Hippocampus	—	—	9.6[104]	5.5[104]
Brainstem	8.6[102]	1.8[102]	8.8[104]	3.0[104]
Cerebellum	8.1[102]	1.2[102]	11.2[104]	3.6[104]
Rest	5.3[102]	1.7[102]	6.8[104]	2.2[104]

[a] The concentrations are given in ng/g of fresh tissue, and the superscripts indicate the source of reference.

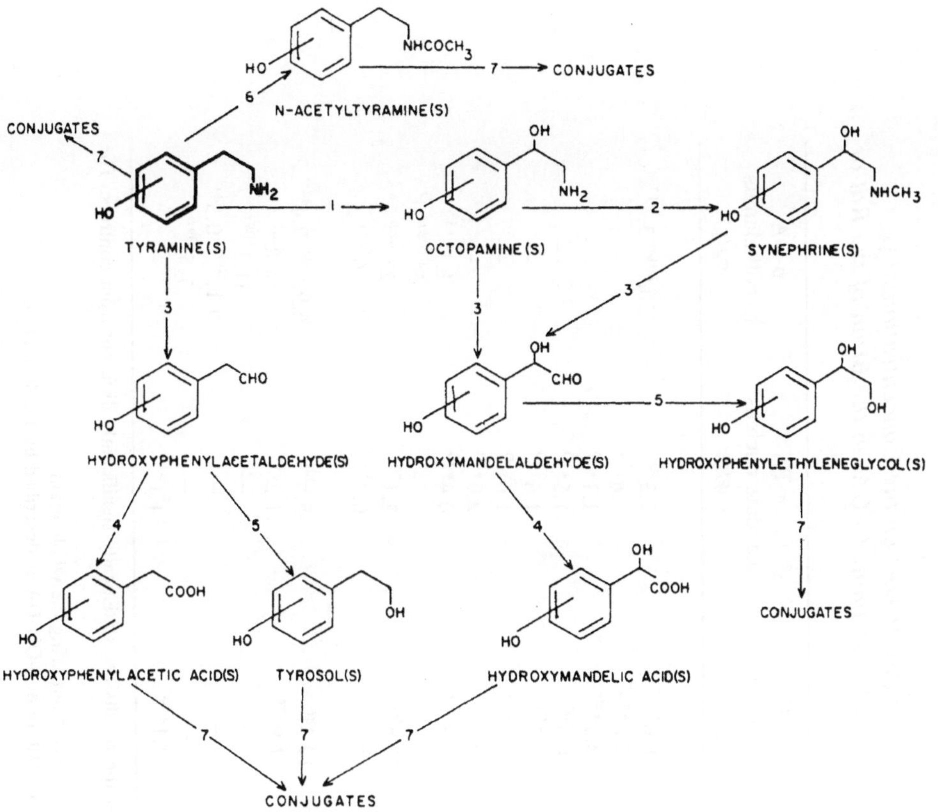

Fig. 4. Metabolism of the phenolic amines. 1, β-hydroxylase; 2, *N*-methyltransferase; 3, monoamine oxidase; 4, aldehyde oxidase; 5, alcohol (de)hydrogenase; 6, arylalkylamine acetyltransferase; 7, conjugation with glycine, sulfate, and/or glucuronic acid. Taken from Boulton[61] and reprinted by permission of University Park Press.

have been observed following the administration of reserpine and tetrabenazine (see Table VII). A cerebral intraventricular injection of 6-hydroxydopamie (250 μg) produced a significant reduction in the rat striatal levels of both *p*- and *m*-TA (see Table VII) after 10 days; no changes were observed when the animals were injected with solvent only (see Table VII). These results suggest that *p*- and *m*-TA, at least in part, may be stored by an intraneuronal reserpine- or tetrabenazine-sensitive storage mechanism. Alternatively, the TAs may replace some of the catecholamines from their storage granules and be coreleased along with the catecholamines. It is also possible that the observed changes in TA levels might reflect the fact that these amines are metabolically related to another amine that is stored in reserpine-sensitive granules.[50]

4.5. Uptake and Release

The presence of saturable high- and low-affinity uptake systems for *p*-TA and *m*-TA has been demonstrated in slices of rat caudate nucleus and hypothalamus; the systems seem similar to those used by DA[111-114] when the usual uptake blockers and metabolic poisons such as dinitrophenol, low sodium concentrations, ouabain, and cocaine are used.[115,116] Differences between the

Table VII

The Effect of Reserpine, Tetrabenazine, and 6-Hydroxydopamine (6-OH-DA) on the Concentration of β-Phenylethylamine (PE), p-Tyramine (p-TA), m-Tyramine (m-TA), and p-Octopamine (p-OA) in the Brain of the Rat or Mouse[a]

	Dose (mg/kg)	Time (h)	PE hypothalamus (ng/g)	p-TA caudate nucleus (ng/g)	m-TA caudate nucleus (ng/g)	p-OA hypothalamus (ng/g)
Rat						
Controls	—	—	2.5[50]	8.0[50]	2.3[50]	20,[b108] 3.4[13]
Reserpine	0.4	3	2.0[50]	3.8*[50]	1.6*[50]	—
	0.4	6	2.7[50]	3.1*[50]	1.1*[50]	—
	0.4	12	3.2[50]	6.5[50]	1.5[50]	—
	0.4	24	3.2[50]	6.3[50]	1.6[50]	—
	0.4	96	1.8[50]	5.4*[50]	1.6[50]	—
	0.4	456	—	10.5[50]	2.0[50]	—
	1	24	2.8[50]	1.4*[50]	0.4*[50]	1.1*[13]
	10	24	—	—	—	5*[108b]
Controls	—	—	—	10.6[50]	2.1[50]	23[108b]
6-OH-DA[c]	—	—	—	3.6*[50]	0.9[50]	6[108b]
Mouse						
Controls	—	—	—	19.3[86]	5.8[86]	6.9,[109b] 9.4[110]
Reserpine	1	24	—	1.9*[86]	1.2*[86]	4.8*[d]
	5	2	—	—	—	11.1[109b]
	5	8	—	—	—	9.1,[109b] 0.3*[d]
	5	18	—	—	—	7.5[109b]
Tetrabenazine	20	1	—	3.1*[86]	1.1*[86]	—

[a] The concentrations are given in ng/g fresh tissue; the asterisk (*) indicates that the results are statistically significantly different from controls. The superscript indicates the source of reference.
[b] p-Octopamine and m-OA determined together. Determinations from ref. 109 were carried out using the whole brain.
[c] Rats were killed 10 days after cerebral intraventricular administration of solvent of 250 μg or 6-OH-DA as described by Boulton et al.[50]
[d] A. V. Juorio, unpublished observations.

TA uptake mechanisms and that for DA become apparent, however, when other blocking agents such as benztropine are used.[117]

Rat striatal slices release preloaded labeled *p*-TA in a manner similar to that exhibited by DA in the presence of the usual depolarizing agents (K⁺, amphetamine, other amines, etc.).[115,118-121] Marked differences between the release of DA and of the TAs become apparent, however, when slices previously incubated with radioactive *p*-TA, *m*-TA, or DA are treated with methylphenidate or other stimulants such as cocaine, nomifensine, and amfonelic acid[122,123] (Fig. 5) or the aminotetralins.[125] In these latter cases, both isomers

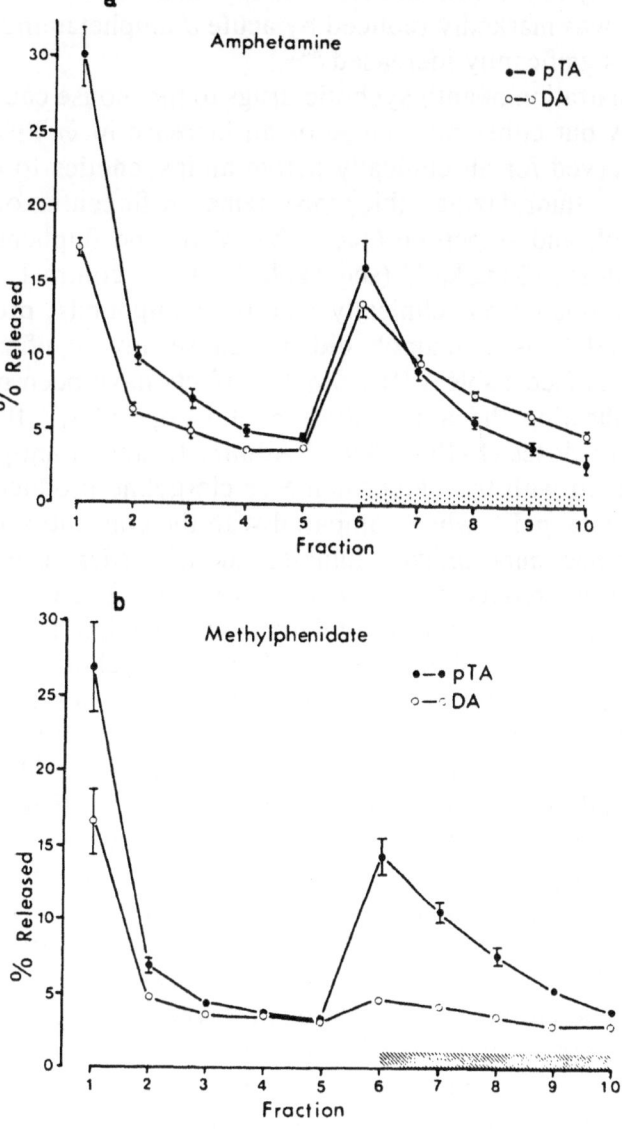

Fig. 5. (a) Effect of amphetamine on the release of dopamine (DA) and *p*-tyramine (pTA) from rat striatal slices; the dotted bar indicates the addition of *d*-amphetamine, 10⁻⁵ M. (b) Effect of methylphenidate on the release of dopamine (DA) and *p*-tyramine (pTA) from rat striatal slices; the dotted bar indicates the addition of methylphenidate, 10⁻⁵ M. These results are from Dyck *et al.*[122] and are published with permission of the *European Journal of Pharmacology*.

of TA are released, but only a very small amount of DA (see Fig. 5). From this, we are able to conclude that the TAs, although able to mimic DA, certainly possess uptake and release mechanisms that are uniquely tyraminergic.

4.6. Drug Effects

Acute and chronic treatment with *d*-amphetamine in rats caused a significant reduction in *p*-TA levels in the striatum and olfactory tubercles.[49] No changes were observed in the hypothalamus or hippocampus.[49] *m*-Tyramine was increased in both the striatum and the olfactory tubercles after acute injections but only in the striatum after chronic administration.[49] In the mouse striatum, *p*-TA was markedly reduced by acute *d*-amphetamine injections, but *m*-TA was not significantly increased.[84]

The administration of antipsychotic drugs to the mouse caused a reduction in striatal *p*-TA but either no change or an increase in *m*-TA. These effects have been observed for all clinically active antipsychotics to date, including chlorpromazine, thioridazine, thioproperazine, α-flupenthixol, (+)-butaclamol, haloperidol, and spiperone (see Table VIII) and fluphenazine and molindone in low doses (2 mg/kg)[87] (see Table VIII). In contrast, treatment with the structurally related but clinically inactive compounds, promethazine, β-flupenthixol, and (−)-butaclamol, did not cause any significant changes in striatal TA levels (see Table VIII). Similar effects have been observed in the rat striatum following the administration of haloperidol,[52] fluphenazine, or molindone in low doses (1–10 mg/kg).[87] Administration of antipsychotic drugs to mice pretreated with tranylcypromine or clorgyline produced a significant reduction in striatal *p*-TA when compared with the concentrations obtained in mice given a monoamine oxidase inhibitor alone[86]; such results suggest that antipsychotic drugs reduce striatal *p*-TA formation. The moderate increases produced by monoamine oxidase inhibitors on striatal *m*-TA were not significantly changed after the administration of an antipsychotic drug.[86]

Because antipsychotic drugs counteract the behavioral effects of *d*-amphetamine,[127,128] the effects on striatal TA levels of *d*-amphetamine injected into mice pretreated with either chlorpromazine or haloperidol[84] was investigated. The initially surprising result was the observation that the reduction in *p*-TA was potentiated although the *m*-TA concentration was increased[84]; this effect is very clearly seen when the *p*-TA/*m*-TA ratio is examined (see Fig. 6). An explanation of this phenomenon could be that *p*-TA and DA are reciprocally related and that both *d*-amphetamine and antipsychotic drugs increase DA turnover,[87,110,124] the former by producing increased release and blockade of uptake,[129] and the latter by blockade of postsynaptic receptors.[130–133] If this explanation is correct, then clearly, increases in DA turnover produced by pharmacological or behavioral manipulations will be accompanied by reductions in striatal *p*-TA. The increase in *m*-TA could be explained if the *m*-TA were formed predominantly from DA or its precursors.

It follows that changes in the opposite direction will occur if the DA turnover[86] is reduced. This predictive state of affairs has been completely confirmed by extensive experimentation[86,124] and is summarized in Table IX.

Table VIII

The Effect of the Administration of Some Antipsychotic Drugs on Mouse Cerebral p-Tyramine (p-TA), m-Tyramine (m-TA), p-Octopamine (p-OA), m-Octopamine (m-OA), and Homovanillic Acid (HVA)[a]

	Dose (mg/kg)	Time (h)	Striatum		Hypothalamus		Striatum
			p-TA (ng/g)	m-TA (ng/g)	p-OA (ng/g)	m-OA (ng/g)	HVA (ng/g)
Controls	—	—	19.1,[85] 21.8,[86] 20.6,[87] 18.1,[126]	5.1,[85] 6.2,[86] 5.7[87]	9.4[87]	1.2[87]	920[87]
Chlorpromazine	2	2	8.3[85]	5.6[85]	—	—	—
Thioridazine	2	2	12.5*[126]	6.7[126]	—	—	—
Thioproperazine	2	2	2.6*[126]	5.9[126]	—	—	—
Promethazine	20	2	15.8[85]	4.1[85]	—	—	—
Fluphenazine	0.1	2	7.4*[87]	6.1[87]	7.9[87]	1.4[87]	2460*[87]
Fluphenazine	1	2	8.0*[87]	10.5*[87]	5.5*[87]	1.1[87]	3280*[87]
α-Flupenthixol	2	2	4.9*[85]	5.8[85]	—	—	—
β-Flupenthixol	2	2	18.2[85]	6.2[85]	—	—	—
(+)-Butaclamol	1	2	7.4*[85]	5.8[85]	—	—	—
(−)-Butaclamol	1	2	17.3[85]	4.7[85]	—	—	—
Molindone	2	2	13.2*[87]	10.3*[87]	7.7[87]	3.4[87]	2140*[87]
Molindone	20	2	23.7[87]	8.4*[87]	25.6*[87]	3.5*[87]	1510*[87]
Molindone	100	2	46.3*[87]	3.7*[87]	26.4*[87]	2.7*[87]	1300*[87]
Haloperidol	0.2	2	6.8*[85]	6.7[85]	—	—	—
Spiperone	0.2	2	6.0*[86]	7.5[86]	—	—	—

[a] The concentrations are given in ng/g of fresh tissue, and the asterisk indicates that the results are statistically significantly different from controls. The superscript indicates the source of reference.

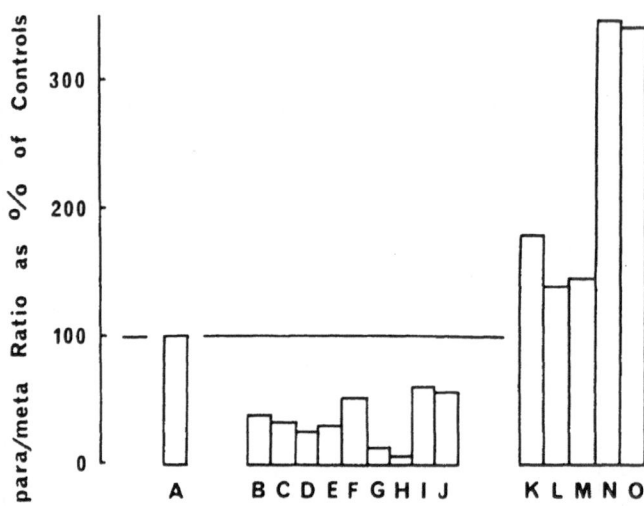

Fig. 6. Effects of various treatments on the p-tyramine/m-tyramine (*para/meta*) ratio in the mouse striatum. The ordinates indicate the results expressed as percentages of the control ratio. In the scale, the different treatments are identified by letters as follows: A, controls, mean value from refs. 84–86, 110; B, chlorpromazine, 2 mg/kg, 2 h[84]; C, fluphenazine, 0.1 mg/kg, 2 h[110]; D, haloperidol, 0.2 mg/kg, 2 h[84]; E, molindone, 2 mg/kg, 2 h[87]; F, d-amphetamine, 5 mg/kg, 0.5 h[84]; H, haloperidol, 0.2 mg/kg, 2 h + d-amphetamine, 5 mg/kg, 0.5 h[84]; I, l-DOPA, 50 mg/kg, 2 h[124]; J, crowding stress[124]; K, apomorphine, 20 mg/kg, 1 h[86]; L, peribedil, 10 mg/kg, 1 h[86]; M, lergotrile, 10 mg/kg, 1 h[86]; N, α-methyl-p-tyrosine, 200 mg/kg, 2 h[86]; O, molindone, 100 mg/kg, 2 h.[87]

The increase in DA turnover that follows the administration of its precursor, l-DOPA, or subjection of mice to some types of stress yielded a significant reduction in striatal p-TA and either no change or an increase in m-TA (see Table X). Similarly, a reduction in DA turnover induced by DA receptor activation following administration of apomorphine, lergotrile, or piribedil (Table XI)[86,134–136] or inhibition of DA synthesis with α-methyl-p-tyrosine[137] produced

Table IX
Correlations between Changes in Brain Dopamine (DA) Turnover and p-Tyramine
(p-TA) and m-Tyramine (m-TA) Concentrations[a]

	DA turnover	p-TA concentration	m-TA concentration
Dopamine receptor blockers (chlorpromazine, haloperidol, and related compounds	Increase	Reduction	Tendency to increase
Dopamine releasor and uptake blocker (d-amphetamine)	Increase	Reduction	Tendency to increase
Dopamine receptor agonists (lergotrile, apomorphine, piribedil)	Reduction	Increase	No change
Dopamine synthesis inhibitor (α-methyl-p-tyrosine)	Reduction	Increase	Reduction
Dopamine synthesis increase (by l-DOPA administration)	Increase	Reduction	No change
Stress	Increase	Reduction	Increase

[a] Modified from ref. 124.

Table X

The Effect of Aggregation Stress, Cold Strees, and Administration of L-DOPA on the Concentration of p-Tyramine (p-TA), m-Tyramine (m-TA), and Homovanillic Acid (HVA) in the Mouse Striatum[a]

	Time (h)	p-TA (ng/g)	m-TA (ng/g)	HVA (ng/g)
Controls	—	21.5	6.2	930
Aggregation stress	—	18.2*	9.2*	1200*
Cold stress	—	16.9*	8.3*	880*
Controls	—	26.7	5.0	970
L-DOPA (50 mg/kg)	2	16.2*	7.9	1760*
L-DOPA (200 mg/kg)	2	10.4*	3.9	9140*

[a] The concentrations are given in ng/g of fresh tissue, and the asterisk indicates that the results are statistically significantly different from controls. The results are from Juorio.[124]

an increase in striatal p-TA and either no change or a decrease in m-TA.[86] It is not yet possible to decide whether a TA controls DA or *vice versa*, or whether the situation is more complex.

An examination of rat urine for p- and m-TA following treatment with chlorpromazine or haloperidol alone or in association with d-amphetamine did not reveal the profound changes observed in the brain.[138] This emphasizes the difficulties of trying to correlate brain changes with changes in urinary levels of p- or m-TA and thus the considerable difficulties that obtain in biological psychiatric research when the only material available for analysis is urine.

5. PHENYLETHANOLAMINE

Phenylethanolamine (PEOH) has been claimed to be present in human and rat brain at levels of 4–6 ng/g determined by radioenzymatic and mass spec-

Table XI

The Effect of the Administration of Apomorphine, Piribedil, or Lergotrile on p-Tyramine (p-TA), m-Tyramine (m-TA), and 3,4-Dihydroxyphenylacetic Acid (DOPAC) in the Mouse Striatum[a]

	Dose (mg/kg)	Time (h)	p-TA (ng/g)	m-TA (ng/g)	DOPAC (ng/g)
Controls	—	—	22.9	6.2	694
Apomorphine	20	0.5	31.7*	5.8	—
	20	1	37.9*	5.5	301*
	20	2	22.1	5.1	—
Piribedil	1	1	30.3*	6.9	—
	10	1	37.0*	7.3	399*
Lergotrile	10	1	40.7*	7.6	339*

[a] The concentrations are given in ng/g fresh tissue, and the asterisk indicates that the results are statistically significantly different from controls. The results are from Jurio.[86]

trometric analytical procedures.[10,139,140] Within the brain, the highest levels of PEOH were reported to exist in the hypothalamus (8–25 ng/g) and striatum (6–10 ng/g), and the lowest in the brainstem (3–7 ng/g) and cerebellum (3 ng/g) (Table III). Phenylethanolamine has also been claimed to exist in various ganglia isolated from the *Aplysia* nervous system[141] and to act in that species as a neurotransmitter. These initial claims, however, have not been confirmed by more recent publications. In the first of these subsequent reports,[13] PEOH was analysed after conversion to *N*-methylphenylethanolamine followed by dansylation and separation twice by thin-layer chromatography. The radio-activity associated with the separated and isolated DNS-*N*-methylphenyle-thanolamine zone revealed that the amounts present in the rat hypothalamus, striatum, and brainstem were below the limits of detection (<0.5 ng/g), even though this modified procedure is very sensitive and specific.[13] Only after inhibition of monoamine oxidase with pargyline (75 mg/kg, 3 h) could PEOH be measured in the hypothalamus where it was present at 1 ng/g and in the striatum and brainstem at 0.6 and 0.7 ng/g, respectively.[13] Somewhat higher concentrations of PEOH have recently been reported in the hypothalamus of the Roman strains of rats where it ranged from 0.6 to 1.3 ng/g.[15] In a similar manner, McCaman and McCaman[14] could not detect any PEOH in various ganglia of the *Aplysia* nervous system. It appears, therefore, that although PEOH probably does exist in the mammalian brain or *Aplysia* ganglia, the amount present is much lower than earlier reports indicated.

6. OCTOPAMINES

6.1. Presence and Regional Distribution

Octopamine [presumably its *p*-isomer, *p*-hydroxyphenylethanolamine (*p*-OA)] was first observed in the posterior salivary gland of the octopus[142,143] and then in human urine and in mammalian organs during monoamine oxidase blockade.[144] Following the development of a sensitive radioenzymatic assay,[30,33] it became possible to estimate the combined *p*- and *m*- isomers of OA and demonstrate their presence in the vertebrate brain and invertebrate ganglia. The improvement of this method by adding a derivatization step (reaction with dansyl chloride) followed by purification by thin-layer chromatography[13] has permitted the separate detection and quantitation of both *p*- and *m*-OA.

In the whole brain of the rat (Table I), guinea pig, mouse, or rabbit, *p*-OA plus *m*-OA range in concentration from 3 to 14 ng/g.[30,33,109] Their highest concentrations in the central nervous system were observed in the rat hypo-thalamus and spinal cord (13–26 and 24 ng/g, respectively)[30] (Table III), the guinea pig midbrain and hypothalamus (24 and 13 ng/g, respectively),[30] and the human hypothalamus (Table XII). Their lowest levels were in the rat cortex, striatum, and cerebellum[30,108] and guinea pig striatum and cortex.[30] The con-centration of *m*-OA in the hypothalamus of the rat or the mouse is <0.4 or 1.3 ng/g, respectively (Table XII), whereas in the domestic fowl hypothalamus, *p*-OA is 2.7 and *m*-OA <0.5 ng/g (Table XII).

Table XII
The Concentration of p- and m-Octopamine Determined Together (OA) or
Separately (p-OA), (m-OA) in the Mammalian Hypothalamus

	OA (ng/g)	p-OA (ng/g)	m-OA (ng/g)
Rat	26,[30] 11,[108] 24[108]	3.4,[13] 4.9,[15] 12.2[15]	<0.4,[13] 0.6,[15] 1.3[15]
Mouse	—	9.0,[145] 9.4[95]	1.2[95]
Guinea pig	13[30]	1.2[145]	—
Domestic fowl	—	2.7[145]	<0.5[145]
Human	80[30]	—	—

[a] The concentrations are given in ng/g of fresh tissue. The superscripts indicate the source of reference.

In the ganglia of invertebrate species so far studied, the concentrations of OAs are considerably higher than those obtaining in the mammalian brain. The circumesophageal ganglia of the snail, for example, contains about 75 ng/g (calculated from ref. 146), whereas the buccal, cerebral, and pedal ganglia of *Aplysia* contain 130 to 200 ng/g, and the abdominal and pleural ganglia 31–35 ng/g.[147] Other workers[14] have found considerably higher values for the buccal, cerebral, and pedal ganglia (530 to 1840 ng/g) and none in the abdominal or pleural ganglia. These different results appear to originate as a result of differences in the dissection of the tissues. The optic lobes of the octopus and other cephalopods contain values of OA that range from 220 to 2210 ng/g[148] (Table I). The arthropods also contain quite high levels, with the brain or optic lobes of the locust containing 2430 and 3910 ng/g, respectively,[47] whereas the brain of the lobster contained 226 ng/g.[149] In the arm nerve of an echinoderm (sunflower starfish), the levels of OA are about 260 ng/g.[48] These high concentrations of OA in the invertebrate nervous system suggest that it may possess a neurotransmitter role there. Such levels, however, really place *p*-OA outside the definition of a trace amine. For further discussion of its functional significance in invertebrates, see the review by Robertson and Juorio.[150]

In assessing *m*- and *p*-OA levels in mammalian brain regions, the direct radioenzymatic method, as opposed to the modified one by Danielson *et al.*,[13] yields values 3–11 times too high depending on the region. As with PEOH, it is assumed that the reason for this is the presence of some contaminant that methylates and separates with *p*-OA. In analyzing trace quantities of any metabolite, therefore, it is clear that unambiguous and sensitive methods must be used.

6.2. Neurophysiological Studies

The iontophoretic administration of OA (presumably the *p*- isomer) to cortical neurons of the rat produced depression in 48% of the cells, excitation in 42%, and no effect on the rest[151]; similar results were observed on dorsal horn neurons of the spinal cord.[152] It has been suggested that the effects of OA are pharmacologically different from those exhibited by NA and DA, and it is concluded that specific OA receptors may exist in central neurons of the rat. No studies have yet been performed using *m*-OA.

6.3. Synthesis and Catabolism

The possible biosynthetic routes for the formation of m- and p-OA can be seen in Figs. 3 and 4. The horizontal route in Fig. 4 following the hydroxylation of phenylalanine has been demonstrated.[92,108,109] If the existence of TA is a prerequisite for the formation of OA then the possibilities outlined in Section 4.3 are equally possible for the OAs, and indeed, its formation from l-DOPA has been demonstrated.[92] It remains at least possible that hydroxylation of PEOH and dehydroxylation of NA could occur.

The main pathway for the degradation of p-OA is deamination by mono-amine oxidase to form the corresponding aldehyde which is then, in turn, oxidized by aldehyde dehydrogenase or reduced by aldehyde reductase to form p-hydroxymandelic acid or p-hydroxyphenylglycol, respectively; both compounds have been detected in brain tissue.[153] These catabolic routes, as indicated in Fig. 5, are further supported by the large increases in cerebral m-OA and p-OA that follow administration of the monoamine oxidase inhibitors pheniprazine and pargyline, respectively.[92,108,109] In the rat and mouse, the hypothalamus exhibited marked increases in both p-OA and m-OA following administration of pargyline, phenelzine, and some of its p-halogenated derivatives.[13,101] Similarly, the hypothalamic concentration of p-OA in the domestic fowl was markedly increased after pargyline and tranylcypromine treatment.[145]

6.4. Storage

Reserpine (1–10 mg/kg) produces a substantial reduction (68–99%) in the rat hypothalamic concentration of p- and m-OA determined together or of p-OA determined by itself (Table VII). In other experiments using mouse whole brain,[109] no changes were observed in the concentration of p and m-OA determined together up to 18 h after the administration of 5 mg/kg of reserpine (Table VII); these results are different from those obtained by Danielson et al.[13] for the rat hypothalamus or for p-OA in the mouse hypothalamus[95,108] (Table VII). The domestic fowl hypothalamic p-OA level is also reduced by reserpine (1 mg/kg) 3 and 24 h after drug treatment.[145] The differences observed between the effects of reserpine in whole brain as opposed to the hypothalamus suggest that reserpine either exhibits a differential effect in some brain regions or that the determinations on the whole brain (carried out using the direct radioenzymatic method) included some contaminants.

A cerebral intraventricular injection of 6-hydroxydopamine (250 µg) produced 2 days later a 74% reduction in the concentration of p- plus m-OA.[108] The above results suggest that p- and m-OA are kept in a neuronal reserpine-sensitive compartment.

6.5. Uptake and Release

Radioactive OA (presumably p-OA) is accumulated in vitro by rat brain homogenates; the uptake is more active in striatal than in cortical tissue and decreases in the sequence NA > TA > OA.[154] Octopamine uptake was inhibited by a decrease in temperature, a decrease in glucose concentration, cyanide, dinitrophenol, or the neuronal uptake blockers desmethylimipramine and cocaine.[154] Tritiated *p*-OA concentrated by various brain regions can be released *in vitro* by electrical stimulation in the following order: caudate nucleus > hypothalamus > cerebral cortex (in the ratio 100:77:52, respectively).[155]

6.6. Drug Effects

By using the direct radiochemical method for OA ($m + p$), it has been shown[109] that chlorpromazine (10 mg/kg, 2 h), chlordiazepoxide (10 mg/kg, 5 h), imipramine (10–50 mg/kg, 18 h), and iprindol (20 mg/kg, 18 h) exhibit no effect on their concentration in mouse brain. Using the modified radiochemical method,[13] we have shown that low doses of fluphenazine (0.1 mg/kg, 2 h) similarly exert no significant effect on mouse brain *p*-OA but that larger doses (1–5 mg/kg, 2 h) produce significant reductions in *p*-OA and no changes in *m*-OA.[87] Small doses of haloperidol (0.1 mg/kg, 2 h) or *d*-amphetamine (5 mg/kg, 0.5 h) were without effect on rat hypothalamic *p*-OA.[52] In contrast, the administration of *d*-amphetamine to chlorpromazine-treated rats produced a significant reduction in hypothalamic *p*-OA. This can most likely be explained as the consequence of a diminished availability of *p*-TA.[52,61,86,95] Molindone is an antipsychotic drug that exhibits some atypical effects[87]: at low doses (2 mg/kg, 2 h), it produced no change in either *p*- or *m*-OA; at higher doses, however, increases in both OAs were observed (Table VIII), and in this case their changes paralleled those obtaining for striatal *p*- and *m*-TA.

7. SYNEPHRINE

A mass spectrometric method for the determination of synephrine (the combined *p*- and *m*- isomers) has recently been developed.[156,157] Using this procedure *p*- plus *m*-synephrine in the mouse hypothalamus was found to be below the limit of sensitivity of the method (presently about <3.8 ng/g). In an additional preliminary study, only very small amounts of *p*- plus *m*-synephrine could be observed in the domestic fowl diencephalon or the turtle brain.[156]

8. TRYPTAMINE

8.1. Presence and Regional Distribution

By a fluorimetric procedure, T was found to be below the limits of sensitivity of the method for the rat, guinea pig, or dog brain.[158,159] Although it was claimed that T could be detected following the administration of tryptophan or tryptophan plus a monoamine oxidase inhibitor,[158] it later transpired that the substantial carry-over of tryptophan into the extracted tryptamine fraction produced false indications of T content.[159] when an ion-exchange chromatographic step was incorporated into the procedure, T could only be found after pretreatment of the animal (guinea pig) with tryptophan and a monoamine oxidase inhibitor.[159]

More recently, the presence of T in bovine, rat, and human brain has been reported using more advanced separative and analytical methods.[34,35,37,39,160] As is usual, however, the less precise methods (radiochemical and spectrofluorimetric) produce values that are higher (20.9–69 ng/g) than those obtained by mass spectrometry (0.5 ng/g). The mass spectrometric determination of T in the whole brain of the domestic fowl, the optic lobes of the octopus, and ganglia of the locust indicated that the levels were below the limits of sensitivity of the method (see Table I). The only tissue so far examined that exhibits high concentrations of T is the arm nerve of the starfish (1251 ng/g). Since there is very little 5-HT in the nervous system of the starfish, however, it may be that T rather than 5-HT is the important indolealkylamine in echinoderms.[48] The highest concentration of T in the rat brain was observed in the caudate nucleus followed by the hypothalamus (Table III); lower values were found in all other regions.

8.2. Neurophysiological Studies

Whereas T decreased in a most profound manner the firing rate of most neurons, 5-HT excited and depressed almost equal numbers of neurons.[161] The depressant effects of 5-HT were most substantially enhanced by very weak applications of T, amounts that by themselves did not alter in any way the base-line firing rate (Fig. 7a). Excitatory responses to 5-HT were consistently reversed into depressant responses during subthreshold applications of T (Fig. 7b). These findings support the proposal that T acts as a neuromodulator in 5-HT-mediated neurotransmission in the central nervous system.

8.3. Synthesis and Catabolism

The administration of tryptophan (800 mg/kg, 0.3–2 h) produces no detectable change in the concentration of brain T in the guinea pig,[159] whereas in the presence of iproniazid as well (150 mg/kg, 16 h), brain T increased from an undetectable level to 50–270 ng/g.[159] Similar increases (from <0.5 to 133 ng/g) have been observed in the rat brain following the administration of tryptophan (100 mg/kg, 1 h) and pargyline (75 mg/kg, 2 h).[11] The inhibition of monoamine oxidase results in large increases (from 2 to 300 times their control levels) in brain concentrations of T. This effect has been demonstrated for the rat, rabbit, dog, cat, and domestic fowl.[11,37,39,40,45,59] In a detailed study carried out using the rat striatum, it was found that inhibition with the type A monoamine oxidase inhibitor clorgyline caused a significant increase in the concentration of T but that no increase followed inhibition with deprenyl, the type B monoamine oxidase inhibitor (Table IV). The other drugs used inhibited both type A and type B monoamine oxidase and produced significant increases.[59] The turnover rate for T as measured by its loss after intraventricular injection[162] or its rate of accumulation following treatment with pargyline[3] was 0.24 nmol/g per h, which is about six to ten times lower than that of 5-HT (Table XIII).

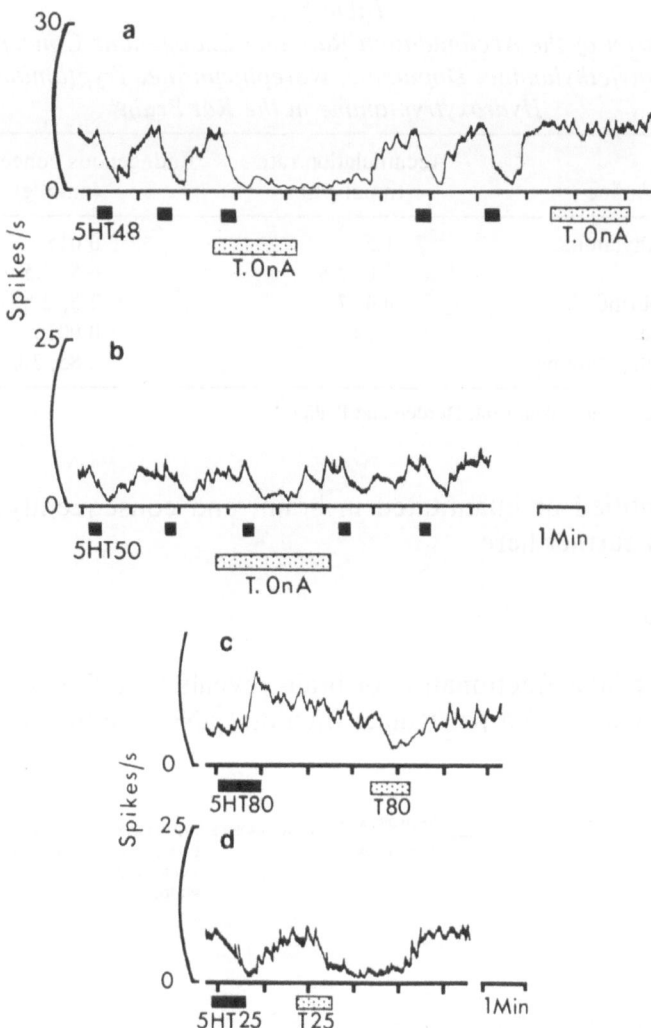

Fig. 7. Excerpts from the ratemeter records of two cortical neurons. The bars beneath the trace indicate periods of amine application, and the figures are the intensities of the ejecting current in nanoamperes. (Details as Fig. 2.) (a) The depressant response of this cell to 5-HT was profoundly potentiated when T was allowed to diffuse out of the electrode. Tryptamine allowed to diffuse from the electrode in the absence of 5-HT had little effect on neuronal firing rate. (b) Again, diffusion of T from the electrode potentiates the 5-HT response. On this occasion, the response following termination of the T application was also potentiated. (c) This cortical cell was excited by 5-HT but depressed by both 5-HT and T. Note the larger response to T. (d) This cell was depressed by both 5-HT and T. Note larger response to T. These results are from Jones and Boulton[161] and are published with permission of *Life Sciences*.

As indicated in Fig. 8, the principal breakdown product of T is indoleacetic acid. This acid has been identified and quantitated in rat brain,[163] with the largest amounts being located in the caudate nucleus and hypothalamus, and in human cerebrospinal fluid.[164] Although an enzyme capable of methylating T to the hallucinogenic substances *N*-methyltryptamine and *N,N*-dimethyltryptamine has been located in mammalian tissues, these substances have not

Table XIII
*Comparison of the Accumulation Rate and Endogenous Concentrations
of β-Phenylethylamine, Dopamine, Norepinephrine, Tryptamine, and 5-
Hydroxytryptamine in the Rat Brain*[a]

Amine	Accumulation rate (nmol/g/h)	Endogenous concentration (nmol/g)
β-Phenylethylamine	1.5	0.015
Dopamine	1.8, 2.8	6.5, 7.5
Norepinephrine	0.4, 7	2.5, 2.8
Tryptamine	0.24	0.0016
5-Hydroxytryptamine	2.5	2.85, 3.0

[a] These values were taken from Durden and Philips.[3]

yet been identified or quantitated in brain, and consequently, they are not discussed any further here.

8.4. Storage

The subcellular fractionation of brain reveals that T is found to be associated in the main with a particulate fraction.[63] No significant changes in dog

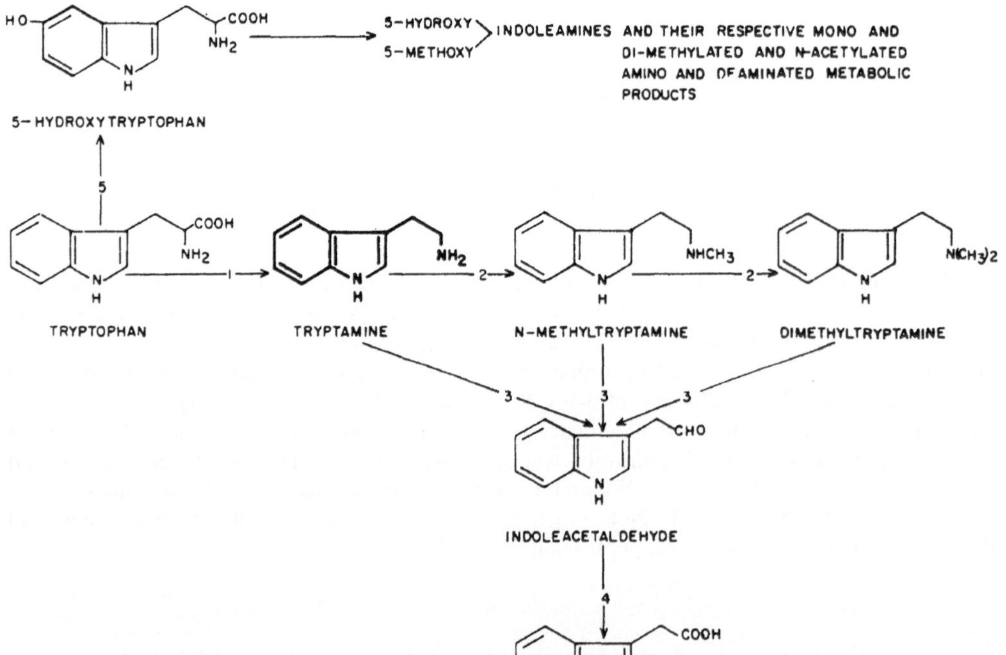

Fig. 8. Metabolism of tryptamine. 1. decarboxylase; 2, *N*-methyltransferase; 3, monoamine oxidase; 4, aldehyde oxidase; 5, tryptophan hydroxylase. Taken from Boulton[61] and reprinted with the permission of University Park Press.

brain T have been observed over 1–4 days following the administration of 0.5–1 mg/kg of reserpine.[37] These results should be interpreted with caution, however, since the control levels for T were quite high; the values obtained may therefore include some contaminant substances.

8.5. Uptake and Release

The ability of the brain to take up T appears to lie somewhere between that of PE which passes easily through membranes and that of 5-HT which passes only with great difficulty.[62]

9. CLINICAL RELEVANCE

Many of the trace amines have been implicated in a variety of mental and neurological disorders, and it is a fact that PE, the TAs, and T are behaviorally active and will mimic the stereotypies of amphetamine,[24,127,165] a drug which when ingested chronically is said to produce a clinical state resembling paranoid schizophrenia.[166,167] As can be seen from Section 4, Table VIII, and Fig. 6, the concentration of *m*-TA and *p*-TA in the caudate nucleus and mesolimbic system are definitely affected by amphetamine treatment as well as by all clinically efficacious neuroleptic drugs. Phenylethylamine and T, however, are not so affected. It has been argued, on somewhat tenuous grounds, that DA and NA may be involved in the etiology of schizophrenia; if this turns out to be so, we would submit that the reciprocal and direct relationships (see Table IX) that obtain among *p*-TA and *m*-TA and DA turnover, respectively, provide a reasonably convincing argument for their involvement as well. All of the trace amines are affected quite profoundly by the monoamine oxidase inhibitors, and it is now possible to manipulate any one of them (see Tables IV, XIII, and refs. 46,59,101) upwards to levels that render the term "trace" quite anachronistic. Since some of the monoamine oxidase inhibitors are useful as antidepressants, it becomes possible to argue, therefore, that some of the trace amines may also be involved in the etiology of the uni- or bipolar affective disorders.

Both Sandler and Wyatt and co-workers [168,169] have referred to PE as an endogenous amphetamine, and both of them[168–171] have proposed a fairly direct role for PE in schizophrenia; they point to the reportedly high levels of phenylacetic acid in blood and cerebrospinal fluid as evidence in support of their suggestion. Sabelli and Mosnaim.[172] have in a similar manner proposed that PE may be involved in the affective disorders. For reasons outlined earlier, however, scepticism must be exercised in this latter case, since hypotheses constructed on the basis of inaccurate tissue levels of PE arising from inappropriate techniques cannot be given the same weight as those based on more acceptable methodologies. In a more recent study, Sandler *et al.*[173] have proposed that an overproduction of PE, as measured as phenylacetic acid in blood plasma, occurs in aggressive psychopaths. This intriguing observation awaits confirmation. The increase in PE[23,174,175] that is found in phenylketonuria presumably arises as a consequence of the increase in its substrate, phenylalanine. Phenylethylamine, along with p-TA, has also been proposed to be the dietary trigger in the induction of migraine.[176,177]

p-Tyramine is excreted in urine in an abnormal manner in patients suffering with parkinsonism,[178-180] schizophrenia,[181] and depression,[182,183] whereas its infusion or ingestion is claimed to cause a pressor response in depression, a hypertensive crisis in patients consuming monoamine oxidase inhibitors,[184-186] and the precipitation of migraine and an activation of the EEG in migrainous and epileptic patients.[177,187,188] In a general psychiatric population, PE, p-TA, and T are all excreted abnormally by a proportion of patients.[189,190] In their unconjugated form, PE, p-TA, m-TA, and T are excreted in a remarkably stable manner that indicates that they probably arise exclusively from an endogenous origin[190-192]; this is not the case with respect to their conjugated forms or their principal acidic metabolites,[193] however, where it is apparent that exogenous as well as endogenous sources contribute to their urinary concentration. Among the metabolites of p-TA or tyrosine, p-hydroxyphenylacetic acid is excreted abnormally in hypertyrosinemia,[194] cystic fibrosis,[195,196] and spina bifida.[197]

The β-hydroxylated metabolite of p-TA, p-OA, has been implicated in hepatic and renal encephalopathies, where its mechanism of action is proposed to be that of a false neurotransmitter.[198-203] In view of the obvious neurophysiological properties of some of the trace amines, however, as outlined in Sections 3, 4, 6, and 8, it is at least worth asking the question, is it possible for them to exert "false" roles?

Tryptamine has been claimed to be abnormally excreted in schizophrenic urine,[204,205] whereas its N-methylated derivatives have been claimed to be causal agents.[206-210] These latter metabolites have not yet been unambiguously identified as tissue metabolites, but they do represent one of the many endogenous psychotogens that have, over the years, been proposed as causal agents of schizophrenia. In cerebrospinal fluid, Young[211,212] and in urine, Domino et al.[213] have claimed that indoleacetic acid, the principal metabolite of T, is increased in patients suffering with depression, schizophrenia, and hepatic coma.

Because urine represents the sink into which all body metabolites are poured, and because it is not easily possible to separate what proportion of a particular metabolite arises from the central nervous system as opposed to the periphery, attention has been turning in recent years to blood and cerebrospinal fluids as well as to the principal metabolites of the trace amines (named for consistency's sake "trace acids") and those enzymes associated with the monoamines. It is of interest to note that the trace amines are usually the substrates used when reductions (or increases) in some of the enzymes (MAO and DβH) are claimed. On the basis of the best possible objective assessment at this time, the trace amines and their associated metabolites and enzymes (along with their respective activators and inhibitors[214,215]) appear to be implicated in schizophrenia, depression, hyperkinesis, parkinsonism, aggression, renal and hepatic comas, as well as some of the subnormality conditions, so further studies using the best available methodologies appear warranted.

10. HYPOTHESES AND CONCLUSIONS

The concept of the trace amines acting as so-called synaptic activators was first advanced in 1975[4,63]; at that time it was proposed that a trace amine would interact in the synapse in such a way as to cause hyperpolarization or depolarization, not at a level sufficient to create significant postsynaptic events, but separated spatially and temporally in such a way as to maintain the synapse in a state of readiness. This state of readiness, it was claimed, was a requirement for the propagation of impulses. It was further suggested that this could occur either directly by the trace amine itself interacting with the synaptic membrane or indirectly by release of discrete quantities of other monoamines or even other putative neurotransmitters. This notion was somewhat further refined[5,216] following Dismukes[217] to provide a definition for substances active in neurotransmission. In the case of the trace amines, this produced the following definition: neurohumor (cytosolic, pre- and/or post-synaptic, synaptic, secondary, direct).

As we can see from the described iontophoretic studies,[57,161] such a role for the trace amines is really quite plausible. Of course, such a role does not preclude any one of them acting in a more classical manner in a particular location, and indeed, this seems perhaps to be the case for the two TAs.[86,218] Clearly, if a trace amine is required to maintain "neuronal homeostasis" and to hold the synaptic area in a "state of readiness," it is not difficult to see how any perturbation caused by too much or too little could lead to cerebral dysfunction and thus some of the behavioral and functional manifestations seen in certain of the psychiatric and neurological disorders. As we have seen in Section 9, there are already indications that some of the trace amines may be implicated in some clinical disorders, and it has been further established that several stimulants, neuroleptic drugs, and dopamine agonists and antagonists, as well as stressful conditions, will cause differential changes to occur with respect to m- and p-TA in the caudate nucleus and mesolimbic system. Whether the TAs or any other of the trace amines control catecholamine synthesis or function, maintain synaptic readiness, or are implicated in any of the major psychiatric of neurological illnesses will require much further work. What is now apparent, however, is that the trace amines have moved away from being curiosities or metabolic accidents to become important and functional substances in their own right.

ACKNOWLEDGMENTS. We thank our colleagues in the Psychiatric Research Division for their helpful comments and criticism and Saskatchewan Health and the Medical Research Council of Canada for continuing financial support.

REFERENCES

1. Usdin, E., and Sandler, M. (eds.), 1976, *Trace Amines and the Brain,* Marcel Dekker, New York.
2. Boulton, A. A., 1974, *Lancet* 2:7871.

3. Durden, D. A., and Philips, S. R., 1980, *J. Neurochem.* **34:**1725–1732.
4. Boulton, A. A., 1976, *Trace Amines and the Brain,* (E. Usdin and M. Sandler, eds.), Marcel Dekker, New York, pp. 21–39.
5. Boulton, A. A., 1979, *Behav. Brain Sci.* **2:**418.
6. Durden, D. A., Davis, B. A., and Boulton, A. A., 1974, *Biomed. Mass Spectrom.* **1:**83–95.
7. Durden, D. A., 1978, *Research Methods in Neurochemistry,* Volume 4 (N. Marks and R. Rodnight, eds.), Plenum Press, New York, pp. 205–250.
8. Durden, D. A., and Boulton, A. A., 1979, *Techniques in the Life Sciences* B2/11 (H. L. Kornberg, H. Metcalfe, D. Northcote, C. Pogson, and K. Tipton, eds.), Elsevier/North Holland, Amsterdan, pp. B214 1–25.
9. Davis, B. A., Durden, D. A., and Boulton, A. A., 1980, *Recent Advances in Mass Sprectrometry and Medicine,* Volume 6 (D. Frigerio and M. McCamish, eds.), Elsevier/North Holland, Amsterdam, pp. 83–93.
10. Willner, J., LeFevre, H. F., and Costa, E., 1974, *J. Neurochem.* **23:**857–859.
11. Warsh, J. J., Godse, D. D., Stancer, H. C., Chan, P. N., and Coscina, D. V., 1977, *Biochem. Med.* **18:**10–20.
12. Karoum, F., Nasrallah, H., Potkin, S., Chuang, L., Moyer-Schwing, J., Phillips, I., and Wyatt, R. J., 1979, *J. Neurochem.,* **33:**201–212.
13. Danielson, T. J., Boulton, A. A., and Robertson, H. A., 1977, *J. Neurochem.* **29:**1131–1135.
14. McCaman, M. W., and McCaman, R. E., 1978, *Brain Res.* **141:**347–352.
15. David, J. C., and Delacour, J., 1980, *Brain Res.* **195:**231–235.
16. Philips, S. R., 1981, *Advances in Cellular Neurobiology,* Volume 2 (L. Hertz and S. Federoff, eds.), Academic Press, New York pp. 355–391.
17. Saavedra, J. M., 1974, *J. Neurochem.* **22:**211–216.
18. Suzuki, O., and Yagi, K., 1976, *Anal. Biochem.* **75:**192–200.
19. Martin, I. L., and Baker, G. B., 1976, *J. Chromatogr.* **123:**45–50.
20. Baker, G. B., Coutts, R. T., and LeGatt, D. F., 1980, *Can. J. Neurol. Sci.* **7:**235.
21. Barger, G., and Dale, H. H., 1910, *J. Physiol. (Lond.)* **41:**19–59.
22. Asatoor, A. M., and Dalgliesh, C. E., 1959, *Biochem. J.* **73:**26P.
23. Jepson, J. B., Lovenberg, W., Zaltzman, P., Oates, J. A., Sjoerdsma, A., and Udenfriend, S., 1960, *Biochem. J.* **74:**5P.
24. Randrup, A., and Munkvad, I., 1966, *Acta Psychiat. Scand. [Suppl.]* **191:**193–199.
25. Fischer, E., Ludmer, R. I., and Sabelli, H. C., 1967, *Acta Physiol. Lat. Am.* **17:**15–21.
26. Mantegazza, P., and Riva, M. J., 1963, *J. Pharm. Pharmacol.* **15:**472–478.
27. Philips, S. R., Rozdilsky, B., and Boulton, A. A., 1978, *Biol. Psychiatry* **13:**51–57.
28. Mosnaim, A. D., and Luwang, E. E., 1973, *Anal. Biochem.* **54:**561–577.
29. Durden, D. A., Philips, S. R., and Boulton, A. A., 1973, *Can. J. Biochem.* **51:**995–1002.
30. Saavedra, J. M., 1974, *Anal. Biochem.* **59:**628–633.
31. Philips, S. R., Durden, D. A., and Boulton, A. A., 1974a, *Can. J. Biochem.* **52:**366–373.
32. Philips, S. R., Davis, B. A., Durden, D. A., and Boulton, A. A., 1975, *Can. J. Biochem.* **53:**65–69.
33. Molinoff, P. B., and Axelrod, J., 1969, *Science* **164:**428–429.
34. Philips, S. R., Durden, D. A., and Boulton, A. A., 1974, *Can. J. Biochem.* **52:**447–451.
35. Saavedra, J. M., and Axelrod, J., 1972, *J. Pharmacol. Exp. Ther.* **182:**363–369.
36. Tallman, J. F., Saavedra, J. M., and Axelrod, J., 1976, *J. Pharmacol. Exp. Ther.* **199:**216–221.
37. Sloan, J. M., Martin, W. R., Clements, T. H., Buchwald, W. F., and Bridges, S. R., 1975, *J. Neurochem.* **24:**523–532.
38. Fischer, E., Spatz, H., Heller, B., and Reggiani, H., 1972, *Experientia* **28:**307–308.
39. Snodgrass, S. R., and Horn, A. S., 1973, *J. Neurochem.* **21:**687–696.
40. Boulton, A. A., Juorio, A. V., Philips, S. R., and Wu, P. H., 1975, *Brain Res.* **96:**212–216.
41. Borison, R. L., Mosnaim, A. D., and Sabelli, H. C., 1974, *Life Sci.* **15:**1837–1848.
42. Faraj, B. A., Mu, J.-Y., Lewis, M. S., Wilson, J. P., Israili, Z. H., and Dayton, P. G., 1975, *Proc. Soc. Exp. Biol. Med.* **149:**664–669.
43. Mosnaim, A. D., Inwang, E. E., and Sabelli, H. C., 1974, *Biol. Psychiatry* **8:**227–234.
44. Gunne, L.-M., and Jonsson, J., 1965, *Acta Physiol. Scand.* **64:**434–438.
45. Juorio, A. V., 1976, *Brain Res.* **111:**442–445.

46. Juorio, A. V., and Philips, S. R., 1976, *Neurochem. Res.* 1:501–509.
47. Robertson, H. A., 1976, *Experientia* 32:552–553.
48. Juorio, A. V., and Robertson, H. A., 1977, *J. Neurochem.* 28:573–579.
49. Danielson, T. J., Wishart, T. B., and Boulton, A. A., 1976, *Life Sci.* 18:1237–1244.
50. Boulton, A. A., Juorio, A. V., Philips, S. R., and Wu, P. H., 1977, *Br. J. Pharmacol.* 59:209–214.
51. Philips, S. R., 1978, *Noncatecholic Phenylethylamines* (A. D. Mosnaim, and M. E. Wolf, eds.), Marcel Dekker, New York and Basel, pp. 113–138.
52. Juorio, A. V., and Danielson, T. J., 1978, *Eur. J. Pharmacol.* 50:79–82.
53. Spector, S., Melmon, K., Lovnberg, W., and Sjoerdsma, A., 1963, *J. Pharmacol. Exp. Ther.* 140:229–235.
54. Wu, P. H., and Boulton, A. A., 1975, *Can. J. Biochem.* 53:42–50.
55. Giardina, W. J., Pedemonte, W. A., and Sabelli, H. C., 1973, *Life Sci.* 12:153–161.
56. Henwood, R. W., Boulton, A. A., and Phillis, J. W., 1979, *Brain Res.* 164:247–351.
57. Jones, R. S. G., and Boulton, A. A., 1980, *Can. J. Physiol. Pharmacol.* 58:222–227.
58. Snodgrass, S. R., 1974, *J. Pharm. Pharmacol.* 26:931–936.
59. Philips, S. R., and Boulton, A. A., 1979, *J. Neurochem.* 33:159–167.
60. Yang, H.-Y. T., and Neff, N. H., 1973, *J. Pharmacol. Exp. Ther.* 187:365–371.
61. Boulton, A. A., 1979, *Int. Rev. Biochem.* 26:179–206.
62. Oldendorf, W. H., 1971, *Am. J. Physiol.* 221:1629–1639.
63. Boulton, A. A., and Baker, G. B., 1975, *J. Neurochem.* 25:477–481.
64. Shore, P. A., Silver, S. L., and Brodie, B. B., 1955, *Science* 122:284–285.
65. Holzbauer, M., and Vogt, M., 1956, *J. Neurochem.* 1:8–11.
66. Bertler, A., 1961, *Acta Physiol. Scand.* 51:75–83.
67. Fuxe, K., Grobecker, H., and Jonsson, J., 1967, *Eur. J. Pharmacol.* 2:202–207.
68. Jonsson, J., Grobecker, H., and Holtz, P., 1966, *Life Sci.* 5:2235–2246.
69. Jackson, D. M., and Smythe, D. B., 1973, *Neuropharmacology* 12:663–668.
70. Baker, G. B., Raiteri, M., Bertollini, A., Del Carmine, R., Keane, P. E., and Martin, I. L., 1976, *J. Pharm. Pharmacol.* 28:456–457.
71. Schmitt, R., and Nasse, O., 1865, *Liebigs Ann. Chem.* 133:211–216.
72. Barger, G., 1909, *J. Chem. Soc.* 95:1123–1129.
73. Henze, M., 1913, *Hoppe Seylers Z. Physiol. Chem.* 87:51–58.
74. Nishimura, T., and Gjessing, L. R., 1966, *Scand. J. Clin. Lab. Invest.* 18:217–220.
75. King, G. S., Goodwin, B. L., Ruthven, C. R. J., and Sandler, M., 1974, *Clin. Chim. Acta* 51:105–107.
76. Boulton, A. A., Dyck, L. E., and Durden, D. A., 1974, *Life Sci.* 15:1673–1683.
77. Boulton, A. A., and Dyck, L. E., 1974, *Life Sci.* 14:2497–2506.
78. Mitoma, C., Posner, H. S., Bogdanski, D. F., and Udenfriend, S., 1957, *J. Pharmacol. Exp. Ther.* 120:188–194.
79. Majer, J. R., and Boulton, A. A., 1970, *Nature* 225:658–660.
80. Boulton, A. A., and Majer, J. R., 1970, *J. Chromatogr.* 48:322–327.
81. Boulton, A. A., and Majer, J. R., 1971, *Can. J. Biochem.* 49:993–998.
82. Faraj, B. A., Dayton, P. G., Camp, V. M., Wilson, J. P., Malvean, E. J., and Schlant, R. C., 1977, *J. Pharmacol. Exp. Ther.* 200:384–393.
83. Tallman, J. C., Saavedra, J. M., and Axelrod, J., 1976, *J. Neurochem.* 27:465–469.
84. Juorio, A. V., 1977, *Brain Res.* 126:181–184.
85. Juorio, A. V., 1977, *Life Sci.* 20:1663–1668.
86. Juorio, A. V., 1979, *Br. J. Pharmacol.* 66:377–384.
87. Juorio, A. V., 1980, *Br. J. Pharmacol.* 70:475–480.
88. Bevan, P., Bradshaw, C. M., Pun, R. Y. K., Slater, N. T., and Szabadi, E., 1978, *Br. J. Pharmacol.* 63:651–657.
89. Boulton, A. A., and Wu, P. H., 1973, *Can. J. Biochem.* 51:428–435.
90. Fellman, J. H., Roth, G. S., and Fujita, T. S., 1976, *Arch. Biochem. Biophys.* 174:562–567.
91. Boulton, A. A., and Quan, L., 1970, *Can. J. Biochem.* 48:1287–1291.
92. Brandau, K., and Axelrod, J., 1972, *Naunyn-Schmiedebergs Arch. Pharmacol.* 273:123–133.
93. Silkaitis, R. P., and Mosnaim, A. D., 1976, *Brain Res.* 114:105–115.

94. Mosnaim, A. D., Edstrand, D. L., Wolf, M. E., and Silkaitis, R. P., 1977, *Biochem. Pharmacol.* **26**:1725–1728.

95. Davis, B. A., and Boulton, A. A., 1980, *Eur. J. Mass Spectrom. Med. Envir. Res.* **1**:149–153.

96. Ewins, A. J., and Laidlaw, P. P., 1910, *J. Physiol.*, **41**:78–87.

97. Blaschko, H., Richter, D., and Schlossmann, H., 1937, *Biochem. J.* **31**:2187–2196.

98. Yu, P. H., and Boulton, A. A., 1980, *J. Neurochem.* **35**:255–257.

99. Wu, P. H., and Boulton, A. A., 1974, *Can. J. Biochem.* **52**:374–381.

100. Philips, S. R., Baker, G. B., and McKim, H. R., 1980, *Experientia* **36**:241–242.

101. Juorio, A. V., Davis, B. A., and Boulton, A. A., 1980, *Res. Commun. Psychol. Psychiatry Behav.* **5**:255–264.

102. Durden, D. A., and Boulton, A. A., 1981, *J. Neurochem.* **36**:129–135.

103. Davis, B. A., Durden, D. A., Li, P. P., and Boulton, A. A., 1977, *J. Chromatogr.* **142**:517–522.

104. McQuade, P. S., Juorio, A. V., and Boulton, A. A., 1981, *J. Neurochem.* **37**:735–739.

105. Karoum, F., Gillin, J. C., Wyatt, R. J., and Costa, E., 1975, *Biomed. Mass Spectrom.* **2**:183–189.

106. Karoum, F., Wyatt, R. J., and Majchrowicz, E., 1976, *Br. J. Pharmacol.* **56**:403–411.

107. Yu, P. H., and Boulton, A. A., 1979, *Can. J. Biochem.* **57**:1204–1209.

108. Buck, S. H., Murphy, R., and Molinoff, P. B., 1977, *Brain Res.*, **122**:281–297.

109. Harmar, A. J., and Horn, A. S., 1976, *J. Neurochem.*, **26**:987–993.

110. Juorio, A. V., 1980, *Catecholamines and Stress: Recent Advances* (E. Usdin, R. Kvetnansky, and I. J. Kopin, eds.), Pergamon Press, Oxford, pp. 423–426.

111. Commarato, M. A., Brody, T. M., and McNeil, J. H., 1969, *Can. J. Physiol. Pharmacol.* **47**:511–514.

112. Steinberg, M. I., and Smith, C. B., 1971, *J. Pharmacol. Exp. Ther.* **176**:139–148.

113. Ross, S. B., and Renyi, A. L., 1971, *J. Pharm. Pharmacol.*, **23**:276–279.

114. Lentzen, H., and Phillippu, A., 1977, *Naunyn Schmiedebergs Arch. Pharmacol.*, **300**:25–30.

115. Dyck, L. E., 1978, *Neurochem. Res.*, **3**:775–791.

116. Petrali, E. H., Boulton, A. A., and Dyck, L. E., 1979, *Neurochem. Res.* **4**:633–642.

117. Petrali, E. H., 1979, *Neurochem. Res.* **5**:297–300.

118. Stoof, J. C., Liem, A. L., and Mulder, A. H., 1976, *Arch. Int. Pharmacodyn. Ther.* **220**:62–71.

119. Dyck, L. E., and Boulton, A. A., 1980, *Res. Commun. Psychol. Psychiatry Behav.*, **5**:61–78.

120. Boulton, A. A., and Dyck, L. E., 1980, *Res. Commun. Psychol. Psychiatry Behav.* **5**:79–94.

121. Arbilla, S., Kamal, L. A., and Langer, S. Z., 1981, *Br. J. Pharmacol.* **72**:499P–500P.

122. Dyck, L. E., Boulton, A. A., and Jones, R. S. G., 1980, *Eur. J. Pharmacol.* **68**:33–40.

123. Dyck, L. E., 1981, *Eur. J. Pharmacol.* **69**:371–374.

124. Juorio, A. V., 1979, *Brain Res.* **179**:186–189.

125. Dyck, L. E., 1981, *Neurochem. Res.* **6**:365–375.

126. Boulton, A. A., and Juorio, A. V., 1979, *Biol. Psychol.* **14**:413–419.

127. Randrup, A., Munkvad, I., and Usden, P., 1963, *Acta Pharmacol. Toxicol. (Kbh.)* **20**:143–157.

128. Espelin, D. E., and Done, A. K., 1968, *N. Eng. J. Med.* **278**:1361–1365.

129. Costa, E., Groppetti, A., and Naimzada, M. K., 1972, *Br. J. Pharmacol.* **44**:742–751.

130. Carlsson, A., and Lindquist, M., 1963, *Acta Pharmacol. Toxicol.* **20**:140–144.

131. Andén, N.-E., Roos, B.-E., and Werdinius, B., 1964, *Life Sci.* **3**:149–158.

132. Laverty, R., and Sharman, D. F., 1965, *Br. J. Pharm.*, **24**:759–772.

133. O Keeffe, R., Sharman, D. F., and Vogt, M., 1970, *Br. J. Pharmacol.* **38**:287–304.

134. Roos, B.-E., 1969, *J. Pharm. Pharmacol.* **21**:263–264.

135. Corrodi, H., Fuxe, K., and Ungerstedt, U., 1971, *J. Pharm. Pharmacol.* **23**:989–991.

136. Fuller, R. W., and Perry, K. W., 1978, *J. Neural Transm.* **42**:23–36.

137. Spector, S., Sjoerdsma, A., and Udenfriend, S., 1965, *J. Pharmacol.* **147**:86–97.

138. Huebert, N. D., and Boulton, A. A., 1978, *Res. Commun. Chem. Path. Pharmacol.* **22**:73–82.

139. Inwang, E. E., Mosnaim, A. D., and Sabelli, H. C., 1973, *J. Neurochem.* **20**:1469–1473.

140. Saavedra, J. M., and Axelrod, J., 1973, *Proc. Natl. Acad. Sci. U.S.A.* **70**:769–772.

141. Saavedra, J. M., Ribas, J., Swann, J., and Carpenter, D. O., 1977, *Science* **195**:1004–1006.

142. Erspamer, V., 1948, *Acta Pharmacol. Toxicol. (Kbh.)* **4**:224–227.

143. Erspamer, V., 1952, *Nature* **169**:375–376.

144. Kakimoto, Y., and Armstrong, M. D., 1962, *J. Biol. Chem.* **237**:422–427.

145. Juorio, A. V., 1978, *Experientia* **34**:1329–1330.
146. Walker, R. J., Ramage, A. G., and Woodruff, G. N., 1972, *Experientia* **28**:1173–1174.
147. Saavedra, J. M., Brownstein, M. J., Carpenter, D. O., and Axelrod, J., 1974, *Science* **185**:364–365.
148. Juorio, A. V., and Molinoff, P. B., 1974, *J. Neurochem.* **22**:271–280.
149. Barker, D. L., Molinoff, P. B., and Kravitz, E. A., 1972, *Nature [New Biol.]* **236**:61–63.
150. Robertson, H. A., and Juorio, A. V., 1976, *Int. Rev. Neurobiol.* **19**:173–224.
151. Hicks, T. P., and McLennan, H., 1978, *Br. J. Pharmacol.* **64**:485–491.
152. Hicks, T. P., and McLennan, H., 1980, *Brain Res.* **157**:402–406.
153. Karoum, F., Gillin, J. C., and Wyatt, R. J., 1975, *J. Neurochem.* **25**:653–658.
154. Baldessarini, R. J., and Vogt, M., 1971, *J. Neurochem.* **18**:2519–2533.
155. Baldessarini, R. J., and Vogt, M., 1972, *J. Neurochem.* **19**:755–761.
156. Durden, D. A., Juorio, A. V., and Davis, B. A., 1978, *Quantitative Mass Spectrometry in Life Sciences II, Proceedings of the Second International Symposium* (A. P. de Leenheer, R. R., Roncucci, and C. van Peteghem, eds.), Elsevier Scientific Publishing Company, Amsterdam, Oxford, New York, pp. 389–397.
157. Durden, D. A., Juorio, A. V., and Davis, B. A., 1980, *Anal. Chem.* **52**:1815–1820.
158. Hess, S. M., Redfield, B. G., and Udenfriend, S., 1959, *J. Pharmacol. Exp. Ther.* **127**:178–181.
159. Eccleston, D., Ashcroft, G. W., Crawford, T. B. B., and Loose, R., 1966, *J. Neurochem.* **13**:93–101.
160. Martin, W. R., Sloan, J. W., Christian, S. T. and Clements, T. H., 1972, *Psychopharmacologia* **24**:331–346.
161. Jones, R. S. G., and Boulton, A. A., 1980 *Life Sci.* **27**:1849–1856.
162. Wu, P. H., and Boulton, A. A., 1973, *Can. J. Biochem.* **51**:1104–1112.
163. Warsh, J. J., Chan, P. W., Godse, D. D., Coscina, D. V., and Stancer, H. C., 1977, *J. Neurochem.* **29**:955–958.
164. Young, S. N., Anderson, G. M., Gauthier, S., and Purdy, W. C., 1980, *J. Neurochem.* **34**:1087–1092.
165. Faurbye, A., 1968, *Compr. Psychiatry* **9**:155–177.
166. Connell, P. H., 1958, *Amphetamine Psychosis*, Chapman and Hall, London.
167. Snyder, S. H., 1973, *Am. J. Psychiatry* **130**:61–67.
168. Sandler, M., and Reynolds, G. P., 1976, *Lancet* **1**:70–71.
169. Potkin, S. G., Karoum, F., Chuang, L. W., Cannon-Spoor, H. E., Phillips, I., and Wyatt, R. J., 1979, *Science* **206**:470–471.
170. Wyatt, R. J., Gillis, J. C., Stoff, A. M., Majo, E. A., and Tinklenborg, J. R., 1977, *Neuroregulators and Psychiatric Disorders* (E. Usdin, D. A. Hamburg, and J. D. Barchas, eds.), Oxford University Press, New York, pp. 31–45.
171. Sandler, M., Ruthven, C. R. J., Goodwin, B. L., King, G. S., Pettit, B. R., and Reynolds, G. P., 1978, *Commun. Psychopharmacol.* **2**:199–202.
172. Sabelli, H. C., and Mosnaim, A. D., 1974, *Am. J. Psychiatry* **131**:695–699.
173. Sandler, M., Ruthven, C. R. J., Goodwin, B. C., Field, H., and Mathews, R., 1978, *Lancet* **2**:1269–1270.
174. Edwards, D. J., and Blau, K., 1973, *Biochem. J.* **132**:95–100.
175. Reynolds, G. P., Seakins, J. W. T., and Gray, D. O., 1978, *Clin. Chim. Acta* **83**:33–39.
176. Sandler, M., Youdim, M. B. H., and Hanington, E., 1974, *Nature* **250**:335–337.
177. Hanington, E., 1967, *Br. Med. J.* **2**:550–551.
178. Boulton, A. A., Majer, J. R., and Pollitt, R. J., 1967, *Nature* **215**:132–134.
179. Smith, I., and Kellow, A. H., 1969, *Nature* **221**:1261.
180. Boulton, A. A., and Marjerrison, G. L., 1972, *Nature* **236**:76–78.
181. Boulton, A. A., Marjerrison, G. L., and Majer, J. R., 1971, *J. Acad. Sci. (USSR)* **5**:68–70.
182. Bonham-Carter, S., Sandler, M., Goodwin, B. L., Sepping, P., and Bridges, P. K., 1978, *Br. J. Psychiatry* **132**:125–132.
183. Sandler, M., Ruthven, C. R. J., Goodwin, B. L., Reynolds, G. P., Rao, V. A. R., and Coppen, A., 1979, *Nature* **278**:357–358.
184. Ghose, K., Turner, P., and Coppen, A., 1975, *Lancet* **1**:1317–1318.
185. Pickar, A., Cohen, R. M., Murphy, D., and Fried, A., 1979, *Am. J. Psychiatry* **136**:1460–1463.

186. Blackwell, B., and Marley, E., 1964, *Lancet* **1**:530–531.
187. Scott, D. F., Moffett, A., and Swash, M., 1972, *Epilepsia* **13**:365–375.
188. Swash, M., Moffett, A. M., and Scott, A. F., 1975, *Nature* **258**:769–750.
189. Slingsby, J. M., and Boulton, A. A., 1976, *J. Chromatogr.* **123**:51–56.
190. Slingsby, J. M., 1975, *Factors affecting urinary excretion of arylalkylamines in a randomly selected psychiatric population*, Ph.D. Thesis, University of Saskatchewan, Saskatoon.
191. Boulton, A. A., 1968, *Prog. Neurogenet.* **1**:437–441.
192. Huebert, N., and Boulton, A. A., 1979, *J. Chromatogr. Biomed. Appl.* **162**:169–176.
193. Davis, B. A., and Boulton, A. A., 1981, *J. Chromatogr. Biomed. Appl.* **222**:161–169.
194. Bremer, H. J., Jaenicke, U., and Lupold, D., 1969, *Clin. Chim. Acta* **23**:244–246.
195. Gjessing, L. R., and Lindeman, R., 1967, *Lancet* **2**:47–48.
196. Gibbons, I. S. E., Seakins, J. W. T., and Ersser, R. S., 1967, *Lancet* **1**:877–878.
197. Duckworth, T., Bond, P. A., McKibbin, B., and Mitchell, E. C., 1973, *Clin. Chim. Acta* **43**:249–255.
198. Fischer, J. E., and Baldessarini, R. J., 1971, *Lancet* **2**:75–80.
199. Lam, K. C., Tall, A. R., Goldstein, G. B., and Mistilis, S. P., 1973, *Scand. J. Gastroenterol.* **8**:465–472.
200. Manghani, K. K., Lanzer, M. R., Billing, B. H., and Sherlock, S., 1975, *Lancet* **2**:943–946.
201. Rossi-Fanelli, F., Cangiano, C., Attili, A., Angelico, M., Cascino, A., Capocaccia, L., Strom, R., and Crifo, C., 1976, *Clin. Chim. Acta* **67**:255–261.
202. Capocaccia, L., Cangiano, C., Attili, A. F., Angelico, M., Cascino, A., and Fanelli, F. R., 1977, *Clin. Chim. Acta* **75**:99–105.
203. Baldessarini, R. J., and Fischer, J. E., 1977, *Arch. Gen. Psychiatry* **34**:958–964.
204. Berlet, H. H., Pscheidt, G. R., Spaid, J. K., and Himwich, H. E., 1964, *Nature* **203**:1198–1199.
205. Sullivan, J. L., Cottey, C. E., Basuk, B., Cavenar, J. D., Maltbie, A. A., and Zung, W. K., 1980, *Biol. Psychiatry* **15**:113–120.
206. Saavedra, S., and Axelrod, J., 1972, *Science* **175**:1365–1366.
207. Morgan, M., and Mandell, A. J., 1969, *Science* **165**:492–493.
208. Morgan, M., and Mandell, A. J., 1971, *Nature [New Biol.]* **230**:85.
209. Mandell, L. R., Rosenweig, S., and Kuehl, F., 1971, *Biochem. Pharmacol.* **20**:712–716.
210. Mandell, L. R., Ahn, A. S., Vandenheuvel, N. J. A., and Walker, R. W., 1972, *Biochem. Pharmacol.* **21**:1197–1200.
211. Young, S. N., and Lal, S., 1980, *J. Neural Transm.* **47**:153–161.
212. Young, S. N., 1980, *Enzymes and Neurotransmitters in Mental Disease* (E. Usdin, T. L. Sourkes, and Q. M. B. H. Youdim, eds.), John Wiley & Sons, New York, pp. 247–260.
213. Domino, E. F., Matthews, B. N., and Tait, S. K., 1979, *Biomed. Mass Spectrom.* **6**:331–334.
214. Yu, P. H., O'Sullivan, K. S., Keegan, D., and Boulton, A. A., 1980, *Psychiatry Res.* **3**:205–210.
215. Yu, P. H., and Boulton, A. A., 1979, *Life Sci.* **25**:31–36.
216. Boulton, A. A., 1980, *Neurochemistry and Clinical Neurology* (L. Battiskin, G. Hashin, and A. Lajtha, eds.), Alan R. Liss, New York, pp. 291–303.
217. Dismukes, R. K., 1979, *Behav. Brain Sci.* **2**:409–448.
218. Juorio, A. V., and Jones, R. S. G., 1981, *J. Neurochem.* **36**:1898–1903.

Polyamines

N. Seiler

1. INTRODUCTION

A group of simple aliphatic diamines and the compounds that derive from them by addition of one or two aminopropyl moieties are commonly called polyamines. This obviously erroneous designation, together with such unsavory names as putrescine and cadaverine for some of the compounds, did not help to put these compounds into the focus of general attention until recently. However, the history of the polyamines is even older than that of the nucleic acids, and a number of famous names are attached to their discovery, among which those of A. van Leeuwenhoek, N. Vauquelin, and, more recently, O. Rosenheim are most prominent.[1]

It is one of the outstanding features of these amines that they occur in all living organisms. They share this ubiquity with amino acids, proteins, and nucleic acids. This is perhaps the strongest evidence for basically important functions of these amines at all levels of organizational complexity, in prokaryotic as well as eukaryotic cells, including the cells of the mammalian nervous system. In the latter, only putrescine, spermidine, and spermine are of significance. Therefore, only these three amines will be discussed in this chapter.

Two physicochemical attributes of the polyamines are predictable from their chemical structure (Fig. 1). At physiological pH, all nitrogen atoms are protonated, so these amines represent structures in which positive charges are distributed along carbon chains of certain lengths. The second typical feature is their conformational flexibility which means that polyamines can attain various forms, more or less bent or extended.

The positive charges allow them to interact with a variety of negatively charged molecules among which nucleic acids are the most important,[2] but anionic sites of membranes[3] and proteins may also be binding sites. Because of their polycationic nature, polyamines show many features of inorganic polycations such as Mg^{2+}. However, the distribution of the charges along a

---

N. Seiler • Centre de Recherche Merrell International, 67100 Strasbourg. France.

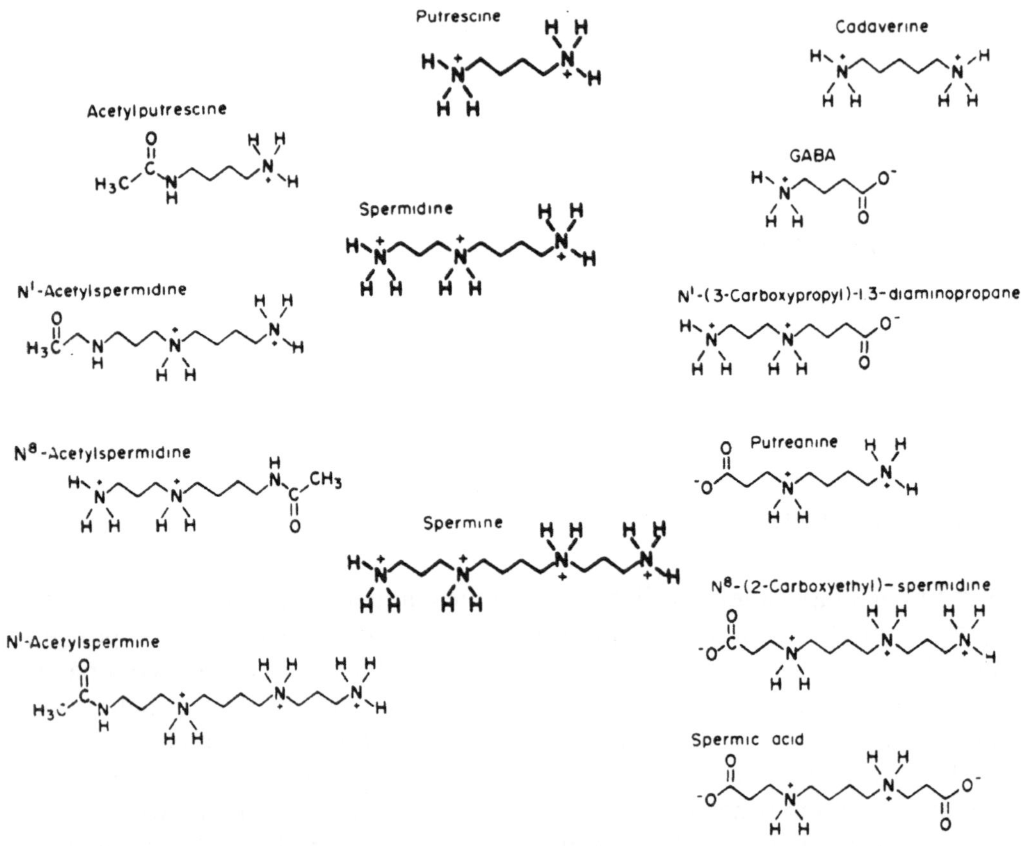

Fig. 1. Structural formulas of the polyamines and related compounds.

flexible chain allows them not only to bind, for instance, to DNA but to bridge distances such as the small groove that is formed by the two strands of the double helix.[4] A similar clamp function is also known in the helical parts of the single-stranded RNAs[2] and may play a role at other binding sites as well. It has been shown that the helical parts of the nucleic acids can be stabilized by polyamines against thermal denaturation and enzymic degradation.[1]

A key distinction between inorganic cations and the organic polycations should be especially emphasized: cellular Mg^{2+} and other metal cations ultimately derive from the circulation. Their levels are dependent on their diffusion or transport through membranes. Even more important, ion–ion interactions between metallic polycations and anionic sites can be interrupted either by displacement by other cations or by removal of the negatively charged binding site by a metabolic reaction. In contrast, the organic polycations can be formed, conjugated, or degraded within the cell, and their formation and degradation are regulated. This permits adjustment of intracellular concentrations to physiological needs in a more subtle way than via transport.

One could consider the polyamines as a kind of sophisticated Mg^{2+}, and indeed, many of the effects the polyamines exert *in vitro* on enzymes can be mimicked by Mg^{2+} or by other metal cations, although higher concentrations of cations are usually needed than those of the polyamines. There are, however, examples that indicate that polyamines have functions that cannot be fulfilled by metal cations.[1] On the basis of a large number of experiments, one can now state with considerable certainty that polyamines affect practically all enzymic reactions *in vitro* in which nucleic acids are involved.[1,4,5] These reactions comprise all steps in the transcription and translation of genetic information.

Some of the reported *in vitro* effects of the polyamines might be experimental artifacts and may not necessarily support cause-and-effect relationships between DNA, RNA, or protein synthesis and polyamines. Nonetheless, it is striking that the only types of mammalian cells that are low in polyamines are those that have no nuclei, like erythrocytes and platelets, and those in which nuclear DNA replication and RNA transcription as well as extramitochondrial protein synthesis are irreversibly switched off, as in ejaculated spermatozoa.[6]

Increasing attention was paid to the polyamines during the last decade when it became evident that polyamine metabolism was enhanced in tumors[7,8] and that certain levels of polyamines in eukaryotic cells were essential for normal cellular growth and proliferation. The latter findings were possible because of the development of specific inhibitors of polyamine biosynthesis.[9,10] Consequently, growth rates of tumors could be slowed by inhibition of polyamine biosynthesis.[11]

The present interest is perhaps best documented by the fact that numerous aspects of polyamine biochemistry and function were reviewed during the last decade in books,[1,4,12] proceedings of meetings,[7,13-18] and review articles.[2,5,6,9,11,19-44] The number of original publications is now barely numerable.

Brain aspects of polyamine biochemistry have occasionally been summarized,[22,38,42,45-51] but no comprehensive review exists. Evidence is accumulating that the metabolism of polyamines in the mammalian brain differs in several respects from that in visceral organs. It might therefore be useful to survey the literature in order to see whether our present knowledge allows us to deduce brain-specific functions of the polyamines.

2. SYNTHETIC AND CATABOLIC PROCESSES

Figure 2 gives an overview of the reactions involved in the formation, interconversion, and catabolism of the polyamines in mammalian cells. Spermidine and spermine can be regarded as metabolic derivatives of L-ornithine and L-methionine. Putrescine is formed by decarboxylation of ornithine. The aminopropyl residues that combine with putrescine to form spermidine and spermine originate from *S*-adenosylmethionine which forms *S*-adenosyl-*S*-methylhomocysteamine by decarboxylation. From the latter compound, the aminopropyl moiety is transferred to putrescine to form spermidine, and to spermidine to form spermine.[1,4,6,19,25,28,29] 5'-Methylthioadenosine is the other product of these reactions. It is rapidly converted to 5'-methylthioribose-1-phosphate and adenine. The 5'-methylthioribose-1-phosphate is hydrolytically split by methylthiolase to ribose-1-phosphate and methylthiol.[4,6]

In comparison with the synthetic reactions, the steps of polyamine catabolism are less well established. Many of the reactions shown in Fig. 2 have only recently been demonstrated to exist. For the conversion of spermine into

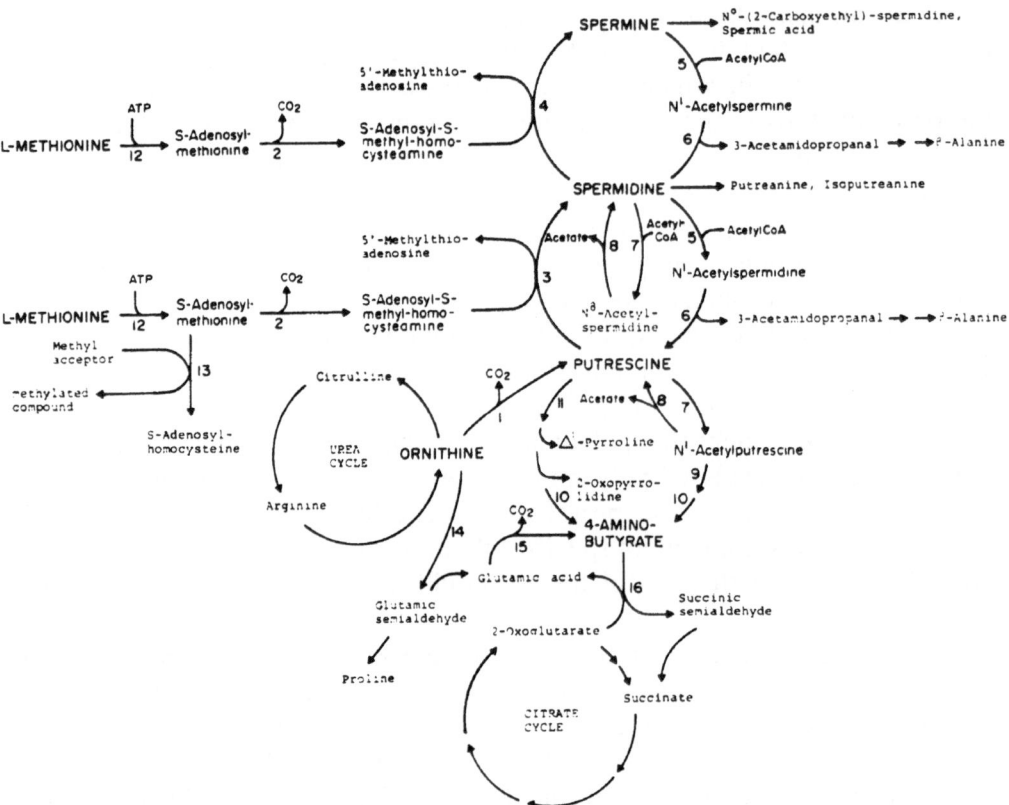

Fig. 2. Reactions involved in the biosynthesis and degradation of the polyamines and their precursors. The reactions are catalyzed by the following enzymes: 1, ornithine decarboxylase (E.C. 4.1.1.17); 2, S-adenosyl-L-methionine decarboxylase (E.C. 4.1.1.50); 3, spermidine synthase; 4, spermine synthase; 5, acetyl-CoA:polyamine acetyltranferase (cytoplasmic); 6, polyamine oxidase; 7, acetyl-CoA:polyamine acetyltransferase (nuclear); 8, acetylpolyamine deacetylase; 9, monoamine oxidase (E.C. 1.4.3.4); 10, aldehyde dehydrogenase; 11, diamine oxidase (E.C. 1.4.3.6); 12, ATP:L-methionine-S-adenosyltransferase (E.C. 2.5.1.6); 13, S-adenosylmethionine:methylacceptor methyltransferases; 14, L-ornithine:2-oxoglutarate aminotransferase (E.C. 2.6.1.13); 15, glutamate decarboxylase (E.C. 4.1.1.15); 16, 4-aminobutyrate:2-oxoglutarate aminotransferase (E.C. 2.6.1.19).

spermidine and of spermidine into putrescine within the cells, monoacetylation seems necessary.[52-55] N^1-Acetylspermine and N^1-acetylspermidine are substrates of polyamine oxidase which forms spermidine and putrescine, respectively, and 3-acetamidopropanal.[53]

In principle, this type of interconversion could also be achieved from the nonacetylated amines because they are also substrates of polyamine oxidase, although poorer substrates than the acetyl derivatives.[53] Within the circulation, the plasma amine oxidase can convert spermine and spermidine to aldehydes by oxidation of the primary amino groups of the C_3 moieties. These aldehydes can then spontaneously eliminate acrolein to form spermidine and putrescine, respectively;[1,4,19,56] i.e., the net result is almost the same as that shown in Fig. 2 for the intracellular catabolism. The mode of formation of the acidic metabolites putreanine,[57,58] isoputreanine (3-carboxypropyl)-1,3-diaminopropane,[59,61] N^8-(2-carboxyethyl)-spermidine[58,59] and spermic acid[58,59] has been recently clarified. They are formed from the abovementioned aldehydes by further oxidation by an aldehyde dehydrogenase i.e., only the second step occurs intracellularly, whereas the precursor aldehyde is formed within the circulation.[62]

The catabolism of putrescine is analogous to that of spermidine and spermine.[51] In organs that are rich in diamine oxidase and in circulation, 4-aminobutyraldehyde is formed and can either spontaneously cyclize to Δ^1-pyrroline, or the aldehyde can be further oxidized to GABA by an aldehyde dehydrogenase.[63-65] In organs such as the adult mammalian brain with low diamine oxidase activity, acetylation to monoacetyl putrescine may occur first. This is a substrate of monoamine oxidase and can be transformed to GABA in a three-step reaction.[66] 2-Oxopyrrolidine can also derive from oxidative degradation of putrescine, but it seems not to be formed by brain tissue.[67]

N^1-Acetylspermidine, N^8-acetylspermidine, and acetylputrescine are normal urinary excretion products.[59] In human urine they comprise about 80% of the total polyamines excreted.[68-71]

2.1. Enzymes of Polyamine Biosynthesis

2.1.1. Enzymes Involved in Putrescine Formation

The fact that arginase (L-arginine amidinohydrolase; E.C. 3.5.3.1) is present in mammalian brains[72] suggests that a major part of ornithine for putrescine formation is provided by hydrolysis of arginine, but it is presently not known how far ornithine abstracted as such from blood is utilized for polyamine synthesis in the CNS and to what extent ornithine formation from L-glutamate is essential.

In contrast to plants and microorganisms, the mammalian organism has only one putrescine-forming reaction, the decarboxylation of ornithine.[1,4,6,37] The responsible enzyme, ornithine decarboxylase (L-ornithine 1-carboxy-lyase; E.C. 4.1.1.17) has not been purified from brain tissue. There is some evidence for differences in the kinetic properties of the brain and liver enzymes,[73] but it is unlikely that these differences are significant.

The liver enzyme, which has been extensively purified,[74] is pyridoxal and thiol requiring.[6,37] In the absence of dithiothreitol or similar thiols, it appears to undergo polymerization, leading to molecular species devoid of enzymic activity.[75] Its apparent K_m for L-ornithine is 0.09 mM. It can attack L-lysine, but its affinity for this substrate is much lower (K_m 9 mM).[76]

Ornithine decarboxylase activity is low in all nondividing tissues including the brain[77] (Table I). Its most conspicuous characteristic is its very short biological half-life. Because of methodological difficulties, exact figures are not available for tissues,[6,37] but one can assume that it is of the order of 20 min or less depending on the tissue and its physiological status.

Ornithine decarboxylase is induced in brain by a large variety of factors, among which nerve growth factor has been most extensively studied.[78-82] This hormonelike protein is necessary for the development and maintenance of peripheral sympathetic neurons; it also has functions in the brain, although these are not presently understood.[83,84] Electrical stimulation,[85] certain

Table I

Distribution of Enzymes Involved in Polyamine Biosynthesis and Degradation in the Tissues of Adult Rats

Tissue	Ornithine decarboxylase[a,c] (pmol $^{14}CO_2$/mg protein per h)	S-adenosyl-L-methionine decarboxylase[a,c] (pmol $^{14}CO_2$/mg protein per h)	Spermidine synthase[a,c] (pmol/mg protein per h)	Spermine synthase[a,c] (pmol/mg protein per h)	Methionine adenosyl-transferase[a,c] (pmol/mg protein per h)	N^8-Acetyl-spermidine deacetylase[a,d] (pmol/mg protein per h)	Polyamino oxidase[b,e] (nmol/g tissue per h)
Brain	28	592 ± 22	6,180 ± 300	6,440 ± 480	4,740 ± 600	191 ± 47	3,780 ± 1,260
Heart	74	10	2,400	540	2,460	—	1,200 ± 180
Kidney	150	148	2,900	1,018	18,400	755 ± 18	8,940 ± 1,320
Liver	14	136 ± 16	6,440 ± 400	532 ± 66	336,600 ± 15,600	1,967 ± 47	12,120 ± 1,680
Lung	40	56	3,960	788	2,280	513 ± 9	2,580 ± 1,080
Pancreas	14	578	75,600	1,126	33,500	—	22,860 ± 2,200
Ventral prostate	1,882	2,340	36,200	5,480	8,460	—	4,140 ± 480
Skeletal muscle	16	78	500	554	840	302 ± 38	600 ± 180
Small intestine	402	302	11,820	1,332	4,680	—	6,480 ± 960
Spleen	24	172	8,160	1,778	3,900	1,135 ± 9	9,120 ± 1,560
Testes	198	380 ± 42	4,360 ± 460	414 ± 158	4,200	—	5,880 ± 480
Thymus	126	246	11,420	2,180	12,700	—	5,580 ± 1,200

[a] These values were obtained from tissue extracts with appropriate buffers.
[b] Total enzyme activity as measured in tissue homogenates.
[c] Data from Raina et al. ref. 77.
[d] Data from Blankenship, ref. 126.
[e] Data from Seiler et al., ref. 125.
[f] Mean ± S.D.

drugs,[47,86-90] hypoxia,[91] and most probably brain injuries as well cause an increase in the activity of ornithine decarboxylase. In newborn rats even the early interruption of mother–infant interaction significantly changes brain ornithine decarboxylase activity,[42,92,93] and cortisol, thyroxine, and many other hormones influence its developmental pattern.[42,94-97]

The high turnover rate of ornithine decarboxylase and its inducibility suggest that this enzyme is a trigger for the initiation of adaptive changes and other physiological events. However, its activity is not controlled by allosteric behavior but most probably by *de novo* enzyme protein synthesis and degradation, wherein an inactive form of the protein molecule may be formed in the first steps. But other mechanisms are currently being considered as well. Since it has not been studied in the brain, the regulation of ornithine decarboxylase will not be discussed in detail in this chapter. The reader is referred to pertinent reviews of the subject.[5,6,28,37,39,98,99]

Tissue levels of putrescine have a significant role in the regulation of ornithine decarboxylase activity. This amine does not exert its effect by direct interaction with the enzyme protein but rather by influencing enzyme formation. High putrescine concentrations can induce the formation of an endogenous inhibitor called antizyme.[37,99,100] Whether antizyme can be formed in brain seems not to be known.

2.1.2. Enzymes Involved in the Formation of S-Adenosyl-S-methylhomocysteamine

S-Adenosylmethionine is formed from L-methionine and ATP. The reaction is catalyzed by methionine adenosyltransferase[101] (ATP:L-methionine S-adenosyltransferase; E.C. 2.5.1.6), an enzyme normally abundantly available in most tissues[77] (Table I), so the availability of S-adenosylmethionine is normally not rate limiting in polyamine formation. There is evidence that tissue methionine levels are rate limiting in S-adenosylmethionine formation.[102]

Decarboxylation of S-adenosylmethionine by S-adenosylmethionine decarboxylase (S-adenosyl-L-methionine 1-carboxy-lyase; E.C. 4.1.1.50) provides S-adenosyl-S-methylhomocysteamine ("decarboxylated S-adenosylmethionine"). This compound contains the "active aminopropyl" moiety that is transferred to putrescine and spermidine (Fig. 2). Tissue levels of S-adenosyl-S-methylhomocysteamine are very low under normal conditions.[103]

The mammalian enzyme has pyruvate covalently linked.[104] Pyridoxal phosphate, although suspected,[105] is not a cofactor.

Putrescine markedly enhances enzyme activity at physiological concentrations; its K_m value as effector is 5–30 μM. Stimulation by putrescine increases the affinity of the enzyme for S-adenosylmethionine,[106] but the kinetics are not classically allosteric. There is some evidence that putrescine interacts in some way with the carbonyl group of pyruvate. Spermine antagonizes the stimulating effect of putrescine.[28]

A second mode of control is exerted at the level of turnover. The biological half-life of S-adenosylmethionine decarboxylase in liver—that of the brain enzyme is not known—is almost as short as that of ornithine decarboxylase, namely, around 40 min.[28] Decrease of spermidine levels in cultured cells by inhibition of ornithine decarboxylase produces a very marked increase in the activity of S-adenosylmethionine decarboxylase.[107,108] Spermidine and possibly spermine levels are, therefore, the most likely candidates for indirect regulation of the tissue levels of this enzyme.

In adult brain, S-adenosylmethionine decarboxylase activity is high[77] (Table I) compared with its activity in most other tissues and especially in relation to ornithine decarboxylase activity. The possible physiological significance of this enzyme pattern will be discussed in a subsequent paragraph (see Section 4.3).

2.1.3. Spermidine and Spermine Synthases

In contrast to the two decarboxylases, ornithine decarboxylase and S-adenosylmethionine decarboxylase, which are easily inducible enzymes with rapid turnover, the two enzymes that form spermidine and spermine from putrescine and spermidine, respectively, by transfer of aminopropyl residues from S-adenosyl-S-methylhomocysteamine (Fig. 2) seem to be stable enzymes[28] and abundantly present in tissues.[77]

Brain tissue has high spermidine synthase (S-adenosyl-S-methylhomocysteamine: putrescine 3-aminopropyl transferase) and exceptionally high spermine synthase (S-adenosyl-S-methylhomocysteamine: spermidine 3-aminopropyl transferase) activities (Table I). These enzymes have no requirement for cofactors.

Spermine synthase has recently been purified to apparent homogeneity[109] from bovine brain. It consists of two identical subunits with molecular weights of approximately 44,000. Its apparent K_m for S-adenosyl-S-methylhomocysteamine is about 0.6 μM and for spermidine 60 μM. Both reaction products, spermine and 5'-methylthioadenosine, compete with S-adenosyl-S-methylhomocysteamine for the enzyme's active site. Putrescine, on the other hand, is competitive with respect to spermidine. It is expected that purified spermidine synthase has analogous properties.

2.2. Enzymes Involved in the Interconversion and Catabolism of Polyamines

The interconversion of the polyamines, i.e., the formation of spermidine from spermine and of putrescine from spermidine, has been shown to occur in the brain of mouse,[110] rabbit,[111] and fish[46,112] and also in the retina of rat and fish.[113]

There is little doubt that these reactions occur not only as a result of nonphysiological intraventricular or intracerebral injections of the labeled precursors[46,111,113] but also under physiological conditions. This was revealed by following the specific radioactivities of the polyamines over periods of 2 and 3 months after administration of labeled putrescine.[110,112] The reactions involved in these interconversions have only very recently been suggested[52,53] and need further elucidation. However, increasing evidence shows that the pathways, as formulated in Fig. 1, are in principle correct; i.e., acetylation initiates the reaction sequence. The acetylated polyamines are then either degraded to the lower analogue or they may be excreted.

2.2.1. Acetylcoenzyme A:Amine Acetyltransferases

Acetylputrescine has been found to be a normal brain constituent.[114,115] It is also a constituent of human and rat urine.[71,116] N^1-Acetylspermidine and N^8-acetylspermidine seem also to be present in low concentrations in tissues (N. Seiler, unpublished observation), and these compounds are also urinary excretion products.[71,116]

N^1-Acetylspermine has not yet been detected in mammalian tissues or as an excretory product, but it can be formed by both bacterial[117] and mammalian acetylases.[55,118,119] However, the enzymes that transfer the acetyl residue from acetyl-CoA to the various polyamines have not been well characterized.

Acetylation of putrescine by brain tissue was demonstrated first. The main activity of the enzyme was localized in the cell nuclei, but microsomal preparations were also active.[120] With acetyl-CoA as cofactor, liver cell nuclei acetylated putrescine as actively as brain cell nuclei; however, liver homogenates did not form acetylputrescine.[120] Starting from these observations, it was shown[118,119] that the nuclear enzyme resided with the chromatin. It was purified 5000-fold.[121] This purified preparation acetylated both polyamines and histones.

The K_m values of the nuclear enzyme are between 0.5 and 2.6 mM for spermidine, spermine, and putrescine as substrates.[118-120]

Chromatin[119] or purified nuclei (N. Seiler, unpublished observation) with spermidine as substrate produce N^8-acetylspermidine almost exclusively. It was shown very recently that liver cytosol is also capable of acetylating spermidine and spermine, although not putrescine. The reaction product of this enzyme with spermidine is exclusively N^1-acetylspermidine (I. Matsui and A. E. Pegg, unpublished observation). It derives from these observations that the interconversion of the polyamines is initiated by the cytosolic acetyl-CoA:spermidine-N^1-acetyltransferase. The function of the nuclear spermidine N^8-acetyltransferase is presently unknown, notwithstanding the fact that acetylputrescine and N^8-acetylspermidine are normal excretory products.[59,68-71] The total activity of the latter enzyme changes in the developing rat brain inversely with total putrescine content.[122] Regulation of ornithine decarboxylase activity should be sufficient for controlling brain putrescine concentrations. The developmental pattern of the acetylase indicates, however, that this enzyme may have an additional role in the limitation of brain putrescine levels.

2.2.2. Polyamine Oxidase

Plasma amine oxidases are probably not involved in the active regulation of the intracellular polyamine levels, although they may have a role in the formation of putreanine and related compounds, as was pointed out earlier. Their function is presumably the metabolism of the polyamines within the circulation. The polyamines in the circulation are either of alimentary or bacterial origin from the gastrointestinal tract or polyamines that were liberated from cells under physiological conditions, by cell damage, or by cell death. It seems important for the organism to avoid accumulation of polyamines in the circulation. Plasma polyamine oxidases are part of the system that keeps circulating polyamines low. Whether the aldehydes formed by plasma amine oxidases have additional functions is not known.

Only two enzymes have been claimed to be involved in intracellular metabolism of polyamines. Caldarera and co-workers noted an amine oxidase in chick embryo brain.[123] Its inhibition with iproniazid caused an increase of polyamines and nucleic acids in the embryo. Unfortunately, this enzyme has never been characterized. It has not been ruled out that it is identical with the polyamine oxidase purified by Hölttä[124] from rat liver. This flavin-containing amine oxidase in many respects resembles monomamine oxidase, although it is not localized in mitochondria but is found in the cytoplasm and the peroxisomes.

Polyamine oxidase is abundantly present in all tissues, including the mammalian brain[125] (Table I). In contrast to the cytosolic acetyltransferase, it seems not to be easily inducible. Therefore, it is presumed that the acetyltransferase limits the formation of N^1-acetylspermidine (and probably also of N^1-acetylspermine) and controls the amount of substrate for the conversion of spermine into spermidine and spermidine into putrescine (Fig. 2).

2.2.3. Acetylpolyamine Deacetylase

Whereas N^1-acetylspermidine can be oxidatively degraded to putrescine, spermidine can be regenerated from N^8-acetylspermidine, and putrescine from acetylputrescine. These deacetylations are catalyzed by a cytosolic enzyme the activity of which is highest in liver and spleen, although brain also contains measurable activities[126] (Table I). The functional significance of this reaction is presently unknown. Since acetylpolyamines are less polar and carry less positive charges than the nonacetylated derivatives, acetylation may be a suitable reaction to displace polyamines from anionic binding sites and to move them through membranes[37,52,119]; i.e., by successive acetylation and deacetylation, polyamines could be carried from one cellular compartment into another.

3. POLYAMINE CONCENTRATION IN BRAIN

Spermidine and spermine have been determined in the brains of a variety of vertebrate species: mouse,[127] rat,[128-135] golden hamster,[132] rabbit,[130,136] guinea pig,[132] sheep,[129,130] dog,[130] monkey,[137,138] man,[129,130,139,140] fowl,[132] duck,[132] pigeon,[132] parrot,[132] and some fish.[46,141] Fewer data are available of brain putrescine concentrations.[46,131-134,140,141]

The published values show relatively small differences among various mammals and birds.[132] Whole-brain putrescine concentrations are of the order of 10–15 nmol/g net weight of tissue. Spermidine varies between 300 and 600 nmol/g, and spermine between 280 and 410 nmol/g net tissue weight. Guinea pig brain is exceptional with regard to its putrescine concentration (90 $\pm$ 3 nmol/g). In fish brain, the putrescine concentration is even higher than that of the polyamines, around 620 nmol/g; spermidine and spermine concentrations are somewhat lower (440 nmol/g and 170 nmol/g)[141] than those in the brains of mammals and birds.

Brain levels of the polyamines are of the same order of magnitude as those in other tissues. One cannot derive, therefore, specific functional roles from the average brain amine levels. However, polyamine concentrations are high compared with those of biogenic amines such as dopamine, norepinephrine, and serotonin which function as neurotransmitters.

3.1. Developmental Aspects

3.1.1. The Polyamines in the Growing Fish Brain

The brains of many fish species grow in parallel to the increase in body size until a very late stage of their lives. Cellular growth and proliferation of the various structures proceeds in such a way that brain functions are not significantly altered during growth. This is demonstrated by the fact that the basic behavioral pattern of small and large individuals of the same species is essentially the same except for the sexual behavior of mature animals. The fish brain is therefore a model for a slowly growing functionally constant organ with a high degree of cellular complexity.

Analyses of trout brains weighing between 60 and 1000 mg showed linear increases of putrescine, spermidine, protein, and RNA content with brain weight.[141] In contrast, the rate of spermine and DNA increase slowed with increasing brain weight. Spermidine and putrescine were consequently directly proportional to the RNA and protein content, and spermine showed a linear correlation with DNA. Since the DNA content of the various cell types of brain is with few exceptions the same, the cellular spermine concentration is apparently constant during growth, whereas the content of the presumed cytoplasmic constituents increases with increasing cell size. These findings are good, although necessarily indirect, evidence for the preferential localization of spermidine and spermine in different cellular compartments. One would, of course, like to suggest that spermine is mainly localized within the cell nucleus and can function there as a structural component of DNA. However, direct experimental evidence for this is lacking, and autoradiographic attempts to localize polyamines within the cells have failed. Usually, both cytoplasm and nucleus were tagged.[112,142]

3.1.2. Polyamine Concentrations and Activities of Enzymes of Polyamine Biosynthesis during Brain Development in Mammals and Birds

In contrast to the steadily growing fish brain, the mammalian brain grows in thrusts. Periods of rapid cell proliferation are followed by periods of cellular differentiation. Functional maturation occurs at a time when neuronal proliferation has practically ceased. However, because of the cellular heterogeneity of the brain and the differences in the time course of maturation of the various parts of the brain, it is not possible to correlate biochemical findings on whole brain with specific events in development. In general, however, enzymes that are involved in the process of cell proliferation and growth should be found in earlier stages of life. Enzymes concerned with specific functional activity tend to develop later, as physiological function matures. Therefore, even whole-brain data reflect major events taking place at a certain period.[72] With regard to the polyamines, this means that one expects high ornithine decarboxylase activities and high putrescine concentrations in embryonal brains, i.e., at periods of rapid cell proliferation. Indeed, the experimental data fully support this expectation: ornithine decarboxylase activity is highest in rat brain at a fetal age of 14–16 days, i.e., 7–9 days before birth. It declines gradually to adult levels which are reached around day 20 of postnatal life. The gradual decline is only interrupted at the time of birth by a burst of ornithine decarboxylase activity. This massive increase of enzyme activity seems to be typical for brain since it does not occur in heart, skeletal muscle, or liver.[143] In the cerebellum, two maxima of ornithine decarboxylase activity were observed, one about 2 days before birth and the second around day 7 of postnatal life, with a rapid decline thereafter.[143,144] Hence, the periods of greatest ornithine decarboxylase activity parallel those of maximal neuronal proliferation. In agreement with this statement are findings in rhesus monkey brain where significant activities of ornithine decarboxylase were only measurable in embryos until a gestational age of 70 and 75 days. In both rat[133] and monkey brain[138] putrescine concentrations declined gradually with maturation and paralleled ornithine decarboxylase activity. The burst of this enzyme at birth was not reflected in a similar increase of putrescine levels. Neither glial proliferation nor the postnatal development of cerebellum was associated with corresponding changes in whole-brain putrescine levels. Relationships similar to those in brain seem to hold in rat retina,[145] and declining putrescine levels in parallel with embryonal maturation were also observed in chicken brain[146,147] and retina.[148]

A study of human pre- and postnatal development shows a more or less constant brain putrescine concentration until a fetal age of 35 weeks. Then an increase in putrescine level takes place within a few weeks which is followed by a gradual decrease over the course of 30 months after birth to adult levels. The rate differences in the development of the various parts of the brain are reflected in the putrescine levels: whereas in forebrain and cerebellum maximal putrescine concentrations were observed before birth, brainstem showed maximal values only around 10 months post-partum.[140] The changes in putrescine concentrations are considerably smaller than those observed, for instance, in rat[133] or monkey brain.[138] This is demonstrated by the fact that total putrescine in human brain increases during the first 30 months of postnatal life,[140] whereas a dramatic decrease in putrescine is observed in rat brain from day 10 post-partum, although the brain weight is nearly doubled between day 10 and 60 in the latter species.

High brain putrescine levels at birth indicate functionally immature brains, whereas autophagous animals, i.e., those that are born with a quasimature behavioral pattern, such as guinea pig and fowl, are born with an adultlike brain polyamine pattern, although the brain weight at birth may only be a fraction of that of the adult animal.[132]

In primary cultures of chick embryo brain cells, an age-dependent decrease of ornithine decarboxylase activity was observed and was similar to that seen in the optic lobe, cerebral hemispheres, and midbrain–diencephalon of the brain[149] if the results were expressed in terms of protein content. In pure neuronal cultures from 8-day-old chick embryo hemispheres, cellular ornithine decarboxylase activity increased during the first 3 days of incubation, i.e., at a nominal embryonal age of 8–11 days. Thereafter, it remained practically constant until a time when the neurons had formed numerous outgrowths and synaptic contacts.[150] In other words, ornithine decarboxylase activity and putrescine concentration seem not to change significantly once the nerve cells have reached a certain stage of maturity. Completion of their cytoplasmic constituents, dendritic arborization, and establishment of synaptic contacts take place at constant cellular putrescine concentrations.

In contrast with ornithine decarboxylase activity, S-adenosylmethionine decarboxylase activity is low at early stages of embryonal and cellular maturation.[138,144,151] In rat brain, the activity of this enzyme starts to increase from about day 10 of postnatal life. Maximum levels are not reached before 2 months. During this period, a very large number of processes gain biological and functional maturation, and consequently, many enzymes involved in myelin formation and synaptogenesis are increasing markedly.[72] Since glial proliferation takes place during the same phase of development, it was suggested[151] that S-adenosylmethionine decarboxylase may be mainly localized in glial cells. In pure neuronal cultures, the activity of this enzyme increased during the period of cellular differentiation,[150] although the increase was only twofold, in comparison with the very dramatic increase of this enzyme's activity in whole brain.[151]

Data about developmental changes of the remaining two enzymes of polyamine biosynthesis, spermidine and spermine synthase, are scarce and somewhat contradictory: a parallel increase of spermidine synthase and S-adenosylmethionine decarboxylase activity was reported for rhesus monkey brain,[138] whereas a rapid decrease of this enzyme during the first days post-partum was observed for rat brain.[77] Spermine synthase seems to be fairly constant during postnatal development.[77] Since these enzymes are present in brain in large excess, as compared with the two decarboxylases (Table I), it is somewhat surprising that neither spermidine nor spermine concentrations increase concomitantly with S-adenosylmethionine decarboxylase activity. In fact, both spermidine and spermine concentrations decrease slightly during maturation, if the data are based on the tissue wet weight. This is true for all species so far studied.[123,127,128,132,133,135,140,145,147,148] The ratios of spermidine and spermine to RNA and DNA are usually very similar in brains of newborn and adult animals regardless of whether they are auto- or heterophagous.[132] Total brain spermidine and spermine increase linearly with brain weight until about day

20 of postnatal life, as do RNA and DNA.[133] The nucleic acids and spermine do not thereafter change very significantly until old age,[152] but spermidine increases. This implies certain functions of spermidine during the periods of slow maturation that are independent of those of spermine.

These findings resemble those in cultured chick embryo neurons.[150] Putrescine and spermine concentrations become constant at an early stage of cellular maturation, whereas spermidine increases per cell even at a time when RNA concentration remains practically constant.

Another finding that indicates different functional roles of spermidine and spermine concerns the spermidine content of microsomes. In preparations obtained from 5-day-old rats, the spermidine content per milligram RNA was significantly higher than that in preparations from brains of adult rats. The microsomal spermine content was, however, the same in both preparations.[153] This finding is of interest in view of the presumed role of the polyamines as structural components of nucleic acids[1,2] and their role in protein biosynthesis.[34]

3.1.3. Developmental Changes Related to Polyamines

3.1.3a. Putreanine. Putreanine is a metabolite of spermidine (see Fig. 2). It seems to be a normal constituent of brain but was found in other tissues[57,58,154,155] as well, and its formation in neuroblastoma cell culture has been shown.[156]

It was not possible to detect putreanine in monkey brains before birth.[157] Its postpartum increase was gradual and was similar to that in rat brain.[57] Since gray and white matter contained about the same amount of putreanine, and since no simple relationships between developmental putreanine and spermidine changes could be observed, no conclusions were reached as to what cell type might form it. In view of the recent finding that putreanine is formed in a reaction sequence involving plasma spermine oxidase and tissue aldehyde dehydrogenase,[62] one can assume that a correlation may exist between the activity of these enzymes and the concentration of putreanine in brain.

3.1.3b. Histamine. The developmental pattern of histamine is different from that of neural transmitter amines. Its concentration is high in fetal rat brain at 17 days of gestation but decreases sharply before birth. At days 5 to 10 of postnatal life, a second maximum of brain histamine is observed with a decline to adult values by the time of weaning.[135] Spermidine follows a similar pattern but with a 24- to 48-h lag.[135] It was felt that changes in brain histamine correlated best with periods of rapid cell proliferation, and there is some evidence that ornithine decarboxylase and histidine decarboxylase activities are changed in parallel in situations that induce DNA synthesis. Interrelationships between histamine and polyamine metabolism, therefore, deserve attention.

3.1.3c. Formation of GABA from Putrescine. The metabolic interrelationships between putrescine and GABA have been briefly mentioned. A full account of this topic is given in ref. 51. Glutamate decarboxylase activity is low at early stages of brain development and increases only in parallel with synaptogenesis.[72,158] This is also true for the developing nerve cell in culture.[150] However, it is known that GABA is found in chick[158,159] and rat[160] embryo brain by early stages of development, when glutamate decarboxylase was not detectable. The finding that GABA formation from putrescine is independent of glutamate decarboxylase activity prompted a comparative study of GABA formation from labeled glutamate and putrescine in the developing chick embryo brain.[147] In accordance with expectations, a significant transformation of glutamate into GABA was observed only in the brains of chick embryos older than 8 days. The maximal rate of GABA formation from putrescine was, however, found between days 6 and 8, with low rates after 12 days of incubation. The time of maximum rate of GABA formation from putrescine corresponds perhaps best with the time of early differentiation of the neuroblasts into neuronal cells. In a prior study the retina of 6-day-old chick embryos was shown to form as much as 20% of the total GABA from putrescine.[148]

It is not presently known along which pathway (see Fig. 2) putrescine is transformed into GABA in the chick embryo brain and retina, i.e., whether GABA is the result of oxidative deamination of putrescine by diamine oxidase or of monoamine oxidase with monoacetylputrescine as intermediate. In the brain of newborn rats the specific activity of the putrescine-acetylating enzyme is higher than at any later stage.[161] This can be taken as an indication that this pathway might be active in embryonal brains.

The finding that transformation of putrescine into GABA is maximal at the time of neuronal differentiation suggests a role of GABA in a process related to cellular development. Evidence in the same direction is the fact[162] that GABA formation from putrescine in neuroblastoma cells was low during the logarithmic phase of growth but was dramatically increased during the stationary phase or in cultures with low serum concentration.

3.2. Regional Distribution

The regional distribution of spermidine and spermine has been described for rat,[47,128,130,131,152] rabbit,[111,130,136] sheep,[129,130] dog,[130] monkey,[137,163] and human[45,129,130] brain. Data for putrescine are only available from rat,[131,164] cat,[164] and human.[45,164]

There is no obvious correlation between spermidine and spermine concentrations and *S*-adenosylmethionine decarboxylase activity.[47,165] As a general feature, it appears that spermine concentrations are relatively higher in structures rich in nerve cells. In the gray matter they exceed spermidine concentrations; the highest spermine concentrations were observed in the olfactory bulb of the rat.[131] In contrast, spermine is low, and putrescine and spermidine are extremely high in white matter and in peripheral nerves. If the data are based on the DNA content, putrescine and spermidine show very similar patterns of regional distribution. Lowest cellular concentrations are found in the olfactory bulb and the cerebellum of the rat, highest concentrations in the spinal cord. The patterns resemble those of protein and RNA[131] with the difference that spinal cord and peripheral nerves contain relatively much more putrescine and spermidine than RNA and protein. Another unusual feature is the high putrescine concentration in the hypothalamus of the rat[131] (in contrast to cat[164]).

In striking contrast, the average cellular spermine concentration is practically constant in all telencephalic and diencephalic regions—around 0.1 μmol/μmol DNA-P—and about half this value, but again constant, in the lower midbrain, medulla, spinal cord, cerebellum, and peripheral nerves. This observation is again in support of different functional roles of these two narrowly related polycations.

The unusually high spermidine and putrescine concentrations in structures rich in myelin gave rise to the hypothesis that spermidine might have a function in clamping adjacent myelin sheaths.[166–168] Indeed, spermidine and putrescine have been localized within the myelin sheath of peripheral nerves by electron microscopic autoradiography,[142] and it was shown that it has a lower turnover rate in myelin than in other structures.[167]

In myelin-deficient mutants (quaking and jimpy mutant mice), deficiencies of spermidine were seen in the hindbrain and spinal cord.[169] This could be taken as further support for the above hypothesis.

Whether the high hypothalamic putrescine concentrations are related to neurosecretion is a matter of speculation.

3.3. Subcellular Localization

3.3.1. Biosynthetic Decarboxylases

Ornithine decarboxylase and *S*-adenosylmethionine decarboxylase activities have been determined in subfractions of various tissues after homogenization in aqueous media. There is little doubt that in most tissues the two decarboxylases are mainly cytoplasmic, although very recently it was possible to demonstrate a certain portion of ornithine decarboxylase in a crude nuclear fraction of rat liver which was resistant to inhibition and was not changed by measures that induced the cytoplasmic enzyme.[170] In rat brain homogenates, however, around 43% of ornithine decarboxylase and about 7% of the *S*-adenosylmethionine decarboxylase activities were associated with the crude nuclear fraction.[151] A similarly high percentage of nuclear ornithine decarboxylase activity was reported for chick embryo.[171] Attempts to purify the nuclei resulted in a loss of enzyme activity.

More recently, it was stated[172] that in nuclei isolated from rat brain tumor cells the specific activity of *S*-adenosylmethionine decarboxylase is 7.4 times higher than in the cytoplasm, whereas ornithine decarboxylase was mainly cytoplasmic. One has to conclude from these somewhat incongruent findings that a certain part of the decarboxylases is located within the cell nuclei; however, they are osmotically sensitive and tend to leak out unless homogenization and separation of the subcellular fractions are very carefully controlled.

Two to three percent of the decarboxylases seemed to reside with the synaptosomal fraction of rat brain homogenates.[151] It is, however, likely that the major portion of the activity that produced $^{14}CO_2$ from [1-^{14}C]-L-ornithine was not ornithine decarboxylase but ornithine-δ-transaminase. This enzyme can initiate a reaction sequence leading to CO_2 formation from ornithine. Evidence for this notion comes from the fact that γ-acetylenic GABA, an inhibitor of GABA and ornithine aminotransferases,[173] completely blocked $^{14}CO_2$ formation from [1-^{14}C]-L-ornithine by synaptosomes.[174] Thus, ornithine decarboxylase activity in nerve endings is at least very low.

Data for other enzymes involved in polyamine biosynthesis are missing.

3.3.2. Polyamines

Because of their polycationic nature, the polyamines can bind to various anionic sites. After disruption of the cells by homogenization, it is therefore not possible to distinguish unambiguously between physiological and nonphysiological binding sites. Isolation of subcellular fractions in nonaqueous solvents can prevent redistribution of polyamines. With this technique it was shown that in the nuclei from rat brain tumor cells spermidine and spermine concentrations were, respectively, 8.9 and 5.3 times higher than in the cytoplasm.[172]

Methods for the isolation of subcellular elements of brain homogenates in nonaqueous media are not available. Partial results can be obtained by comparison of the subcellular distribution of the endogenous amines with that of labeled amines added to the aqueous homogenization medium. The following findings were obtained by this approach.[48,175]

3.3.2a. Distribution in Primary Cell Fractions. Fractionation of rat liver homogenates showed that in a medium of low ionic strength spermidine and spermine are distributed with RNA; 60% of total spermidine and spermine and of RNA were found in the microsomal fraction.[176] In surprising contrast, putrescine and the polyamines were not distributed along with RNA in the primary subfractions of rat brain cortex homogenates; rather, distribution followed the distribution of proteins. It is of especial interest that around 50% of the polyamines were found in the crude mitochondrial fraction, 20–30% resided with the nuclear fraction, and about the same percentage in the supernatant fraction which contained the microsomes as well. The exogenous radioactive amines were not completely equilibrated with the endogenous amines.

Similar results were reported from the measurement of the distribution of exogenous amines in guinea pig brain cortex, but up to 60% of the added spermidine was found attached to the nuclear fraction of rat hypothalamus.[177] This underlines the presumed importance of the polyamines in nuclear metabolism. From autoradiographic studies it is known that spermidine and spermine are more or less evenly distributed over nucleus and cytoplasm of brain cells.[112,142]

3.3.2b. Distribution in Subfractions of the Crude Mitochondrial Fraction. Substantial amounts of the endogenous polyamines were found together with myelin, but a significantly higher proportion was in the fraction that contained the sheared-off nerve endings. The exogenous amines showed the opposite distribution; i.e., the labeled amines that were added to the homogenization medium were preferentially attached to myelin fragments and were only partially equilibrated with the synaptosomal polyamine content. This supports the view that polyamines may be entrapped within the nerve endings. As one would expect, putrescine was liberated from the synaptosomes after hypoosmotic disruption. In contrast, spermidine and spermine remained attached to the synaptosomal membranes under these conditions. These amines were more easily displaced by Mg^{2+} from the binding sites of the myelin fraction than from the synaptosomal membranes, thus demonstrating a high affinity of the polyamines for membranes of the synaptic region.

Although no conclusions can be drawn with regard to the function of the polyamines within the synaptic region, a few observations should be mentioned that may help in elucidating physiological roles at this site.

There is no evidence for an energy-requiring high-affinity uptake system for the polyamines in nerve endings, but ornithine is actively taken up by rat cortex synaptosomes.[49,175] For neurotransmitterlike functions, such a system would be expected, but acetylation or other conjugation reactions could also fulfill the role of physiological inactivation processes. Moreover, temperature- and inhibitor-sensitive low-affinity uptake systems in the cerebral hemispheres of rodents have been described for putrescine[178] and spermidine,[179] and those could also function as inactivating transport routes.

Some evidence is available for binding of spermine to synaptic vesicles.[175] Among the many compounds that have been tested, spermidine and spermine were most effective in inhibiting binding of acetylcholine, but not of GABA, to synaptic vesicles.[180]

The high affinity of spermine and spermidine for binding sites of acetylcholine is further demonstrated by the fact that the activity of acetylcholinesterase can be modulated by these amines at micromolar concentration.[146,181,182] A physiological function as modulators at cholinergic synapses is therefore not unlikely, although direct evidence is presently lacking.

Polyamines inhibit[183] or, at low potassium concentrations,[184] activate Na^+,K^+-ATPase, and they counteract the disorganizing effect of imipramine on this enzyme.[185] It is tempting to speculate that polyamines could take part in the regulation of ion transport by influencing membrane ATPases.[186]

3.4. Transport

3.4.1. Cerebrospinal Fluid

Labeled putrescine, if injected into the lateral ventricles of mice[187] or fish,[112] is taken up by the brain, but maximum labeling is found by quantitative autoradiography in various brain parts only between 5 and 10 days after injection, with a subsequent slow decline of radioactivity. Normally, the nerve-cell-rich structures are much more heavily tagged than the surrounding white matter, but some tracts, such as the anterior commissure of mice, show accumulation of radioactivity. This suggests that oligodendrocytes are capable of accumulating putrescine.

Injections of labeled spermidine and spermine showed very similar distribution patterns in the brain to those obtained with radioactive putrescine. However, in fish brain, the circumventricular structures (meninges, ependymal cell layer) were very heavily labeled even 21 days after administration of [³H] spermine.[112] It is likely that similar observations can be made in mammalian brain as well. Studies of the dynamics of polyamine metabolism (and presumably also of other natural and nonnatural compounds) after intraventricular injections of labeled material may, therefore, give misleading results.

Recent studies with ventriculocisternal perfusion of rabbits indicated a role of the choroid plexus in the active accumulation from CSF of spermidine and spermine.[188] These share the same uptake system (K_m 21 μM and 24 μM, respectively). Part of the injected polyamines is removed from CSF via circulation.[111]

3.4.2. Axonal Transport

Axonal transport of the polyamines seems not to occur in the intact visual systems of mature rats[189] and goldfish[190] or in the R2 neuron of *Aplysia californica*.[191] However, putrescine seems to be transported in the growing axons of the embryonic zebra fish[192] and the regenerating optic axons of goldfish.[190,193] It appears that axonal transport of polyamines is a phenomenon related to axonal growth. It may be that the inability of neurons to regenerate following damage is linked to their inability to transport compounds critical to growth, and putrescine may belong to these compounds.[194]

4. TURNOVER

For any compound that regulates cellular function, concentration changes near the site of its action are essential. Local concentration changes can be achieved by several mechanisms. One possibility is a change of the relative rates of biosynthesis and degradation. High turnover rates allow rapid achievement of newly required high or low concentrations, whereas low turnover rates indicate that rapid concentration changes cannot occur by synthetic and degradative processes.

In previous attempts to measure *in vivo* rates of polyamine turnover in brain,[58,110,112,171,195,196] the specific radioactivities of the polyamines were determined over a certain period of time, and biological half-lives were calculated from the semilogarithmic plots of the specific radioactivity–time relationships.

This approach relies on the following assumptions: (1) the labeled precursor equilibrates rapidly with the endogenous pools whose sizes are not influenced by the material administered; (2) the polyamine pools of all cellular compartments are labeled to a comparable extent; and (3) at the time when the decline of the specific radioactivity is measured, no significant amount of the labeled precursor is present in the organ or cell studied.

4.1. The Experimental Approach

Repeated intraperitoneal injections of trace amounts of [1,4-^{14}C]putrescine were used for the labeling of the endogenous polyamine pools in mice,[110] instead of single intracerebroventricular injections,[58,112,171,195,196] in order not to label only the pools with a high turnover rate. The specific radioactivities of the polyamines spermidine and spermine were measured during a subsequent period of more than 2 months. The most pertinent results of these measurements were as follows. Spermidine and spermine showed identical apparent half-lives of about 42 days in the brain. In the other tissues, the half-lives of the polyamines were considerably shorter, in the range of 10 to 16 days. With [2-^{14}C]methionine instead of labeled putrescine as precursor, the observed half-life of brain spermidine was only 16 days, and that of spermine was 18 days. The corresponding values were 13 and 18 days, respectively, with [2-^{3}H]methionine as precursor. These latter values are close to those observed for liver polyamines and also similar to the half-lives of polyamines in other visceral organs.[110] If this difference did not result from an incomplete labeling of the brain polyamine pools by methionine—and there is no reason to assume this, since the experiment with [^{14}C]methionine was extended over a period of 129 days—it indicates an exceptional characteristic of polyamine metabolism in vertebrate brain and requires an explanation.

4.2. The Reutilization of Putrescine and Spermidine

It has been previously shown in mouse[187] and trout brain[112] that labeled putrescine can be detected over a long period of time after a single intraperitoneal or intracerebral dose of [^{14}C]putrescine. In trout brain, the specific radioactivity of putrescine paralleled in the late phase that of spermidine and spermine. These findings allow one to conclude that putrescine is formed from spermidine. The analogous formation of spermidine from injected spermine has been shown in trout brain,[46] goldfish and rat retina,[113] and rabbit brain.[111] The reaction sequence most probably responsible for these transformations is shown in Fig. 2. The aminopropyl part of spermidine and spermine, which originates from methionine, is irreversibly eliminated from the cycle during the degradation step.

If putrescine is formed from spermidine within a cellular compartment that contains the polyamine-synthesizing machinery, or if putrescine is transported into such a compartment before it is exposed to conjugation or oxidative degradation, it could reenter the cycle. If this type of reutilization of putrescine and spermidine for polyamine biosynthesis occurs to a great extent in brain, it could explain the observed differences in the apparent biological half-lives of spermidine and spermine in brain as compared with other organs. Putrescine formation from ornithine would have the function of replenishing irreversible losses of putrescine carbon from the system that occur by degradation and elimination of putrescine itself and of the polyamines (Fig. 2).

The actual synthesis and degradation rates of spermidine and spermine in such a system could be much higher than is suggested by the rate of putrescine formation. Assuming only small losses of putrescine carbon, the actual rate-limiting step could be the rate of production of decarboxylated S-adenosylmethionine.

4.3. The Physiological Significance of Putrescine and Spermidine Reutilization

Putrescine reutilization could be a general phenomenon active to different degrees in different cells and during different physiological states of the same cell. It could be an essential part of a system for the subtle regulation of polyamine biosynthesis rates and provide the possibility of an enormous increase of spermidine and spermine synthesis rates by induction of ornithine decarboxylase. The reasoning is as follows.

In a system with a low equilibrium concentration of putrescine and high S-adenosylmethionine decarboxylase and spermidine and spermine synthase activities as is present in the mature rat brain (Table I), a certain rate of polyamine turnover can be maintained even at low ornithine decarboxylase activity. The rate of spermidine synthesis in such a system is limited by the equilibrium concentration of putrescine. S-Adenosylmethionine decarboxylase would be only partially activated. A relatively small increase of ornithine decarboxylase activity and a concomitant local increase of putrescine concentration, however, would significantly increase the synthesis rate of both spermidine and spermine, since S-adenosylmethionine decarboxylase would be activated further and thus increase the amount of decarboxylated S-adenosylmethionine. However, a system such as an embryonal brain cell, with high ornithine and low S-adenosylmethionine decarboxylase activities presumably has a close to maximally activated S-adenosylmethionine decarboxylase. The further increase in the putrescine concentration from induction of ornithine decarboxylase would enhance the synthesis rate only by increasing the substrate concentration. The rate of spermidine synthesis would initially be increased with little effect on the rate of spermine formation.

Putrescine reutilization is closely analogous to reutilization of spermidine for spermine biosynthesis. The equilibration of the radioactive carbon between spermidine and spermine as suggested by the finding of identical specific radioactivities of the two polyamines in brain is a good argument for the reentry of spermidine formed by spermine degradation into the synthetic cycle.

The regulatory influence of cerebral putrescine concentrations on polyamine biosynthesis rates was also concluded from observations made on mutant mice with deficient S-adenosylmethionine decarboxylase activity.[197]

4.4. The Validity of Turnover Rate Determinations

It is not possible to design experimental conditions that would permit the determination of actual turnover rates of the polyamines under physiological conditions. Even disregarding various experimental problems, the active interconversion of the polyamines is an obstacle difficult to overcome. The finding that the two parts of the polyamine structure deriving from ornithine and methionine, respectively, may have widely differing biological life-spans necessitates the definition of life-span of polyamines as the time from the first incorporation of a certain carbon atom into the spermidine molecule until it is part of a molecule that cannot reenter the polyamine structure. The reentry of putrescine carbon into the polyamine cycle once putrescine has been degraded to GABA or the reentry of a β-alanine carbon deriving from the C_3 moiety of the polyamines is negligibly low.

The difference in the life-spans of the putrescine moiety and of the aminopropyl moieties of the polyamines may be zero in the case where putrescine formed by degradation of spermidine is quantitatively degraded or conjugated and excreted from the cell. This situation should prevail in cells with high ornithine decarboxylase and low S-adenosylmethionine decarboxylase activities, i.e., in nondifferentiated, rapidly proliferating cells. From the available data, one can conclude that putrescine may reenter the polyamine cycle several times before it is finally eliminated. This should occur preferentially in highly differentiated, nongrowing cells with low ornithine decarboxylase and high S-adenosylmethionine decarboxylase activities, such as cells of the central nervous system.

5. POLYAMINES AND BRAIN TUMORS

Increased rates of polyamine biosynthesis have frequently been reported to be interrelated with the enhancement of cell growth and proliferation.[1,4–8] Examples related to the nervous system are the degenerating peripheral nerve, the scrapie-affected mouse brain, and brain tumors.

During the so-called Wallerian degeneration of injured peripheral nerves, Schwann cell proliferation, among others, is greatly enhanced, and the uptake of [1,4-^{14}C]putrescine parallels the rate of cell proliferation.[198]

Inoculation of scrapie agent into the brain of mice induces a fourfold increase of spermidine formation. This and the levels of polyamines correlate well with the pattern of astrocyte hypertrophy.[199]

In a systematic comparison, putrescine levels of astrocytomas were found to be enhanced in proportion to the malignancy of the tumor as determined by histopathological criteria. In slowly growing tumors, putrescine levels were not significantly different from those in normal brain.[200]

In recent years the excretion of polyamines in the urine has been studied as a means of diagnosis of proliferative diseases[5,7,16] based on the above concept that enhanced cell proliferation is accompanied by an increased rate of polyamine biosynthesis.

This concept was applied to the study of brain tumors. Some of the tumors of human brain showed elevated putrescine levels (astrocytoma,[200,201] glioblastoma,[45,201] meningioma,[45] craniopharyngioma[45]). In some tumors, spermine concentration was elevated (epidural carcinoma,[45] lymphoma[45]). Nitrosomethyl-urea-induced brain tumors of rats[122] and tumors produced by inoculation of cultured brain tumor cells that were originally obtained by nitrosomethylurea injections[202] showed dramatically elevated putrescine concentrations with little change in spermidine and spermine concentrations. Ornithine decarboxylase activity was greatly enhanced.[202] The nuclear polyamine acetylase activity was also enhanced in nitrosomethylurea-induced brain tumors.[122]

These observations prompted the analysis of the polyamines in the cerebrospinal fluid of patients with various glial tumors. From the extensive systematic work of Marton and collaborators,[203-205] the following generalizations can be made: increased levels of putrescine and spermidine in the CSF clearly indicate the presence of gliomas and astrocytomas, but it appears that CSF polyamine determinations may be of little use for the monitoring of tumor progression in patients with glioblastoma multiforme or anaplastic astrocytoma. The reason is that the amounts of spermidine and putrescine appearing in the CSF are dependent not only on tumor growth rate and mass but also on the situation of the tumor within the brain. The polyamines undergo metabolism, and its extent is dependent on the proximity of the tumor to the ventricular system.

The situation is more favorable in the case of medulloblastoma. These are tumors of the fossa posterior that frequently seed along the spinal cord, so that one obtains high putrescine levels in the lumbar CSF.

More important than tumor diagnosis seems the use of polyamine analysis as a means of following the effects of tumor therapy. In the case of medulloblastoma, increased polyamine levels in the CSF are predictive of recurrence of the tumor.

6. EFFECTS OF EXOGENOUS POLYAMINES

The pharmacological properties of spermidine and spermine are qualitatively the same, but spermine usually acts at much lower concentrations. Putrescine also produces some of the effects of the polyamines, but very large doses are usually needed. Thus, for instance, putrescine seems not to produce acute toxicity if given intraperitoneally. However, spermine·4HCl in doses exceeding 0.1 mmol/kg produces curarelike paralysis of the musculature, and the animals die from respiratory arrest.[206,207] Amounts of spermidine at least 20 times higher are necessary to achieve the same effect.

The interaction of the polyamines with cholinergic receptors is also documented by the fact that they selectively antagonize the contractions of the guinea pig ileum produced by nicotine.[207] Toxic intraperitoneal doses of spermine produced sedation and hypothermia,[207] and intravenous doses of spermine (10–30 mg/kg) and spermidine (50–90 mg/kg) produced hyperglycemia in rats and rabbits.[208] However, all neuropharmacological properties of the polyamines can be demonstrated in much smaller doses if the blood–brain barrier is avoided by intraventricular administration of the compounds.[206,207] The effects on spontaneous motor activity and body temperature were long lasting, and prolongation of pentobarbital-induced sleep developed slowly.[209]

A few micrograms of intraventricularly administered spermine produced convulsions. These developed over a period of several hours. The animals became extremely hyperexcitable, and convulsions were precipitated by sound or touch.[210] Intraventricular spermidine in high doses produces a pathological syndrome. Ataxia appeared at 24 h. The animals showed anorexia and adipsia and became moribund after about 5 days. Focal lesions were found in the ventral medulla and in the pyramidal tract. Superficial lesions were also seen in the spinal cord.[210] Whether this lesioning bears any relationship to human disease is not known.

Putrescine injected in 100-μg amounts into the third cerebral ventricle of chicks produced, 30 min later, a significant depletion of GABA in the diencephalon but not in the cerebral hemispheres. This effect appeared to be dependent on the inhibition of glutamate decarboxylase.[211]

The central effects of putrescine in mice[210] have been suggested to result from the increased formation of spermidine and spermine in the brain and were described as occurring with a latency of 24 h. However, behavioral excitation, marked postural changes, and epileptogenic cortical discharges were seen in the chicken after only a 5- to 10-min period of sedation. Since convulsions and electrocortical epileptogenic discharges are known to occur after depletion of brain GABA,[212] it was suggested that the effects of intraventricular putrescine were caused by the impairment of GABA neurons.

It has previously been shown that electrophoretically applied polyamines depress spontaneous firing of neurons in the brainstem of rat and cat; excitatory effects were more common in the rat.[213] This suggests that polyamines may be involved in the control of neuronal activity. However, in contrast to GABA, perfusion of the isolated nerve ring of the snail with up to 10^{-4} M putrescine, spermidine, or spermine had no effect on the spontaneous activity of the neurons.[214]

It was claimed that mice made aggressive by isolation showed an increase in brain spermidine that paralleled the development of aggressivity. This change was not observed in nonaggressive isolated mice,[215] but return of aggressive mice to grouped housing, which rendered them nonaggressive, reversed the increase of spermidine.[216] Intracerebroventricularly administered spermidine and spermine had little effect on the sleep–wakefulness cycle in rats.[217]

It is not possible to infer a functional role of the polyamines in a distinct group of neurons from their pharmacological properties. Most of the evidence points to cholinergic systems, although doses of spermidine and spermine that produced central effects were without effect on brain acetylcholine content (or on catecholamines, serotonin, or GABA).[218] Turnover studies of the neurotransmitter amines under these conditions are lacking.

The stimulated release of polyamines has been reported in rhesus monkey brain[163] but could not be shown in guinea pig cortex.[175] It remains, therefore, an open question whether the polyamines can participate in functions that require rapid local concentration changes.

7. MODULATION OF POLYAMINE METABOLISM

It was pointed out above (see Section 2.1.1) that several eucaryotic cell types respond to a variety of hormones, growth-promoting agents, and drugs by a rapid induction of ornithine decarboxylase activity. (For general reviews of this topic see refs. 5,37,98,219).

A number of hormones, when administered to neonatal rats, are capable of altering the developmental pattern of brain ornithine decarboxylase activity.[42] Thyroxine accelerates neuronal development and consequently compresses the time course in which ornithine decarboxylase passes through its ontogenic pattern, whereas corticosteroids prolong the pattern.[94] Drugs such as morphine, methadone, reserpine, or guanethidine similarly prolong the developmental pattern of the brain and of ornithine decarboxylase.[42]

The interruption of the normal mother–infant interaction in rats caused an immediate decrease in brain ornithine decarboxylase activity which was accompanied by a decline in putrescine levels. It could be shown that returning the pup to the mother restored ornithine decarboxylase activity to normal or supranormal values. Studies with nipple-ligated and with anesthetized mothers showed, furthermore, that the changes were not caused by interruption of feeding but resulted from interruption of active maternal behavior.[42,92,93] Thus, ornithine decarboxylase turned out to be a sensitive tool for studies of normal and perturbed ontogeny of the central and peripheral nervous system.

However, very little is known of the biochemical and biological consequences directly related to the induction of ornithine decarboxylase activity. For example, a clonal line of rat adrenal pheochromocytoma cells, when treated with nerve growth factor, acquires certain features of sympathetic neurons including the formation of extensive neurites. These events are preceded by a rapid increase in ornithine decarboxylase activity. From the fact that inhibition of the induced ornithine decarboxylase activity with 5-hexyne-1,4-diamine and 1,3-diaminopropane did not prevent the outgrowth of neurites, it was concluded that ornithine decarboxylase induction was not causally related to these events but may be a "dead end" or terminal consequence of other events related to the interaction of nerve growth factor with its target cells.[82] On the other hand, it was postulated that there is a direct relationship between the delay in the postnatal fall-off of brain ornithine decarboxylase activity that was induced by prenatal reserpine administration and subsequent deficits in tyrosine hydroxylase activity.[220,221] A critical period in prenatal development seems to exist in which reserpine can produce long-lasting altered brain monoamine biosynthesis.

Treatment of adult rats with reserpine also resulted in the induction of ornithine decaboxylase activity in both brain and adrenal medulla.[87] Complete inhibition of the induced ornithine decarboxylase activity by α-difluoromethyl ornithine, a specific, irreversible inhibitor of ornithine decarboxylase,[11] did not change the induction of tyrosine hydroxylase in the adrenal medulla (N. Seiler and M. J. Jung, unpublished results). This example could again be used as an argument against the direct causal interrelationship between induction of ornithine decarboxylase activity and later events such as the induction of tyrosine hydroxylase. However, at the present stage of our knowledge, it may be premature to draw such far-reaching conclusions. It cannot be excluded, for example, that even the enzyme protein that is catalytically inactivated by the inhibitor may fulfill functions such as the activation of RNA polymerase I[222] and thus act as a trigger for following events. On the other hand, it is not unlikely that induced ornithine decarboxylase may fulfill certain functions only at a certain time in a specific cellular environment but may be without any long-lasting physiological consequence at any other time. It remains a matter of future research to clarify whether the known alterations of the activities of ornithine decarboxylase and S-adenosylmethionine decarboxylase and of putrescine concentration in adult mammalian brain are indicative of physiological events or whether they are only a sign of short-lasting imbalances of brain polyamine metabolism. Examples for this type of change are the effects of ablation of the pituitary, adrenal, or thyroid glands,[223] administration of inhibitors of GABA transaminase with consequential long-lasting increases of brain GABA levels,[160,224] or treatment with allylglycine,[90,225] β-p-chlorophenyl-GABA,[88,90] L-DOPA,[89] or 6-hydroxydopamine.[86] In none of these studies were major changes in spermidine or spermine levels observed. However, rats exposed to hyperbaric oxygen and consequently convulsing are reported to undergo a fall in brain polyamine levels by 40% within less than 4 h.[226] Further elucidation of this model could produce insights into an unexpected lability of brain polyamine levels and biochemical processes related to convulsive states.

8. FUNCTIONAL CONSIDERATIONS

It is now well established that polyamines are essential for the normal growth and replication of dividing cells,[5,8,227] but what role can they play in a nondividing, fully differentiated, highly organized system such as the mammalian central nervous system?

Naturally, the normal functioning of basic metabolic reactions such as biosynthesis of RNA, protein, and lipid are essential for the brain as well. Many of the *in vitro* effects of the polyamines originally demonstrated in bacterial systems or in liver have now been established in subcellular preparations from brain tissue. For example, spermidine is capable of stimulating nuclear RNA polymerase II in preference to RNA polymerase I, suggesting that DNA-like RNA formation may be stimulated by polyamines.[228] In cell-free protein-synthesizing systems from rat cerebral cortex and cerebellum, polyamines stimulate binding of template RNA and aminoacyl RNA to ribosomes.[229] Furthermore, *in vitro* stimulation of phosphorylation and acetylation of histones of the cerebral cortex has been observed,[230] suggesting an involvement of the polyamines in the alteration of gene expression during development. Finally, some regulatory involvement in cerebral glycerolipid biosynthesis has been evidenced.[231] This finding and the fact that enhancement of brain phospholipids by feeding rats with a diet rich in phospholipids or oleic acid increased brain spermidine and spermine levels[232] are perhaps the most direct indications for a function of the polyamines in membrane structure.

Structural interrelations of spermidine and spermine with RNAs, RNA-containing cellular elements, and DNA are suggested from the simple quantitative interrelations among these cellular components during growth,[123,132,133,141] and structural interrelations of spermidine with myelin[166-168] from the abundance of this amine[131] and its slow turnover[167] in myelin-rich structures.

Other general functions, such as involvement in ion transport, are evidenced from the effects of the polyamines on Na^+,K^+-ATPase and on the neuronal firing rate.[213]

Usually, the pharmacological properties suggest involvement in certain functions, but in the case of the polyamines very little has been learned in this respect. A modulatory role, presumably of the cholinergic neuronal system, is not unlikely[48,49]; however, direct evidence for this is still lacking. The evidence for the interrelation of the polyamines with the GABA-ergic system, other than a metabolic interrelationship, is also very preliminary[51] and needs further exploration. The finding that spermidine and spermine are reversible, noncompetitive inhibitors of succinic semialdehyde dehydrogenase (E.C. 1.2.1.16) is suggestive of a participation of the polyamines in the regulation of GABA degradation.[233]

It appears that we are not in a position to pinpoint a specific functional role of the polyamines in brain, a function distinct from those presumed in other organs. The unusually high activities of *S*-adenosylmethionine decarboxylase and spermidine synthase reflect, perhaps mainly, the very active RNA and protein biosynthesis of the brain. The now increased interest in these simple, multifunctional polycations may, however, generate new insights into brain-specific functions, especially if one succeeds in modulating brain polyamines in a more specific manner than was possible hitherto.

REFERENCES

1. Cohen, S. S., 1971, *Introduction to the Polyamines*, Prentice-Hall, Englewood Cliffs, New Jersey.
2. Cohen, S. S., 1978, *Nature* **274**:209–210.
3. Silver, S., Wendt, L., Bhatthacharyya, P., and Beauchamp, R. S., 1970, *Ann. N.Y. Acad. Sci.* **171**:838–862.
4. Bachrach, U., 1973, *Function of the Naturally Occurring Polyamines*, Academic Press, New York.
5. Jänne, J., Pösö, H., and Raina, A., 1978, *Biochim. Biophys. Acta* **473**:241–293.
6. Williams-Ashman, H. G., and Canellakis, Z., 1979, *Perspect. Biol. Med.* **22**:421–453.
7. Russell, D. H. (ed.), 1973, *Polyamines in Normal and Neoplastic Growth*, Raven Press, New York.
8. Raina, A., Eloranta, T., Pajula, R. L., Mäntyjärvi, R., and Tuomi, K., 1980, *Polyamines in Biomedical Research* (J. M. Gaugas, ed.), John Wiley & Sons, Chichester, New York, Brisbane, Toronto, pp. 35–50.
9. Mamont, P. S., Duchesne, M.-C., Joder-Ohlenbusch, A.-M., and Grove, J., 1978, *Enzyme-Activated Irreversible Inhibitors* (N. Seiler, M. J. Jung, and J. Koch-Weser, eds.), Elsevier/North Holland Biomedical Press, Amsterdam, New York, Oxford, pp. 43–54.

10. Heby, O., and Jänne, J., 1981, *Polyamines in Biology and Medicine* (D. R. Morris and L. H. Marton, eds.), Marcel Dekker, New York, pp. 243–310.

11. Seiler, N., Danzin, C., Prakash, N. J., and Koch-Weser, J., 1978, *Enzyme-Activated Irreversible Inhibitors* (N. Seiler, M. J. Jung, and J. Koch-Weser, eds.), Elsevier/North Holland Biomedical Press, Amsterdam, New York, Oxford, pp. 55–72.

12. Gaugas, J. M. (ed.), 1980, *Polyamines in Biomedical Research*, John Wiley & Sons, Chichester, New York, Brisbane, Toronto.

13. Herbst, E. J., and Bachrach, U., 1970, *Ann. N.Y. Acad. Sci.*, **171**:691–1009.

14. Caldarera, C. (ed.), 1976, *A Tribute to Giovanni Moruzzi, Professor of Biochemistry at the University of Bologna, on the Occasion of his 70th Birthday, Ital. J. Biochem.* **25**:1–114.

15. Campbell, R. A., Morris, D. R., Bartos, D., Daves, G. D., Jr., and Bartos, F. (eds.), 1978, *Advances in Polyamine Research*, Volumes 1 and 2, Raven Press, New York.

16. Russell, D. H., and Drurie, B. G. M., 1978, *Polyamines as Biochemical Markers in Normal and Malignant Growth, Progress in Cancer Research*, Volume 8, Raven Press, New York.

17. Morris, D. R., and Marton, L. H. (eds.), 1981, *Polyamines in Biology and Medicine*, Marcel Dekker, New York.

18. Daves, G. D., Jr., (ed.), 1980, *Polyamine Metabolites and Conjugates in Man and Higher Animals, Physiol. Chem. Phys.* **12**(5):387–480.

19. Tabor, M., and Tabor, C. W., 1964, *Pharmacol. Rev.* **16**:245–300.

20. Stevens, L., 1970, *Biol. Rev.* **45**:1–27.

21. Bachrach, U., 1970, *Annu. Rev. Microbiol.* **24**:109–134.

22. Kremzner, L. T., 1970, *Fed. Proc.* **29**:1583–1588.

23. Herbst, E. J., and Tanguay, R. B., 1971, *Progress in Molecular and Subcellular Biology* Volume 2, (F. E. Halm, ed.), Springer-Verlag, Berlin, Heidelberg, New York, pp. 166–180.

24. Smith, T. A., 1971, *Biol. Rev.* **46**:201–241.

25. Williams-Ashman, H. G., Jänne, J., Coppoc, G. L., Geroch, M. E., and Schenone, A., 1972, *Adv. Enzyme Regul.* **10**:225–245.

26. Russell, D. H., 1973, *Life Sci.* **13**:1635–1647.

27. Seiler, N., 1975, *Research Methods in Neurochemistry* Volume 3, (N. Marks and R. Rodnight, eds.), Plenum Press, New York, pp. 409–441.

28. Raina, A., and Jänne, J., 1975, *Med. Biol.* **53**:121–147.

29. Tabor, C. W., and Tabor, H., 1976, *Annu. Rev. Biochem.* **45**:285–306.

30. Sakai, T. T., and Cohen, S. S., 1976, *Prog. Nucleic Acid Res. Mol. Biol.* **17**:15–42.

31. Seiler, N., 1977, *Clin. Chem.* **23**:1519–1526.

32. Inoue, H., and Takeda, Y., 1977, *J. Jpn. Biochem. Soc.* **49**:411–428.

33. Karpetsky, T. P., Hieter, P. A., Frank, J. J., and Levi, C. C., 1977, *Mol. Cell. Biochem.* **17**:89–99.

34. Algranati, I. D., and Goldenberg, S. M., 1977, *Trends Biochem. Sci.* **2**:272–274.

35. Russell, D. H., 1977, *Clin. Chem.* **23**:22–27.

36. Cohen, S. S., 1977, *Cancer Res.* **37**:939–942.

37. Canellakis, E. S., Vicepts-Madore, D., Kyriakidis, D. A., and Heller, J. S., 1979, *Curr. Topics Cell. Regul.* **15**:155–200.

38. Shaw, G. G., 1979, *Biochem. Pharmacol.* **28**:1–6.

39. Maudsley, D. V., 1979, *Biochem. Pharmacol.* **28**:153–161.

40. Quash, G., Roch, A. M., 1979, *Ann. Biol. Clin. (Paris)* **37**:317–325.

41. Stevens, L., and Winther, M. D., 1979, *Adv. Microbiol.* **19**:63–148.

42. Slotkin, T. A., 1979, *Life Sci.* **24**:1623–1629.

43. Russell, D. H., 1980, *Pharmacology* **20**:117–129.

44. Milano, G., Viguier, E., Cassuto, J. P., Schneider, M., Namer, M., Boublil, J. L., Lesbats, G., Cambon, P., Krebs, B. P., and Lalanne, C. M., 1980, *Pathol. Biol. (Paris)* **28**:328–334.

45. Kremzner, L. T., 1973, *Polyamines in Normal and Neoplastic Growth* (D. H. Russell, ed.), Raven Press, New York, pp. 27–40.

46. Seiler, N., 1973, *Polyamines in Normal and Neoplastic Growth* (D. H. Russell, ed.), Raven Press, New York, pp. 137–156.

47. Snyder, S. H., Shaskan, E. G., and Harik, S. I., 1973, *Polyamines in Normal and Neoplastic Growth* (D. H. Russell, ed.), Raven Press, New York, pp. 199–213.

48. Seiler, N., and Deckardt, K., 1978, *Advances in Polyamine Research*, Volume 2 (R. A. Campbell, D. R. Morris, D. Bartos, G. D. Daves, Jr., and F. Bartos, eds.), Raven Press, New York, pp. 145–159.

49. Seiler, N., and Deckhardt, K., 1978, *Advances in Polyamine Research*, Volume 2 (R. A. Campbell, D. R. Morris, D. Bartos, G. D. Daves, Jr., and F. Bartos, eds.), Raven Press, New York, pp. 161–167.

50. Ingoglia, N. A., and Sturman, J. A., 1978, *Advances in Polyamine Research*, Volume 2 (R. A. Campbell, D. R. Morris, D. Bartos, G. D. Daves, Jr., and F. Bartos, eds.), Raven Press, New York, pp. 169–182.

51. Seiler, N., 1980, *Physiol. Chem. Phys.* **12**:411–429.

52. Seiler, N., 1981, *Polyamines in Biology and Medicine* (D. R. Morris, and L. H. Marton, eds.), Marcel Dekker, New York, pp. 127–150.

53. Bolkenius, F., and Seiler, N., 1981, *Int. J. Biochem.* **13**:287–292.

54. Seiler, N., Bolkenius, F. N., and Knödgen, B., 1980, *Biochim. Biophys. Acta* **633**:181–190.

55. Matsui, I., and Pegg, A., 1980, *Biochem. Biophys. Res. Commun.* **92**:1009–1015.

56. Morgan, D. M. L., 1980, *Polyamines in Biomedical Research* (J. M. Gaugas, ed.), John Wiley & Sons, Chichester, New York, Brisbane, Toronto, pp. 285–302.

57. Kakimoto, J., Nakajima, T., Kumon, A., Matsuoka, Y., Imaoka, N., and Sano, I., 1969, *J. Biol. Chem.* **244**:6003–6007.

58. Nakajima, T., Noto, T., and Kato, N., 1980, *Physiol. Chem. Phys.* **12**:401–410.

59. Aigner-Held, R., and Daves, G. D., Jr., 1980, *Physiol. Chem. Phys.* **12**:389–400.

60. Nakajima, T., 1973, *J. Neurochem.* **20**:735–742.

61. Asatoor, A. M., 1979, *Biochim. Biophys. Acta* **586**:55–62.

62. Seiler, N., Knödgen, B., Gittos, M. W., Chan, W. Y., Griesmann, G., and Rennert, O. M., 1981, *Biochem. J.* **200**:123–132.

63. Seiler, N., and Eichentopf, B., 1975, *Biochem. J.* **152**:201–210.

64. Tsuji, M., and Nakajima, T., 1978, *J. Biochem. (Tokyo)* **83**:1407–1412.

65. Konishi, H., Nakajima, T., and Sano, I., 1977, *J. Biochem. (Tokyo)* **81**:355–360.

66. Seiler, N., and Al-Therib, M. J., 1974, *Biochem. J.* **144**:29–35.

67. Lundgren, D. W., and Hankins, J., 1978, *J. Biol. Chem.* **253**:7130–7133.

68. Seiler, N., Koch-Weser, J., Knödgen, B., Richards, W., Tardif, C., Bolkenius, F. N., Schechter, P., Tell, G., Mamont, P., Fozard, J., Bachrach, U., and Grosshans, E., 1981, *Advances in Polyamine Research*, Volume 3, (C. M. Caldarera, V. Zappia, and U. Bachrach, eds.), Raven Press, New York, pp. 197–211.

69. Seiler, N., and Knödgen, B., 1979, *J. Chromatogr. Biomed. Appl.* **164**:155–168.

70. Abdel-Monem, M. M., and Ohno, K., 1977, *J. Pharm. Sci.* **66**:1195–1197.

71. Abdel-Monem, M. M., and Ohno, K., 1978, *J. Pharm. Sci.* **67**:1671–1673.

72. Seiler, N., 1969, *Handbook of Neurochemistry*, Volume 1, (A. Lajtha, ed.), Plenum Press, New York, pp. 325–468.

73. Butler, S. R., and Schanberg, S. M., 1976, *Life Sci.* **18**:759–762.

74. Obenrader, M. K., and Prouty, W. F., 1977, *J. Biol. Chem.* **252**:2860–2865.

75. Jänne, J., and Williams-Ashman, M. G., 1971, *J. Biol. Chem.* **246**:1725–1732.

76. Pegg, A. E., and McGill, S., 1979, *Biochim. Biophys. Acta* **568**:416–427.

77. Raina, A., Pajula, R.-L., and Eloranta, T., 1976, *FEBS Lett.* **67**:252–255.

78. MacDonnell, P. C., Nagaiah, K., Lakshmanan, J., and Guroff, G., 1977, *Proc. Natl. Acad. Sci. U.S.A.* **74**:4681–4684.

79. Lewis, M. E., Lakshmanan, J., Nagaiah, K., MacDonnell, P., and Guroff, G., 1978, *Proc. Natl. Acad. Sci. U.S.A.* **75**:1021–1023.

80. Otten, U., Hatanaka, H., and Thoenen, H., 1978, *Experientia* **34**:951.

81. Ikeno, T., MacDonnell, P. C., Nagaiah, K., and Guroff, G., 1978, *Biochem. Biophys. Res. Commun.* **82**:957–963.

82. Greene, L. A., and McGuire, J. C., 1978, *Nature* **276**:191–193.

83. Harper, G. P., and Thoenen, H., 1980, *J. Neurochem.* **34**:5–16.

84. Greene, L. A., and Shooter, E. M., 1980, *Annu. Rev. Neurobiol.* **3**:353–402.

85. Pajunen, A. E. I., Hietala, O. A., Virransalo, E. L., and Piha, R. S., 1978, *J. Neurochem.* **30**:281–283.

86. Belin, M.-F., and Pujol, J.-F., 1977, *Biochem. Pharmacol.* **26**:2473–2475.

87. Deckardt, K., Pujol, J.-F., Belin, M.-F., Seiler, N., and Jouvet, M., 1978, *Neurochem. Res.* **3**:745–753.

88. Pajunen, A. E. I., Virransalo, E. L., Hietala, O. A., and Piha, R. S., 1978, *Acta Chem. Scand.* [*B*] **32**:322–326.

89. Harik, S. I., 1979, *Eur. J. Pharmacol.* **54**:235–242.

90. Pajunen, A., 1979, Polyamine Metabolism in Mouse Brain *Acta Universitatis Ouluensis*, Ser. A. Sci. Rer. Nat. No. 79, pp. 5–47.

91. Kleihues, P., Hossmann, K. A., Pegg, A. E., Kobayashi, K., and Zimmermann, V., 1975, *Brain Res.* **95**:61–75.

92. Butler, S. R., and Schanberg, S. M., 1977, *Life Sci.* **21**:877–884.

93. Butler, S. R., Suskind, M. R., and Schanberg, S. M., 1978, *Science* **199**:445–447.

94. Anderson, T. R., and Schanberg, S. M., 1975, *Biochem. Pharmacol.* **24**:495–501.

95. Roger, L. J., and Fellows, R. E., 1977, *Fed. Proc.* **36**:589.

96. Butler, S. R., Hurley, T. W., Schanberg, S. M., and Handwerger, S., 1978, *Life Sci.* **22**:2073–2078.

97. Bartolomé, J., Lau, C., and Slotkin, T. A., 1977, *J. Pharmacol. Exp. Ther.* **202**:510–518.

98. Bachrach, U., 1980, *Polyamines in Biomedical Research* (J. M. Gaugas, ed.), John Wiley & Sons, Chichester, New York, Brisbane, Toronto, pp. 81–107.

99. McCann, P. P., 1980, *Polyamines in Biomedical Research* (J. M. Gaugas, ed.), John Wiley & Sons, Chichester, New York, Brisbane, Toronto, pp. 109–123.

100. Heller, J. S., and Canellakis, E. S., 1980, *Polyamines in Biomedical Research* (J. M. Gaugas, ed.), John Wiley & Sons, Chichester, New York, Brisbane, Toronto, pp. 135–145.

101. Cantoni, G. L., and Durell, J., 1957, *J. Biol. Chem.* **225**:1033–1048.

102. Eloranta, T., 1977, *Biochem. J.* **166**:521–529.

103. Hibasami, H., Hoffman, J. L., and Pegg, A. E., 1980, *J. Biol. Chem.* **255**:6675–6678.

104. Pegg, A. E., 1977, *FEBS Lett.* **84**:33–36.

105. Sturman, J. A., and Kremzner, L. T., 1974, *Biochim. Biophys. Acta* **372**:162–170.

106. Hanonnen, P., 1975, *Acta Chem. Scand.* [*B*] **29**:295–299.

107. Alhonen-Hongisto, L., 1980, *Biochem. J.* **190**:747–754.

108. Mamont, P. S., Joder-Ohlenbusch, A. M., Nüssli, M., and Grove, J., 1981, *Biochem. J.* **196**:411–422.

109. Pajula, R.-L., Raina, A., and Eloranta, T., 1979, *Eur. J. Biochem.* **101**:619–626.

110. Antrup, H., and Seiler, N., 1980, *Neurochem. Res.* **5**:123–143.

111. Halliday, C. A., and Shaw, G. G., 1976, *J. Neurochem.* **26**:1199–1206.

112. Seiler, N., Al-Therib, M.-J., Fischer, H. A., and Erdmann, G., 1979, *Int. J. Biochem.* **10**:961–974.

113. Sturman, J. A., Ingoglia, N. A., and Lindquist, T. D., 1976, *Life Sci.* **19**:719–724.

114. Perry, T. L., Hansen, S., and MacDougall, L., 1967, *J. Neurochem.* **14**:775–782.

115. Seiler, N., Al-Therib, M.-J., and Knödgen, B., 1973, *Hoppe Seylers Z. Physiol. Chem.* **354**:589–590.

116. Noto, T., Tanaka, T., and Nakajima, J., 1978, *J. Biochem. (Tokyo)* **83**:543–552.

117. Dubin, D. T., and Rosenthal, S. M., 1960, *J. Biol. Chem.* **235**:776–782.

118. Blankenship, J., and Walle, T., 1975, *Arch. Biochem. Biophys.* **179**:235–242.

119. Blankenship, J., and Walle, T., 1978, *Advances in Polyamine Research*, Volume 2, (R. A. Campbell, D. R. Morris, D. Bartos, G. D. Daves, Jr., and F. Bartos, eds.) Raven Press, New York, pp. 97–110.

120. Seiler, N., and Al-Therib, M.-J., 1974, *Biochim. Biophys. Acta* **354**:206–212.

121. Libby, P. R., 1978, *J. Biol. Chem.* **253**:233–237.

122. Seiler, N., Lamberty, U., and Al-Therib, M.-J., 1975, *J. Neurochem.* **24**:787–800.

123. Caldarera, C. M., Moruzzi, M. S., Rossoni, C., and Barbiroli, B., 1969, *J. Neurochem.* **16**:309–316.

124. Hölttä, E., 1977, *Biochemistry* **16**:91–100.

125. Seiler, N., Bolkenius, F. N., Knödgen, B., and Mamont, P., 1980, *Biochim. Biophys. Acta* **615**:480–488.

126. Blankenship, J., 1978, *Arch. Biochem. Biophys.* **189**:20–27.

127. Shimizu, H., Kakimoto, Y., and Sano, I., 1965, *Nature* **207**:1196–1197.

128. Shaskan, E. G., Haraszti, J. H., and Snyder, S. H., 1973, *J. Neurochem.* **20**:1443–1452.

129. Kremzner, L. T., Barrett, R. E., and Terrano, M. J., 1970, *Ann. N.Y. Acad. Sci.* **171**:735–748.

130. Shaw, G. G., and Pateman, A. J., 1973, *J. Neurochem.* **20**:1225–1230.

131. Seiler, N., and Schmidt-Glenewinkel, T., 1975, *J. Neurochem.* **24**:791–795.

132. Seiler, N., and Lamberty, U., 1975, *Comp. Biochem. Physiol. [B]* **52**:419–425.

133. Seiler, N., and Lamberty, U., 1975, *J. Neurochem.* **24**:5–13.

134. Marton, L. J., Heby, O., Wilson, C. B., and Lee, P. L. Y., 1974, *FEBS Lett.* **41**:99–103.

135. Pearce, L. A., and Schanberg, S. M., 1969, *Science* **166**:1301–1303.

136. Shimizu, H., Kakimoto, Y., and Sano, I., 1964, *J. Pharmacol. Exp. Ther.* **143**:199–204.

137. Michaelson, I. A., Coffman, P. Z., and Vedral, D. F., 1968, *Biochem. Pharmacol.* **17**:2435–2441.

138. Sturman, J. A., and Gaull, G. E., 1975, *J. Neurochem.* **25**:267–272.

139. Sturman, J. A., and Gaull, G. E., 1974, *Pediatr. Res.* **8**:231–237.

140. McAnulty, P. A., Yusuf, H. K. M., Dickerson, J. W. T., Hey, E. N., and Waterlow, J. C., 1977, *J. Neurochem.* **28**:1305–1310.

141. Seiler, N., and Lamberty, U., 1973, *J. Neurochem.* **20**:709–717.

142. Fischer, H. A., Schröder, J. M., and Seiler, N., 1972, *Z. Zellforsch.* **128**:393–405.

143. Anderson, T. R., and Schanberg, S. M., 1972, *J. Neurochem.* **19**:1471–1481.

144. Gilad, G. M., and Kopin, I. J., 1979, *J. Neurochem.* **33**:1195–1204.

145. Macaione, S., and Calatroni, A., 1978, *Life Sci.* **23**:683–690.

146. Heinrich-Hirsch, B., Ahlers, J., and Peter, H. W., 1977, *Enzyme* **22**:235–241.

147. Sobue, K., and Nakajima, T., 1978, *J. Neurochem.* **30**:277–279.

148. De Mello, F. G., Bachrach, U., and Nirenberg, M., 1976, *J. Neurochem.* **27**:847–851.

149. Parker, K., and Vernadakis, A., 1980, *J. Neurochem.* **35**:155–163.

150. Seiler, N., Sarhan, S., and Roth-Schechter, B. F., 1981, *Dev. Neurosci.* **4**:181–187.

151. Schmidt, G. L., and Cantoni, G. L., 1973, *J. Neurochem.* **20**:1373–1385.

152. Shaskan, E. G., 1977, *J. Neurochem.* **28**:509–516.

153. Giorgi, P. P., 1971, *Brain Res.* **34**:199–202.

154. Shiba, T., and Kaneko, T., 1969, *J. Biol. Chem.* **244**:6006–6007.

155. Perry, T. L., Hansen, S., and Kloster, M., 1972, *J. Neurochem.* **19**:1395–1397.

156. Kremzner, L. T., Hiller, J. M., and Simon, E. J., 1975, *J. Neurochem.* **25**:889–894.

157. Kremzner, L. T., and Sturman, J. A., 1979, *J. Neurochem.* **33**:1115–1117.

158. Van den Berg, C. J., Van Kempen, G. M. J., Schadé, J. P., and Veldstra, H., 1965, *J. Neurochem.* **12**:863–869.

159. Roberts, E., and Kuriyama, K., 1968, *Brain Res.* **8**:1–35.

160. Seiler, N., Bink, G., and Grove, J., 1980, *Neuropharmacology* **19**:251–258.

161. Seiler, N., Lamberty, U., and Al-Therib, M.-J., 1975, *J. Neurochem.* **24**:797–800.

162. Sobue, K., and Nakajima, T., 1977, *J. Biochem. (Tokyo)* **82**:1121–1126.

163. Russell, D. H., Gfeller, E., Marton, L. J., and LeGendre, S. M., 1974, *J. Neurobiol.* **5**:349–354.

164. Harik, S. I., and Snyder, S. H., 1974, *Brain Res.* **66**:328–331.

165. Kremzner, L. T., and Cote, L. J., 1980, *Trans. Am. Soc. Neurochem.* **11**:126.

166. Mugnaini, E., 1978, *Proc. Eur. Soc. Neurochem.* **1**:3–31.

167. Giorgi, P. P., 1978, *Neurosci. Lett.* **10**:335–340.

168. Schmidt-Glenewinkel, T., 1979, Untersuchung der regionalen und zellulären Verteilung von Polyamines und Nukleinsäuren im Nervensystem der Ratte, Ph.D. Thesis, Johann Wolfgang Goethe University, Frankfurt am Main, Germany.

169. Russell, D. H., and Meier, H., 1975, *J. Neurobiol.* **6**:267–275.

170. Bartholeyns, J., 1981, *Life Sci.* (in press).

171. Snyder, S. H., Kreuz, D. S., Medina, V. J., and Russell, D. H., 1970, *Ann. N.Y. Acad. Sci.* **171**:749–771.

172. Heby, O., 1977, *Proc. Am. Assoc. Cancer Res.* **18**:79.

173. Jung, M. J., and Seiler, N., 1978, *J. Biol. Chem.* **253**:7431–7439.

174. Seiler, N., and Sarhan, S., 1980, *Neurochem. Res.* **5**:97–100.

175. Seiler, N., and Deckardt, K., 1976, *Neurochem. Res.* **1**:469–499.

176. Raina, A., and Teloranta, T., 1967, *Biochim. Biophys. Acta* **138**:200–203.
177. Michaelson, I. A., and Smithson, H. R., 1971, *Biochem. Pharmacol.* **20**:2091–2094.
178. Lajtha, A., and Sershen, H., 1974, *Arch. Biochem. Biophys.* **165**:539–547.
179. Pateman, A. J., and Shaw, G. G., 1975, *J. Neurochem.* **25**:341–345.
180. Kuriyama, K., Roberts, E., and Vos, J., 1968, *Brain Res.* **9**:231–252.
181. Kossorotow, A., Wolf, H. U., and Seiler, N., 1974, *Biochem. J.* **144**:21–27.
182. Anand, R., Gore, M. G., and Kerkut, G. A., 1976, *J. Neurochem.* **27**:381–385.
183. Peter, H. W., Gies, A., Neumeier, M., Schädler, R., and Wegener, I., 1979, *Gen. Pharmacol.* **10**:123–141.
184. Tashima, Y., Hasegawa, M., and Mizunuma, H., 1978, *Biochem. Biophys. Res. Commun.* **82**:13–18.
185. Nag, D., and Gosh, J. J., 1973, *J. Neurochem.* **20**:1021–1027.
186. Peter, H. W., Wolf, H. U., and Seiler, N., 1973, *Hoppe Seylers Z. Physiol. Chem.* **354**:1146–1148.
187. Fischer, H. A., Korr, H., Seiler, N., and Werner, G., 1972, *Brain Res.* **39**:197–212.
188. Halliday, C. A., and Shaw, G. G., 1978, *J. Neurochem.* **30**:807–812.
189. Siegel, L. G., and McClure, W. O., 1975, *Brain Res.* **93**:543–547.
190. Ingoglia, N. A., Sturman, J. A., and Eisner, R. A., 1977, *Brain Res.* **130**:433–445.
191. Kremzner, L. T., and Abron, R. T., 1979, *Trans. Am. Soc. Neurochem.* **10**:208.
192. Fischer, H. A., and Schmatolla, E., 1972, *Science* **176**:1327–1329.
193. Ingoglia, N. A., Jaggard, P., Perez, C., and Sturman, J. A., 1979, *Neurosci. Abstr.* **10**:2309.
194. Ingoglia, N. A., and Sturman, J. A., 1980, *Neurochem. Res.*, **5**:913–914.
195. Shaskan, E. G., and Snyder, S. H., 1973, *J. Neurochem.* **20**:1453–1460.
196. Shaw, G. G., 1979, *Neurochem. Res.* **4**:269–275.
197. Shaskan, E. G., and Marshall, J. M., 1973, *Fed. Proc.* **32**:1166.
198. Seiler, N., and Schröder, J. M., 1970, *Brain Res.* **22**:81–103.
199. Giorgi, P. P., Field, E. J., and Joyce, G., 1972, *J. Neurochem.* **19**:255–264.
200. Harik, S. I., and Sutton, C. H., 1979, *Cancer Res.* **39**:5010–5015.
201. Mirzoyan, P. A., and Promyslov, M. Z., 1979, *Ukr. Biokhim. Zh.* **51**:474–476.
202. Marton, L. J., and Heby, O., 1974, *Int. J. Cancer* **13**:619–628.
203. Marton, L. J., Heby, O., Levin, V. A., Lubich, W. P., Crafts, D. C., and Wilson, C. B., 1976, *Cancer Res.* **36**:973–977.
204. Marton, L. J., Edwards, M. S., Levin, V. A., Lubich, W. P., and Wilson, C. B., 1979, *Cancer Res.* **39**:993–997.
205. Seidenfeld, J., and Marton, L. J., 1979, *J. Natl. Cancer Inst.* **63**:919–931.
206. Buss, J., 1963, Die pharmakologischen Wirkungen von Gamma-Aminobutyrylcholin und von Spermin, Spermidin und Putrescin, M.D. Thesis, Freie Universität, Berlin.
207. Shaw, G. G., 1972, *Arch. Int. Pharmacodyn. Ther.* **198**:36–48.
208. Anderson, D. J., and Shaw, G. G., 1974, *Br. J. Pharmacol.* **52**:205–211.
209. Sakurada, T., Kohno, H., Tadano, T., and Kisara, K., 1977, *Jpn. J. Pharmacol.* **27**:453–460.
210. Anderson, D. J., Crossland, J., and Shaw, G. G., 1975, *Neuropharmacology* **14**:571–577.
211. Nistico, G., Ientile, R., Rotiroti, D., and Di Giorgio, R. M., 1980, *Biochem. Pharmacol.* **29**:954–957.
212. Meldrum, B. S., 1978, *Lancet* **2**:304.
213. Wedgwood, M. A., and Wolstencroft, J. H., 1977, *Neuropharmacology* **16**:445–446.
214. Gould, R. M., and Cottrell, G. A., 1976, *Comp. Biochem. Physiol.* **483**:591–597.
215. Tadano, T., Onoki, M., and Kisara, K., 1974, *Folia Pharmacol. Jpn.* **70**:9–18.
216. Tadano, T., 1974, *Folia Pharmacol. Jpn.* **70**:457–464.
217. Sakurada, T., and Kisara, K., 1978, *Jpn. J. Pharmacol.* **28**:125–132.
218. Shaw, G. G., 1977, *Biochem. Pharmacol.* **26**:1450–1451.
219. Russell, D. H., Byus, C. V., and Manen C.-A., 1976, *Life Sci.* **19**:1297–1306.
220. Lau, C., Bartolomé, J., Seidler, F. J., and Slotkin, T. A., 1977, *Neuropharmacology* **16**:799–809.
221. Bartolomé, J., Seidler, F. J., Anderson, T. R., and Slotkin, T. A., 1976, *J. Pharmacol. Exp. Ther.* **197**:293–302.
222. Manen, C.-A., and Russell, D. H., 1977, *Science* **195**:505–506.

223. Harik, S. I., 1979, *J. Neurochem.* **33**:1131–1133.
224. Seiler, N., Bink, G., and Grove J., 1979, *Neurochem. Res.* **4**:425–435.
225. Pajunen, A. E. I., Hietala, O. A., Baruch-Virransalo, E.-L., and Piha, R. S., 1979, *J. Neurochem.* **32**:1401–1408.
226. Tsvetnenko, E. Z., and Shugalev, V. S., 1978, *Biull. Eksp. Biol. Med.* **85**:28–30.
227. Heby, O., and Andersson, G., 1980, *Polyamines in Biochemical Research* (J. M. Gaugas, ed.), John Wiley & Sons, Chichester, New York, Brisbane, Toronto, pp. 17–34.
228. Singh, V. K., and Sung, S. C., 1972, *J. Neurochem.* **19**:2885–2888.
229. Goertz, B., 1979, *Brain Res.* **173**:125–135.
230. Das, R., and Kanungo, M. S., 1979, *Biochem. Biophys. Res. Commun.* **90**:708–714.
231. Zilliken, F., Gerken, U., Sokolis, U., and Giesing, M., 1977, *Hoppe Seylers Z. Physiol. Chem.* **358**:328–329.
232. Heger, H. W., and Peter, H. W., 1979, *Pharmacology* **10**:433–435.
233. Lapinjoki, S. P., Pajunen, A. E. I., Hietala, O. A., and Piha, R. S., 1980: *FEBS Lett.* **112**:289–292.

10

Cyclic Nucleotides in the Central Nervous System

Kenneth A. Bonnet

1. INTRODUCTION

1.1. Scope of the Chapter

This chapter is intended to present perspectives on the role of cyclic nucleotides in central nervous system function and pathology. The coverage is representative rather than comprehensive and extends to July, 1980. The particular focus of the present chapter is on the cyclic nucleotides themselves. The synthesizing and degradative enzymes in the cyclic nucleotide systems and the protein kinases activated by the cyclic nucleotides in the central nervous system are the subjects of individual chapters of subsequent volumes of this series.

1.2. The Role of Cyclic Nucleotides as Second Messengers in the Nervous System

The cyclic nucleotides have figured prominently in a substantial number of hormone and neurotransmitter systems. Generally, several classes of neurotransmitters and neuropeptidyl hormones are known to act through receptors that are coupled to the enzymes that synthesize intracellular cyclic nucleotides. The intracellular cyclic nucleotides, in turn, regulate the activity of a number of protein-phosphorylating enzymes, the kinases, that reversibly alter the levels of enzyme activities and ongoing protein synthesis. It is believed that through this process the specificity of agents acting on the cell membrane receptors is translated to very selective effects on the internal state of the receiving cell. This second-messenger role of the cyclic nucleotides is somewhat complex and provides for a high degree of "fine tuning" and interactive effects of several

---

Kenneth A. Bonnet • Department of Psychiatry, New York University School of Medicine, New York, New York 10016.

hormones or neurotransmitters acting in concert on the same neuron or glial cell.

The second-messenger role of the cyclic nucleotides cannot be taken as a given in every cell type or in every brain region under study. There are neurotransmitters that have little direct effect on cyclic nucleotide metabolism. There have been a number of studies that have attempted to bypass the neurotransmitter receptor by injecting stable analogues of cyclic AMP but have become confounded by substantially differing effects of different analogues injected into the same brain sites. Therefore, as with the criteria for the classification of hormones as neurotransmitters, there are stringent criteria by which one can define the role of a cyclic nucleotide as a second messenger in a particular system under study. These criteria were defined by Siggins[1] and serve well in the evaluation of work in the literature as well as in the design of systematic research efforts.

There are four major criteria, adapted from the criteria used for hormones, to establish that the action of a neurotransmitter is mediated by a cyclic nucleotide.

1. Exogenous neurotransmitter or neuromodulator substance and the activation of the synaptic pathway directly must each regulate intracellular levels of cyclic nucleotide in the target postsynaptic cell.
2. The change in intracellular nucleotide content should precede "the biological event" triggered by the transmitter or nerve pathway.
3. Responses to the neurotransmitter, neuromodulator, or nerve pathway should be logically altered by drugs that specifically interact with the nucleotide cyclase or phosphodiesterase.
4. Exogenous cyclic nucleotides (and analogues that activate protein kinases) should elicit the biological event caused by the neurotransmitter, neuromodulator, or pathway stimulation.

The attempt to satisfy all of these criteria is difficult, particularly from the technical standpoint. The rapidity of synaptic events far exceeds the speed of our sampling techniques. The measurement of nucleotide levels may represent an integral of cell types and of interactive effects of several transmitters at any given time. The application of exogenous cyclic nucleotides is met with limited penetration into the target cells, and the phosphodiesterase inhibitors most commonly used have little selectivity for the potentiation of one cyclic nucleotide over the other.

The following sections are oriented toward the exposition of the general areas of study involving the cyclic nucleotides in the function of specific nervous system functions. Emphasis is placed on critical evaluation of methodology or interpretation rather than on presentation of a comprehensive review.

The complex nature of the nucleotide cyclases and phosphodiesterases requires that these topics be treated in a subsequent chapter in Volume 4 of this series. Likewise, the protein kinases that are stimulated by the cyclic nucleotides in the central nervous system are discussed individually in subsequent volumes as well.

2. METHODS FOR DETERMINING CYCLIC NUCLEOTIDE LEVELS

The determination of the cyclic nucleotide levels in brain has been difficult and requires specialized techniques. The primary difficulty lies in the requirement for rapid fixation of tissues to circumvent the rapid postmortem changes that occur as a result of tissue ischemia and anoxia.[2] If the tissue is rapidly fixed, the determination of cyclic nucleotide levels proceeds as in any other tissue. The extraction of the tissue is followed by an assay based on use of cyclic nucleotide binding proteins or on radioimmunoassays; these are described in Section 2.4.

2.1. Problems Associated with Rapid Postmortem Changes in Brain Cyclic Nucleotide Levels

The problems of rapid postmortem changes in brain cyclic nucleotides were specified in early studies.[2] Decapitation and delayed fixation lead to considerable changes in brain glycogen, glucose, lactic acid, and the products of the Embden–Meyerhof pathway.[3-5] Accompanying these changes are considerable alterations in the forebrain cyclic AMP levels which are not necessarily uniform in all brain regions.[6] Several methods for rapid fixation have been studied. Each has definite advantages and definite liabilities that one must consider before the use of the technique or the interpretation of the data derived from tissues fixed by that technique.

2.2. Rapid Fixation Procedures to Stabilize Postmortem Levels for Measurement

The cyclic nucleotides are very stable in preparations of the tissue that are devoid of enzyme activity. The inactivation of enzymes that alter cyclic nucleotide, has been the subject of considerable technology in the last decade.

Microwave fixation has provided the most rapid and thorough method used for brain fixation to date. The focusing of up to 4 kW on the head of the immobilized animal provides inactivation of rapidly changing *in vivo* metabolic systems.[7-10] However, microwave fixation must be applied with full knowledge of the relative efficiency of stabilization of cyclic nucleotide levels in specific brain regions as a function of head position, delivered beam power, and time.

The most carefully delineated contribution of each of these factors was reported by Lenox and co-workers.[11] The inactivation of enzymes as a result of regional heating varies with perturbations of microwave distribution by differences in regional tissue composition and orientation of the head of the animal. "Hot spots" can occur that lead to vacuolization and diffusion of materials into adjacent regions; other areas may be insufficiently heated to stabilize enzyme activity.[12] Hypothalamus, for example, is among the least heated regions. Regional nonuniformity in microwave fixation has been reported by others as well.[13]

Lenox and co-workers emphasize the necessity for attention to study designs that allow exposure durations that are relevant or standardized for specific regions of interest. In addition, it is imperative to minimize line voltage fluctuations. The target brain regions must be inspected for structural integrity with the completion of a standardized fixation procedure. A comparison of these methods of sacrifice indicates that 3.5 kW forward power on the immobilized head for 3 s was optimal, whereas 3.0 kW for 5 s was not. This was assessed both by regional cyclic nucleotide levels and by residual phosphodiesterase activity.[11]

One consideration seldom addressed in procedures of rapid brain fixation is the stress engendered by the immobilization of the animal in preparation for fixation. In early studies attempting rapid freezing of the animal in liquid nitrogen or Freon®, it was common practice to insert the animal into a wire cage prior to lowering it into the liquid phase. In focused microwave fixation, the animal is inserted into a holder so that the head is immobilized, and the holder is locked in place in the microwave apparatus. Such stressful handling substantially elevates the levels of cyclic AMP and cyclic GMP in brains of mice and rats.[14-16] Therefore, the levels of cyclic nucleotides found in most standard microwave fixation studies are not resting levels but represent elevations of cyclic AMP that are about 1.5- to twofold higher than resting levels (as determined by the same measurements in animals that were repeatedly habituated in the device for several days prior to sacrifice).[15] Moreover, the changes caused by the stress of prefixation handling may not be uniform over all brain regions prior to microwave or freeze-fixing of the animal. The level of cyclic GMP in the cerebellum, for example, is reduced by 80% in animals that are adapted to the device for several days prior to sacrifice.

The levels of cyclic AMP and cyclic GMP elevated in several brain regions in response to short exposure to a variety of stresses.[9,14-17] The effects on cyclic nucleotide levels of sedating or anxiolytic drugs must be interpreted with caution. Such drugs have the effect of reducing susceptibility of the animal to handling stress in preparation for sacrifice. The lowering of cerebellar cyclic GMP by such drugs is similar to the 80% lower cerebellar cyclic GMP levels in handling-adapted animals.[15] Furthermore, when drug effects are studied, it is necessary to insure that results are specific to the selective action of the drug rather than caused by procedurally induced changes in the levels of regional cyclic nucleotides.

Rapid freezing techniques provide an alternative means of stabilizing brain cyclic nucleotide levels. Freezing mixtures usually consist of liquid nitrogen,[16,18] Freon®,[19] or a mixture.[20] The freezing liquid must be of the liquid nitrogen or Freon® temperature range to effect rapid freezing. Freezing techniques have the distinct advantages of reproducibility, economy, and independence of head size, orientation, or movement during fixation.[21] The cost of rapid freezing is comparatively low and requires minimal apparatus. In addition, the frozen tissue can be removed in the cold room but requires the use of cooled metal tools for removal or dissection. The unthawed tissue can be mounted for histochemical techniques or homogenized for assay of tissue constituents or enzyme activity. Rapid freezing offers an additional advantage of maintaining tissues with substantially intact membranes and associated enzymes that can be activated at will by temperature elevation under the control of the experimenter.[22-24] A disadvantage of the freezing procedures is the necessity for dissection in the cold (thawing the frozen head to $-24°C$ renders the skull sufficiently pliable to permit easier exposure of the intact brain). The immobilization of the animal presents the same problems as discussed above for microwave fixation. However, with the freezing procedure, there is a specific method of circumventing this problem in mice and rats.[25]

Several variations on the freezing method have been studied. Whole-body immersion of mice is very effective and results in brain cyclic nucleotide levels that are very close to the levels found in the most careful microwave fixation studies. Whole-body immersion of the rat is potentially effective as well, but only if the animal is small.[26] The larger the animal, the greater the risk of rapid outer body freezing that insulates the inner tissue from rapid freezing. Rats 200 g or smaller may be fixed effectively by whole-body immersion so that cyclic nucleotide levels and other substances that are sensitive to anoxia and ischemia compare well with the levels determined in microwave-fixed animal brains.[27,28] Decapitation into liquid nitrogen is very poor in that the core of the brain cools very slowly and the effects of anoxia are reflected in the same substantial elevations in cyclic nucleotides that are seen with simple decapitation.[27,28]

An effective variation on the rapid freezing procedure has been described[25] for use with animals up to 500 g. The animal is picked up by the posterior third of the tail, immersed rapidly to midabdomen in liquid nitrogen for 1.5 s, and elevated so that only the head and upper third of the thorax are immersed; the immersed head is kept in that position for a total time of 2 min. The heart continues to beat through the early part of the period, and the circulating blood in the brain acts as heat exchanger until that portion of the brain tissue freezes. Brain core temperature (at the centroid of the brain) reaches $5°C$ by 7 to 8 s, $0°C$ by 15 s, where it plateaus for a brief period before descending rapidly in temperature. Thus, the brain (in even the most insulated area) reaches temperatures below $10°C$ within seconds in a procedure that does not require prior immobilization and that does not produce elevated brain temperatures prior to enzyme inactivation. A disadvantage may be that brain tissue remains substantially metabolically intact on thawing, and the processing must be carried out at very cold temperatures until the tissue has been deproteinized or otherwise processed. An advantage of the procedure, however, is that the enzymes contributing to the levels of cyclic nucleotides remain accessible for subsequent *in vitro* assay as well. In addition, constituents that are heat labile are preserved. Generally, this procedure circumvents the problems described for the whole-body immersion in animals larger than mice.[28]

Another freezing method, the freeze-blowing technique, is very rapid. The head is rapidly pierced by two probes through one of which 25 psi air pressure is delivered, blowing the brain out through the other probe into a liquid-nitrogen-cooled metal chamber.[29] The tissue is frozen very rapidly, but the brain is not available for regional dissection. Studies of light-induced changes in retinal cyclic GMP have employed the use of isolated, dark-adapted eyes that are stimulated with light for discrete times and then quickly crushed by the use of liquid-nitrogen-cooled hammers.[30]

Visual inspection after fixation often provides quality assurance in fixation of tissues. Microwave-fixed tissue should appear tan and is not fixed if some pink color remains.[11] Conversely, adequately frozen tissues will appear white, with red lines demarcating frozen blood vessels; slight gray or tan color in brain tissue indicates inadequate freezing (not below $-20°C$) or results from partial thawing by contact with warm instruments or surfaces or with ungloved hands.

Dissection of microwave-fixed tissue is relatively straightforward if the specimen has not been overheated. Cooling of the heads in ice renders the tissue firmer for dissection. Dissection of frozen tissues is best done with mounting on a cryostat for sectioning and freeze-punching with cooled stainless steel tubes of appropriate diameters.[25,31]

2.3. Tissue Extraction Procedures for Cyclic Nucleotide Measurement

Cyclic nucleotides may be prepared from most tissue sources by the same procedures. Heat and acid stability of cyclic nucleotides permit preparation by boiling or by acid precipitation of associated proteins and enzymes. Generally, tissue can be dropped into boiling water or buffer for 5 min, cooled, and centrifuged to remove coagulated protein. The supernatant contains the cyclic nucleotides and is compatible for direct use in many assays if sufficiently concentrated. Alternatively, fresh or freeze-fixed tissue is homogenized in the cold in five or ten volumes of 6% trichloroacetic acid or acidic ethanol. The supernatant is washed with ether extensively in the case of the trichloroacetic acid and neutralized, taken to dryness, and reconstituted at an appropriate concentration for the assay to be used. Trichloroacetic acid interferes slightly with the protein binding assay for cyclic AMP. If the radioimmunoassay is to be used, trichloroacetic acid extracts can be neutralized with $CaCO_3$, but the Ca^{2+} interferes in the protein binding assay.[32]

2.4. Methods of Quantitating Cyclic Nucleotides in Samples

The assay of cyclic nucleotides is efficient by radioimmunoassay or by protein binding assay. The protein binding assay for cyclic AMP utilizes protein kinase protein from beef muscle,[33] and the protein binding assay for cyclic GMP uses a protein from lobster muscle. Competitive displacement of [³H] cyclic AMP or [³H]cyclic GMP from the corresponding high-affinity protein binding sites is the basis for the protein binding and for the radioimmunoassays. The advantage of higher sensitivity with the radioimmunoassays (0.05 to 10 pmol) is offset by their higher cost and shorter shelf life compared to the protein binding assays (0.5 to 20 pmol). The standard curve for the saturation of the protein binding assays is not linear. Therefore, the reading of sample values may be problematic. It appears that the nonlinear nature of the curves results from the presence of two binding sites on the protein, and reading of sample concentrations from the standard curve can be resolved with good accuracy by use of an equation derived specifically to accomodate this assay.[34] An excellent comparison of the various assay systems and the extraction procedures optimal for each have been reported.[35]

Immunohistochemical studies are very sensitive to the presence and localization of specific cyclic nucleotides. Freeze-fixing of brain tissue permits freeze sectioning for immunohistochemical processing.[25] The specificity of the immunohistochemical procedure is dependent on the antibodies used and may sometimes react with ATP or GTP in sufficient quantity to give spurious results. The specifity can be defined in immunohistochemical studies by the use of gaseous nitrous acid or acetylation to control for the contribution of ATP or GTP.[36]

3. CYCLIC NUCLEOTIDES IN THE CENTRAL NERVOUS SYSTEM

Cyclic nucleotides are integrally involved in the mediation of the actions of several neurotransmitters and in the actions of peptides and hormones in a variety of areas of the central nervous system. The cyclic nucleotides are involved in the synthesis and release of neurotransmitters as well as in the mediation of postsynaptic effects. In addition, the release of many peptidyl hormones from central nervous system sites and from pituitary is regulated by factors that act through the cyclic nucleotides.

3.1. Cyclic Nucleotide Structure, Levels, and Distribution in Nervous System Tissue

The structures of cyclic AMP, cyclic GMP, and cyclic IMP are shown in Figure 1.

The levels of the cyclic nucleotides in a specific brain region represent the net contribution of the cyclase and of degradation by phosphodiesterases. The levels may thus reflect net activation of the cyclase through some mechanism

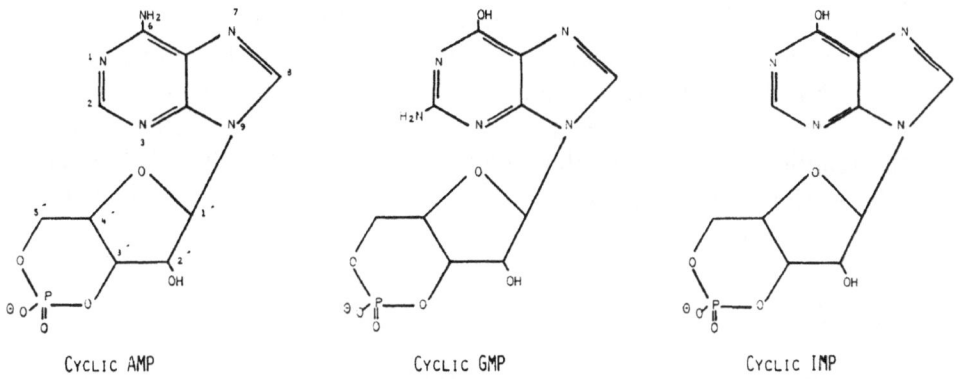

CYCLIC AMP   CYCLIC GMP   CYCLIC IMP

Fig. 1. Structure of the cyclic nucleotides.

such as a neurotransmitter receptor binding an agonist ligand or net activation of the phosphodiesterase (i.e., reduced cyclase activity relative to increased phosphodiesterase activity). The levels of one cyclic nucleotide may reflect the primary consequence of some event such as catecholamine release for postsynaptic stimulation. Alternatively, the levels of that cyclic nucleotide may reflect the secondary consequences of the initiating event (e.g., decreased release of dopamine as a result of a presynaptic α-adrenergic agonist diminishing dopamine release).

The levels of cyclic nucleotides in the central nervous system vary across brain regions. Cyclic AMP levels range from about 6 to 15 pmol/mg protein throughout the brain, and cyclic GMP levels are 0.2 to 0.5 pmol/mg protein except in brainstem (0.8 to 1.0) and cerebellum (4 to 8).[37]

Levels of cyclic nucleotides are higher in brain than in most peripheral tissues. The levels of cyclic nucleotides in brain do not correlate very well with reported enzyme activities; the best correlation is with the phosphodiesterase activity.[38] The regulation of levels by phosphodiesterase activity makes it unlikely that excretion from the cell is important to the removal of the cyclic nucleotides. Thus, the release of cyclic nucleotides to extracellular space or cerebrospinal fluid may not provide direct reflections of cellular responses to neurotransmitter activity.

3.2. Enzymes Synthesizing Cyclic Nucleotides

A number of factors regulate the synthesis of cyclic nucleotides. Synthesis of cyclic AMP is accomplished by a single enzyme, adenylate cyclase (E.C. 4.6.1.1). Adenylate cyclase requires the presence of guanosine triphosphate (GTP) and magnesium ion in association with the ATP substrate.[39] In addition, the coupling to hormone or neurotransmitter receptor systems requires additional factors for the inhibition or stimulation of the enzyme. Generally, receptors mediating hormonal stimulation require the presence of calcium ions in the membrane, calcium-binding protein (calmodulin), and possibly metal ions among other possible factors.[40–42]

Guanylate cyclase (E.C. 4.6.1.2) occurs as a soluble and as a membrane-bound enzyme in brain, is hydrophobic, and is stimulated by free calcium ions and by fatty acids.[43] The enzyme is activated by sodium azide, sodium arachidonate, and sodium nitroprusside, but not by sodium fluoride.[44]

Notwithstanding our concept of the purity of product from one cyclase or another, it has been demonstrated that partially purified soluble guanylate cyclase can produce cyclic AMP from ATP.[45] The agents that effectively stimulate the reaction are all those that selectively increase cyclic GMP levels but not cyclic AMP levels in various tissues. The nucleotide cyclases will be treated in detail in Volume 4 of this series.

3.3. The Phosphodiesterases

There are several 3′,5′-cyclic nucleotide phosphodiesterase isoenzymes (E.C. 3.1.4.17) that convert cyclic nucleotides to 5′-nucleotides. The isoenzymes for cyclic AMP vary in their affinity (high and low K_m), whereas the affinity for the cyclic GMP enzyme varies little.[43,45] Subcellular distribution of cyclic AMP phosphodiesterase equally divided between membrane-bound and soluble fractions, whereas the cyclic GMP enzyme is largely in the soluble fraction.[46]

Several factors are known to regulate the activity of the phosphodiesterases. Calcium-stimulated phosphodiesterase is stimulated in the presence of calmodulin. The calcium-stimulated forms of phosphodiesterase appear to utilize cyclic GMP at a fourfold greater velocity at low substrate concentrations and cyclic AMP at a threefold greater velocity than cyclic GMP at high substrate concentrations.[47] There are also several endogenous factors, still uncharacterized, that inhibit the stimulation of the calcium-sensitive phosphodiesterases.[48]

A number of agents stimulate the activity of phosphodiesterases. Best characterized among these agents is imidazole.[49] Brain gangliosides also stimulate the activity of the calcium-stimulated and of the calcium-independent phosphodiesterases.[50]

Potassium ions at physiological levels inhibit the calcium-stimulated phosphodiesterase of brain.[51] Ro20-1724 [4-(3-butoxy-4-methoxybenzyl)-2-imidazolidinone] given systemically increases cyclic AMP levels in almost all brain regions with little effect on cyclic GMP levels. The cerebellum is exceptional in that Ro20-1724 increases both cyclic nucleotide levels in that region.[52] Another selective phosphodiesterase inhibitor is ZK 62711 [4-(3-cyclopentyloxy-4-methoxyphenyl)-2-pyrrolidone].[53] Although different from the imidazolidinone compounds, ZK 62711 is also able to increase cyclic AMP levels selectively.

Various derivatives of the methylxanthines are known to be effective inhibitors of phosphodiesterases. Theophylline and 3-isobutyl-1-methylxanthine (IBMX) are effective in potentiating cyclic nucleotide accumulation but are not selective and are more potent in blocking adenosine stimulation of cyclic AMP formation than they are in blocking phosphodiesterase activity.[54,55] A novel variant of these compounds, 1-isoamyl-3-isobutylxanthine, is a potent inhibitor of phosphodiesterase activity and potentiates norepinephrine, histamine, and adenosine stimulation of cyclic AMP formation.[56] Generally, the methylxanthines are not selective for inhibiting the degradation of one nucleotide and generally lead to accumulations of elevated levels of both cyclic AMP and cyclic GMP. It remains to be determined if this applies to the isoamyl derivative.

4. HORMONAL AND NEUROTRANSMITTER REGULATION OF CYCLIC NUCLEOTIDE METABOLISM

Since the earliest studies by Sutherland and colleagues,[57] the stimulation of adenylate cyclase activity by catecholamines and other hormones has become the pervasive model of transduction of the hormonal message.[38,58] The transduction process amounts to an amplification process. A recent review of this process has detailed the putative steps from the initiation by agonist–receptor binding through the eventual degradation of the cyclic nucleotide generated.[38]

The system first observed to involve the cyclic AMP system was the stimulation of liver cells by epinephrine.[57] The catecholamines have each been observed to stimulate cyclic AMP accumulation, as have a large number of other neurotransmitter and hormonal systems. It is now apparent that many of the neurotransmitter systems exhibit receptor heterogeneity and that within each of these systems not all receptor subtypes are coupled to cyclic nucleotide metabolism.

The β-adrenergic system appears to consist of two subtypes of receptors.[59] The β_1 receptor population in brain appears to be adenylate cyclase linked, varies 20-fold across various brain regions, and changes density of binding sites in response to chronic under- or overstimulation. The β_2 receptor population shows very even distribution across brain regions and does not respond to frequency of ligand encounter. This receptor population is suggested to be associated with cerebral blood vessels or nonneural elements.[60] α-Adrenergic receptors have also been reported to mediate accumulation of cyclic AMP and cyclic GMP in brain tissues. 6-Fluoronorepinephrine and 2-fluoronorepinephrine are selective α- and β-receptor agonists. The use of these compounds has made it evident that β_1-receptor-mediated accumulations of cyclic AMP are the result of direct coupling to adenylate cyclase, whereas accumulation of cyclic AMP through the α-adrenergic system is by potentiation of the accumulation of cyclic AMP mediated by the adenosine, histamine, or β_1 receptor systems.[61]

Agonists working primarily at the postsynaptic α-adrenergic receptors increase cerebellar cyclic GMP levels, whereas putative antagonists to these postsynaptic sites promote decreases in the levels of cyclic GMP.[60] Clonidine, an agonist for both pre- and postsynaptic α receptors, effects decreases in cyclic GMP at low doses, presumably at presynaptic sites. In these studies, β-adrenergic agonists or antagonists had no effect on cerebellar cyclic GMP levels.[62]

Like the adrenergic receptor systems, it has been proposed that the histamine receptors are of at least two types. The H_2 system is coupled to adenylate cyclase regulation, whereas the H_1 system resembles the α-adrenergic system by apparently potentiating cyclic AMP accumulations stimulated by H_2 or other such coupled receptor sites.[63–65]

The dopamine systems in the forebrain contain receptor subtypes coupled to adenylate cyclase.[66] The classical dopamine receptor antagonists are the antipsychotic drugs.[66] However, the antagonists are not selective for the D_1 receptor and have relatively poor blocking potency compared to their potency in receptor binding studies.[66] It is interesting that antipsychotic drugs of several types show potent binding to calmodulin, the calcium-binding protein that is the putative intermediary in the coupling of the dopamine D_1 receptor to adenylate cyclase.[67] We demonstrated that chronic blockade of the dopamine receptor results in increased adenylate cyclase stimulation that can be reversed by subsequent treatment with L-DOPA.[68]

A serotonin-receptor-mediated stimulation of cyclic AMP accumulation has been described which shows selective blocking by serotonin receptor blockers but not by high-affinity dopamine blocking agents. The cortex and the hippocampus are particularly rich in this activity in comparison to other brain regions.[69]

Adenosine is a potent stimulator, through the adenosine receptor, of cyclic AMP accumulation. Through local metabolic conversion, 5'-AMP, ATP, and ADP are capable of stimulating these receptors as well.[70] The alkylxanthines are potent inhibitors of this activity and are thus very effective in assays where the aim is to increase cyclic nucleotide accumulation by simultaneously blocking phosphodiesterase activity and acting through mechanisms other than the adenosine system.

The prostaglandins, of the E series in particular, can stimulate the accumulation of cyclic AMP in some brain regions.[71] Both adenosine and prostaglandin stimulation of cyclic AMP accumulation can be inhibited in a stereospecific manner in some types of neural cell populations by opiate narcotics and opioid peptides.[72-74] The central role of the calcium ion in the actions of the prostaglandins as displaced by opioid agonists at the cell membrane calcium binding sites and as inhibitor of adenosine release suggests that the calcium ion may be central to the action of these agents on cyclic AMP metabolism.[75,76]

Ion channels can be manipulated to alter cyclic nucleotide metabolism. Depolarizing agents such as veratridine or potassium ions at high concentrations increase cyclic GMP levels in several brain areas in a calcium-dependent manner.[77-79] Sea anemone toxin II is known to keep activated sodium channels open and raises cyclic GMP levels in cerebellum 35-fold.[80] This effect was blocked by lithium ion.

Cholinergic effects on cyclic nucleotide levels are most prominent in the elevation of cyclic GMP levels. The specific responses to cholinergic activation in select brain regions reveal a region-specific response in cyclic nucleotide metabolism. Six brain regions show cyclic GMP elevations with cholinergic activation, and six regions of the 18 studied show cholinergic agonist-mediated increases in cyclic AMP as well.[81]

In cerebellum, glutamate, glycine, GABA, apomorphine, and depolarizing agents all increase cyclic GMP levels.[82] In addition, histamine acting through the H_1 receptor increases cyclic GMP levels in sympathetic ganglia when optimal levels of calcium ion are present in the medium.[83] It is noteworthy that increases in guanylate cyclase activity induced by hormone stimulation have not yet been reported for cell-free preparations[84] but are demonstrable for slice preparations.

A number of neurotransmitter and hormone systems in brain tissue effect an inhibition of adenylate cyclase activity. Muscarinic acetylcholine receptors mediate inhibition.[85] The epinephrine interaction with the α-adrenergic receptor and opioids interacting with the enkephalinergic opioid receptor inhibit the basal and prostaglandin E_1-stimulated accumulation of cyclic AMP.[86,87] It is apparent that not all of the effects of narcotic agonists are to effect decreased cyclic AMP levels, however, since the effects of various classes of narcotic agonists and antagonists vary with the brain region studied.[25,88] The membranes of secretory vesicles of the adrenal medulla contain a cyclic-AMP-generating system that is inhibited by *l*-isoproterenol or epinephrine.[89] In all of these inhibitory actions, the requirement for GTP seems to be general. The specific role of GTP in inhibitory actions on adenylate cyclase activation is not yet known. Calcium ions are inhibitory to many types of adenylate cyclase activity in brain but are stimulatory to calmodulin-linked adenylate cyclase.[90] Lead ions potently inhibit adenylate cyclase in brain[91,92] as do lithium and lanthanum ions in some brain regions.[93] Bacterial endotoxin (*Bacillus thuringiensis*) inhibits adenylate cyclase activity, whereas cholera toxin irreversibly maximally activates the same enzyme.[94,95]

Many of the differences seen in the literature that create controversy often result from the use of differing brain regions, poor duplication of procedure or reagents in the assay, or the use of different strains or species of animals.[38] Many of these difficulties can be resolved if one consults the cataloguing by Daly[37] of the effects on neurotransmitters on cyclic nucleotide accumulations. Although this material is becoming chronologically dated, the usefulness of the materials will undoubtedly serve an instructive role for some time to come. Daly has classified effects within a number of animal species and within each species with respect to specific neurotransmitter effects in homogenate or slice preparations.

5. CYCLIC NUCLEOTIDE REGULATION OF PROTEIN PHOSPHORYLATION

The cyclic nucleotides are pervasive throughout most cell types and tissues. The specificity of the messages they impart to the internal cellular metabolism is conveyed by the stimulation of specific protein kinases that catalyze the phosphorylation of specific protein substrates in the cell membrane, ribosomes, nuclear histones, etc., to alter the local activity of the cell. These complex and highly specific effects will be discussed in detail in other chapters in this series. The phosphoprotein phosphatases remove the phosphate groups from the protein substrates of the kinases, thus providing a mechanism for the reversible alteration of the conformation and activity of the enzyme or other protein.[96,97] This reversible regulation of specific proteins is exemplified by the phosphorylation of tyrosine hydroxylase for the synthesis of catecholamine neurotransmitters as discussed in Section 6.1.

The cyclic-nucleotide-binding proteins are primarily associated with the protein kinases. There are two types of cyclic-AMP-binding proteins of this type.[98] RI is about 47,000 daltons and is more often found in free, dissociated form, whereas RII is almost always found associated only with the type II protein kinase activity and may be a portion of a type II protein kinase holoenzyme.

The affinity for protein-kinase-associated binding sites for cyclic AMP is about 0.7 nM for RI and about 2 nM for RII, with an apparent capacity of about two sites per molecule and little evidence of cooperativity.[99,100] Binding sites for cyclic-GMP-stimulated protein kinase evidence positive cooperativity ($_n$H = 1.6) and forms a dimeric complex consisting of two kinase molecules coupled to two cyclic GMP molecules.[101]

6. CYCLIC NUCLEOTIDE REGULATION OF NEUROTRANSMITTER SYNTHESIS AND RELEASE

6.1. Cyclic Nucleotide Regulation of Neurotransmitter Synthesis

The regulation of the primary synthetic enzyme in catecholamine synthesis, tyrosine hydroxylase (E.C. 1.14.16.2), has been convincingly shown to be cyclic AMP stimulated through a protein kinase.[102-105] The resulting phosphorylation of the tyrosine hydroxylase molecule effects a covalent modification that elevates the V_{max} of the enzyme about twofold without altering its affinity for substrate.[106]

The regulation of the synthesis of tyrosine hydroxylase molecules has also been shown to be under the influence of cyclic-AMP-mediated enzyme-induction mechanisms.[107] The role of cyclic AMP in the induction of increased synthesis of the enzyme molecules has been demonstrated in the adrenal medulla and in the superior cervical ganglion.[108]

Other neurotransmitter-synthesizing enzymes have been shown to be stimulated by cyclic AMP. Dopamine-β-hydroxylase (DBH) in the superior cervical ganglion, adrenal medulla, and neuroblastoma cells shows increased activity in the presence of dibutyryl cyclic AMP.[109-111] The activity of acetylcholinesterase and that of choline acetyltransferase have been demonstrated to increase in the presence of dibutyryl cyclic AMP in fetal brain tissues and in neuroblastoma cells.[112-114] However, as Nathanson points out,[58] these effects are not reported in adult, differentiated tissues, and it is probable that the increased activity of these enzymes is the result of cyclic-AMP-induced growth or differentiation of the cells or tissues rather than of a direct regulation of the enzymes themselves.

Glutamic acid decarboxylase, the enzyme synthesizing γ-aminobutyric acid in brain, is increased in cultured brain cells by cyclic AMP.[115] Tryptophan hydroxylase, the enzyme synthesizing serotonin, is unaffected by cyclic AMP, but effects on the transport of tryptophan may account for cyclic-AMP-mediated increases in serotonin synthesis by this enzyme.[116,117]

Melatonin synthesis in the pineal gland is well characterized as being responsive to norepinephrine stimulation of adenylate cyclase. The elevated cyclic AMP levels result in increases in *N*-acetyltransferase.[118] This elegant model system responds to environmental light flux with bidirectional changes in norepinephrine receptor sensitivity, cyclic-AMP-mediated regulation of enzyme induction, and the final yield of dramatic alterations in the rate of melatonin production in direct correlation with environmental light–dark cycles.

6.2. Cyclic Nucleotide Regulation of Neurotransmitter Release

Presynaptic levels of cyclic nucleotides have been implicated in neurotransmitter and peptidyl hormone release in brain tissues and in pituitary cells. Generally, cyclic AMP level elevation in the nerve ending results in facilitated release of norepinephrine, acetylcholine, or vasopressin, whereas relative elevations in cyclic GMP attenuate the release of these substances.[119-122] Presynaptic release is modulated by cyclic AMP in relation to a depolarizing event that facilitates the uptake of calcium that is required for the release of stored vesicular neurotransmitter. It has recently been demonstrated that vesicular storage of neurotransmitter is in association with a protein bearing substantial resemblance to the calmodulin that figures prominently in the postsynaptic action of the catecholamine neurotransmitters' activating adenylate cyclase.[123] This extends the coordinated activity of cyclic AMP and calcium ion to the nerve ending and the stimulus-secretion process.

The role of cyclic GMP is contrary to that of cyclic AMP in the processes of stimulus–secretion-coupled release of neurotransmitter. In several systems, cyclic GMP level elevation in the nerve ending results in attenuation of the stimulus–secretion coupling.[124–126] Cyclic GMP reverses the potassium (depolarization)-induced release of norepinephrine. Whereas the increase of cyclic AMP levels in the presynaptic area potentiates the release of acetylcholine, increases in presynaptic cyclic GMP levels are mediated by intrasynaptic acetylcholine binding to presynaptic muscarinic receptors.[127,128] The presynaptic α-adrenergic receptors are thought to truncate the release of norepinephrine, and the actions of the α-receptor agonist clonidine are thought to occur at this site. Norepinephrine or clonidine, acting at the presynaptic site, stimulates increased levels of cyclic GMP in the presynaptic area.[129,130] These effects are blocked by the specific α-adrenergic receptor antagonist piperoxane.

Serotonin-sensitive cyclic AMP accumulation has also been shown to facilitate presynaptic activity in sensory neurons of *Aplysia*.[131]

6.3. Cyclic Nucleotide Regulation of Neurotransmitter Receptor State

The number of acetylcholinergic receptors in developing tissues has been reported to be increased by the presence of high levels of cyclic AMP or of phosphodiesterase inhibitors that increase cyclic AMP levels preferentially.[96]

7. NEUROPHARMACOLOGY OF CYCLIC NUCLEOTIDES IN THE CENTRAL NERVOUS SYSTEM

The direct application of the cyclic nucleotides or their analogues to nervous tissues can mimic the effects of several of the neurotransmitters or neurohormones that act through cyclic nucleotide systems. Early studies in this vein utilized high concentrations of cyclic nucleotide, often injected into the ventricles.[58,132] The effects resulting from such procedures are of limited importance, since the levels required for such studies are extremely high and often approximate the cyclic nucleotide content of the entire animal brain. Moreover, the effects in such procedures are exerted largely on those cells bordering on the ventricles themselves.

The use of cyclic nucleotides in cell culture or by iontophoresis onto distinct cell populations in brain regions provides enlightening information. Cyclic AMP can induce cellular differentiation and slow the growth of neuroblastoma or glioma tumors.[133,134] The differentiation of these cells and of cells among many other types in other tissues can be induced by several analogues of cyclic AMP or by use of phosphodiesterase inhibitors that preferentially increase cyclic AMP levels.[135,136] These effects require high concentrations of cyclic AMP. Adenosine monophosphate, ADP, ATP, and cyclic GMP also inhibit neuroblastoma growth but do not stimulate neurite outgrowth as does cyclic AMP.[134]

The outgrowth of neurites and the elongation of microtubules in the neuroblastoma cell exposed to elevated levels of cyclic AMP are accompanied by increases in tyrosine hydroxylase activity, acetylcholinesterase and choline acetyltransferase activities, increased prostaglandin production, and morphological signs of nearly normal differentiation.[138-140] The elongation of microtubules and microfilaments induced in this way appears to be effected locally and can be induced in enucleated cells.[141]

The use of dibutyryl cyclic AMP or GMP provides a complication not often addressed. Butyric acid itself, or when liberated from the dissociation of dibutyryl cyclic nucleotide, induces elevation in levels of cyclic AMP in the cell and initiates alterations in RNA metabolism as well.[142] Butyrate alone can induce the type of differentiation seen with cyclic AMP additions and has even been tested clinically for the arrest of human neuroblastoma.

8. CYCLIC NUCLEOTIDE MEDIATION OF BEHAVIOR AND PSYCHOENDOCRINOLOGY

8.1. Regulation of Hormone Release

Cyclic nucleotides are now known to be integrally involved in the mechanisms of the release of the neurotransmitters and in the mediation of the regulation of the release of peptidyl hormones in central and in hypophyseal sites. Prototypic of this role of cyclic nucleotides is the regulation of the release of prolactin. Prolactin-inhibiting factor has been shown to be identical with dopamine.[143] This action of dopamine may not be mediated by receptors coupled to adenylate cyclase.[65] Cyclic AMP, or a number of synthetic derivatives, promotes the release of prolactin. The release is effected in the presence of cyclic AMP only in the presence of membrane-bound calcium. A different type of mechanism is evident in the cyclic-AMP-mediated inhibition of release of vasopressin from the posterior pituitary.[144] Microinjection of dibutyryl cyclic AMP results in an increased ACTH release and consequent elevations in plasma corticosterone as well.[145]

8.2. Cyclic Nucleotides in Behavioral Systems

The vegetative effects of the application of the cyclic nucleotides are fairly well characterized. Cyclic AMP produces hyperthermia with intraventricular injection that is reversed by antipyretics.[146] This is also seen in the intrahypothalamic injection of cyclic AMP. Predictably, these effects were potentiated by theophylline, a phosphodiesterase inhibitor, but were not affected by the aspirinlike compounds, since the exogenous cyclic nucleotide bypasses the prostaglandin-receptor-mediated phase of production of hyperthermia.[147] Intracerebral dibutyryl cyclic AMP increases food intake and growth in the rat when implanted in the mammilary body of the forebrain.[148] Lateral hypothalamic sites are sensitive to cyclic-AMP-induced water intake.[149] Intraventricular cyclic AMP increases blood pressure and heart rate.[150–152] Generally, these vegetative effects are highly site or brain region specific and are indicative of the probable involvement of neurotransmitter-coupled adenylate cyclase systems that have highly localized regulatory effects on these functions.

Behavioral effects of exogenous cyclic nucleotides have been studied in a large number of behavioral systems and species. The injection of cyclic nucleotides into the cerebrum results in diverse and global behavioral changes. Dibutyryl cyclic AMP injected intraventricularly results in an increase in spontaneous motor activity and exploration in several species.[37,153–155] Injection of dibutyryl cyclic AMP into specific brain regions, however, results in differing behavioral alterations. Cyclic AMP in the hypothalamus or in the cerebellum results in sleep rather than activation.[153] Injection of dibutyryl cyclic AMP into the amygdala results in convulsions in cats.[156,157] However, there appears to be a genetic factor in most of these effects, since the injection of dibutyryl cylic AMP into mice of some strains, and in guinea pigs, results in sedation rather than behavioral activation.[158,159]

The specificity of the effects of intracerebral or even intraregional injection of cyclic nucleotides or nucleotide derivatives presents problems in that the effects may result from alterations in cyclic nucleotide levels or balance in a great many cells in the vicinity of the injection. Yet, many of these cells may not participate in the same manner in the normal regulation of the behavioral endpoint or of interactive cell network activities in that brain region. A specific example is seen in the effects of stimulation of the locus coeruleus. Direct stimulation of the locus coeruleus or self-stimulation by electrodes implanted in the locus coeruleus results in a substantial elevation of cyclic AMP in the hippocampus pyramidal neurons but not in other brain regions.[160,161] The general effects of cyclic AMP on these cell types is inhibitory, as is also true of the effects of catecholamines on Purkinje cells in the cerebellum and pyramidal cells in the rostral brain regions.[162] Although the "yin–yang" hypothesis of opposing effects of cyclic AMP and cyclic GMP does apply in many instances, the administration of cyclic GMP to forebrain sites does not always attenuate or oppose the effects of cyclic AMP. Moreover, the effects of cyclic AMP do not always correspond to the effects produced by the dibutyryl derivatives or the other metabolically stable derivatives of cyclic AMP (reviewed in detail by Daly[37]).

It is necessary, then, that such studies employ a second means of confirming that the effect resembles the physiological effects naturally regulated by the cyclic nucleotides in that tissue. A concerted program of research of this type is the study of the effects of locus coeruleus-derived noradrenergic effects on a variety of forebrain regions by electrical stimulation of the locus coeruleus and by local administration of iontophoretically applied norepinephrine or cyclic AMP in the target regions receiving afferents that originate in the locus coeruleus. Further, the use of phosphodiesterase inhibitors that specifically potentiate the levels of the particular cyclic nucleotide of interest should also produce the net effect seen with the more local and specific application of norepinephrine or cyclic AMP.

It is generally the case that the catecholamine effects that are mediated by the cyclic AMP systems are of long latency and long duration.[163] Moreover, those effects that are mimicked with the injection of cyclic AMP are also of rather long duration.

9. CYCLIC NUCLEOTIDES IN BLOOD AND OTHER BODY FLUIDS: CONTRIBUTIONS FROM THE CENTRAL NERVOUS SYSTEM

Considerable interest was generated by the description of the very high levels of cyclic nucleotides in urine and of measurable levels of cyclic nucleotides in the plasma.[164,165] The subsequent search for pathological conditions that are reflected in these levels has been somewhat disappointing. The levels in the various body fluids are contributed to from several sources that remain poorly understood. Even cerebrospinal fluid is not directly representative of the action of neurotransmitter flux in specific central nervous systems sites as had been hoped in the early pursuit of such studies.

9.1. Method for Determining Cyclic Nucleotide Levels in Body Fluids

Cyclic nucleotide levels in urine are measurable in a direct manner without the necessity for separation of the nucleotides or for the extraction from interfering substances in urine.[166] The levels of urinary cyclic nucleotides are sufficiently high to require dilution of the sample in buffer prior to inclusion in the assay. If the assay is sufficiently specific, the diluted sample can be added directly to the assay with excellent recovery and reproducibility.[167]

The plasma cyclic nucleotide levels are substantially lower than those in urine, but the changes in plasma levels are more dramatic as a result of autonomic nervous system (or other) manipulation that increases adenylate cyclase activity or free calcium levels. A modification of the commonly used radioimmunoassay may provide greater sensitivity to levels in plasma.[168] Sensitivity can be increased to the point of enabling determination of the levels of both cyclic AMP and cyclic GMP in samples of plasma of less than 0.05 ml.[169]

The cerebrospinal fluid contains measurable levels of both cyclic nucleotides that are about the same as those in plasma. However, the levels are not contributed to substantially by the plasma.[170]

9.2. Levels of Cyclic Nucleotides in Body Fluids

The levels of cyclic nucleotides in plasma are about 20 pmol/ml for cyclic AMP and about 1 pmol/ml for cyclic GMP.[168,171,172] The levels in cerebrospinal fluid are similar, with cyclic AMP at about 30 pmol/ml and cyclic GMP at about 3 pmol/ml.[173-175] Urinary levels are about 1000 times higher than in other body fluids. Urinary cyclic AMP level is about 2.5 μmol/g creatinine, and cyclic GMP about 0.55 μmol/g creatinine.[165] Normally, the ranges of these levels among individuals can vary about twofold in any of the above body fluids. Cyclic nucleotides are found in other body fluids such as saliva, breast milk, bile, and amniotic fluid.[165]

9.3. Changes in Cyclic Nucleotide Levels that Reflect Acute States

The levels of urinary cyclic nucleotides are rather stable and are not substantially affected by the level of physical activity or by the ingestion of methylxanthines.[175] There are daily rhythms of both cyclic AMP and cyclic GMP that seem to peak at about midnight and to ebb at about 9:00 a.m.[176] The excretion rate in urine increases with age from childhood to adulthood.[165] Most of the cyclic nucleotides in the urine are from plasma sources filtered through the kidney glomeruli (about 50–75%); the remainder is contributed by local synthesis and secretion from the kidney itself (about 25–50%). This latter source is stimulated substantially by parathyroid hormone.[177]

Plasma cyclic nucleotides are contributed by the autonomic nervous system, by platelet synthesis, and by excretion from a variety of tissues. In contrast to urinary levels, the plasma levels of cyclic nucleotides respond to vigorous exercise, cold stress, or changing bodily postures.[168] The increases in cyclic AMP can be abolished by propanolol (the β-adrenergic antagonist), and the increases in cyclic GMP can be blocked by atropine (a muscarinic cholinergic antagonist). Intravenous epinephrine increases the plasma cyclic AMP levels more than the cyclic GMP levels, and the epinephrine effect is blocked by β-adrenergic antagonists. The small elevations in plasma cyclic GMP induced by epinephrine can be reversed by an α-adrenergic antagonist.[168,178,179] Methacholine-induced hypotension significantly increases the plasma cyclic GMP levels with relatively little effect on cyclic AMP levels.[180]

9.4. Changes in Cyclic Nucleotides in Body Fluids with Pathological States of Central Nervous System Function

Early studies of cyclic nucleotide levels in plasma and urine suggested that elevations in cyclic AMP accompanied manic states, whereas decreases in cyclic AMP levels accompanied depressive states.[181,182] The plasma levels appear to be more sensitive to altered states, however, and the cerebrospinal fluid is potentially of value but less accessible than plasma.

Lithium has been shown to inhibit a number of adenylate cyclase activities and is demonstrated to inhibit the epinephrine-induced increase in plasma cyclic AMP. There is some specificity, however, in that the elevation in plasma cyclic AMP induced by glucagon is not blocked by lithium.[183] The modest epinephrine-induced elevation in plasma cyclic GMP was partially blocked by lithium at therapeutic levels.[183] A series of studies indicates that the plasma cyclic nucleotide levels may have poor correlation with mood state in spite of early expectations.

A number of studies have been reported in which cyclic nucleotides in the cerebrospinal fluid appeared to correlate with psychiatric conditions or with psychopharmacological intervention. The cerebrospinal fluid in the ventricles may provide a more direct representation of local cyclic nucleotide excretion than the more distal lumbar samples. Nonetheless, it is difficult to know what cell populations are actually contributing to the levels of cyclic nucleotides in the cerebrospinal fluid.[184,185] Intraventricular administration of norepinephrine or dopamine in animals leads to elevations in cerebrospinal fluid cyclic AMP levels, and the dopamine-induced effect is blocked by administration of haloperidol.[186,187] However, the norepinephrine-induced effect may not be blocked by propanolol.[188,189]

The early studies demonstrating that cerebrospinal fluid levels of cyclic AMP are elevated with manic states and lowered with depressive states have not been replicated.[190] Nor have these cyclic nucleotide levels correlated with other psychiatric conditions.[191] However, Belmaker and colleagues appear to have identified a subpopulation of schizophrenics, from among drug-free patients, who present elevated levels of cerebrospinal fluid cyclic AMP levels.[175,192]

Neuroleptic drug administration may lower the cyclic AMP in cerebrospinal fluid and elevate cyclic GMP levels.[193,194] However, it is suggested that the depression of cyclic AMP levels may be a response to those neuroleptics that have a relatively high potency in blocking the dopamine receptor systems coupled to adenylate cyclase.[137] It is currently difficult to understand the mechanism by which apparent neurotransmitter-specific agents effect changes in cyclic nucleotide levels in cerebrospinal fluid in light of the studies of Belmaker and colleagues.[175] Parkinsonism patients have cyclic AMP levels and cyclic GMP levels that are significantly below those of schizophrenics; however, these levels do not respond to L-DOPA or to propanolol administration. Recently, we have found urinary levels of cyclic AMP in schizophrenic children to be about four times normal levels observed in age-matched controls (K. Bonnet, S. Gusik, and M. Campbell, unpublished observations).

REFERENCES

1. Siggins, G. R., 1979, *Adv. Exp. Med. Biol.* **116**:41–64.
2. Schmidt, M., Schmidt, D., and Robison, G., 1971, *Science* **173**:1142–1143.
3. Hutchkins, D., and Rodgers, K., 1970, *Br. J. Pharmacol.* **39**:9–25.
4. Mark, J., Godin, G., and Mandel, P., 1963, *J. Neurochem.* **15**:141–143.
5. Lowry, O., Passonneau, J., Hasselberger, F., and Schultz, D., 1964, *J. Biol. Chem.* **239**:18–30.
6. Kakiuchi, S., and Rall, T. W., 1968, *Mol. Pharmacol.* **4**:379–388.
7. Stavinoha, W., Weintraub, S., and Modak, T., 1973, *J. Neurochem.* **20**:361–371.
8. Schmidt, M. J., Schmidt, D. E., and Robison, G. A., 1971, *Science* **173**:1142–1143.
9. Mao, C., Guidotti, A., and Costa, E., 1974, *Mol. Pharmacol.* **10**:736–745.
10. Balcom, G., Lenox, R. H., and Meyerhoff, J., 1975, *J. Neurochem.* **24**:609–613.
11. Lenox, R. H., Meyerhoff, J., Ghandi, O., and Wray, H., 1977, *J. Cyclic Nucleotide Res.* **3**:367–379.

12. Lenox, R. H., Balcom, G., and Meyerhoff, J., 1979, *Neurosci. Abstr.* **6**:788.
13. Butcher, R. W., and Butcher, S., 1976, *Life Sci.* **19**:1079–1087.
14. Biggio, G., and Guidotti, A., 1976, *Brain Res.* **107**:365–373.
15. Corda, M., Biggio, G., and Gessa, G., 1980, *Brain Res.* **188**:287–290.
16. Delapaz, R. L., Dickman, S. R., and Grosser, B. I., 1975, *Brain Res.* **85**:171–175.
17. Dinnendahl, V., 1975, *Brain Res.* **100**:716–719.
18. Bonnet, K., 1975, *Life Sci.* **16**:1877–1882.
19. Steiner, A., Ferendelli, J., and Kipnis, D., 1972, *J. Biol. Chem.* **247**:1121–1124.
20. Cramer, H., Paul, M., Silbergold, S., and Forn, J., 1971, *J. Neurochem.* **18**:1605–1608.
21. Lenox, R., Brown, P., and Meyerhoff, J., 1979, *Trends Neurosci.* **2**:106–109.
22. Bonnet, K., Branchey, L., Friedhoff, A., and Ehrlich, Y., 1978, *Life Sci.* **22**:2003–2008.
23. Ehrlich, Y., Bonnet, K., Davis, L., and Brunngraber, F., 1979, *Mechanisms Regulation and Special Functions of Protein Synthesis in Brain* (S. Roberts, ed.), Plenum Press, New York, pp. 273–277.
24. Ehrlich, Y., Bonnet, K., Davis, L., and Brunngraber, E., 1978, *Life Sci.* **23**:137–145.
25. Bonnet, K. A., 1975, *Life Sci.* **16**:1877–1882.
26. Lust, W., Passonneau, J., and Veech, R., 1973, *Science* **181**:280–282.
27. Lenox, R., Meyerhoff, J., Ghandi, O., and Wray, H., 1977, *J. Cyclic Nucleotide Res.* **3**:367–379.
28. Schmidt, M., Schmidt, D., and Robison, G., 1971, *Science* **173**:1142–1143.
29. Veech, R., Harris, R., Veloso, D., and Veech, E., 1973, *J. Neurochem.* **20**:183–188.
30. Kilbride, P., and Ebrey, T., 1980, *J. Gen. Physiol.* **74**:415–426.
31. Palkovits, M., 1973, *Brain Res.* **59**:449–450.
32. Meurs, H., Kauffman, H., Koeter, G., and DeVries, K., 1980, *Clin. Chim. Acta* **106**:91–97.
33. Gilman, A., 1970, *Proc. Natl. Acad. Sci. U.S.A.* **67**:305–312.
34. Hughes, R., and Ayad, S., 1979, *Clin. Chim. Acta* **97**:197–204.
35. Vossenberg, J., 1979, *J. Clin. Chem. Clin. Biochem.* **17**:581–585.
36. Rosenberg, E., LaVallee, G., Weber, F., and Tucci, S., 1979, *J. Histochem. Cytochem.* **27**:913–923.
37. Daly, J., 1977, *Cyclic Nucleotides in the Nervous System*, Plenum Press, New York.
38. Bartfai, T., 1980, *Curr. Topics Cell. Regul.* **16**:225–269.
39. Rodbell, M., Lin, M. C., and Salomon, Y., 1974, *J. Biol. Chem.* **249**:59–65.
40. Rasmussen, H., and Goodwin, D. B. P., 1977, *Physiol. Rev.* **57**:421–509.
41. Hegstrand, L., Minneman, K., and Molinoff, P. B., 1979, *J. Pharmacol. Exp. Ther.* **210**:215–221.
42. Rodbell, M., Lin, M., Salomon, Y., and Londos, C., 1975, *Adv. Cyclic Nucleotide Res.* **5**:3–29.
43. Goldberg, N. D., and Haddox, M. K., 1977, *Annu. Rev. Biochem.* **46**:823–896.
44. Severin, E., Gulayaev, N., Bulargina, T., and Kochetkova, M., 1979, *Adv. Enzyme Regul.* **17**:251–282.
45. Mittal, C., Braughler, J., Ichihara, K., and Murad, F., 1980, *Biochim. Biophys. Acta* **585**:333–342.
46. Appelman, M. M., Thompson, W., and Russell, T., 1973, *Adv. Cyclic Nucleotide Res.* **3**:66–98.
47. Brostrom, C. O., and Wolff, D. J., 1976, *Arch. Biochem. Biophys.* **172**:301–311.
48. Ho, R. J., and Sutherland, E. W., 1975, *Adv. Cyclic Nucleotide Res.* **5**:533–548.
49. Cheung, W. Y., 1971, *Biochim. Biophys. Acta* **242**:395–409.
50. Davis, C. W., and Daly, J. W., 1980, *Mol. Pharmacol.* **17**:206–211.
51. Davis, C. W., and Daly, J. W., 1978, *J. Biol. Chem.* **253**:8683–8686.
52. Kant, G. J., Meyerhoff, J. L., and Lenox, R. H., 1979, Biochem. Pharmacol. **29**:369–373.
53. Schwabe, U., Miyake, M., Ohga, Y., and Daly, J. W., 1976, *Mol. Pharmacol.* **12**:900–910.
54. Smellie, F. W., Davis, C. W., Daly, J. W., and Well, J., 1979, *Life Sci.* **24**:2475–2480.
55. Rall, T. W., 1978, *Neuronal Information Transfer* (A. Karlin, V. Tennyson, and H. Vogel, eds.), Academic Press, New York, pp. 323–338.
56. Smellie, F. W., Daly, J. W., and Wells, J. N., 1979, *Life Sci.* **25**:1917–1924.
57. Sutherland, E. W., and Rall, T. W., 1957, *J. Am. Chem. Soc.* **79**:3608.

58. Nathanson, J., 1977, *Physiol. Rev.* **57**:157–256.

59. Ahlquist, R. P., 1948, *Am. J. Physiol.* **153**:586–600.

60. Minneman, K. P., Dibner, M. D., Wolfe, B. B., and Molinoff, P. B., 1979, *Science* **204**:866–868.

61. Daly, J. W., Padgett, W., Nimitkitpaisan, Y., Creveling, C. R., Cantacuzene, D., Kirk, K. L., 1980, *J. Pharmacol. Exp. Ther.* **212**:382–389.

62. Haidamous, M., Kouyoumdjian, J., Briley, P., Gonnard, P., 1980, *Eur. J. Pharmacol.* **63**:287–294.

63. Daly, J. W., McNeal, E., Partington, C., Neuwirth, M., and Creveling, C. R., 1980, *J. Neurochem.* **35**:326–337.

64. Daly, J. W., McNeal, E. T., and Creveling, C. R., 1979, *Histamine Receptors* (T. O. Yellin, ed.), SP Medical and Scientific Books, New York, pp. 299–323.

65. Kanoff, P. O., and Greengard, P., 1979, *Mol. Pharmacol.* **15**:445–461.

66. Kebabian, J. W., and Calne, D. B., 1979, *Nature* **277**:93–96.

67. Weiss, B., Prozialeck, W., and Cimino, M., 1980, *Adv. Cyclic Nucleotide Res.* **12**:213–225.

68. Friedhoff, A. J., Bonnet, K. A., and Rosengarten, H., 1977, *Res. Commun. Chem. Pathol. Pharmacol.* **16**:411–423.

69. Pagel, J., Christian, S. T., Quayle, E. S., and Monti, J. A., 1976, *Life Sci.* **19**:819–824.

70. Mah, H., and Daly, J., 1976, *Pharmacol. Res. Commun.* **8**:65–79.

71. Traber, J., Reiser, G., Fischer, K., and Hamprecht, B., 1975, *FEBS Lett.* **52**:327–332.

72. Bonnet, K. A., 1979, *Modulators, Mediators and Specifiers in Brain Function* (Y. Ehrlich, J. Volavka, L. Davis, and E. Brunngraber, eds.), Plenum Press, New York, pp. 247–259.

73. Lambert, A., Nirenberg, M., and Klee, W. A., 1976, *Proc. Natl. Acad. Sci. U.S.A.* **73**:3165–3167.

74. Hamprecht, B., 1980, *J. Neurochem.* **33**:999–1005.

75. Ross, D. H., and Cardenas, H. L., 1980, *Adv. Biochem. Psychopharmacol.* **20**:301–317.

76. Bonnet, K. A., Branchey, L., Friedhoff, A. J., and Ehrlich, Y., 1978, *Life Sci.* **22**:2003–2008.

77. Kinscherf, D. A., Chang, M., Rubin, E., Schneider, D., and Ferendelli, J. A., 1976, *J. Neurochem.* **26**:527–530.

78. Ferendelli, J. A., Kinscherf, D., and Chang, M., 1973, *Mol. Pharmacol.* **9**:445–454.

79. Shimizu, H., Creveling, C. R., and Daly, J. W., 1970, *Adv. Biochem. Psychopharmacol.* **3**:135–154.

80. Ahnert, G., Glossmann, H., and Habermann, E., 1980, *Arch. Pharmacol.* **307**:151–157.

81. Lenox, R. H., Kant, G., and Meyerhoff, J. L., 1980, *Life Sci.* **26**:2201–2209.

82. Kebabian, J. W., Steiner, A. L., and Greengard, P., 1975, *J. Pharmacol. Exp. Ther.* **193**:474–488.

83. Daly, J. W., McNeal, E., and Creveling, C., 1979, *Histamine Receptors* (T. O. Yellin, ed.), SP Medical Books, New York, pp. 299–323.

84. Bartfai, T., Study, R. E., and Greengard, P., 1977, *Cholinergic Mechanisms and Pharmacology* (D. J. Jenden, ed.), Plenum Press, New York, p. 285.

85. Nathanson, N. M., Klein, N. L., and Nirenberg, M., 1978, *Proc. Natl. Acad. Sci. U.S.A.* **75**:1788–1791.

86. Haidamous, M., Kouyoumdjian, J., Briley, P., and Gonnard, P., 1980, *Eur. J. Pharmacol.* **63**:287–294.

87. Klee, W. A., and Nirenberg, M., 1974, *Proc. Natl. Acad. Sci. U.S.A.* **71**:3474–3477.

88. Watanabe, A. M., McConnaugh, M. M., Strawbridge, R. A., Fleming, J. W., Jones, L. A., and Besch, H. R., 1978, *J. Biol. Chem.* **253**:4833–4836.

89. Nikodijevic, O., Nikodijevic, B., Zinder, O., Yi, M., Guroff, G., and Pollard, H. B., 1976, *Proc. Natl. Acad. Sci. U.S.A.* **73**:771–774.

90. Westcott, K. R., LaPorte, D., and Storme, D. R., 1979, *Proc. Natl. Acad. Sci. U.S.A.* **76**:204–208.

91. Nathanson, J. A., and Bloom, F. E., 1975, *Nature* **255**:419–420.

92. LeMay, A., and Jurett, L., 1975, *J. Cell Biol.* **65**:39–50.

93. Walker, J. B., 1974, *Biol. Psychiatry* **8**:245–251.

94. Woolf, C. J., Willies, G. H., and Rosendorff, C., 1976, *Naturwissenschaften* **63**:94–95.

95. Miller, R. J., and Kelly, P. H., 1975, *Nature* **255**:163–166.

96. Blosser, J., and Appel, S., 1980, *J. Biol. Chem.* **255**:1235–1238.
97. Ehrlich, Y., 1979, *Modulators, Mediators and Specifiers in Brain Function* (Y. Ehrlich, J. Volavka, L. Davis, and E. Brunngraber, eds.), Plenum Press, New York, pp. 75–102.
98. Walter, U., Costa, M. R., Breakfield, X. O., and Greengard, P., 1979, *Proc. Natl. Acad. Sci. U.S.A.* **76**:3251–3255.
99. Gabibov, A., Kochetkov, S., Sashchenko, L., and Severin, E., 1979, *Biochim. Biophys. Acta* **569**:145–152.
100. Builder, S. E., Beavo, J. A., and Krebs, E. G., 1980, *J. Biol. Chem.* **255**:2350–2354.
101. Yagura, T., Sigman, C. C., Sturm, P., Reist, E., Johnson, H. I., and Miller, J., 1980, *Biochem. Biophys. Res. Commun.* **92**:463–467.
102. Hanbauer, I., and Guidotti, A., 1975, *Arch. Pharmacol.* **287**:213–217.
103. Lovenberg, W., Bruckwick, E., and Hanbauer, I., 1975, *Proc. Natl. Acad. Sci. U.S.A.* **72**:2955–2958.
104. Morgenroth, V. H., Hegstrand, L., Roth, R., and Greengard, P., 1975, *J. Biol. Chem.* **250**:1946–1948.
105. Joh, T., Park, D., and Reis, D., 1978, *Proc. Natl. Acad. Sci. U.S.A.* **75**:4744–4748.
106. Vulliet, P., Langan, T., and Weiner, N., 1980, *Proc. Natl. Acad. Sci. U.S.A.* **77**:92–96.
107. Costa, E., Chang, D., Guidotti, A., and Uzunov, P., 1975, *Clinical Tools in Catecholamine Research* (T. Malmfors, ed.), North Holland, Amsterdam, pp. 283–292.
108. Guidotti, A., Zivkovic, B., Pfeiffer, R., and Costa, E., 1973, *Arch. Pharmacol.* **278**:195–206.
109. Keen, P., and McLean, W., 1974, *J. Neurochem.* **22**:5–10.
110. Hamprecht, B., and Traber, J., 1974, *FEBS Lett.* **42**:221–226.
111. Kvetnansky, R., Gewirty, G., Weise, V., and Kopin, I., 1971, *Endocrinology* **89**:50–55.
112. Furmanski, P., Silverman, D. J., and Lukin, M., 1971, *Nature* **233**:413–415.
113. Werner, I., Peterson, G., and Shuster, L., 1971, *J. Neurochem.* **18**:141–151.
114. Prasad, K., and Mandel, 1972, *Exp. Cell Res.* **74**:532–534.
115. Schrier, B., and Shapiro, D., 1973, *Exp. Cell Res.* **80**:459–462.
116. Forn, J., Tagliamonte, A., Tagliamonte, P., and Gessa, G., 1972, *Nature* **237**:245–247.
117. Tagliamonte, A., Tagliamonte, P., Forn, J., Perez-Cruet, J., Krishna, G., and Gessa, G., 1970, *J. Neurochem.* **18**:1191–1196.
118. Axelrod, J., 1974, *Science* **184**:1341–1348.
119. Ichida, S., Yonehara, N., Watanabe, Y., and Yoshida, H., 1980, *Brain Res.* **192**:487–494.
120. Berridge, M., 1975, *Adv. Cyclic Nucleotide Res.* **6**:1–98.
121. Miyamoto, D., and Breckenridge, B., 1974, *J. Gen. Physiol.* **63**:609–624.
122. Wilson, D., 1974, *J. Pharmacol. Exp. Ther.* **188**:447–452.
123. DeLorenzo, R., Freedman, S., Yoke, W., and Mauren, S., 1979, *Proc. Natl. Acad. Sci. U.S.A.* **76**:1838–1842.
124. Ichida, S., Yonehara, N., Watanabe, Y., and Yoshida, H., 1980, *Brain Res.* **192**:487–494.
125. Pelayo, F., Dubocovich, M. L., and Langer, S. Z., 1978, *Nature* **274**:76–78.
126. Yonehara, N., Matsuda, T., Saito, K., and Yoshida, H., 1981, *Brain Res.* (in press).
127. Haidamous, M., Kouyoumdjian, J. C., Briley, P., and Gonnard, P., 1980, *Eur. J. Pharmacol.* **63**:287–294.
128. Greengard, P., 1979, *Fed. Proc.* **38**:2208–2217.
129. Wikberg, J., 1978, *Nature* **273**:164–166.
130. Starke, K., Endo, T., and Taube, H., 1975, *Arch. Pharmacol.* **291**:55–78.
131. Daszuta, A., Pons, F., and Cadhilac, J., 1979, *Eur. J. Pharmacol.* **56**:397–401.
132. Daly, J. W., 1977, *Cyclic Nucleotides in the Nervous System*, Plenum Press, New York, pp. 211.
133. Daniels, M., and Hamprecht, B., 1974, *J. Cell Biol.* **63**:691–699.
134. Prasad, K., and Vernadakis, A., 1972, *Exp. Cell Res.* **70**:27–32.
135. Klier, G., Schubert, D., and Heinemann, S., 1975, *Neurobiology* **5**:1–7.
136. Prasad, K., Mandel, B., Waymire, J., Lees, G., Vernadakis, A., and Weiner, N., 1973, *Nature* **241**:117–119.
137. Clement-Courmier, Y., 1980, *Adv. Biochem. Psychopharmacol.* **22**:256–269.
138. Hamprecht, B., Joffe, B., and Philpott, G., 1973, *FEBS Lett.* **36**:193–198.
139. Furmanski, P., Silverman, D., and Lubin, M., 1971, *Nature* **233**:413–415.

140. Prasad, K., Kumar, S., Gilmer, K., and Vernadakis, A., 1973, *Biochem. Biophys. Res. Commun.* **50**:973–977.
141. Miller, R. J., and Ruddle, F., 1974, *J. Cell Biol.* **63**:295–299.
142. Glazer, R., and Schneider, F., 1975, *J. Biol. Chem.* **250**:2745–2749.
143. Thorner, M., 1977, *J. Clin. Endocrinol. Metab.* **6**:201–202.
144. Yancey, S., Weiner, R., and Schechter, J., 1980, *Am. J. Anat.* **157**:345–356.
145. Mathison, R., and Lederis, K., 1980, *Endocrinology* **106**:842–848.
146. Clark, R. B., Crumby, H., Davis, H., 1974, *J. Physiol. (Lond.)* **240**:493–504.
147. Woolf, C., Willies, G., and Rosendorff, C., 1976, *Naturwissenschaften* **63**:94–95.
148. Breckenridge, B., and Lisk, R., 1969, *Proc. Soc. Exp. Biol. Med.* **131**:934–935.
149. Rindi, G., Sciorelli, G., Poloni, M., and Acanfora, F., 1972, *Experientia* **28**:1047–1049.
150. Beani, L., Bianchi, C., and Bertelli, A., 1975, *Pharmacol. Res. Commun.* **7**:347–359.
151. Walland, A., 1975, *Arch. Pharmacol.* **290**:419–423.
152. Saxena, P., and Bahargava, K., 1974, *Pharmacol. Res. Commun.* **6**:347–355.
153. Gessa, G., Krishna, G., Forn, J., Tagliamonte, A., and Brodie, B., 1970, *Adv. Biochem. Psychopharmacol.* **3**:371–381.
154. Brus, R., Herman, Z., and Kostman, J., 1974, *Pharmacol. Biochem. Behav.* **2**:719–724.
155. Clark, R., Crumby, H., and Davis, H., 1974, *J. Physiol. (Lond.)* **240**:493–504.
156. Brus, R., Herman, Z., and Kostman, J., 1974, *Pharmacol. Biochem. Behav.* **2**:719–724.
157. Purpura, D., and Schoffer, R., 1974, *Brain Res.* **38**:179–181.
158. Henion, W. F., Sutherland, E., and Paternak, T., 1967, *Biochim. Biophys. Acta* **148**:106–113.
159. Vagaric, V., and Bedeslin, D., 1973, *Brain Res.* **57**:252–254.
160. Siggins, G., 1978, *Neuronal Information Transfer* (A. Karlin, V. Tennyson, and H. J. Vogel, eds.), Academic Press, New York, pp. 339–359.
161. Siggins, G., 1979, *Modulators, Mediators and Specifiers in Brain Function* (Y. Ehrlich, J. Volavka, L. Davis, and E. Brunngraber, eds.), Plenum Press, New York, pp. 41–64.
162. Bloom, F., 1979, *Catecholamines: Basic and Clinical Frontiers* (E. Usdin, I. Kopin, and J. Barchas, eds.), Pergamon Press, New York, pp. 609–618.
163. Nathanson, J., 1977, *Physiol. Rev.* **57**:157–256.
164. Butcher, R., and Sutherland, E., 1962, *J. Biol. Chem.* **240**:4515–4523.
165. Murad, F., 1973, *Adv. Cyclic Nucleotide Res.* **3**:355–383.
166. Aurbach, G., 1980, *Adv. Cyclic Nucleotide Res.* **12**:1–9.
167. Murad, F., 1973, *Adv. Cyclic Nucleotide Res.* **3**:355–383.
168. Ui, M., Honma, M., Kunitada, S., Okada, F., Ide, H., Hata, S., and Satoh, T., 1980, *Adv. Cyclic Nucleotide Res.* **12**:25–35.
169. Honma, M., Satoh, T., Takezawa, J., and Ui, M., 1977, *Biochem. Med.* **18**:257–273.
170. Brooks, B., Engel, W., and Sode, J., 1977, *Arch. Neurol.* **34**:468–469.
171. Broadus, A., 1977, *Adv. Cyclic Nucleotide Res.* **8**:509–548.
172. Honma, M., and Ui, M., 1978, *Eur. J. Pharmacol.* **47**:1–10.
173. Cramer, H., Ng, L., and Chase, T., 1972, *J. Neurochem.* **19**:1601–1602.
174. Ebstein, R., Kara, T., and Belmaker, R., 1977, *Acta Pharmacol. Toxicol. (Kbh.)* **41**:80–83.
175. Belmaker, R., Zohar, J., and Ebstein, R., 1980, *Adv. Cyclic Nucleot. Res.* **12**:187–198.
176. Murad, F., and Pak, C., 1972, *N. Engl. J. Med.* **286**:1382–1387.
177. Pak, C., 1980, *Adv. Cyclic Nucleotide Res.* **12**:393–403.
178. Ebstein, R. P., Belmaker, R., Gunhaus, L., and Rimon, R., 1976, *Nature* **259**:411–413.
179. Guttler, R., Croxson, M., DeQuatro, V., Warren, D., Otis, C., and Niroloff, J., 1977, *Metabolism* **26**:1155–162.
180. Hamet, P., Franks, D., Adnot, S., and Coquil, J., 1980, *Adv. Cyclic Nucleotide Res.* **12**:11–23.
181. Abdullah, Y., and Hamadah, K., 1970, *Lancet* **1**:378–381.
182. Brown, B., Salway, J., Albano, J., Hublin, R., and Elkins, R., 1972, *Br. J. Psychiatry* **120**:405–408.
183. Belmaker, R., Kon, M., Ebstein, R., and DasBerg, H., 1981, *Biol. Psychiatry* **15**:3–8.
184. Robison, G. A., 1980, *Adv. Cyclic Nucleotide Res.* **12**:239–242.
185. Tsang, D., Lal, S., Sourkes, T., Ford, R., Aranoff, A., 1976, *J. Neurol. Neurosurg. Psychiatry* **39**:1186–1190.
186. Sebens, J., Koff, J., 1975, *Exp. Neurol* **46**:333–344.

187. Korf, J., Boer, P., and Felkes, D., 1976, *Brain Res.* **113**:551–561.

188. Kiessling, M., Lindl, T., and Cramer, H., 1975, *Arch. Psychiatr. Nervenkr.* **220**:325–333.

189. Belmaker, R., Ebstein, R., Biederman, J., Stern, R., Berman, M., and van Praag, H., 1978, *Psychopharmacology* **58**:307–310.

190. Post, R., Cramer, H., and Goodwin, F., 1977, *Psychol. Med.* **7**:599–605.

191. Kadouch, R., Belmaker, R., Ebstein, R., and Perez, L., 1977, *Neuropsychobiology* **3**:250–255.

192. Biederman, J., Rimon, R., Ebstein, R., Belmaker, R., and Davidson, J., 1977, *Br. J. Psychiatry* **130**:64–67.

193. Biederman, J., Rimon, R., Ebstein, R., Zohar, J., and Belmarker, R., 1976, *Neuropsychobiology* **2**:324–327.

194. Kebabian, J., Petzold, G., and Greengard, P., 1972, *Proc. Natl. Acad. Sci. U.S.A.* **69**:2145–2149.

11

Biopterin

E. Martin Gál

1. INTRODUCTION

Three generations ago, F. G. Hopkins[1] drew the attention of scientists to the compounds responsible for the multicolored pigments of butterfly wings. These compounds were named pteridines from the Greek word *pteros* meaning wing. The term "pterin" is now used solely for the derivatives of 2-amino-4-hydroxypyrimido(4,5-b)pyrazines. The pterins, like folic acid, occur either unconjugated or conjugated. However, mammalian tissue does not catabolize folate to biopterin.[2]

This chapter will deal with the results of the last 10 years as they relate to the biochemistry of unconjugated pterins in the mammalian nervous system. For the story of the long and exciting road of research that led from the pterins of butterfly wings to those of the human brain, readers are referred to other sources.[3]

Two of the landmarks of the chemistry and biology of pterins as they influenced neurochemical research need to be mentioned. In 1963, Rembold and his colleagues[4] reported the first indirect evidence for the biosynthesis of biopterin in the rat by demonstrating that a biopterin-free diet had no effect on the continuous urinary output of biopterin. At the same time, Kaufman[5] described the structure of the cofactor of phenylalanine hydroxylase as tetrahydrobiopterin.

2. OCCURRENCE OF PTERINS IN MAMMALIAN BRAIN

During the last 10 years, the presence of pterins in mammalian brains has been confirmed. The pterins of biochemical importance described herein all possess C-6 alkyl chains (Fig. 1).

Although the flagellate *Crithidia fasciculata* is dependent on biopterin for

E. Martin Gál • Neurochemical Research Laboratories, Department of Psychiatry, University of Iowa, Iowa City, Iowa 52242.

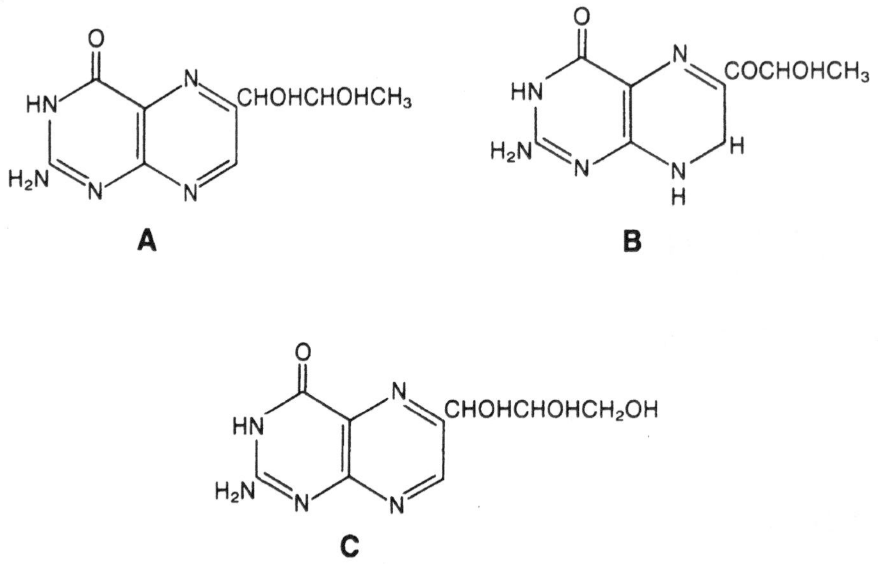

Fig. 1. Structures of (A) biopterin, (B) sepiapterin, and (C) neopterin.

its growth, at high concentrations, folic acid was found to be a good substitute because of the ability of this protozoan to cleave folic acid to yield pterin.

The first reported quantitation of pterins in the brain relied on the *Crithidia*[6] and *Pseudomonas* phenylalanine hydroxylase assays.[7] This latter test permitted quantitation of tetrahydrobiopterin (BH₄). Recent improvements in methodology of high-pressure liquid chromatography (HPLC)[8,9] combined with gas chromatography (GC)[8] allowed assessment of picomole amounts of pterins and their reduced forms in brain. The occurence of biopterins in the brain is briefly illustrated in Table I. Obviously, determination of pterin content by various tests has led to a range of values. In general, the concordance of values is greatest between those obtained by *Pseudomonas* assay and HPLC–GC techniques.

Table I
Occurrence of Pterins in Mammalian Brain

Species	Pterin	μg/g wet weight	Assay	References
Human adult	Biopterin (B)	0.035	*Crithidia*	10
	B	0.11	HPLC–GC	11
	BH₂	0.24	HPLC–GC	11
	BH₄	0.21	*Crithidia*	11
Rat	B	0.20	*Crithidia*	6
	B	0.08	*Crithidia*	12
	BH₂, BH₄	0.75	*Pseudomonas*	7
	B	0.12	HPLC–GC	8
	BH₂	0.20	HPLC–GC	8
	BH₄	0.33	HPLC–GC	8
	B	0.09	HPLC	9

Table II
Regional Levels of Biopterins in Human and Rat Brain[a]

Area	Human B	Rat B	Human BH$_2$	Rat BH$_2$	Human BH$_4$	Rat BH$_4$[b]
Cerebellum	0.07	0.29(0.07)[c]	0.17	0.08	0.11	0.12
Colliculi	0.25	0.19	0.65	0.19	0.62	—
Cortex	0.11	0.07	0.20	0.13	0.16	0.062
Hippocampus	—	(0.06)[c]	—	—	—	0.10
Hypothalamus	0.05	—	0.14	—	0.25	0.53
Pons–medulla	0.10	0.09(0.12)[c]	0.60	0.79	0.23	0.21
Septum	0.20	0.19	0.12	—	0.24	0.29
Striatum	0.24	(0.26)[c]	0.17	—	0.17	0.51
Tectum	—	—	—	—	—	0.23
Thalamus	0.08	(0.11)[c]	0.19	—	0.16	0.23

[a] Values are expressed as μg/g (wet weight).
[b] Calculated from the values by Levine *et al.*[15] assuming that 1 g of rat brain (wet weight) contains 186 mg of protein.
[c] Values in parentheses are from reference 8.

The levels of biopterins in various areas of the human and rat brains (Table II) permit an insight into the locations of their major cerebral pools. It seems that there is a somewhat similar distribution of biopterins in the brains of the two species. Analysis of rat brain areas for pool sizes of dihydrobiopterin (BH$_2$), BH$_4$, and B revealed that the largest amounts of reduced pterins were recoverable in the pontine region, whereas in the cerebellum, biopterin predominated.[16]

The major form of biopterin in the brain appeared to be BH$_4$ following intraperitoneal injection of [^{14}C]biopterin or reduced biopterins. The cerebral turnover rates of biopterins are fairly rapid (Table III), amounting to replacement of the cerebral pool in half an hour. This turnover rate implied an active cerebral synthesis of dihydrobiopterin. This was eventually confirmed by experimental evidence.[17]

The hypophysis also contains appreciable amounts of tetrahydrobiopterin (2.0 μg/g),[15] perhaps indicative of other functions than its cofactorial role for monooxygenases. The presence of biopterin was established in sheep pineal gland,[18] and tetrahydrobiopterin (22.5 μg/g) was measured in rat pineals.[18] The biopterin content of several cell lines of human neuroblastoma was assayed by the *Crithidia* growth test.[19]

Table III
Data on Cerebral Efflux Rates and Turnover by Isotope Dilution

Parameter	BH$_2$	B	BH$_4$
$t_{\frac{1}{2}}$ (h)	1.29	1.26	1.73
K	0.54	0.55	0.40
Pool (μg/g)	0.19	0.11	0.32
Turnover (nmol/g per h)	0.43	0.25	0.53

The presence of biopterins, neopterin, sepiapterin, and pterin-6-carboxylic acid, as well as their catabolic products, was demonstrated in liver. In addition, 3-hydroxysepiapterin[20] and various pteridines were also present in urine.[21] Hitherto, only traces of dihydroneopterin triphosphate (2 ng/g) had been detected in rat brain.[22] The heaviest concentration of biopterins are found in the postmitochondrial fractions. Crude mitochondria subjected to sucrose density gradient fractionation retained only 34% of their pterins, whereas 66% was recoverable from the synaptosomes.[16] The synaptosomal biopterins thus constituted 6% of the total cerebral pool.[16]

3. CEREBRAL METABOLISM

3.1. Synthesis of Pterins in Vivo

The chemical similarity between guanosine and pterin rings has led to speculation about biochemical interversion. Studies with microbes, insects, and lower vertebrates confirmed conversion of purines into pterins. These results have been competently reviewed.[23] Eventually, in 1973, the first communication appeared that reported the urinary excretion of [^{14}C]biopterin from [U-^{14}C]guanosine triphosphate (GTP) injected into rats and mice.[24-26] However, demonstration of the synthesis of biopterin in tissues was unsuccessful, since technical difficulties combined with large pool sizes of GTP and biopterin precluded isolation of [^{14}C]biopterins. Clones of mouse neuroblastoma given [2-^{14}C]- or [8-^{14}C]guanosine were capable of synthesizing labeled pterins in sufficient quantities to enable their isolation. There was no radioactivity incorporated from [8-^{14}C]guanosine.[25] These results were consistent with an earlier suggestion[27] that guanosine nucleotides could be considered precursors for biopterin synthesis.

None of the above studies have irrefutably demonstrated the origin of the 6-alkyl side chain of biopterin. One of the problems was to explain the conversion of D-erythroneopterin to the L-erythro configuration of the 1',2'-hydroxypropyl side chain of biopterin. This transformation and the chemistry of its intermediates were not elucidated by the experiments with cell cultures. Eventually, experiments with homogenates of different tissues from hamsters led to the demonstration of the synthesis of D-erythrodihydroneopterin triphosphate (NPTH$_2$-P$_3$) and of a "*Crithidia*-active" substance from guanosine nucleotides.[28]

It was soon established that B, BH$_2$, and BH$_4$ did not penetrate the brain in amounts sufficient to account for their cerebral pool and turnover.[16] These findings implied the existence of cerebral synthesis of dihydrobiopterin. This was confirmed by experiments in which intracerebral injection [U-^{14}C]-GTP was converted to [^{14}C]-BH$_2$, -BH$_4$, and -biopterin.[17] The identity of [^{14}C]-BH$_2$ was also confirmed by mass fragmentography. Advances in methodology enabled the isolation of some of the intermediates of the pathway from [U-^{14}C]-GTP to [^{14}C]-BH$_2$ such as the ^{14}C-labeled 2-amino-4-hydroxy-5-(or 6-)formamido-6-ribosylamino pyrimidine triphosphate (FPyd-P$_3$) and D-erythro-7,8-

dihydroneopterin triphosphate. [Incidentally, dGTP was also cleaved to the corresponding 2-amino-4-hydroxy-5-(or 6-)formamido-6-(2')-deoxyribosylamino pyrimidine triphosphate (dFPyd-P$_3$).] The FPyd-P$_3$ content of the rat brain was about 2 ng/g, and NPTH$_2$-P$_3$, measured as neopterin, was 6 ng/g. These compounds were isographic with their synthetic samples by gas chromatography. Of the guanosine nucleotides, GTP is the best precursor for BH$_2$ synthesis. Two hours after intracerebral administration of 0.55 μCi of GTP, 0.26% of the label was in BH$_2$. The rate of [^{14}C]-BH$_2$ synthesis was linear for 2 hr and then rapidly declined because of the decrease in the [U-^{14}C]-GTP pool in the brain. The rate of BH$_2$ synthesis was 0.53 nmol/g wet brain per h.[17] This rate of cerebral synthesis of BH$_2$ agrees well with its turnover rate and pool size (Table III).

The accumulation *in vivo* of the above compounds was achieved by intraventricular administration of selective inhibitors of the GTP-to-BH$_2$ pathway. For instance, 2,4-diamino-6-hydroxypyrimidine (DAOPyr) brought about 80% inhibition of [U-^{14}C]-GTP conversion to [^{14}C]-BH$_2$ by blocking the pathway at the FPyd-P$_3$ level. Dietary DAOPyr (1%) given to rats for 2 weeks did not appreciably inhibit the synthesis of BH$_2$.[17] Another inhibitor was dGTP which is enzymatically converted to dFPyd-P$_3$. This compound, unlike FPyd-P$_3$, cannot undergo cyclization. According to the results *in vitro*, the K_I of dFPyd-P$_3$ is 5×10^{-7} M,[29] and its cerebral half-life is about 14 min.[17] Intraventricular injection of 70 g of dFPyd-P$_3$ in repeated doses over a period of 3 h caused an almost complete arrest of [^{14}C]-BH$_2$ synthesis from [U-^{14}C]-GTP[30,31] (Table IV) with a tenfold increase in [^{14}C]-FPyd-P$_3$. There is little doubt that the synthesis of BH$_2$ from GTP *in vivo* progresses from FPyd-P$_3$ through NPTH$_2$-P$_3$ to BH$_2$. Several metabolic intermediates, such as sepiapterin, could be envisaged, but their presence in the brain has not yet been proven.

The enzymes involved in the cerebral synthesis of BH$_2$ catalyze irreversible reactions. The specific activities in [^{14}C]-FPyd-P$_3$, NPTH$_2$-P$_3$, and BH$_2$ revealed that the carbon skeleton of both the purine (except for C-8) and ribose

Table IV
Inhibition of Cerebral Synthesis of BH$_2$ in Vivo

	BH$_2$		
	Total dpm	dpm/nmol	Inhibition (%)
Control[a]	3405	2481	0
dFPyd-P$_3$	75	38	99
2,4-Diamino-6-hydroxypyrimidine	561	421	83
2,4-Diamino-6-hydroxypyrimidine[b]	1486	1666	33
2,4,6-Triaminopyrimidine	1809	1887	24
8-Mercaptoguanosine	3030	2273	8
6-Mercaptoguanosine	1884	1413	43

[a] Injected intraventricularly with [U-^{14}C]-GTP (0.54 μg or 1.2×10^6 dpm). Inhibitors (300 μg) except dFPyd-P$_3$ (210 μg) were administered in three doses by the same route in the solution containing [U-^{14}C]-GTP.
[b] Inhibitor was given with the folate-deficient diet.

of [U-^{14}C]-GTP remained unaltered along the biochemical pathway to BH$_2$. The concentration of GTP in rat brain is about 0.66 mol/g wet weight, and only a fraction of it is needed to sustain the synthesis of FPyd-P$_3$. The rate of this reaction was 0.05 nmol/g per h *in vivo*, and it is the rate-limiting step of the pathway to BH$_2$. Competitive amounts of other nucleotides or deoxyribonucleotides will affect this rate as these compounds are available to the brain cells. The rate of conversion of [U-^{14}C]-FPyd-P$_3$ to [^{14}C]-BH$_2$ had an estimated rate of 1.17 nmole/g wet weight per h *in vivo* and could significantly be affected by the degree of phosphorylation of the catalyzing enzyme[32] or by the presence of interfering amounts of dGTP or dFPyd-P$_3$. The rate of conversion of intraventricularly injected [U-^{14}C]-D-erythro-7,8-dihydroneopterin triphosphate into [^{14}C]-BH$_2$ was calculated to be 1.4 nmole/g wet weight per h. This rapid conversion of NPTH$_2$-P$_3$ explains its small pool observed *in vitro*.

3.2. Synthesis of Pterins in Vitro

Dialyzed extracts of hamster brain homogenates revealed the synthesis of biopterin *in vitro*.[33] The activity of organ extracts was markedly stimulated by the addition of a mixture of pyridine nucleotides (NADPH$_2$ and NADP).[34,35] These experiments also demonstrated that the 6-alkyl side chain of biopterin derived from the alkyl chain of NPTH-P$_3$. This dependence was not observed with a 96% pure enzyme preparation (by technical amino acid analysis) from rat, guinea pig, and beef brain.[22] However, the Ultro® gel AcA-34 enzyme from hamster kidneys or the rat brain fraction[35] were crude preparations. The question of participation of pyridine nucleotides in the conversion of NPTH$_2$-P$_3$ to dihydrobiopterin thus remains unresolved. The failure to obtain conversion to biopterin with nonphosphorylated dihydroneopterin suggested that phosphates were essential for the enzyme-catalyzed reaction to proceed.

Contrary to a recent statement,[35] incubation of postmitochondrial fractions from brains of various species with [U-^{14}C]-GTP led to the isolation of [^{14}C]-BH$_2$[36] (Table V). This synthetic ability of the brain appears to be markedly area dependent (Table VI). Subcellular fractionation solubilizes the enzymes

Table V
Cerebral Synthesis of BH$_2$ in Various Species^a

Brain	BH$_2$ (dpm/g per h)	BH$_2$ (μg)	BH$_2$ (dpm/nmol)
Bovine	1245 ± 266	0.46	2588
Porcine	305 ± 49	0.22	1323
Human	70 ± 23	0.15	455
Guinea pig	78 ± 19	0.20	372
Rat	43 ± 21	0.12	366
Rabbit	60 ± 23	0.20	283

^{a} 12,000g Tris-acetate supernatants, equivalent to 2 g of tissue, were incubated with 1.2 × 10^6 dpm [U-^{14}C]-GTP at 38°C for 2 h. Whole brain was used except for the human sample where parietal neocortex was the source of the enzyme.

Table VI
Regional Biosynthetic Capacity

Area[a]	BH$_2$ (dpm/g per h)
Cerebellum	12,110 ± 2440
Pons	5,760 ± 1480
Colliculi	1,550 ± 525
Thalamus	952 ± 76
Globus pallidus	851 ± 266
Hypothalamus	483 ± 33
Neocortex	194 ± 21

[a] Areas incubated with 110,000 dpm (49 ng) [^{14}C]-GTP for 1 h in supernatant obtained after homogenization of the area in 2 volumes of 0.05 M Tris-Ac, pH 8.0, and centrifugation at 16,000g. Values are mean ± SD from three rats.

of the GTP-to-BH$_2$ pathway. Only 10–15% of the original activity was retained in the crude mitochondrial pellet. The presence of these enzymes was also detected in the synaptosomes.[17] The choice of fractionation technique has a quantitative influence on the synaptosomal enzyme activity. Synaptosomes prepared by sucrose gradient (with appreciable mitochondrial contamination) synthesized 12 pmol/mg per h BH$_2$, whereas those isolated by Ficoll density gradient produced very low yields of BH$_2$ (0.8 pmol/mg per h) displayed a very active uptake of [^{14}C]-BH$_2$ biopterin. This and the retention of [^{14}C]-BH$_2$ speak for the functional integrity of the synaptosomes isolated by Ficoll gradient fractionation.

Finally, both glial and neuronal fractions from neocortices of rabbits synthesized BH$_2$ from GTP.[36]

4. ENZYMOLOGY OF CEREBRAL BIOPTERIN SYNTHESIS

4.1. Conversion of GTP to FPyd-P$_3$

The enzymatic cleavage of GTP at its imidazole carbon 8 produces a formamidated pyrimidine which is ultimately hydrolyzed to formic acid and Pyd-P$_3$. It is an irreversible reaction. The same reaction takes place with dGTP as substrate. These reactions are catalyzed by GTP cyclohydrolases I and II in the mammalian cells. The mammalian GTP cyclohydrolases, unlike those of microbial or plant origin,[37,38] selectively catalyze the reaction to FPyd-P$_3$ and are unable to catalyze ring closure to NPTH$_2$-P$_3$.[29] Both enzymatic and chemical breakdown of GTP to formamido pyrimidine require the presence of phosphate groups. The phosphate (α, β, or γ) in [^{32}P]-GTP reappeared in all three phosphoric acids of FPyd-P$_3$.

The GTP cyclohydrolase I from mammalian brain was purified 2000-fold.[29] This enzyme is inhibited by its substrate, GTP, at 5×10^{-5} M or above and competitively inhibited by dGTP, dATP, and ATP. It is thermostable with $T_{1/2}$

of 11 min for inactivation at 80°C. It is not activated by Mg^{2+}. Its estimated molecular weight is 135,000 as obtained by gel filtration, and it has a pI of 4.95. It shows a K_m of 1.2×10^{-6} M for GTP and a V_{max} of 7.1 nmol/mg per h.[29]

The GTP cyclohydrolase II is Mg^{2+} dependent and resembles the enzyme from *E. coli*. It is thermolabile; its molecular weight is 34,000, and it has a pI of 6.65. Its K_m for GTP is 5.8×10^{-5} M, and its V_{max} is 12.2 nmol/mg per h in the presence of 6.7×10^{-4} Eq of Mg^{2+}.[29] Magnesium pyrophosphate is a strong inhibitor of this enzyme.

If one compares these two enzymes according to their stoichiometries and consistent with the pool sizes of the substrate *in vivo*, GTP cyclohydrolase I seems to be the preferred enzyme in the mammalian tissue for producion of FPyd-P_3. This implies that about 6.6×10^{-6} M GTP of its 6.6×10^{-4} M cerebral pool would be cleaved to yield 0.53 nmol/g per h of cerebral BH_2, whereas GTP cyclohydrolase II would require 1.3×10^{-3} M GTP for the same amount.[22]

4.2. D-*Erythro-q-dihydroneopterin Triphosphate Synthetase*

The first experimental evidence for the synthesis of neopterins from GTP was obtained with an extract of *Pseudomonas*.[39] There is a well-defined basic protein in mammalian brain and other tissues which performs the sole function of converting FPyd-P_3 into D-erythro-NPTH$_2$-P_3 and is unable to use GTP as substrate. This enzyme is 9177 daltons and is a single strand of 68 amino acids except the enzyme purified from rat brain which contains an additional aspartic acid. The structure has been determined by manual Edman degradation.[40] This protein has three active sulfhydryl groups and, in its most active form, is phosphorylated at the serine residue 66 (or 67). The catalytic rate of phosphorylated D-erythro-*q*-dihydroneopterin triphosphate synthetase is about 0.1 μmol/mg per h of *q*-NPTH$_2$-P_3[32] (very close to its rate *in vivo*). The antiserum against the enzyme from beef brain cross reacted with its samples from brains and livers of other mammals.[40] This enzyme is different from the one isolated from avian tissue[41] in that the latter utilizes GTP as substrate to produce NPTH$_2$-P_3. The mammalian enzyme releases its product as the quinonoid (*q*-) tautomer of D-erythrodihydroneopterin triphosphate.[42]

4.3. L-*Erythro-q-dihydrobiopterin Synthetase*

This enzyme, which converts to *q*-NPTH$_2$-P_3 to *q*-BH_2 in brain and other organs, has been purified to electrophoretic homogeneity.[29] The enzyme is an acidic, heat-labile protein of about 240,000 daltons with a pI of 4.9. It catalyzes an irreversible reaction and has an apparent K_m of 1.7×10^{-7} M and a V_{max} of 20 nmol/mg per hr. A purified preparation of this enzyme did not depend on the presence of pyridine nucleotides for its action,[42] although in another report the need for them was emphasized.[35] Essential mechanisms of the reaction catalyzed by this enzyme are not yet established. Dihydrobiopterin may be formed through epimerization and dephosphorylation by a metal-catalyzed hydrogenolytic cleavage[29] or through a postulated intermediate, 6-(1′,2′-dioxopropyl)-7,8-dihydropterin.[43] This latter compound, through reduction to sepiapterin in the presence of cerebral sepiapterin reductase,[44] would result in L-erythrodihydrobiopterin.[43] It is possible that this enzyme is multifunctional or a complex of several proteins with catalytic properties. A simplified pathway of the conversion of GTP to q-BH$_4$ is presented in Fig. 2.

4.4. Quinonoid–Dihydropteridine Reductase

The mechanism of conversion of 7,8-BH$_2$ to 5,6,7,8-BH$_4$ was described several years ago.[45] In a coupled oxidative reaction, BH$_4$ was oxidized to q-BH$_2$. Quinonoid–dihydropterin reductase (q-DHPR) (E.C. 1.6.99.7) catalyzed reduction of q-BH$_2$ back to BH$_4$ in the presence of NADH$_2$ or NADPH$_2$. This permits a continuous shuttle of a pair of hydrogen ions to reach the enzymatic sites of product formation. Quinonoid–dihydropterin reductase is widely distributed in brain.[46–49] The molecular weight of bovine cerebral q-DHPR was found to be 47,000, with the enzyme comprising two monomers of 27,400 daltons[47,48] with K_m for NADH$_2$ of 5.5×10^{-5} M and 4×10^{-4} M for NADPH$_2$.[48,50] The apparent[46] K_m of BH$_4$ is 1.1×10^{-6} M, and V_{max} is 2.0–1.8 mol/mg per min. The regional and subcellular distribution of q-DHPR in rat brain was measured.[48,49] The enzyme of the soluble fractions of the brain displayed the highest specific activity. Its regional distribution was highest in midbrain and lowest in the cortices (Table VII). The assays were done with DMPH$_4$ or 6-MPH$_4$ instead of BH$_4$; consequently, the data may not convey the true biological values. Irrespective of the choice of cofactor, these data do not support the suggestion that most of the DHPR are extraneuronal.[51]

Intraventricularly injected FPyd-P$_3$ was enzymatically converted to the quinonoid D-erythrodihydroneopterin triphosphate from which the quinonoid dihydrobiopterin was formed. Additionally, intraventricularly injected meth-

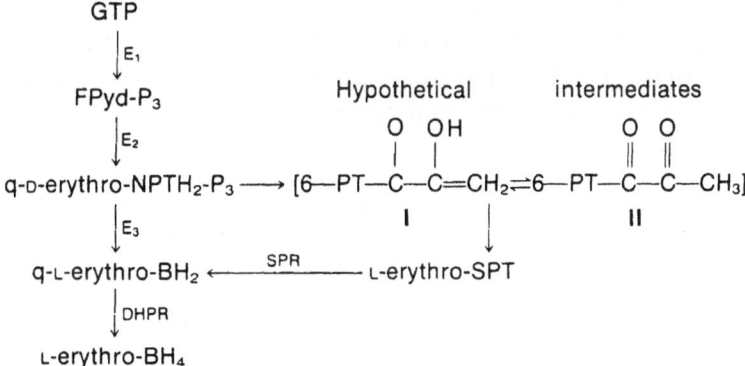

Fig. 2. Pathway of L-erythro-BH$_4$ biosynthesis from GTP. GTP, guanosine triphosphate; FPyd-P$_3$, formamidopyrimidine ribotide; q-D-erythro-NPTH$_2$-P$_3$, quinonoid-D-erythrodihydroneopterin triphosphate; q-L-erythro-BH$_2$, quinonoid dihydrobiopterin; L-erythro-BH$_4$, L-erythrotetrahydrobiopterin; L-erythro-SPT, sepiapterin. Hypothetical intermediates: I, 6-(1′-keto-2′-hydroxyallyl)dihydrobiopterin; II, 6-(1′.2′-diketopropyl)dihydropterin. Enzymes involved are: E$_1$, GTP cyclohydrolase I and II; E$_2$, q-D-erythrodihydroneopterin triphosphate synthetase; E$_3$, q-L-erythrodihydrobiopterin synthetase; SPR, sepiapterin reductase; DHPR, dihydropteridine reductase.

Table VII
Quinonoid Dihydropterin Reductase Activity
in Various Regions of Adult Rat Brain

| Region | Enzyme activity nmol NADH oxidized/mg per min. | |
	Ref. 48	Ref. 49
Caudate nucleus	620	—
Cerebellum	1040	109
Cortices	140	72
Globus pallidus	—	112
Hippocampus	330	66
Hypothalamus	1250	—
Inferior colliculi	2350	184
Locus coeruleus	2210	266
Mammillary bodies	—	121
Nucleus accumbens	—	77
Olfactory bulb	1300	108
Raphe nuclei	1730	196
Substantia nigra	1160	180
Superior colliculi	1340	144
Thalamus	930	121

otrexate at 5×10^{-6} M exerted no inhibition of the reduction of the biosynthetic q-BH$_2$ to BH$_4$ *in vivo* (Table VIII), yet it completely inhibited the reduction of intraventricularly injected tritiated dihydrofolate to tetrahydrofolate[42] (Table IX). Corollary evidence for the absence of synthesis of BH$_4$ through DHFR came from experiments *in vivo* with trimethoprim [2,4-diamino-5-(3′,4′,5′-tri-methoxybenzyl)pyrimidine], a strong competitive inhibitor of DHFR. Injection of 18.5–185 mg/kg of this inhibitor had no effect on the mean hepatic concentrations of BH$_4$ in the rat. This finding implied either that the drug was unavailable to cellular DHFR or that the reduction of 7,8-dihydrobiopterin to BH$_4$ did not contribute to the maintenance of BH$_4$ concentration.[52] It was thus

Table VIII
Effect of Methotrexate on Cerebral Synthesis of
Reduced Biopterins from [U-^{14}C]-GTP^a

| | Total recovered (dpm ± S.D.) | |
	[^{14}C]-BH$_2$	[^{14}C]-BH$_4$
Control (4)	1615 ± 71	456 ± 42
Methotrexate (4)	1570 ± 69	464 ± 58

^{a} Animals received 0.54 μg (1.2×10^6 dpm) of [^{14}C]-GTP intraventricularly 1½ h before sacrificing. Methotrexate (67 μg) was given simultaneously to the experimental group. Number of animals in brackets. From ref. 42.

Table IX
Effect of Methotrexate on Inhibition of Cerebral
Dihydrofolate Reductase in Vivo[a]

	Recovered [^{3}H]-FH$_4$ ± S.D.	Inhibition (%)
Control (4)	3,177,000 ± 193,000	—
Methotrexate (4)	37,000 ± 16,000	98.8

[a] Eight rats were injected with 10 μCi of [^{3}H]dihydrofolic acid in 20 μl water. Four of these animals also received 67 μg of methotrexate interventricularly dissolved in the solution containing the dihydrofolic acid. Animals were killed 1½ h later, and the levels of FH$_4$ were measured by scintillation counting. Number of animals in brackets. From ref. 42.

concluded that "this would indicate that there is no significant isomerization of quinonoid dihydrobiopterin to 7,8-dihydrobiopterin *in vivo* and that the biosynthesis of tetrahydrobiopterin in the rat is unlikely to involve reduction of 7,8-dihydrobiopterin."[52] The data (Table VIII and IX) are consistent with this statement.

5. CATABOLISM OF PTERINS IN THE BRAIN

The control of catabolism of a compound will often depend on its origin, i.e., whether the compound is exogenous or endogenous to its biological system. Injected [^{14}C]biopterin appeared in the urine as 20% biopterin, 45% pterin-6-carboxylic acid, 16% pterin, 0.4% isoxanthopterin, etc.[16,53] Elimination of reduced biopterins was usually slower. The main product of neopterin and BH$_4$ catabolism *in vivo* and *in vitro* was 6-hydroxylumazine, lumazine, and 7,8-dihydroxanthopterin. There are at least three major reactions in the catabolism of biopterin and neopterin,[54] i.e., the cleavage of the side chain, the deamination of the pterin ring, and the appearance of oxygen at C-6 of the ring.

The nature of products obtained by enzyme catabolism of pterins is dependent on the oxygen tension available to the tissues. It could therefore be rewarding to look for cerebral catabolites of biopterins at varying oxygen levels. To date, this reviewer is not aware of any catabolic studies of pterins involving the mammalian brain.

6. COFACTORIAL ROLE OF PTERINS IN THE BRAIN

6.1. Monooxygenases

Enzymatic reactions relevant to folic acid and its derivatives are numerous.[55] The participation of unconjugated pterins, by contrast, is only known for tetrahydrobiopterin which was recognized about 20 years ago as the natural cofactor for phenylalanine hydroxylase[5] (E.C. 1.14.3.1). Phenylalanine hydroxylase is well defined in most of its properties and thus could serve as a prototype for the other BH_4-dependent monoxygenases such as L-tyrosine and L-tryptophan hydroxylase. However, it is absent from the brain, and its kinetic and regulatory nature *vis-à-vis* BH_4 has been reviewed.[56]

Tyrosine hydroxylase (E.C. 1.14.16.2) catalyzes the conversion of L-tyrosine to L-3,4-dihydroxyphenylalanine (L-DOPA)[57] and of L-tryptophan (L-TRP) to L-5-hydroxytryptophan (L-5-HTP).[58] The dependence of this enzyme on reduced pterins for its function was known for some time.[59,60] Addition of reduced biopterin instead of 6-MPH_4 led to a sixfold increase of DOPA synthesis. L-Erythrotetrahydroneopterin ($NPTH_4$) (10^{-3} M) was less effective than 6-MPH_4 in catalyzing the conversion of L-tyrosine to L-DOPA, producing 24 nmol of DOPA versus 72 nmol by 6-MPH_4.[61] These studies also indicated that the reduced cofactors at 1 or 2×10^{-2} M were strongly inhibitory.[61] The response of tyrosine hydroxylase, the stoichiometry of the reaction to various reduced pterins, and the substrate specificity are also dependent on the stereochemistry of BH_4[62] This accords well with the data obtained from phenylalanine hydroxylase. Here again, tyrosine hydroxylase was inhibited by its substrate in the presence of BH_4 and, when phosphorylated, responded to 6-MPH_4 with a change in its affinity for this cofactor.[63]

Studies with tyrosine hydroxylase in rat brain homogenates suggested that, unlike the enzyme in adrenal tissue, the cerebral enzyme was particle bound and was insensitive to 1 to 30×10^{-4} M $DMPH_4$ in the presence of 6×10^{-2} M mercaptoethanol.[64] This observation was consistent with the report that $DMPH_4$ and $NADPH_2$ had no effect on the particle-bound tryptophan hydroxylase of rat brain.[65] Recently, solubilized tyrosine hydroxylase from brain tissue had the same affinity as the particulate for $DMPH_4$ (K_m 1.5×10^{-5} M). In contrast to the solubilized enzyme, the untreated soluble enzyme had a K_m of 7.4×10^{-4} M.[66]

Another pterin-dependent enzyme is tryptophan-5-hydroxylase (E.C. 1.14.16.4) which converts L-tryptophan and α-methyltryptophan to the corresponding 5-hydroxytryptophans. Tryptophan-5-hydroxylase from rabbit hindbrain can also hydroxylate L-phenylalanine to L-tyrosine in the presence of BH_4.[67] The dependence of this enzyme on the presence of reduced biopterin was soon recognized.[65,68] Addition of $NADPH_2$ to $DMPH_4$ enhanced the activity of purified soluble tryptophan-5-hydroxylase[69] but did not affect the activity of the particle-bound enzyme.[65] Similarly to other pterin-dependent monooxygenases, tryptophan-5-hydroxylase had a much lower apparent K_m for BH_4 than for its analogues. Also, at 2×10^{-4} M of L-tryptophan, this substrate became inhibitory only in presence of the natural cofactor *in vitro*[70] and *in vivo*.[58,71] Phospholipids did not stimulate this enzyme in the presence of BH_4.[67] In the search for a stimulating factor, a pteridinelike compound was found in 30,000g supernatant of rat brain homogenates. This factor would stimulate cerebral hydroxylation of L-tyrosine and L-tryptophan.[72] Unlike phenylalanine hydroxylase stimulating protein (PHS), this factor is not a protein. It was thermostable, alkali labile, dialyzable, and light sensitive and lost its activity after a few days of storage at $-27°C$. The presence of BH_4 was an absolute requirement for its effect.

L-Erythro-5,6,7,8-tetrahydrobiopterin (BH_4) is the natural cofactor of the three hydroxylases discussed in this section. It is the natural cofactor by virtue of its demonstrated synthesis from GTP *in vivo* and by its unique role *in vivo* as the active hydrogen carrier for the enzymes it serves.

The natural cofactor, BH_4, of the three hydroxylases meets the criteria suggested[73]; i.e., it (1) is specific, (2) has catalytic functions, and (3) is utilized during a substrate-dependent enzymatic reaction of hydroxylation. The apparent K_m, of BH_4 for phenylalanine, tyrosine, and tryptophan hydroxylases is about the same order of magnitude.

6.2. Oxygenases and Other Enzymes

Because of their autoxidizable nature, tetrahydropterins were found to stimulate indoleamine 2:3 dioxygenase,[74] an enzyme also found in the brain.[65] This enzyme, like hepatic tryptophan dioxygenase, is a heme protein which oxidatively cleaves the pyrrole ring of L-tryptophan to L-kynurenine, but unlike the liver enzyme, it also attacks D-tryptophan,[65] DL-5-HTP, 5-HT, tryptamine, and melatonin.[75,76] It is induced by tryptophan but not by cortisol.[77] Instead of oxygen, it requires superoxide O_2^- for its catalysis. Tetrahydropterins generate superoxide during their autoxidation and together with catalase promote synthesis of L-kynurenine. The presence of catalase may prevent nonenzymatic oxidation of tetrahydropterins by H_2O_2 which is generated during their autoxidation. Tetrahydrobiopterin was found to be the most effective among the tetrahydropterins in catalyzing the breakdown of tryptophan. The maximal activity of the enzyme was attained at 2×10^{-4} M of BH_4, and half-maximal activity at 7×10^{-5} M, which implied[74] that BH_4 may serve as the natural cofactor of indoleamine-2:3-dioxygenase but could be inhibited by another enzyme, superoxide dismutase, which catalyzed the dissipation of O_2^-. At present there is no evidence that BH_4 is a cofactor of indoleamine-2:3-dioxygenase *in vivo*.

Recently, it was suggested that pteridines may participate in the light-dependent regulation of the hydroxyindole-*O*-methyltransferase (HIOMT) of the pineal gland and retina.[78]

Another example of the participation of tetrahydropterins was revealed in studies on the oxidation of glyceryl ethers to fatty acids and glycerol, with long chain aldehydes as intermediates. This reaction was demonstrated to require oxygen and an unconjugated tetrahydropterin.[79] L-Erythrotetrahydroneopterin (BH_4 was not available to the experimenters) was six times as active as $DMPH_4$. Tetrahydrofolate (*l*-L-enantiomorph) at 5×10^{-5} M was ten times less active than 6-MPH_4 in catalyzing the oxidation of glyceryl ethers to their hemiacetals which would spontaneously break down to glycerol and an aldehyde. The aldehyde is then oxidized by NAD^+ to the acid. The oxidative step could be accomplished by well-washed microsomes, O_2, and a tetrahydropterin. In the presence of *l*-L-tetrahydrofolate and DFPR, the reaction was inhibited by amethopterin (MTX).

Tetrahydropterins were also implicated in the function of other oxygenases such as the one involved in 17α hydroxylation of progesterone.[80] The synthesis of prostaglandins from 8,11,14-eicosatrienoic acid was stimulated by BH_4 *in vitro*.[81] Preliminary experiments *in vivo* did not show any change in the cerebral levels of prostaglandins $PGF_{1\alpha}$ and PGE_2 in rats whose cerebral synthesis of BH_2 and BH_4 was completely arrested and whose pool sizes of reduced biopterins were decreased by 50%.[30] Purified lipoyl dehydrogenase from pig brain was inhibited by pterins, but its transhydrogenase and diaphorase activities were unaffected.[82]

6.3. Mitochondrial Respiration

It was noted that mitochondrial respiration at the level of cytochrome *c* was enhanced by addition of tetrahydropterin without this compound having any effect on the ATP-generating system.[83] A more recent report has indicated that a 0.9 P:O ratio was obtained with rat liver mitochondria on addition of $DMPH_4$. This implies phosphorylation at site 3 of the electron transport system.[84] An interesting feature of this report was that the reduction of cytochrome *c* by the tetrahydropterin proceeded anaerobically as well and was not inhibited by superoxide dismutase.

7. CLINICAL ASPECTS OF DEFECTS IN BIOPTERIN SYNTHESIS

A recent review of the biochemical mechanism of PKU[85] has dealt with the relevant terminology and definitions. Therefore, only a short overview will be given of the deficiencies in the level or synthesis of BH_4 as they relate to the function of phenylalanine hydroxylase. Human liver phenylalanine hydroxylase has been partially purified and studied for some of its properties.[86] Its instability and low activity, even with BH_4 in the absence of lysolecithin, precluded accurate estimates of the K_m values; nevertheless, the K_ms found were 3×10^{-6} M for BH_4 (with lysolecithin added) and 4×10^{-5} M and 5×10^{-5} M for 6-MHP_4 and $DMPH_4$, respectively. Tetrahydrobiopterin and lysolecithin failed to stimulate the synthesis of tyrosine by a sample of PKU liver beyond 0.27% of the rate in normal liver.[87]

A few years ago, PKU was reported in a male infant whose liver, brain, and skin fibroblast samples revealed the absence of dihydropterin reductase. The sum of the concentrations of BH_2 and BH_4 in this liver sample was half of that of normal subjects (6 nmol/g liver wet weight), and less than one-sixth of the BH_4 content of the normal liver was present.[88] The brain biopsy of the frontal cortex from a DHPR-deficient patient revealed less than 1% activity of a control. Another case of DHPR deficiency was reported as a variant of phenylketonuria.[89] The patient was a severely retarded 17-month-old girl excreting abnormally elevated amounts of 7,8-dihydrobiopterin. Her serum level of total biopterins was also high. Intravenous injection of BH_4 in graded doses from 1 to 100 mg produced a transient decrease in the serum phenylalanine level of the patient. This experiment lent further support to the idea that BH_4

functions as a cofactor of phenylalanine hydroxylase *in vivo*. It also pinpointed the defect at the DHPR site.

It is well to note that although human serum has been found capable of effective binding of BH_4, possibly by α_2-macroglobulins,[90] this binding, as well as the distribution and synthesis of BH_4, is greatly altered in diseases.[12]

Little is known of the effect of DHPR deficiency on tryptophan and tyrosine hydroxylases. Dihydrofolate reductase may have substituted to a limited extent for DHPR, although 7,8-dihydrobiopterin as a substrate for DHFR greatly decreases the rate of BH_4 synthesis. In some patients there was a total absence of BH_4 in spite of the presence of reducing substances such as glutathione, ascorbic acid, etc., which could conceivably have converted some $7,8-BH_2$ to BH_4. A trial of ascorbate therapy, 5 g daily for 19 days, failed to be remedial in the patients with peripheral and neurological symptoms.[88] These observations seem to strengthen the data which indicated that: (1) dihydrofolate has no significant functional role in reduction *in vivo* of dihydrobiopterin which is synthesized as the quinonoid-BH_2; (2) the cerebral activity of DHPR must fall below 5–10% before the synthesis of L-DOPA and 5-HTP becomes critically impaired. Such impairment might lead to the demise of the patient. Dihydropterin reductase-deficient patients revealed no observable defect in the biosynthetic route to $7,8-BH_2$ as manifested by excessive accumulation of $7,8-BH_2$. Appearance of urinary 7,8-dihydroxanthopterin and xanthopterin was also noted in these patients.[91] This defect might be construed as a mutation in the DHPR gene.

Another type of hyperphenylalaninemia was diagnosed as resulting from low levels of reduced biopterins.[88,92] The patient, a child, had normal phenylalanine DHPR, DFPR, and PHS protein as indicated by assays of his biopsied liver. However, reduced biopterins constituted only 5% of the normal level. This case represented a block along the biosynthetic pathway of quinonoid BH_2 from GTP. Recently, additional reports have appeared on patients with atypical phenylketonuria and normal liver dihydropteridine reductase and phenylalanine-4-hydroxylase activities who excreted neopterin but minimal biopterin or dihydrobiopterin in urine[93–95] (Table X). Attempts were made to

Table X
Urinary Levels of Biopterin and Neopterin in Patients with Deficiencies of the
GTP-to-BH_4 Pathway

	Biopterin (μmol/24 h urine)[a]		Neopterin (μmol/24 h urine)[a]	
	Ref. 94	Ref. 95	Ref. 94	Ref. 95
Controls	11.4 (25)	9.07 (3)	11.8 (25)	9.2 (3)
Typical PKU	45.3 (49)	28.9 (3)	44.6 (49)	8.6 (3)
DHPR deficiency	129.0 (4)	147.5 (1)	20.9 (4)	11.5 (1)
BH_2 deficiency	0.2 (7)	0.6 (2)	161.2 (7)	17.8 (2)

[a] Values were calculated from data in references by taking 920 μmol of creatinine excreted in an average volume of 1.5 liters of 24-h urine collection. Number of probands is in parentheses.

treat the patients with BH_4, BH_2, and sepiapterin, and the results appeared promising.[13,14]

REFERENCES

1. Hopkins, F. G., 1889, *Nature* **40**:335.
2. Pabst, W., and Rembold, H., 1966, *Z. Physiol. Chem.* **344**:107–112.
3. Gál, E. M., 1981, *Advances in Neurochemistry*, Volume 4 (B. Agranoff and M. Aprison, eds.), Plenum Press, New York, pp. 83–148.
4. Kraut, H., Pabst, W., Rembold, H., and Wildemann, L., 1963, *Z. Physiol. Chem.* **332**:101–108.
5. Kaufman, S., 1963, *Proc. Natl. Acad. Sci. U.S.A.* **50**:1085–1092.
6. Rembold, H., and Metzger, H., 1967, *Z. Naturforsch.* **22**:827–830.
7. Guroff, G., Rhoads, C. A., and Abramowitz, A., 1967, *Anal. Biochem.* **21**:273–278.
8. Gál, E. M., and Sherman, A. D., 1977, *Prep. Biochem.* **72**:155–164.
9. Fukishima, T., and Nixon, J., 1980, *Anal. Biochem.* **102**:176–188.
10. Leeming, R. J., Blair, J. A., Green, A., and Raine, D. N., 1976, *Arch. Dis. Child.* **51**:771–777.
11. Gál, E. M., and Sherman, A. D., 1981, (unpublished results).
12. Baker, H., Frank, O., Bacchi, D. J., and Hunter, S. H., 1974, *Am. J. Clin. Nutr.* **27**:1247–1253.
13. Curtius, H.-C., Niederwieser, A., Viscontini, M., Otten, A., Schaub, J., Scheibenreiter, S., and Schmidt, H., 1979, *Clin. Chim. Acta.* **93**:251–262.
14. Kapatos, G., and Kaufman, S., 1981, *Science* **212**:955–956.
15. Levine, R. A., Kuhn, D. A., and Lovenberg, W., 1979, *J. Neurochem.* **32**:1575–1578.
16. Gál, E. M., Hanson, G., and Sherman, A., 1976, *Neurochem. Res.* **1**:511–523.
17. Gál, E. M., and Sherman, A. D., 1976, *Neurochem. Res.* **1**:627–639.
18. van der Have-Kirchberg, M. L. L., de Morée, A., van Laar, J. F., Gerwig, G. J., Versluis, C., Ebels, I., Haus-Citharel, A., L'Heritier, A., Roseau, S., Zurburg, W., and Moszkowska, A., 1977, *J. Neural Transm.* **40**:205–220.
19. Albrecht, A. M., Biedler, J. L., Baker, H., Frank, O., and Hutner, S., 1978, *Res. Commun. Chem. Pathol. Pharmacol.* **19**:377–380.
20. Niederwieser, A., Matasovic, A., and Curtius, H.-C., 1980, *FEBS Lett.* **118**:299–302.
21. Rembold, H., 1970, *Chemistry and Biology of Pterins* (K. Iwai, M. Akino, M. Goto, and Y. Iwanami, eds.) International Academy, Tokyo, pp. 163–178.
22. Gál, E. M., and Sherman, A. D., 1977, *Structure and Function of Monoamine Enzymes* (E. Usdin, N. Weiner, and M. Youdim, eds.), Marcel Dekker, New York, pp. 23–42.
23. Shiota, T., Jackson, R., and Baugh, C. M., 1970, *Chemistry and Biology of Pterins* (K. Iwai, M. Akino, M. Goto, and Y. Iwanami, eds.), International Academy, Tokyo, pp. 264–269.
24. Sugiura, K., and Goto, M., 1973, *Experientia* **29**:1481–1482.
25. Buff, K., and Dairman, W., 1975, *Chemistry and Biology of Pteridines* (W. Pfleiderer, ed.), W. de Gruyter, Berlin, pp. 273–284.
26. Buff, K., and Dairman, W., 1975, *Mol. Pharmacol.* **11**:87–93.
27. Kaufman, S., 1967, *Annu. Rev. Biochem.* **36**:171–184.
28. Fukushima, F., Eto, I., Saliba, D., and Shiota, T., 1975, *Biochem. Biophys. Res. Commun.* **65**:644–651.
29. Gál, E. M., Nelson, J. M., and Sherman, A. D., 1978, *Neurochem. Res.* **3**:69–88.
30. Sherman, A. D., and Gál, E. M., 1979, *Life Sci.* **23**:1675–1680.
31. Gál, E. M., and Whitacre, D. H., 1981, *Neurochem. Res.* **6**:233–241.
32. Gál, E. M., and Sherman, A. D., 1978, *Biochem. Biophys. Res. Commun.* **83**:593–598.
33. Eto, L., Fukushima, K., and Shiota, T., 1976, *J. Biol. Chem.* **251**:6505–6512.
34. Fukushima, K., Eto, I., Mayumi, T., Richter, W., Goodson, S., and Shiota, T., 1975, *Chemistry and Biology of Pteridines* (W. Pfleiderer, ed.), W. de Gruyter, Berlin, pp. 247–263.
35. Lee, C., Fukushima, T., and Nixon, J. C., 1979, *Chemistry and Biology of Pteridines* (R. L. Kisliuk and G. M. Brown, eds.), Elsevier/North Holland, Amsterdam, pp. 125–128.
36. Gál, E. M., Henn, F. A., and Sherman, A. D., 1978, *Neurochem. Res.* **3**:493–499.
37. Burg, A. W., and Brown, G. M., 1968, *J. Biol. Chem.* **243**:2349–2358.

38. Cone, J., and Guroff, G., 1971, *J. Biol. Chem.* **24**:979–985.
39. Guroff, G., and Strenkoski, C. A., 1966, *J. Biol. Chem.* **241**:2220–2227.
40. Gál, E. M., Dawson, M. R., Dudley, D. T., and Sherman, A. D., 1979, *Neurochem. Res.* **4**:605–625.
41. Fukushima, K., Richter, W. E., Jr., and Shiota, T., 1977, *J. Biol. Chem.* **252**:5750–5755.
42. Gál, E. M., Bybee, J. A., and Sherman, A. D., 1979, *J. Neurochem.* **32**:179–186.
43. Tanaka, K., Akino, M., Hagi, Y., and Shiota, T., 1979, *Chemistry and Biology of Pteridines* (R. L. Kisliuk and G. M. Brown, eds.), Elsevier/North Holland, Amsterdam, pp. 147–152.
44. Katoh, S., 1971, *Arch. Biochem. Biophys.* **146**:202–214.
45. Kaufman, S., 1964, *J. Biol. Chem.* **239**:332–338.
46. Craine, J. E., Hall, S. E., and Kaufman, S., 1972, *J. Biol. Chem.* **247**:6082–6091.
47. Cheema, S., Soldin, S. J., Knapp, A., Hofmann, T., and Scrimgeour, K. G., 1973, *Can. J. Biochem.* **51**:1229–1239.
48. Snady, H., and Musacchio, J. M., 1978, *Biochem. Pharmacol.* **27**:1939–1953.
49. Bullard, W. P., Guthrie, P. B., Russo, V., and Mandell, A. J., 1978, *J. Pharmacol. Exp. Ther.* **206**:4–20.
50. Scrimgeour, K. G., and Cheema, S., 1971, *Ann. N.Y. Acad. Sci.* **186**:115–118.
51. Turner, A. J., 1977, *Biochem. Pharmacol.* **26**:1009–1014.
52. Stone, K. J., 1976, *Biochem. J.* **157**:105–109.
53. Rembold, H., 1964, *Pteridine Chemistry* (W. Pfleiderer, and E. C. Taylor, eds.), Pergamon Press, Oxford, pp. 465–484.
54. Rembold, H., 1970, *Chemistry and Biology of Pteridines* (K. Iwai, M. Akino, M. Goto, and Y. Iwanami, eds.), International Academy, Tokyo, pp. 163–178.
55. Blakley, R. L., 1969, *The Biochemistry of Folic Acid and Related Pteridines*, North-Holland, Amsterdam.
56. Kaufman, S., 1971, *Adv. Enzymol.* **32**:245–319.
57. Ikeda, M., Levitt, M., and Udenfriend, S., 1965, *Biochem. Biophys. Res. Commun.* **18**:482–488.
58. Gál, E. M., 1975, *Pavlov. J. Biol. Sci.* **10**:145–160.
59. Nagatsu, T., Levitt, M., and Udenfriend, S., 1964, *J. Biol. Chem.* **239**:2910–2917.
60. Brenneman, A. R., and Kaufman, S., 1964, *Biochem. Biophys. Res. Commun.* **17**:177–183.
61. Ellenbogen, L., Taylor, R. J., and Brundage, G. B., 1965, *Biochem. Biophys. Res. Commun.* **19**:708–715.
62. Numata (Sudo), Y., Kato, T., Nagatsu, T., Sugimoto, T., and Matsuura, S., 1977, *Biochim. Biophys. Acta* **480**:104–112.
63. Lovenberg, W., Bruckwick, E. A., Hanbauer, I., 1975, *Proc. Natl. Acad. Sci. U.S.A.* **72**:2955–2958.
64. McGeer, E. G., Gibson, S., and McGeer, P. L., 1967, *Can. J. Biochem.* **45**:1557–1563.
65. Gál, E. M., Armstrong, J. C., and Ginsberg, B., 1966, *J. Neurochem.* **13**:643–654.
66. Kuczenski, R., and Mandell, A. J., 1972, *J. Biol. Chem.* **247**:3114–3122.
67. Tong, J. H., and Kaufman, S., 1975, *J. Biol. Chem.* **250**:4152–4158.
68. Gál, E. M., 1965, *Fed. Proc.* **24**:580.
69. Nakumara, S., Ichiyama, A., and Hayaishi, O., 1965, *Fed. Proc.* **24**:604.
70. Friedman, P. A., Kappelman, A. H., and Kaufman, S., 1972, *J. Biol. Chem.* **247**:4165–4173.
71. Gál, E. M., Young, R. B., and Sherman, A. D., 1978, *J. Neurochem.* **31**:237–244.
72. Gál, E. M., and Roggeveen, A. E., 1973, *Science* **179**:809–811.
73. Kaufman, S., 1973, *Frontiers in Catecholamine Research* (E. Usdin, and S. Snyder, eds.), Pergamon Press, Oxford, pp. 53–60.
74. Nishikimi, M., 1975, *Biochem. Biophys. Res. Commun.* **63**:92–98.
75. Tsuda, H., Noguchi, T., and Kido, R., 1972, *J. Neurochem.* **19**:887–890.
76. Hirata, F., Hayaishi, O., Tokuyama, T., and Senoh, S., 1974, *J. Biol. Chem.* **249**:1311–1313.
77. Gál, E. M., 1974, *J. Neurochem.* **22**:861–863.
78. Cremer-Bartels, G., and Hollwich, F., 1978, in *Abstr. of the 6th Internat. Symp. on the Chemistry and Biology of Pteridines*, Elsevier/North Holland, Amsterdam, p. 24.
79. Tietz, A., Lindberg, M., and Kennedy, E. P., 1964, *J. Biol. Chem.* **239**:4081–4090.
80. Hagerman, D. D., 1964, *Fed. Proc.* **23**:480.

81. Samuelsson, B., 1972, *Fed. Proc.* **31**:1442–1450.
82. Millard, S. A., Kubose, A., and Gál, E. M., 1969, *J. Biol. Chem.* **244**:2511–2515.
83. Rembold, H., and Buff, K., 1972, *Eur. J. Biochem.* **28**:579–585.
84. Taylor, D., and Hochstein, P., 1975, *Biochem. Biophys. Res. Commun.* **67**:156–162.
85. Kaufman, S., 1977, *Advances in Neurochemistry*, Volume 2 (B. W. Agranoff, and M. H. Aprison, eds.), pp. 1–116.
86. Friedman, P. A., and Kaufman, S., 1973, *Biochim. Biophys. Acta* **293**:56–61.
87. Friedman, P. A., Fisher, D. B., Kang, E. S., and Kaufman, S., 1973, *Proc. Natl. Acad. Sci. U.S.A.* **70**:552–556.
88. Kaufman, S., 1975, *Chemistry and Biology of Pteridines* (W. Pfleiderer, ed.), W. du Gruyter, Berlin, pp. 291–303.
89. Danks, D. M., Cotton, R. G., and Schlesinger, P., 1975, *Lancet* **2**:1043.
90. Rembold, H., Buff, K., and Hernings, G., 1977, *Clin. Chim. Acta* **76**:329–338.
91. Watson, B. M., Schlesinger, P., and Cotton, R. G. H., 1977, *Clin. Chim. Acta* **78**:417–423.
92. Milstien, S., Orloff, S., Spielberg, S., Berlow, S., Schulman, J. D., and Kaufman, S., 1977, *Pediatr. Res.* **11**:460.
93. Niederwieser, A., Curtius, H.-C., Bettoni, O., Bieri, J., Schircks, B., Viscontini, M., and Schaub, J., 1979, *Lancet* **2**:131–133.
94. Niederwieser, A., Curtius, H.-C., Gitzelmann, R., Otten, A., Baerlocher, K., Blehova, B., Berlow, S., Grobe, H., Rey, F., Schaub, J., Scheibenreiter, S., Schmidt, H., and Viscontini, M., 1980, *Helv. Paediatr. Acta* **35**:335–342.
95. Nixon, J. C., Lee, C.-L., Milstient, S., Kaufman, S., Bartholome, K., 1980, *J. Neurochem.* **35**:898–904.

12

Neurons

Bert Csillik

1. BUILDING BLOCKS OF THE NERVOUS SYSTEM

Each of the $\sim 10^{11}$ neurons that constitute the human nervous system, $\sim 10^{10}$ of which are in the cerebral cortex, is surrounded by and, to some extent, symbiotic with satellite cells[1] (glial cells in the CNS; Schwann cells, amphicytes, and teloblasts in the peripheral NS) but is nonetheless an independent cellular unit, functionally as well as metabolically. Up-to-date techniques of electrophysiology and morphology (especially histochemistry and electron microscopy) have refuted "reticularist" and "continuist" theories that culminated in the delusion of the "Terminalretikulum"[2] and have proved the full validity of the classical concept formulated by Waldeyer, Lenhossék, and Cajal according to which every neuron represents an independent entity in terms of anatomy (dendrites and axon are cytoplasmic appendages of the perikaryon; each neuron is an independent cellular unit surrounded by an uninterrupted cell membrane; neurons are contiguous but not continuous with each other), physiology (the nerve impulse proceeds unimpeded within the domain of the neuron but has to "jump" to the next neuron at the synapse), genesis (all portions of the neuron, i.e., perikaryon, dendrites, and axon derive from the same single neuroblast of ectodermal origin), and trophism (the distal stump of a transected axon undergoes Wallerian degeneration; regeneration starts from the proximal stump; the trophic center of the neuron is the perikaryon).

A relatively novel extension of the classical tetralogy, Dale's principle, postulates that every neuron transmits impulses by means of one (and only one) chemical mediator substance at every one of its axon terminals and contains the enzyme systems responsible for synthesis and inactivation of this transmitter. Neurochemical implications of Dale's principle, known also as the cytochemical entity of the neuron, are obvious.

Bert Csillik • Department of Anatomy, University Medical School, Szeged, Hungary.

2. *STRUCTURAL OVERVIEW*

Within an apparently homogeneous hyaloplasm, the cell body, or peri-karyon, of the neuron contains microorganelles as does any cell; some of these can readily be traced cytochemically on the basis of genuine marker enzymes. These include mitochondria, the Golgi apparatus, lysosomes, peroxisomes, as well as smooth and rough (granulated) endoplasmic reticulum. Cisterns of the rough endoplasmic reticulum, associated with the Golgi apparatus, and lysosomes deriving from them are referred to as the Golgi-associated endo-plasmic reticulum plus lysosomes (GERL)[3] (Table I).

Dendrites, protruding from several micrometers to several hundred mi-crometers and usually equipped with spines, contain mitochondria and rough endoplasmic reticulum; the Golgi apparatus, however, extends only, if at all, into the most proximal portion of the dendrite. In contrast, the axon, often several centimeters long, is entirely devoid of both Golgi apparatus and rough endoplasmic reticulum. Neurofilaments, microfilaments, and microtubuli are arranged, as a rule, paralled to the longitudinal axis of the axon; mitochondria and cisterns of the smooth endoplasmic reticulum, often referred to as axo-plasmic reticulum, are also included and move along proximodistally at a rate of 1–10 mm/day in the axoplasm (slow transport). Microtubules seem to take an active part in the mechanism of fast axoplasmic transport, 100–1000 mm per day.[5] A retrograde transport mechanism (see Vol. 5) conveys material back to the perikaryon.

According to the principle of histodynamic polarity, dendrites (receptive processes) conduct impulses to the perikaryon, whereas the axon, an effector process, conducts impulses to its ending, the nerve terminal. Terminals of other neurons impinge on the perikaryon and the dendrites; every spine rep-resents at least one synapse. Axoaxonic and dendrodendritic synapses are less frequent contact types. At synaptic sites, condensation of the receptor mole-cules and association of other proteins anchored to the interior of the cell result in postsynaptic thickenings of the neuronal surface membrane; an intrinsic protein kinase mediates the transfer of phosphate from ATP to endogenous

Table I
Cytochemical Marker Enzymes of Neuronal
Microorganelles

Organelle	Marker enzyme
Mitochondria	Succinic dihydrogenase
	Cytochrome oxidase
Golgi apparatus	Thiamine pyrophosphatase
Lysosomes	Acid phosphatase
Rough endoplasmic reticulum	Acetylcholinesterase[a]
GERL	Acid phosphatase
	Nucleoside phosphatase
Peroxisomes	D-Amino acid oxidase

[a] Not only in cholinergic nerve cells but also in many noncholinergic neurons.[4]

membrane proteins of the postsynaptic density.[6] An element of the rough endoplasmic reticulum called the subsurface cistern is often seen subjacent to the postsynaptic thickening. Roles of the subsurface cistern in the supply of high-energy phosphate produced by mitochondria, in the active transport of metabolites into or out of the neuron, and in the synthesis of receptor proteins have been suggested.[7] In many cases, a spine apparatus consisting of flattened sacs and lamellae is subjacent to the postsynaptic membrane. Suggestions on the role of the spine apparatus in memory and learning await further corroboration.[8]

The axon terminal contains mitochondria and synaptic vesicles associated with the transmitter substance and a ring of microtubules within a seemingly homogeneous terminal axoplasm. Demonstrability of microtubules within the terminal is dependent on the albumin content of the fixative used prior to the processing of tissue for electron microscopy.[9] The axonal surface membrane of the terminal, called the presynaptic membrane, is adjacent to and coupled by an "intersynaptic organelle" to the postsynaptic surface membrane of the next cell. Synaptic vesicles are clustered at the active sites of the presynaptic membrane which are characterized by presynaptic protrusion(s) representing calcium channels or calcium gates.[10]

Nuclei of the neurons are spherical, large, and light, as compared to those of other cells, since the DNA content is dispersed in a huge volume. As a rule, large projective neurons contain tetraploidic amounts of DNA, whereas small interneurons are equipped only with a diploidic set of DNA.[11] The nucleolus is dense, with a high RNA content. Indentations of the nuclear membrane intrude into the nucleoplasm and reach the vicinity of the nucleolus. The nuclear membrane is built up of a perinuclear cistern of the rough endoplasmic reticulum.

3. NEURONAL PROTEINS

In addition to enzymes specific for cell organelles and to those participating in the production and breakdown of transmitter substances, structural proteins characterizing neurons have been identified. The biochemistry of proteins in the nervous system will be dealt with extensively in Volume 5 of this *Handbook*. However, a taxonomic enumeration of structural proteins specific for nerve cells, as opposed to glial proteins such as S-100, GFA, NS-1, etc. can properly be included in this chapter (Table II).

Although not characteristic of nerve cells per se (since it occurs in nearly every cell of the animal kingdom), tubulin, the microtubule subunit protein, is present in virtually every type of neuron. This ubiquitous dimeric protein comprises a large percentage of the protein in the mammalian brain. Microtubules measure 20–24 nm in diameter and have a well-defined lumen of about 15 nm; they are equipped with short side arms. Microtubules fulfill important functions as neuronal cytoskeleton and participate in the mechanism of axoplasmic transport, although earlier suggestions of an exclusive role of microtubules in transport have recently been challenged. Tubulin consists of two

Table II
Nerve-Cell-Specific Proteins

Protein	Characteristics
14-3-2	Specific for neurons of mammals and birds. Some neurons contain it only in early phases of development; other neurons contain it throughout life.[12] Highest concentration in synaptic terminals.[13]
GP-350	Brain-specific sialoglycoprotein.[14]
Synaptin	Present in nerve terminals.[14]
D1, D2, D3	Brain-specific membrane proteins.[14]
P-400	Membrane-bound protein present only in the molecular layer of the cerebellum (probably in dendrites or Purkinje cells).[15]

monomers; α tubulin has a molecular weight of 56,000, whereas β tubulin is a 53,000-dalton monomer. In addition, the side arms of neurotubules contain a high-molecular-weight (HMW) protein, also referred to as microtubule-associated protein (MAP), that consists of 350,000-, 300,000-, 150,000-, and 70,000-dalton components.[16] Earlier studies performed with labeled leucine indicated a half-life of approximately 4 days for of brain tubulin[17]; recent studies with labeled $NaHCO_3$ suggest a half-life more than twice as long.[18] This is considerably longer than the mean half-life of brain proteins (6.6 days).

Other fibrous proteins present in neurons are actin[19] and myosin.[20] F-actin seems to be involved in the force-generating mechanism in axoplasmic transport.[21] Actin forms microfilaments* with a diameter of 2–5 nm[16] or 5–7 nm[22]; these are prominent in axonal growth cones.[16] Distinctly different are neurofilaments* or intermediate filaments, the width of which varies between 8–11 nm[22] or 8–13 nm.[16] Recent studies suggest that the major neurofilament subunit is a 17,000-dalton protein[16] that is capable of forming a variety of stable aggregates. These might include filarin, the 70,000-dalton neurofilament protein isolated by Davison,[23] as well as the ~50,000-dalton neurofilament proteins isolated from human, bovine, rodent, and lagomorph brains[22] and the 70,000- and 160,000-dalton complexes in giant axons.[22] According to Lasek and Hoffmann,[24] the 68,000-, 145,000-, and 200,000-dalton components are parts of the "triplet" that constitutes the filamentous cytoskeleton of the axon.

The dynamics of protein turnover provide important insight into neuronal function. In this context, it should be noted that with respect to nontubulin proteins, labeling with $NaH^{14}CO_3$ indicates considerably shorter half-lives than earlier studies performed with [4,5-^{3}H]leucine or [1-^{14}C]leucine.[25] Obviously, leucine is reutilized by nerve cells. Some relevant results are summarized in Table III.

4. DALE'S PRINCIPLE

On the basis of the transmission mechanisms, cholinergic, nomoaminergic, aminacidergic, and peptidergic neurons can be distinguished.[26] Neurochemical

---

* Differences in values given by various authors reflect the effects of different preparative techniques used prior to the electron microscopic procedure. It seems safe to say that neurofilaments are twice as thick as microfilaments.

Table III
Half-Lives of Neuronal Proteins[25]

Type of protein	Type of labeling	Half-life
Tubulin	$[^3H]$- or $[^{14}C]$leucine	4 days
	$NaH^{14}CO_3$	9.8 days
Gross brain protein	$[4,5-^3H]$Leucine	8.4 and 16.5 days
	$[1-^{14}C]$Leucine	8.9 and 14.2 days
	$NaH^{14}CO_3$	3.3 and 8.7 days

types of neurons and the transmitter substances involved are summarized in Table IV.

Histochemical localization of acetylcholinesterase (AChE) at light and electron microscopic levels proved to be an important contribution to the concept of the cytochemical entity.[4] In cholinergic motor nerve cells (of the spinal cord and the brainstem), AChE is located in the perikaryon, in dendrites and in both neuromuscular and recurrent axon terminals. Within the perikaryon, the AChE reaction is confined to the endoplasmic reticulum, clusters of which, at the light microscope level, constitute Nissl bodies; high-power electron microscopy reveals that within the lumina of endoplasmic cisterns globular units measuring ~35 nm in diameter are devoid of the enzyme reaction.

It has been postulated that polypeptide chains synthesized on the surface of ribosomes and transported into the cisternal lumina acquire their tertiary configuration on the surfaces of such "endoplasmic units."[37] This seems to account for the well-known fact that esterolytic activity is exerted only by intracisternally located AChE molecules.[37] In large dendrites, the situation is similar, yet some of the enzyme may leak out of the dendritic compartments and into capillaries by a process called dendritic secretion.[38] In axons, AChE activity is confined to the axolemmal membrane; both histochemical and biochemical studies prove a proximodistal migration of the enzyme. In axon terminals, AChE is concentrated in the presynaptic membrane and is accompanied by various amounts of postsynaptic AChE activity. (The postsynaptic AChE seems to derive, both ontogenetically and functionally, from the postsynaptic element; recent studies prove profound biochemical differences between the molecular structures of neural and postsynaptic AChEs.)[39]

Double (pre- and postsynaptic) membrane localization of AChE is evident in neuromuscular junctions of type α subneural apparatuses in which the postsynaptic membrane is thrown into multiple folds; thus, its surface area surpasses that of the presynaptic membrane by a factor of about ten. Accordingly, type α subneural apparatuses are characterized by an overwhelmingly postsynaptic AChE activity. Since this type was initially studied almost exclusively, this peculiar situation led to an erroneous generalization with regard to the postsynaptic localization of AChE in cholinergic synapses. In type γ neuromuscular junctions, however, which are equipped with less numerous and shallow folds, the postsynaptic contingent of AChE is only slightly higher than the presynaptic one.[40] Finally, in axon terminals of recurrent collaterals synapsing with bulbous dendrites of Renshaw cells, the overwhelmingly presynaptic localization of AChE is clearly evident.[41]

Table IV
Neurochemical Types of Neurons and Chemically Labeled Pathways

Neuron type	Transmitter involved	Example
Cholinergic	Acetylcholine (ACh)	Spinal motoneurons
		Preganglionic autonomic neurons
		Postganglionic parasympathetic neurons
		Vestibulo–cerebellar neurons (in rat)
	ACh-like substances:	Neurons in cerebral cortex (?)
	Acetylcarnitine[27]	
	Acetates of mono-, di-, and triethylcholine[27]	
	Betaine-Coenzyme A[27]	
	N-Acetyl aspartate[27]	
	Homocholine[28]	
Monoaminergic	Catecholamines	
	Norepinephrine (NA)	Postganglionic sympathetic neurons
		Neurons in locus coeruleus
	Dopamine (DA)	Neurons in substantia nigra
		SIF cells in autonomic ganglia
		Mesolimbic and tubero–infundibular neurons
	Octopamine[29]	Scattered neurons in rat brain
	Indoleamines:	
	5-Hydroxytriptamine (serotonin)	Neurons in raphe nuclei
Aminacidergic	Excitatory amino acids[30]	
	Glutamate	Excitatory neurons in cerebral and cerebellar cortex
	Aspartate	
	β-Alanine	
	Proline	
	Inhibitory amino acids[30]	Inhibitory neurons in cerebral and cerebellar cortex
	GABA	
		Limbic system
		Basal ganglia
		Spinal cord
	Taurin	Retina
Peptidergic[a]	Enkephalin[32]	Neurons in sympathetic ganglia
	Substance P[33]	Primary nociceptive neurons
	Somatostatin[33]	
	Neurotensin[34]	Neurons in hypothalamus
	Neurophysin-I[35]	Neurons in hypothalamus
	Oxytocin[35]	Neurons in hypothalamus
	Vasoactive intestinal polypeptide (VIP)[36]	Neurons innervating the gut

[a] For a compilation of the immunohistochemical studies on VIP, substance P, enkephalin, and somatostatin, see Schultzberger *et al.*[32]

Thus, histochemistry proves that, in accord with Dale's principle, AChE is present constantly in every component of the cholinergic neuron (Fig. 1).

Two other proteins characteristic of cholinergic neurons, the enzyme choline acetyltransferase (choline acetylase, CAT) and the acetylcholine receptor (AChR), are also distributed throughout the neuron, being synthesized in the perikaryon and transported to the axon terminal via axoplasmic flow.[42] The transmitter molecule itself, ACh, is also present throughout the cholinergic neuron; recent studies prove that it is subject to axoplasmic transport.[42]

In accord with Dale's principle, most, if not all, classes of neurons are genuinely labeled by transmitter-specific enzymes and receptors, some of them subject not only to neurochemical but also to cytochemical demonstration. Tyrosine hydroxylase, the marker enzyme of noradrenergic neurons, can be demonstrated at light and electron microscope levels by means of immunohistochemistry[43]; presence of the transmitter itself, NA, can be proved by means of fluorescence microscopy based on the formaldehyde condensation of monoamines.[44,45]

Immunohistochemistry is a powerful tool to locate other enzymes involved in the production of transmitter substances, including DOPA decarboxylase,[46] phenylethanolamine-*N*-methyltransferase,[47] etc. Also, the demonstration of peptide transmitters such as substance P,[33] somatostatin,[33] neurotensin,[34] neurophysin-I,[35] oxytocin,[35] and VIP[36] by means of immunohistochemistry proves the validity of Dale's principle in many neuronal classes. Among these, peptidergic primary nociceptive neurons are labeled by a histochemically readily demonstrable fluoride-resistant acid phosphatase (FRAP)[48] probably related to the production of the transmitter peptide(s). Related to FRAP, a calcium-activated cytidine triphosphate phosphohydrolase activity has been shown by recent biochemical studies in dorsal roots.[49] In the domain of aminacidergic neurons, high-affinity uptake of labeled amino acids such as [^{3}H]glutamate, [^{3}H]aspartate, and [^{3}H]-GABA have been used for localization; this approach has also been widely used for the study of monoaminergic systems.

Apparently, however, not all neurons obey the law of cytochemical entity.[50] The number of neurons identified as releasing more than one transmitter substance is steadily increasing.[51] On the other hand, sensitivity and efficiency of neurons are regulated by numerous nonsynaptically liberated chemical substances: modulators, neurohormones, and neurohumors.[52]

In addition to releasing more than one transmitter substance at their abundantly ramifying axon terminals, a fair number of interneurons (local circuit neurons[53]) are also distinguished by their ability to transmit impulses by specialized regions of their dendritic arborization. Such presynaptic dendrites, which had been thought to represent unique exceptions to the law of histodynamic polarization, were found in amazing numbers in recent years in nearly every region of the central nervous system.[54] Thus, what seemed to be a sporadic exception is now regarded a general rule of the network organization of interneurons. By means of presynaptic dendrites, interneurons are often engaged in triadic arrangements with an adjacent (usually projective) axon terminal and a neighboring conventional dendrite.[55] Triads are thought to be involved in information processing, with presynaptic dendrites exerting inhibitory influence on the adjacent conventional dendrite.[56]

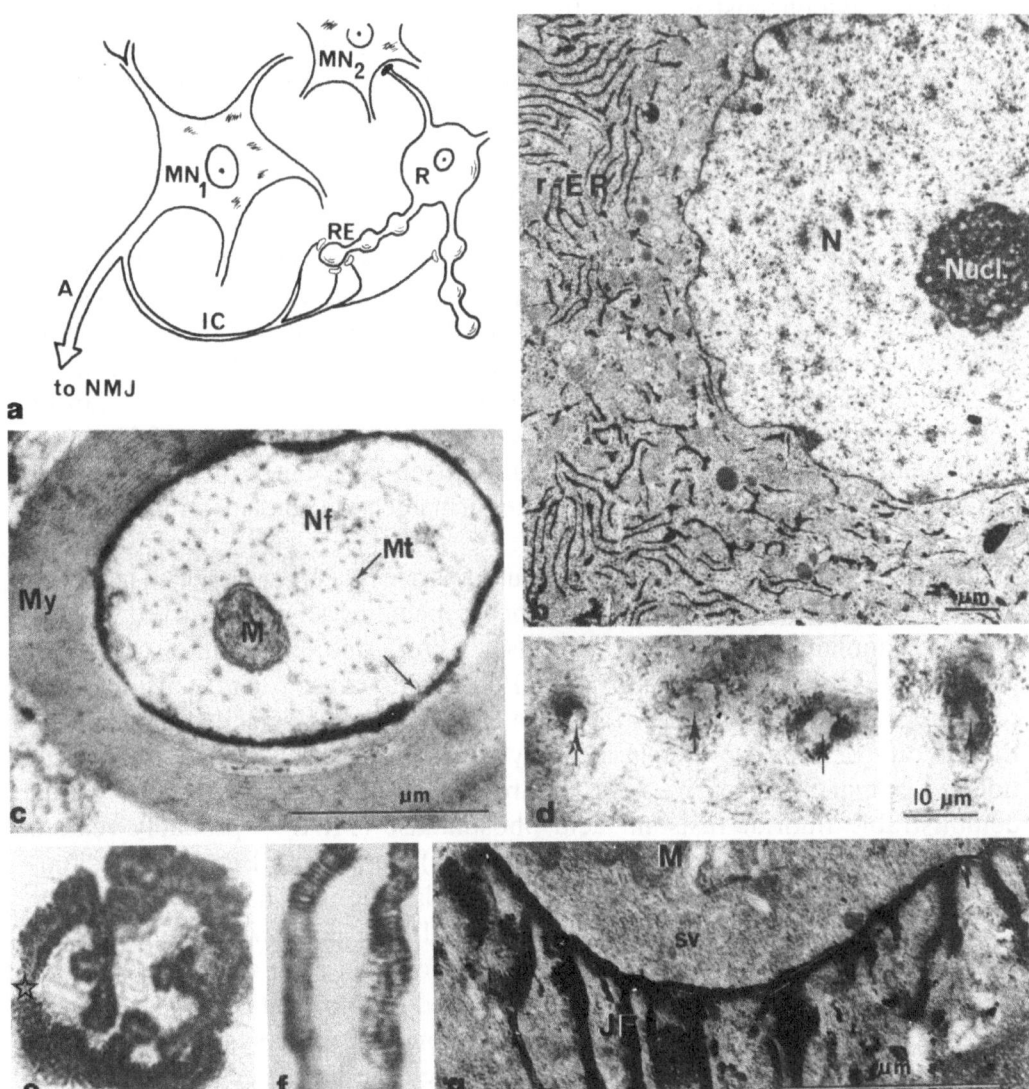

Fig. 1. Distribution of AChE in the spinal motoneuron (Dale's principle). (a) Diagram of cholinergic motoneurons (MN$_1$ and MN$_2$) in the ventral horn. Axon of the motoneuron (A) proceeds via the ventral root to cholinergic neuromuscular junctions (NMJ). Initial collateral of the axon (IC) establishes cholinergic synapses with bulbous dilatations of Renshaw cell (R) dendrites; these latter are known as "Renshaw elements" (RE). Axon of the Renshaw cell inhibits adjacent motoneuron (MN$_2$). (b) Acetylcholinesterase activity in the cytoplasm of a motoneuron at the level of electron histochemistry. Note abundant cisterns of rough endoplasmic reticulum (rER) filled with the reaction product of the AChE reaction. Other components of the cytoplasm as well as the nucleus (N) and nucleolus (Nucl) are devoid of any AChE activity. (c) Cross section of a motoneuron axon; electron histochemistry. Note intense AChE reaction of the axolemmai membrane (arrow). Mitochondrion (M), myelin sheath (My), neurofilaments (Nf), and microtubules (Mt) are devoid of any AChE reaction. (d) Acetylcholinesterase activity of Renshaw elements; light microscopy. These consist of a nonreacting core (arrow), corresponding to a bulbous dilatation of the Renshaw cell dendrite, and of numerous AChE-active cholinergic terminals of motoneuronal axon collaterals surrounding the core. (e) Acetylcholinesterase activity of a neuromuscular junction (type α subneural apparatus) in the rat gastrocnemius muscle. Light microscopy. Axon terminal enters the gutter formed by semicircular folds of the postsynaptic membrane at the site marked by an asterisk. (f) High-power light micrograph of the junctional folds in an amphibian neuromuscular junction

5. GENERATION AND DEGENERATION OF NEURONS

Generation of nerve cells in the embryonic CNS starts with a highly intensive cytokinetic activity of neuroepithelial cells in the ventricular layer of the neural tube, followed by lateral migration of neuroblasts, and accomplished by their settling in areas forming the prospective anlage of the cell groups that form the mature CNS.

[3H]Thymidine labeling, introduced by Sidman *et al.*[57] in the investigation of early development of the central nervous system, proved to be an exceptionally useful marker for studying neurogenesis. Since [3H]thymidine is incorporated into DNA of cells preparing for division, the radiochemical marker remains in such cells during and after mitosis as well as in their progeny and can be located by radioautography.

Mitotic figures in the neural tube are exclusively found adjacent to the central canal. New nuclei produced by mitosis at the lumen move away and enter the peripheral zone before synthesizing DNA. Interphase nuclei remain in the peripheral zone for at least 4 h or even longer before returning to the luminal border to divide.

During this pendulumlike movement, neuroepithelial cells are anchored to the luminal surface. Thus, while the perikaryon moves away from the lumen, the luminal portion of the cytoplasm remains attached to adjacent neuroepithelial cells by the terminal bars; this results, after interphase, in a "bouncing back" of the cell to its original position. Finally, when the cytokinetic capacity of the cell is exhausted, and the final migration into position takes place, this luminal anchoring provides the basis of the trailing process. On the other hand, another expansion, called the leading process, grows peripherally and anchors the cell to the external surface of the neural tube.

Trailing and leading processes give rise to the first apparent bipolarity of neuroblasts. For example, in motoneurons, the leading process is transformed into the axon, whereas the trailing process (transient dendrite) is resorbed and disappears when the cell starts to migrate in the mantle zone. In other nerve cells, e.g., the Purkinje cells of the cerebellum and in pyramidal cells of the cerebral cortex, the leading process gives rise to the dendritic arborization, and the trailing process is transformed into the axon. In the activity of axonal growth cones, actin seems to be involved.[58]

Generation of neurons follows rostrocaudal and ventrodorsal gradients. In addition, and sometimes overruling these gradients, large projective cells are generated earlier than small interneurons in every given structural entity. Neuroblasts giving rise to interneurons undergo several postmigratory mitoses that result in the reduction of the DNA content of their nuclei, whereas large projective neuroblasts maintain their original tetraploidicity[11] since they are not subjected to postmigratory mitoses.

---

stained by the AChE technique. (g) Electron microscopic histochemistry of AChE in the neuromuscular junction; rat diaphragm. The axon terminal contains mitochondria (M) and synaptic vesicles (sv). Acetylcholinesterase activity is seen in pre- and postsynaptic membranes of the neuromuscular junction; because of the intensity of the enzyme reaction, the synaptic gap also contains the enzyme reaction product. Junctional folds (JF) exert an intense AChE reaction.

Developing neurons are subjected to effects of various hormones produced by embryonic or maternal endocrine glands.[59] Differentiation of the dendritic arborization and the generation of synapses are largely dependent on such hormonal influences.[60]

Epinephrine and several peptide hormones are known to bind to specific receptors at the cell surface membrane; by activating nucleotide cyclase, these hormones stimulate the conversion of ATP to cyclic AMP which, in turn, as a "second messenger," stimulates anabolic processes in the developing neuroblast or neuron.[61]

Steroid hormones and thyroxin, on the other hand, seem to penetrate the neuroblast and bind to cytoplasmic receptor proteins that carry the hormone molecules to the nucleus of the cell where they stimulate synthesis of specific RNA. Thyroxin acts at the level of transcription, in all probability by increasing the activity of RNA polymerase.[62] In addition, thyroxin increases the activity of thymidine kinase.[63] Binding of glucocorticoids to nuclear and cytoplasmic receptors has been demonstrated by means of radioautography.[64]

Although the growth hormone somatotropin obviously influences the ontogeny of neurons, it is difficult to decide whether this is related to direct hormonal effects or is secondary to general effects on metabolism. On the other hand, the direct effect of nerve growth factor (NGF) on developing neurons is beyond question. Nerve growth factor seems to be taken up selectively at nerve endings of adrenergic sympathetic nerve cells and at those of primary sensory neurons by specific receptors at the cell surface of the terminal. It is then transported via retrograde transport to the perikaryon, where it exerts its effect on RNA and protein synthesis, thus stimulating axonal outgrowth.[61] Neurochemical correlations of NGF will be analyzed in detail by Guroff in Volume 5.

During embryonic life, nerve cells are generated in surplus (e.g., in dorsal root ganglia, three times more than necessary[65]); they are decimated in the course of subsequent periods characterized by peaks of the cell death rate. The mechanism by which and reason why nerve cells die are essentially unknown; also, the factors by which individual cells are selected for death is a mystery. Phylogenetic, morphogenetic, and histiogenetic types of embryonic nerve cell death have been distinguished.[66] Phylogenetic cell death results in regression of vestigial organs, e.g., the paraphysis in the CNS of mammalian embryos. Morphogenetic cell death is a programmed obsolescence of neurons that have completed their function in the formation of the CNS, especially in areas subjected to folding or bending of the neural tube.[67]

Histiogenetic cell death related to the lack of appropriate peripheral connections of the axons seems to be understood best. According to Jacobson[61] "histiogenetic cell death of nerve cells occurs before the cells have become functionally mature, and it appears to be a means of selection of cells whose functions are most effectively adapted to the final ensemble." Lack of oxygen and the action of thyroid hormone have been considered responsible for histiogenetic cell death.[68] Many motoneuroblasts and dorsal root ganglion cells in the thoracic region of the spinal cord (between limb bud zones) die this way. Another example of histiogenetic embryonic nerve cell death is that that occurs in ventrodorsal rows within the alar plate (the area of the prospective dorsal horn), which suggests the effect of locally accumulated toxic metabolic products in an early period when the trophism of the neuroblasts in the developing spinal cord (which lacks blood vessels) is wholly dependent on diffusion of tissue fluids. Although there is no convincing evidence for lysosomal acid hydrolases being involved in the mechanism of cell death,[69] compliance to a preexisting program that ultimately results in the sculpturing of the definite structure of the CNS by disposing of superfluous elements has been repeatedly suggested.[70] In this context, DNA cross linking has been postulated as a working hypothesis.[71]

In the course of extrauterine life, cell death in the nervous system proceeds at an amazingly high rate, amounting to 2.5×10^9 cells in a human life span of 80 years, i.e., each fourth nerve cell in the cerebral cortex will be discarded. Cell death is usually ascribed to age-related changes of the affected neurons and/or to their microenvironment. Although this is undoubtedly the case in senescence,[72] it is difficult to envisage such a mechanism in young individuals. It seems more probable that the life-span of the individual neurons is genetically programmed.

Not only neurons but, in all probability, also their axon terminals and the synapses brought about by them are produced in the course of embryonic life in a redundant number. Recent studies on the plasticity of the spinal cord suggest that a fair number of synapses (perhaps also their parent nerve cells) are nonfunctioning under normal conditions; after injuries of vicinal input channels, however, such connections may become viable.[73] Unfortunately, for the time being, neither morphological nor cytochemical techniques are capable of distinguishing such silent synapses from active ones. By increasing the resolving power of the 2-D-deoxyglucose labeling technique by which functionally active regions of the brain can be distinguished from resting ones using a semimicroscopical autoradiography,[74] the prospect of differentiating active synapses from silent ones at the fine-structural level might be brought closer to reality.

6. THE INTRACELLULAR SIGNALING SYSTEM OF THE NEURON

Neurons are specialized to generate, conduct, and transmit impulses that are integrated into the regulatory function of the nervous system. In addition to this overt and ostensible function, however, the extreme geometry of the neuron renders indispensable an intracellular mechanism of internal signaling which provides information on the state of the axon, which may be 10^3 times longer than the diameter of the cell body and the volume of which may be $10^3–10^4$ times larger than that of the perikaryon. Such a signaling system is responsible for the switchover of perikaryal protein-synthesizing machinery whenever the axon is injured and production of large new masses of axoplasmic material is needed to replace the lost process by a new, regenerating one.

Microscopic signs of such a switchover in the perikaryal protein-synthesizing machinery (chromatolysis, "primäre Reizung") include swelling and displacement of the nucleus and an apparent dissolution of Nissl bodies—in fact, a rearrangement of the rough endoplasmic reticulum.[75] Simultaneously, chromatolytic cells undergo dramatic metabolic changes: production of neurotransmitter-specific enzymes and organelles is reduced,[76] and synthetic priorities are reorganized to provide material for axonal regeneration.[77] Ironically, chromatolysis is often referred to as "retrograde degeneration," although, as a rule, chromatolytic neurons do not die except for those in very young animals (Gudden's effect) or if the lesion was formed too close to the cell body or was made with blunt instruments. Chromatolysis also leads to cell degeneration if axonal regeneration is unsuccessful, e.g., in the majority of neurons the axons of which terminate in the CNS except for those that possess a "sustaining collateral." In all other cases, however, chromatolysis ("tigrolysis") is a preparatory step for full restoration of the lost process and of lost synaptic connectivity of the neuron.

What is the signal for chromatolysis? How does the cell body know that its axon has been damaged? In his classical paper, Cragg[78] enumerated ten possible answers to this question, and, although some of the original possibilities can now be ruled out with certainty, some of the factors listed there still remain. These include:

1. A blood-borne substance that is released by the transected end of the nerve and reaches the cell body via circulation.
2. Depolarization of the parent cell by the cut end of the axon.
3. Loss of action potential in the axon.
4. Proliferation of neurofilaments.
5. Proliferation of Schwann cells.
6. Breakdown of a blood–nerve barrier.
7. Increased loss of fast moving particles from the cell body.
8. Increased loss of axoplasm and mitochondria.
9. Loss of a hypothetical trophic substance coming from the periphery.
10. Loss of a repressor substance normally produced by the cell body and transported down the axon which represses that same neuron's RNA production.

An additional possibility would be a genetically programmed time course of chromatolysis.[77]

Although possibilities 1, 2, and 3 can be excluded with certainty, there is still only circumstantial evidence in favor of 9 and 10. The involvement of the retrograde axoplasmic transport mechanism in the signal triggering chromatolysis seems undoubtable. Such a retrograde movement of intraaxonal material was first seen 60 years ago[79] and has been rediscovered since then several times. Prestige[80] called the retrogradely transported material responsible for maturation and survival of neurons the "maintenance factor." According to Jacobson,[61] "the possibility must be entertained that materials are transferred from the peripheral tissues to the tips of the axons, and a message is then transmitted up the axon to the neuron soma, resulting in growth and maintenance of the neuron."

Thoenen and his group published convincing evidence that nerve growth factor is transported retrogradely in sensory neurons at a rate of 13 mm/h[81] and in sympathetic postganglionic neurons at a rate of 2.5 mm/h.[82] In the latter case, nerve growth factor induces synthesis of tyrosine hydroxylase and dopamine-β-hydroxylase, the enzymes that catalyze the rate-limiting steps in the synthesis of norepinephrine. However, it may not only be nerve growth factor that is involved in the signaling system, since, e.g., in spinal motoneurons NGF fails to be transported retrogradely. Therefore, other substances transported retrogradely have to be taken into consideration; these include various rapidly migrating regulatory proteins[83] and glycoproteins that contain exposed *N*-acetyl-D-glucosamine.[84]

The role of axoplasmic transport in communicating neurotrophic information has been conclusively proved both in sensory and motor neurons. Inhibition of fast retrograde axonal transport by colchicine mimics many of the neuronal responses observed after axotomy including chromatolysis[85,86] as well as sprouting of neighboring uninjured nerves.[87] Recent studies suggest that the chemical composition of retrogradely transported material after axotomy differs from that under normal conditions. It seems that after axotomy the same kinds of rapidly transported proteins that were transported orthogradely return to the cell body without the transformation that would normally occur in the terminal.[88] In other words, retrograde axonal transport is a positive information feedback that "continuously informs the cell body about the status of the axon and elicits the appropriate metabolic response."[76]

Another, more sophisticated theory (Diamond *et al.*[89]) postulates a negative feedback system in which agents that produce axonal sprouting are released from the target tissue and transported by retrograde flow to the cell body. These agents are postulated to induce synthesis in the perikaryon of antisprouting agents which would, in turn, be transported orthogradely. After axotomy, sprouting agents would no longer be transported to the cell body; in consequence, no antisprouting agent would be synthesized. In fact, after axotomy, the chemical composition of orthogradely transported proteins undergoes profound alterations: a new type with a molecular weight of 42,000 appears at 5–15 days after axotomy, while simultaneously, another protein with a molecular weight of 19,000 disappears from the complement of orthogradely transported proteins.[90] This latter would fit the properties of the hypothetical antisprouting agent.

A novel aspect of the intraneuronal signaling system is the transganglionic degenerative atrophy of central terminals of bipolar primary sensory neurons that ensues after lesion of the peripheral axonal branch.[91] Transganglionic degenerative atrophy results in a marked decrease of fluoride-resistant acid phosphatase activity of central terminals of primary nociceptive neurons which are labeled by this enzyme under normal conditions.[92] In accord with the above theories regarding the role of retrograde transport in the signaling system, local (perineural) application of microtubule inhibitors (colchicine, vinblastine, vincristine, leurosine) around the peripheral axonal branch induces, without Wallerian degeneration of the peripheral axon, degenerative atrophic alterations of the central terminal that are similar to those observed after peripheral ax-

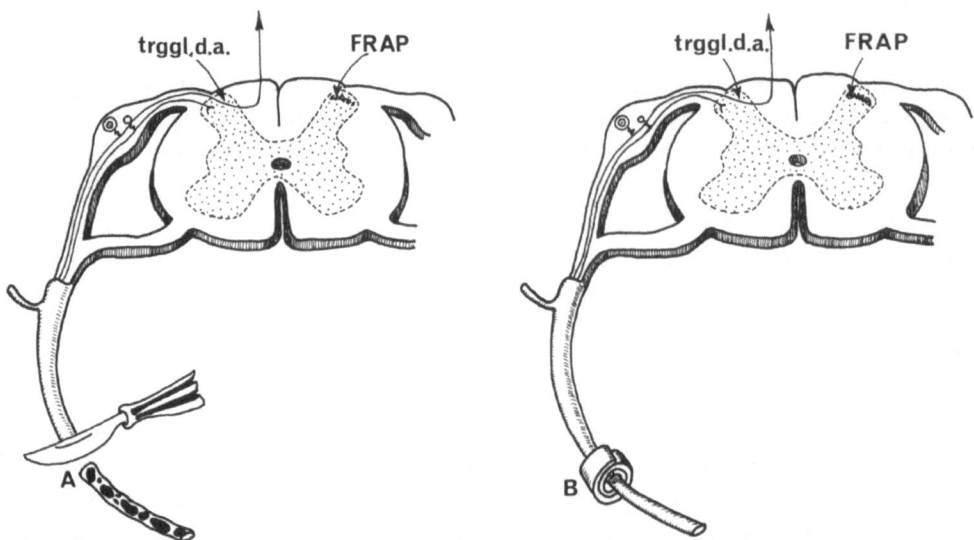

Fig. 2. Transganglionic degenerative atrophy (trggl.d.a.) of central terminals of primary sensory neurons after transection (A) and local microtubule inhibitor treatment (B) of the peripheral axonal branch. Note that the peripheral stump undergoes Wallerian degeneration after transection (A) whereas, after short-term application of a cuff soaked in vinblastine (or other vinca alcaloids), the peripheral branch remains structurally intact (B). In spite of this, central terminals in the corresponding area in the spinal cord undergo trggl.d.a. in both cases, accompanied by a simultaneous disappearance of the marker enzyme FRAP (fluoride-resistant acid phosphatase) from the Rolando substance.

otomy[93] (Fig. 2). Transganglionic degenerative atrophy of central terminals of primary sensory neurons obviously results from the lack of flow of an infrastructural signal from the periphery; it seems that in this case, the lacking "maintenance factor" is nerve growth factor itself.

7. THE NEURON: A SYNAPTOCHEMICAL ENTITY OF PROTEIN METABOLISM

The powerful protein-synthesizing activity of the neuron is kept in balance with a similarly intense proteolytic activity.[94] The site of protein synthesis is the perikaryon (and the dendrites); also, protein breakdown is thought to be restricted mainly to the perikaryon which contains large numbers of lysosomes. Yet, striking differences between chemical compositions of orthogradely and retrogradely transported axoplasmic material[76] suggest that additional protein breakdown may take place *en route*. The main site of this additional protein turnover seems to be the axon terminal; the intensity of this axoterminal proteolysis appears to be function dependent as follows from histochemical experiments (Fig. 3).

Axon terminals are, as a rule, devoid of lysosomes; thus, the proteolytic activity in nerve endings has to be caused by an extralysosomal enzyme. In this respect, it should be recalled that more than 15 years ago, Guroff described a Ca^{2+}-activated neutral protease in the soluble fraction from rat brain; this enzyme was inactivated by sulfhydryl inhibitors.[95] Calcium ions were shown to change the viscosity of the axoplasm[96]; this effect is accompanied by degradation of axoplasmic proteins which is blocked by 1 mM *para*-chlormercuribenzoate.[97] A divalent-cation-activated nucleoside triphosphatase (substrate: ATP) is involved in the transport of divalent cations across neuronal cell membranes.[98] A divalent-cation-dependent neutral protease active at physiological pH and inhibited by sulfhydryl-blocking agents was found in nerve fibers.[98] Finally, a Ca^{2+}-activated protease is thought to be responsible for the disassembly of the axonal cytoskeleton in nerve terminals.[24]

Thus, it seems that Ca^{2+} ions, essential both in the release of neurochemical transmitter substances and in excitation–performance coupling,[100] also play an important role in the axoterminal protein metabolism of the neuron. A Ca^{2+}-binding protein with a molecular weight of 15,000 was found in the brain,[101] in the axoplasm of giant nerve fibers,[102] and in the electric organ[103] which is, in fact, a modified neuromuscular synapse. The calcium-binding protein is transported down the axons in mammalian nerves at a daily rate of 410 mm.[104]

The dynamic condition of synaptic proteins in axon terminals has been reviewed by Droz.[105] Figure 4, redrawn after his paper, summarizes the main neurochemical events that possibly occur presynaptically. According to Marks

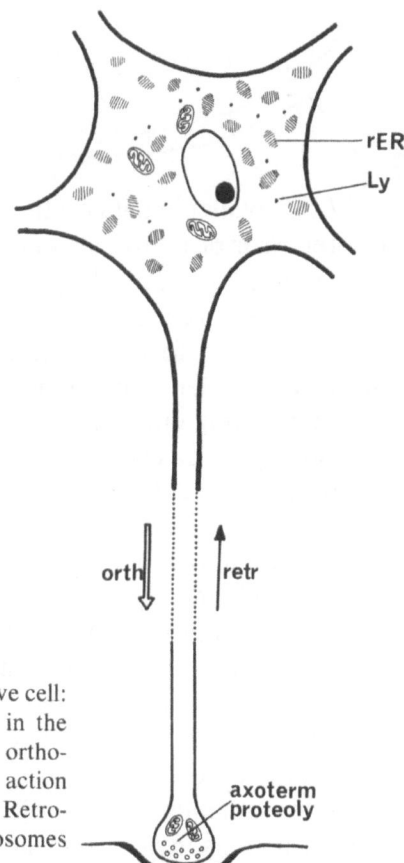

Fig. 3. Highly simplified scheme of protein metabolism in a nerve cell: synaptochemical entity of the neuron. Proteins synthesized in the perikaryal rough endoplasmic reticulum (rER) are transported orthogradely to the axon terminal where they are subjected to the action of a Ca^{2+}-linked, function-dependent axoterminal proteolysis. Retrogradely transported proteins are degraded by perikaryal lysosomes (Ly).

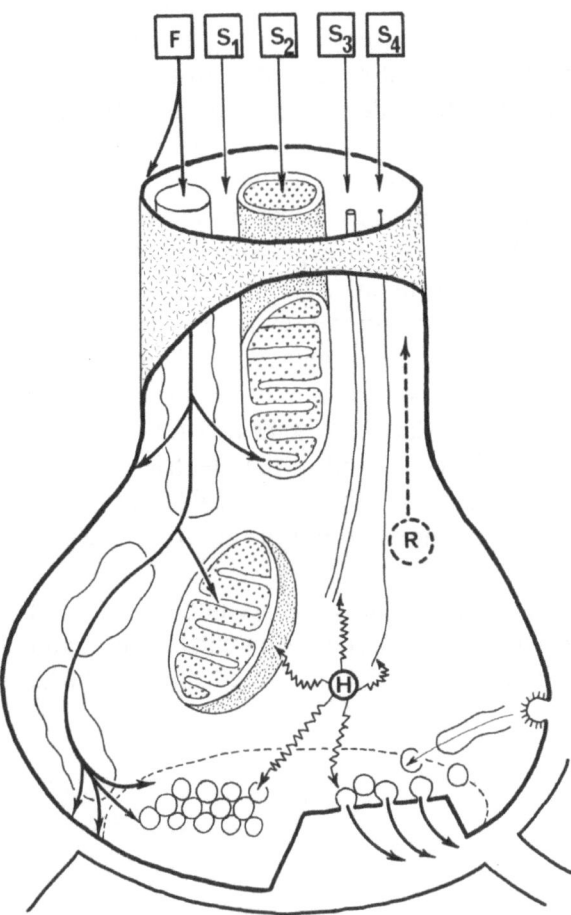

Fig. 4. Dynamic condition of synaptic proteins in an axon terminal. F, fast orthograde transport of axolemma and axoplasmic reticulum; S, slow orthograde transport (S_1, axoplasm; S_2, mitochondria; S_3, microtubules; S_4, neurofilaments); R, retrograde transport; H, hydrolytic enzymes (proteinases and peptide hydrolases). Redrawn with slight modifications after Droz.[101]

et al.,[106] available data on the formation of neuropeptides favor the breakdown of precursor proteins as a major pathway rather than *de novo* synthesis of such transmitter peptides. The formation of peptides is mediated by proteolytic enzymes and related hydrolases. For hormone peptides, multiple macromolecular protein forms are known, and in many cases a common mechanism activates smaller active peptides. This common mechanism involves the action of a converting enzyme that cleaves bonds adjacent to a paired basic residue. According to Lasek and Black,[107] proteins of the axonal cytoskeleton, i.e., neurofilament and microtubule proteins, which are transported continuously to the axon terminal are turned over as soon as they arrive there. The regulatory factor that activates proteases in the terminal seems to be Ca^{2+}.

Histochemical experiments prove that ionic Ca^{2+} is released in axon terminals after supramaximal stimulation.[108,109] At the same time, neutral protease activity of the terminal, which cannot be demonstrated histochemically under resting conditions, becomes highly active[110]; increased protease activity of axon terminals persists for about 12 h after a train of supramaximal impulses.[111]

Function-dependent, Ca^{2+}-linked axoterminal protease activity seems to be a fundamental feature of neurons from at least two viewpoints. First, it stands to reason that the modification of retrogradely transported proteins, brought about to different extents by axoterminal proteolysis, may inform the cell body about the extent of neurochemical transmission processes that were previously ongoing in the axon terminal, thus providing a material basis for neurocellular memory.[112] Second, it may serve as a basis for a more general interpretation of neurochemical transmission. Breakdown products of the axoterminal proteolysis, like excitatory and inhibitory amino acids and various peptide transmitters, can be regarded as the very original, ancient types of intercellular communication in the nervous system. In this respect, cholinergic transmission seems to be a relatively novel achievement of phylogenesis. Acetylcholine, known to decrease the surface tension of membranes,[113] seems to have originally played the role of an adjuvant promoting the efflux of protein breakdown products from the terminal; in the course of nervous system evolution, it may have become preponderant in a variety of nerve cells. In conclusion, Dale's principle, i.e., that each neuron makes and utilizes a single neurotransmitter, seems to be a marginal case of a more general rule.

REFERENCES

1. Varon, S., and Somjen, G. G., 1979, Neuron–glia interactions, *Neurosci. Res. Program Bull.* **17:**3–239.
2. Stöhr, P., Jr., 1957, *Mikroskopische Anatomie des Vegetativen Nervensystems, Möllendorff's Handbuch der Mikroskopischen Anatomie des Menschen,* Volume IV, Part 5, Springer, Berlin.
3. Novikoff, P. M., Novikoff, A. B., Quintana, N., and Hauw, J. J., 1971, *J. Cell Biol.* **50:**859–886.
4. Csillik, B., 1975, *Int. Rev. Neurobiol.* **18:**69–140.
5. Heiwall, P.-O., Larsson, P.-A., and Dahlstöm, A., 1978, *Acta Physiol. Scand.* **104:**156–166.
6. Ng, M., and Matus, A., 1979, *Neuroscience* **4:**169–180.
7. Takahashi, K., and Wood, R. L., 1970, *Z. Zellforsch.* **110:**311–320.
8. Westrum, L. E., Jones, D. H., Gray, E. G., and Barron, J., 1980, *Cell Tissue Res.* **208:**171–181.
9. Gray, E. G., 1978, *Proc. R. Soc. (Lond.) [Biol.]* **203:**219–227.
10. Csillik, B., and Knyihár-Csillik, E., 1980, *Acta Biol. Acad. Sci. Hung.* **31:**49–56.
11. Herman, C. J., and Lapham, L. W., 1968, *Science* **160:**537.
12. Moore, B. W., 1972, *Int. Rev. Neurobiol.* **15:**215–225.
13. Grasso, A., and Pirazzi, R., 1975, *Brain Res.* **90:**324–328.
14. Bock, E., 1978, *J. Neurochem.* **30:**7–14.
15. Mikoshiba, K., Mallet, J., and Changeux, J. P., 1977, *Proc. Int. Soc. Neurochem.* **6:**283.
16. Iqbal, K., Grundke-Iqbal, I., Terry, R. D., and Wisniewski, H. M., 1977, *Mechanism, Regulation and Special Function of Protein Synthesis in the Brain* (S. Roberts, A. Lajtha, and W. H. Gispen, eds.), Elsevier/North Holland Biomedical Press, Amsterdam, pp. 171–179.
17. Hemminki, K., 1973, *Biochim. Biophys. Acta* **310:**285–288.
18. Forgue, S. T., and Dahl, J. L., 1978, *J. Neurochem.* **31:**1289–1297.
19. Bray, D., 1977, *Biochimie* **59:**1–6.
20. Unsicker, K., Drenckhahn, D., Gröschel-Stewart, U., Schumacher, U., and Griesser, G. H., 1978, *Neuroscience* **3:**301–306.
21. Isenberg, G., Schubert, P., and Kreutzberg, G. W., 1980, *Brain Res.* **194:**588–593.
22. Shelanski, M. L., Liem, R., and Yen, S.-H., *Mechanisms, Regulation and Special Function of Protein Synthesis in the Brain* (S. Roberts, A. Lajtha, and W. H. Gispen, eds.), Elsevier/North Holland Biomedical Press, Amsterdam, pp. 137–152.
23. Davison, P. F., and Winslow, B., 1974, *J. Neurobiol.* **5:**119–133.
24. Lasek, R. J., and Hoffmann, P., 1976, *Cell Motility,* Volume C (R. Goldman, T. Pollard, and J. Rosenbaum, eds.), Cold Spring Harbor Laboratory, New York, pp. 1021–1050.

25. Chee, P. Y., and Dahl, J. L., 1978, *J. Neurochem.* **30**:1485–1493.
26. Csillik, B., 1980, *Neurotransmitters. Comparative Aspects* (J. Salánki and T. M. Turpaev, eds.), Akadémiai Kiadó, Budapest, pp. 149–189.
27. Hosein, E. A., Proulx, P., and Ara, R., 1962, *Biochem. J.* **83**:341–346.
28. Collier, B., Lovat, S., Ilson, D., Barker, L. A., and Mittag, T. W., 1977, *J. Neurochem.* **28**:331–339.
29. Saavedra, J. M., Coyle, J. T., and Axelrod, J., 1974, *J. Neurochem.* **23**:511–515.
30. Curtis, D. R., and Johnston, G. A. R., 1974, *Ergeb. Physiol.* **29**:97–188.
31. Baskin, S. I., Leibman, A. J., and Cohn, E. M., 1976, *Adv. Biochem. Psychopharmacol.* **15**:153–164.
32. Schultzberger, M., Hökfelt, T., Terenius, L., Elfvin, L.-G., Lundberg, J. M., Brandt, J., Elde, R. P., and Goldstein, M., 1979, *Neuroscience* **4**:249–270.
33. Hökfelt, T., Elde, R., Johansson, O., Luft, R., Nilsson, G., and Arimura, A., 1976, *Neuroscience* **1**:131–136.
34. Uhl, G. R., and Snyder, S. H., 1977, *Life Sci.* **19**:1827–1832.
35. Swanson, L. W., and McKeller, S., 1979, *J. Comp. Neurol.* **188**:87–106.
36. Fuxe, K., Hökfelt, T., Said, S. I., and Mutt, V., 1977, *Neurosci. Lett.* **5**:241–246.
37. Csillik, B., and Knyihár, E., 1968, *Acta Biochim. Biophys. Acad. Sci. Hung.* **3**:165–170.
38. Kreutzberg, G. W., Tóth, L., and Kaiya, H., 1975, *Adv. Neurol.* **12**:269–281.
39. Fernandez, H. L., Duell, M. J., and Festoff, B. W., 1979, *J. Neurobiol.* **10**(5):441–454.
40. Csillik, B., 1965, *Functional Structure of the Post-Synaptic Membrane in the Myoneural Junction*, Academy Publishing House, Budapest.
41. Csillik, B., and Tóth, L., 1972, *J. Histochem. Cytochem.* **20**:385–387.
42. Heiwall, P.-O., Larsson, P. A., and Dahlström, A., 1978, *Acta Physiol. Scand.* **104**:156–166.
43. Joh, T. H., and Reis, D. J., 1975, *Brain Res.* **85**:146–151.
44. Falck, B., Hillarp, N.-A., Thieme, G., and Torp, A., 1962, *J. Histochem. Cytochem.* **10**:348–354.
45. Bloom, F. E., and Battenberg, E. L. T., 1976, *J. Histochem. Cytochem.* **24**:561–571.
46. Hökfelt, T., Fuxe, K., and Goldstein, M., 1973, *Brain Res.* **62**:461–469.
47. Goldstein, M., Anagnoste, B., Freedman, L. S., Roffman, M., Ebstein, R. P., Park, D. H., Fuxe, K., and Hökfelt, T., 1973, *Frontiers in Catecholamine Research* (E. Usdin and S. Snyder, eds.), Pergamon Press, New York, pp. 69–78.
48. Knyihár, E., 1971, *Experientia* **27**:1205–1207.
49. Collingridge, G. L., and Keen, P., 1978, *J. Neurochem.* **31**:681–684.
50. Burnstock, G., 1976, *Neuroscience* **1**:239–248.
51. Dismukes, R. K., 1979, *Behav. Brain* **2**:409–418.
52. Daly, J. W., Hoffer, B. J., and Dismukes, R. K., 1980, *Neurosci. Res. Program Bull.* **18**(3):325–456.
53. Rakic, P., 1976, *Local Circuit Neurons*, MIT Press, Cambridge, Massachusetts.
54. Ralston, H. J. III., 1971, *Nature* **230**:585–587.
55. Harding, B. N., and Powell, T. P. S., 1977, *Phil. Trans. R. Soc. Lond [Biol.]*, **279**:357–412.
56. Jahr, C. E., and Nicoll, R. A., 1980, *Science* **207**:1473–1475.
57. Sidman, R. L., Miale, I. L., and Feder, N., 1959, *Exp. Neurol.* **1**:322–333.
58. Santerre, R., and Rich, A., 1976, *Dev. Biol.* **54**:1–12.
59. Balázs, R., 1974, *Br. Med. Bull.* **30**:126–134.
60. Rebière, A., and Legrand, J., 1972, *C. R. Acad. Sci [D] (Paris)* **274**:3581–3584.
61. Jacobson, M., 1978, *Developmental Neurobiology*, 2nd ed. Plenum Press, New York.
62. Pitot, H. C., and Yatvin, M. D., 1973, *Physiol. Rev.* **53**:228–325.
63. Weichsel, M. E., Jr., 1974, *Brain Res.* **78**:455–465.
64. Stumpf, W. E., 1971, *J. Neuro-Visceral Relations [Suppl.]* **10**:51–64.
65. Prestige, M. C., 1967, *J. Embryol. Exp. Morphol.* **17**:453–471.
66. Glücksmann, A., 1951, *Biol. Rev.* **26**:59–86.
67. Cowan, W. M., 1973, *Development and Aging in the Nervous System* (M. Rockstein and M. L. Sussman, eds.), Academic Press, New York, London, pp. 19–41.
68. Prestige, M. C., 1965, *J. Embryol. Exp. Morphol.* **13**:63–72.
69. O'Connor, T. M., and Wyttenbach, C. R., 1974, *J. Cell Biol.* **60**:448–459.
70. Saunders, J. W., 1966, *Science* **154**:604–612.

71. Webster, D. A., and Gross, J.. 1970, *Dev. Biol.* 22:157–184.
72. Bondareff, W., and Liu-Liu, S., 1977, *Am. J. Anat.* 148:57–64.
73. Wall, P. D., 1977, *Phil. Trans. R. Soc. Lond.* [*Biol.*] 278:361–372.
74. Plum, F., Gjedde, A., and Samson, F. E., 1976, *Neurosci. Res. Program Bull.* 14(4):457–518.
75. Lieberman, A. R., 1974. *Essays on the Nervous System* (R. Bellairs and E. G. Gray, eds.), Calderon Press, Oxford, pp. 71–105.
76. Bulger, V. T., and Bisby, M. A., 1978, *J. Neurochem.* 31:1411–1418.
77. Grafstein, B., 1975, *Exp. Neurol.* 48:32–51.
78. Cragg, B. G., 1970, *Brain Res.* 23:1–21.
79. Matsumoto, T., 1920, *Bull. Johns Hopkins Hosp.* 30:91–93.
80. Prestige, M. C., 1970. *The Neurosciences. Second Study Program* (F. O. Schmitt, ed.), Rockefeller University Press, New York, pp. 73–82.
81. Stoeckel, K., Schwab. M., and Thoenen, H., 1975, *Brain Res.* 89:1–14.
82. Paravicini, V., Stoeckel, K., and Thoenen, H., 1975, *Brain Res.* 84:279–291.
83. McClure, W. O., 1972. *Adv. Pharmacol. Chemother.* 10:185–220.
84. Karlsson, J.-O., 1979. *J. Neurochem.* 32:491–494.
85. Albuquerque, E. S., Warmick, J. E., Tasse, J. R., and Sansone, F. M., 1972, *Exp. Neurol.* 37:607–634.
86. Pilar, G., and Landmesser. L., 1972, *Science* 177:1116–1118.
87. Aguilar, C. E., Bisby, M. A., Cooper, E., and Diamond, J., 1973, *J. Physiol. (Lond.)* 234:449–464.
88. Bisby, M. A., and Bulger, V. T., 1977, *J. Neurochem.* 29:313–320.
89. Diamond, J., Cooper, E., Turner, C., and McIntyre, L., 1976, *Science* 193:371–377.
90. Theiler, R. F., and McClure, W. O., 1978, *J. Neurochem.* 31:433–447.
91. Knyihár, E., and Csillik, B., 1976, *Exp. Brain Res.* 26:73–87.
92. Csillik, B., and Knyihár, E., 1976, *Neuron Concept Today* (J. Szentágothai, J. Hámori and E. S. Vizi, eds.) Akadémiai Kiadó, Budapest, pp. 27–38.
93. Csillik, B., Knyihár, E., Jójárt, I., Elshiekh, A. A., and Pór, I., 1978. *Res. Commun. Chem. Pathol. Pharmacol.* 21:467–484.
94. Marks, N., and Lajtha, A., 1971, *Handbook of Neurochemistry*, Volume 5A (A. Lajtha, ed.), Plenum Press, New York, pp. 49–139.
95. Guroff, G., 1964, *J. Biol. Chem.* 239:149–155.
96. Hodgkin, A. L., and Katz, B., 1949, *J. Exp. Biol.* 26:292–294.
97. Gilbert, D. S., Newby, B. J., and Anderton, B. H., 1975, *Nature* 256:586–589.
98. Stefanovic, V., Ledig, M., and Mandel, P., 1976, *J. Neurochem.* 27:799–805.
99. Pant, H. C., Terekawa, S., and Gainer, H., 1979, *J. Neurochem.* 32:99–102.
100. Llinás, R. R., and Heuser, J. E., 1977, Depolarization–release coupling systems in neurons. *Neurosci. Res. Program Bull.* 15(4):557–687.
101. Wolff, D. J., and Siegel, F. L., 1972. *J. Biol. Chem.* 247:4180–4185.
102. Alema, S., Calissano, P., Rusca, G., and Giuditta, A., 1973, *J. Neurochem.* 20:681–689.
103. Childers, S. R., and Siegel, F. L., 1975, *Biochim. Biophys. Acta* 405:99–108.
104. Iqbal, Z., and Ochs, S., 1978, *J. Neurochem.* 31:409–418.
105. Droz, B., 1973, *Brain Res.* 62:383–394.
106. Marks, N., Benuck, M., and Grynbaum, A., 1977, *Mechanism, Regulation and Specific Functions of Protein Synthesis in the Brain* (S. Roberts, A. Lajtha, and W. H. Gispen, eds.), Elsevier/North-Holland Biomedical Press, Amsterdam, pp. 355–372.
107. Lasek, R. J., and Black. M. M., 1977, *Mechanism, Regulation and Specific Functions of Protein Synthesis in the Brain* (S. Roberts, A. Lajtha, and W. H. Gispen, eds.), Elsevier/North-Holland Biomedical Press, pp. 161–169.
108. Csillik, B., and Sávay, G., 1963, *Nature*, 198:399.
109. Palánkai, G., Sávay, G., and Poberai, M., 1978, *Acta Histochem. (Jena)* 62:170–175.
110. Poberai, M., Sávay, G., and Csillik, B., 1972, *Neurobiology* 2:1–7.
111. Poberai, M., and Sávay, G., 1976, *Acta Histochem. (Jena)* 57:44–48.
112. Csillik, B., 1974, *J. Neural Trans.* [*Suppl.*] 11:13–42.
113. Jain, M. F., 1972, *The Bimolecular Lipid Membrane*. Van Nostrand-Reinhold, Princeton, New Jersey.

13

Astrocytes

L. Hertz

1. INTRODUCTION

During the decade since the first appearance of this *Handbook,* few, if any, major areas of neurobiology have developed at a faster rate than our knowledge of astrocytic biochemistry and function. This development is undoubtedly related to the methodologies that became available during the 1960s and early 1970s, i.e., microdissection of samples consisting primarily of glial cells,[1] bulk separation by gradient centrifugation of a fraction highly enriched in astroglia,[2-6] and culturing of glial cells either in cell lines, e.g., the C-6 glioma cell line,[7] or in primary cultures of astrocytes.[8,9] Each of these methodologies has its own sources of error and uncertainties.[6,10-13] These include cross contamination between isolated cell fractions, dedifferentiation of transformed cell lines, and the possibility of a deficient maturation in primary cultures that are prepared from biochemically immature tissue. These factors may not only delete or reduce astrocytic characteristics from the astrocytic preparations but may also add nonastrocytic qualities. In this connection, it should be kept in mind that neurons, astrocytes, and oligodendrocytes develop from common precursor cells (subventricular cells), meaning that cells that remain at, or revert to, immature stages may display phenomena that, in the adult brain, are only displayed by the other cell types. Because of these potential sources of error, great caution should be exerted before any particular characteristic is accepted as astrocytic, although features that have been established using different preparations of astrocytes are likely to represent true characteristics of astrocytes.

Studies of astrocytic biochemistry and function are facilitated by the fact that, in a few cases, studies on whole brain have allowed the identification of compounds or properties that seem to be either absent from or confined to astrocytes and may thus serve as "negative" or "positive" markers for this cell type. The latter group includes glial fibrillary acidic (GFA) protein[14,15] and glutamine synthetase.[16] The detailed function of GFA remains unknown, which

---

L. Hertz • Department of Pharmacology, University of Saskatchewan, Saskatoon, Saskatchewan S7N 0W0, Canada.

does not detract from the value of this compound as a macromolecular astrocytic marker,[17] and the presence of the GFA protein in retinal Müller cells and in the C-6 glioma cell line[15] indicates the astrocytic character of these cells. The function of the glutamine synthetase obviously is to catalyze the formation of glutamine from ammonia and glutamate, a process which in the normal brain seems to account quantitatively for the incorporation of ammonia from blood into the brain.[18] The glutamine synthetase activity is of value not only as a qualitative or quantitative[9,13] marker but also by indicating that glutamine synthesis exclusively or predominantly occurs in astrocytes, a finding that is of major importance in connection with the well-established compartmentation of metabolism in brain.[19–21]

For a genuine understanding of astrocytic biochemistry and function, it is essential to possess quantitative data on whether the characteristics are expressed to an extent that is physiologically meaningful, how they compare to those in whole brain, and whether estimates can be made of the relative importance of neuronal, oligodendrocytic, microglial, vascular, and astrocytic events. In general, it seems possible to obtain such quantitative information with greater reliability from studies of primary cultures of astrocytes or from bulk-separated astroglia than from studies of cell lines (e.g., refs. 22, 23). For an interpretation of the extent to which specific processes occur in astrocytes, it is a prerequisite to know the astrocytic contribution to total cell number and to total volume of the brain cortex. This varies among species, but in large animals (brain weight > 100 g), glial cells outnumber neurons,[24,25] and the contribution of astrocytes to the total cellular volume of the brain cortex in man is about one-third.[25] Since astrocytes have an extremely large surface–volume ratio,[26] the astrocytic fraction of the total surface area is even larger, and in the normal rat neocortex, protoplasmic astrocytes form about 30% of the total inner surface of the neuropil.[27] The distribution of astrocytes is not uniform at different areas, but the astrocytes seem to be concentrated at strategic positions, such as around synapses, at the periphery of capillaries (astrocytic endfeet), and beneath the pial surface.[28] These locations, in connection with the large surface–volume ratio, make astrocytes singularly well suited to regulate concentrations of neuroactive compounds in extracellular spaces including synaptic clefts.

The detailed structural relationships between astrocytes and neurons at the electron microscopic level were recently discussed by S. L. Palay. According to this description,[29] astrocytes have a large number of relatively sparsely branched processes which penetrate the neuropil and/or terminate on other cells, on blood vessels, or at the subpial surface of the brain. In contrast to previous postulates of a relative paucity of mitochondria in astrocytes, the processes appear to be preferential locations for mitochondria. The thin astrocytic processes break up the neuropil into a mosaic of small regions or "parcels," each containing a synaptic field. A similar "parcel" is found surrounding clusters of synaptic terminals. Although neurons and oligodendrocytes may be directly apposed, astrocytic processes often intervene between the two other cell types. Membrane specializations such as gap junctions and assemblies are abundant between astrocytes,[29,30] and like other morphological features[27] they seem to undergo continuous, dynamic alterations. Electronic coupling, which has been described both in invertebrate glia[31] and in cultured rat astrocytes,[32,33] and metabolic cooperation between cells may to a large extent take place through the gap junctions,[30] and "the concept of ionic and metabolic communications via gap junctions in neural tissues deserves considerable attention even though factual information is so far limited."[29]

In this review, several aspects of astrocytic neurochemistry will be discussed. I shall not attempt to cover these areas comprehensively but shall give references to recent, more complete literature reviews by various investigators. Along the same lines, the evolution of ideas concerning glial function will not be dealt with, but "historic notes" are included in the recent review by Varon and Somjen[29] on neuron–glia interactions, and Glees[34] and Kuffler and Nicholls[31] have also described the early history of glial research. Emphasis will be placed on evidence that a major astrocytic function is to remove neuroactive compounds (e.g., certain transmitters, potassium ions) from the extracellular space and thus participate in the regulation of activity in the nervous system. The functional implications of this activity may be anywhere between a garbage-collecting role for astrocytes and a modulation and selective transmission of neuronal activity which could be of fundamental importance for "higher mental functions."[13,35,36] It is in agreement with the latter point of view that astrocytes possess receptors for certain transmitter compounds, i.e., react to signaling from specific neurons.[13,37–40]

Drugs with specific therapeutic action on schizophrenia (antipsychotic drugs), manic–depressive illness (antidepressant drugs), or anxiety states (antianxiety drugs) are also bound to specific binding sites on astrocytes and seem to interact with separate, specific transmitter systems.[13,37,40–42] The ways in which astrocytic function is affected by the transmitter signals in the absence or presence of drugs is largely unknown, but it seems likely that several aspects of astrocytic function are altered and that this may, in turn, modify neuronal–astrocytic interactions. Invaluable information about the role of astrocytes in brain function may, in all likelihood, be obtained from further knowledge in this area combined with other information on the mechanism of action for these groups of drugs and the symptomatology and pathophysiology of the diseases against which they are therapeutically effective. For this reason, information about acute or chronic effects of drugs on astrocytes will be dealt with in some detail.

2. ENERGY METABOLISM

2.1. Basal Conditions

2.1.1. Oxygen Consumption

In view of the resurgence of interest in energy metabolism of the brain during different functional states, it seems appropriate initially to discuss energy metabolism of astrocytes. In contrast to previous concepts (but in agreement with the recent realization of the high mitochondrial density in astrocytic processes described in the previous section), the metabolic activity of astrocytes is now known to be quite high, and it is probably roughly comparable to that of neurons. Thus, there is little doubt that the rate of oxygen uptake by rodent astrocytes amounts to 1.5–5 μmol/min per 100 mg protein.[43–47].* The lower of these values is similar to the average oxygen consumption in human brain, and the higher is at least as high as the average rate of oxygen uptake in the rat brain (for further details and references see review by Hertz[48]). The respiratory rate in cultured astrocytes is independent of the oxygen concentration down to near anoxic levels,[46] which might explain the fact that astrocytes seem to react to graded hypoxic insults with an augmentation of energy metabolism as evidenced by an increase in mitochondrial density,[49] a reaction that may be of relevance for the pathology and symptomatology of ischemic–hypoxic brain damage.[50]

2.1.2. Glucose Utilization

Further perfection of the ingenious 2-deoxyglucose method developed by Sokoloff and his co-workers[51] is eagerly awaited to allow determination of glucose utilization *in vivo* by the various cell types of the brain. Such work, focusing on replacement of ^{14}C with the less energy-rich ^3H isotope and improvements of fixation techniques, is well under way (M. H. Des Rosiers and L. Descarries, personal communication), and some previously reviewed evidence[48] suggests that the rates of glucose utilization are roughly identical in nerve cell bodies and in neuropil. This is consistent with the recent observation of Kao-Jen and Wilson[52] that the hexokinase activity in normal brain is as high in astrocytic as in neuronal cell bodies. An additional, potentially extremely important finding by these authors is that the cell processes of both astrocytes and neurons have very low, if any, hexokinase activity. For this reason, it was suggested that they might utilize other substrates than glucose, e.g., glutamate.[52] If glutamate is provided by metabolic processes in other parts of the cells, this will not affect the *R.Q.* of brain tissue (1.00) which generally is taken as evidence that exclusively or mainly glucose is metabolized in brain. Further evidence that astrocytes may metabolize glutamate is discussed in Section 3.2.2.

The hexokinase is one of the enzymes (e.g., hexokinase, glucose-6-phosphate dehydrogenase, succinate dehydrogenase, and glutamate dehydrogenase) that Roth-Schechter and Mandel[53–55] have found to increase in activity in pentobarbital-dependent astrocytes which also show an augmentation in size and number of mitochondria and a dilation of the rough endoplasmic reticulum. Further information about activities of glucose-metabolizing enzymes or metabolite levels in microdissected glial cells, bulk-separated astrocytic fractions, or cultured astrocytes can be obtained from reviews by Hertz and Schousboe,[56] Passonneau *et al.*,[57] Nagata and Tsukada,[4] or Pevzner.[58]

---

* In this review most values will be expressed per 100 mg protein, which roughly equals 1 g wet weight.

2.1.3. Carbon Dioxide Production

The end products of aerobic metabolism are carbon dioxide and water. Carbon dioxide readily passes cell membranes and can leave the cells as such, but only as long as the CO_2 tension within the cells is higher than at their outside. Carbonic anhydrase, which converts CO_2 to carbonic acid, seems to be highly enriched in astroglia,[59,60] although other authors claim an exclusively oligodendrocytic localization.[61] At physiological pH, most of the carbonic acid will dissociate to hydrogen ions and bicarbonate. The latter is probably, as in erythrocytes,[62] removed in exchange with chloride,[50,63] whereas the hydrogen ions are exchanged with other cations, i.e., conceivably either sodium or potassium. These transport processes are discussed in Section 5. Reasonably high activities of carbonic anhydrase have been observed in primary cultures of astrocytes.[64,65] The enzyme is absent in C-6 glioma cells[66] but present in retinal Müller cells.[67]

2.2. Effects of Ionic Deviations

2.2.1. Potassium

2.2.1a. Oxygen Consumption. The effect of an increase in the extracellular potassium concentration on oxygen consumption by astrocytes may be of special interest because of the fluctuations of extracellular potassium that occur in the central nervous system *in vivo* during normal and abnormal conditions (Section 5.1.1). A transient increase of 50% or more in respiration of microdissected glial cells, bulk-prepared astroglia, or cultured astrocytes during exposure to excess potassium (Table I) has been observed by several different investigators.[43,45,71–73] However, the cultured cells must be at least 3–4 weeks old (E. Hertz and L. Hertz, unpublished experiments) which probably reflects the absence of a potassium-induced stimulation of oxygen uptake by brain slices from neonatal rodents.[74] An obvious question is the correlation between this effect of excess potassium on oxygen consumption in astrocytes and on their function and metabolism in the central nervous system *in vivo*. Fluorometric monitoring *in vivo* of the concentrations of NADH and other coenzymes

Table I

Oxygen Uptake Rates, Na^+,K^+-ATPase Activities and Rates of Net Potassium Uptake into Astrocytes during Exposure to Normal and Elevated (50 mM) Extracellular Potassium Concentrations[a]

	Oxygen uptake	Na^+,K^+-ATPase activity	Potassium uptake
Normal potassium	5.4	9.7	25–30
Elevated potassium	9.2	21.7	100–125

[a] Rates for oxygen uptake[68] and potassium uptake[69] were measured in primary cultures of astrocytes; those for Na^+,K^+-ATPase activity[70] in bulk-separated astroglia. A comparable activity (up to 12 µmol/min per 100 mg protein during exposure to a normal potassium concentration) is, however, found in astrocytes in primary cultures (P. H. Wu and L. Hertz, unpublished experiments). All values expressed as µmol/min per 100 mg protein.

of the respiratory chain have shown that the elevated extracellular potassium concentration that accompanies increased functional activity is accompanied by an increase in metabolism. During normal excitation there is a very good correlation between the external potassium concentration (rising to a "ceiling" of 12 mM) and the metabolic intensity,[75,76] suggesting that the increased potassium concentration per se might trigger an energy-requiring process; but during certain pathological conditions, e.g., seizures,[76-78] the metabolic activity becomes greater than could be expected from the normal correlation. Whether two different types of metabolic responses to extracellular potassium (or related stimuli) exist in astrocytes or whether other cell types or cell constituents are also involved is unknown. Although most authors have not observed any potassium-induced stimulation of oxygen consumption in neurons (for a review, see refs. 48, 56), such an effect was reported by Hultborn and Hydén,[79] and synaptosomal preparations also show an increased rate of oxygen uptake when exposed to elevated concentrations of potassium.[80] The significance of the latter finding may be reduced by the fact that synaptosomal fractions are known to be contaminated to a certain extent with astrocytes.[81]

The potassium-induced stimulation of oxygen uptake in astrocytes is abolished in the presence of barbiturates[68,82] which, under the conditions used, had no effect on the oxygen uptake of astrocytes incubated in a medium with a normal concentration of potassium. This effect may be a major reason that the rate of oxygen uptake by whole brain is diminished by almost one-half under anesthesia[83] and that its metabolic activity, measured by monitoring of alterations in the concentrations of respiratory coenzymes, is affected by barbiturates (for references, see ref. 84). Neuronal respiration is also reduced by barbiturates, but the effect on neurons is at least as large in the presence of a normal potassium concentration as in the presence of excess potassium.[68] The frequently reported observation that respiration by brain slices is more inhibited by barbiturates (and some other drugs) in media with high potassium concentrations than in media with a normal content of potassium[84] therefore suggests a large astrocytic contribution to the total oxygen consumption by the slices.

2.2.1b. Glucose Utilization. It is consistent with the enhancement of oxidative metabolism by excess potassium that elevated potassium concentrations cause an increase of glucose uptake in astrocytes in primary cultures and in isolated retinal Müller cells.[85,86] Excess potassium also evokes a substantial augmentation of the utilization of phosphoenolpyruvate in bulk-separated astroglia,[4] possibly reflecting the stimulation by excess potassium of several enzymes including the phosphoenolpyruvate kinase, which has been observed in brain slices,[87] or the Na^+,K^+-ATPase.

2.2.1c. ATP and Na^+,K^+-ATPase Activity. It has been suggested that the potassium-induced stimulation of oxygen uptake in brain slices should be a metabolic manifestation of an increased breakdown of ATP by Na^+,K^+-ATPases or related enzymes involved in active transport of sodium and potassium (for references, see ref. 56). It is in keeping with this concept that the ATP content is cultured astrocytes is substantially reduced by elevated potassium concentrations, whereas the ATP content in cultured neurons is unaffected.[88]

Biochemical measurements in preparations of isolated cells have repeatedly demonstrated that the activity of the Na^+,K^+-ATPase is higher in glial cells than in neurons,[12,70,89,90] although both cell types do show Na^+,K^+-ATPase activity. A considerable increase in activity during postnatal development has been demonstrated in bulk-separated astroglia[70,91,92] as well as in astrocytes in primary cultures.[93] These findings are in keeping with the demonstration that inhibition of Na^+,K^+-ATPase activity *in vivo* by intracranially injected or superfused ouabain leads to vacuolization of astrocytes as well as swelling of neuronal elements.[94,95] They are, however, at variance with histochemical observations in intact gray matter that have led to the conclusion that the Na^+,K^+-ATPase activity is considerably higher in neurons than in glial cells.[96] However, evidence has been found that this is caused by a largely specific inhibition of astrocytic ATPase activity under the experimental conditions used in the histochemical assay.[70,97]

The characteristics of the Na^+,K^+-ATPase in astrocytic, neuronal–perikaryal, and synaptosomal preparations have recently been studied in detail by Grisar[70,98,99] who has presented convincing evidence that the neuronal–perikaryal and synaptosomal enzymes are identical and are little, if at all, stimulated by excess (>5 mM) potassium, which, in contrast, decreases the affinity of the enzyme for ATP. However, the activity of the astrocytic enzyme is increased by elevated potassium concentrations (Table I), a response that is absent in cells from immature animals. This potassium effect on Na^+,K^+-ATPase activity is consistent with the potassium-induced stimulation of oxygen uptake although quantitatively not as marked as could have been expected on the basis of a P/O_2 ratio of 6 (Table I). Such a quantitative discrepancy might partly be explained on the basis of the different preparations used and the fact that the Na^+,K^+-ATPase is a hysteretic enzyme the initial velocity of which is higher than the steady-state values shown in the table.[70,98] It should, however, also be kept in mind that the potassium-induced stimulation of oxygen uptake may not only (or mainly) be a reflection of an increase in Na^+,K^+-ATPase activity but might also be the result of a stimulation of other energy-requiring processes in astrocytes (Section 2.2.1b) and that in the brain *in vivo* evidence was found for two different types of response to excess potassium.

An even more dramatic effect of excess potassium on the astrocytic Na^+,K^+-ATPase is to increase the affinity to ATP, an effect which is the opposite of that found for the neuronal enzyme.[70,99] The differences between the two enzymes are so profound that it was concluded that they must constitute two different molecular entities, a conclusion in agreement with other indications for two different Na^+,K^+-ATPases in brain,[100] one of which is located in astrocytes.[101] The functional implications[70] of these findings are (1) that the development of a specific type of enzyme in astrocytes which is greatly enhanced by excess potassium is a qualitative, not only quantitative, indication of the involvement of astrocytes in active removal of excess potassium from the extracellular space of the mammalian central nervous tissue, and (2) that it is especially under conditions when the extracellular potassium concentration is elevated above its normal level that the potassium uptake preferentially occurs into astrocytes.

It has been suggested that norepinephrine, a mainly inhibitory transmitter, owes its action to a hyperpolarization of neuronal cells brought about by a stimulation of Na^+,K^+-ATPase activity and, thus, increased active transport of sodium and potassium (e.g., ref. 102). This effect does not require that potassium be accumulated into the neurons but only that either the intracellular level of potassium in the neurons be increased or that the extracellular potassium concentration be reduced. It is an indication of a possible astrocytic involvement in the response to norepinephrine *in vivo* that the Na^+,K^+-ATPase activity in astrocytes in primary cultures is stimulated by norepinephrine (for details, see Section 4.4.3).

2.2.2. Sodium

Respiration by astrocytes in primary cultures is remarkably resistant to respiratory inhibition in media of increased osmolality, since no inhibition occurs even at 1600 mosmol/liter, which reduces oxygen uptake by at least some other cell types.[103] This is in agreement with the finding that mitochondria, or a subpopulation of mitochondria, from whole brain during maturation become resistant to respiratory inhibition in media of increased osmolality,[103,104] an ontogenetic alteration which appears simultaneously with astrocytic maturation. Absence of sodium leads, on the other hand, to a considerable reduction (50%) in the rate of oxygen uptake by microdissected glial cells or by astrocytes in primary cultures[43,46] and an even larger decrease (85%) in brain slices.[105] This effect might be related to the decreases in enzyme activities in cultured astrocytes that have been demonstrated histochemically after lowering of the sodium concentration of the medium[106] or, possibly, to a deficient uptake of glutamate into astrocytes in the absence of sodium (Section 3.2.1) which could impair astrocytic utilization of glutamate (Section 3.2.2).

3. UPTAKE AND METABOLISM OF AMINO ACIDS INCLUDING AMINO ACID TRANSMITTERS

3.1. γ-Aminobutyric Acid

3.1.1. Uptake

The concept that astrocytes accumulate and thus inactivate GABA represents one of the earliest hypotheses of glial cell function. It was originally developed on the basis of studies on metabolic compartmentation in whole brain that suggested that GABA was formed in one, presumably neuronal, metabolic compartment and at least partly degraded in a different compartment which could be astrocytic (e.g., ref. 107). The finding that GABA is accumulated into bulk-separated astroglia, glioma cell lines, and glial cells in peripheral ganglia yielded good support to this concept, but quantitatively, the uptake into the cell lines and the peripheral ganglia are unimpressive, and GABA is known also to be taken up into neuronal constituents (for references, see refs. 22, 23, 108, 110). This is illustrated in Table II which shows that synaptosomes, cerebellar glomeruli, and cultured neurons accumulate GABA by a high-affinity uptake with a V_{max} comparable to that in brain slices, whereas the uptake rates into glial cell lines and peripheral ganglia are much less intense. An uptake of GABA also occurs, however, into astrocytes in primary cultures. These cells show a higher uptake rate than any of the other glial preparations, although even this uptake is less intense than the rate of GABA uptake into synaptosomes, cerebellar glomeruli, or cultured neurons.[109] This suggests that GABA uptake in the cerebral cortex *in vivo* occurs into both neuronal and astrocytic constituents, a concept in agreement with histochemical observations that have shown that there is accumulation of radioactive label over nerve terminals, neuronal cell bodies, and glial cells after exposure of brain cortical tissue to [³H]-GABA. See ref. 22. Similar observations have been made in organotypic cultures from the CNS.[111] The intensity of the uptake, as estimated by V_{max} values, is several times higher into the neuronal cells and maybe especially into nerve endings.

Drugs seem to affect neuronal and astrocytic GABA uptake to various extents. Thus, pentobarbital has a more potent effect on the uptake in astrocytes,[82] whereas ketamine seems to inhibit neuronal uptake more than astrocytic uptake.[112] Several conformationally restricted GABA analogues are relatively specific in inhibiting astrocytic uptake of GABA.[113] The concept that β-alanine is a potent and specific inhibitor of GABA uptake into glial cells, which is mainly based on experiments with peripheral ganglia,[114] is not supported by observations in primary cultures of astrocytes.[109,115]

Table II
Kinetics for Uptake of GABA and Glutamate into Brain Slices and Isolated Glial and Neuronal Preparations[a]

	GABA		Glutamate	
	K_m (μM)	V_{max} (μmol/min per 100 mg)	K_m (μM)	V_{max} (μmol/min per 100 mg)
Brain slices	11–31	0.03–0.17	20–480	0.06–4.8
Synaptosomes	0.4–13	0.11–0.22	1.9–30	0.1–0.6
Cerebellar glomeruli	10–15	0.15–0.16	5	0.02
Cultured neurons	0.3–26	0.02–0.08	27–50	0.6–3.2[b]
Bulk-prepared astrocytes	0.6		10–12	0.06
Glial cell lines	0.2–50	0.001–0.002	12–66	0.02–0.4
Peripheral ganglia	7–10	0.001–0.003	20–21	0.01
Cultured cortical astrocytes	40–45	0.04	10–220	0.4–7.5
Rat retina	38–40	0.007	21	3.5

[a] From ref. 108. Reference to individual results can be obtained from ref. 23 and 108.
[b] The higher of these values is only observed in glutamatergic neurons; furthermore, these cultures do contain some astrocytes which may contribute to the uptake. All values are expressed per 100 mg protein.

A cellular high-affinity uptake of GABA can be of importance for termination of transmitter activity only if it represents a net accumulation rather than a 1:1 homoexchange. Such a homoexchange may occur in some glial preparations, but uptake of GABA into astrocytes in primary cultures is by net accumulation.[109,116] So is the uptake into rat retina in which GABA is localized in the Müller cells.[117] The uptake of GABA into cultured astrocytes is, as into many other cells, virtually abolished in the absence of sodium in the medium, and the energy for the GABA accumulation may reside in the sodium gradient.[108,118,119]

3.1.2. Formation

Based on experiments with bulk-separated astroglial cells, glioma cell lines, and peripheral ganglia, it has been claimed that glial cells possess glutamate decarboxylase (GAD) activity and are able to synthetize GABA (e.g., ref. 29). However, the peripheral ganglion preparation contains both neurons and glial cells, the bulk-separated fractions are contaminated to some extent with synaptosomes, and the glioma cell lines are more or less dedifferentiated, and primary cultures of astrocytes have no measurable GAD activity (J.-Y. Wu, L. Hertz, and A. Schousboe, unpublished experiments) and only produce GABA in exceedingly slight amounts.[120] I therefore support the view (Fig. 1) that GABA formation does not occur to a significant extent in glial cells.[21,121]

3.1.3. Release

There is no doubt that astrocytes contain GABA, probably acquired by accumulation from the extracellular fluid,[12] and that they release this GABA when exposed to GABA-deficient media (e.g., ref. 116), but the release rates are low (Fig. 1). It has repeatedly been claimed that stimulation with excess potassium causes a GABA release from glial cells[29] and the stimulated release of GABA in brain slices occurs from more than one pool.[122] There is, however, no compelling evidence that a potassium-induced release of GABA occurs from normal astrocytes in the central nervous system and is of functional importance.[108]

3.1.4. Metabolism

The presence of GABA transaminase activity has been demonstrated in astrocytes prepared by gradient centrifugation, peripheral ganglia, C-6 glioma cells, and astrocytes in primary cultures (for references see refs. 22, 23). Although the activity is at least twice as high in the primary cultures of astrocytes as in other glial preparations, it is still only about one-third of the activity found in mouse brain homogenates. Such a limited but not negligible capacity for GABA metabolism in astrocytes is in perfect agreement with the immunohistochemical demonstration[123,124] of GABA transaminase activity in both glial and neuronal cell bodies. It is also compatible with the uptake of GABA into both neurons and glial cells (Section 3.1.1) and with the conclusion by Van

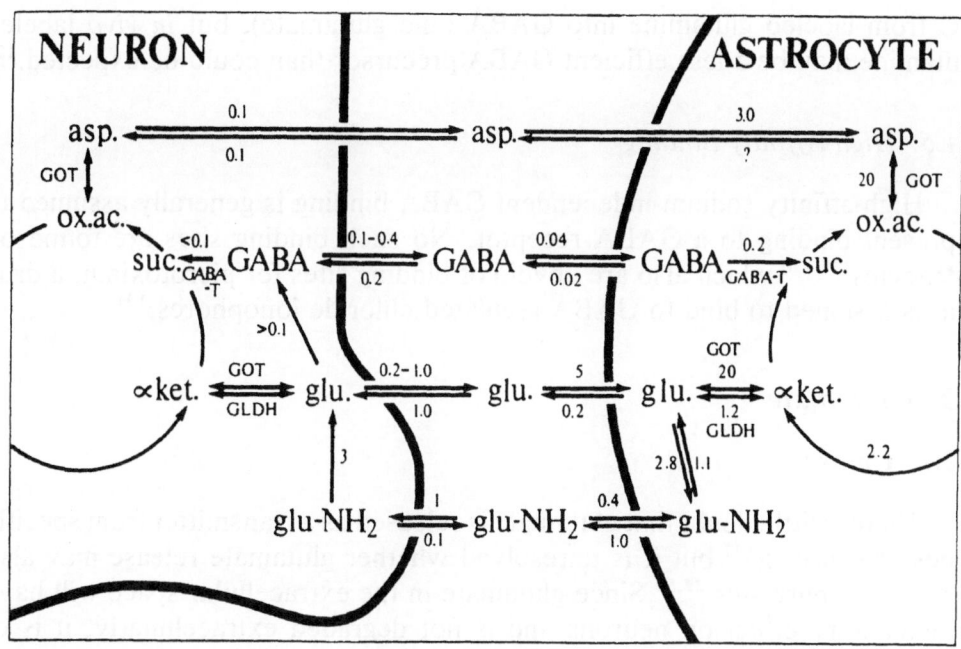

Fig. 1. Diagrammatic representation of rates of uptake and release (μmol/min per 100 mg protein) for glutamate, GABA, glutamine, and aspartate in neurons and astrocytes, together with rates of metabolic interconversion between these amino acids or between individual amino acids and corresponding tricarboxylic acid cycle intermediates. No attempt has been made to cover the literature comprehensively in this figure, but the rates indicated are assumed to be representative. With the exception of glutamine, the uptake rates are V_{max} values; the rates for glutamine uptake are those estimated to occur at a concentration of 0.5 mM. Slightly modified from ref. 22 where references are given for individual results.

den Berg *et al.*[125] that "there are two metabolic spaces involved in the degradation of GABA . . . one of the GABA degradation metabolic spaces leads to a high labeling of glutamine, the other not." The astrocytic localization of glutamine synthetase activity indicates that the former of these "metabolic spaces" is in astrocytes. It is completely unknown which factor(s) decides whether GABA molecules that have been released from a certain neuron under a certain condition are accumulated into one or the other cell type. However, this difference may be of major importance for the further fate of the accumulated GABA: the molecules accumulated into neurons may be reutilized as a transmitter or metabolized, whereas those accumulated into astrocytes may be released in a non-transmitter-related fashion, degraded to CO_2 and water, or, via the tricarboxylic acid cycle, be converted to glutamine.

Uptake and metabolism of neuronally released GABA in astrocytes means a depletion of TCA constituents from the neurons, and it has been suggested that glutamine, which has no transmitter activity, may be released from the astrocytes to the extracellular space in quantitatively corresponding amounts, taken up by the neurons, and used as a precursor for GABA (for references see, e.g., refs. 22, 23, 56). Support for this concept can be obtained from the histochemical demonstration of glutaminase activity in GABAergic nerve endings[126] and from experiments with synaptosomes[127] which rapidly incorporate

[14]C from labeled glutamine into GABA (and glutamate), but *in vivo* labeled glutamine may be a less efficient GABA precursor than could be expected.[128]

3.1.5. High-Affinity Binding

High-affinity sodium-independent GABA binding is generally assumed to represent binding to a GABA receptor. No such binding sites are found on astrocytes[129,130] which also are devoid of binding sites for picrotoxinin, a drug that is assumed to bind to GABA-regulated chloride ionophores.[131]

3.2. Glutamate

3.2.1. Uptake

There is little doubt that glutamate is released as a transmitter from specific types of neurons,[132] but it is unresolved whether glutamate release may also serve other purposes.[22,56] Since glutamate in the extracellular space will have an excitatory effect on neurons and is not degraded extracellularly, it is of crucial importance that efficient high-affinity uptake mechanisms exist. From Table II it can be seen that the glutamate uptake into astrocytes is much more (100 times) intense than that of GABA and also much more pronounced than the glutamate uptake into neurons. A high rate of glutamate uptake into astrocytes is supported by the very avid accumulation into rat retina (Table II) where the glutamate uptake, like the GABA uptake, predominantly occurs into Müller cells and by the autoradiographic demonstration that glutamate *in vivo* is almost exclusively accumulated into glial cells.[133] As in the case of GABA, the uptake into the astrocytes represents a net uptake, not a homoexchange, and it requires the presence of both sodium and potassium.[134] Recent experiments by Schousboe and Divac[135] have demonstrated that there appears to be a correlation between the extent of astrocytic glutamate uptake in a specific brain region and the presumed quantitative importance of glutamatergic transmission in that particular brain area. Glutamatergic neurons themselves seem, however, also to have a more intense uptake of glutamate than nonglutamatergic neurons (A. Yu and L. Hertz, unpublished data).

In contrast to GABA, which is found in very small amounts in cultured astrocytes unless they have been exposed to GABA-containing medium, the content of glutamate is high in all preparations of astrocytes.[12]

3.2.2. Metabolism

The confinement of glutamine synthetase activity in the brain *in vivo* to astrocytes[16] and the high activity of this enzyme in primary cultures of astrocytes[65,136,137] are consistent with the concept that glutamate that has been accumulated into astrocytes is to a large extent converted to glutamine. However, the two other glutamate-metabolizing enzymes, glutamate dehydrogenase and glutamate oxaloacetate transferase, which convert glutamate to α-ketoglutarate, are also present in astrocytes at high activities (Fig. 1).[22,23,65] This suggests that glutamate accumulated into astrocytes may be converted via α-ketoglutarate to TCA cycle constituents and thus possibly serve as a metabolic substrate. This concept is supported by the finding that radioactivity from labeled glutamate is incorporated not only into glutamine but also into aspartate and even, to a considerable extent, into CO_2.[138] The use of glutamate as a substrate for metabolic oxidation in astrocytes is further supported by the finding that cultured astrocytes seem to maintain their rates of oxygen uptake better during incubation in a medium that contains glutamate and no glucose than in a glucose-free medium that does not contain any glutamate (E. Hertz and L. Hertz, unpublished experiments). These observations are compatible with the previously mentioned observation by Kao-Jen and Wilson[52] that astrocytic processes may utilize a different substrate than glucose. They are, however, not in agreement with the concept[139] that the glutamate transfer from neurons to astrocytes, which is brought about when glutamate is accumulated into astrocytes, is compensated for by an equivalent transport of glutamine from astrocytes back to neurons where it is hydrolyzed to glutamate. Nor are the observations that glutamine is not preferentially accumulated into neurons[140] (Fig. 1), that a considerable glutaminase activity is found in astrocytes,[141–143] that no glutaminase activity is found in glutamatergic nerve endings,[126] and that the glutamine uptake is not enhanced in cultured glutamatergic neurons (A. Yu and L. Hertz, unpublished experiments) compatible with this concept.

A major part of the glutamate uptake into astrocytes thus seems to represent a net transfer from neurons, and two likely roles for this process are (1) to provide glutamate as an acceptor for ammonia in the detoxification of this compound by formation of glutamine and (2) to serve as a metabolic substrate for astrocytes. Evidence for the latter of these roles has already been given and is further discussed by Hertz.[22] If astrocytes are, indeed, dependent on a supply of glutamate, the rather large amounts of this amino acid and of glutamine (and glutaminase activity) in astrocytes[12] may simply, like glycogen, constitute a nutritional reservoir. The former role, i.e., detoxification of ammonia, is a process that is not yet fully understood,[144] although it has been demonstrated that under normal conditions virtually all fixation of ammonia in the brain *in vivo* occurs into the amide group of glutamine.[18] The histochemical demonstration that glutamine synthetase in the brain *in vivo* is confined to astrocytes[16] has led to general acceptance that this process exclusively occurs in astrocytes, although it should be mentioned as a *caveat* that neither bulk-prepared neurons[132] nor cultured neurons (E. Hertz, S. Mukerji, and L. Hertz, unpublished experiments) are as devoid of glutamine synthetase activity as they should be according to this concept. This could, however, be because of cross contamination of the bulk-prepared cells and a deficient metabolic differentiation of the cultured neurons. If the glutamate required for glutamine synthesis is mainly produced in neurons, and no comparable return of glutamine occurs, there must be a considerable *de novo* synthesis of α-ketoglutarate, requiring fixation of both CO_2 and ammonia. Carbon dioxide fixation does take place quite avidly in the brain *in vivo*[145] but not as well in brain slices[144] (which might be part of the reason that glutamine was found to be a preferred precursor for glutamate production in slices from the dentate gyrys,[132]) and the cellular location of this process is unknown. Intense formation of glutamate in neurons from α-ketoglutarate and ammonia is not easily compatible with the observation that ammonia in the brain is almost exclusively incorporated into glutamine.[18] However, the concept of a compensatory return of glutamine to the neurons and subsequent hydrolysis also creates a problem with ammonia which in that case would have to be transferred in large amounts from astrocytes to neurons. Further studies of the fate of ammonia (and of CO_2) at the cellular level are therefore urgently needed[22,144] (see also the chapter by A. M. Benjamin in this *Handbook*).

3.3. Other Amino Acids

Aspartate is accumulated almost as avidly as glutamate into astrocytes (Fig. 1),[22] and, as previously mentioned, it is also formed in astrocytes from accumulated glutamate.

Glycine is an inhibitory transmitter at the spinal level but not in the brain. It is another indication of differences between astrocytes from different regions that gradient-separated astrocytes from spinal cord, but not those from frontal cortex, show a high-affinity uptake of glycine.[146]

A quite efficient high-affinity uptake for taurine has been observed both in astrocytes in primary cultures and in glial cell lines, whereas the uptake rate into synaptosomes is low.[134] This may seem quite surprising in view of the fact that metabolic degradation of taurine is supposed to occur very slowly.

Incorporation of amino acids into proteins is discussed in Section 6.1. This obviously requires uptake of at least all the essential amino acids. Information is available on uptake kinetics for some of these into C-6 glioma cells.[147]

4. UPTAKE, RECEPTOR BINDING, AND EFFECTS OF OTHER TRANSMITTERS AND DRUGS

4.1. Neuronal Signaling to Astrocytes

Uptake into astrocytes of transmitters that have been released from excited neurons represents one facet of neuronal–astrocytic interactions. A different aspect is that transmitters may serve as specific signals that can modify astrocytic function. This fascinating possibility was underscored when Henn *et al.*[37] demonstrated that bulk-prepared astrocytes from the caudate nucleus have a higher density of high-affinity binding sites for dopamine than do synaptosomes and that the clinical efficacy of antipsychotic drugs was better correlated with their ability to displace dopamine from its astrocytic binding sites than with their ability to displace the transmitter from its synaptosomal binding sites. Another equally intriguing observation made at about the same time by, e.g., Descarries *et al.*,[148,149] Chan-Palay,[150] and Tennyson *et al.*[151] is that in noradrenergic, serotonergic, and dopaminergic pathways in the brain only a small number of the axonal varicosities are connected to postsynaptic membranes, i.e., form classical synapses. The remainder, which have all the classical morphological characteristics of synaptic boutons, lack the close cellular apposition and postsynaptic thickening, with the implication "that the majority of amine-containing varicosities release neurotransmitter which must diffuse to remote receptors, a model of operation intermediate between the private addressing of classical synaptic messengers and the broadcasting of neuroendocrine secretion."[152] Such effects might equally well be exerted on glial cells as on neurons.

The presence of receptor sites for transmitters on astroglia does not preclude that these transmitters might also be accumulated (after binding to transport sites constituting a different population of binding sites) and metabolized in astrocytes. In this chapter evidence for both astrocytic receptor sites and uptake mechanisms for transmitters other than GABA and glutamate will be discussed.

4.2. Uptake and Metabolism

4.2.1. Monoamines

4.2.1a. Uptake. Bulk-prepared astroglia, C-6 glioma cells, and glial cells in the frog *filum terminale* preparation are able to accumulate the three monoamines norepinephrine, dopamine, and serotonin,[153–157] although the uptake into the bulk-separated cells is considerably less intense than into synaptosomes. The catecholamines[158] and serotonin[159] are also accumulated by intact astrocytes in primary cultures, but the serotonin uptake rate into these cells is low, only slightly concentrative, not saturable within the range 1–200 μM, and apparently not sodium-dependent (L. Hertz and A. Schousboe, unpublished experiments; cf. ref. 157), and no serotonin uptake into glial cells could be observed in organotypic cultures.[111] It might, therefore, appear questionable whether a meaningful net accumulation of serotonin does, indeed, occur into astroglia, but the autoradiographic demonstration of an *in vivo* accumulation of tritiated serotonin into both neuronal constituents and glial cells[160] does support this view, as does the fact that the accumulation into C-6 cells is a high-affinity uptake with a V_{max} comparable to that for GABA uptake (Table II) into the same cell type.[161]

4.2.1b. Metabolism. Primary cultures of astrocytes have monoamine oxidase (MAO) activity and metabolize norepinephrine and dopamine at rates of 0.01–0.1 μmol/100 mg protein per h,[158] an activity that is rather low compared to that of the enzymes metabolizing amino acid transmitters (Fig. 1) but not compared to the MAO activity in brain; even higher activities have been reported in bulk-prepared astroglia[162] and C-6 glioma cells.[163–165] The enzyme in the latter cells metabolizes serotonin with somewhat higher affinity than norepinephrine and almost the same V_{max} as in brain.[164,165]

Based on specificities for different substrates, it has long been claimed that different types of monoamine oxidase, i.e., MAO A and MAO B, exist, and recently differences in the molecular entities have been revealed by limited proteolysis of hepatoma cell monoamine oxidase followed by peptide mapping.[166] C-6 glioma cells, and most other glial (or neuronal) cell lines[154] contain only the A form, whereas astrocytes in primary cultures contain not only the A,[167] but also the B form—especially after culturing with dibutyryl cyclic AMP (P. H. Yu and L. Hertz, unpublished experiments).

Catechol-O-methyl transferase (COMT) activity has been observed in primary cultures of astrocytes[158] and C-6 glioma cells,[165] but it is lower than that of the monoamine oxidase, and no COMT activity could be demonstrated in bulk-separated astrocytes.

4.2.2. Adenosine

4.2.2a. Uptake. Adenosine is accumulated into both glioma cell lines[169,170] and astrocytes in primary cultures[171] by a high-affinity uptake which in the latter preparation is as intense as that of GABA (Table II) but of higher affinity. Such an accumulation may conceivably serve one or both of two purposes, i.e., termination of transmitter activity or uptake of a precursor for formation of nucleotides. Brain slices accumulate adenosine less intensely,[171] which may suggest that adenosine *in vivo* is mainly accumulated into astrocytes, and autoradiographic studies have shown uptake of adenosine into both neurons and glial cells *in vivo*.[172] The uptake into cultured astrocytes is competitively and efficiently inhibited by papaverine[171] and certain other drugs, including diazepam (Section 4.4.4).

4.2.2b. Metabolism. At low concentrations adenosine is mainly converted to nucleotides, but at higher concentrations it is to a large extent metabolized to inosine and hypoxanthine,[170] which is consistent with the presence of adenosine deaminase in astrocytoma cells.[173] The incorporation of exogenous adenosine into cyclic AMP is far too small to explain the increase in intracellular cyclic AMP that occurs when the cells are exposed to adenosine (Section 4.3.4b). The further metabolic fate of the nucleotides is unknown, but they are to some extent released from the cells.[170] Extracellular nucleotides, some of which thus may originate from astroglia, can be hydrolyzed by an extracellular nucleotidase mainly found on glial cells.[174]

4.2.3. Acetylcholine and Choline

4.2.3a. Uptake. Choline uptake, which has often been claimed to be rate limiting for the synthesis of acetylcholine, has been demonstrated in astrocytoma cells[154] and primary cultures of astrocytes.[175] The accumulated choline may also serve as a precursor for the synthesis of lipids (Section 6.3).

4.2.3b. Metabolism. On the basis of histochemical examination, it has often been concluded that neurons show acetylcholineesterase (AChE) activity whereas glial cells (including astrocytes), as first demonstrated by G. B. Koelle more than 25 years ago, have butyrylcholinesterase (BuChE) activity. It is in accordance with this that Giacobini[176] found higher activity of BuChE in spinal cord astrocytes than in neurons and no AChE in the glial cells. Bulk-prepared astrocytes have analogously been found by some investigators to contain less acetylcholinesterase than corresponding neuronal samples,[177] but the difference is less striking than could be expected, and other authors have reported almost identical activities in the two cell types (for review, see refs. 58, 178).

4.3. Receptor Binding

4.3.1. Norepinephrine and Antidepressant Drugs

4.3.1a. Binding. C-6 glioma cells and human astrocytoma cells possess high-affinity binding sites (K_D 0.1–7 nM; B_{max} 40–500 fmol/mg protein) for dihydroalprenolol and iodohydroxybenzylpindolol, two β-adrenergic ligands.[179–184] Intact astrocytes in primary cultures resemble these cells in binding dihydroalprenolol, although the affinity is considerably lower and the maximum binding considerably higher. The joint effect of these kinetic differences is that the binding at 1.8 nM amounts to 1500–2000 fmol/mg protein,[40] a value that is larger than that in the glioma cells and also larger than the binding of about 100 fmol/mg protein in whole brain.[185,186] Since dihydroalprenolol is notorious for binding not only to β-adrenergic sites but also to nonspecific sites, it is of importance that part of the binding can be displaced by propranolol in a stereospecific manner. However, the potency of propranolol is relatively modest, and isoproterenol has no effect. A similar lack of sensitivity to isoproterenol has been observed in intact C-6 cells where the affinity for this compound was found to increase 1000 times after disruption of the cells,[182] a phenomenon that might suggest that the β-adrenergic binding sites are mainly located intracellularly.[187] The binding is inhibited by all groups of antidepressant drugs tested, in therapeutically effective concentrations, but not by antianxiety drugs or antipsychotics.[40] This may be of considerable pharmacological relevance since long-term administration of antidepressant drugs *in vivo* evokes a subsensitivity of β-adrenergic function, a phenomenon that may be closely connected to the mechanism of action of the antidepressants. A similar response has been observed in astrocytes in primary cultures after chronic exposure to antidepressant drugs.[40,188]

4.3.1b. Adenylyl Cyclase Activity. An overwhelmingly large amount of evidence shows that a variety of astroglial preparations, including C-6 cells, human astrocytoma cells, and primary cultures of astrocytes, respond to norepinephrine or isoproterenol with a large increase in the intracellular level of cyclic AMP.[39,189] It is consistent with the concept of an effect by antidepressant drugs on β-adrenergic receptors that the response to isoproterenol is partly inhibited by tricyclic antidepressants.[40,190]

In primary cultures of astrocytes the maximum stimulation of cyclic AMP production by norepinephrine (an α and β agonist) is less pronounced than that by isoproterenol (a relatively pure β agonist), and α-adrenergic agonists decrease the effect of isoproterenol. For these reasons it has been concluded that the cells possess both adrenergic β receptors which stimulate adenylyl cyclase activity and α receptors which depress this activity.[39,191] The biological role of this arrangement is not known.

4.3.2. Dopamine and Antipsychotic Drugs

4.3.2a. Binding. Binding of dopamine and of the antipsychotic drug halo-peridol (140 fmol/mg protein at 1 nM) to bulk-separated astroglial cells from the caudate nucleus was first described by Henn and co-workers[13,37] and has been confirmed in astrocytes in primary cultures.[167] However, the expected interaction between dopamine and antipsychotic drugs could not be established in the cultured astrocytes, perhaps because they originated from whole cerebral hemispheres rather than specifically from the caudate nucleus (see below).

4.3.2b. Adenylyl Cyclase Activity. Several reports have appeared that do-pamine increases the level of cyclic AMP in primary cultures of astrocytes, C-6 cells, and human astrocytoma cells, but these effects might well result from a weak β-adrenergic activity of dopamine, since they are more effectively antagonized by propranolol than by dopamine antagonists and since cyclic AMP accumulation in glioma cells is increased to a much smaller extent by dopamine than by norepinephrine.[39]

Henn and co-workers[13] observed a stimulation by dopamine of adenylyl cyclase activity in bulk-separated astroglia from the caudate nucleus but not in a similar preparation from neocortex. This response seems, in contrast, to be mediated via genuine dopamine receptors, since dopamine was more potent than norepinephrine and since the effect was antagonized by dopamine receptor blockers but not by propranolol. This is, again (see Sections 3.2.1 and 3.3), a demonstration of functional differences between astrocytes from different regions of the central nervous system. For further discussion of dopamine receptors on astrocytes, see the recent review by Henn and Henn.[13]

4.3.3. Serotonin and Serotonin-Binding Protein

4.3.3a. Binding. A relatively modest binding (50 fmol/mg protein at 11 nM) of serotonin to astrocytes in primary cultures was reported by Hertz *et al.*[38] and has been confirmed in bulk-separated astroglia by Fillion *et al.*[192]

The characteristics of a neuronal serotonin-binding protein (found in brain and certain other tissues) have been studied in detail by Tamir and co-workers.[193,194] The serotonin-binding proteins in neurons and in astrocytes are kinetically different and are affected in different ways, or to a different extent, by compounds enhancing or inhibiting the binding (Table III).[195] This strongly suggests two different molecular entities of the binding proteins.

4.3.3b. Adenylyl Cyclase Activity. The kinetics for the serotonin binding to the protein from astrocytes fit reasonably well with those for a stimulation of adenylyl cyclase activity which, although modest (about a doubling), is undoubtedly present in both astrocytes in primary cultures[159] and bulk-prepared astroglia.[192]

Table III
Characteristics of Neuronal and Astrocytic Serotonin-Binding
Proteins[195]

	Astrocytic	Neuronal
K_D for specific binding	3×10^{-7} M	4×10^{-10} M
		3×10^{-8} M
Specific binding (% of total)	70 ± 2.2	86 ± 3.1
Enhancement by Fe^{2+}	2–3 times	15 times
Inhibition by sodium phosphate	32%	80%
Inhibition by Tris	32%	90%
Enhancement by gangliosides	0 or ↓	4–5 times
Inhibition by reserpine	70%	50%
Inhibition by fluoxetine	30%	0%

4.3.4. Other Transmitters

4.3.4a. Acetylcholine. No binding of quinuclidinyl benzilate, a muscarinic cholinergic ligand, to astrocytes in primary cultures was found by Braestrup *et al.*[167] However, Schwann cell membranes do have nicotinic cholinergic receptors,[196] and some evidence is also found suggesting the presence of such sites on astrocytes (for details and references see ref. 39). It was previously mentioned that choline is accumulated into astroglia.

4.3.4b. Adenosine. At a low concentration adenosine inhibits the isoproterenol-induced stimulation of cyclic AMP production in primary cultures of astrocytes, but at higher concentrations it is also capable of stimulating cyclic AMP accumulation in astrocytoma cells and in primary cultures of astrocytes.[39,197] That this results from a direct effect on an extracellular receptor and not from conversion of the applied adenosine to cyclic AMP was concluded by Clark *et al.*[198] (see also ref. 39) on the basis of the finding that the response was increased in the presence of an uptake inhibitor.

4.3.4c. Peptides. Vasopressin affects, to a moderate extent, the level of cyclic AMP in primary cultures of astrocytes, even at extremely low concentrations (e.g., 10^{-10} M) (L. Hertz and J. R. McNeill, unpublished experiments). The response seems, like that to adenosine, to be quite complex. The possible presence of receptors for other peptide hormones on glial cells has been discussed by Van Calker and Hamprecht.[39] No enkephalin receptors seem to occur on astrocytes in primary cultures as evidenced by the lack of specific binding of naloxone.[167]

4.3.5. Benzodiazepines and Barbiturates

There is no doubt that bulk-separated astroglia, C-6 glioma cells, and astrocytes in primary cultures have specific binding sites for benzodiazepines,[41,42,199,200] a group of drugs that probably interact with an unknown endogenous ligand. What remains controversial is the characteristics of this binding compared to that to brain membranes as well as possible regional differences. The latter have been explored by Henn and co-workers who reached the conclusion that the binding in the frontal cortex mainly occurs to astroglia, whereas in the cerebellum there is an undisputable binding to Purkinje cells.[13] With respect to binding characteristics it has been postulated that the binding to astroglia is to a "peripheral-type" receptor of little biological relevance.[167,199] It is likely that the benzodiazepine receptor, like the Na^+,K^+-ATPase and the serotonin-binding protein may be found in different molecular forms in neurons and astrocytes and that these may have different drug specificities. It is, however, in support of the neurobiological relevance of the astrocytic binding that therapeutically relevant concentrations of clonazepam or flurazepam cause a distinct displacement of diazepam from its binding sites on astrocytes in primary cultures[42,200] and that the limited number of other benzodiazepines that have been tested cause a similar displacement in a rank order that matches their pharmacological potency (L. Hertz, unpublished experiments). This is in contrast to the absence of such a displacement in the C-6 glioma cells.[199,200]

Since the pharmacological actions of benzodiazepines and barbiturates are similar in many respects, it may be of functional importance that diazepam binding to astrocytes in primary cultures is inhibited by barbiturates.[42,82] Relatively high concentrations are required, but this is in agreement with the rather low clinical potency of barbiturates.

4.4. Neurochemical Consequences of Receptor Occupancy

4.4.1. Phospholipid Metabolism

Recent results from Axelrod's and Tallman's laboratories have shown that the reaction of benzodiazepines with receptor sites in C-6 glioma cells, as evidenced by displacement of labeled diazepam, is mirrored by an enhancement of the incorporation of methyl groups into phosphatidyl choline. As can be seen in Fig. 2, these two events have almost exactly the same quantitative correlation with varying concentrations of three different benzodiazepines of widely different potencies.[201,202,*] β-Adrenergic agonists have a similar effect,[203] and incubation of astrocytoma cells with both benzodiazepines and β-adrenergic agonists causes an additive increase in methylation, indicating activation of separate pools of phospholipid-methylating enzymes by each drug.[202]

These alterations of phospholipid methylation are related to alterations in membrane fluidity which may be involved in the coupling between receptor occupancy and activation of adenylyl cyclase activity.[202] In addition, they might conceivably affect a multitude of other biochemical processes, e.g., membrane transport.

* It is generally assumed that R05-4864 has no clinical effect.

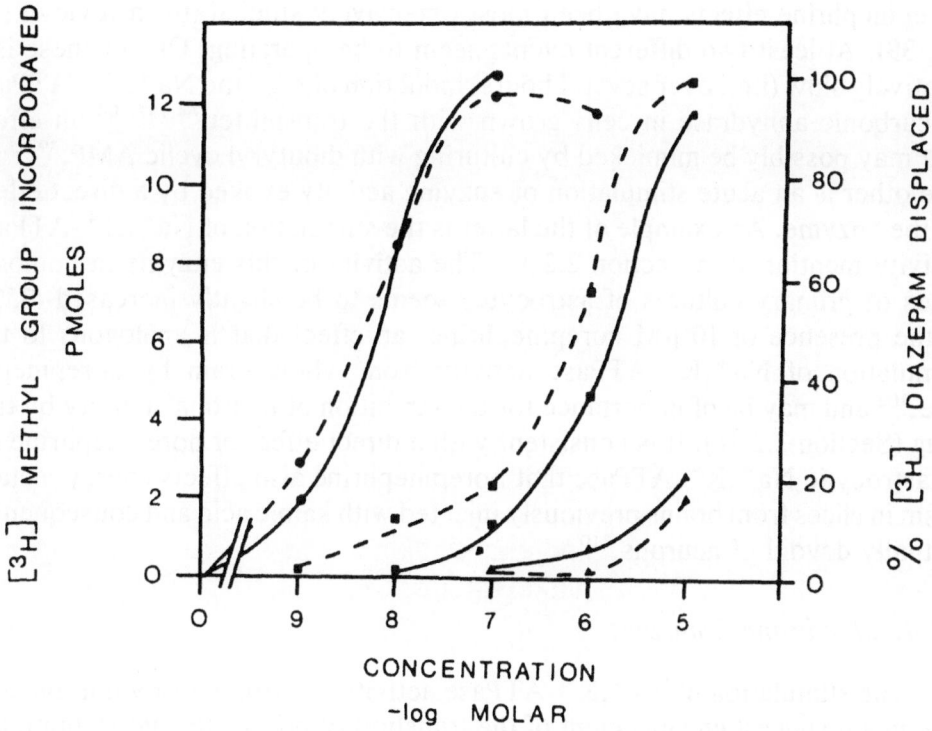

Fig. 2. Phospholipid methylation (fully drawn lines) and [³H]diazepam displacement in C-6 glioma cells in the presence of the benzodiazepines R05-4864 (circles, left), clonazepam (squares, middle), or chlordiazepoxide (triangles, right). From ref. 202 where experimental details are described.

4.4.2. Protein Phosphorylation

Part of the action of transmitter agents on glial cells may, as in synaptic membranes, occur via activation of a protein kinase, catalyzing phosphorylation of proteins. Such a system which is sensitive to dopamine has been demonstrated in bulk-prepared astroglia from the caudate nucleus.[13]

4.4.3. Energy Metabolism

It is consistent with the known stimulation of liver phosphorylase by β-adrenergic stimulation and the presence of large glycogen stores in astrocytes that norepinephrine stimulates glycogenolysis in C-6 glioma cells[204,205] and increases the release of radioactivity from cells preloaded with labeled glucose.[206] The rapid disappearance of the glycogen in astrocytes under hypoxic conditions[29] might well be a result of an increase in adrenergic activity, since brain hypoxia is known to increase the level of cyclic AMP.[207]

Dopamine causes an analogous increase of phosphorylase activity in tissue from the caudate nucleus. The cellular localization of this phenomenon is unknown, but the astrocytic localization of glycogen might suggest an action on astrocytes.[13]

Certain enzyme activities in glial cells are increased by transmitters, and norepinephrine effects have been most extensively studied (for a review, see ref. 39). At least two different events seem to be operating. One of these is a relatively slow (i.e., over several hours) induction of e.g., the Na^+,K^+-ATPase or carbonic anhydrase in cells grown with the transmitter,[64,197,208] an effect that may possibly be mimicked by culturing with dibutyryl cyclic AMP.[39,65,209] The other is an acute stimulation of enzyme activity evoked by a direct effect on the enzyme. An example of the latter is the stimulation of Na^+,K^+-ATPase activity mentioned in Section 2.2.1c. The activity of this enzyme in homogenates of primary cultures of astrocytes seems to be slightly increased (35%) in the presence of 10 μM norepinephrine, an effect that is analogous to the stimulation of Na^+,K^+-ATPase activity from whole brain by norepinephrine,[102] and may be of importance for the inhibition of neuronal activity by this drug (Section 2.2.1c). It is consistent with a direct effect of norepinephrine on an astrocytic Na^+,K^+-ATPase that norepinephrine also affects energy metabolism in slices from brains previously injected with kainic acid and consequently virtually devoid of neurons.[210]

4.4.4. Membrane Transport

The stimulation of Na^+,K^+-ATPase activity in astrocytes by norepinephrine might suggest enhancement of the transport of potassium and sodium, but no investigations exploring this possibility seem to have been carried out. The only effects on transport in astroglia by a transmitter or a transmitter-related drug reported so far may therefore be an increase in methionine uptake by norepinephrine[85] and an inhibition of adenosine uptake by diazepam.[211] The latter effect may, in the brain *in vivo*, lead to a delayed termination of the inhibitory effect of adenosine, an effect that Phillis *et al.*[212] have suggested could be causally related to the mechanism of action of the benzodiazepines.

5. TRANSPORT OF INORGANIC IONS

5.1. Potassium

5.1.1. Need for Potassium Homeostasis in the Central Nervous System in Vivo

After accumulation into astrocytes, glutamate may be converted to glutamine or metabolized to α-ketoglutarate and CO_2, adenosine may be converted to its nucleotides or deaminated, catecholamines may be methylated and/or oxidized, and serotonin may also be oxidized. In contrast, potassium ions obviously remain potassium ions. It is therefore remarkable that a neuronal–astrocytic interaction also seems to be operating in the maintenance of potassium homeostasis at the cellular level of the brain.

The first similarity is that potassium, like the transmitter compounds, is released from excited neurons to the extracellular space of the central nervous system, leading to a measureable increase in potassium concentration during physiological activity (i.e., from the resting level of 3.0 to about 4.0 mM), a larger increase (up to about 12 mM) during seizures, and an enormous, completely reversible increase (up to >50 mM) during "spreading depression" (a peculiar neurophysiological phenomenon observed after massive local stimulation) or brief exposure to anoxia.[29,50,84,213] These increases in extracellular potassium concentration have all been measured *in vivo* by different groups of neurophysiologists utilizing potassium-sensitive microelectrodes. If anything, they are underestimates because of the enlargement of the extracellular space at the location of the electrode (1–2 μm) and the distance between the excited cells and the electrode which, on the average, might be as much as 35–40 μm.[214] The *in vivo* experiments have also convincingly demonstrated that the subsequent reduction to a normal (or in some cases even subnormal) level of potassium in the mammalian central nervous system is predominantly brought about by active, energy-requiring reaccumulation into adjacent cells.[215,216] The maximum transport rate probably amounts to 1 μmol/sec or about 60 μmol/min per g wet wt. of the brain.[29] Ultimately, potassium ions that were originally released from neurons must obviously be reaccumulated into neurons in order not to deplete these cells of potassium. Since the relative alteration of the extracellular potassium concentration in the narrow extracellular clefts must be much more pronounced than in the neurons, a reestablishment of normal extracellular potassium concentration will, however, be more urgent than a correction of the small decrease in intracellular potassium content in the neurons. The stimulation of Na^+,K^+-ATPase activity and oxygen uptake in astrocytes, but not in neurons, when the potassium concentration is elevated (Section 2.2.1) suggests that astroglia may play an active role in the immediate cellular reaccumulation of extracellular potassium. In this section the characteristics for potassium uptake into astroglia will therefore be discussed.

5.1.2. Potassium Uptake

5.1.2a. Uptake Kinetics. The kinetics for active uptake of potassium (measured as net uptake or ouabain-sensitive uptake) into astrocytes are probably more disputed than those of any other compound, as evidenced by the fact that Hertz,[69] using primary cultures of mouse astrocytes, reached the conclusion that such an uptake is intense enough and sufficiently stimulated by elevated potassium concentrations[217] (Table I) to be the major mechanism involved in the immediate clearing of extracellular potassium *in vivo*. The uptake rate in the presence of an elevated potassium concentration (>100 μmol/min per 100 mg protein, which approximately equals 1 g wet wt.) is two times higher than the estimated potassium uptake of 60 μmol/min per g wet wt. into cells in the brain *in vivo*. With astrocytes constituting one-third of the cellular volume (Section 1), this is enough to account for the major part of the *in vivo* uptake, and the transport rates shown in Table I were measured in cells that had been exposed to cold, potassium-free medium for 30 min, which may have reduced their uptake capacity. Other authors[218–221] have, however, found that the uptake rates are relatively low and not increased by an elevation in the extracellular potassium concentration beyond physiological levels, which does not suggest a major role of astrocytes in potassium homeostasis. Recently, Moonen[33] has reinvestigated this problem. Although the uptake rates observed by him were somewhat lower than those observed by Hertz, the affinity for potassium, i.e., the correlation between the potassium concentration and the uptake rates, was identical in the two studies. Moreover, Moonen[33] could also demonstrate that the alterations in uptake rate as a function of the external potassium concentration match the changes in the electrogenic component of the membrane potential, estimated as the differences between the membrane potentials in the presence and absence of ouabain.

Besides the system described above, astrocytes may possess an even more powerful uptake mechanism which is not called into action until the extracellular potassium concentration exceeds 50 mM,[217] but it has not been determined to which extent this represents a net uptake.

5.1.2b. Uptake into Astrocytes versus Uptake into Other Cell Types. The potassium uptake into cultured neurons in primary cultures is much slower than that into astrocytes and is barely, if at all, increased when the potassium concentrations exceed 5 mM.[50] It can, however, not be excluded that the cultured neurons may not have attained maximum functional maturation, and in C-6 glioma cells the uptake rate is only twice as high as in neuroblastoma cells.[220] Synaptosomes also accumulate potassium much less intensely,[222,223] but this may partly be a result of damage to the preparation during the isolation. The potassium uptake into brain slices is also slower than that into astrocytes in primary cultures, but it has the same K_m,[224] supporting the concept that a similar mechanism is involved, i.e., that the uptake in the slices predominantly occurs into the contained astrocytes. Finally, and maybe most important, Coles and Tsacopoulos[225] have, in the drone retina, shown that physiological stimulation (light flashes) leads to a decreased potassium activity in the neuronal photoreceptors, an increased potassium activity in the extracellular space (analogous to that described in the mammalian central nervous system by several authors), and an increase in the potassium activity inside the glial cells, i.e., that glial cells in a cellularly intact preparation and thus in the presence of nearby neurons participate substantially in potassium homeostasis at the cellular level by accumulating potassium ions.

A key difficulty for the acceptance of the view that an initial active uptake of potassium occurs into astrocytes is that, somehow, the potassium must be returned to the neurons, a process requiring additional energy. No exact information is available about the mechanisms(s) for such an uptake, but in C-6 glioma cells a lowering of the temperature inhibits the transport of potassium out of the cells,[218] and Latzkovits et al.[226] have deduced from a computer analysis of the shape of the influx curve for potassium into mixed cultures of neurons and astrocytes that in addition to a direct uptake into both cell types there must be a transfer from one of the two types into the other. If such a transport mechanism can be regulated, it might be of major importance for modulation of activity in the central nervous system.[36] Evidence for a transcellular transport of potassium in the mammalian brain *in vivo* was recently obtained by Gardner-Medwin[227] who imposed a known current on the intact cerebral cortex and measured the amount of potassium accumulating under the cathode. The fraction of the current that was carried by potassium ions was too large to be explainable by the amount of potassium in the extracellular fluid, for which reason it was concluded that a significant part of the current moved through an intracellular compartment.[29] This observation is consistent with the concept of a potassium transport through astrocytes but obviously does not give any information about the cell type(s) involved.

5.1.2c. Phylogenetic and Ontogenetic Development. The concept of an active accumulation of potassium ions into astrocytes followed by a transfer to neurons is in many ways analogous to that of a current-carried redistribution of potassium ions through glial cells which has been suggested[228] on the basis of glial cell characteristics in non-mammalian nervous systems,[31,228–230] a mechanism that is not as directly dependent on metabolic energy. For this reason, Hertz and Nissen[231] suggested that an active transport of potassium ions by astrocytes might represent a more highly developed, more efficient, and more regulable version of a passive, less efficient, and less regulable mechanism operating in other types of nervous systems with glial cells more or less similar to those in the vertebrate nervous system.[232,233]

Like other phenomena in which potassium is involved (potassium effect on oxygen consumption and on Na^+,K^+-ATPase activity), the potassium transport rates are altered during development, and the potassium uptake is considerably faster in 4-week-old cultures of astrocytes than in 2-week-old cultures.[33]

5.1.2d. Correlation with Potassium Effects on Energy Metabolism. The potassium uptake into astroglia is substantially reduced, although not abolished, in the presence of ouabain[33,50,218,220] which inhibits the Na^+,K^+-ATPase and thus the exchange between potassium and sodium (Fig. 3). The correlation between potassium net uptake rates and Na^+,K^+-ATPase activity can be estimated from Table I which shows that the ratio between the number of potassium ions accumulated and the number of phosphate bonds hydrolyzed is higher than the expected ratio of 2.[234] This might be explained by the fact that the Na^+,K^+-ATPase is a hysteretic enzyme (Section 2.2.1c), but it could also indicate that some of the potassium ions might be accumulated by processes other than stimulation of Na^+,K^+-ATPase activity. It is compatible with this view that the carbonic anhydrase inhibitor acetazolamide, which inhibits chloride uptake into an astrocytic cell line,[235] reduces the potassium uptake into astrocytes in primary cultures by about 30%; the chloride transport inhibitor ethacrynic acid has, in contrast, no effect.[50]

The ratio between the amounts of potassium accumulated and oxygen consumed are well within generally accepted ranges for active ion transport (e.g., ref. 236). Increased levels of potassium enhance not only Na^+,K^+-ATPase activity but also the activities of several other enzymes involved in glucose oxidation in brain (Section 2.2.1b). The stimulation of metabolism might, in turn, lead to the potassium-induced swelling of astrocytes (Section 5.5.2) and possibly enhance the uptake of potassium. For further discussion

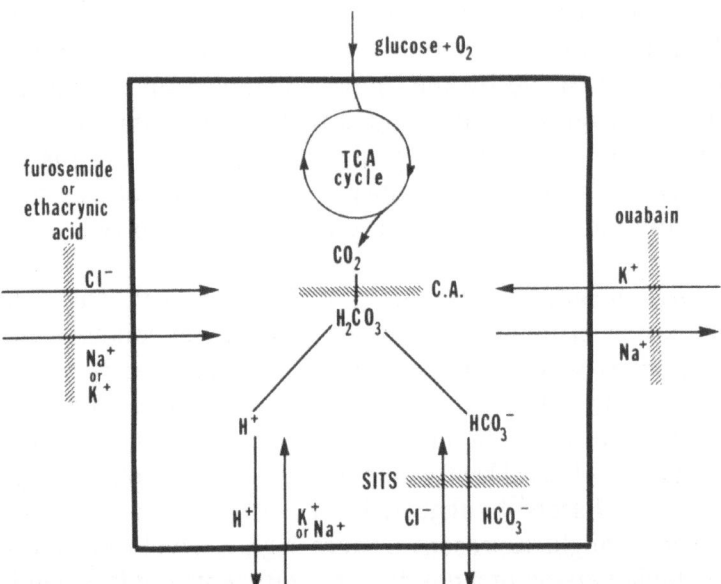

Fig. 3. Schematic representation of mechanisms in astrocytes that may lead to accumulation of potassium, sodium, and chloride as well as extrusion of hydrogen ions and bicarbonate. The shaded bars indicate inhibition by specific drugs (C.A., carbonic anhydrase inhibitor). Potassium uptake in exchange for sodium can not per se be expected to lead to a swelling of the cells, whereas both cation uptake together with chloride and exchange of chloride with bicarbonate can be expected to lead to cellular swelling (for details see text). Conceivably, stimulation of astrocytic oxygen uptake by potassium may, by increasing the amount of bicarbonate available for exchange with chloride, be directly connected with the astroglial swelling evoked by excess potassium. From ref. 50.

of the correlation (or lack of correlation) between potassium effects on energy metabolism and on potassium transport, see the review by Hertz and Chaban.[237]

5.2. Sodium

Relatively little information is available about sodium transport in astrocytes.[84] In contrast to neuroblastoma cell lines, but in accordance with the concensus that glial cells are nonexcitable, C-6 glioma cells do not show an increased sodium uptake during exposure to veratridine.[238]

5.3. Lithium

Lithium transport in astroglia might be of relevance in connection with the therapeutic and prophylactic effect of this ion on the manic-depressive psychosis. Lithium uptake has been described in both glioma cells and primary cultures of astrocytes[239,240] but seems to be less than that into neurons.[240]

5.4. Calcium

There is little difference between calcium uptake in bulk-separated astroglia and in bulk-separated neurons, but the uptake rate is higher in synaptosomes.[241] It is a further indication of a more intense calcium uptake in neuronal constituents that neurons in primary cultures take up ^{45}Ca five times faster (0.3 μmol/min per 100 mg protein) than do corresponding cultures of astrocytes and that only the neuronal uptake is further enhanced by exposure to elevated concentrations of potassium.[242] The decrease in extracellular calcium concentration resulting from neuronal uptake during excitation may have profound effects on astroglia.[243]

5.5. Chloride

5.5.1. Chloride Transport in Other Tissues

Chloride transport has been studied quite extensively in astroglia and seems to be of major importance in connection with regulation of fluid content. From studies of other cell types, it is known that at least two mechanisms exist for chloride uptake, i.e., an accumulation together with a cation (sodium or potassium) and an exchange with bicarbonate (Fig. 3). The existence of a transport process together with a cation is well established in the kidney and provides the basis for the action of loop diuretics, e.g., furosemide and ethacrynic acid, which block the active reuptake of chloride from renal tubuli and thus the establishment of a hypertonic environment in the medulla (up to 1200 mosmol/liter). An exchange between chloride and bicarbonate has mainly been studied in erythrocytes where it is responsible for the rapid efflux of bicarbonate required during the passage of blood through pulmonary capillaries (e.g., ref. 62). The bicarbonate is formed together with hydrogen ions by dissociation of carbonic acid which, in turn, results from hydration of CO_2, a process requiring the presence of carbonic anhydrase in order to proceed at sufficient speed (Section 2.1.3).

5.5.2. Chloride Uptake into Astrocytes

The presence of carbonic anhydrase activity in astrocytes (Section 2.1.3) suggests that the CO_2 that results from aerobic metabolism is hydrated to carbonic acid and dissociated to hydrogen ions and bicarbonate ions, the latter of which are removed from the cells in a process analogous to that described above. It is in favor of this concept that SITS, an inhibitor of chloride–bicarbonate exchange, inhibits chloride uptake into astrocytes in primary cultures, and Kimelberg et al.[63] have suggested a transport model according to which a simultaneous exchange of intracellular bicarbonate with extracellular chloride and intracellular hydrogen with extracellular sodium serves to remove the bicarbonate and hydrogen ions formed. One reason that sodium was specifically suggested as the cation exchanging with hydrogen is that addition of sodium to a sodium-free medium in which primary cultures of astrocytes are incubated increases the rate of acidification of the medium.[63] This effect could equally well be caused by metabolic actions of sodium[63] (see also Section 2.2.2), and the previously mentioned finding that uptake of ^{42}K into astrocytes in primary cultures is inhibited by acetazolamide does suggest that some potassium is accumulated into astrocytes by this mechanism. However, it cannot be excluded that hydrogen is exchanged primarily with sodium which subsequently stimulates sodium–potassium exchange catalyzed by the Na^+,K^+-ATPase. Such a process would energetically be less advantageous. Since one molecule of glucose gives rise to six molecules of CO_2 which are exchanged with six molecules of sodium and/or potassium chloride, the process will lead to a cellular hypertonicity which, for osmotic reasons, induces an uptake of water. This may provide an explanation for the potassium-induced swelling of astrocytes which may be a direct consequence of the potassium-induced stimulation of oxygen uptake and thus of CO_2 production (Fig. 3), a concept that explains the many ontogenetic and quantitative correlations that have been described between potassium-induced stimulation of oxygen uptake and of swelling in brain slices.[56,74,84,]* One may wonder if astrocytic permeability to water might occasionally be low (as is the case in the kidney) so that the cells might be able to accumulate more salt than fluid. This would provide a reasonable explanation for the resistance to respiratory inhibition in hypertonic media (Section 2.2.2) and might facilitate the return of potassium to neurons.

In addition to the bicarbonate–chloride exchange, astrocytes may also take up chloride together with a cation, since furosemide, an inhibitor of this process, decreases uptake of ^{36}Cl into astrocytes in primary cultures.[244]

6. MACROMOLECULAR CONSTITUENTS

6.1. Proteins

6.1.1. Rate of Synthesis

Many different cell types are responsible for the overall protein synthesis in brain (neurons, oligodendrocytes, astrocytes, microglial cells, pericytes, and endothelial cells). The quantitative contribution by each cell type to the total rate of synthesis has been found to vary both with the age of the animal and the preparation used to study the synthesis. The rate of protein synthesis in the brain is high in newborn animals and declines over the first month of growth (for references, see ref 245). Johnson and Sellinger[246,247] determined the *in vivo* labeling of neuronal perikarya and bulk-separated astrocytes over short incorporation times as a function of age and found that protein synthesis declines considerably more in the neurons than in the glial cells, so that protein synthesis in the mature brain is almost twice as intense in glial cells as in

---

* It should be very much emphasized that a different reason for astrocytes (and most other cells) to swell in potassium-rich media is that replacement of sodium with the more easily diffusible potassium ion in itself enhances swelling for purely passive reasons, a swelling that will continue until the concentrations of the osmotically active cellular constituents approach that of the reduced extracellular sodium concentration.[12,50] The swelling described in the text is a phenomenon over and above that resulting from the replacement of sodium with potassium, and it can be observed in astrocytes in brain slices incubated in a potassium-rich medium with a maintained sodium concentration.

neurons. High rates of protein synthesis in astrocytes in primary cultures regardless of age support this point of view.[245] Conflicting results have, however, been reported by other workers who, using slightly different techniques than Johnson and Sellinger, found that in mature animals neurons synthesize protein several times faster than glial cells. The more recent results (e.g., refs. 248, 249) indicate, however, that the differences are relatively small, i.e., less than a factor of 2. The concensus of these studies therefore seems to be that developing neurons have higher protein synthesis rates than developing astrocytes but that the quantitative differences in the mature central nervous system are relatively slight.

6.1.2. Qualitative Characteristics

Several qualitative differences between glial and neuronal cells with respect to protein composition have been reported,[250,251] and certain proteins, e.g., GFA are exclusively found in astrocytes.[14,15] In the present review it has been noted at several places that differences in molecular entities seem to exist between proteins in neurons and in astrocytes that may have analogous functions, e.g., the Na^+,K^+-ATPase. For further discussion of glial (not necessarily astrocytic) proteins and of indications suggesting functionally important interactions between neurons and glial cells in protein turnover, refer to recent reviews by Lange,[252] Stewart and Rosenberg,[251] Varon and Somjen,[29] and Pevzner.[58]

6.2. Nucleic Acids

In the 1960s it was observed by Hydén[253] and Pevzner[58] that during certain forms of activity RNA content and base composition change in reciprocal ways in glial cells and in neurons, and a transfer between the two cell types was suggested.[29,58,253] It is unknown to what extent astrocytes may specifically be involved. According to Watson,[254] nucleic acids are similar in neurons and glial cells, but recently Sellinger and Der[255] have obtained evidence for an apparently cell-specific pattern of tRNA synthesis and methylation in astrocytes in primary cultures.

High rates of RNA synthesis in bulk-prepared astroglia have been demonstrated by Yanagihara,[256] but different results have been reported by other investigators and may be related to differential membrane permeabilities or to differences in cytoplasmic precursor pool size.[12]

6.3. Lipids

Lipid composition and metabolism have been studied in different preparations of bulk-separated astroglia[257-260] and in glioma cell lines.[261] The findings have been authoritatively reviewed by Norton *et al.*[257] who concluded that "neuronal perikarya and astrocytes have very similar lipid profiles. Gangliosides, long thought to be specific for neuronal cell membranes, are present in all cells. . . . The galactolipids, cerebroside and sulfatide, once thought to be specific to myelin and, by inference, to oligodendroglia, are present in all brain cell types."

In agreement with the large surface–volume ratio in astrocytes,[26] their lipid content is high,[12,257] and the incorporation of fatty acids is faster than into neuronal perikarya.[262–263] Glioma cell lines have a much lower lipid content, the composition of which is very different from that in bulk-separated astroglia; the cell lines therefore do not seem to be suitable for studies of lipid composition or metabolism in astroglia.[257] Primary cultures of brain cells may, however, be much better for this purpose,[264] and differences between ganglioside composition in astrocytes and neurons in primary cultures have been observed by Dreyfus *et al.*[265] In view of transmitter effects on membrane fluidity (Section 4.4.1) and the large surface area and high lipid content in astrocytes, further information about lipid metabolism in these cells is urgently needed.

6.4. Carbohydrates

It is well established that glycogen is present in astrocytes but under normal conditions absent from neurons. For further details and references see, e.g., refs. 12, 29. In Section 4.4.3. it was mentioned that the utilization of glycogen can be regulated by external factors.

7. CONCLUDING REMARKS

Any review of a rapidly expanding, controversial area of research is of necessity to some extent subjective and selective. The present author has tried to emphasize dynamic properties of astrocytes that seem to be of crucial importance in neuronal–astrocytic interactions. Most of the properties that have been discussed are, in the opinion of the author, relatively well established, but a few hypotheses with much less experimental foundation were included. One aspect of astrocytic neurochemistry that should be stressed is the fact that regional variations do occur and that astrocytes undergo a very considerable postnatal maturation. This maturation may to a large extent explain neurochemical alterations of the brain occurring during development, and it calls for great caution in selecting astrocytic preparations that have attained a sufficient degree of metabolic differentiation.

REFERENCES

1. Hydén, H., 1959, *Nature* **184**:433–435.
2. Sellinger, O. Z., and Azcurra, J. M., 1974, *Research Methods in Neurochemistry*, Volume 2 (N. Marks and R. Rodnight, eds.), Plenum Press, New York, pp. 3–38.
3. Hamberger, A., Hansson, H. A., and Sellström, Å., 1975, *Exp. Cell Res.* **92**:1–10.
4. Nagata, Y., and Tsukada, Y., 1978, *Reviews of Neuroscience*, Vol. 3 (S. Ehrenpreis and I. Kopin, eds.), Raven Press, New York, pp. 195–221.
5. Farooq, M., and Norton, W. T., 1978, *J. Neurochem.* **31**:887–894.

6. Henn, F. A., 1980, *Advances in Cellular Neurobiology*, Vol. 1 (S. Fedoroff and L. Hertz, eds.), Academic Press, New York, pp. 373–403.
7. Benda, P., 1978, *Dynamic Properties of Glial Cells* (E. Schoffeniels, G. Franck, L. Hertz, and D. B. Tower, eds.), Pergamon Press, Oxford, pp. 66–81.
8. Booher, J., and Sensenbrenner, M., 1972, *Neurobiology* 2:97–105.
9. Hertz, L., Juurlink, B. H. J., Fosmark, H., and Schousboe, A., 1981, *Neuroscience Approached Through Cell Culture*, Vol. 1 (S. E. Pfeiffer, ed.), CRC Press, Boca Raton (in press).
10. Johnston, P. V., and Roots, B. I., 1970, *Int. Rev. Cytol.* 29:265–280.
11. Johnston, P. V., and Roots, B. I., 1978, *Neuroscience* 3:649–650.
12. Hertz, L., 1977, *Cell, Tissue and Organ Cultures in Neurobiology* (S. Fedoroff and L. Hertz, eds.), Academic Press, New York, pp. 39–71.
13. Henn, F. A., and Henn, S. W., 1980, *Prog. Neurobiol.* 15:1–17.
14. Eng, L. F., Vanderhaeghen, J. J., Bignami, A., and Gerstl. B., 1971, *Brain Res.* 28:351–354.
15. Bignami, A., Dahl, D., and Rueger, D. C., 1980, *Advances in Cellular Neurobiology* (S. Fedoroff and L. Hertz, eds.), Academic Press, New York, pp. 285–310.
16. Norenberg, M. D., and Martinez-Hernandez, A., 1979, *Brain Res.* 161:303–310.
17. Varon, S., 1978, *Dynamic Properties of Glia Cells* (E. Schoffeniels, G. Franck, L. Hertz, and D. B. Tower, eds.), Pergamon Press, Oxford, pp. 93–103.
18. Cooper, A. J. L., McDonald, J. M., Gelbard, A. S., Gledhill, R. F., and Duffy, T. E., 1979, *J. Biol. Chem.* 254:4982–4992.
19. Berl, S., Lajtha, A., and Waelsch, H., 1961, *J. Neurochem.* 7:186–197.
20. Berl, S., Clarke, D. D., and Schneider, D. (eds.), 1974, *Metabolic Compartmentation and Neurotransmission. Relation to Brain Structure and Function*, Plenum Press, New York.
21. Balazs, R., and Cremer, J. E., 1972, *Metabolic Compartmentation in the Brain*, Macmillan Press, London, pp. 1–383.
22. Hertz, L., 1979, *Prog. Neurobiol.* 13:277–323.
23. Schousboe, A., 1981, *Int. Rev. Neurobiol.* 22:1–45.
24. Tower, D. B., 1978, *Dynamic Properties of Glia Cells* (E. Schoffeniels, G. Franck, L. Hertz, and D. B. Tower, eds.), Pergamon Press, Oxford, pp. 443–460.
25. Pope, A., 1978, *Dynamic Properties of Glia Cells* (E. Schoffeniels, G. Franck, L. Hertz, and D. B. Tower, eds.), Pergamon Press, Oxford, pp. 13–20.
26. Wolff, J. R., 1970, *VIth International Congress of Neuropathology, Paris*, Masson Cie, Paris, pp. 327–333, 352.
27. Wolff, J. R., and Güldner, F.-H., 1978, *Dynamic Properties of Glia Cells* (E. Schoffeniels, G. Franck, L. Hertz, and D. B. Tower, eds.), Pergamon Press, Oxford, pp. 115–118.
28. Palay, S. L., and Chan-Palay, V., 1977, *Handbook of Physiology*. Section 1: *The Nervous System*, Volume I, *Cellular Biology of Neurons*, Part 1 (J. M. Brookhart, V. B. Mountcastle, E. R. Kandel, and S. R. Geiger, eds.), American Physiological Society, Bethesda, pp. 5–37.
29. Varon, S. S., and Somjen, G. G., 1979, *Neurosci. Res. Program Bull.* 17:1–239.
30. Brightman, M. W., Anders, J. J., and Rosenstein, J. M., 1980, *Advances in Cellular Neurobiology*, Volume 1, (S. Fedoroff and L. Hertz, eds.), Academic Press, New York, pp. 3–29.
31. Kuffler, S. W., and Nicholls, J. G., 1966, *Ergeb. Physiol.* 57:1–90.
32. Moonen, G., and Nelson, P. G., 1978, *Dynamic Properties of Glia Cells* (E. Schoffeniels, G. Franck, L. Hertz, and D. B. Tower, eds.), Pergamon Press, Oxford, pp. 389–393.
33. Moonen, G., 1980, *Aspects Structuraux et Fonctionnels de la Croissance et de la Différenciation d'Astrocytes et de Neurones Cultivés in Vitro*, Thesis, University of Liège, Liège.
34. Glees, P., 1955, *Neuroglia Morphology and Function*, Charles C Thomas, Springfield, Illinois.
35. Galambos, R., 1961, *Proc. Natl. Acad. Sci. U.S.A.* 47:129–136.
36. Hertz, L., 1965, *Nature* 249:663–664.
37. Henn, F. A., Anderson, D. J., and Sellström, Å., 1977, *Nature* 266:637–638.
38. Hertz, L., Baldwin, F., and Schousboe, A., 1979, *Can. J. Physiol. Pharmacol.* 57:223–226.
39. Van Calker, D., and Hamprecht, B., 1980, *Advances in Cellular Neurobiology*, Vol. 1 (S. Fedoroff and L. Hertz, eds.), Academic Press, New York, pp. 31–67.
40. Hertz, L., Mukerji, S., and Richardson, J. S., 1982, *Psych. Res.*, submitted.

41. Henn, F., and Henke, D. J., 1978, *Neuropharmacology* **17**:985–988.
42. Hertz, L., and Mukerji, S., 1980, *Can. J. Physiol. Pharmacol.* **58**:217–220.
43. Hertz, L., 1966, *J. Neurochem.* **13**:1373–1387.
44. Roth-Schechter, B., Sensenbrenner, M., and Mandel, P., 1976, *C. R. Acad. Sci. [D] (Paris)* **283**:1333–1335.
45. Hertz, E., and Hertz, L., 1979, *In Vitro* **15**:429–436.
46. Olson, J. E., and Holtzman, D., 1981, *J. Neurosci. Res.* **5**:497–506.
47. Olson, J. D., and Holtzman, D., 1981, *Neurochem. Res.* (in press).
48. Hertz, L., 1978, *Dynamic Properties of Glia Cells* (E. Schoffeniels, G. Franck, L. Hertz, and D. B. Tower, eds.), Pergamon Press, Oxford, pp. 121–132.
49. Petito, C., 1981, *Abstracts Second Beaune Conference, Scientific Internat. Research*, Montrouge, France.
50. Hertz, L., 1981, *J. Cerebral Blood Flow and Metab.*, **1**:143–153.
51. Sokoloff, L., Reivich, M., Kennedy, C., Des Rosiers, M. H., Patlak, C. S., Pettigrew, K. D., Sakurada, O., and Shinohara, M., 1977, *J. Neurochem.* **28**:897–916.
52. Kao-Jen, J., and Wilson, J. E., 1980, *J. Neurochem.* **35**:667–678.
53. Roth-Schechter, B. F., and Mandel, P., 1978, *Dynamic Properties of Glia Cells* (E. Schoffeniels, G. Franck, L. Hertz and D. B. Tower, eds.), Pergamon Press, Oxford, pp. 403–412.
54. Roth-Schechter, B. F., Tholey, G., and Mandel, P., 1979, *Neurochem. Res.* **4**:83–97.
55. Roth-Schechter, B. F., Winterith, M., Tholey, G., Dierich, A., and Mandel, P., 1979, *J. Neurochem.* **33**:669–676.
56. Hertz, L., and Schousboe, A., 1975, *Int. Rev. Neurobiol.* **18**:141–211.
57. Passonneau, J. V., Schwartz, J. P., and Lust, W. D., 1978, *Dynamic Properties of Glia Cells* (E. Schoffeniels, G. Franck, L. Hertz, and D. B. Tower, eds.), Pergamon Press, Oxford, pp. 133–142.
58. Pevzner, L. Z., 1979, *Functional Biochemistry of the Neuroglia*, Consultants Bureau, Plenum Press, New York.
59. Giacobini, E., 1962, *J. Neurochem.* **9**:169–177.
60. Roussel, G., Delaunoy, J.-P., Nussbaum, J.-L., and Mandel, P., 1979, *Brain Res.* **160**:47–55.
61. Ghandour, M. S., Langley, O. K., Vincendon, G., and Gombos, G., 1979, *J. Histochem. Cytochem.* **27**:1634–1637.
62. Wieth, J. O., 1979, *J. Physiol. (Lond.)* **294**:521–539.
63. Kimelberg, H. K., Biddlecome, S., and Bourke, R. S., 1979, *Brain Res.* **173**:111–124.
64. Kimelberg, H. K., Narumi, S., and Bourke, R. S., 1978, *Brain Res.* **153**:55–77.
65. Schousboe, A., Nissen, C., Bock, E., Sapirstein, V., Juurlink, B. H. J., and Hertz, L., 1980, *Tissue Culture in Neurobiology* (E. Giacobini, A. Vernadakis, A. Shahar, eds.), Raven Press, New York, pp. 397–409.
66. De Vellis, J., and Brooker, G., 1973, *Tissue Culture of the Nervous System* (S. Gato, ed.), Plenum Press, New York, pp. 231–246.
67. Sarthy, P. V., and Lam, D. M. K., 1978, *J. Cell Biol.* **78**:675–684.
68. Yu, A. C. H., Hertz, E., and Hertz, L., 1982, *Trans. Amer. Soc. Neurochem.* **13** (in press).
69. Hertz, L., 1979, *Neuropharmacology* **18**:629–633.
70. Grisar, T., 1979, *L'Astroglie dens les Méchanismes de Régulation des Ions K⁺ au Sein des Espaces Extracellulaires du Système Nerveux Central*, Thesis, University of Liege, Liege.
71. Aleksidze, N. G., and Blomstrand, C., 1969, *Dokl. Akad. Nauk. SSSR* **186**:1429–1430; Translated in: *Proc. Acad. Sci. USSR Biochem. Series*.
72. Haljamae, H., and Hamberger, A., 1971, *J. Neurochem.* **18**:1903–1912.
73. Hertz, L., Dittmann, L., and Mandel, P., 1973, *Brain Res.* **60**:517–529.
74. Hertz, L., 1976, *Transport Phenomena in the Nervous System, Physiological and Pathological Aspects, Advances of Experimental Medicine and Biology*, Volume 69 (G. Levi, L. Battistin, and A. Lajtha, eds.), Plenum Press, New York, pp. 371–383.
75. Lewis, D. V., and Schuette, W. H., 1975, *J. Neurophysiol.* **38**:405–417.
76. Lothman, E., La Manna, J., Cordingley, G., Rosenthal, M., and Somjen, G., 1975, *Brain Res.* **88**:15–36.
77. Lewis, D. V., O'Connor, M. J., and Schuette, W. H., 1974, *Electroencephalogr. Clin. Neurophysiol.* **36**:347–356.

78. Jobsis, F., Rosenthal, M., La Manna, J., Lothman, E., Cordingley, G., and Somjen, G., 1975, *Brain Work, Alfred Benzon Symposium*, Volume 8 (D. H. Ingvar and N. A. Lassen, eds.), Munksgaard, Copenhagen, pp. 185–196.
79. Hultborn, R., and Hydén, H., 1974, *Exp. Cell. Res.* **87**:346–350.
80. Whittaker, V. P., 1969, *Handbook of Neurochemistry*, 1st. ed., Vol. 2 (A. Lajtha, ed.), Plenum, New York, pp. 327–364.
81. Sieghart, W., Sellström, Å., and Henn, F., 1978, *J. Neurochem.* **30**:1587–1589.
82. Hertz, L., Sastry, B. R., Hertz, E., Larsson, O. M., Krogsgaard-Larsen, P., and Schousboe, A., 1980, *Brain Res. Bull.* **5**(Suppl. 2):653–658.
83. Kety, S. S., 1957, *Metabolism of the Nervous System* (D. Richter, ed.), Pergamon Press, Oxford, pp. 221–237.
84. Hertz, L., 1977, *Pharmacol. Rev.* **29**:35–65.
85. Cummins, C. J., Glover, R. A., and Sellinger, O. Z., 1979, *Brain Res.* **170**:190–193.
86. Trachtenberg, M. C., and Packey, D. J., 1980, *Abstr. Soc. Neurosci.* **6**:325.
87. Takagaki, G., 1972, *J. Neurochem.* **19**:1737–1751.
88. Schousboe, A., Fosmark, H., and Hertz, L., 1975, *J. Neurochem.* **25**:909–911.
89. Cummins, J., and Hydén, H., 1962, *Biochim. Biophys. Acta.* **60**:271–283.
90. Henn, F. A., Haljamae, H., and Hamberger, A., 1972, *Brain Res.* **43**:437–443.
91. Medzihradsky, F., Sellinger, D. Z., Nandhasri, P. S., and Santiago, J. C., 1972, *J. Neurochem.* **19**:543–545.
92. Nagata, Y., Mikoshiba, K., and Tsukada, Y., 1974, *J. Neurochem.* **22**:493–503.
93. Moonen, G., and Franck, G., 1977, *Neurosci. Lett.* **4**:263–267.
94. Bignami, A., and Palladini, G., 1966, *Nature* **209**:413–414.
95. Lowe, D. A., 1978, *Brain Res.* **148**:347–363.
96. Stahl, W. L., and Broderson, S. H., 1976, *Fed. Proc.* **35**:1260–1265.
97. Grisar, T., Franck, G., and Schoffeniels, E., 1978, *Dynamic Properties of Glia Cells* (E. Schoffeniels, G. Franck, L. Hertz, and D. B. Tower, eds.), Pergamon Press, Oxford, pp. 359–369.
98. Grisar, T., Frere, J. M., Charlier-Grisar, J., Franck, G., and Schoffeniels, E., 1978, *FEBS Lett.* **89**:173–176.
99. Grisar, T., Frere, J. M., and Franck, G., 1979, *Brain Res.* **165**:87–103.
100. Hansen, O., 1976, *Biochim. Biophys. Acta* **433**:383–397.
101. Sweadner, K. J., 1979, *J. Biol. Chem.* **254**:6060–6067.
102. Sastry, B. S. R., and Phillis, J. W., 1977, *Can. J. Physiol. Pharmacol.* **55**:170–179.
103. Olson, J. D., and Holtzman, D., 1980, *Abstr. Soc. Neurosci.* **6**:547.
104. Holtzman, D., Herman, M. M., Desautel, M., and Lewiston, N., 1979, *J. Neurochem.* **33**:453–460.
105. Hertz, L., and Schou, M., 1962, *Biochem. J.* **85**:93–104.
106. Friede, R. L., 1964, *J. Cell. Biol.* **20**:5–15.
107. Van den Berg, C. J., and Garfinkel, D., 1971, *Biochem. J.* **123**:211–218.
108. Hertz, L., and Schousboe, A., 1980, *Brain Res. Bull.* **5**(Suppl. 2):389–395.
109. Balcar, V. J., Mark, J., Borg, J., and Mandel, P., 1979, *Neurochem. Res.* **4**:339–354.
110. Cohen, J., Balazs, R., and Woodham, P. C., 1980, *Neurochem. Res.* **5**:963–981.
111. Hösli, L., and Hösli, E., 1978, *Rev. Physiol. Biochem. Pharmacol.* **81**:135–188.
112. Wood, J. D., and Hertz, L., 1980, *Neuropharmacology* **19**:805–808.
113. Schousboe, A., Hertz, L., Larsson, O. M., and Krogsgaard-Larsen, P., 1980, *Brain Res. Bull.*, Suppl. 2, **5**:403–409.
114. Kelly, J. S., and Dick, F., 1978, *Dynamic Properties of Glia Cells* (E. Schoffeniels, G. Franck, L. Hertz, and D. B. Tower, eds.), Pergamon Press, Oxford, pp. 183–192.
115. Schousboe, A., Krogsgaard-Larsen, P., Svenneby, G., and Hertz, L., 1978, *Brain Res.* **153**:623–626.
116. Hertz, L., Wu, P. H., and Schousboe, A., 1978, *Neurochem. Res.* **3**:313–323.
117. Lake, N., and Voaden, M. J., 1976, *J. Neurochem.* **27**:1571–1573.
118. Martin, D. L., 1976, *GABA in Nervous System Function* (E. Roberts, T. N. Chase, and D. B. Tower, eds.), Raven Press, New York, pp. 347–386.
119. Larsson, O. M., Hertz, L., and Schousboe, A., 1980, *J. Neurosci. Res.* **5**:469–477.

120. Wu, P. H., Durden, D. A., and Hertz, L., 1979, *J. Neurochem.* **32**:379–390.
121. McLaughlin, B., Wood, J. G., Saito, K., Barber, R., Vaughn, J. E., Roberts, E., and Wu, J.-Y., 1974, *Brain Res.* **76**:377–391.
122. Szerb, J. C., 1979, *J. Neurochem.* **32**:1565–1573.
123. Barber, R., and Saito, K., 1976, *GABA in Nervous System Function* (E. Roberts, T. N. Chase, and D. B. Tower, eds.), Raven Press, New York, pp. 113–132.
124. Chan-Palay, V., Wu, J.-Y., and Palay, S. L., 1979, *Proc. Natl. Acad. Sci. U.S.A.* **76**:2067–2071.
125. Van den Berg, C. J., Matheson, D. F., and Ronda, G., 1974, *Metabolic Compartmentation and Neurotransmission* (S. Berl, D. D., Clarke, and D. Schneider, eds.), Plenum Press, New York, pp. 515–540.
126. McGeer, E. G., and McGeer, P. L., 1979, *J. Neurochem.* **32**:1071–1075.
127. Bradford, H. F., Ward, H. K., and Thomas, A. J., 1978, *J. Neurochem.* **30**:1453–1459.
128. Van den Berg, C. J., Nijenmanting, W. C., Bruntink, R., and Matheson, D. F., 1979, *Brain Res. Bull.* **4**:687.
129. Schousboe, A., 1980, *Cell. Mol. Biol.* **26**:505–513.
130. Ossola, L., DeFeudis, F. V., and Mandel, P., 1980, *J. Neurochem.* **34**:1026–1029.
131. Ticku, M. K., 1980, *Eur. J. Pharmacol.* **65**:135–136.
132. Hamberger, A., Cotman, C. W., Sellstrom, Å., and Weiler, C. T., 1978, *Dynamic Properties of Glia Cells* (E. Schoffeniels, G. Franck, L. Hertz, and D. B. Tower, eds.), Pergamon Press, Oxford, pp. 163–172.
133. McLennan, H., 1976, *Brain Res.* **115**:139–144.
134. Schousboe, A., 1978, *Dynamic Properties of Glia Cells* (E. Schoffeniels, G. Franck, L. Hertz, and D. B. Tower, eds.), Pergamon Press, Oxford, pp. 173–182.
135. Schousboe, A., and Divac, I., 1979, *Brain Res.* **177**:407–409.
136. Hertz, L., Bock, E., and Schousboe, A., 1978, *Dev. Neurosci.* **1**:226–238.
137. Juurlink, B. H. J., Schousboe, A., Jørgensen, O. S., and Hertz, L., 1981, *J. Neurochem.* **36**:136–142.
138. Hertz, L., Yu, A., Potter, R. L., Nicklas, W. J., and Schousboe, A., 1981, *Abstracts 8th Meeting of the Internat. Soc. for Neurochem.*, Nottingham, p. 174.
139. Quastel, J. H., 1978, *Dynamic Properties of Glia Cells* (E. Schoffeniels, G. Franck, L. Hertz, and D. B. Tower), Pergamon Press, Oxford, pp. 153–162.
140. Hertz, L., Yu, A., Svenneby, G., Kvamme, E., Fosmark, H., and Schousboe, A., 1980, *Neurosci. Lett.* **16**:103–109.
141. Utley, J. D., 1964, *Biochem. Pharmacol.* **13**:1383–1392.
142. Promyslov, M. S., and Andreeva, T. V., 1967, *Ukr. Biokhim. Zh.* **39**:590–592. Quoted in ref. 58.
143. Schousboe, A., Hertz, L., Svenneby, G., and Kvamme, E., 1979, *J. Neurochem.* **32**:943–950.
144. Berl, S., Nicklas, W. J., and Clarke, D. D., 1978, *Dynamic Properties of Glia Cells* (E. Schoffeniels, G. Franck, L. Hertz, and D. B. Tower, eds.), Pergamon Press, Oxford, pp. 143–149.
145. Waelsch, H., Berl, S., Rossi, C. A., Clarke, D. D., and Purpura, D. P., 1964, *J. Neurochem.* **11**:717–728.
146. Henn, F. A., 1980, *Advances in Cellular Neurobiology* (S. Fedoroff and L. Hertz, eds.), Academic Press, New York, pp. 373–403.
147. Pfeiffer, S. E., Betschart, B., Cook, J., Mancini, P., and Morris, R., 1977, *Cell, Tissue and Organ Cultures in Neurobiology* (S. Fedoroff, and L. Hertz, eds.), Academic Press, New York, pp. 287–346.
148. Descarries, L., Baudet, A., and Watkins, K. C., 1975, *Brain Res.* **100**:563–588.
149. Descarries, L., Watkins, K., and Lapierre, Y., 1977, *Brain Res.* **133**:197–222.
150. Chan-Palay, V., 1976, *Brain Res.* **102**:103–130.
151. Tennyson, V. M., Heikkila, R., Mytilineon, C., Cote, L., and Cohen, G., 1974, *Brain Res.* **82**:341–348.
152. Dismukes, K., 1977, *Nature* **269**:557–558.
153. Henn, F., and Hamberger, A., 1971, *Proc. Natl. Acad. Sci. U.S.A.* **68**:2686–2691.
154. Haber, B., and Hutchison. H. T., 1976, *Transport Phenomena in the Nervous System, Phys-*

iological and Pathological Aspects, Advances of Experimental Medicine and Biology, Volume 69 (G. Levi, L. Battistin, and A. Lajtha, eds.), Plenum Press, New York, pp. 179–198.

155. Vernadakis, A., and Nidess, R., 1976, *Neurochem. Res.* **1**:385–402.
156. Ritchie, T., Glusman, S., and Haber, B., 1978, *Trans. Am. Soc. Neurochem.,* Volume 9.
157. Ritchie, T., Glusman, S., and Haber, B., 1981, *Neurochem. Res.* **6**:441–452.
158. Pelton, E. W., Kimelberg, H. K., Shipherd, S. V., and Bourke, R. S., 1981, *Life Sci.* **14**:1655–1663.
159. Schousboe, A., Tamir, H., Schousboe, I., Mukerji, S., and Hertz, L., 1981, *Trans. Am. Soc. Neurochem.* **12**:135.
160. Ruda, M. A., and Gobel, S., 1980, *Brain Res.* **184**:57–83.
161. Suddith, R. L., Hutchison, H. T., and Haber, B., 1978, *Life Sci.* **22**:2179–2188.
162. Hazama, H., Ito, M., Hirano, M., and Uchimura, H., 1976, *J. Neurochem.* **26**:417–419.
163. Silberstein, S. D., Shein, H. M., and Berv, K. R., 1972, *Brain Res.* **41**:245–248.
164. Murphy, D. L., Donelly, C. H., and Richelson, E., 1976, *J. Neurochem.* **26**:1231–1235.
165. Skaper, S. D., Adelson, G. L., and Seegmiller, J. E., 1976, *J. Neurochem.* **27**:1065–1070.
166. Cawthon, R. M., and Breakefield, X. O., 1979, *Nature* **281**:692–694.
167. Braestrup, C., Nissen, C., Squires, R. F., and Schousboe, A., 1978, *Neurosci. Lett.* **9**:45–49.
168. Arbogast, B. W., and Arsenis, C., 1974, *Neurobiology* **4**:21–37.
169. Schultz, J., Hamprecht, B., and Daly, J. W., 1972, *Proc. Natl. Acad. Sci. U.S.A.* **69**:1266–1270.
170. Lewin, E., and Bleck, V., 1979, *J. Neurochem.* **33**:365–367.
171. Hertz, L., 1978, *J. Neurochem.* **31**:55–62.
172. Schubert, P., and Kreutzberg, G. W., 1976, *The Cerebral Vessel Wall* (J. Cervais-Navarro, E. Betz, F. Matakas, and R. Wüllemweber, eds.), Raven Press, New York, pp. 207–213.
173. Trams, E. G., and Lauter, C. J., 1975, *Biochem. J.* **152**:681–687.
174. Kreutzberg, G. W., Barrow, K. D., and Schubert, P., 1978, *Brain Res.* **158**:247–257.
175. Massarelli, R., Sensenbrenner, M., Ebel, A., and Mandel, P., 1974, *Neurobiology* **4**:414–418.
176. Giacobini, E., 1964, *Morphological and Biochemical Correlates of Neural Activity* (M. M. Cohen and R. S. Snider, eds.), Harper & Row, New York, pp. 15–38.
177. Vernadakis, A., and Gibson, D. A., 1973, *Abstracts, Fourth International Meeting of the International Society for Neurochemistry,* p. 138.
178. Vernadakis, A., and Arnold, E. B., 1980, *Advances in Cellular Neurobiology,* Volume 1 (S. Fedoroff, and L. Hertz, eds.), Academic Press, New York, pp. 229–283.
179. Maguire, M. E., Wiklund, R. A., Anderson, H. J., and Gilman, A. G., 1976, *J. Biol. Chem.* **251**:1221–1231.
180. Schmitt, H., and Pochet, R., 1977, *FEBS Lett.* **76**:302–305.
181. Lucas, M., and Bockaert, J., 1976, *Mol. Pharmacol.* **13**:314–329.
182. Terasaki, W. L., and Brooker, G., 1978, *J. Biol. Chem.* **253**:5418–5425.
183. Johnson, G. L., Wolfe, B. B., Harden, T. K., Molinoff, P. B., and Perkins, J. P., 1978, *J. Biol. Chem.* **253**:1472–1480.
184. Dolphin, A., Adrien, J., Hamon, M., and Bockaert, J., 1979, *Mol. Pharmacol.* **15**:1–15.
185. Bylund, D. B., and Snyder, S. H., 1976, *Mol. Pharmacol.* **12**:568–580.
186. Wolfe, B. B., Harden, K. T., Sporn, J. R., and Molinoff, P. B., *J. Pharmacol. Exp. Ther.* **207**:446–457.
187. Schneck, D. W., Pritchard, J. F., and Hayes, A. H., Jr., 1977, *J. Pharmacol. Exp. Ther.* **203**:621–629.
188. Hertz, L., Mukerji, S., and Richardson, J. S., 1981, *Eur. J. Pharmacol.* **72**:267–268.
189. Gilman, A., and Nirenberg, M., 1971, *Proc. Natl. Acad. Sci. U.S.A.* **68**:2165–2168.
190. Hertz, L., Richardson, J. S., and Mukerji, S., 1980, *Can. J. Physiol. Pharmacol.* **58**:1515–1519.
191. McCarthy, K. D., and de Vellis, J. D., 1979, *Life Sci.* **24**:639–649.
192. Fillion, G., Beaudoin, D., Rousselle, J. C., and Jacob, J., 1980, *Brain Res.* **198**:361–374.
193. Tamir, H., Klein, A., and Rapport, M. M., 1976, *J. Neurochem.* **26**:871–878.
194. Tamir, H., Bebivian, R., Muller, F., and Casper, D., 1980, *J. Neurochem.* **35**:1033–1044.
195. Hertz, L., and Tamir, H., 1981, *J. Neurochem.* 1331–1334.
196. Villegas, J., 1978, *Dynamic Properties of Glia Cells* (E. Schoffeniels, G. Franck, L. Hertz, and D. B. Tower, eds.), Pergamon Press, Oxford, pp. 207–215.

197. Narumi, S., Kimelberg, H. K., and Bourke, R. S., 1978, *J. Neurochem.* **31**:1479–1490.

198. Clark, R. B., Cross, R., Su, Y.-F., and Perkins, J. P., 1974, *J. Biol. Chem.* **249**:5296–5303.

199. Syapin, P. J., and Skolnick, P., 1979, *J. Neurochem.* **32**:1047–1051.

200. Gallager, D. W., Mallorga, P., Oertel, W., Henneberry, R., and Tallman, J., 1981, *J. Neurosci.* **1**:218–225.

201. Strittmatter, W., Hirata, F., Axelrod, J., Mallorga, P., Tallman, J. F., and Henneberry, R. C., 1979, *Nature* **282**:857–859.

202. Tallman, J. F., Gallager, D. W., Mallorga, P., Thomas, J. W., Strittmatter, W., Hirata, F., and Axelrod, J., 1979, *Receptors for Neurotransmitters and Peptide Hormones, Advances in Biochemical Psychopharmacology,* Volume 21 (G. Pepe, M. J. Kuhar, and S. J. Enna, eds.), Raven Press, New York, pp. 277–283.

203. Hirata, F., Tallman, J. F., Jr., Henneberry, R. C., Mallorga, P., Strittmatter, W. J., and Axelrod, J., 1979, *Receptors for Neurotransmitters and Peptide Hormones, Advances in Biochemical Psychopharmacology,* Volume 21 (G. Pepe, M. J. Kuhar, and S. J. Enna, eds.), Raven Press, New York, pp. 91–97.

204. Opler, L. A., and Makman, M. M., 1972, *Biochem. Biophys. Res. Commun.* **46**:1140–1145.

205. Browning, E. T., Schwartz, J. P., and Breckenridge, B. McL., 1974, *Mol. Pharmacol.* **10**:162–174.

206. Newburgh, R. W., and Rosenberg, R. N., 1973, *Biochem. Biophys. Res. Commun.* **52**:614–619.

207. Kogure, K., Scheinberg, P., Matsumoto, A., Busto, R., and Reinmuth, O. M., 1975, *Arch. Neurol.* **32**:21–24.

208. Church, G. A., Kimelberg, H. K., and Sapirstein, V. A., 1980, *J. Neurochem.* **34**:873–879.

209. Hertz, L., 1981, *Eleventh International Congress of Anatomy: Glial and Neuronal Cell Biology* (S. Fedoroff, ed.), Alan R. Liss, New York, pp. 45–58.

210. Segal, M., Sagie, D. B., and Mayevsky, A., 1980, *Brain Res.* **202**:387–399.

211. Hertz, L., Wu, P. H., and Phillis, J. W., 1979, *Soc. Neurosci. Abstr.* **5**:404.

212. Phillis, J. W., Bender, A. S., and Wu, P. H., 1980, *Brain Res.* **195**:494–498.

213. Somjen, G. G., 1975, *Annu. Rev. Physiol.* **37**:163–190.

214. Vyklický, L., Syková, E., and Kriz, N., 1975, *Brain Res.* **87**:77–80.

215. Vern, B. A., Schuette, W. H., and Thibault, L. E., 1977, *J. Neurophysiol.* **40**:1015–1023.

216. Cordingley, G. E., and Somjen, G. G., 1978, *Brain Res.* **151**:291–306.

217. Hertz, L., 1978, *Brain Res.* **145**:202–208.

218. Kukes, G., De Vellis, J., and Elul, R., 1976, *Brain Res.* **104**:93–105.

219. Kukes, G., Elul, R., and De Vellis, J., 1976, *Brain Res.* **104**:71–92.

220. Kimelberg, H. K., 1974, *J. Neurochem.* **22**:971–976.

221. Kimelberg, H. K., Bowman, C., Biddlecome, S., and Bourke, R. S., 1979, *Brain Res.* **177**:533–550.

222. Marchbanks, R. M., and Campbell, C. W. B., 1976, *J. Neurochem.* **26**:973–980.

223. Pastuszko, A., Wilson, D. F., Erecinska, M., and Silver, I. A., 1981, *J. Neurochem.* **36**:116–123.

224. Hertz, L., and Franck, G., 1978, *Dynamic Properties of Glia Cells* (E. Schoffeniels, G. Franck, L. Hertz, and D. B. Tower, eds.), Pergamon Press, Oxford, pp. 383–388.

225. Coles, J. A., and Tsacopoulos, M., 1979, *J. Physiol. (Lond.)* **290**:525–549.

226. Latzkovits, L., Sensenbrenner, M., and Mandel, P., 1974, *J. Neurochem.* **23**:193–200.

227. Gardner-Medwin, A. R., 1977, *J. Physiol. (Lond.)* 32P–33P.

228. Orkand, R. K., 1969, *Basic Mechanisms of the Epilepsies* (H. H. Jasper, A. A. Ward, and A. Pope, eds.), Little, Brown, Boston, pp. 737–746.

229. Abbott, N. J., 1979, *Trends Neurosci.* **2**:91–93.

230. Schlue, W. R., and Deitmer, J. W., 1980, *J. Exp. Biol.* **87**:23–43.

231. Hertz, L., and Nissen, C., 1976, *Brain Res.* **110**:182–188.

232. Roots, B., 1978, *Dynamic Properties of Glia Cells* (E. Schoffeniels, G. Franck, L. Hertz, and D. B. Tower, eds.), Pergamon Press, Oxford, pp. 45–54.

233. Radojcic, T., and Pentreath, V. W., 1979, *Progr. Neurobiol.* **12**:115–179.

234. Sweadner, K. J., and Goldin, S. M., 1975, *J. Biol. Chem.* **250**:4022–4024.

235. Gill, T. H., Young, O. M., and Tower, D. B., 1974, *J. Neurochem.* **23**:1011–1018.

236. Zerahn, K., 1956, *Acta. Physiol. Scand.* **36**:300–318.
237. Hertz, L., and Chaban, G., 1981, *Neuroscience Approached Through Cell Culture*, Vol. 1 (S. E. Pfeiffer, ed.), CRC Press, Boca Raton (in press).
238. Catterall, W. A., and Nirenberg, M., 1973, *Proc. Natl. Acad. Sci. U.S.A.* **70**:3759–3763.
239. Gorkin, R. A., and Richelson, E., 1979, *Brain Res.* **171**:365–368.
240. Janka, Z., Szentistvanyi, I., Juhasz, A., and Rimanoczy, A., 1980, *Neuropharmacology* **19**:827–830.
241. Lazarewicz, J. W., Kanje, M., Sellström, Å., and Hamberger, A., 1977, *J. Neurochem.* **29**:495–502.
242. Barnes, E. M., and Mandel, P., 1981, *J. Neurochem.* **36**:82–85.
243. Latzkovits, L., 1978, *Dynamic Properties of Glia Cells* (E. Schoffeniels, G. Franck, L. Hertz, and D. B. Tower, eds.), Pergamon Press, Oxford, pp. 327–336.
244. Kimelberg, H. K., and Biddlecome, S., 1980, *Soc. Neurosci. Abstr.* **6**:548.
245. White, F. P., and Hertz, L., 1981, *Neurochem. Res.* **6**:353–364.
246. Johnson, D. E., and Sellinger, O. Z., 1971, *J. Neurochem.* **18**:1445–1460.
247. Johnson, D. E., and Sellinger, O. Z., 1973, *Neurobiology* **3**:113–123.
248. Rose, S. P. R., 1975, *J. Neurosci. Res.* **1**:19–30.
249. Hamberger, A., and Sourander, P., 1975, *Neurochem. Res.* **3**:535–547.
250. Packman, P. M., Blomstrand, C., and Hamberger, A., 1971, *J. Neurochem.* **18**:1–9.
251. Stewart, R. C., and Rosenberg, R. N., 1979, *Int. Rev. Neurobiol.* **21**:275–309.
252. Lange, P. W., 1978, *Dynamic Properties of Glia Cells* (E. Schoffeniels, G. Franck, L. Hertz, and D. B. Tower, eds.), Pergamon Press, Oxford, pp. 231–245.
253. Hydén, H., 1967, *The Neurosciences* (G. C. Quarton, T. Melnechuk, and F. O. Schmitt, eds.), The Rockefeller University Press, New York, pp. 248–266.
254. Watson, W. E. 1974, *Physiol. Rev.* **54**:245–271.
255. Sellinger, O. Z., and Der, O., *J. Neurochem.* **35**:1436–1445.
256. Yanagihara, T., 1979, *J. Neurochem.* **32**:169–177.
257. Norton, W. T., Abe, T., Poduslo, S. E., and DeVries, G. H., 1975, *J. Neurosci. Res.* **1**:57–75.
258. Francescangeli, E., Govacci, G., Piccinin, G. L., Mozzi, R., Woelk, H., and Porcellati, G., 1977, *J. Neurochem.* **28**:171–176.
259. Abe, T., Miyatake, T., Norton, W. T., and Suzuki, K., 1979, *Brain Res.* **161**:179–182.
260. Marggraf, W. D., Wang, H., and Kanfer, J. N., 1979, *J. Neurochem.* **32**:353–361.
261. Dawson, G., 1979, *Complex Carbohydrates of Nervous Tissue* (R. U. Margolis and R. K. Margolis, eds.), Plenum Press, New York, pp. 291–325.
262. Cohen, S. R., and Bernsohn, J., 1973, *Brain Res.* **60**:521–525.
263. Morand, O., Baumann, N., and Bourre, J. M., 1979, *Neurosci. Lett.* **13**:177–181.
264. Siegrist, H. P., Bologa-Sandru, L., Burkart, T., Wiesmann, U., Hofmann, K., and Herschkowitz, N., 1981, *J. Neurosci. Res.* **6**:293–301.
265. Dreyfus, H., Louis, J. C., Harth, S., and Mandel, P., 1980, *Neuroscience* **5**:1647–1655.

14

Oligodendrocytes

Leonid Pevzner

1. INTRODUCTION

Although the discovery of the neuroglia in general by Rudolph Virchow[1] was an achievement of the mid-19th century, it was not until the first quarter of this century that Pio Del Rio Hortega[2] described a particular kind of neuroglia. He suggested for it the term *oligodendroglia,* originating from the Greek words ὀλίγος (little, few), δένδρον (tree), and γλια (glue), to emphasize differences in size and in the number of processes between the oligodendrocytes (ODC) and astrocytes (AC).

Historically, biochemical as well as morphological characterization of ODC began later than that of AC. True, the first microchemical and quantitative cytochemical studies of the neuroglia dealt with the perineuronal glia which consist mainly, although not completely, of ODC. However, the most abundant and reliable information about glial metabolism was not amassed until the bulk isolation technique had appeared.[3] This technique, developed initially to compare chiefly neuronal and glial biochemical features, consisted of a combination of brain tissue disaggregation, filtration through meshes, and gradient centrifugation. The combination has resulted in the separation of neuronal- and glial-cell-enriched fractions.[4-8] The subsequent morphological and immunochemical characterization of the glial-enriched fraction showed that it contained more AC than ODC[7,9,10] (see also the chapter by L. Hertz in this volume).

Another promising approach, glial cell lines cultured *in vitro,* also initially provided cells that most often turned out to be of AC nature.[10,11]

Therefore, it was of great value that several groups subsequently developed methods of bulk isolation of sufficiently intact and viable ODC.[12-18] Based on these methods, among which the original procedure by Poduslo and Norton[13] was the most widely adopted, much data has been amassed on the chemical composition and metabolic properties of ODC. These data will be the main topics of consideration in this chapter.

Leonid Pevzner • The Saul R. Korey Department of Neurology, Albert Einstein College of Medicine, Bronx, New York 10461. Present address: Laboratory of Developmental Neurobiology, National Institute of Child Health and Human Development, National Institutes of Health, Bethesda, Maryland 20205.

2. MAIN MORPHOLOGICAL FEATURES OF
OLIGODENDROCYTES

Light and electron microscopic findings have shown that ODC have a spherical or polygonal cell body with a few delicate radiating processes.[19-27] The cell soma ranges from 10 to 20 μm in diameter; the margin of the cell is irregular and somewhat compressed against the adjacent neuropil. The nucleus is usually ovoid and lies, as a rule, in an eccentric position (Figs. 1-3).

Two chief classes of ODC can be distinguished. Perineuronal satellite ODC in the gray matter often are in close association with neuronal bodies; the plasma membranes of these ODC sometimes lie in direct apposition (Fig. 1). Although about 90% of perineuronal satellites are actually ODC, both AC and microglia cells can also be found.[23,28,29] The AC, though located far from the neuronal body, frequently spread their processes between the neuron and the perineuronal ODC (Fig. 2).

The second class of ODC is the interfascicular ODC of the white matter (Fig. 3). These are larger and lighter than satellite ODC and may be aligned in rows between the nerve fibers. These cells play a unique role in the formation of myelin in the central nervous system.[30-32] They are actively dividing cells.[33-37]

Ultrastructures of the gray matter and white matter ODC are rather similar.[33] Within the cytoplasm, parallel cisternae of rough endoplasmic reticulum and a widely dispersed Golgi apparatus are found. Ribosomes associated with membranes and free ribosomes occur, the latter scattered among occasional multivesicular bodies, mitochondria, and coated vesicles. The mitochondria do not show any unusual morphological features. Dense bodies also are common in the cytoplasmic matrix; most of them are related to lipid droplets, lysosomes, or pigment granules.[23,33]

A particular feature of the ODC that distinguishes them from the AC is the constant presence of microtubules, which are about 25 nm in diameter and are very abundant both in perikaryal cytoplasm and in the processes. On the other hand, the ODC, unlike the AC, have few glial filaments or glycogen granules.[23,33]

A number of publications have dealt with the morphological criteria that distinguish large ODC from AC and neurons, and small ODC from microglial cells.[23] The possibility of such confusion should be kept in mind during studies on the biochemical characterization of ODC.

3. MAIN PHYSIOLOGICAL FEATURES OF
OLIGODENDROCYTES

In electrophysiological studies of brain and spinal cord, standing negative potentials were described which, unlike the membrane potentials of neurons, could not be induced to generate impulses or synaptic potentials. These potentials were shown to be generated across the membrane of glial cells.[38-41]

The glial cells seem to possess resting membrane potentials comparable

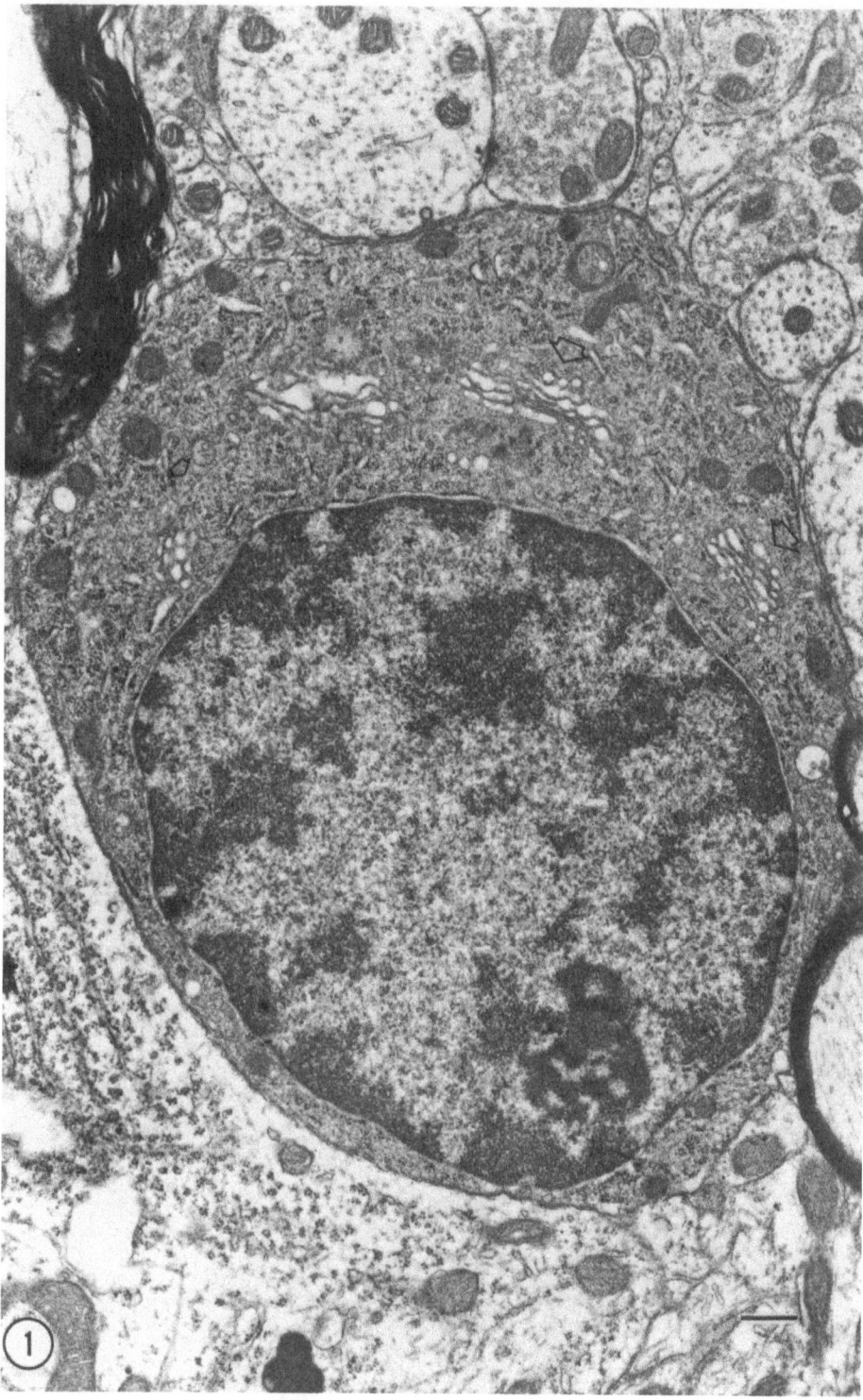

Fig. 1. Perineuronal satellite oligodendrocyte from anterior horn of segment L7 of the rat spinal cord. Note Golgi apparatus (large arrowheads) and microtubules (small arrowheads), important distinguishing features of oligodendrocytes. Below the body of the oligodendrocyte, neuronal perykaryon with the Nissl substance is seen. Note apposition of plasma membranes of the neuron and oligodendrocyte. ×14,400. Bar = 0.5 μm. (Courtesy of Dr. C. S. Raine.)

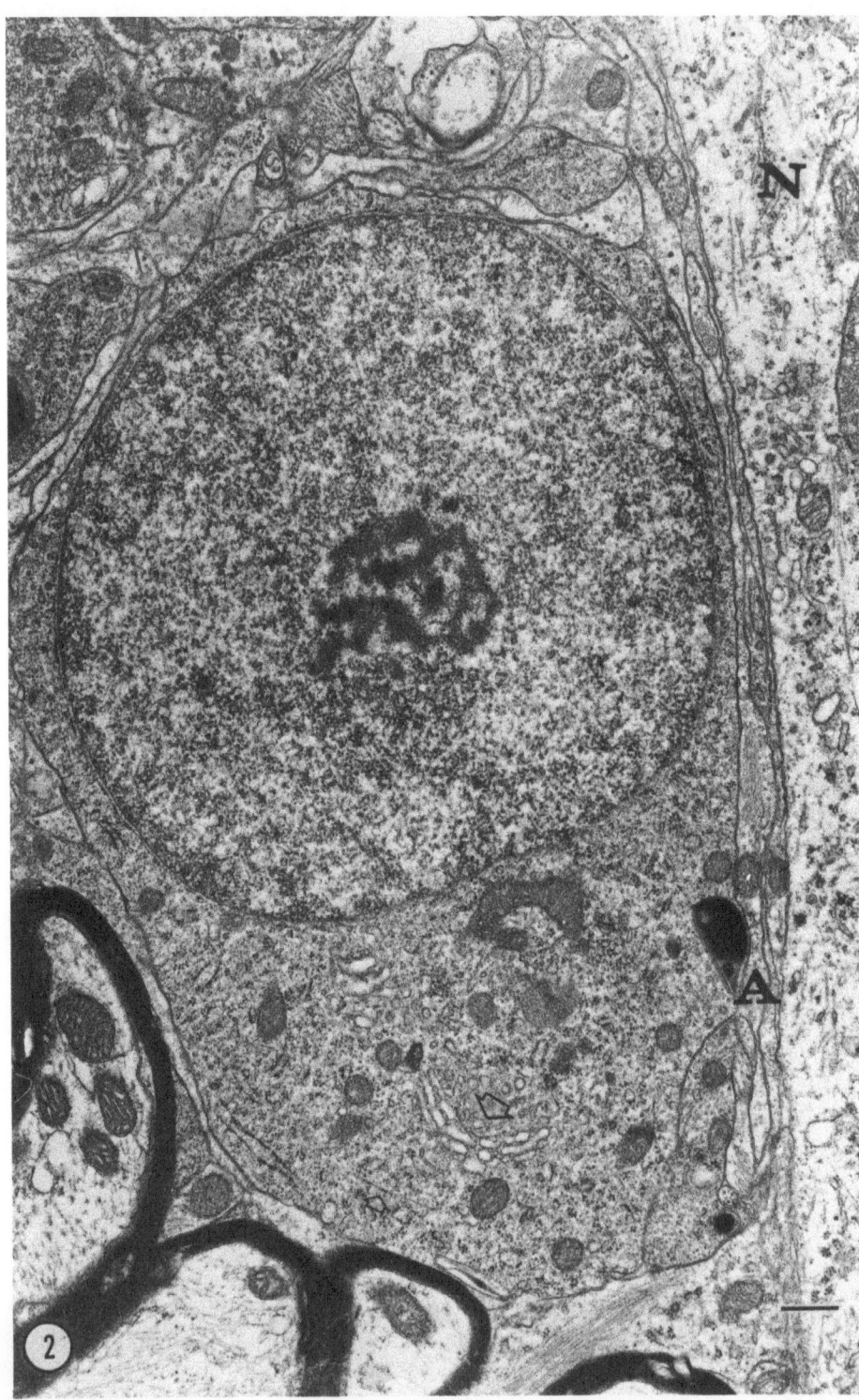

Fig. 2. Perineuronal oligodendrocyte from anterior horn of segment L7 of the dog spinal cord. Golgi apparatus (large arrowhead) and microtubules (small arrowhead) are evident. N, neuron; A, astrocyte, thin processus, with microfilaments, intervening between the perikarya of the neuron and oligodendrocyte. ×14,600. Bar = 0.5 μm. (Courtesy of Dr. C. S. Raine.)

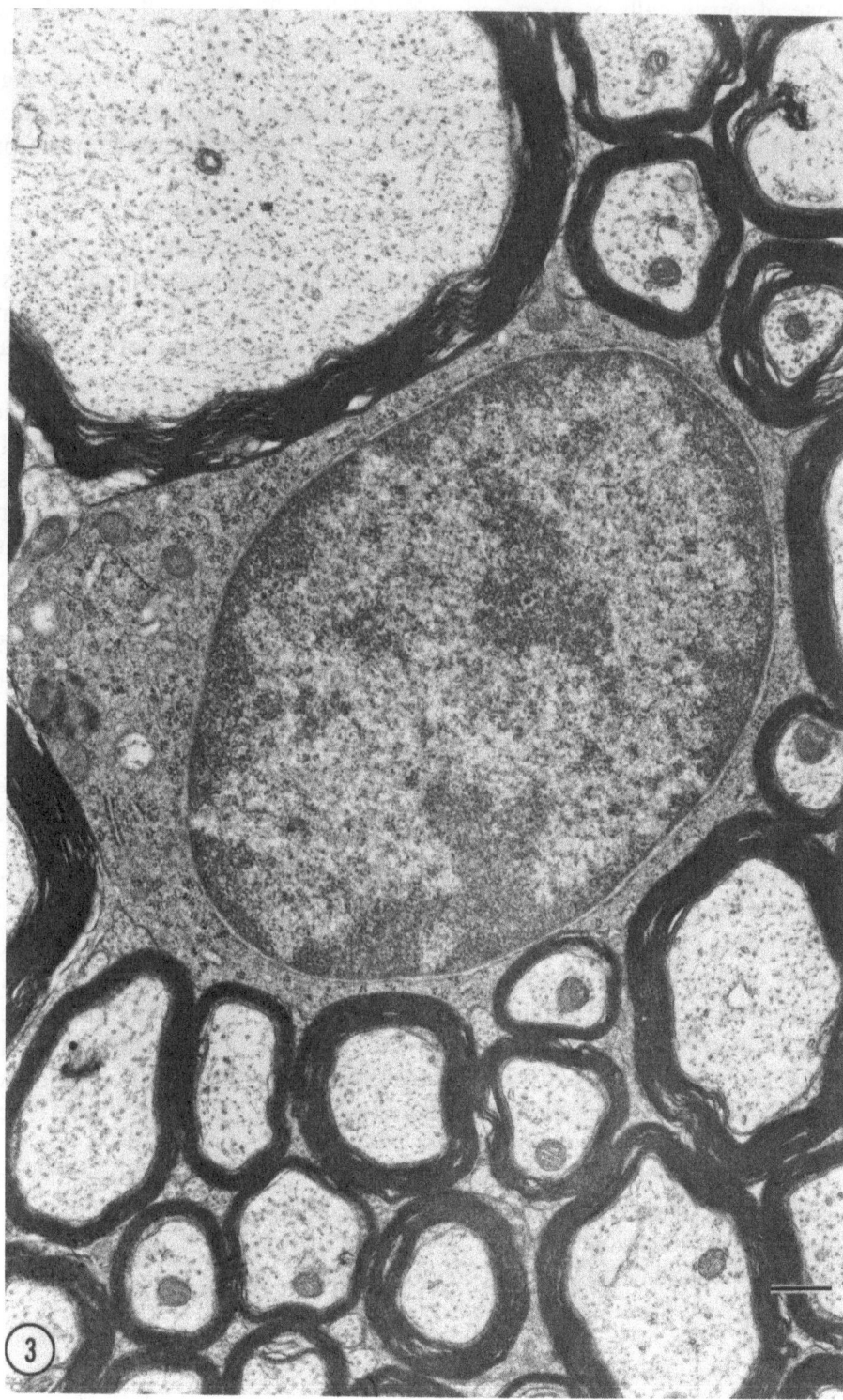

Fig. 3. Interfascicular oligodendrocyte from white matter of segment L7 of the rat spinal cord. Oligodendrocyte is surrounded by cross-sectioned myelinated axons. ×15,600. Bar = 0.5 μm. (Courtesy of Dr. C. S. Raine.)

to those of nerve cells or even somewhat higher. They usually vary between −50 mV and −80 mV in mammalian brain or spinal cord.[38,42-45] On occasion higher values, up to −95 mV, have been reported.[41,46] However, the highest values are considered as coming from the technically best experiments, i.e., from cases of the least damage to the glial cells by the microelectrode puncture.[41,47]

The input resistance of the glial cell membrane fluctuates from 1–12 MΩ in spinal cord[47] to 5–50 MΩ in neocortex.[43,45,48] In tissue culture, this value is usually lower, i.e., 0.5–5 MΩ.[49,50] The specific resistance of glial cells has been found to be 190–480 Ω·cm² in cerebral cortex[41,43] and 3–300 Ω·cm² in tissue culture.[42,50] Time constant values, 200–600 μs, turned out to be rather similar in the cerebral cortex and in tissue culture.[41,43,45,50]

An essential feature of the neuroglial cells is their capability to undergo slow waves of depolarization when the nerve cells of which they are satellites are excited.[30,40,47,51,52] This depolarization is thought to be caused by potassium ions released by the neurons during their activity and accumulated in intercellular clefts. However, the input resistance of glial cells did not change during K^+-induced depolarization.[47] In an intriguing experiment,[53] capillary microelectrodes filled with cation solutions were inserted into cerebral cortex glial cells. Such iontophoretic injection of Na^+ or Li^+, but not of K^+, caused the high-frequency discharge of neurons in the vicinity. Injection of any of these cations into cerebral cortex neurons did not induce their discharge.

All the data mentioned above were obtained with the aid of the microelectrode technique, i.e., under conditions in which the exact morphological characteristics of the recorded glial cells in the cerebral cortex or spinal cord were very difficult to evaluate. Most authors assume that these cells are, as a rule, AC.[51,54] In some papers, the ODC and AC, both in neocortex and in culture, are said to have similar electric properties.[47,50,55]

However, Vernadakis and Berni[56] reported that in 2-day culture of the 15-day chick embryo brain, most glial cells exhibited resting membrane potentials in the range −35 mV to −40 mV, whereas at 5 days in culture, such potentials were seen in only about 15% of cells, the remainder having potentials from −10 mV to −25mV. At 2 days, the bulk of the glial cells in the culture were ODC, but at 5 days, all cells possessed the morphological characteristics of AC.

In gliomas, ODC-like cells were recently shown to have a resting membrane potential of −53 mV, i.e., the same (−50 mV) as AC-like cells.[57]

Thus, the physiological features of ODC as compared with those of AC still remain to be elucidated.

4. METHODS OF ISOLATION OF OLIGODENDROCYTES

Since, as mentioned above, ODC are present in both the gray and white matter of the brain, it is not possible to assess indirectly the chemical composition of ODC by comparing different areas of the brain, as can be done, for instance, for neurons[58-60] (see also the chapter by B. Csillik in this volume).

The direct methods of selective biochemical analysis of ODC can be subdivided into four main approaches.

4.1. Microdissection of Perineuronal Satellite Oligodendrocytes

This approach is usually applied to the perineuronal capsule of large neurons such as spinal cord motoneurons, neurons of dorsal root ganglia,[58,61-64] and particularly Deiters' cells of the lateral vestibular nucleus.[65-72]

With the aid of a steel needle under the stereomicroscope, the neuronal body and the adjacent glial capsule can be dissected out separately from slices of nerve tissue. Because of the very low mass of the sample, in the picogram range, ultramicromethods had to be developed for measurements of substrate utilization, macromolecular composition and turnover, individual enzyme activities, etc. (for literature, see Refs. 3, 64, 70, 73, 74 and the chapter by B. Csillik in this volume).

The microdissection approach was, in fact, the first to be used for functional–biochemical analysis of the neuroglia. It enabled publication of initial data[67,71] which stimulated interest in biochemical investigations of the neuroglia and their interrelation with the neuron. At the same time, it should be kept in mind that this approach provides a glial fraction highly contaminated by surrounding structures. In Hydén's earlier work it was stated[71] that the Deiters' neuron was surrounded by 35–40 ODC, the proportion of AC being not more than 10%. Subsequently, however, the same group reported[72,75] that the sample of perineuronal glia contained eight cells, seven ODC and one AC. Electron microscopy has shown that even when the Deiters' neurons seem light microscopically to be surrounded by satellite ODC, the latter are often separated from the neuronal bodies by thin sheets of AC processes.[76]

4.2. Bulk Isolation

Among several schemes of bulk isolation of ODC,[12,13,15,77] the most widely adopted procedure has been that developed by Poduslo and Norton.[13,78,79] In this method, bovine brain white matter is minced and incubated in trypsin solution at pH 6.0. Then the tissue pieces are disrupted by passing them through nylon and steel meshes to form a cell suspension which is applied to a sucrose density gradient and centrifuged at low speed. The interface between 1.40 and 1.55 M sucrose contained ODC of better than 90% purity. The overall yield amounted to 10×10^6 cells/g tissue. Electron microscopy of the ODC showed them to be morphologically intact, with plasma membranes and well-preserved cytoplasm.[80-82]

This method has been applied successfully by many authors for bovine white matter[83-85] and even for rat brain[86] and spinal cord.[87] Recently, a new procedure has been developed by Norton's group.[16] It uses more physiological medium and a balanced salt solution at pH 7.2 throughout, and it can be effectively applied to undissected rat forebrain and to the nerve tissue of other species as well. Because of the selective lysis of neurons and AC, 90% of the cells in the preparation are ODC, the rest being red blood cells, phase-dark cells, and nuclei, ependyma, and capillary fragments. Cell yield has amounted, depending on the age of rats, to 4–7 × 10⁶/g or 4–5 × 10⁶/brain. Electron microscopy has confirmed that the majority of cells are well preserved and have, for the most part, intact plasma and nuclear membranes.[16]

A rather sophisticated addition to the bulk-isolation approach is electronic cell sorting.[88] A dissociated cell suspension passes through a capillary with an inner diameter such that cells move one by one. The capillary section is illuminated with a laser light. From the light scatter frequency and the fluorescence frequency distribution analyses, a homogeneous cell population can be studied. Specific fluorescent antisera make it possible to analyze only the cells binding some marker antibodies; for instance, anti-corpus-callosum serum thereby labels glial cells.[88] At present, this procedure has been applied in only a few studies. Perhaps further modifications will make this method ever more useful for glial cell investigations.

Of more limited use are methods of bulk isolation of brain cell nuclei.[89] With respect to ODC, these methods possess the substantial defect of providing fractions that are, as a rule, a mixture of ODC and microglial cell nuclei. Microglia, however, are of quite a different origin; histogenetically they are much closer to mesenchymal elements of the nerve tissue.[21–24,33,90] Thus, biochemical properties of the microglial cells can be expected to differ significantly from those of ODC.

The initial scheme of separation of neuronal and glial cell nuclei suggested by Løvtrup-Rein and McEwen[91] included centrifugation of a brain tissue homogenate on a sucrose concentration gradient. This scheme was further modified: a zonal centrifugation was applied on a discontinuous gradient of sucrose. Five fractions were obtained, one of them containing ODC nuclei, although contamination by microglial nuclei was noted.[92]

4.3. Glial Cell Lines

This approach provides viable, genetically and functionally homogeneous populations of glial cells in quantities sufficient for most biochemical investigations.[11] Earlier, glial cell lines derived from experimental or natural glial tumors were used.[93,94] Of particular interest are experimental gliomas induced by various carcinogens such as methylcholanthrene[95] and methyl-[96] or ethyl-nitrosourea.[97,98] Subsequent cultivation and cloning of the tumor cells led to the establishment of several stable glial cell lines.[94]

Unfortunately for those interested in ODC, these lines, as a rule, give cells with morphological, biochemical, and immunological properties of AC rather than of ODC[11,99–102] (see also the chapter by L. Hertz in this volume). Oligodendrogliomas and cell lines derived from them are rare.[94,103] Besides, their tumorlike features bring about some biochemical differences between glioma cells and ODC.[103]

Recently, increasing attention has been paid to attempts at cultivating normal ODC from the brain. The first papers reported a failure to keep long-term cultures of the ODC bulk-isolated from adult brain.[104,105] Perhaps mature ODC, although generated histogenetically from precursor cells much later than neurons and AC,[36,106,107] cannot grow in tissue culture. However, it is also possible that the procedure of bulk isolation induces some irreversible alterations in segregated ODC. At any rate, several successful attempts at maintaining cultures of the ODC bulk-isolated from fetal[108] or even adult brain[17,108] for some months have recently been published. The most promising technique seems to be the use of primary cultures of dissociated brain cells. According to the procedure developed by Sensenbrenner and her co-workers,[109,110] dissociated brain cells from newborn rats and mice cultivated on a plastic surface provide for a pure glial cell population. The neuronal cells degenerate during the first 2 days, and some oligodendroblasts survive for about a week. Subsequently, however, the flat polygon-shaped cells multiply actively to form, within 2 weeks, a monolayer. Morphological and immunologic analysis of the cells of this layer indicates their AC nature.[110] Furthermore, by adding brain extract to the nutrient medium, Sensenbrenner's group has recently managed to enrich the culture of astroblasts in ODC.[111]

Labourdette *et al.*[112,113] maintained cultures enriched in ODC (or, at least, ODC-like cells) from newborn rat brain up to 5 months without signs of degeneration. The authors seeded trypsinized cells from newborn brain hemispheres in a medium supplemented with 10% calf serum. This cultivation led to a great number (almost half of the total cell number) of small cells identified immunologically as ODC-like cells on top of a monolayer of AC.

A novel method has recently been developed by McCarthy and deVellis[114] to separate nearly pure cultures of AC and ODC. Dissociated cells from brain hemispheres of 1- or 2-day-old rat pups were cultivated in Eagle's medium supplemented with 15% fetal calf serum. On the tenth day, the flasks containing the culture were shaken on an orbital shaker overnight. The suspended cells were then collected, pooled, and filtered. The purity of the ODC fraction was initially about 95%, but after another filtration and an additional cultivation of the cells for 24 h, it could be increased to more than 99%. The separated ODC could be maintained in culture for several weeks.

4.4. Histochemistry In Situ

Achievements in histochemistry have provided a great number of color reactions that enable visualization, under the microscope, of many cellular and subcellular biochemical components.[115–117] Immunohistochemical procedures have greatly improved the selectivity of the histochemical approach.[118]

It is possible, in some cases, to determine quantitatively the amount of certain, mainly macromolecular, components. For this purpose, visual cytospectrophotometry is most often applied. Sections of smears of nerve tissue are stained for some chemical components with selective stoichiometric dyes, i.e., dyes that stain only one chemical substance in cells and are bound to it quantitatively. Unfortunately, the number of such color reactions is rather limited. The most reliable among them are Feulgen reactions for DNA, gallocyanin chrome alum for total nucleic acids, periodic acid-Schiff reaction for glycogen, and fast green FCF for total or basic (depending on pH of the staining solution) protein.[119] Recently, some new staining reactions have appeared, among them the reactions for cytospectrophotometric study of glial cells, such as amido black 10B for total protein,[120] toluidine blue for acidic proteins,[121] heparin–alcian blue for basic proteins,[121] ammoniac silver for the ratio of arginine-rich to lysine-rich histones,[122] tetrazolium tetranitroblue with urea or extra substrate for H and M subunits of lactate dehydrogenase,[123] etc.

Under standard conditions of histochemical procedure, with certain precautions taken into account, absorption of the monochromatic light passed through the glial cell within a nerve tissue histochemical preparation can be a source of information about the quantity of substance per cell.[119,124–126] In addition to visual cytospectrophotometry, ultraviolet cytospectrophotometry of unstained sections is used, as are some other methods of quantitative histochemistry such as cytofluorimetry, radioautography, interference microscopy,[127] etc. Their general principles and particular application to studies of ODC are set forth in the monograph by Pevzner.[3]

5. BIOCHEMICAL MARKERS FOR OLIGODENDROCYTES

Achievements in immunology have made it possible to use a number of antisera against individual morphological structures. Thus, antisera against ODC have been successfully raised.[82,98,103,128–132] However, their use and the use of some cell surface antigens as well as markers for ODC are handicapped by the fact that the ODC share several antigens with AC, Schwann cells, or, particularly, myelin. Advantages and disadvantages of this approach are competently discussed in previous reviews.[10,94,133–135] In this chapter, only the biochemical components that are hoped to be more or less reliable markers for ODC will be considered.

5.1. Galactocerebrosides

Oligodendrocytes have consistently been found to contain a large amount of galactocerebrosides[13,17,104,136–141] which are synthesized in large amounts' in ODC that are bulk isolated[140] and maintained in culture.[17,104] Antisera to galactocerebroside reacted specifically with ODC in brain tissue sections,[142] in bulk-isolated ODC,[140] and in ODC cell cultures.[132,135,143–146] No other classes of central nervous system cells (neurons, AC, microgliocytes, meningeal cells) were immunostained for galactocerebrosides.[132,135] Schwann cells also bound antigalactocerebroside serum; however, after 3 days in culture they lost this capability.[132] Schwannoma cells did not react with this antiserum, but oligodendroglioma cells did.[146]

Similar selective localization in ODC was immunohistochemically demonstrated for sulfatide.[132] This agrees with the data indicating extensive synthesis of sulfatide in ODC clonal cell lines[146] and a particularly high activity of cerebroside sulfotransferase, the sulfatide-synthetic enzyme, in bulk-isolated ODC[84] and in ODC maintained in culture.[85]

5.2. Myelin Proteins

Since ODC play an essential role in elaborating myelin,[30-33] an antigenic similarity between myelin and ODC can be expected. Nevertheless, there is some controversy on this point. Antisera to bulk-isolated ODC reacted specifically with ODC in corresponding enriched fractions[140,147] and in brain tissue sections[131]; this activity was blocked by absorption by ODC but not by absorption by purified myelin.[147]

Perhaps basic protein as an ODC marker is valuable only for actively myelinating glial cells. Indeed, Sternberger and her co-workers could detect the myelin basic protein immunohistochemically in brain ODC only in newborn rats, whereas in 25-day-old animals the protein was not observed by this method.[148,149] Later on, it appeared again in adult brain[150]; in chick brain, it was first observed on the 14th day of incubation.[150] In ODC in culture, a quite definite positive reaction was found with antiserum to the myelin basic protein.[17,112]

Immunostaining of ODC in brain sections was reported with antisera to the major CNS myelin glycoprotein[151] and to proteolipid protein[150]. The latter was also clearly visible in the cytoplasm and processes of actively myelinating ODC (in the corpus callosum of 10-day-old rats), but in adult animals, all ODC stains were completely negative.[152,153]

Evidence of a selective ODC localization of myelin Wolfgram protein has been presented[154-156] In 18-day-old rats, the Wolfgram protein was immunohistochemically located only in ODC,[154] mainly in polysomes but not in mitochondria or nuclei.[155] The presence of this protein in the cells of primary ODC cultures has also been reported.[112,113]

5.3. 2',3'-Cyclic Nucleotide 3'-Phosphohydrolase

2',3'-Cyclic nucleotide 3'-phosphohydrolase (CNPase) activity appears to reflect a portion of myelin in close proximity to the glial membrane.[30] At the same time, it is present at relatively high specific activity in a number of nonmyelinating glial cell lines[128,157,158] and in oligodendroglioma tissue.[159] By applying the bulk isolation technique, several groups obtained a high CNPase activity in isolated ODC.[13,18,86,87,138,140] Data by Poduslo[138,140] have indicated that CNPase is concentrated in the plasma membrane of ODC. Such CNPase activity has frequently been observed in ODC and in cultivated ODC cell lines.[17,105,110,112,113,128,134] Convincing data were recently reported by Vernadakis and her co-workers[159]; at early (21 to 26) cell passages, when cells of C-6 line, 2B clone, possessed small, round, dark nuclei, scanty cytoplasm, and very short cytoplasmic processes, their CNPase activity was markedly high. But at late (82 to 88) passages, when the cell nuclei became larger and paler, the cytoplasm more abundant, and prominent elongated cytoplasmic processes were evident, the activity of CNPase decreased 3–4 times.

Recently, evidence has been presented that a form of CNPase is a component of W1 Wolfgram protein.[160,161]

5.4. Carbonic Anhydrase

The presence of carbonic anhydrase in perineuronal glial cells in much higher amount than in the corresponding neuronal bodies was first found using microdissected Deiters' neurons and the surrounding glial capsule.[65] Subsequently, high carbonic anhydrase activity was shown in an isolated myelin fraction of the brain tissue[162,163] (see also chapter by V. S. Sapirstein in Volume 4 of this *Handbook*).

This enzyme has been detected in cultivated glial cell types; some of them were, morphologically and immunochemically, closer to AC, others to ODC.[94,128,147,164] High activity of carbonic anhydrase was also revealed in primary AC[165] and in the glial-enriched fractions containing mainly AC.[165–167] Recently, however, immunochemical and immunohistochemical techniques provided convincing evidence of the localization of carbonic anhydrase in ODC.[156,168–170]

5.5. Glycerol-3-Phosphate Dehydrogenase

The function of this enzyme is quite general, so it can be expected to be present in many animal tissues. It has been shown to be immunologically similar in brain, liver, and skeletal muscle.[171] However, deVellis's group has amassed considerable proof that one specific property of this dehydrogenase, its inducibility by hydrocortisone, is peculiar only to the nervous system.[172,173] Initially, the induction of glycerol phosphate dehydrogenase by hydrocortisone was demonstrated in a number of cell lines some of which seemed of AC, rather than ODC, nature.[174,175] Nevertheless, both indirect comparisons[176] and direct immunohistochemical observations[173,177] subsequently showed the ODC localization of this enzyme.

6. BIOCHEMISTRY OF OLIGODENDROCYTES

As shown above, there are several methods for the isolation of ODC. These methods differ in their basic principles, in the degree of injury sustained by ODC in the course of their isolation, in the contamination by other classes of cells or other structures of the nervous tissue, etc. These may be reasons why the data of various authors, reviewed below, are at variance.

6.1. Oxidative Metabolism

As seen from Table I, endogenous respiration in microdissected ODC (or, more exactly, in the whole perineuronal capsule) around rabbit Deiters' neurons was one order of magnitude higher than that in the microdissected bodies of the Deiters' neurons.[178] Addition of the substrate (particularly of the mixture pyruvate plus malate) stimulated the respiration markedly.

With succinate as a substrate, regional variations in the O_2 consumption have been demonstrated in the perineuronal ODC microdissected from various

Table I

Utilization of Various Substrates by Microdissected Perineuronal Oligodendrocytes of Rabbit Lateral Vestibular Nucleus

Substrate	Concentration of the substrate in the incubation medium	O_2 consumption (10^{-4} μl O_2/h) per glial clump[a]	Ratio of glia to neuron	O_2 consumption per g wet weight as calculated by Hertz[164]		Reference
				pl O_2 per h	μmol O_2 per min	
No added substrate (endogeneous respiration)	0	0.3 ± 0.06	7.5	4	0.03	178
Succinate	25 mM	4.5	2.05	60	4.2	71
Succinate	25 mM	4.2 ± 0.5	1.91	70	5.1	72
Succinate	25 mM	2.0 ± 0.3	0.53	20	1.5	179
α-Ketoglutarate	13 mM	2.1 ± 0.2	0.95	40	2.9	72
α-Ketoglutarate	13 mM	0.9 ± 0.19	0.82	10	0.8	178
Glutamate	15 mM	1.1 ± 0.2	0.50	20	1.3	72
Glutamate	15 mM	0.6 ± 0.14	0.27	10	0.6	178
Cytochrome c	0.1 mM	11.5 ± 0.8	2.74			71
Pyruvate + malate	13 mM + 0.5 mM	1.9 ± 0.47	2.37			178
Glucose	6 mM	40[b]	2[b]			180

[a] Clump of the perineuronal glial capsule of a size equal to the size of the Deiters' neuronal body, approximately 10^5 μm³. The glial clump seems to contain eight glial cells: seven oligodendrocytes and one astrocyte.[72,75]

[b] Taken from the authors' graph; O_2 consumption is expressed in arbitary units (unit is a change in the manometer pressure, mm H_2O).

areas of the CNS, such as lateral vestibular nucleus of medulla,[72,179] gigantocellular[75] or oral nucleus of the reticular formation,[181] nucleus of the trigeminal[181] or hypoglossal nerve.[182] However, these variations are not great.

Oligodendrocytes were found to possess a very high level of cytochrome *c* oxidase, its activity in microdissected ODC being almost three times higher than that in the bodies of the Deiters' neurons.[183]

6.2. Carbohydrate and Phosphorus Metabolism

The rate of anaerobic glycolysis in microdissected perineuronal ODC of the rabbit vestibular nucleus was almost identical with that in the Deiters' neurons of this nucleus.[183] However, the activity of hexokinase, the initial step of glycolysis, has been shown to be present immunohistochemically in all types of CNS cells except ODC.[184,185] As mentioned in the discussion of the morphological features of ODC, they contain, unlike AC, few glycogen granules.

In the bulk-isolated ODC from the calf brain, the content of heparan sulfate, chondroitin sulfate, and hyaluronic acid was lower than in bulk-isolated neurons and AC.[186] At the same time, the concentrations of N-acetylglucosamine, N-acetylgalactosamine, N-acetylneuraminic acid, galactose, and mannose isolated from glycoproteins were much higher in the ODC. The only exception was fucose, its concentration being equal in ODC and in neurons but lower than in AC.[186]

Passoneau's group[187] compared the concentrations of glucose, phosphocreatine, and ATP in the microdissected bodies of mouse cerebral cortex pyramidal neurons and in cerebellar Purkinje cells, and also in adjacent neuropil. The last contained, apart from perineuronal glial cell bodies, a number of other components of the cortex tissue. However, no difference between neurons and the perineuronal tissue samples was seen in either case. X-ray film autoradiography of rat brain and spinal cord sections after administration of [^{3}H]-2-deoxy-D-glucose showed[188] that the intensity of glucose metabolism was approximately the same or only slightly higher in neuronal perikarya than in the immediately surrounding neuropil. In brain gray matter, the glucose consumption was similar in glial cell nuclei and in the neuropil, whereas in the spinal cord, it was several times less in the white matter ODC nuclei than in the gray matter or perineuronal perikarya.

In microdissected perineuronal ODC of the rabbit vestibular nucleus, the concentration of ATP was two orders of magnitude lower than in the Deiters' neurons.[189] The ATPase activity was somewhat higher in the glial capsule than in the neuronal bodies and showed a sharp pH optimum at 8.0, whereas the neuronal pH optimum had a broad peak around pH 7.4.

Activity of acid phosphatase in bulk-isolated ODC from rabbit brain was much lower than in AC and in neurons. In beef brain, activities in ODC and AC were almost equal and only a little lower than in neurons.[190]

Electron microscopic histochemistry showed[191] that thiamine pyrophosphatase activity in ODC was rather low. Products of the reaction were seen in perineuronal satellite ODC only at the smooth endoplasmic reticulum or

Golgi apparatus. In the other kinds of glial cells, these products were abundant in various cell organelles such as mitochondria, lysosomes, etc.

As mentioned above, the activity of one particular phosphatase, 2',3'-cyclic nucleotide 3'-phosphohydrolase, was much higher in ODC than in other CNS cells.[13,18,86,138,192] It progressively increased during the postnatal development of the rat and was more than two times higher in 29-day-old than in 16-day-old rats.[86]

6.3. Mineral and Amino Acid Composition

Data on mineral composition of glial cells are not numerous. Almost all of them deal with AC (see chapter by L. Hertz in this volume). Only a single paper reports the assay of potassium concentration in microdissected ODC of rabbit vestibular nucleus.[193] With the aid of an X-ray fluorescence microanalysis, it was found that the value of K^+ content in these cells and in the bodies of Deiters' neurons was practically the same. If the microdissected ODC sample was incubated for 1 h in a glucose-containing medium at 4°C, the K^+ content was reduced; when incubated at 20°C, it remained unchanged, but at 37°C it increased. In the Deiters' neurons, there was a lowering of the potassium concentration at all temperatures.

Amino acid metabolism and transport systems in glial cells are subjects of an intensive study. However, these systems have been examined either in AC or in perineuronal satellite glial cells of sympathetic and dorsal root ganglia.[3,164,194–196] The latter cells, although analogous in a way to ODC, are formations of peripheral nervous system, with a number of morphological peculiarities.[21,23,197] Electron microscopic autoradiography provided evidence of similar uptake systems for a number of amino acids in glial cells of CNS.[196,198] However, these systems appear to be peculiar both to AC and ODC.

Analysis of 19 free amino acids in seven different cell lines demonstrated some individual fluctuations in the amino acid content. At the same time, no particular differences were revealed that correlated with the neuronal or glial origin of the cell line.[199] Different cell lines, whether neuronal or glial in origin, also failed to differ in their capability of uptake of [^{3}H]-GABA.[200]

6.4. Lipid Metabolism

The main source of lipid assay in ODC is bovine or rat brain white matter from which bulk-isolated ODC have been obtained (Table II). An essential feature is the rather high amount of galactolipid, several times higher than in AC or neurons.[137]

Analysis of these data shows[201] that the general pattern of lipid composition in ODC is more similar to that in myelin and axon but different from the lipid pattern of bulk-isolated neurons and AC. Oligodendrocytes have much higher levels of galactoceramide and sulfatide than other classes of brain cells. The concentration of glucosylceramide is greater than, whereas that of ceramide dihexoside is rather similar to, their concentration in myelin or the whole white matter. However, when expressed as percent of total lipid, ODC are ten times

Table II
Main Lipid and Phospholipid Fractions in Bulk-Isolated
Bovine White Matter Oligodendrocytes

	References	
Fractions	136	13
Total lipid (% dry weight)	20.8	29.5
Cholesterol (% lipid weight)	30.6	14.1
Cerebroside (% lipid weight)	10.5	7.3
Sulfatide (% lipid weight)	4.6	1.5
Total phospholipid (% lipid weight)	48.2	62.2
Individual phospholipids (mol/100 mol lipid P)		
Choline phosphoglycerides	29	48
Ethanolamine phosphoglycerides	36	24
Serine phosphoglycerides	21	8
Phosphatidylinositol	11	9
Total plasmalogen	9	16

richer in glucosylceramide and two times richer in ceramide dihexoside than myelin or white matter.[201] Among phospholipid fractions, the major ODC phosphatide (65–66% of the total) was phosphatidylcholine.[18]

Sphingolipids in ODC contained fatty acids with a much longer chain length than that in neurons and AC. On the other hand, a small percentage of β-hydroxy fatty acids was found in ODC.[137]

Bulk-isolated ODC had a higher capacity than other kinds of rat brain cells for accumulation and enzymatic synthesis of monogalactosyl diglyceride. They also had a higher rate of biosynthesis of galactosylceramide but not of glucosylceramide. In 16-, 19-, and 29-day-old rats there was, with age, a progressive increase, per milligram of protein, in the content of monogalactosyldiglyceride and a decrease in the [^{14}C]galactose incorporation into the ODC mono- and digalactosyl diglycerides.[86]

The lipid composition of plasma membranes isolated from calf white matter ODC was characterized by a higher percentage of cholesterol and galactolipids and a lower percentage of phospholipids than that of the whole ODC. The ODC membranes differed from isolated neuronal plasma membranes and also from synaptic membranes.[138,140]

In bulk-isolated ODC from the human brain (frozen autopsied material), the total content of gangliosides per milligram protein was about three times lower than in calf brain ODC.[141] This difference might be partially dependent on species specificity but most probably can be accounted for by poor morphological preservation and worse purification of the cells isolated from frozen brain tissue. Thin-layer chromatographic analysis of the gangliosides has shown the ganglioside pattern of human ODC to be quite complex, with a number of differences from that of myelin. A high concentration of sialosylgalactosylceramide (ganglioside M$_4$) was found in the ODC in addition to G$_{M1}$, G$_{D1a}$, G$_{D1b}$, and G$_{T1b}$. The usual minor brain gangliosides G$_{M3}$, G$_{M2}$, and G$_{D3}$ were also enriched in the human ODC.[141]

Rates of *in vitro* CDP-diglyceride synthesis, an essential step in phospholipid synthesis, from [^{3}H]-CDP and phosphatidic acid were about seven times less in small dark nuclei than in large pale cell nuclei isolated from rabbit cerebral cortex. The latter nuclei are considered most probably neuronal ones, whereas the former seem to be the nuclei of ODC substantially contaminated with the nuclei of microgliocytes and perhaps of small neurons.[202]

In vitro uptake of [1-^{14}C] fatty acids by bovine white matter ODC was the greatest for palmitic acid and decreased with decreasing chain length. Fatty acids were incorporated into both cerebrosides and phospholipids, preferably into ethanolamine phosphoglyceride, and underwent chain elogation, desaturation, and oxidation.[203] The specific activity of total phospholipid after *in vitro* incubation with [1-^{14}C]linoleate was found to be 23 times higher in bulk-isolated rat ODC than in neurons and 13 times higher than in AC.[83]

In vivo use of [1-^{14}C]linoleate has demonstrated its much active incorporation into phospholipid of bulk-isolated rat ODC than into that of myelin and other cell types. At the same time, no direct precursor–product relationship was revealed between the lipids of ODC and of myelin in the rat brain.[204]

Calf brain white matter ODC maintained in culture for 2 days were capable of *in vitro* synthesis of the lipids that predominate in brain, particularly those enriched in myelin. With [^{3}H]- or [^{14}C]galactose as a precursor, cerebrosides had the most radioactivity; with [^{3}H]acetic acid, cholesterol, and [^{3}H]glycerol, the phospholipids were predominant, especially phosphatidylcholine.[104,140]

In ODC bulk isolated from calf brain white matter and maintained for 48 h in suspension culture, ^{35}SO$_4$ or [1-^{14}C]acetate were actively incorporated into sulfatide and unesterified sterol, respectively, of the cultivated ODC. The activity of 3-hydroxy-3-methylglutaryl CoA reductase, the microsomal enzyme rate limiting in cholesterol synthesis, was more than three times higher in the isolated ODC than in the whole white matter.[85]

An intense labeling of unesterified sterol has been shown[205] after the incubation of bulk-isolated calf ODC with ^{3}H$_2$O. Ninety per cent of the radioactivity was in cholesterol and desmosterol, in a ratio of 5 : 1. Labeled ketone bodies, D-[3-^{14}C]-(−)-β-hydroxybutyrate or [3-^{14}C]acetoacetate, also were actively incorporated into the unesterified sterol of ODC. The specific activity of acetoacetyl CoA synthetase in the ODC turned out to be more than 20 times greater than that of whole brain. Evidence has been presented in support of an extramitochondrial pathway in ODC for the conversion of acetoacetate but not of hydroxybutyrate to acetyl CoA[205].

A very low content of ganglioside was found in microdissected ODC from ox brain Deiters' nucleus.[206]

In cultured human oligodendroglioma cells, Yu and co-workers[207] failed to demonstrate the presence of the ganglioside G$_{M4}$. In the bulk-isolated ODC from human brain, Yu and Igbal[141] found a substantial concentration of this ganglioside, which they suggest as a specific marker for human ODC perikarya.

Among 19 cell lines of neurological origin, only two of them, G26-20 and G26-21, were characterized by high levels of ^{35}SO$_4$ incorporation into sulfatide. Tumors in mice produced by subcutaneous injections of these lines contained a particularly high content of sulfogalactocerebroside and of galactosylceramide. These cell lines have been considered to be of ODC or, perhaps, of Schwann cell origin.[146] Their glycosphingo- and sialoglycosphingolipid composition was shown to be simpler than the composition of neuroblastoma cell lines but more complex than that of astrocytoma cell lines. Oligodendroglioma cell lines G26-15, G26-19, G26-20, and G26-24, as compared with neuroblastoma cell lines NB2a, NB18, and NB41A, possessed lower activity of N-acetylneuraminyltransferase and G_{M1}-ganglioside-β-D-galactosidase but higher activity of arylsulfatase A and galactosylceramide-β-galactosidase.[208] The later fact agrees well with the data presented by Norton's group[209] concerning bulk-isolated ODC (see below).

Addition of hydrocortisone to the lines of presumed ODC origin (G26 series) led to a marked increase in $^{35}SO_4$ incorporation into cell sulfogalactosylceramide *in vitro* and to a three- to fourfold activation of galactosylceramide sulfotransferase. Similar changes were observed with cortisone and dexamethasone but not with sex hormones. The hydrocortisone effect was not observed in neuroblastoma cell lines.[210,211]

In bulk-isolated ODC from bovine white matter, the activity of acid lipase with U-methylumbelliferyl oleate as a substrate was much higher than in an initial sample of white matter and, particularly, than in the isolated myelin fraction.[212]

Activities of galactosylceramide-β-galactosidase and arylsulfatase A were the highest in ODC when compared to neurons and AC bulk-isolated from calf brain. Four other glycolipid hydrolases, asialo-G_{M1}-gangliosidase, G_{M1}-gangliosidase, β-galactoside-β-galactosidase, and N-acetyl-β-glucosaminidase, were not very active in any particular kind of brain cells.[209]

In the bodies of anterior horn neurons microdissected from human lumbar spinal cord (autopsy material obtained 3.5 h after death), hexosaminidase activity was one order of magnitude higher than in the perineuronal neuropil. The neuronal enzyme contained a much smaller proportion of the heat-stable isoenzyme (hexosaminidase B) than did the perineuronal tissue sample.[213]

Plasmalogenase activity turned out to be several times greater in bulk-isolated ODC of the bovine brain than in AC or neurons.[214] The activities of phospholipases A_1 and A_2 in bulk-isolated ODC were 2–3 times lower than in neurons and only a little lower (phospholipase A_2) or even somewhat higher (phospholipase A_1) than in AC from bovine brain. It was only ODC phospholipases that were activated by Ca^{2+} (10 μM), whereas the neuronal enzymes were not changed, and AC enzymes were, in contrast, inhibited by Ca^{2+} ions.[215].

In bulk-isolated ODC from beef and particularly from rabbit brain, the activities of arylsulfatases A and B were significantly lower than in corresponding neurons and AC. In both animal species, the activity of β-galactosidase in ODC was only slightly higher than in AC but one order of magnitude lower than in neurons.[190]

6.5. DNA Metabolism

In bulk-isolated ODC the mean amounts of DNA per cell determined by various authors were rather similar (Table III). This amount seems to corre-

Table III
DNA and RNA Content in Bulk-Isolated Oligodendrocytes

Source	Age	DNA content (pg/cell)	RNA content (pg/cell)	Reference
Human autopsied brain	Adult	5.3 ± 0.6	1.8 ± 0.1	219
Bovine white matter	Calf	5.14	1.95	13
	Adult	6.7 ± 0.5	1.6 ± 0.14	14
		5.43 ± 0.77	1.8 ± 0.16	18
		8.50 ± 0.25	1.90 ± 0.02	220
	2 years	6.44 ± 0.45	1.44 ± 0.28	220
	1 year	11.45 ± 1.6	2.24 ± 0.11	220
	8 months of gestation	9.38 ± 0.29	2.29 ± 0.13	220
Rat brain	60 days	7.9 ± 1.1	2.5 ± 0.5	16
	30 days	6.2 ± 1.5	2.7 ± 1.2	16
	10 days	7.2 ± 0.4	2.9 ± 0.4	16
	Adult	7.4 ± 0.3	2.5 ± 0.1[a]	92
	10 days	5.9 ± 0.4		

[a] Nuclear RNA (pg/nucleus).

spond to the diploid chromosome set, which confirms earlier cytospectropho-tometric data.[3,216]

The average length of nucleosomal DNA was determined to be 200 base pairs in cerebral cortex cell fractions enriched in small dark nuclei, presumed to be mainly ODC. In large pale nuclei considered as neuronal, the length was 160 base pairs.[217] The molecular weight of chromosomal DNA in neuronal nuclei-enriched and in ODC nuclei-enriched fractions from guinea pig cerebral cortex was 6.2×10^5 and 8.5×10^5, respectively, before incubation of the nuclei with all four nucleotides in the medium for the synthesis of DNA. After the incubation, the molecular weight of the neuronal DNA increased 1.4-fold, whereas glial DNA increased twofold, to result in a more than twofold higher molecular weight of DNA in the nuclei from ODC.[218]

Proliferation and, hence, synthesis of DNA occurs in ODC ontogenetically much longer than in AC and neurons.[106,107] Indeed, the autoradiographic technique has provided evidence for [³H]thymidine incorporation in ODC of post-natal rat brain.[34,35,222,223]

In bulk-isolated ODC nuclei from the brains of 10-day-old rats injected with [³H]thymidine, incorporation of the label into DNA was three times as much as that in neuronal or AC nuclei.[221] It cannot be ruled out, however, that the ODC nuclear fraction was contaminated with nuclei from proliferating spongioblasts.

In a fraction from rat brain enriched in the nuclei of ODC and microglio-cytes, DNA polymerase β, resistant to N-ethylmaleimide, was somewhat more active than in neuron- or AC-enriched fractions. Activities of N-ethylmaleim-ide-sensitive DNA polymerase α and terminal deoxynucleotidyl transferase were higher in ODC nuclei than in neuronal but lower than in AC nuclei.[224] These data are difficult to interpret because ODC were in a mixture with microgliocytes, cells of a mesenchymal origin with quite different properties of proliferation, phagocytosis, etc.

In isolated ODC nuclei from adult guinea pig cerebral cortex, *in vitro* synthesis of DNA from [³H]-TTP was three times lower than in bulk-isolated neuronal nuclei greatly contaminated with AC nuclei. In the same nuclear fraction, DNA polymerase activity, with calf thymus DNA added, was only slightly lower in ODC than in neuronal plus AC nuclei.[218] At the same time, the activity of DNA ligase in ODC nuclei was 11-fold lower than in the neuronal nuclei. Apparent K_m for the substrate [³²P]phosphoryl DNA was similar in ODC and in neuronal nuclei. Glial nuclei contained about five times fewer nicks available as substrate in chromatin DNA (per microgram of DNA) than did the neuronal nuclei.[218,225]

It was calculated that 3.4% of neuronal chromatin DNA, but only 1.32% of glial chromatin DNA, was utilized as template.[218] The DNase activity in isolated ODC nuclei from adult rat brain was several times lower than that in the neuronal and AC nuclei. In the ODC nuclei of 10-day-old animals, this activity was much higher; it reached the activity level of the AC nuclei but was still lower than the activity in neuronal nuclei.[221] However, in ODC nuclei bulk-isolated from adult guinea pig cerebral cortex, DNase activity was about 1.5 times greater than that in a mixture of the neuronal and AC nuclei.[226] These data are difficult to compare because, apart from possible species differences, substantial differences were present in the bulk isolation procedures used.

6.6. RNA Metabolism

Data summarized in Table III show that bulk-isolated ODC from brain contain as little as several picograms of RNA. This is in good agreement with results of ultraviolet cytospectrophotometry that showed that the perineuronal ODC from spinal cord anterior horns contained 3–4 pg of RNA.[3,227]

The average molecular weight of RNA in ODC-enriched nuclear fraction from guinea pig cerebral cortex was found to be 71,000; it was 79,000 in the neuronal (and, perhaps, AC) nuclei and 42,000 in the liver cell nuclei.[228]

Comparison of RNA fractions in bulk-isolated cell nuclei showed similarities rather than differences: in the AC, neuronal, and ODC plus microglial nuclei, the main fractions of RNA were 38 S and 45 S. In the neuronal nuclei, however, 35 S RNA was also detected.[229]

Little regularity in RNA base ratios has been found when large neurons have been compared with their perineuronal glial capsules (Table IV). As mentioned above, the perineuronal capsule of the Deiters' neurons is found to contain seven ODC and one AC.[72,75] If the amount of RNA in AC is 30 pg per cell,[13] and the maximal amount of RNA in ODC is 4 pg per cell,[3] the total RNA content in the perineuronal sample should not be higher than 60 pg. Meanwhile, a value of 120 pg of RNA was reported by Hydén's group,[67,71,230] which suggests that half of the RNA in the perineuronal sample is of nonglial origin, probably axonal and microglial. The *in vitro* synthesis of RNA, as judged from [³H]uridine incorporation, was as low in isolated ODC nuclei as in neuronal nuclei, whereas in AC nuclei, this synthesis was much more intensive. The same relationships were seen *in vivo* 30 min after intracisternal injection of [³H]uridine.[235]

Table IV
Nucleotide Composition of RNA in Microdissected Neuronal Bodies and
Perineuronal Glia from Various Areas of the Central Nervous System

Area	Nitrogen base	Neuronal RNA[a]	Glial RNA[a]	Difference in glia compared to neurons[b]	Reference
Lateral vestibular nucleus of rabbit	Adenine	19.7 ± 0.37	20.8 ± 0.28		230
	Guanine	33.5 ± 0.39	28.8 ± 0.64	− 14%	
	Cytosine	28.8 ± 0.36	31.8 ± 0.27	+ 10%	
	Uracil	18.0 ± 0.18	18.6 ± 0.55		
Lateral vestibular nucleus of rat	Adenine	20.5 ± 0.54	25.3 ± 0.16	+ 23%	231
	Guanine	33.7 ± 0.33	29.0 ± 0.24	− 14%	
	Cytosine	27.4 ± 0.34	26.5 ± 0.43		
	Uracil	18.4 ± 0.26	19.2 ± 0.27		
Nucleus of the hypoglossal nerve of rabbit	Adenine	21.1 ± 0.63	28.1 ± 1.30	+ 34%	232
	Guanine	24.8 ± 0.60	23.5 ± 1.47		
	Cytosine	31.9 ± 0.53	21.8 ± 1.15	− 32%	
	Uracil	22.2 ± 0.53	26.6 ± 1.83	+ 20%	
Spinal cord anterior horn of rat	Adenine	21.8 ± 0.70	26.5 ± 0.93	+ 22%	233
	Guanine	31.1 ± 0.55	25.7 ± 0.93	− 17%	
	Cytosine	25.4 ± 0.95	23.2 ± 0.59		
	Uracil	21.7 ± 0.47	24.8 ± 0.87	+ 14%	
Globus pallidus of human	Adinine	18.3 ± 0.42	19.0 ± 0.78		234
	Guanine	30.5 ± 0.44	29.1 ± 0.15		
	Cytosine	35.3 ± 0.60	33.7 ± 0.72		
	Uracil	15.9 ± 0.36	18.2 ± 0.36	+ 14%	

[a] Nucleotide composition is expressed as molar proportions from the total base sum.
[b] Differences between neurons and glia are shown only when they are statistically significant ($P < 0.05$).

Less intensive *in vitro* incorporation of various precursors such as orotic acid, uridine, cytosine, UTP, or GTP into the nuclear RNA of ODC than of neurons was observed by several authors.[236-238]

Injections of a powerful nicotinamide antagonist, 6-aminonicotinamide, inhibited the *in vitro* incorporation of [³H]-UMP into the RNA of isolated ODC nuclei of the rat brain. Such injections reduced both the number of RNA initiation sites on chromatin and the [³H]acetate uptake into chromatin-bound histones. The effect of aminonicotinamide on the neuronal RNA transcription was just the opposite.[239]

In isolated ODC nuclei from rat brain[92] or from guinea pig cerebral cortex,[228] the activity of DNA-dependent RNA polymerase was several times lower than in neuronal nuclei. Sheep white matter ODC nuclei had an active DNA-dependent RNA polymerase with a sharp maximum at pH 7.8; actinomycin D markedly inhibited its activity.[240]

In bulk-isolated ODC nuclei from 15- to 18-day-old rat brain probably contaminated with microglial cell nuclei, transcription from unique DNA sequences was markedly less effective than in a fraction containing a mixture of neuronal and AC nuclei. Transcription from the repeated DNA sequences was similar in all kinds of the nuclei.[241]

With the aid of ultracentrifugation in a discontinuous sucrose gradient, two classes of rat brain cell nuclei were isolated, large and small, and were considered neuronal plus AC and ODC nuclei, respectively. Analysis of transcription in a cell-free system demonstrated twice the number of RNA initiation sites on the neuronal chromatin than on the ODC chromatin.[242,243] This can be accounted for, at least partly, by a twofold higher acetylation of neuronal nuclear histones; this fact, in turn, can be attributed to a higher acetate uptake by neuronal than by glial chromatin-bound histones.[243,244] Under conditions that allowed repeated initiation of RNA chains at the same initiation site, rat brain RNA polymerase utilized the neuronal initiation sites more frequently than the ODC sites.[245]

The ODC nuclear chromatin was found to contain a much lower proportion of transcribable chromatin than the neuronal chromatin.[228] The DNase-sensitive, Mg^{2+}-soluble chromatin represented only 4.6% of the total chromosomal DNA in bulk-isolated rat brain ODC nuclei but 15–20% of the DNA in the neuronal nuclei.[245]

By applying [^{3}H]-DNA–RNA saturation hybridization it has been shown that the unusually high unique DNA expression value (16%) in the rat brain nuclear RNA is of neuronal origin only. Oligodendrocyte nuclei contained RNA with the same low degree of complexity (10–11%) as cell nuclei from liver and spleen.[246]

If bulk-isolated ODC nuclei and the nuclei of neurons and of AC from adult brain were incubated with yeast RNA, RNase activity per nucleus was found to be virtually the same in all three classes of cell nuclei.[92]

6.7. Protein Metabolism

As seen from Table V, the total protein content per cell found by various authors applying different schemes of bulk isolation of ODC is in reasonable agreement. The protein amount values per cell fluctuate on the average from 30 to 60 pg, and the values per cell nucleus from 20 to 30 pg.

Comparison of the ODC isolated from the white matter of fetal (8 months of gestation), calf (1-year-old), baby (2-year-old), and adult (5-year-old) bovine brain has shown a higher protein content in fetal and calf ODC. Surprisingly few ontogenetic changes have been revealed electrophoretically in the composition of the protein fractions from the bovine ODC.[220]

Bulk-isolated ODC from rat brainstem and spinal cord actively incorporated [U-^{14}C]leucine *in vitro*. This incorporation was markedly reduced by the addition of cyanide, cycloheximide, or chloramphenicol to the incubation medium.[87] The *in vitro* incorporation of [^{14}C]leucine into proteins of a mixture of bulk-isolated ODC and microglial cell nuclei was somewhat lower than that into proteins of the AC nuclear fractions and much lower than that into proteins of the neuronal fraction.[247] *In vivo*, after intraventricular injection of [^{3}H]leucine, electron microscopic autoradiography of rat hippocampus demonstrated the high intensity of ODC labeling, which equaled or even exceeded that of pyramidal neurons.[248]

The addition of K^+ (105 mM) or glutamate (10 mM) to the bulk-isolated

Table V
Protein Content in Bulk-Isolated Oligodendrocytes

Source	Age	Protein content (pg/cell)	Reference
Human autopsied brain	Adult	47 ± 3	219
Bovine white matter	Calf	30–39	82
	Adult	40 ± 5	201
		46.1 ± 4.0	14
		32.9 ± 4.4	18
	5 years	38.6 ± 2.4	220
	2 years	33.0 ± 2.4	220
	1 year	44.2 ± 4.5	220
	8 months of gestation	45.1 ± 3.9	220
Rat brain	Adult	42.0 ± 5.1	15
	60 days	57	16
	30 days	56	16
	10 days	60	16
Rat brain cell nuclei	Adult	31.0 ± 2.0[a]	202
		17.6 ± 1.8[a]	92
	10 days	26.2 ± 0.3[a]	221

[a] Expressed as pg/nucleus.

glia from bovine white matter significantly reduced the *in vivo* incorporation of uniformly labeled [^{14}C]leucine.[77]

Bovine ODC that were bulk-isolated from the white matter and maintained as suspension cultures for 2 days in the presence of [^{3}H]proline exhibited active incorporation of the precursor into cell protein. Polyacrylamide gel disk electrophoresis revealed the spectrum of proteins synthesized by the ODC; mostly high-molecular-weight proteins were labeled with this amino acid.[139]

Protein kinase and histone methyltransferase activities in the ODC-enriched nuclear fraction were lower than in the neuronal nuclear fraction from mouse brain.[249]

Analysis of individual proteins in ODC has been restricted mainly to myelin proteins. As mentioned above, a high content of myelin basic protein[17,112,148–150,219] and a selective localization of the Wolfgram proteins[112,113,154–156] have been demonstrated in bulk-isolated ODC or glial cell lines. The ratio of myelin basic protein to proteolipid protein in the bulk-isolated ODC from human autopsied brain was found to be 6.9 ± 0.3; in the myelin fraction, this ratio was only 2.3 ± 0.3. Two-dimensional peptide maps of a tryptic digest of the basic protein isolated from the ODC and from the white matter myelin were identical.[219]

The discovery by Hydén and McEwen[250] of the presence of the brain-specific S-100 protein in the glial capsule around Deiters' neurons was confirmed immunohistochemically by other authors using both microdissected perineuronal glia[251] and cell lines derived from oligodendrogliomas.[97,103] However, this protein could not be detected in bulk-isolated ODC from bovine white matter.[14]

In transcriptionally inactive chromatin of bulk-isolated rat brain neuronal and ODC nuclei, the patterns of histones were shown electrophoretically to be practically identical. In the template-active chromatin, histone fraction H1 was reduced in ODC and almost nonexistent in neurons. Nonhistonal chromosomal proteins from the ODC nuclei, when separated by SDS-polyacrylamide gel electrophoresis, showed three peaks associated with the template-active chromatin but absent in the template-inactive chromatin. The same kind of observation was made in the neuronal nuclei, but four peaks were present in the template-active chromatin, and these peaks were different from the three ODC histones.[245] Acetylation of the nuclear ODC histones and, correspondingly, acetate uptake by the chromatin-bound ODC histones were two times lower than in the neuronal nuclei.[243,244]

7. FUNCTIONAL CHANGES IN THE OLIGODENDROCYTE METABOLISM

Changes in the functional states of the nervous system have been shown by many authors to be accompanied by metabolic changes in the corresponding neurons (for literature see refs. 70, 126, 252 and the chapter by B. Csillik in this volume). Data summarized in Table VI indicate that in many cases the metabolism of ODC also changes under the effects of various experimental conditions such as hypoxia, convulsions, forced muscular activity, electrostimulation, and injections of various drugs. Wherever possible, these metabolic changes in the ODC are compared in this table with the accompanying changes in the neuronal metabolism.

Interpretation of the data presented in Table VI is difficult for several reasons. First of all, only a few metabolic processes were followed in these studies. Many important properties of carbohydrate, phosphorus, and lipid metabolism, turnover of macromolecules, and activities of enzymes have not been analyzed at all with respect to functional changes in the ODC.

Furthermore, the methods of the ODC isolation that have been used in the papers referred to in Table VI are not always reliable. As mentioned above, microdissected perineuronal samples contain, apart from the ODC, a number of other elements, the whole mass of which should be at least equal to the mass of the ODC. It is quite possible that many synaptic structures are also included in the microdissected perineuronal sample. Their metabolism is, perhaps, the most sensitive to stimulation, pharmacological agents, behavioral factors, etc.

Quantitative cytochemistry (in situ histochemistry) cannot distinguish with certainty the ODC from other perineuronal cells such as the AC or microgliocytes. Besides, assays in fixed, dehydrated tissue sections have their specific limitations (for critical evaluation of this approach, see refs. 3, 70).

At present, the most reliable approach for the preparation of ODC is bulk isolation. This method, however, has been used mainly for characterization of normal ODC. Its contribution to the functional biochemistry of ODC has been limited to few papers.

Therefore, the comparison of the metabolic changes in the neurons and ODC presented in Table VI should be evaluated with great caution. As a

Table VI

Metabolic Changes in Oligodendrocytes during Changes of the Functional State of the Nervous System

Biochemical component and experimental conditions	Animal species	Area of the nervous system	ODC isolation	Changes in ODC	Changes in neurons	Reference
Anaerobic glycolysis						
Hypoxia (12-h exposure at 8% O_2)	Rabbit	LVN[a]	MD[a]	Moderate activation	Marked activation	183
Rotation for 7 days, 25 min/day	Rabbit	LVN	MD	Activation	Reduction	183
Glucose uptake						
Swimming for 45 min	Rat	Cerebral cortex and spinal cord	QC[a]	Increase	Increase	188
Oxidation of succinate						
Regeneration of the sectioned hypoglossal nerve	Rabbit	Nucleus of hypoglossal nerve	MD	Increase 6 days and then 48 days after the section	Increase during all postoperational period (48 days)	182
Injection of urea (3 g/kg, i.v.)	Rabbit	LVN	MD	Increase	Increase	253
Injection of GABA (20 mg/kg, i.p.)	Rabbit	LVN	MD	Increase 1 h after the injection	Decrease 1 h after the injection	179
Injection of hydroxylamine (4.8 mg/kg, i.v.)	Rabbit	LVN	MD	Decrease 1 h after the injection	Increase 1 h after the injection	179
Injection of thiosemicarbazide (8 mg/kg, i.v.)	Rabbit	LVN	MD	None	Increase 1 h after the injection	179
Barbiturate sleep (sodium pentobarbital, 27 mg/kg, i.v.)	Rabbit	Gigantocellular nucleus of reticular formation	MD	None	Decrease 30 min after the barbiturate injection	75
Natural sleep	Rabbit	Gigantocellular nucleus of reticular formation	MD	None	Activation after the 1.5-h sleep	181
Forced insomnia	Rabbit	Gigantocellular nucleus of reticular formation	MD	Increase after the 7-day worrying (5 hr/day)	Decrease	181

(Continued)

Table VI. (Continued)

Biochemical component and experimental conditions	Animal species	Area of the nervous system	ODC isolation	Changes in ODC	Changes in neurons	Reference
Forced insomnia	Rabbit	Oral nucleus of reticular formation	MD	Decrease after the 7-day worrying (5 h/day)	Decrease	181
Oxidation of α-ketoglutarate						
Injection of tricyanoaminopropene (20 mg/kg, i.v.)	Rabbit	LVN	MD	Increase 1 h after the injection	None	72
Lactate dehydrogenase activity						
Injection of KCN (10 mg/kg, i.p.)	Mouse	Occipital cortex	QC	None	Activation of H-forms 15 min after the injection	123
Injection of KCN (10 mg/kg, i.p.)	Mouse	SCAH[a]	QC	None	Activation of M-forms 15 min after the injection	123
Audiogenic convulsions caused by sound of 96 dB	Rat	SCAH	QC	Activation of H-forms after 2-min convulsions	None	123
Cytochrome oxidase activity						
Hypoxia (12-h exposure at 8% O₂)	Rabbit	LVN	MD	None	Increase	183
Rotation for 25 min	Rabbit	LVN	MD	Slight decrease	None	71
Rotation for 7 days (25 min/day)	Rabbit	LVN	MD	Marked decrease	Decrease	71
Injection of tricyanoaminopropene (20 mg/kg, i.v.)	Rabbit	LVN	MD	None	Increase 1 h after the injection	183
Injection of tranylcypromine (0.3 mg/kg, i.v.)	Rabbit	LVN	MD	None 1 h after the injection; progressive decrease in the following 5 days	Increase 1 h after the injection; gradual return to the norm in the following 5 days	254
RNA content per cell						
Ontogenetic development:						
Comparison of fetal, young, and adult animals	Bovine	Brain white matter	BI[a]	Decrease in the course of development		220

	Animal	Forebrain	BI			16
Comparison of 10-, 30- and 60-day-old animals	Rat			None		
Seasonal changes in hibernating animals	Ground squirrel	Supraoptic nucleus of hypothalamus	QC	Increase during the winter	Decrease during the winter	255
Diurnal changes	Rat	Supraoptic nucleus of hypothalamus	QC	Zenith in the light, nadir in the dark period of the 24-h cycle	Zenith in the dark, nadir in the light period of the 24-h cycle	256
Diurnal changes	Rat	SCAH	QC	Zenith in the dark, nadir in the light period of the 24-h cycle	Zenith in the light, nadir in the dark period of the 24-h cycle	256
Hypoxic hypoxia for 1 h	Cat	Cerebral cortex, layer II	QC	Decrease in visual, motor, and auditory cortices	Decrease in visual and motor cortices	257
Ischemic hypoxia for 1h	Cat	Cerebral cortex, layer II	QC	Decrease in motor and auditory cortices	Decrease in visual and motor cortices	257
Hypoxic hypoxia for 1 h	Rat	SCAH	QC	None	Increase	258
Hypothermia	Rat	SCAH	QC	None	Increase	259
Cooling	Rat	Hypothalamus	QC	Increase during the cold adaptation	Increase during the cold adaptation	260
Gamma irradiation	Mouse	SCAH	QC	Increase	Increase	261
REM-sleep[a] deprivation	Rat	Supraoptic nucleus of hypothalamus	QC	None	Decrease	262
Hyperoxic hyperbaric convulsions	Rat	Motor cortex, layer V	QC	Decrease	Decrease	263
Hyperoxic hyperbaric convulsions	Rat	SCAH	QC	None	Decrease	263
Convulsions induced by anticholinesterase drug (0.6 mg/kg, i.p.)	Rat	SCAH	QC	Decrease	Decrease	264
Convulsions induced by pentylenetetrazole (45–55 mg/kg, i.p.)	Rat	SCAH	QC	Decrease during the convulsions, rapid increase afterwards	Decrease during the convulsions, slow increase afterwards	265
Convulsions induced by transcorneal electrostimulation	Rat	Hippocampus	QC	None	Increase	266

(Continued)

Table VI. (Continued)

Biochemical component and experimental conditions	Animal species	Area of the nervous system	ODC isolation	Changes in ODC	Changes in neurons	Reference
Foot shock	Rat	SCAH	QC	Decrease during the foot shock, rapid increase afterwards	Decrease during the foot shock, slow increase afterwards	258
Electrostimulation	Rat	SCAH	QC	Decrease, then return to normal	None, then decrease	267
Light stimulation for 2 hr	Rat	Visual cortex	QC	None	Increase	268
Rotation for 7 days (25 min/day)	Rabbit	LVN	MD	Decrease	Slight Increase	68
Uninterrupted rotation for 1, 7, and 30 days	Rat	LVN	MD	Decrease	Decrease	269
Running in a wheel for 1 h	Mouse	Motor cortex, layer V	QC	None	Increase	270
Swimming for 3 h	Mouse	SCAH	QC	None	Increase	227
Injection of tricyanoaminopropene (20 mg/kg, i.v.)	Rabbit	LVN	MD	Increase 1 h after the injection; changes in nucleotide composition of cellular RNA	Decrease 1 h after the injection; nucleotide changes reciprocal to those in the ODC	71
Injection of phenylcyclopropylamine (0.3 mg/kg, i.v.) or of imipramine (4 mg/kg, i.v.)	Rabbit	LVN	MD	Increase 1 h after the injection	Decrease 1 h after the injection	68
Injection of tranylcypromine (0.3 mg/kg, i.v.)	Rabbit	LVN	MD	Decrease; changes in RNA nucleotide composition	Increase; nucleotide composition changes more marked than in the ODC	254
Injections of epinephrine daily for 2 weeks (0.15 mg/kg per day)	Rat	Spinal cord lateral horns	QC	Increase	None	258
Injections of epinephrine daily for 2 weeks (0.15 mg/kg per day)	Rat	SCAH	QC	Increase	Decrease	258
Adrenalectomy	Rat	Supraoptic nucleus of hypothalamus	QC	Decrease; return to normal after cortisol injections	None	271

Protein content per cell
Ontogenetic development:

Item	Species	Region	Method			Ref.
Comparison of fetal, young and adult animals	Bovine white matter	Brain white matter	BI	Decrease in the course of development		220
Comparison of 10-, 30- and 60-day-old animals	Rat	Forebrain	BI	None		16
Comparison of 5–45-day-old animals	Rat	Corpus callosum	QC	Increase between the fifth and eighth day, then decrease		272
Diurnal changes	Rat	Supraoptic nucleus of hypothalamus	QC	Zenith at the evening hours, nadir at the morning hours	Zenith at the evening and night hours, nadir at the morning hours	256
Diurnal changes	Rat	SCAH	QC	Zenith at the evening hours, nadir at the morning hours	Zenith at the day hours, nadir at the night hours	256
Cooling	Rat	Supraoptic nucleus of hypothalamus	QC	Decrease, then increase during the initial cooling; return to normal at the cold adaptation	Increase, then decrease during the initial cooling; return to normal at the cold adaption	273
Cooling	Rat	Medial preoptic area of hypothalamus	QC	Decrease during the initial cooling; return to normal at the cold adaptation	Increase, then decrease during the initial cooling; return to normal at the cold adaptation	273
REM-sleep deprivation	Rat	Supraoptic nucleus of hypothalamus	QC	Moderate decrease	Moderate decrease	274
REM-sleep deprivation	Rat	Red nucleus of midbrain	QC	Moderate, then marked decrease	Marked decrease	274
Convulsions induced by pentylenetetrazole (45–55 mg/kg, i.p.)	Rat	SCAH	QC	Decrease during the convulsions; rapid increase afterwards	Decrease during the convulsions; slow increase afterwards	265
Anticipation stress (daily foot shocks for 7 days)	Rat	SCAH	QC	None	Increase in the nucleus and cytoplasm	275

(Continued)

Table VI. (Continued)

Biochemical component and experimental conditions	Animal species	Area of the nervous system	ODC isolation	Changes in ODC	Changes in neurons	Reference
Foot shock-motivated passive avoiding	Rat	CA$_3$ area of hippocampus	QC	None during the acquisition; increase by the moment of consolidation	Increase during the acquisition; return to normal by the moment of consolidation	276
Injection of amphetamine (6 mg/kg, every 3 h for 96 h, s.c.)	Rat	Supraoptic nucleus of hypothalamus	QC	Marked decrease	Marked decrease	274
Injection of amphetamine (6 mg/kg, every 3 h for 96 h, s.c.)	Rat	Red nucleus of midbrain	QC	Marked stable decrease	Moderate temporary decrease	274
Injections of sodium amobarbital (70 or 100 mg/kg, s.c.)	Rat	Supraoptic nucleus of hypothalamus	QC	None or moderate decrease	Marked decrease	3
Injections of sodium amobarbital (70 or 100 mg/kg, s.c.)	Rat	Red nucleus of midbrain	QC	Decrease	Decrease	3
Injections of morphine	Mouse	Brain	BI	Increase in the phosphorylation of acidic chromatin proteins		277

^{a} Abbreviations: BI, bulk isolation; LVN, lateral vestibular nucleus; MD, microdissection; REM, rapid eye movements; SCAH, spinal cord anterior horns; QC, quantitative cytochemistry.

preliminary conclusion, the concept of Holger Hydén[67-72] can be accepted: the perineuronal ODC are capable of reacting actively to changes in the functional state of the nervous system. Indeed, in many cases, as seen from Table VI, metabolic responses in the perineuronal ODC and in corresponding neurons are similar. At the same time, specific differences are often seen in the neuronal and ODC metabolic responses to the same experimental conditions; sometimes these responses are reciprocal (Table VI). When these conditions no longer affect the nervous system, recovery of the ODC metabolism proceeds more rapidly than that of the neuronal metabolism.

8. CONCLUSION

The ODC are small in size, and their content of RNA, protein, and total lipid per cell is lower than the corresponding amounts in neurons or AC.[3,13,15,16,18]

Oligodendrocytes are closely connected with myelin structures, and their chemical composition is similar to that of myelin or axons[13,18,86,87,112,113,136-141, 150,151,154-156,168-170]; however, differences between them have also been revealed with more sensitive methods of analysis.[137,140,147-151]

The biological role of the ODC remains obscure. The only function that can be definitely accepted is their participation in the formation of myelin.[30-32]

The participation of the ODC in the physiological activity of neurons is less well defined.[3,9,10] The hypothesis of Roitbak[52] tries to combine the myelin-forming function of the ODC with their functional role. A facilitating effect is postulated in the axonal pathways in which the percentage of myelinated presynaptic terminals is augmented through the activation of the myelinating ODC. However, this hypothesis lacks several proofs, particularly that depolarization of the ODC membranes triggers myelin formation and that this formation is rapid enough to provide for short-term synaptic events.

The data summarized in Table VI indicate that changes in the functional activity of the nervous system are accompanied in many cases by metabolic changes not only in related neurons but also in their perineuronal ODC; the direction of the neuronal and ODC changes can be either the same or opposite. However, it is not clear whether these ODC changes are specific and, even if so, what part they play in the physiological activity of the corresponding neurons.

From some indirect proofs, Hydén and Lange[69,70] put forward a hypothesis about an intercellular transfer of RNA from the perineuronal ODC to the activated neurons. The mechanism of pinocytosis is suggested by the authors to explain such macromolecular transfer. Subsequently, some data were reported both in favor[278-280] and against this hypothesis.[281] Even if one accepts the fact of the migration of the glial RNA to neurons,[3,10,252] no explanation has been presented about the possible role of this exogenous RNA (see Table IV) in the biochemical machinery of the neuron.

Perineuronal localization of the satellite ODC has made it tempting to speculate about the ODC as a morphological substrate for exchange of materials between the neuronal body and the capillary network of the nerve tissue.[3,19.]

[20,24,25] To an extent, this is, perhaps, true. However, the credibility of this idea has been lessened, on the one hand, by the findings of an intensive transport of material through intercellular space[46,47,54] and, on the other hand, by the description of AC that, apart from their vascular end-feet, possessed processes intervening between the neuronal bodies and the perineuronal ODC[23,76] (see Fig. 2).

Undoubtedly peculiar to the glial cells are regulatory functions such as supplying the neuron with some materials, homeostatic maintenance of the ion and acid–base balance, modulating effects on synapses through uptake and release of physiologically active substances, etc.[3,10,164,194,282] It remains to be seen which of these functions, and to what degree, are properties of the ODC and which of the AC.

ACKNOWLEDGMENTS. Supported by Grant RG-1089 from the National Multiple Sclerosis Society. I sincerely thank Dr. William T. Norton and the whole staff of the Departments of Neurology and Pathology of the Albert Einstein College of Medicine for constant help and friendly encouragement. The excellent secretarial assistance of Mrs. Marion Levine and Miss Renée Sasso is gratefully acknowledged. Thanks are due to Dr. Abel Lajtha and Dr. Gordon Guroff for their fruitful criticism of the manuscript and to Dr. Cedric S. Raine for his elegant microphotographs of the oligodendrocytes.

REFERENCES

1. Virchow, R., 1846, *Allgem. Z. Psychiatr.* **3**:242–250.
2. Del Rio Hortega, P., 1921, *Bol. R. Soc. Espan. Hist. Nat.* **21**:63–92.
3. Pevzner, L. Z., 1979, *Functional Biochemistry of the Neuroglia,* Plenum Press, New York.
4. Rose, S. P. R., 1967, *Biochem. J.* **102**:33–43.
5. Freysz, L., Bieth, R., Judes, C., Sensenbrenner, M., Jacob, M., and Mandel, P., 1968, *J. Neurochem.* **15**:307–313.
6. Blomstrand, C., and Hamberger, A., 1969, *J. Neurochem.* **16**:1401–1407.
7. Norton, W. T., and Poduslo, S. E., 1970, *Science* **167**:1144–1145.
8. Sellinger, O. Z., Azcurra, J. M., Johnson, D. E., Ohlsson, W. G., and Lodin, Z., 1971, *Nature* **230**:253–256.
9. Schoffeniels, E., Franck, G., Hertz, L., and Tower, D. B. (eds.), 1978, *Dynamic Properties of Glia Cells,* Pergamon Press, Oxford.
10. Varon, S., and Somjen, G. G., 1979, *Neurosci. Res. Progr. Bull.* **17**:1–239.
11. Fedoroff, S., and Hertz, L. (eds.), 1977, *Cell, Tissue, and Organ Cultures in Neurobiology,* Academic Press, New York.
12. Fewster, M. E., Scheibel, A. B., and Mead, J. F., 1967, *Brain Res.* **6**:401–408.
13. Poduslo, S. E., and Norton, W. T., 1972, *J. Neurochem.* **19**:727–736.
14. Fewster, M. E., Blackstone, S. C., and Ihrig, T. J., 1973, *Brain Res.* **63**:263–271.
15. Chao, S.-W., and Rumsby, M. G., 1977, *Brain Res.* **124**:347–351.
16. Snyder, D. S., Raine, C. S., Farooq, M., and Norton, W. T., 1980, *J. Neurochem.* **34**:1614–1621.
17. Szuchet, S., Stefansson, K., Wollmann, R. L., Dawson, G., and Arnason, B. G. W., 1980, *Brain Res.* **200**:151–164.
18. Farooq, M., Cammer, W., Snyder, D. S., Raine, C. S., and Norton, W. T., 1981, *J. Neurochem.* **36**:431–440.
19. Glees, P., 1955, *Neuroglia; Morphology and Function,* Blackwell Scientific Publications, Oxford.

20. Nakai, J. (ed.), 1963, *Morphology of Neuroglia*, Igaku Shoin, Tokyo.
21. Kuhlenbeck, H., 1970, *The Central Nervous System of Vertebrates*, Volume 3, Part I. *Structural Elements: Biology of Nervous System*, S. Karger, Basel.
22. Johnston, P. V., and Roots, B. I., 1972, *Nerve Membranes. A Study of the Biological and Chemical Aspects of Neuron–Glia Relationship*, Pergamon Press, Oxford.
23. Peters, A., Palay, S. L., and Webster, H. deF., 1976, *The Fine Structure of the Nervous System. The Neurons and Supporting Cells*, W. B. Saunders, Philadelphia.
24. Penfield, W., 1932 (1965 facsimile edition), *Cytology and Cellular Pathology of the Nervous System*, Volume II (W. Penfield, ed.), Hafner Publishing Co., New York, pp. 421–479.
25. Polak, M., 1965, *Prog. Brain Res.* **15**:12–34.
26. Vaughn, J. E., and Peters, A., 1971, *Cellular Aspects of Neural Growth and Differentiation* (D. C. Pease, ed.), University of California Press, Berkeley, pp. 103–134.
27. Roots, B. I., 1978, *Dynamic Properties of Glia Cells* (E. Schoffeniels, G. Franck, L. Hertz, and D. B. Tower, eds.), Pergamon Press, Oxford, pp. 45–54.
28. Cammermeyer, J., 1966, *Am. J. Anat.* **118**:227–247.
29. King, J. S., 1968, *Anat. Rec.* **161**:111–124.
30. Norton, W. T., 1976, *Basic Neurochemistry*, 2nd ed. (G. J. Siegel, R. W. Albers, R. Katzman, and B. W. Agranoff, eds.), Little, Brown, Boston, pp. 74–99.
31. Morell, P. (ed.), 1977, *Myelin*, Plenum Press, New York.
32. Palo, J. (ed.), 1978, *Myelination and Demyelination*, Plenum Press, New York.
33. Raine, C. S., 1976, *Basic Neurochemistry*, 2nd ed. (G. J. Siegel, R. W. Albers, R. Katzman, and B. W. Agranoff, eds.), Little, Brown, Boston, pp. 5–33.
34. Mori, S., and Leblond, C. P., 1970, *J. Comp. Neurol.* **139**:1–30.
35. Paterson, J. A., Privat, A., Ling, E. H., and Leblond, C. P., 1973, *J. Comp. Neurol.* **149**:83–102.
36. Privat, A., 1977, *Int. Rev. Cytol.* **40**:281–323.
37. Bondar, R. L., 1978, *Dynamic Properties of Glia Cells* (E. Schoffeniels, G. Franck, L. Hertz, and D. B. Tower, eds.), Pergamon Press, Oxford, pp. 3–11.
38. Kelly, J. S., Krnjević, K., and Yim, G. K. W., 1967, *Brain Res.* **6**:767–769.
39. Grossman, R. G., and Hampton, T., 1968, *Brain Res.* **11**:316–324.
40. Sugaya, E., Karahashi, Y., Sugaya, A., and Haruki, F., 1971, *Jpn. J. Physiol.* **21**:149–157.
41. Orkand, R. K., 1977, *Handbook of Physiology*, Section 1: *The Nervous System*, Volume I. *Cellular Biology of Neurons*, Part 2 (J. M. Brookhart, V. B. Mountcastle, E. R. Kandel, and S. R. Geiger, eds.), American Physiological Society, Bethesda, pp. 855–875.
42. Hild, W., and Tasaki, I., 1962, *J. Neurophysiol.* **25**:277–304.
43. Trachtenberg, M. C., and Pollen, D. A., 1970, *Science* **167**:1248–1252.
44. Krnjević, K., and Morris, M. E., 1972, *Can. J. Physiol. Pharmacol.* **50**:1214–1217.
45. Glötzner, F. L., 1973, *Brain Res.* **55**:159–171.
46. Kuffler, S. W., Nicholls, J. G., and Orkand, R. K., 1966, *J. Neurophysiol.* **29**:768–787.
47. Somjen, G. G., 1975, *Annu. Rev. Physiol.* **37**:163–190.
48. Krnjević, K., and Schwartz, S., 1967, *Exp. Brain Res.* **3**:306–319.
49. Wardell, W. M., 1966, *Proc. R. Soc. (Lond.) [Biol.]* **165**:326–361.
50. Trachtenberg, M. C., Kornblith, P. L., and Häuptli, J., 1972. *Brain Res.* **38**:279–298.
51. Pape, L. G., and Katzman, R., 1972, *Brain Res.* **38**:71–92.
52. Roitbak, A. I., and Fanardzhyan, V. V., 1973, *Proc. USSR Acad. Sci. [Biol. Sci.] (Engl. Transl.)* **211**:340–343.
53. Grossman, R. G., and Seregin, A., 1977, *Science* **195**:196–198.
54. Grossman, R. G., 1978, *Dynamic Properties of Glia Cells* (E. Schoffeniels, G. Franck, L. Hertz, and D. B. Tower, eds.) Pergamon Press, Oxford, pp. 105–113.
55. Dennis, M. J., and Gerschenfeld, H. M., 1969, *J. Physiol. (Lond.)* **203**:211–222.
56. Vernadakis, A., and Berni, A., 1973, *Brain Res.* **57**:223–228.
57. Picker, S., Pieper, C. F., and Goldring, S., 1980, *Soc. Neurosci. Abstr.* **6**:326.
58. Lowry, O. H., Roberts, N. R., Leiner, K. Y., Wu, M.-L., Farr, A. L., and Albers, R. W., 1954, *J. Biol. Chem.* **207**:39–49.
59. Kety, S., and Elkes, J. (eds.), 1961, *Regional Neurochemistry*, Pergamon Press, Oxford.
60. Friede, R., 1966, *Topographic Brain Chemistry*, Academic Press, New York.

61. Lowry, O. H., 1955, *Biochemistry of the Developing Nervous System* (H. Waelsch, ed.), Academic Press, New York, pp. 350–357.
62. Lowry, O. H., Roberts, N. R., and Chang, M.-L. W., 1956, *J. Biol. Chem.* **222**:97–107.
63. Lowry, O. H., 1962, *Bull. N.Y. Acad. Med.* **38**:789–798.
64. Lowry, O. H., and Passonneau, J. V., 1972, *A Flexible System of Enzymatic Analysis,* Academic Press, New York.
65. Giacobini, E., 1962, *J. Neurochem.* **9**:169–177.
66. Giacobini, E., 1964, *Morphological and Biochemical Correlates of Neural Activity* (M. M. Cohen and R. S. Snider, eds.), Harper & Row, New York, pp. 15–38.
67. Hydén, H., 1960, *The Cell,* Volume IV (J. Brachet and A. E. Mirsky, eds.), Academic Press, New York, pp. 215–323.
68. Hydén, H., 1964, *Recent Adv. Biol. Psychiatry* **6**:34–54.
69. Hydén, H., 1967, *Neuron* (H. Hydén, ed.), Elsevier, Amsterdam, pp. 179–217.
70. Hydén, H., 1973, *Macromolecules and Behaviour* (G. B. Ansell and P. B. Bradley, eds.) Macmillan, London, pp. 3–75.
71. Hydén, H., and Pigon, A., 1960, *J. Neurochem.* **6**:57–72.
72. Hamberger, A., 1963, *Acta Physiol. Scand. [Suppl.]* **203**:1–58.
73. Roots, B. I., and Johnston, P. V., 1972, *Research Methods in Neurochemistry,* Volume 1 (N. Marks and R. Rodnight, eds.), Plenum Press, New York, pp. 3–17.
74. Osborn, N. N., 1974, *Microchemical Analysis of Nervous Tissue,* Pergamon Press, Oxford.
75. Hamberger, A., Hydén, H., and Lange, P. W., 1966, *Science* **151**:1394–1395.
76. Sotelo, C., and Palay, S. L., 1968, *J. Cell Biol.* **36**:151–179.
77. Takahashi, Y., Hsu, C. S., and Honma, S., 1970, *Brain Res.* **23**:284–287.
78. Poduslo, S. E., and Norton, W. T., 1972, *Research Methods in Neurochemistry,* Volume 1 (N. Marks and R. Rodnight, eds.), Plenum Press, New York, pp. 19–32.
79. Poduslo, S. E., and Norton, W. T., 1975, *Methods in Enzymology* (S. P. Colowick and N. O. Kaplan, eds.), Volume XXXV, *Lipids,* Part B (J. M. Lowenstein, ed.) Academic Press, New York, pp. 561–579.
80. Raine, C. S., Poduslo, S. E., and Norton, W. T., 1971, *Brain Res.* **27**:11–24.
81. Trapp, B. D., Dwyer, B., and Bernsohn, J., 1975, *Neurobiology* **5**:235–248.
82. Raine, C. S., Traugott, U., Iqbal, K., Snyder, D. S., Cohen, S. R., Farooq, M., and Norton, W. T., 1978, *Brain Res.* **142**:85–96.
83. Cohen, S. R., and Bernsohn, J., 1973, *Brain Res.* **60**:521–525.
84. Benjamins, J. A., Garnieri, M., Miller, K., Sonneborn, M., and McKhann, G. M., 1974, *J. Neurochem.* **23**:751–756.
85. Pleasure, D., Abramsky, O., Silberberg, D., Quinn, B., Parris, J., and Saida, T., 1977, *Brain Res.* **134**:377–382.
86. Desmukh, D. S., Flynn, T. J., and Pieringer, R. A., 1974, *J. Neurochem.* **22**:479–485.
87. Banik, N. L., and Smith, M. E., 1976, *Neurosci. Lett.* **2**:235–238.
88. Campbell, G. LeM., Schachner, M., and Sharrow, S. O., 1977, *Brain Res.* **127**:69–86.
89. McEwen, B. S., and Zigmont, R. E., 1972, *Research Methods in Neurochemistry,* Volume 1 (N. Marks and R. Rodnight, eds.), Plenum Press, New York, pp. 139–161.
90. Zubzhitskaya, L. B., and Pevzner, L. Z., 1973, *Proc. USSR Acad. Sci. [Biol. Sci.] (Engl. Transl.)* **210**:249–252.
91. Løvtrup-Rein, H., and McEwen, B. S., 1966, *J. Cell Biol.* **30**:405–416.
92. Austoker, J., Cox, D., and Mathias, A. P., 1972, *Biochem. J.* **129**:1139–1155.
93. Pontén, J., 1975, *Human Tumor Cells In Vitro* (J. Fogh, ed.), Plenum Press, New York, pp. 175–206.
94. Pfeiffer, S. E., Betschart, B., Cook, J., Mancini, P., and Morris, R., 1977, *Cell, Tissue, and Organ Cultures in Neurobiology* (S. Federoff and L. Hertz, eds.), Academic Press, New York, pp. 287–346.
95. Zimmerman, H. M., 1955, *Am. J. Pathol.* **31**:1–29.
96. Wechsler, W., Pfeiffer, S. E., Swenberg, J. A., and Koestner, A., 1973, *Acta Neuropathol.* **24**:287–303.
97. Schubert, D., Heinemann, S., Carlisle, W., Tarikas, H., Kimes, B., Patrick, J., Steinbach, J. H., Culp, W., and Brandt, B. L., 1974, *Nature* **249**:224–227.

98. Fields, K. L., Gosling, C., Medson, M., and Stern, P. L., 1975, *Proc. Natl. Acad. Sci. U.S.A.* **72:**1296–1300.

99. Henn, F. A., Anderson, D., and Rustad, D., 1976, *Brain Res.* **101:**341–344.

100. Vernadakis, A., and Nidess, R., 1976, *Neurochem. Res.* **1:**385–402.

101. Parker, K. K., Norenberg, M. D., and Vernadakis, A., 1980, *Science* **208:**179–181.

102. Eng, L. F., 1980, *Proteins of the Nervous System*, 2nd ed. (R. A. Bradshaw and D. M. Schneider, eds.), Raven Press, New York, pp. 85–117.

103. Sundarraj, N., Schachner, M., and Pfeiffer, S. E., 1975, *Proc. Natl. Acad. Sci. U.S.A.* **72:**1927–1931.

104. Poduslo, S. E., and McKhann, G. M., 1977, *Neurosci. Lett.* **5:**159–163.

105. Varon, S., 1977, *Cell, Tissue, and Organ Cultures in Neurobiology* (S. Fedoroff and L. Hertz, eds.), Academic Press, New York, pp. 237–261.

106. Federoff, S., 1977, *Cell, Tissue, and Organ Cultures in Neurobiology* (S. Fedoroff and L. Hertz, eds.), Academic Press, New York, pp. 265–286.

107. Fedoroff, S., 1978, *Dynamic Properties of Glia Cells* (E. Schoffeniels, G. Franck, L. Hertz, and D. B. Tower, eds.), Pergamon Press, Oxford, pp. 83–92.

108. Fewster, M. E., and Blackstone, S. C., 1975, *Neurobiology* **5:**316–328.

109. Booher, J., and Sensenbrenner, M., 1972, *Neurobiology* **2:**97–105.

110. Sensenbrenner, M., 1977, *Cell, Tissue, and Organ Cultures in Neurobiology* (S. Fedoroff and L. Hertz, eds.), Academic Press, New York, pp. 191–213.

111. Pettmann, B., Delaunoy, J. P., Courageot, J., Devilliers, G., and Sensenbrenner, M., 1980, *Dev. Biol.* **75:**278–287.

112. Labourdette, G., Roussel, G., Ghandour, M. S., and Nussbaum, J. L., 1979, *Brain Res.* **179:**199–203.

113. Labourdette, G., Roussel, G., and Nussbaum, J. L., 1980, *Neurosci. Lett.* **18:**203–209.

114. McCarthy, K. D., and deVellis, J., 1980, *J. Cell Biol.* **85:**890–902.

115. Pearse, A. G. E., 1968, *Histochemistry, Theoretical and Applied*, 3rd ed. Williams & Wilkins, Baltimore.

116. Burstone, M. S., 1962, *Enzyme Histochemistry*, Academic Press, New York.

117. Gersch, I. (ed.), 1973, *Submicroscopic Cytochemistry*, Academic Press, New York.

118. Sternberger, L. A., 1979, *Immunocytochemistry*, 2nd ed. Wiley, New York.

119. Wied, G. L. (ed.), 1966, *Introduction to Quantitative Cytochemistry*, Academic Press, New York.

120. Klenikova, V. A., and Pevzner, L. Z., 1979, *Microsc. Acta* **82:**207–214.

121. Brumberg, V. A., and Pevzner, L. Z., 1978, *Microsc. Acta* **80:**323–330.

122. Pevzner, L. Z., Raygorodskaya, T. G., and Agroskin, L. S., 1978, *Microsc. Acta* **81:**9–14.

123. Brumberg, V. A., and Pevzner, L. Z., 1976, *Acta Histochem.* **55:**1–7.

124. Caspersson, T., 1950, *Cell Growth and Cell Function*, Norton, New York.

125. Caspersson, T., 1955, *Experientia* **11:**45–60.

126. Pevzner, L. Z., 1966, *Macromolecules and Behavior* (J. Gaito, ed.), Appleton-Century-Crofts, New York, pp. 43–70.

127. Hale, A. J., 1958, *The Interference Microscope in Biological Research*, Livingstone, Edinburgh.

128. Varon, S., 1975, *Exp. Neurol.* **48:**93–134.

129. Raine, C. S., Wisniewski, H. M., Iqbal, K., Grundke-Iqbal, I., and Norton, W. T., 1977, *Brain Res.* **120:**269–286.

130. Traugott, U., Snyder, D. S., Norton, W. T., and Raine, C. S., 1978, *Ann. Neurol.* **4:**431–439.

131. Abramsky, O., Lisak, R. P., Pleasure, D., Gilden, D. H., and Silberberg, D. H., 1978, *Neurosci. Lett.* **8:**311–316.

132. Raff, M. C., Fields, K. L., Hakomori, S.-I., Mirsky, R., Pruss, R. M., and Winter, J., 1979, *Brain Res.* **174:**283–308.

133. Bock, E., 1977, *Cell, Tissue, and Organ Cultures in Neurobiology* (S. Fedoroff and L. Hertz, eds.), Academic Press, New York, pp. 407–422.

134. Varon, S., 1978, *Dynamic Properties of Glia Cells* (E. Schoffeniels, G. Franck, L. Hertz, and D. B. Tower, eds.), Pergamon Press, Oxford, pp. 93–103.

135. Fields, K., 1979, *Current Topics in Developmental Biology*, Volume 13 (A. A. Moscona and A. Monroy, eds.), Academic Press, New York, pp. 237–257.

136. Fewster, M. E., and Mead, J. F., 1968, *J. Neurochem.* **15**:1041–1052.
137. Norton, W. T., Abe, T., Poduslo, S. E., and DeVries, G. H., 1975, *J. Neurosci.. Res.* **1**:57–75.
138. Poduslo, S. E., 1975, *J. Neurochem.* **24**:647–654.
139. Poduslo, S., Miller, K., and McKhann, G. M., 1978, *J. Biol. Chem.* **253**:1592–1597.
140. Poduslo, S. E., 1978, *Myelination and Demyelination* (J. Palo, ed.), Plenum Press, New York, pp. 71–94.
141. Yu, R. K., and Iqbal, K., 1979, *J. Neurochem.* **32**:293–300.
142. Johnson, A. B., and Bornstein, M. B., 1978, *Brain Res.* **159**:173–182.
143. Raff, M. C., Mirsky, R., Fields, K. L., Lisak, R. P., Dorfman, S. H., Silberberg, D. H., Gregson, N. A., Leibowitz, S., and Kennedy, M. C., 1978, *Nature* **274**:813–816.
144. Lisak, R. P., Abramsky, O., Dorfman, S. H., George, J., Manning, M. C., Pleasure, D. E., Saida, T., and Silberberg, D. H., 1979, *J. Neurol. Sci.* **40**:65–73.
145. Mirsky, R., Winter, J., Abney, E. R., Pruss, R. M., Gavrilovic, J., and Raff, M. C., 1980, *J. Cell Biol.* **84**:483–494.
146. Dawson, G., Sundarraj, N., and Pfeiffer, S. E., 1977, *J. Biol. Chem.* **252**:2777–2779.
147. Poduslo, S. E., McFarland, H. F., and McKhann, G. M., 1977, *Science* **197**:270–272.
148. Sternberger, N. H., Itoyama, Y., Kies, M. W., and Webster, H. deF., 1978, *J. Neurocytol.* **7**:251–263.
149. Sternberger, N. H., Itoyama, Y., Kies, M. W., and Webster, H. deF., 1978, *Proc. Natl. Acad. Sci. U.S.A.* **75**:2521–2524.
150. Hartman, B. K., Agrawal, H. C., Kalmbach, S., and Shearer, W. T., 1979, *J. Comp. Neurol.* **188**:273–290.
151. Sternberger, N. H., Quarles, R. H., Itoyama, Y., and Webster, H. deF., 1979, *Proc. Natl. Acad. Sci. U.S.A.* **76**:1510–1514.
152. Agrawal, H. C., Hartman, B. K., Shearer, W. T., Kalmbach, S., and Margolis, F., 1977, *J. Neurochem.* **28**:495–508.
153. Agrawal, H. C., and Hartman, B. K., 1980, *Proteins of the Nervous System*, 2nd ed. (R. A. Bradshaw and D. M. Schneider, eds.), Raven Press, New York, pp. 145–169.
154. Roussel, G., Delaunoy, J. P., Nussbaum, J. L., and Mandel, P., 1977, *Neuroscience* **2**:307–313.
155. Roussel, G., Delaunoy, J. P., Mandel, P., and Nussbaum, J. L., 1978, *J. Neurocytol.* **7**:155–163.
156. Mandel, P., Roussel, G., Delaunoy, J.-P., and Nussbaum, J.-L., 1978, *Dynamic Properties of Glia Cells* (E. Schoffeniels, G. Franck, L. Hertz, and D. B. Tower, eds.), Pergamon Press, Oxford, pp. 267–274.
157. Pfeiffer, S. E., and Wechsler, W., 1972, *Proc. Natl. Acad. Sci. U.S.A.* **69**:2885–2889.
158. Zanetta, J. P., Benda, P., Gombos, G., and Morgan, I. G., 1972, *J. Neurochem.* **19**:881–883.
159. Kurihara, T., Kawakami, S., Ueki, K., and Takahashi, Y., 1974, *J. Neurochem.* **22**:1143–1144.
160. Drummond, R. J., and Dean, G., 1980, *J. Neurochem.* **35**:1155–1165.
161. Sprinkle, T. J., Wells, M. R., Garver, F. A., and Smith, D. B., 1980, *J. Neurochem.* **35**:1200–1206.
162. Cammer, W., Bieler, L., Fredman, T., and Norton, W. T., 1977, *Brain Res.* **138**:17–28.
163. Yandrasitz, J. R., Ernst, S. A., and Salganicoff, L., 1976, *J. Neurochem.* **27**:707–715.
164. Hertz, L., 1977, *Cell, Tissue, and Organ Cultures in Neurobiology* (S. Fedoroff and L. Hertz, eds.), Academic Press, New York, pp. 39–71.
165. Kimelberg, H. K., Narumi, S., Biddlecome, S., and Bourke, R. S., 1978, *Dynamic Properties of Glia Cells* (E. Schoffeniels, G. Franck, L. Hertz, and D. B. Tower, eds.), Pergamon Press, Oxford, pp. 347–457.
166. Murai, S., 1973, *Nihon Univ. Med. J.* **32**:583–605.
167. Nagata, Y., Mikoshiba, K., and Tsukada, Y., 1974, *J. Neurochem.* **22**:493–503.
168. Roussel, G., Delaunoy, J.-P., Nussbaum, J.-L., and Mandel, P., 1979, *Brain Res.* **160**:47–55.
169. Ghandour, M. S., Langley, O. K., Vincendon, G., and Gombos, G., 1979, *J. Histochem. Cytochem.* **27**:1634–1637.
170. Ghandour, M. S., Langley, O. K., Vincendon, G., Gombos, G., Filippi, D., Limozin, N., Dalmasso, C., and Laurent, G., 1980, *Neuroscience* **5**:559–571.
171. McGinnis, J. F., and deVellis, J., 1977, *Arch. Biochem. Biophys.* **179**:682–691.

172. DeVellis, J., and Inglish, D., 1973, *Neurobiological Aspects of Maturation and Aging* (D. H. Ford, ed.), Elsevier, Amsterdam, pp. 321–330.

173. DeVellis, J., McGinnis, J. F., Breen, G. A. M., Leveille, P., Bennett, K., and McCarthy, K., 1977, *Cell, Tissue, and Organ Cultures in Neurobiology* (S. Fedoroff and L. Hertz, eds.), Academic Press, New York, pp. 485–511.

174. Breen, G. A. M., and deVellis, J., 1974, *Dev. Biol.* **41**:255–266.

175. Claisse, P., and Roscoe, J. P., 1976, *Brain Res.* **109**:423–425.

176. Hirsch, H. E., Blanco, C. E., and Parks, M. E., 1980, *J. Neurochem.* **34**:760–762.

177. Leveille, P. J., McGinnis, J. F., Maxwell, D. S., and deVellis, J., 1980, *Brain Res.* **196**:287–305.

178. Hamberger, A., 1961, *J. Neurochem.* **8**:31–35.

179. Aleksidze, N. G., and Blomstrand, C., 1968, *Brain Res.* **11**:717–719.

180. Aleksidze, N. G., and Blomstrand, C., 1969, *Proc. USSR Acad. Sci. [Biochem.] (Engl. Transl.)* **186**:140–141.

181. Hydén, H., and Lange, P. W., 1965, *Science* **149**:654–658.

182. Hamberger, A., and Sjöstrand, J., 1966, *Acta Physiol. Scand.* **67**:76–88.

183. Hamberger, A., and Hydén, H., 1963, *J. Cell Biol.* **16**:521–525.

184. Wilkin, G. P., and Wilson, J. E., 1977, *J. Neurochem.* **29**:1039–1051.

185. Kao-Jen, J., and Wilson, J. E., 1980, *J. Neurochem.* **35**:667–678.

186. Margolis, R. U., and Margolis, R. K., 1974, *Biochemistry* **13**:2849–2852.

187. McCandless, D. W., Feussner, G. K., Lust, W. D., and Passonneau, J. V., 1979, *Proc. Natl. Acad. Sci. U.S.A.* **76**:1482–1484.

188. Sharp, F. R., 1976, *Brain Res.* **110**:127–139.

189. Cummins, J., and Hydén, H., 1962, *Biochim. Biophys. Acta* **60**:271–283.

190. Freysz, L., Farooqui, A. A., Adamczewska-Goncerzewicz, Z., and Mandel, P., 1979, *J. Lipid Res.* **20**:503–508.

191. Kumamoto, T., Nakagawa, S., Yata, Y., and Shimizu, E., 1976, *Histochemistry* **47**:101–109.

192. Carruthers, A., and Carey, E. M., 1979, *Biochem. Soc. Trans.* **7**:418–419.

193. Hamberger, A., and Röckert, H., 1964, *J. Neurochem.* **11**:757–760.

194. Quastel, J. H., 1978, *Dynamic Properties of Glia Cells* (E. Schoffeniels, G. Franck, L. Hertz, and D. B. Tower, eds.), Pergamon Press, Oxford, pp. 153–162.

195. Minchin, M. C. W., and Beart, P. M., 1975, *Brain Res.* **83**:437–449.

196. Kelly, J. S., and Dick, F., 1978, *Dynamic Properties of Glia Cells* (E. Schoffeniels, G. Franck, L. Hertz, and D. B. Tower, eds.), Pergamon Press, Oxford, pp. 183–192.

197. Oksche, A. (ed.), 1980, *Handbuch der Mikroscopischen Anatomie des Menschen*, Band 4, Teil 10, *Neuroglia I*, Springer-Verlag, Berlin.

198. Leach, M. J., Riddall, D. R., and Winkley, C. M., 1976, *J. Neurochem.* **27**:1281–1282.

199. Drummond, R. J., and Phillips, A. T., 1977, *J. Neurochem.* **29**:101–108.

200. Schubert, D., 1975, *Brain Res.* **84**:87–98.

201. Abe, T., and Norton, W. T., 1979, *J. Neurochem.* **32**:823–832.

202. Thompson, B. J., 1977, *J. Neurochem.* **29**:387–391.

203. Fewster, M. E., Ihrig, T., and Mead, J. F., 1975, *J. Neurochem.* **25**:207–213.

204. Cohen, S. R., and Bernsohn, J., 1978, *J. Neurochem.* **30**:661–669.

205. Pleasure, D., Lichtman, C., Eastman, S., Lieb, M., Abramsky, O., and Silberberg, D., 1979, *J. Neurochem.* **32**:1447–1450.

206. Derry, D. M., and Wolfe, L. S., 1967, *Science* **158**:1450–1452.

207. Manuelidis, L., Yu, R. K., and Manuelidis, E. E., 1977, *Acta Neuropathol. (Berl.)* **38**:129–135.

208. Dawson, G., 1979, *J. Biol. Chem.* **254**:155–162.

209. Abe, T., Miyatake, T., Norton, W. T., and Suzuki, K., 1979, *Brain Res.* **161**:179–182.

210. Dawson, G., and Kernes, S. M., 1978, *J. Neurochem.* **31**:1091–1094.

211. Dawson, G., and Kernes, S. M., 1979, *J. Biol. Chem.* **254**:163–167.

212. Hirsch, H. E., Wernicke, J. F., Myers, L. W., and Parks, M. E., 1977, *J. Neurochem.* **29**:979–985.

213. Hirsch, H. E., 1972, *J. Neurochem.* **19**:1513–1517.

214. Dorman, R. V., Toews, A. D., and Horrocks, L. A., 1977, *J. Lipid Res.* **18**:115–117.

215. Fu, S. C., and Horrocks, L. A., 1978, *Fed. Proc.* **37**:1834.

216. Lapham. L. W., and Johnstone, M. A., 1963, *Arch. Neurol.* **9:**194–202.

217. Thomas, J. O., and Thompson, R. J. 1977, *Cell* **10:**633–640.

218. Inoue, N., Suzuki, O., and Kato, T., 1976, *J. Neurochem.* **27:**113–119.

219. Iqbal, K., Grundke-Iqbal, I., and Wisniewski, H. M., 1977, *J. Neurochem.* **28:**707–716.

220. Fewster, M. E., Einstein, E. R., Csejtey, J., and Blackstone, S. C., 1974, *Neurobiology* **4:**388–401.

221. Stambolova, M. A., Cox, D., and Mathias, A. P., 1973, *Biochem. J.* **136:**685–695.

222. Haas, R. J., Werner, J., and Fliedner, T. M., 1970, *J. Anat.* **107:**421–437.

223. Krauss-Ruppert, R., Laissue, J. Bürki, H., and Odartchenko, N., 1973, *J. Comp. Neurol.* **148:**211–216.

224. Norton, P., and Viola, M. V., 1977, *J. Neurochem.* **29:**299–303.

225. Inoue, N., and Kato, T., 1980, *J. Neurochem.* **34:**1574–1583.

226. Inoue, N., Ono, T., and Kato, T., 1979, *Biochem. J.* **180:**471–480.

227. Brumberg, V. A., 1968, *Proc. USSR Acad. Sci. [Biol. Sci.] (Engl. Transl.)* **182:**158–160.

228. Mizobe, F., Tashiro, T., and Kurokawa, M., 1974, *Eur. J. Biochem.* **48:**25–33.

229. Løvtrup-Rein, H., and Grahn, B., 1970, *J. Neurochem.* **17:**845–852.

230. Egyhazi, E., and Hydén, H., 1961, *J. Biophys. Biochem. Cytol.* **10:**403–410.

231. Hydén, H., and Egyhazi, E., 1963, *Proc. Natl. Acad. Sci. U.S.A.* **49:**618–623.

232. Daneholt, B., and Brattgård, S.-O., 1966, *J. Neurochem.* **13:**913–921.

233. Slagel, D. E., Hartmann, H. A., and Edström, J.-E., 1966, *J. Neuropathol. Exp. Neurol.* **25:**244–253.

234. Gomirato, G., and Hydén, H., 1963, *Brain* **86:**773–780.

235. Austoker, J., Cox, D., and Mathias, A. P., 1973, *Biochem. J.* **132:**813–819.

236. Kato, T., and Kurokawa, M., 1970, *Biochem. J.* **116:**599–609.

237. Thompson, R. J., 1973, *J. Neurochem.* **21:**19–40.

238. Banks-Schlegel, S. P., and Johnson, T. C., 1975, *J. Neurochem.* **24:**947–952.

239. Sarkander, H.-I., Knoll-Köhler, E., and Cervos-Navarro, J., 1978, *J. Pharmacol. Exp. Ther.* **205:**503–514.

240. Slagel, D. E., and Akers, R. D., 1972, *Brain Res.* **44:**245–260.

241. Soga, K., and Takahashi, Y., 1976, *J. Neurochem.* **26:**89–94.

242. Sarkander, H.-I., and Uthoff, C. G., 1976, *Eur. J. Biochem.* **71:**53–56.

243. Sarkander, H.-I., and Dulce, H.-J., 1978, *Exp. Brain Res.* **31:**317–327.

244. Sarkander, H.-I., Fleischer-Lambropoulos, H., and Brade, W. P., 1975, *Eur. J. Biochem.* **52:**40–43.

245. Sarkander, H.-I., and Dulce, H.-J., 1979, *Exp. Brain Res.* **35:**109–125.

246. Ozawa, H., Kushiya, E., and Takahashi, Y., 1980, *Neurosci. Lett.* **18:**191–196.

247. Løvtrup-Rein, H., 1970, *Brain Res.* **19:**433–444.

248. Kiss, J., and Koritsánszky, S., 1975, *Cell Tissue Res.* **159:**267–277.

249. Lee, N. M., and Loh, H. H., 1977, *J. Neurochem.* **29:**547–550.

250. Hydén, H., and McEwen, B. S., 1966, *Proc. Natl. Acad. Sci. U.S.A.* **55:**354–358.

251. Sviridov, S. M., Korochkin, L. I., Ivanov, V. N., Maletskaya, E. I., and Bakhtina, T. K., 1972, *J. Neurochem.* **19:**713–718.

252. Jakoubek, B., 1974, *Brain Function and Macromolecular Synthesis*, Pion, Ltd., London.

253. Hamberger, A., and Løvtrup, S., 1964, *J. Neurochem.* **11:**687–694.

254. Hydén, H., and Egyhazi, E., 1966, *Neurology (Minneap.)* **18:**732–736.

255. Pevzner, L. Z., and Semeshina, T. M., 1976, *Brain Res.* **108:**205–211.

256. Pevzner, L. Z., Litinskaya, L. L., Raygorodskaya, T. G., and Khrust, Yu. R., 1978, *Acta Histochem. (Jena)* **62:**1–11.

257. Pevzner, L. Z., 1979, *Exp. Neurol.* **65:**237–241.

258. Pevzner, L. Z., 1971, *J. Neurochem.* **18:**895–907.

259. Piven', N. V., and Pevzner, L. Z., 1972, *Proc. USSR Acad. Sci. [Biol. Sci.] (Engl. Transl.)* **206:**583–585.

260. Filipchenko, R. E., Pevzner, L. Z., and Slonim, A. D., 1978, *Acta Histochem (Jena)* **61:**23–29.

261. Olkowski, Z. L., and McLaren, J. R., 1975, *Strahlentherapie* **150:**76–79.

262. Demin, N. N., and Rubinskaya, N. L., 1974, *Proc. USSR Acad. Sci. [Biochem.] (Engl. Transl.)* **214:**43–44.

263. Pevzner, L. Z., 1979, *Acta Histochem.(Jena)* **64**:237–242.
264. Doemin, N. N., 1968, *Macromolecules and the Function of the Neuron* (Z. Lodin and S. P. R. Rose, eds.), Excerpta Medica Foundation, Amsterdam, pp. 316–323.
265. Pevzner, L. Z., and Saudargene, E. D., 1971, *Acta Histochem. (Jena)* **39**:101–117.
266. Pevzner, L., 1981, *Acta Histochem. (Jena)* **69**:302–306.
267. Geinisman, Yu. Ya., 1971, *Brain Res.* **28**:251–262.
268. Pevzner, L. Z., and Malinauskaite, O.-L., 1978, *Acta Histochem. (Jena)* **63**:288–291.
269. Grenell, R. G., Hazama, H., Kakazawa, M., and Einberg, E., 1968, *Brain Res.* **9**:115–125.
270. Tiplady, B., Glushchenko, T. S., and Pevzner, L. Z., 1974, *Proc. USSR Acad. Sci. [Biol. Sci.] (Engl. Transl.)* **214**:78–80.
271. Pevzner, L. Z., 1972, *Brain Res.* **46**:329–339.
272. Klenikova, V. A., Pevzner, L. Z., and Mularek, O., 1979, *Neuroscience* **4**:1187–1193.
273. Krichevskaya, A. A., Mogilnitskaya, L. V., and Pevzner, L. Z., 1976, *Proc. USSR Acad. Sci. [Biol. Sci.] (Engl. Transl.)* **226**:85–87.
274. Voronka, G. Sh., Demin, N. N., and Pevzner, L. Z., 1971, *Proc. USSR Acad. Sci. [Biol. Sci.] (Engl. Transl.)* **198**:417–420.
275. Jakoubek, B., Pevzner, L. Z., and Pavlik, A., 1979, *Neuroscience* **4**:1179–1186.
276. Pevzner, L. Z., 1978, *Dynamic Properties of Glia Cells* (E. Schoffeniels, G. Franck, L. Hertz, and D. B. Tower, eds.), Pergamon Press, Oxford, pp. 223–229.
277. Oguri, K., Lee, N. M., and Loh, H. H., 1976, *Biochem. Pharmacol.* **25**:2371–2376.
278. Kolodny, G. M., 1974, *Cell Communication* (R. P. Cox, ed.), John Wiley and Sons, New York, pp. 97–111.
279. Lasek, K. J., Gainer, H., and Barker, J. L., 1977, *J. Cell Biol.* **74**:501–523.
280. Gainer, H., Tasaki, I., and Lasek, R. J., 1977, *J. Cell Biol.* **74**:524–530.
281. Gambetti, P., Autillio-Gambetti, L., and Peck, K., 1980, *Brain Res.* **200**:59–86.
282. Vernadakis, A., Nidess, R., Culver, B., and Arnold, E. B., 1979, *Mech. Ageing Dev.* **9**:553–566.

The Schwann Cell

Robert M. Gould, Dan Matsumoto, and Gary Mattingly

1. INTRODUCTION

Although Schwann cells were not a topic in the previous series of this *Handbook*, their properties were covered in a general review of the peripheral nerve by Porcellati.[1] In addition, aspects of protein and lipid metabolism in peripheral nerve undergoing Wallerian degeneration were reviewed.[2,3] Interest in the peripheral nervous system has grown over the past 10 years, and the discussions in this chapter on Schwann cells and in others on the peripheral nervous system (PNS) including PNS myelin (Vol. 3), peripheral nerve (Vol. 7), and nerve regeneration (Vol. 9) reflect this interest and the advancement made.

This chapter will consider the biochemical properties of Schwann cells as seen in their normal association with both myelinated and unmyelinated axons and with terminal specializations as a means to appreciate the multipotentiality of this dynamic cell. The Schwann cell's role in mature, developing, pathological nerves will also be considered, first, in a morphological context, and then from a biochemical point of view.

2. MORPHOLOGY

2.1. Mature Nerve

Schwann cell morphology has been extensively studied and recently reviewed.[4,5] Each myelin-forming Schwann cell covers an approximate 500- to 1500-μm length of axon, called an internode, which is best displayed in individually teased fibers (Fig. 1A). Consecutive internodes, covering the entire length of axon, are separated by 1-μm-wide nodes of Ranvier (see also Fig. 3A). The oblong nucleus is located in the center of the internode and is su-

Robert M. Gould, Dan Matsumoto: and Gary Mattingly • Laboratory of Membrane Biology, Institute for Basic Research in Mental Retardation, Staten Island, New York 10314.

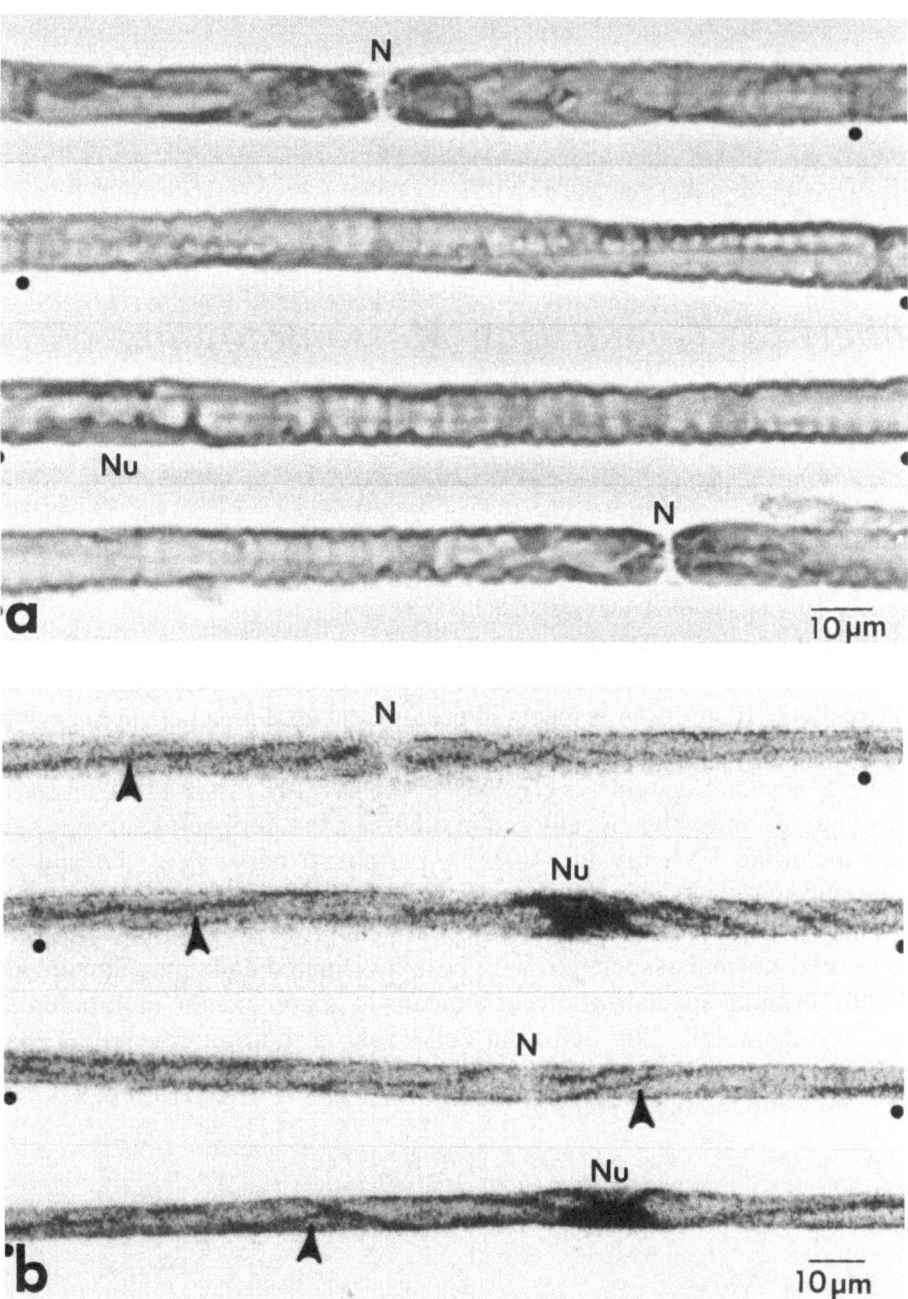

Fig. 1. Micrographs of consecutive segments of individual teased nerve fibers of mouse sciatic nerve (dots represent areas of overlap) showing nodes of Ranvier (N) and Schwann cell perinuclear regions (Nu). (a) Normal fiber, taken with phase-contrast optics. (b) Autoradiograph of a fiber labeled with [^{3}H]choline, demonstrating that synthesis of lecithin occurs in Schwann cell perinuclear regions (Nu) and surface cytoplasmic channels (arrows). Calibration bar indicates 10 μm.

perficial to the myelin sheath. Schwann cell cytoplasm is most abundant in the perinuclear region and in the two paranodal regions at opposing ends of the myelin sheath. These areas are interconnected by cytoplasmic channels (Fig. 1B). Other channels, which penetrate the myelin sheath, are called the Schmidt–Lanterman incisures.

The complex relationship between the Schwann cell cytoplasm and its myelin sheath is illustrated in an "unrolled" schematic (Fig. 2). Narrow rims of Schwann cell cytoplasm not only connect the perinuclear and paranodal regions but also completely surround (in the superficial, paranodal, and adaxonal regions) and invade (in Schmidt–Lanterman clefts) the compact myelin. As shown in this diagram, the compact myelin is the dominant structure, having cellular dimensions on the order of 1 mm (internodal length) by 5 mm (unwrapped sheath).[7,8]

Perinuclear regions, paranodal regions, and Schmidt–Lanterman incisures are readily visualized in longitudinal sections (Fig. 3A). In transverse sections (Fig. 3B), profiles of Schwann cell cytoplasm are crescent shaped in the perinuclear region and, in paranodal regions, make multiple cup-shaped indentations in the myelin.

Also visualized in peripheral nerves are unmyelinated axons (Fig. 3) that

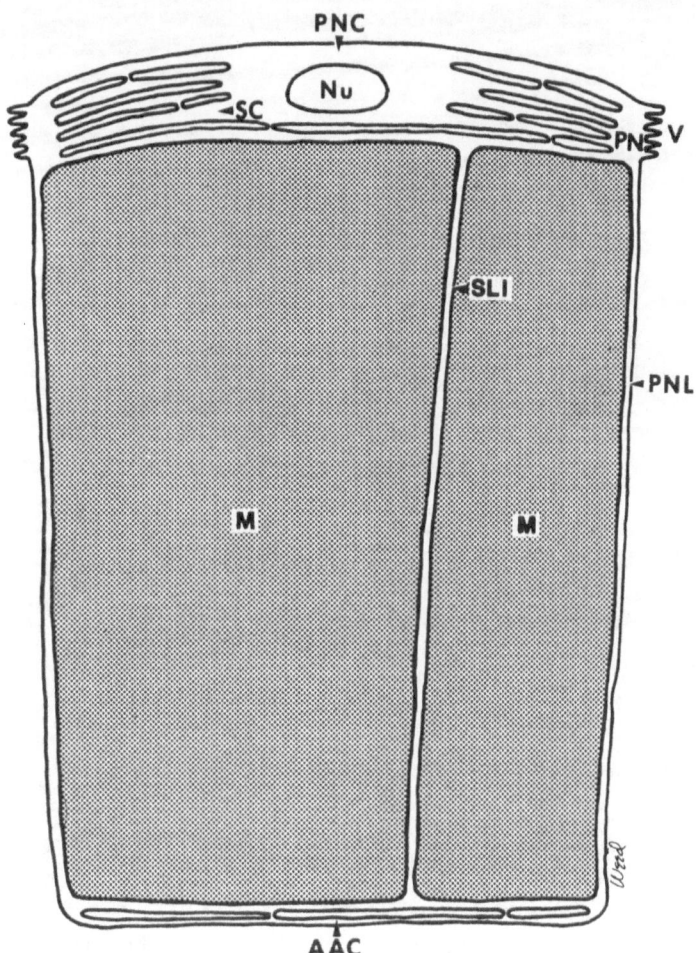

Fig. 2. Schematic of "unrolled" Schwann cell illustrating nucleus (Nu), perinuclear cytoplasm (PNC), cytoplasmic channels (SC) situated between areas of contact with the outermost myelin (represented by ovals), paranodal cytoplasm (PN), paranodal loops (PNL), paranodal villi (V), adaxonal cytoplasm (AAC), Schmidt–Lanterman incisure (SLI), and myelin (M). Modified from Fig. 22 of Mugnaini *et al.*[6]

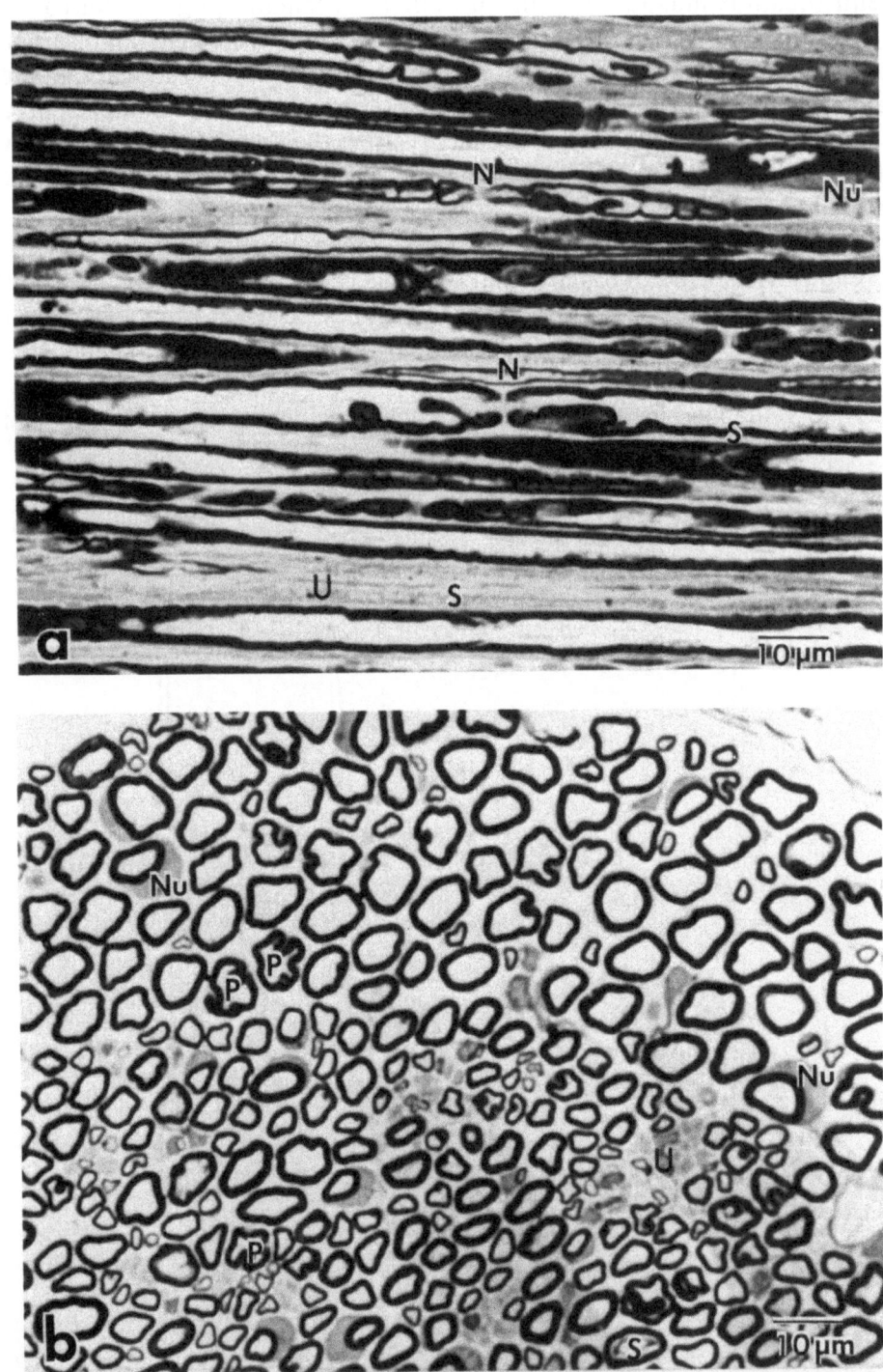

Fig. 3. Sections of mouse sciatic nerve stained with toluidine blue showing areas of Schwann cell nuclei (Nu), nodes of Ranvier (N), Schmidt–Lanterman incisures (S), unmyelinated fiber bundles (U), and paranodal regions (P). (a) Longitudinal section. (b) Transverse section. Calibration bar indicates 10 μm.

are enclosed by Schwann cell processes. Aguayo and his colleagues[9] used serial reconstruction techniques to demonstrate that longitudinal chains of Schwann cells retain an association with multiple unmyelinated axons, although the number of axons enclosed within the Schwann cell processes varies along the length of the nerve. Schwann cells surrounding both myelinated and unmyelinated fibers are enclosed in a basal lamina. Both Schwann cells and fibroblasts secrete collagen fibrils into the endoneurium, the region in which the individual fibers are contained.

Schwann cells are also present in a variety of peripheral terminal regions,[10,11] the best characterized being the neuromuscular junction.[12]

2.2. Development of Peripheral Nerve

The morphological features of peripheral nerve development have recently been reviewed.[13,14] Schwann cells originate in the neural crest and proliferate while migrating along developing neurites, separating axons from each other. The Schwann cells retain a close association with the developing axons. During nerve development, the bipotential nature of the Schwann cell is expressed by its establishing either an unmyelinated or a myelinated condition. In forming myelin sheaths, Schwann cells first limit their contact to a single larger-caliber axon and then enclose it and begin to elaborate extensions of their own plasma membrane which become the multilayered myelin. The process of myelination is continuously repeated in a spatial and temporal framework; proximal sections of an axon are myelinated before distal sections, and the largest-caliber axons are separated from the axon bundles and myelinated before smaller ones. Normally, Schwann cells associated with both myelinated and unmyelinated axons will retain a continued relationship with segments of their axon(s) throughout the lifetime of the animal. The most distally displaced Schwann cells likewise develop and form lasting associations with specialized nerve terminals.

2.3. Response of the Schwann Cell to Injury

Neuropathies are generally subdivided into two types,[15,16] demyelinating and neuronal. In demyelinating neuropathies (e.g., experimental allergic neuritis and diphtheritic neuritis), Schwann cells normally participate in the phagocytosis of their own myelin sheaths, producing denuded axons. The lesions are usually focal, with sections of nerve both proximal and distal retaining normal morphology and physiological function. Schwann cells usually proliferate in response to focal demyelination and remyelinate the affected axon if the injury is not persistent or too extensive. In some cases, axons will degenerate concomitantly with the demise of their myelin sheaths.

In neuronal neuropathies, Schwann cells respond to the injury with autophagic and phagocytic activities directed towards their myelin sheaths and axons, respectively. Early morphological responses, seen also in demyelinating neuropathies, include swelling of Schmidt–Lanterman incisures and widening of nodal gaps. Presumably, these alterations reflect a redistribution of Schwann cell cytoplasm which may aid in its autophagic and phagocytic roles.[17] This response also includes proliferation of the Schwann cells.[18] The regenerative response of the axons depends on the extent of the damage. The axonal sprouts regenerating from sites proximal to the lesion grow within the basal lamina formed by the original Schwann cells. Those axons that reach adequate caliber are myelinated by Schwann cells in a manner similar to the myelination of axons during development. This repair phenomenon, however, results in thinner and shorter internodes than those originally present.[19]

3. BIOCHEMISTRY OF THE SCHWANN CELL

This section will include some of the known biochemical properties of the Schwann cell that are expressed in mature, developing, and pathological peripheral nerves. Emphasis will be placed on the dynamic character of the Schwann cell revealed during its specific interactions with myelinated and unmyelinated axons and with terminal specializations.

3.1. Mature Nerve

3.1.1. Involvement with Myelin

The traditional view that myelin is a metabolically stable structure is no longer tenable. Hendelman and Bunge[20] provided strong evidence that phospholipids are continuously deposited into PNS myelin sheaths. They applied autoradiographic methods (ARG) to nerve fibers in tissue culture and showed that choline-labeled lipids were synthesized and subsequently incorporated into mature myelin. Gould and Dawson[21] performed *in vivo* studies using adult frog and mouse sciatic nerves that demonstrated that choline-labeled lecithin synthesized in Schwann cell cytoplasm was transferred to the myelin sheaths. These results indicated that Schwann cells are continually involved in myelin maintenance.

In the Gould and Dawson study,[21] the synthetic activity was predominantly localized in perinuclear regions. Recently, Gould and Sinatra[22] used teased fiber preparations in conjunction with ARG (Fig. 1B) to demonstrate that phosphatidylcholine synthesis occurs not only in the perinuclear regions but also in superficial channels extending along the entire length of the internode. From these synthetic sites along the internode, lecithin was transferred to the myelin sheath; the outer layers lying closer to these sites were labeled before those further in. Quantitative measurements established that phosphatidylcholine initially deposited in the outermost layers is subsequently redistributed evenly throughout the entire myelin sheath.[21] The time required for this redistribution was dependent on the thickness of the myelin; it took 4 days for lecithin to equilibrate across all layers of the thickest myelin sheaths of mouse sciatic nerve.

One explanation for the movement of the amphipathic phospholipid from outer layers inward would be diffusion within the bilayer along its spiraling course. Another possible route is through the Schmidt–Lanterman incisures and paranodal loops (Fig. 2) which could provide a convenient cytoplasmic pathway for phospholipids to reach inner myelin layers. However, the lack of labeling of these structures is evidence against this latter possibility[21] (R. M. Gould, unpublished observations). Furthermore, the close correlation between diffusion rates of phospholipid in this and other membranes[21] adds evidence to the idea that phosphatidylcholine utilizes a membranous rather than a cytoplasmic pathway. Other phospholipids labeled with tritiated inositol, glycerol, or ethanolamine display similar synthesis and migration patterns to that of phosphatidylcholine.

Besides phospholipid, the only other myelin constituent that has been extensively studied by these techniques is the major peripheral nerve myelin protein, a low-molecular-weight glycoprotein called P_0.[23] Fucose, a highly specific precursor of this protein, has been used in autoradiographic studies that localized the site of synthesis of this glycoprotein in the perinuclear Golgi apparatus.[24] P_0 has also been localized in Golgi regions of the Schwann cell cytoplasm by immunocytochemical methods.[25] In addition to its fucosylation, P_0 is a sulfated protein,[26] and its sulfation may occur in the perinuclear Golgi apparatus.[27]

Following its synthesis, P_0 protein is transported to the superficial layers of the myelin sheath[24]; unlike other myelin constituents, there is a time delay between the processing of P_0 in the Golgi apparatus and its incorporation into myelin.[28] Subsequent movement inward is much slower than that of phospholipid.[24] Even 10 weeks after its formation, this labeled protein is still largely confined to outer layers of the sheath. The much slower diffusion of this protein relative to phospholipids would suggest that these two constituents are not strongly associated in the myelin bilayer.

The continuous deposition of new myelin lipids and proteins requires a compensatory turnover mechanism to remove and degrade existing myelin constituents. The catabolic events could occur either within the sheath, since some hydrolytic enzymes have been shown to be associated with myelin,[29] or in the Schwann cell cytoplasm. Some biochemical evidence suggesting a structural closeness of anabolic and catabolic sites has been obtained by Patsalos *et al.*[30] using a Wallerian degeneration paradigm (see Section 3.3.1).

If normal catabolism occurs largely in perinuclear cytoplasm, then the turnover of protein and lipid would depend on the diffusion of these constituents within the myelin. Those components in the outermost layers of myelin, having a shorter distance to travel, would be more readily catabolized than those present in layers further inward. Thicker myelin sheaths with longer stretches of compact myelin would, therefore, retain their constituents much longer than thinner sheaths, as has been observed for phosphatidylcholine.[21]

These studies demonstrate that phospholipids and the major protein constituent of mature myelin are largely if not exclusively formed (and degraded) in the Schwann cell cytoplasm. It would be of interest to know more about the metabolism of other myelin constituents, e.g., glycolipids, cholesterol, and other proteins, and of those constituents comprising the specialized paranodes and incisures, such as the myelin-associated glycoprotein, MAG (B. D. Trapp, N. Sternberger, R. H. Quarles, and H. deF. Webster, personal communication).

3.1.2. Involvement with Axons

Schwann cells of both myelinated and unmyelinated fibers have membrane surface and regions of their cytoplasm in close apposition to the axon. This intimate relationship between the axon and its entourage of Schwann cells (roughly ten per centimeter of myelinated axon) creates the possibility for active biochemical interactions.

These interactions can proceed in two directions: from Schwann cells to axons and from axons to Schwann cells. Since the metabolic machinery of the myelin-forming Schwann cells lies superficially to its myelin sheath, molecules to be transferred to the axon would have to penetrate the myelin sheath[31] by moving across or along its bilayers or through the cytoplasmic channels provided by the Schmidt–Lanterman incisures and the paranodal loops (see Fig. 2). For the thickest sheaths, the distance from external cytoplasm to the axon would be at least 5 mm.[8] Even though the situation is far simpler for unmyelinated fibers, the molecules would still have to cross two plasma membranes and the intervening gap to reach the axoplasm.

Aguayo and his associates[32,33] provided convincing morphological evidence that Schwann cells influence their axons. In one experiment they interposed a graft from Trembler mutant mouse sciatic nerves into sciatic nerves of normal mice. The host animal's axons regenerated through the graft and were subsequently myelinated by the mutant's Schwann cells in the graft and normal Schwann cells further distally. The calibers of myelinated axons in the graft were smaller than those myelinated by normal Schwann cells in the distal portions. Conversely, when a segment from a normal mouse was grafted into a mutant's nerve, those portions of the axons myelinated by normal Schwann cells in the graft were of larger caliber than those in proximal and distal regions which were myelinated by the Trembler's Schwann cells. These results demonstrated that local Schwann cells influence the size of the axons that they myelinate.

There is biochemical evidence that invertebrate Schwann cells contribute macromolecules to the axon. Lasek, Gainer, and their colleagues[34,35] used the squid giant axon to demonstrate that proteins synthesized in the supporting sheath cells are transported to the axoplasm. These investigators eliminated any potential incorporation of radioactive precursors into protein within the axoplasm. Therefore, they concluded that labeled proteins, localized inside the axon by autoradiographic and biochemical analysis, originated in the glia. Recently, Tytell and Lasek[36] used two-dimensional gels to show that the radioactive proteins synthesized in the Schwann cells and transported into axoplasm form a unique subpopulation of both the total proteins synthesized by the Schwann cells and the intrinsic proteins of the axoplasm.

In mammalian nerves, the potential transfer of macromolecules from glial cells to axons has been studied with autoradiographic methods. Singer and

Salpeter[37] and Singer and Green[38] have presented quantitative evidence that labeled macromolecules, presumably proteins and RNA, are transferred from Schwann cells through the myelin sheath to axons. Because of the difficulty in controlling local incorporation within the axon and the possibility that the grains might represent material that was rapidly transported following synthesis in the perikarya, these results have been criticized.[39] Related studies[21,40] performed with phospholipid precursors have also been inconclusive in demonstrating a transfer of lipid from Schwann cells to their axons. A major problem is the difficulty of distinguishing the lipid synthesized within the axon from that transferred from Schwann cells.

There is autoradiographic evidence from studies of both CNS and PNS that molecules are transferred from the axon to the surrounding glial cells. The appearance of grains over glial cells can result either from a direct transfer of labeled macromolecules or from a transfer of radioactive precursors which are subsequently utilized by the glia. There is strong evidence for direct transfer of phospholipids from axons to the myelin sheaths of supportive Schwann cells[41,42]; however, this pathway is not the major source of myelin phospholipid.[21] There is recent evidence that nucleic acids,[43] phospholipids,[41,42] and proteins[44] are synthesized in glia following the uptake of labeled precursors from the axons. In these studies, a precursor was supplied to the neurons, taken up by the neuronal soma, and transported down the axons either in its original or slightly modified form or as a constitutent of a macromolecule. Since there is no evidence for direct transfer of proteins or nucleic acids to glia, the label incorporated into macromolecules must be liberated by axonal catabolism prior to uptake by glial cells. Taken in concert, the above evidence demonstrates that glial cells receive a continuous supply of material from the axon. Whether this axon–glial transfer plays a nutritional role[45] or provides a means of communication (e.g., informing the Schwann cells of the metabolic state of the axon) is at present unknown.

Another property of axon-associated Schwann cells has been studied by Villegas.[46,47] He has demonstrated that Schwann cells of the squid giant axon have pharmacologically sensitive acetylcholine (ACh) receptors and are hyperpolarized by ACh. He has hypothesized that hyperpolarization of cells is related to conduction of action potentials by the axon. In other studies, Ellisman and his collaborators[48] have proposed that paranodal specializations of mammalian Schwann cells function in regulating the ionic environment during nerve impulse conduction.

3.1.3. Involvement with Terminal Regions

Schwann cells maintain specialized associations with nerve endings by surrounding the terminal portions of the axons but not forming myelin sheaths. One explanation for this absence of myelin at terminals is the lack in these regions of a collagen substrate for the Schwann cells[49] (see also Section 3.2.3). At the endings, the Schwann cells display a variety of forms and functions, depending on the type of terminal, i.e., sensory and motor. Metabolic roles of Schwann cells at sensory endings are not known except for the phagocytic activity accompanying nerve degeneration. At the neuromuscular junction, there is evidence that Schwann cells can take part in the uptake of (putative) transmitters,[50] an analogous function to one expressed by astrocytes.[51] Following nerve degeneration, the Schwann cells are modified to release neurotransmitter.[52] The signal for the Schwann cells to form ACh has been suggested to be the loss of contact with the terminal.[53]

3.2. Development of Peripheral Nerve

3.2.1. Cell Proliferation

A major role of the immature Schwann cells in developing peripheral nerve is to separate the growing axons from each other and ensheath them, thus sealing each axon along its entire length from the extracellular milieu. To accomplish this task, the Schwann cells must greatly increase their numbers to enclose all axons. The axons, which are actively growing to their targets and developing the capacity to propagate electrical impulses, also play a role in controlling Schwann cell numbers. Aguayo and his associates[54] used nerve growth factor antiserum to deplete the number of neurons in the sympathetic ganglia and consequently the number of axons in the sympathetic nerve trunk. Concomitant with the loss of axons, they observed a proportional decrease in the number of Schwann cells in the nerve trunk. In other experiments, McCarthy and Partlow[55] used recombined tissue cultures of neurons and non-neuronal cells (presumably Schwann cells) to show that the proliferation measured by tritiated thymidine incorporation into nonneuronal cells was proportional to the numbers of neurons present in the culture.

The mitogenic properties that neurites exert on Schwann cells has been the focus of many recent investigations.[49,56] Purified Schwann cells, which are normally quiescent in culture,[57,58] can be induced to undergo cell division with the introduction of neurons. It has been shown that proliferation is not stimulated by diffusible mitogens from the neurite but occurs only when Schwann cells come in direct contact with axons,[55,59] a situation mimicking the *in vivo* condition.

Proliferation of cultured Schwann cells is induced by sonicated neurons,[60] neurite membranes,[59,61] and a CNS axolemmal fraction.[62] Homogenates or fractions from other cell types are in general not mitogenic,[63,64] isolated myelin being an exception.[62] Salzer and his associates[59,61] have shown by proteolytic digestion, heating, and aldehyde fixation that surface proteins are an intrinsic part of the mitogenic properties of the neurite membranes. Furthermore, related studies[56] have indicated that carbohydrate moieties on glycoproteins and/or glycolipids that are sensitive to periodate and lectins are required for mitogenic expression. It also appears that neuron-induced mitogens act through a mechanism in which 3',5'-cyclic adenosine monophosphate (cyclic AMP) levels are not increased.[60,64]

In addition to neuron-derived mitogens, a limited number of other factors, most notably cholera toxin[63] and a soluble glial growth factor[65] present in pituitary and brain extracts, have been shown to be mitogenic. The former stimulates adenylate cyclase and causes increased cyclic AMP, whereas the latter factor works independently of cyclic AMP. A large number of compounds have been shown to have no mitogenic activity when tested on Schwann cells in tissue culture.[63,64] A possible role of nonneuritic mitogens would be to stimulate the proliferation of Schwann cells that occurs in response to nerve insult (e.g., crush injury,[18] lysolecithin-induced demyelination[66]; see Section 3.3.2).

3.2.2. Ensheathment

The mitogenic properties of axons must be regulated to establish the proper number of Schwann cells for normal ensheathment and myelination of axons. During development, while some Schwann cells are actively proliferating and separating the growing axons distally, other Schwann cells located more proximally will have stopped proliferating and completed the ensheathment of the axons and started producing myelin. A characteristic feature of the nonproliferating Schwann cells, be they associated with a single (myelinating) or multiple (nonmyelinating) axon(s), is the basal lamina. This Schwann cell covering is not continuous over the entire plasma membrane; those portions of the Schwann cell opposed to the axon (adaxonal) and those forming the mesaxon remain free of basal lamina.

Schwann cells in tissue culture will only form a basal lamina and associated collagen fibrils when appropriate contact with axons is made.[49] If basal lamina is subsequently removed with trypsin treatment, a new basal lamina will form only if contact with the axon is retained.[49] These results suggest that a continuous signal from the axon is needed to stimulate formation and maintenance of the basal lamina. The fibroblast is not required for basal lamina formation; however, when it is present in the culture, basal lamina becomes thicker, suggesting that the fibroblasts influence the development of this structure.[49,67]

When dorsal root ganglion explants are grown in a chemically defined medium, Schwann cell development is arrested prior to ensheathment and myelination[49,68] of the axon. In addition, no basal lamina is formed, and the period of proliferation is protracted. When embryo extract or serum is added, proliferation ceases, and normal development ensues. These results demonstrate that factors present in the extract or serum are needed for continued Schwann cell differentiation. Such factors could act either on the neurites, allowing them to signal the Schwann cells, or directly on the Schwann cells, enabling them to respond to the neurites' signal(s). These studies also demonstrated a relationship between the cessation of the Schwann cells' proliferative activity and the onset of secretion of the basal lamina and collagen.[49] In addition to the elaboration of basal lamina and collagen, it has been suggested that the Schwann cells release substance(s) that mask the neuritic signal, causing Schwann cell proliferation.

It has also been demonstrated that Schwann cells in tissue culture require contact with a collagen substratum secreted by fibroblasts for normal ensheathment and myelination.[69] When Schwann cells are suspended above this substratum, they form an incomplete basal lamina but do not ensheath axons and form myelin. When strips of collagen are layered over the suspended Schwann cells, ensheathment and myelination subsequently proceed. In line with these results it has been proposed that the lack of collagen in nerve rootlets of dystrophic mice may cause the failure of the Schwann cells in the roots to myelinate the axons.[49]

3.2.3. Myelin Formation

The bipotentiality of Schwann cells is expressed at the stage of development at which myelin sheaths are formed. Schwann cells associated with small-caliber axons retain the relationship that they have established. On the other hand, the Schwann cells associated with individual larger caliber fibers begin synthesizing characteristic myelin lipids and proteins and assembling them into the myelin sheath. Myelin-destined phosphatidylcholine and P_0 glycoprotein are formed in Schwann cell cytoplasm surrounding the myelin and are added to the outermost layers of the sheath.[24,25,70] Phosphatidylcholine formed in the Schwann cell cytoplasm is probably redistributed throughout the myelin sheath by lateral diffusion, as is the case in mature nerve. Fucose injected into developing rat sciatic nerve is first incorporated into P_0 protein in Schwann cell cytoplasm and is then transferred to the growing sheath.[24] After a 7-week survival, the labeled protein is localized to the inner layers of myelin, a result that suggests that its diffusion within the myelin is much slower than the rate of myelin deposition.[24] In *in vitro* studies, M. E. Smith (personal communication) showed that tunicamycin blocked the glycosylation of P_0 but not its short-term incorporation into myelin.

Tritiated cholesterol injected into developing mice was shown by ARG to be concentrated in sciatic nerve myelin, and the label was distributed evenly throughout the growing myelin sheaths within 3 h after injection.[71] In a subsequent study, Rawlins demonstrated that cholesterol enters the myelin from both external and adaxonal surfaces.[72]

The myelin sheaths of developing as well as mature nerves cover a uniform length of axon. During development, the size of the axons increases. The fluid character of the myelin membrane, as demonstrated by the mobility of the constituent lipids, would allow it to expand to accommodate the increasing axon caliber. Internodal lengths also increase as the animal gets larger and its nerves correspondingly lengthen. The uniformity of internodal dimensions along each axon might suggest an axonal influence on the growth and expansion of the myelin sheaths.

It is also known that the axons stimulate the Schwann cells to make myelin. Cross-anastomosis[32,73] and nerve-grafting[32] techniques have been used to cause large-caliber myelinated axons to grow into nerves that originally contained mainly unmyelinated axons. The Schwann cells originally associated with the unmyelinated axons proliferated in response to the entry of regenerating fibers and subsequently myelinated these foreign (large-caliber) axons.

The influence exerted on the Schwann cells by axons seems to be continuously required, for when Schwann cells involved in forming myelin are separated from the axons, either *in vivo*[74] or in tissue culture,[75,76] they no longer express characteristic myelin proteins and lipids as detected by biochemical and immunochemical methods. This behavior is different from that of oligodendrocytes which continue to express myelin proteins and lipids in culture.[75] The requirement for the continued axonal influence for the Schwann cells to maintain their myelin-associated metabolism is similar to that noted for basal lamina formation (Section 3.2.2).

3.3. Response to Injury

Another behavior is observed for Schwann cells in damaged peripheral nerve. This behavior is a complex one, involving different phases including phagocytic and autophagic activity, proliferation, the formation of bands of Bungner, and, finally, the formation of new myelin sheaths.[77] Although the Schwann cell responses vary depending on the causative agent, its target, the stage of maturation of the nerve, and other factors,[78,79] emphasis will be the generalized responsiveness of Schwann cells as illustrated in two well-studied models: classical Wallerian degeneration and lysolecithin-induced demyelination.

3.3.1. Autophagic and Phagocytic Behavior

In Wallerian degeneration, the myelin-forming Schwann cells respond rapidly to axonal damage. Within 2 h of nerve transection or crush, hydrolytic enzymes are increased in Schwann cells distal to the site of injury.[29,79] This activity marks the beginning of the destruction of the myelin sheath and axon by the Schwann cells. Retrogradely transported lysosomes[80] may also function in axon degeneration. During degeneration, the clearing of affected axons and myelin sheaths is probably a mandatory step which allows regeneration to occur. In demyelinating diseases in which the axons are spared, destruction of the sheaths may be the only effective means to remove defective components from this slowly metabolizing structure. Myelin can be reformed on the denuded axons and will remain viable if the original injury is acute. On the other hand, a chronic condition can cause repeated demyelination and remyelination, resulting in the classic onion-bulb formation.[15,16]

Early events in degeneration, which are probably related to the changing role of the Schwann cell, include the increase in levels of RNA and DNA,[81] cyclic AMP,[82] and cholesterol ester formation.[83] The retardation of paranodes and swelling of the Schmidt–Lanterman incisures associated with degeneration would redistribute more of the Schwann cells' cytoplasm to the perimeters of compact myelin, giving hydrolytic enzymes in the cytoplasm better access to the degenerating myelin and axons.[17] The findings of Patsalos and his colleagues[30] that those proteins added most recently to the myelin sheaths were the first to be broken down would indicate that catabolic activities occur close to the sites where proteins are originally incorporated into myelin sheath.

3.3.2. Proliferation

Several days after injury (Wallerian degeneration[18,81] or lysolecithin-induced demyelination[66]), the Schwann cells proliferate. This proliferation does not require axonal stimulation but may result from mitogenic effect of myelin debris, because the Schwann cells surrounding unmyelinated fibers do not proliferate following nerve injury.[64,84,85] This injury-related proliferation may be cyclic AMP mediated since Schwann cells isolated from myelinated nerve proliferate when their cyclic AMP levels are increased,[63] whereas Schwann cells isolated prior to myelination do not proliferate in response to increased cyclic AMP.[64]

The proliferative response of the Schwann cells may serve a number of functions. First, in both demyelination and degeneration, the Schwann cells endocytose large amounts of myelin debris, and the increased numbers of Schwann cells would reduce the amount of debris accumulated within each cell.[86] It is known that the addition of isolated myelin inhibits phospholipid exchange activity in an *in vitro* system.[87] One could argue that an excess of myelin debris within the cells might also impair this activity as well as a number of other metabolic processes in much the same way that accumulation of lipids in lipid-storage diseases are thought to affect the cells' metabolic machinery. Second, the myelin-forming Schwann cells are in a highly differentiated state. Proliferation may allow them to return to an undifferentiated state, allowing them to ensheath the axons and form myelin.[66,86] In line with these roles, in both lysolecithin-induced demyelination[66] and Wallerian degeneration,[88] inhibition of local Schwann cell proliferation with mitomycin C was found to stop the Schwann cells that accumulated myelin debris from continuing with events leading to ensheathment and (re)myelination.

3.3.3. Axonal Guidance

During normal nerve development, because the Schwann cells follow the lead axons, they play no role in guiding the axons to their targets. Following axon injury, regenerating axons follow pathways (bands of Bungner)[89] to reach their target tissue. Recent studies, reviewed by Varon and Manthrope,[56] have used tissue cultures to characterize factors that stimulate neurite outgrowth. They hypothesized that the Schwann cells distal to the regenerating nerves can stimulate their outgrowth both by restricting the paths the neurites can take and by secreting neurotrophic factors.

3.3.4. Remyelination

In a manner presumably analogous to myelination in development, daughter Schwann cells, following proliferation, (re)establish contact with denuded axons (remyelination) or regenerating axons, ensheath them, and begin the production of myelin. Those Schwann cells in the injured area involve themselves with the repair activity. Schwann cells in adjacent areas tend not to migrate to aid in the repair, probably because they are inhibited from doing so by Schwann cells intrinsic to the injured site.[86]

Spencer and his colleagues[90] have recently begun to study the formation of myelin proteins during regeneration. They use a new model in which distal degeneration is temporally separated from subsequent regeneration. The cat tibial nerve is transected, and the distal stump is allowed to degenerate. After several weeks, this stump is composed of postmitotic Schwann cells and their progeny, the myelin debris having been digested. The peroneal nerve is then cut, and the proximal end is anastomosed to the degenerated tibial nerve. The regenerating axons from the peroneal nerve grow down bands of Bungner in the tibial nerve, and, after 3 weeks, three distinct zones exist: a distal zone containing no axons, an intermediate zone where contact has been made but no myelin has formed, and a proximal zone in which myelination has begun. At this time, the nerve was cut into segments which were incubated with tritiated amino acids. The radioactive proteins synthesized by each segment were separated on polyacrylamide gels. Analysis of segments in the distal zone showed that the areas to which characteristic myelin proteins would migrate, were unlabeled. In the intermediate zone there was some synthesis of myelin proteins, the basic protein (P_1 and P_2) being formed in greater amounts than P_0. In the proximal zone, all three myelin proteins were formed; however, P_0 was more strongly labeled than the basic proteins. These results demonstrate that Schwann cells in distal stumps of regenerating nerve are induced to form myelin proteins following contact with the axons. It is apparent that the capacity to form the basic proteins and the P_0 protein are not induced simultaneously. It would follow that the synthesis of basic proteins would precede that of P_0 protein in each myelin-producing Schwann cell. Such a pattern, presumably also occurring during development, might suggest different roles of these proteins in the formation of the compact myelin.

4. CONCLUSION

It was the first aim of this review to acquaint the reader with morphological studies demonstrating the various behaviors of Schwann cells during peripheral nerve development, at maturity, and in acutely injured tissue. This framework was constructed so that it would be possible to consider what was known about the biochemistry displayed by Schwann cells during these various conditions. Unfortunately, present knowledge of the biochemical characteristics of Schwann cells is limited and lags far behind knowledge of their dynamic morphological characteristics.

Technology has only recently become available to better study the Schwann cell. Recent advances in tissue culture, immunochemistry, histochemistry, autoradiography, and sensitive biochemical methods should encourage wider-ranging and more directed studies of this fascinating cell.

ACKNOWLEDGMENTS. We wish to thank Drs. Eric Holtzman and Sasha Koulish for their criticism of the manuscript. We are indebted to Drs. Albert Aguayo,

Joyce Benjamins, Jerome Brockes, Richard Bunge, Kay Fields, Richard Peterson, Peter Spencer, Bruce Trapp, Silvio Varon, and Richard Wiggins for sending preprints of their work, to Richard Weed for preparing the figures, and to Ann Cosgove for typing the manuscript. This work was made possible by support from the National Institute of Health (NS13980 and NS16305).

REFERENCES

1. Porcellati, G., 1969, *Handbook of Neurochemistry*, Volume 2 (A. Lajtha, ed.), Plenum Press, New York, pp. 393–422.
2. Porcellati, G., 1972, *Handbook of Neurochemistry*, Volume 7 (A. Lajtha, ed.), Plenum Press, New York, pp. 191–219.
3. Domonkos, J., 1972, *Handbook of Neurochemistry*, Volume 7 (A. Lajtha, ed.), Plenum Press, New York, pp. 93–106.
4. Landon, D. N., and Hall, S., 1976, *The Peripheral Nerve* (D. N. Landon, ed.), Chapman and Hall, London, pp. 1–105.
5. Berthold, C.-H., 1978, *Physiology and Pathobiology of Axons* (S. G. Waxman, ed.), Raven Press, New York, pp. 3–63.
6. Mugnaini, E., Osen, K. K., Schnapp, B., and Friedrich, V. L., Jr., 1977, *J. Neurocytol.* **6**:647–668.
7. Webster, H. deF., 1971, *J. Cell Biol.* **48**:348–367.
8. Friede, R. L., and Bischhausen, R., 1980, *J. Neurol. Sci.* **48**:367–381.
9. Aguayo, A. J., Bray, G. M., Terry, L. C., and Sweezey, E., 1976, *J. Neuropathol. Exp. Neurol.* **35**:136–151.
10. Hubbard, J. I., 1974, *The Peripheral Nervous System*, Plenum Press, New York.
11. Landon, D. N. (ed.), 1976, *The Peripheral Nerve*, Chapman and Hall, London.
12. Heuser, J. E., and Reese, T. S., 1977, *Handbook of Physiology*, Section 1, *The Nervous System*, Volume 1 (J. M. Brookhart and V. B. Mountcastle, eds.), American Physiological Society, Bethesda, pp. 261–294.
13. Webster, H. deF., 1975, *Peripheral Neuropathy*, Volume 1 (P. J. Dyck, P. K. Thomas, and E. H. Lambert, eds.), W. B. Saunders, Philadelphia, pp. 37–61.
14. Spencer, P. S., and Weinberg, H. J., 1978, *The Physiology and Pathobiology of Axons* (S. Waxman, ed.), Raven Press, New York, pp. 389–405.
15. Schroder, J. M., 1975, *Peripheral Neuropathy*, Volume 1 (P. J. Dyck, P. K. Thomas, and E. H. Lambert, eds.), W. B. Saunders, Philadelphia, pp. 337–362.
16. Allt, G., 1976, *The Peripheral Nerve* (D. N. Landon, ed.), Chapman and Hall, London, pp. 666–739.
17. Singer, M., and Steinberg, M. C., 1972, *Am. J. Anat.* **133**:51–84.
18. Asbury, A. K., 1975, *Peripheral Neuropathy*, Volume 1 (P. J. Dyck, P. K. Thomas, and E. H. Lambert, eds.), W. B. Saunders, Philadelphia, pp. 201–212.
19. Bonnaud-Toulze, E. N., and Raine, C. S., 1980, *Neuropathol. Appl. Neurobiol.* **6**:279–290.
20. Hendelman, W. J., and Bunge, R. P., 1969, *J. Cell Biol.* **40**:190–208.
21. Gould, R. M., and Dawson, R. M. C., 1976, *J. Cell Biol.* **68**:480–496.
22. Gould, R. M., and Sinatra, R. S., 1981, *J. Neurocytol.* **10**:161–167.
23. Greenfield, S., Brostoff, S., Eylar, E. H., and Morell, P., 1973, *J. Neurochem.* **20**:1207–1216.
24. Gould, R. M., 1977, *J. Cell Biol.* **75**:326–339.
25. Trapp, B. D., Itoyama, Y., Sternberger, N. H., Quarles, R. H., and Webster, H. deF., 1981, *J. Cell Biol.* **90**:1–6.
26. Matthieu, J. M., Everly, J. L., Brady, R. O., and Quarles, R. H., 1975, *Biochim. Biophys. Acta* **392**:167–174.
27. Young, R. W., 1973, *J. Cell Biol.* **57**:175–189.
28. Rapaport, R. N., and Benjamins, J. A., 1981, *J. Neurochem.* **37**:164–171.
29. Hallpike, J. F., 1976, *The Peripheral Nerve* (D. N. Landon, ed.), Chapman and Hall, London, pp. 605–665.

30. Patsalos, P. N., Bell, M. E., and Wiggins, R. C., 1980, *J. Cell Biol.* **87**:1–5.
31. Singer, M., 1968, *Ciba Foundation Symposium on Growth of the Nervous System* (G. E. W. Wolstensholme and M. O'Connor, eds.), J. and A. Churchill, London, pp. 200–215.
32. Aguayo, A. J., Bray, G. M., Perkins, C. S., and Duncan, I. D., 1979, *Soc. Neurosci. Symp.* **4**:361–383.
33. Aguayo, A. J., Bray, G. M., and Perkins, C. S., 1979, *Ann. N.Y. Acad. Sci.* **317**:512–533.
34. Lasek, R. J., Gainer, H., and Barker, J. L., 1977, *J. Cell Biol.* **74**:501–523.
35. Gainer, H., Tasaki, I., and Lasek, R. J., 1977, *J. Cell Biol.* **74**:524–530.
36. Tytell, M., and Lasek, R. J., 1979, *Soc. Neurosci. Abstr.* **5**:63.
37. Singer, M., and Salpeter, M., 1966, *J. Morphol.* **120**:281–315.
38. Singer, M., and Green, M. R., 1968, *J. Morphol.* **124**:321–343.
39. Gambetti, P., Autilio-Gambelli, L., and Peck, K., 1980, *Brain Res.* **200**:59–68.
40. Gould, R. M., 1976, *Brain Res.* **117**:169–174.
41. Droz, B., DiGiamberardino, L., Koenig, H. L., Boyenval, J., and Hassig, R., 1978, *Brain Res.* **155**:347–353.
42. Droz, B., Brunetti, M., DiGiamberardino, L., Koenig, H. L., and Porcellati, G., 1979, *Soc. Neurosci. Symp.* **4**:344–360.
43. Politis, M. J., and Ingoglia, N. A., 1979, *Brain Res.* **169**:343–356.
44. Matthieu, J. M., Webster, H. deF., DeVries, G. H., Corthay, S., and Koellreutter, B., 1978, *J. Neurochem.* **31**:93–102.
45. Wu, P.-S., and Ledeen, R. W., 1980, *J. Neurochem.* **35**:659–666.
46. Villegas, J., 1975, *J. Physiol. (Lond.)* **249**:679–689.
47. Villegas, J., 1978, *Dynamic Properties of Glia Cells* (E. Schoffeniels, ed.), Pergamon Press, Oxford, pp. 207–216.
48. Ellisman, M. H., Friedman, P. L., and Hamilton, W. J., 1980, *J. Neurocytol.* **9**:185–205.
49. Bunge, R. P., and Bunge, M. B., 1981, in: *Studies in Developmental Neurobiology: Essays in Honor of Victor Hamberger* (W. M. Cowan, ed.), Oxford University Press, New York, pp. 322–353.
50. Salpeter, M. M., and Faeder, I. R., 1971, *Prog. Brain Res.* **34**:103–114.
51. Schousboe, A., 1977, *Cell, Tissue and Organ Cultures in Neurobiology* (S. Federoff and L. Hertz, eds.), Academic Press, New York, pp. 441–446.
52. Dennis, M. J., and Miledi, R., 1974, *J. Physiol. (Lond.)* **237**:431–452.
53. Tucek, S., Zelena, J., Ge, I., and Vyskocil, F., 1978, *Neuroscience* **3**:709–724.
54. Aguayo, A., Peyronnard, J., Terry, L., Romine, J., and Bray, G., 1976, *J. Neurocytol.* **5**:137–155.
55. McCarthy, K. D., and Partlow, L. M., 1976, *Brain Res.* **114**:415–426.
56. Varon, S., and Manthrope, M., 1982, *Advances in Cellular Neurobiology*, Volume 3 (S. Federoff and L. Hertz, eds.), Academic Press, New York (in press).
57. Wood, P. M., 1976, *Brain Res.* **115**:361–375.
58. Brockes, J., Fields, K., and Raff, M., 1979, *Brain Res.* **165**:105–118.
59. Salzer, J. L., Williams, A. K., Glaser, L., and Bunge, R. P., 1980, *J. Cell Biol.* **84**:753–766.
60. Hanson, G. R., and Partlow, L. M., 1978, *Brain Res.* **159**:195–210.
61. Salzer, J. L., Bunge, R. P., and Glaser, L., 1980, *J. Cell Biol.* **84**:767–778.
62. Minier, L. N., and DeVries, G. H., 1980, *Soc. Neurosci. Abstr.* **6**:550.
63. Raff, M., Abney, E., Brockes, J., and Hornby-Smith, A., *Cell* **15**:813–822.
64. Salzer, J. L., and Bunge, R. P., 1980, *J. Cell Biol.* **84**:739–752.
65. Brockes, J. P., Lemke, G. E., and Balzer, D. R., 1980, *J. Biol. Chem.* **255**:8374–8377.
66. Hall, S. M., and Gregson, N. A., 1975, *Neuropathol. Appl. Neurobiol.* **1**:149–170.
67. Bunge, M. B., Williams, A. K., Wood, P. M., Vaitto, J., and Jeffrey, J. J., 1980, *J. Cell Biol.* **84**:184–202.
68. Moya, F., Bunge, M. B., and Bunge, R. P., 1980, *Proc. Natl. Acad. Sci. U.S.A.* **77**:6902–6906.
69. Bunge, R. P., and Bunge, M. B., 1978, *J. Cell Biol.* **78**:943–950.
70. Dawson, R. M. C., and Gould, R. M., 1976, *Function and Metabolism of Phospholipids in the Central and Peripheral Nervous System* (G. Porcellati, ed.), Plenum Press, New York, pp. 95–112.
71. Hedley-Whyte, E. T., Rawlins, F. A., Salpeter, M. M., and Uzman, B. G., 1969, *Lab. Invest.* **21**:536–547.

72. Rawlins, F., 1973, *J. Cell Biol.* **58**:42–53.
73. Spencer, P. S., 1979, *Soc. Neurosci. Symp.* **4**:275–321.
74. McDermott, J. R., and Wisniewski, H. M., 1977, *J. Neurol. Sci.* **33**:81–94.
75. Mirsky, R., Winter, J., Abney, E. R., Pruss, R. M., Gavrilovic, J., and Raff, M. C., 1980, *J. Cell Biol.* **84**:483–494.
76. Brockes, J. P., Raff, M. C., Nishiguchi, D. J., and Winter, J., 1980, *J. Neurocytol.* **9**:67–78.
77. Dyck, P. J., Thomas, P. K., and Lampert, E. H., 1975, *Peripheral Neuropathy,* W. B. Saunders, Philadelphia.
78. Spencer, P. S., and Schaumburg, H. H., 1980, *Experimental and Clinical Neurotoxicology,* Williams & Wilkins, Baltimore.
79. Joseph, B. S., 1973, *Brain Res.* **59**:1–18.
80. Tsukita, S., and Ishikawa, H., 1980, *J. Cell Biol.* **84**:513–530.
81. Oderfeld-Nowak, B., and Niemierko, S., 1969, *J. Neurochem.* **16**:235–248.
82. Appenzeller, O., and Partlow, L. M., 1972, *Brain Res.* **42**:521–524.
83. Belin, J., and Smith, A. D., 1976, *J. Neurochem.* **27**:969–970.
84. Spencer, P. S., and Weinberg, H. J., 1978, *The Physiology and Pathobiology of Axons* (S. Waxman, ed.), Raven Press, New York, pp. 389–405.
85. Romine, J. S., Bray, G. M., and Aguayo, A. J., 1976, *Arch. Neurol.* **33**:49–54.
86. Hall, S. M., 1978, *Neuropathol. Appl. Neurobiol.* **4**:165–176.
87. Miller, E. K., and Dawson, R. M. C., 1972, *Biochem. J.* **126**:823–835.
88. Hall, S. M., and Gregson, N., 1977, *Neuropathol. Appl. Neurobiol.* **3**:65–78.
89. Thomas, P. K., 1974, *Essays on the Nervous System: A Festschrift for Professor J. Z. Young* (R. Bellairs and E. G. Gray, eds.), Oxford University Press, Oxford, pp. 44–70.
90. Spencer, P. S., Politis, M. J., Pellegrino, R. G., and Weinberg, H. J., 1981, *Symposium on Post-Traumatic Peripheral Nerve Regeneration* (A. Gorio, H. Millesi, and S. Mingrino, eds.), Raven Press, New York, pp. 911–929.

16

Physiological Neurochemistry of Cerebrospinal Fluid

James H. Wood

1. INTRODUCTION

The central nervous system is effectively isolated from the rest of the body. This physiological compartmentalization not only provides protection of its delicate function from aberrant peripheral influences but also impedes its diagnostic evaluation. Cerebrospinal fluid (CSF) bathes the brain and spinal cord, is in dynamic equilibrium with its extracellular fluid, and tends to reflect the state of health and activity of the central nervous system. Cerebrospinal fluid examination is the most direct and popular method of assessing the central chemical and cellular environment in the living patient or mammal.

The purpose of this chapter is to provide the sophisticated knowledge of the mechanisms that maintain the physiological composition of CSF and provide an optimal extracellular environment for central neural transmission. This discussion will include the sites of origin, rostrocaudal concentration gradients, and circadian rhythms, as well as age and sex variations of the major constituents in CSF.

2. SOLUTE COMPOSITION

The solute composition of CSF is determined by several factors: (1) metabolism, production, or uptake of solutes by cells of the central nervous system; (2) restriction of intercellular diffusion coupled with special transport mechanisms at both the blood–brain and blood–CSF barriers; (3) rates of CSF production and excretion by bulk flow.[1]

In addition to CSF secretion, the choroid plexus controls the chemical composition of CSF by possessing both a blood–CSF barrier and membrane transport systems for ions, organic acids and bases, hexoses, purines, amino

James H. Wood • Cerebral Blood Flow Laboratories, Division of Neurosurgery, Emory University Clinic, Atlanta, Georgia 30322.

acids, and vitamins.[2-4] These barrier and transport mechanisms have been evaluated employing analysis of the composition of fluid collected at the surface of the plexus,[5] extracorporeal perfusion of the choroid plexus,[6] ventriculocisternal perfusion,[7] and incubation of the choroid plexus *in vitro*.[8]

The influx of nonelectrolytes into CSF may proceed by passive diffusion, pinocytosis, or carrier-mediated transport.[9] In review,[10] *passive diffusion* is regulated by the transcellular concentration gradient, the lipid solubility, and, in the case of electrolytes, the degree of ionization. *Pinocytosis* or vesicular transport pertains more to protein exudation than to solute migration. The most important mode of solute entry into brain and CSF is *carrier-mediated transport*. This process is termed *facilitated transport* if the binding and translocation phases do not require energy to yield equilibration of the intracellular and extracellular solute concentrations. The term *active transport* is reserved for processes utilizing energy to move solutes against a concentration gradient. The transport systems may be saturable and specific for certain classes of solutes, exhibit stereospecificity, or require concurrent exchange for molecules moving in the opposite direction.

Interpretation of CSF solute data may be complicated by pharmacological or disease-related alterations in central metabolism, blood–brain–CSF barriers, CSF circulation, or CSF concentration gradients. The steady-state concentrations of various solutes in plasma and lumbar CSF of man are listed in Table I.

2.1. Ions

Sodium ions are the major cations in both plasma and CSF.[4] The absolute concentration of sodium is slightly greater in plasma than in CSF when sodium levels are corrected for the water content of the two fluids; the CSF/plasma ratio is about 0.93. Contrarily, if sodium concentrations of these fluids are compared employing their actual volumes, then the CSF/plasma ratio is about 1.03.

The Donnan equilibrium produces a 5-mV positive electrical potential in CSF with respect to blood. The active secretion of sodium into CSF prevents a 6–17% sodium excess in plasma that would otherwise result from the inhomogeneity of negatively charged proteins across the blood–CSF barrier.[4,11] Cerebrospinal fluid sodium levels reach a maximum within 1 to 2 h after an acute elevation of plasma sodium concentrations.[12]

Potassium concentrations in CSF are stable despite variations in plasma potassium levels between 3.4 and 5.8 meq/liter.[13] In fact, experimental elevation of plasma potassium to concentrations high enough to cause fatal cardiac arrest (10 meq/liter) may not alter cisternal CSF potassium levels.[14] The potassium concentration in CSF is low compared with that in plasma or in brain cells. The concentration of potassium ions decreases steadily as CSF passes from the cerebral ventricles to the subarachnoid space,[15] suggesting the presence of mechanisms for the removal of potassium as the CSF circulates. The potassium content of CSF rises as a function of time after death,[16] probably because of depletion of sodium–potassium ATPase.

Table I
Composition of Normal Plasma and Lumbar CSF in Man[a]

Constituent	Units	Plasma	CSF	CSF/plasma ratio
Osmolarity	mOsm/liter	295	295	1.0
Water content		93%	99%	
Sodium	mEq/liter	138	138	1.0
Potassium	mEq/liter	4.5	2.8	0.6
Calcium	mEq/liter	4.8	2.1	0.4
Magnesium	mEq/liter	1.7	2.3	1.4
Phosphorus	mg/dl	4.0	1.6	0.4
Chloride	mEq/liter	102	119	1.2
Bicarbonate	mEq/liter	24[b]	22	0.9
P_{CO_2}	mm Hg	41[b]	47	1.1
pH		7.41[b]	7.33	
P_{O_2}	mm Hg	104[b]	43	0.4
Glucose	mg/100 ml	90	60	0.67
Lactate	mEq/liter	1.0[b]	1.6	1.6
Pyruvate	mEq/liter	0.11[b]	0.08	0.73
Lactate/pyruvate ratio		17.6[b]	26	
Total protein	mg/100 ml	7000	35	0.005
Total free amino acids	μmol/100 ml	228	81	0.4
Ammonia	μg/100 ml	37[b]	24	0.6
Urea	mmol/liter	5.4	4.7	0.9
Creatinine	mg/100 ml	1.8	1.2	0.67
Uric acid	mg/100 ml	5.50	0.25	0.05
Iron	μg/100 ml	15.0	1.5	0.01
Putrescine	pmol/ml		184	
Spermidine	pmol/ml		150	
Total lipids	mg/100 ml	750	1.5	0.002

[a] Derived from review of Fishman.[11]
[b] Arterial plasma.

Approximately 50% of plasma calcium exists in an ionized state, but this fraction in CSF is increased to 73% because of the lower protein content in CSF.[17] The majority of calcium enters the CSF by carrier-mediated transport with only minor contributions by passive diffusion.[18] Fortunately, calcium concentrations in CSF appear not to be altered by acute perturbations in plasma calcium levels,[17] especially since arousal and various autonomic functions are modified by changes in CSF calcium content.[19] The concentration of calcium in human ventricular CSF is similar to that in lumbar CSF[20]; however, some animals have been reported to have higher calcium levels in lumbar CSF.[15]

The concentration of magnesium in CSF is about 30% greater than that in plasma, and the fraction in the ionized form is higher in CSF than in plasma.[4] The 5-mV positive electrical potential in CSF would tend to keep the magnesium level in CSF lower than that in plasma; however, the higher CSF levels are maintained by active transport mechanisms.[11] Fortunately, CSF magnesium levels are quite stable despite perturbation of plasma magnesium concentrations,[21,22] thus preventing CSF and extracellular magnesium alterations that would profoundly affect brain excitability. Magnesium concentrations in the ventricles are higher than those in lumbar CSF,[20] and acute elevations occur after death[16] with depletion of energy-dependent transport systems. Recently, correlations between magnesium and protein levels in CSF have been reported in disease states.[23]

The concentration of inorganic phosphorus in normal CSF is about 60% of that in plasma and is relatively stable despite large alterations in plasma phosphorus content.[24] Correlations between phosphorus and protein levels in CSF remain controversial.[18,23]

Chloride is the major anion in CSF, and its CSF concentration exceeds that in plasma.[4] The electrical potential between blood and CSF favors greater levels of chloride in CSF. High elevations of CSF protein content are associated with depression of CSF chloride levels.[11] Although CSF chloride concentrations passively reflect alterations in plasma levels, the changes in CSF chloride are usually less than those in plasma.[13,25] The movement of chloride also appears to be somewhat coupled to the movement of sodium ions,[26] and ventricular CSF chloride levels are marginally higher than the chloride concentrations in lumbar CSF.[15]

2.1.1. Hydrogen Ions, Bicarbonate, and Acid–Base Equilibrium

The CSF–extracellular fluid compartment maintains the pH environment of the central nervous system optimal for the central enzyme activity required for appropriate neurotransmission. The acid–base balance of CSF is determined by the partial pressure of carbon dioxide (P_{CO_2}) and the bicarbonate (HCO_3^-) concentration which together regulate the hydrogen ion (H^+) concentration according to the *Henderson–Hasselbalch equation*:

$$pH = pK_i + \log(HCO_3^-/sP_{CO_2}) \tag{1}$$

where pK_i is the first dissociation constant of carbonic acid (6130 at 38°C at pH 7.30) and s equals the solubility coefficient of CO_2 in CSF (0.0312 nmol/mm Hg at 38°C).[27] Techniques have been described for the accurate determination of pH and P_{CO_2} in CSF despite its very low buffering capacity.[28,29]

Interpretation of acid–base-related data is complicated by the known differences between CSF and arterial blood. The P_{CO_2} in cisternal CSF is higher than that in arterial blood, whereas the HCO_3^- concentrations in these fluids are similar; thus, the CSF pH is somewhat lower than arterial pH.[30,31] Similarly, lumbar CSF P_{CO_2} exceeds cisternal CSF P_{CO_2} and thus, in the absence of a rostrocaudal HCO_3^- concentration gradient, accounts for a lower CSF pH in lumbar CSF. Acute acid–base disturbances are reflected in lumbar CSF more slowly than in cisternal CSF[31,32]; CSF P_{CO_2} depends on the arterial–CSF P_{CO_2} gradient, the cerebral blood flow, and the CO_2 production of neural tissue.[33] Large changes in blood HCO_3^- levels are not acutely reflected in CSF; however, prolonged alterations in blood HCO_3^- concentrations will induce large shifts in CSF HCO_3^- levels.[33] These alterations in the H^+ and HCO_3^- content of CSF may directly affect the respiratory centers located superficially in the ventral medulla.[34] Alteration in the HCO_3^- ion concentration of the CSF also modifies the responsiveness of cerebral arterioles to CO_2[35] and arterial diameter.[36] Specific autonomy of the CSF acid–base system appears to be secondary to the action of regulatory mechanisms at the neuronal level.[36a]

2.1.1a. Investigational Considerations. Normally, arterial blood and CSF pH are about 7.4 and 7.3, respectively. This small pH difference across the blood–CSF barrier influences the distribution of organic acids and bases that occur in solution in both ionized and nonionized forms. The degree of ionization of a drug in CSF or blood is calculated employing the Henderson–Hasselbalch equation.

$$\text{For weak acids:} \quad pH = pK_a + \log(\text{salt/acid}) \tag{2}$$

$$\text{For weak bases:} \quad pH = 14 - pK_b - \log(\text{salt/base}) \tag{3}$$

where the pK or ionic dissociation constant of the drug refers to the pH at which 50% of the drug is ionized. Drugs such as phenytoin ($pK_a = 8.3$), being mostly nonionized in blood, easily enter the CSF, whereas a major portion of drugs such as phenobarbital ($pK_a = 7.4$) with a lower pK_a is in the ionized form and is impeded from entering the CSF unless the blood becomes more acidic, thereby deterring drug ionization.[11]

2.2. Glucose

Glucose in CSF has its origin in plasma, and its CSF concentration is dependent on the blood glucose level.[37] Glucose transfer from plasma to CSF occurs by carrier-facilitated diffusion and exhibits saturation kinetics, stereo-specificity, bidirectionality, competitive inhibition, and counterflow.[11] The CSF/plasma ratio for glucose is normally 0.60–0.80 but may decrease by saturation kinetics following elevations in blood glucose.[37] The glucose content of CSF is relatively increased in both premature and newborn infants, whose CSF/plasma ratio for glucose is 0.8 or greater.[38] These ratios imply that the rate of glucose removal from CSF is greater than that of its entry into CSF. The concentration of glucose in brain is approximately 20 mg/100 ml; thus, the brain serves as a sink for both blood and CSF glucose.[11] The mechanism for the ventriculospinal concentration gradient of glucose in CSF is controversial.[11,37]

Alterations in plasma glucose are reflected in parallel changes in CSF glucose concentrations.[37] However, the peak glucose level in CSF may lag 2 h behind that in blood following rapid intravenous glucose infusions (Fig. 1), and the CSF level may not return to base-line concentrations for 4 to 6 h after the hyperglycemic episode.[11,39] Thus, for investigational purposes, simultaneous CSF and plasma glucose levels should be compared, preferably with patients in the fasting state.

Hypoglycorrhachia may be masked by hyperglycemia and caused by hypoglycemia, neoplastic or inflammatory infiltrations of the meninges, subarachnoid bleeding, or chemical meningitis.[11] The major factors for low CSF glucose levels in meningeal disorders include increased glucose utilization by

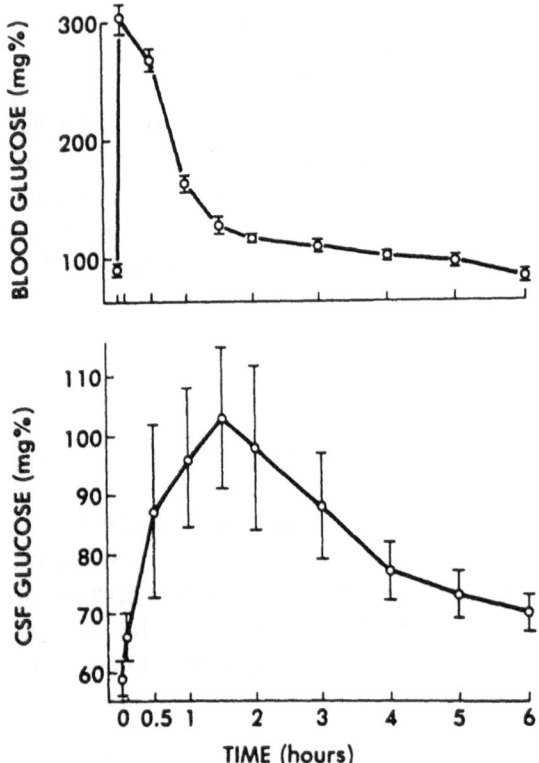

Fig. 1. Alterations of glucose concentration in lumbar CSF following acute hyperglycemia in five patients. Mean ± SD venous glucose level peaked 5 min after intravenous injection of 0.75 g/kg 50% glucose solution and approached base-line level 2 to 3 h thereafter. Glucose concentrations in lumbar CSF rose slowly, beginning within 5 min, peaked after 90–120 min, but did not approach base-line level for 4 to 6 h after injection.[39] Reprinted from Fishman[11] with permission.

nervous tissue or leukocytes and inhibition of glucose transport. An increased CSF lactate level always occurs with hypoglycorrhachia from any cause except hypoglycemia, reflecting increased anaerobic glycolysis.

2.3. Lactic and Pyruvic Acids

The brain content of lactic acid varies with the rate of brain lactate formation but is not dependent on the blood lactate concentration. Intravenous lactic acid infusions do not appear to raise CSF lactate levels,[40] and elevations in CSF lactate concentrations may occur in the presence of normal blood lactate.[40] Thus, the lactate levels in CSF primarily reflect brain lactate content.

The lactate/pyruvate ratio reflects the redox state in the brain and depends in part on the activity of the enzyme lactate dehydrogenase.[41] Whenever central glycolysis is enhanced during hypoxia, subarachnoid bleeding, ischemia, seizures, or nonviral meningitis, both the lactate and pyruvate levels in brain and CSF as well as the lactate/pyruvate ratio are usually elevated.[41]

2.4. Amino Acids

The total free amino acid content of plasma is about three times that of CSF,[42] and similarly, the plasma concentrations of most amino acids except glutamine are 4–16 times those in CSF.[43] Glutamine, the major amino acid, comprises about 73% of the total amino acids in CSF but only 24% of that in plasma.[42] The concentrations of free amino acids in plasma and lumbar CSF of normal fasting individuals are listed in Table II.[44] Several other reports[42,43,45–47] of amino acid profiles in the CSF of control patients have been published. Glycine and serine concentrations appear to be higher, and glutamine levels are lower, in ventricular CSF than in lumbar CSF.[44] Variable ventriculospinal concentration gradients have been observed for other amino acids in CSF.[48]

The CSF contents of most amino acids[45] except glutamine[43] in newborn infants are increased in comparison to those in normal adults. Cerebrospinal fluid concentrations of glutamine, glycine, alanine, citrulline, α-aminobutyric acid, valine, methionine, isoleucine, leucine, phenylalanine, and lysine are high, and those of phosphoethanolamine, aspartate, and homocarnosine are lower in older patients than in younger patients.[43,45,46,49]

2.5. Protein

The proteins present in normal CSF are mostly derived from plasma with the exception of trace proteins originating from brain and some β-globulins which appear to be modified within the brain.[11] Pharmacokinetic studies following intravenous administration have documented that 20 h are required for albumin to reach equilibrium in dog cisternal CSF[50] and that 3 to 6 days are required for γ-globulin to equilibrate in human lumbar CSF.[51]

The low protein content in normal CSF with respect to plasma protein

Table II
Amino Acids in Plasma and Lumbar CSF in Man[a]

Amino acid	Mean plasma concentration (μmol/liter)	Mean CSF concentration (μmol/liter)	CSF/plasma concentration ratio
Alanine–citrulline	488.5	34.3	0.08
2-Aminobutyric acid	29.8	3.5	0.14
Arginine	80.9	22.4	0.31
Asparagine	111.7	13.5	0.12
Glutamic acid	61.3	26.1	0.40
Glutamine	641.0	552.0	0.86
Glycine	282.7	5.9	0.02
Histidine	79.8	12.3	0.16
Isoleucine	76.7	6.2	0.09
Leucine	155.3	14.8	0.10
Lysine	170.7	20.8	0.12
Methionine	27.7	2.5	0.10
Ornithine	73.5	3.8	0.06
Phenylalanine	64.0	9.9	0.17
Phosphoethanolamine	5.1	5.4	1.05
Phosphoserine	8.3	4.2	0.58
Serine	139.7	29.5	0.23
Taurine	77.2	7.6	0.11
Threonine	165.5	35.5	0.25
Tyrosine	73.0	9.5	0.14
Valine	308.6	19.9	0.07

[a] Data from McGale *et al.*[44]

levels suggests the relative exclusion of macromolecules by the blood–brain or blood–CSF barriers.[11] However, the concentrations of prealbumin, transferrin, and γ-trace or β-trace proteins in CSF cannot be entirely explained by an inverse relationship to molecular weight.[52] The CSF/plasma distribution ratios for the various protein components in CSF appear to be better correlated with their hydrodynamic radii than with their molecular weights[53] (Table III), partially accounting for the electrophoretic characteristics of the various CSF proteins that originate in plasma. If pinocytosis of proteins is the major route of entry into CSF, then various-sized macromolecules would be expected to cross these barriers at similar rates.[11] Recently two pathways for protein transfer at the choroid plexus epithelium have been postulated: (1) 117-Å-radius pores that allow the transfer of smaller proteins by diffusion or ultrafiltration or both; and (2) 250-Å-radius pinocytotic vesicles that account for the exchanges of large proteins.[54,55] Although not described in the choroidal epithelium, the 117-Å-radius pores may represent a 0.08% defect in the normally continuous tight junctions that surround and closely connect choroidal epithelial cells.

Since the total protein concentration in CSF is less than 0.5% of that of plasma, the exit of protein is about 200 times the entry rate.

2.5.1. Identification and Quantification

Techniques developed for the quantitative measurement of the total protein content of CSF[56] include ultraviolet spectrophotometry, turbidimetric methods,[57–60] biuret procedures, and the Lowry method.[61] Fractionation of CSF proteins has been accomplished by electrophoresis (using paper or cellulose acetate, agar, agarose, polyacrylamide,[62] and starch gels) and more recently by immunoelectrophoresis, electroimmunodiffusion,[57,63] radioimmunoassay,[64] enzyme immunoassay,[65] and isoelectric focusing techniques.[66–70] Techniques that do not require the preliminary concentration of CSF are preferred to avoid

Table III
Concentrations of Proteins in Plasma and Lumbar CSF in Man[a]

Protein	Molecular weight	Hydrodynamic radius (Å)	Plasma concentration (mg/liter)	CSF concentration (mg/liter)	CSF/plasma ratio
Prealbumin	61,000	32.5	238	17.3	0.07
Albumin	69,000	35.8	36,000	155.0	0.004
Transferrin	81,000	36.7	2040	14.4	0.007
Ceruloplasmin	152,000	46.8	366	1.0	0.003
IgG	150,000	53.4	9,870	12.3	0.001
IgA	150,000	56.8	1,750	1.3	0.0007
α_2-Macroglobulin	798,000	93.5	2,220	2.0	0.0009
Fibrinogen	340,000	108.0	2,964	0.6	0.0002
IgM	800,000	121.0	700	0.6	0.0009
β-Lipoprotein	2,239,000	124.0	3,728	0.6	0.0002

[a] Data derived from Felgenhauer.[53]

technical artifacts caused by protein loss from membrane absorption or protein denaturation.

The characteristic pattern of separation of proteins by electrophoresis depends on their molecular weight and electrical charge. These patterns (pherograms) enable protein separation into albumin, α-globulin, β-globulin, and γ-globulin fractions. Immunoelectrophoresis first employs electrophoresis to separate the protein fractions and then an immunoprecipitation reaction to identify the multiple proteins within each major electrophoretic peak. Proteins identified in normal CSF by immunoelectrophoresis include prealbumin, albumin, α_1-globulin, α_2-globulin, glycoprotein, α_1-antitrypsin, haptoglobins, ceruloplasmin, hemopexin, transferrin, β_1-globulin, IgA, IgM, and IgG. Major CSF proteins identified by electrophoresis and immunoelectrophoresis are listed in Table IV.

Prealbumin accounts for about 5% of total protein in lumbar CSF but constitutes about 10% of that in ventricular CSF. Since the total protein content in lumbar CSF is two to three times that in ventricular CSF, the actual prealbumin concentrations are similar throughout the CSF spaces.[11]

Albumin constitutes about 56–76% of total CSF protein but only 52–67% of total serum protein.[71] The albumin in CSF originates from plasma.[50,51] Albumin elevations are relatively greater when the total CSF protein is increased and leads to a rise in the CSF albumin/total protein ratio.[11]

The α_1-globulins include two discrete α_1-lipoproteins and α_1-glycoproteins. Ceruloplasmin, haptoglobin, erythropoietin, α_2-macroglobulin and α_2-lipoprotein are α_2-globulins. The β-globulins include transferrin, which is the iron-binding β_1-globulin, β_2-transferrin, tau-protein, plasminogen, complement, hemopexin, and β-trace proteins.[11] The normal content of β-trace protein is about 7% of the total CSF protein, but it is almost absent in serum; thus, β-trace proteins may be synthesized by the central nervous system.[72]

The normal source of the γ-globulin content in CSF is plasma γ-globulin.[51] The γ-globulin/total protein ratio in lumbar CSF is about two-thirds the ratio in plasma. The primary immunoglobulins (Ig) in CSF are IgG, IgA, and IgM,[56,71,73,77] although trace amounts of IgD and IgE have been reported.[75] IgG, the major γ-globulin in both plasma and CSF, accounts for 15–18% of total plasma protein but only 5–12% of total CSF protein.[56] Most antibodies against bacteria and viruses are IgG. The IgG response to central inflammations or demyelination must be related to that of other CSF proteins in order to assess whether elevations in CSF IgG represent increased blood–brain barrier permeability of plasma IgG or local central nervous system IgG synthesis. Discrimination between blood–brain barrier dysfunction and inflammation may be accomplished by evaluation of the CSF protein profile.[78] The *IgG–albumin index*[79] may be employed to correct the CSF IgG level for the contribution of plasma IgG that crosses a damaged blood–brain barrier:

$$\text{IgG–albumin index} = \text{IgG}_{\text{CSF}} \times \text{albumin}_{\text{plasma}} / \text{IgG}_{\text{plasma}} \times \text{albumin}_{\text{CSF}} \quad [4]$$

Use of this index requires each laboratory to determine its own normal values. The contribution of locally synthesized IgG to the total CSF IgG content may

Table IV
Cerebrospinal Fluid Proteins Identified with Electrophoresis and Immunoelectrophoresis[a]

Electrophoretic fraction	Prealbumin	Albumin	α_1-Globulin	α_2-Globulin	β-Globulin	γ-Globulin
Immunoelectrophoretic proteins	Prealbumin	Albumin	α_1-Antitrypsin	α_2-Macroglobulin	β-Lipoprotein	IgG
			α_1-Lipoprotein	α_2-Lipoprotein	Transferrin	IgA
			α_1-Glycoprotein (orosomucoid)	Haptoglobulin	Tau-fraction (modified transferrin)	IgM
				Ceruloplasmin	Plasminogen	IgD
					Complement	
				Erythropoietin	Hemopexin	IgE
					β-Trace protein	γ-Trace protein

[a] Reprinted from Fishman[11] with permission. Data from refs. 56,71–76.

be calculated as follows[77,80]:

$$\text{IgG Synthesis} = 5(\text{IgG}_{CSF} - \text{IgG}_{plasma}/369)$$

$$- (\text{albumin}_{CSF} - \text{albumin}_{plasma}/230)(\text{IgG}_{plasma}/\text{albumin}_{plasma})0.43 \quad [5]$$

Percent evaluations of these two corrective methods have found them equally reliable.[81,82]

The IgG molecule may be dissociated into two heavy and two light polypeptide chains. The two types of light chains, κ and λ, are about equal in concentration in both normal CSF and plasma.[83] The CSF κ/λ ratio rises during inflammation or demyelination of the central nervous system.

2.5.2. Variations with Age and Sex

α-Fetoprotein and fetuin are present in relatively high concentrations in fetal CSF.[84] In addition, a greater penetration of proteins from plasma into the CSF occurs early, but not late, in gestation, probably secondary to the immaturity and relatively greater size of the fetal choroid plexus.[84] Low-birth-weight neonates have higher CSF protein levels than neonates of normal birth weight and are more susceptible to subclinical subependymal hemorrhages that may increase CSF protein concentrations.[85] The CSF concentrations of γ-trace proteins and β_2-microglobulins are three and two times higher, respectively, in neonates than in adults, and the CSF/plasma ratios for these proteins were higher in young infants than in older infants.[86] The absolute levels of IgG, IgA, and IgM as well as total protein in CSF positively correlate with age, and the IgG and IgA percentages of total CSF protein also increase with age.[87] The CSF/serum albumin ratio increases with age, whereas the CSF IgG index remains constant in control patients aged six months to 30 years.[87a] Total protein content in lumbar CSF normally varies with the patient's age according to the equation[88]:

$$\text{Total lumbar CSF protein} = 23.8 + 0.39 \times \text{age} \pm 15.0 \text{ mg/100 ml} \quad [6]$$

The concentration of albumin, the largest protein fraction in CSF, appears to be lower in females, but the CSF/plasma albumin ratios are the same for both sexes, suggesting a dynamic equilibrium between CSF and plasma protein in normal individuals.[89]

2.5.3. Ventriculospinal Concentration Gradients

Analysis of protein fractions in ventricular CSF of patients with psychiatric or extrapyramidal disorders undergoing stereotactic operations and in cisternal and lumbar CSF of patients seeking confirmation of suspected neurological disease have demonstrated ventriculospinal protein concentration gradients in CSF[90] (Table V). Lumbar CSF contains approximately 1.6 times more total protein than ventricular CSF, whereas cisternal CSF protein content is

Table V
Protein Concentration Gradients in Human CSF[a]

Protein	Ventricular CSF	Cisternal CSF	Lumbar CSF
Total protein	25.6 ± 1.2 mg/100 ml[b]	31.6 ± 1.0 mg/100 ml	42.0 ± 0.5 mg/100 ml
Albumin	8.3 ± 0.5 mg/100 ml	12.7 ± 0.7 mg/100 ml	18.6 ± 0.6 mg/100 ml
% Albumin	32.4%	40.2%	44.3%
IgG	0.9 ± 0.1 mg/100 ml	1.4 ± 0.1 mg/100 ml	2.3 ± 0.1 mg/100 ml
% IgG	3.5%	4.4%	5.5%
Number of patients	27	33	127

[a] Total protein and protein fractions determined by radioimmunodiffusion (derived from data of Weisner and Bernhardt[90]).

[b] Mean ± standard error.

1.2 times that of ventricular CSF. Generally, total protein concentrations in the ventricles are 6–12 mg/100 ml, whereas those in the cisterna magna and lumbar sac are 15–25 and 20–50 mg/100 ml, respectively.[37] This higher level of protein in lumbar CSF is thought to result from increased permeability of plasma proteins and not from dehydration or poor mixing in the lumbar subarachnoid space.[91]

This mechanism does not appear to affect all protein fractions equally.[90] Lumbar CSF albumin and IgG concentrations were 2.2 and 2.6 times higher, respectively, than their levels in ventricular CSF. Cerebrospinal fluid albumin and IgG concentrations correlate highly in all regions of the subarachnoid space despite the presence of a 10% IgG fraction that does not originate from the serum.[90] Similar variations in IgG and total protein levels have been noted during continuous lumbar CSF drainage.[92] In contrast, the concentration of prealbumin is reduced by a factor of 0.7 in lumbar CSF in comparison to ventricular CSF levels.[90] The CSF levels of IgA are much lower than those of IgG despite the similarity of their molecular weights. This disparity between IgA and IgG concentrations suggests the presence of selective functions of the blood–CSF barrier with respect to protein fractions.

2.6. Ammonia and Glutamine

Ammonia, which is neurotoxic, is combined with α-ketoglutarate in the brain to form glutamine as a defense mechanism against ammonia encephalopathy. Normally, only 10% of blood ammonia is nonionized and capable of crossing the blood–brain barrier. The CSF ammonia concentration is about one-third to one-half of the arterial blood ammonia level,[93] and its distribution between these fluids is unlikely to be explained by passive diffusion.[94]

2.7. Urea, Creatinine, and Uric Acid

The steady-state urea content in CSF is slightly less than that in the plasma, and alterations in plasma urea levels are reflected in CSF.[13] Cerebrospinal fluid creatinine concentrations are about two-thirds that in plasma but may vary when corrected for protein binding.[95]

The CSF/plasma ratio for uric acid is normally about 0.05 and varies with shifts in plasma concentrations.[96] Although the exact mechanism for the removal of uric acid from CSF is unknown, elevated CSF uric acid concentrations have been associated with probenecid administration.[97,98]

2.8. Polyamines

Polycationic polyamines form complexes with polyanionic nucleic acids and are increased in rapidly proliferating cells. Polyamines are present in both free and conjugated forms in plasma and CSF.[99,100] Spermine is not detectable in normal CSF, but low CSF levels of spermidine and their precursor putrescine have been reported.[101,102]

The CSF concentration of spermidine is usually determined by radioimmunoassay,[103] whereas that of putrescine is usually measured by column chromatography with fluorometric quantitation.[102] Cerebrospinal fluid putrescine levels determined by enzymatic–isotopic techniques correlate well with free putrescine concentrations, but this method underestimates total putrescine values.[104] The difference between total and free putrescine concentrations mostly reflects putrescine metabolites such as monoacetylputrescine, which accounts for about 40% of the total putrescine content in CSF.

2.9. Lipids

The total lipid content in CSF is less than 0.2% of the plasma level. Phospholipids in normal CSF amount for only about 1.5% of those in plasma, and the relative distribution of lysolethicin, inositol phosphatide, sphingomyelin, lecithin, phosphatidyl serine, and phosphatidyl ethanolamine in CSF and plasma are similar.[105] A greater resemblance has been noted between the fatty acid compositions of CSF and brain (especially the phospholipids) than between those of CSF and plasma.[105] Approximately 60% of the fatty acids in CSF are saturated.[106] Palmitic acid, the major fatty acid, comprises about 30% of the total fatty acids in CSF. The major polyunsaturated fatty acid in CSF is arachidonic acid.[106] About one-third and two-thirds of the cholestrol in CSF are in the free and esterified forms, respectively.[107] Desmosterol (2,4-dehydrocholesterol) appears in the CSF following the administration of triparanol which blocks the conversion of desmosteral to cholesterol.[108] Small amounts of α_1-lipoprotein are present in normal CSF, and thus, the electrophoretic pattern of CSF lipoprotein differs from that in plasma.[107]

2.10. Prostaglandins

Prostaglandins (PG) are normally present in only small amounts in brain and CSF.[110] Activation of the enzyme phospholipase A_2 by cell membrane perturbation releases arachidonic acid (*cis*-5,8,11,14-eicosatetraenoic acid), the major precursor of PG in man, which is then oxygenated into PG.[111] Although the intraventricular injection of PG produces cerebral vasoconstriction[112] and stupor,[113] the exact function of PG in CSF is unknown.

Prostaglandins formed by the brain parenchyma are released into CSF following stimulation.[114] Most of the PG injected into lateral ventricular CSF can be recovered from cisternal CSF.[115] Prostaglandins are removed from the CSF by active sequestration by the choroid plexus[116] and active transport into pial blood vessels.[117] Labeled $PGF_{2\alpha}$ instilled into cisternal CSF rapidly disappears ($T_{1/2} = 8$ min), and only a small fraction can be found in lumbar CSF.[118] This labeled $PGF_{2\alpha}$ has been retrieved from jugular vein blood, suggesting that PG rapidly egresses from CSF to blood[118] and that the PG concentrations in lumbar CSF may not accurately reflect cisternal levels.[119]

Prostaglandins in CSF have been measured by radioimmunoassay,[120–123] gas chromatography/mass spectrometry (GC/MS),[114,118,124,125] and radioisotope dilution[126] techniques. Prostaglandin values determined by radioimmunoassay and GC/MS generally agree, but the radioisotope dilution method yields approximately 10 to 100 times greater PG values. To date, only $PGF_{2\alpha}$, PGE_2, and thromboxane A_2 have been quantitated in CSF.

Clinical studies suggest that the ''normal'' concentration of $PGF_{2\alpha}$ in lumbar CSF is less than 100 pg/ml in man.[118,121,125] Although similar CSF levels of PG have been reported in dogs,[119] ventricular CSF PG concentrations may vary fivefold in normal dogs.[113] The level of PGE_2 in CSF has been suggested to be a better index of encephalopathy than that of $PGF_{2\alpha}$.[120,123] The CSF concentrations of PGE_2 in men and $PGF_{2\alpha}$ in women increase with age, although the mean levels of these PGs are similar in men and women.[126]

2.11. Adenosine

The adenine nucleoside, adenosine, has been proposed as a regulator of cerebral blood flow,[128] a modulator of synaptic transmission,[127] and a potent stimulator of cyclic adenosine monophosphate production.[129] Minimal quantities of labeled adenosine are present in CSF following its intravascular injection.[130] Intrathecal administration of labeled adenosine results in heavy labeling of brain nucleotides but only limited activity in cerebral venous blood.[130] Thus, a relative blood–brain barrier exists for adenosine,[130,131] and brain uptake is its major mechanism of clearance from CSF.[130] Normally, adenosine is not detectable in CSF.[132]

Adenosine may be incorporated into inosine by brain adenosine deaminase. The uptake of labeled inosine into brain requires its conversion into hypoxanthine by nucleoside phosphorylase; thus, much of its CSF activity is cleared into the sagittal venous system after intrathecal injection.[132] Although inosine is not usually detected in normal CSF, CSF hypoxanthine levels have been reported to be 0.82 nmol/liter in dogs[132] and between 0 and 3 μmol/liter in man.[133]

The levels of exogenous adenosine, inosine, and hypoxanthine are stable in CSF, indicating that their metabolism requires uptake into brain.[132]

3. NEUROTRANSMITTERS, THEIR PRECURSORS AND METABOLITES, AND CYCLIC NUCLEOTIDES

Quantitative determination of neurotransmitter precursors, neurotransmitters, and their respective metabolites in CSF enables the study of the physiological, pathological, and pharmacological alterations of *in vivo* central nervous system activity. Meaningful interpretation of the CSF analysis requires the absence of CSF contamination with peripheral neurotransmitters and their metabolites.[134] In addition, the CSF should reflect the chemical composition of adjacent areas of nervous tissue.[134,135] The reference concentrations of these constituents in the lumbar CSF of control patients and normal volunteers are listed in Table VI.

3.1. Biochemical Physiology

The blood–brain and blood–CSF barriers profoundly affect central nervous system function. Normally, precursors, but not neurotransmitters or their metabolites, cross these barriers, usually by an energy-dependent carrier-mediated transport mechanism.[10] The permeability of neurotransmitter precursors enables manipulation of central neurotransmitter activity if the synthesizing enzymes are not saturated. The relative inability of the neurotransmitters to cross these barriers promotes reuptake and reutilization by somewhat restricting them to the immediate brain and spinal regions of origin,[9] prevents con-

Table VI

Concentration of Neurotransmitters, Their Metabolites and Cyclic Nucleotides in Lumbar CSF in Man

Constituent	CSF concentration[a]	Reference
Acetylcholine	0.07 ± 0.02 μM $(10)^{b,h}$	136
	187 ± 84 pmol/ml $(2)^{b}$	137
γ-Aminobutyric acid	233 ± 12 pmol/ml $(40)^{b}$	138
Norepinephrine	239 ± 17 pg/ml $(9)^{c,d}$	139
	373 ± 38 pg/ml $(18)^{b,e}$	140
3-Methoxy-4-hydroxyphenylethylene glycol	8.1 ± 0.2 ng/ml $(31)^{b}$	141
Homovanillic acid	35.1 ± 2.3 ng/ml $(42)^{b}$	141
5-Hydroxyindoleacetic acid	17.1 ± 1.0 ng/ml $(42)^{b}$	141
Cyclic adenosine 3′,5′-monophosphate	10.4 ± 0.5 pmol/ml $(59)^{c,g}$	97
	19.6 ± 0.7 pmol/ml $(41)^{b,f}$	141
Cyclic guanosine 3′,5′-monophosphate	2.4 ± 0.2 pmol/ml $(41)^{b,f}$	141
	2.4 ± 0.3 pmol/ml $(34)^{c,g}$	97

[a] Mean ± standard error.
[b] Normal volunteers.
[c] Control patients.
[d] CSF collected on ice (4°C) prior to ultracold storage (-70°C).
[e] CSF collected on dry ice (-56°C) prior to ultracold storage (-70°C).
[f] Protein-binding assay.
[g] Radioimmunoassay.
[h] Parentheses indicate number of patients.

tamination from peripheral sources, and increases the significance of the determinations of central neurotransmitter concentrations.[142]

Cerebrospinal fluid concentration gradients aid in our efforts to anatomically localize the origin of constituents found in CSF obtained at distant sampling sites.[135] However, these concentration gradients reflect not only variations in the distribution and metabolism of these constituents within the central nervous system but also their absorption mechanisms and CSF circulation patterns. Knowledge of these gradients is absolutely necessary for the interpretation of precursor, neurotransmitter, and metabolite levels in CSF.[142,143]

Many physiological functions such as sleep–wake activity, level of attention, pain and seizure thresholds, urinary excretory rates, body temperature, and plasma hormone concentrations fluctuate in rhythmic fashion with 24-h periodocity.[144] Most daily biological rhythms originate within the central nervous system,[145] and thus, the levels of many constituents in CSF appropriately follow circadian rhythms as reflections of this periodic central neuronal activity.[144,146] Other normal variations in the levels of CSF constituents may be related to age, sex, and neuroendocrine factors.

3.1.1. Acetylcholine

The quantification of acetylcholine (ACh) in CSF is controversial, and CSF ACh levels appear to reflect the method of analysis. Early investigators employed the cholinesterase inhibitors neostigmine[147] or physostigmine[148] and bioassay techniques.[148] Employing bioassay, the mean ($\pm$S.E.) ACh activity in normal CSF has been reported to be 1.8 $\pm$ 0.5 mg/100 ml and that in lumbar CSF not to be representative of ACh activity in ventricular or cortical subarachnoid CSF.[148] These early studies suggested that ACh activity in intracranial CSF is most likely masked by the rapid degradation of ACh by the plentiful acetylcholinesterase in CSF and brain.

Another study employing GC/MS[149,150] found ACh levels of 0.08 $\pm$ 0.02 μM in normal lumbar CSF.[136] Radiochemical techniques[151] have demonstrated higher ACh levels in ventricular CSF than in lumbar CSF.[137] The concentration of ACh in lumbar CSF has recently been reported to be 187 $\pm$ 84 pmol/ml, and the plasma-to-lumbar CSF ACh ratio appeared to be 9 in patients with minimal pathology.[137] However, the possibility exists that the ACh reported in CSF may have been artifactually introduced during sample preparation or derivatization.[152]

3.1.1a. Choline. Choline is both a precursor and an end product of ACh; however, CSF choline is passively derived,[153] principally from plasma choline rather than from brain ACh.[154] Clinical investigations have indicated that the concentration of choline in lumbar CSF is between 1.5 and 3.5 nmol/ml[155] or 3.35 $\pm$ 0.38 μM[136] or 1745 $\pm$ 709 pmol/ml.[137] The presence of mild erythrocytic contamination has been reported not to affect CSF choline levels. Cerebrospinal fluid levels of choline appear to be between 9 and 25 times higher than those of ACh.[137,148] Choline concentrations in ventricular CSF are higher than in lumbar CSF.[136,137,148] The plasma-to-lumbar CSF choline ratio is approximately 3.5.[137] Choline transport out of CSF occurs by both saturable carrier-mediated and passive, nonsaturable processes.[153]

3.1.2. γ-Aminobutyric Acid

γ-Aminobutyric acid (GABA), a putative inhibitory neurotransmitter, reduces membrane resistance by increasing chloride ion permeability and induces hyperpolarization.[156,157] The plasma-to-lumbar CSF ratio is approximately 0.6 in normal individuals.[158] The GABA clearance from the ventricular CSF by neuronal, glial[159–161] and choroidal tissue is 100 times faster than that of the nonpenetrating solute albumin.[162] γ-Aminobutyric acid is actively accumulated in the adjacent brain tissue[159] against a concentration gradient by a transport system that is inhibited by taurine and β-alanine.[162] Mean lumbar CSF GABA levels of 233 ± 12 and 214 ± 8 pmol/ml, determined by ion-exchange fluorimetry,[163] have been reported in 40[138] and 87[164] normal volunteers, respectively. The CSF GABA levels determined by the radioreceptor assay[165] are approximately 9% higher than, but closely correlate with, those assayed by the ion-exchange fluorimetric method.[166]

Although orally administered L-glutamine has been shown to increase CSF GABA in a patient with Huntington's disease,[167] the ability of GABA to cross the blood–brain or blood–CSF barrier is controversial. Most experimental investigations of systemically administered GABA have failed to demonstrate central GABA accumulations in normal adult animals.[168–171] Recent indirect evidence suggests that minimal quantities of peripheral GABA may cross into the central nervous system after systemic administration of large doses of GABA.[172] Penetration of intravenously injected GABA appears to occur following experimental local breakdown of the blood–brain barrier.[170] High-dose trials of oral GABA have been reported to suppress petit mal and grand mal seizure activity.[173]

The concentration of GABA rises in a dose-dependent manner after the experimental administration of drugs known to elevate brain GABA content.[174,175] Moreover, whole-brain GABA levels have been exponentially correlated with the GABA concentrations in CSF.[174,176] Although pharmacological elevations in serum GABA levels have resulted in some raising of CSF GABA concentrations,[174] plasma GABA levels do not significantly correlate with CSF GABA concentrations.[175] Thus, the vast majority of GABA in CSF appears to have a central origin.[135,142,174,177]

Extrapyramidal nuclei, hypothalamus, base of the pons, and cerebellum constitute the brain regions that contain the highest levels of GABA.[178,179] Small interneurons in the cerebral cortex,[157] hippocampus,[180] and thalamus[181] also release GABA. Although the source of the GABA content in cisternal or ventricular CSF appears to be brain,[174,175] the degree of contribution from the brain and spinal cord to the lumbar CSF GABA content remains to be determined.

The levels of GABA in the gray matter of the spinal cord are generally lower than those in the gray matter of the brain.[182,183] Accordingly, the CSF GABA concentration of the 40th milliliter obtained during continuous lumbar

CSF drainage is approximately 47% higher than that of the eighth milliliter[135,165] (Fig. 2), and the CSF GABA level in the 20th milliliter is approximately 30% higher than that of the first milliliter.[158,163,177] The rate of increase in GABA concentrations in sequential lumbar CSF fractions has been reported to be approximately 2% per ml as compared to the GABA content of the initial lumbar CSF fraction.[158,163,177]

The presence of a significant rostrocaudal GABA concentration gradient in CSF obtained at lumbar puncture may be indirect evidence that lumber CSF GABA levels reflect brain GABA metabolism. Accordingly, low GABA concentrations have been reported in the lumbar CSF of patients with neurological disorders[98,158,186–189] that have been associated with deficiencies in brain GABA.[190–192] However, the low GABA levels in patients with spinal and bulbar atrophy suggest that the spinal cord and brainstem make contributions to the lumbar CSF GABA content.[193,194]

Both experimental[144,195] and clinical[98] investigations have suggested that CSF GABA levels exhibit a daily rhythm with highest concentrations during the daytime. Recent clinical studies[158,164,196] have demonstrated significant negative correlations between lumbar CSF GABA concentrations and age in normal females but not in normal males. More specifically, evaluation of CSF

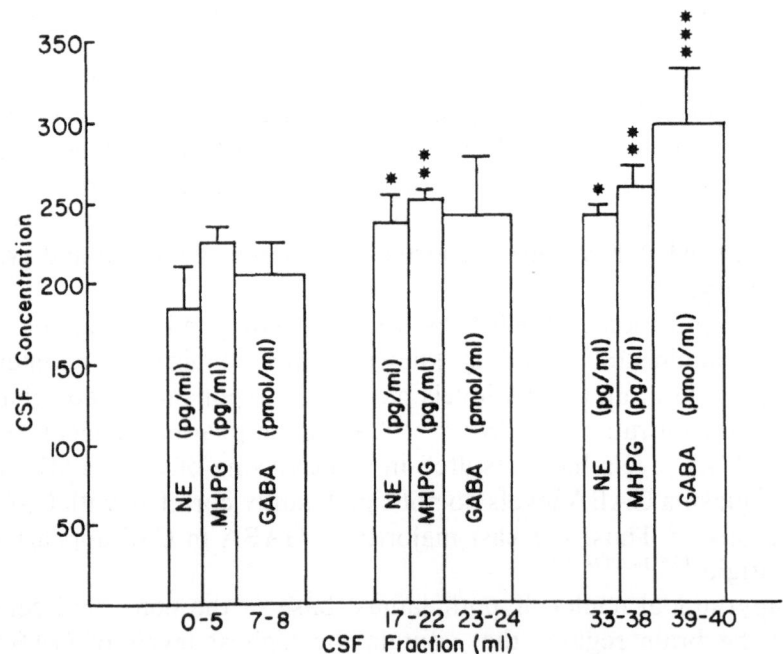

Fig. 2. Norepinephrine (NE), 3-methoxy-4-hydroxyphenylethylene glycol (MHPG), and γ-aminobutyric acid (GABA) concentrations (±SEM) in serial samples of lumbar CSF in man (data taken with permission from Ziegler et al.[184,185] and Enna et al.[165]). Mean incremental increases in *NE and **MHPG levels in 17th–22nd ml and 33rd–38th ml CSF fractions over those respective levels in the first 5-ml CSF aliquot are significant ($P < 0.002$, two-tailed paired Student's t-test). Mean incremental changes in NE and MHPG concentrations in 17th–22nd ml and 33rd–38th ml CSF fractions are not significantly different. ***Mean GABA level in 39th–40th ml CSF fraction is significantly higher ($P < 0.05$) than that in 7th–8th ml CSF aliquot. Reprinted from Wood[135] with permission.

data from homogeneous subgroups of normal individuals reveals that both the propensity of lumbar CSF GABA levels to decrease with age and the magnitude of the rostrocaudal GABA concentration gradient are more pronounced in females (Fig. 3). The significant differences in CSF GABA concentrations between females less than 40 years old and those older than 40 years suggest that this decrease in central GABAergic activity with respect to age may be influenced by hormonal alterations that are known to occur among females in midlife.

Recently GABA has been implicated in the modulation of central neuroendocrine function.[198,453] Although intraventricularly administered GABA promotes growth hormone (GH) release and inhibits secretion of thyrotropin (TSH) by an action on the hypothalamus, its physiological role in the control of these hormones is unknown.[199] Similarly, injection of GABA into ventricular CSF induces release of luteinizing hormone (LH) and its releasing factor (LH-RH), but the effect of GABA on prolactin secretion is dose dependent[200] and may be mediated at both hypothalamic and adenohypophyseal levels.[201] Concentrations of GABA in ventricular CSF not only alter pituitary hormone release but also the release of dopamine and norepinephrine from terminals in the median eminence. The released catecholamines may be important in mediating the effects of GABA on releasing-factor secretion.[202]

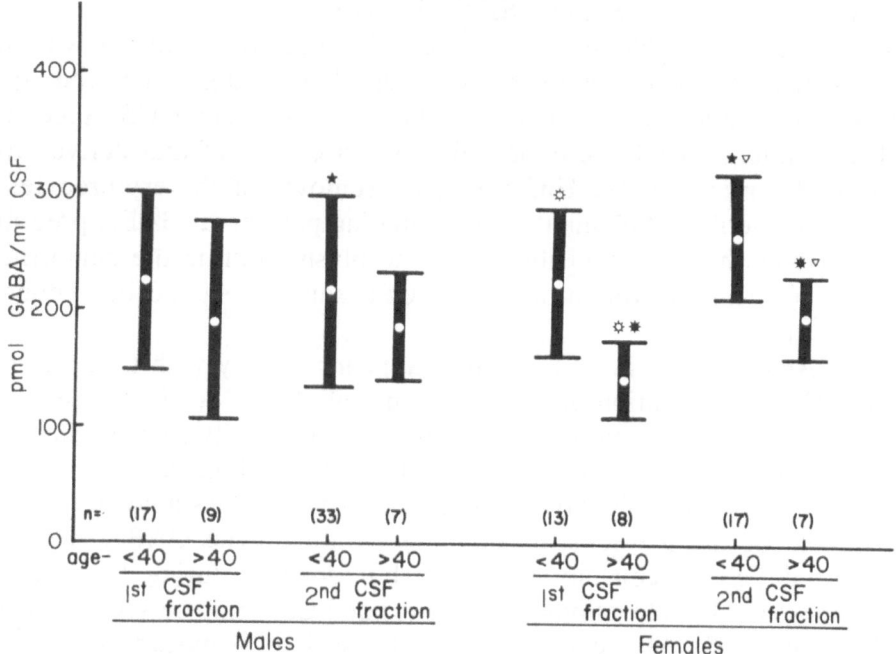

Fig. 3. Mean γ-aminobutyric acid (GABA) concentrations (±SD) in 111 lumbar CSF specimens from population of 87 normal individuals divided in homogenous subgroups with respect to sex, CSF fraction, and age (data taken with permission from Hare *et al.*[164,196]). Significant differences (two-tailed Student's *t*-test) in GABA content were noted (a) between males and females younger than 40 years of age ($P < 0.04$, ★, second CSF fraction), (b) between CSF specimens of females younger than 40 years and those 40 years of age or older ($P < 0.005$, ✿, first CSF fraction; $P < 0.01$, ▽, second CSF fraction), and (c) between first and second CSF aliquots in females 40 years old or older ($P < 0.01$, ✱). Reprinted from Hare[196] with permission.

When administered via ventricular CSF, GABA exerts inhibitory control over the sympathetic activity influencing arterial pressure and heart rate.[203]

3.1.2a. Investigative Considerations. Valid analysis of the GABA content in CSF requires sufficiently sensitive and specific assay systems[163,165,166] in addition to careful protocol formulation to minimize variations resulting from CSF sample procurement, storage, or analysis.[142,143,158,197,204,205] Dietary or pharmacological alterations in catecholamine or indoleamine metabolism should be avoided since CSF GABA concentrations have been reported to correlate positively with CSF levels of norepinephrine and its major central metabolite 3-methoxy-4-hydroxyphenylethylene glycol (MHPG)[206] as well as the serotonin metabolite 5-hydroxyindoleacetic acid (5-HIAA).[98] Obviously, the intake of precursors[167] and glutamic acid decarboxylase or GABA transaminase inhibitors[174-176,190] should be avoided. Anticonvulsant medications variably affect brain GABA content.[207]

Experimental protocols should restrict GABA data comparisons to similar CSF fractions which are collected at the same time of day from age- and sex-matched populations so as to avoid variations secondary to CSF concentration gradients,[135,177] circadian rhythms,[144,195] age, and sex.[158,164,196] Patients being considered for inclusion in clinical studies should be carefully screened for convulsive,[158,187-189] degenerative,[98,158,186,193,194] or demyelinating[98,158,186] disorders which are known to alter CSF GABA content.

Xanthochromic or blood CSF is often obtained from patients with subarachnoid hemorrhage or hemorrhagic cerebral infarctions and from those undergoing traumatic lumbar punctures. The GABA content in CSF specimens from these patients would be expected to be in excess of that derived from central GABAergic activity. Unfortunately, removal of the erythrocytes by centrifugation would not eliminate contaminating peripheral GABA present in the supernatant that was contributed by the plasma during the hemorrhagic event.[158] Thus xanthochromic or bloody CSF samples should be eliminated from most studies.

The GABA content in CSF samples appears to vary with time and temperature. The concentration of GABA in untreated CSF tends to rise significantly as the result of an enzymatic reaction[197] but usually remains stable for 10 min at room temperature or 2 h at 4°C and as long as 40 months at $-70°C$.[158,204,205] Levels of GABA in deproteinized CSF are stable for at least 49 h but double during 3 weeks at room temperature. The GABA content in deproteinized CSF is constant at $-70°c$ but doubles during 11 months at $-20°C$. These GABA elevations in deproteinized CSF may be secondary to *in vitro* degradation of GABA-containing peptides such as homocarnosine[158,204,205] or γ-aminobutyrylcystathionine.[208] Thus, exposure of CSF GABA samples to conditions that permit these artifactual GABA variations should be avoided.

3.1.2b. Homocarnosine and Other GABA-Containing Peptides. The concentration of GABA-containing peptides is 10- to 20-fold higher than that of free GABA in CSF.[49,174] Pharmacological elevations in brain GABA levels in rats are linearly correlated with CSF concentrations of total conjugated GABA and homocarnosine.[174,176] Although several GABA-containing peptides have been identified in brain, only γ-aminobutyrylcystathionine[208] and homocarnosine (γ-aminobutyryl-L-histidine)[49,176] have been detected in CSF.

Although homocarnosine is biosynthetically derived from GABA, no consistent relationship has been shown between GABA content and homocarnosine content in different brain regions.[209] Thus, homocarnosine may be formed from a pool of GABA different from that which serves for inhibitory neurotransmission.[209] Homocarnosine is a major constituent of the pool of GABA-containing peptides in CSF,[176] accounting for about 70% and 27% of the GABA conjugates in the CSF of rats[176] and man,[210] respectively. Homocarnosine levels of 1.69 ± 0.18 nmol/ml have been reported in lumbar CSF obtained from relatively healthy individuals,[210] and appear to be inversely related to age.[49]

3.1.2c. γ-Hydroxybutyric Acid. A metabolite of GABA, γ-hydroxybutyric acid (GHB) has a regional distribution in brain similar to that of GABA with highest quantities in the medulla, basal ganglia, caudate nucleus, hippocampus, cingulate cortex, and colliculi.[211] The concentration ratio of endogenous GHB to GABA in brain is 0.01–0.001, and reported relationships between blood and CSF levels of exogenous GHB suggest some peripheral contamination of CSF following high-dose GHB administration.[212] Pharmacokinetic data obtained following the intravenous administration of GHB suggest the passive diffusion of GHB into CSF, but its rapid entry into brain raises the possibility of active uptake or rapid binding to brain proteins.[213] γ-Hydroxybutyric acid appears to be an active metabolite whose effects on seizure activity,[212,214] behavior,[212] and dopaminergic neurons may be mediated by opioid peptides.[215]

The concentration of GHB in ventricular CSF has been preliminarily reported to be higher than that in lumbar CSF; however, the same study was not able to document the presence of a rostrocaudal concentration gradient in sequential lumbar CSF samples.[214] Although the level of GHB in CSF tends to rise *in vitro* over time if samples are not properly stored, the relative stability of GHB in CSF[208] compared to that of GABA[158,204,205] suggest that GHB in CSF is not derived from CSF GABA metabolism.[214] The CSF concentration of GHB appears to decrease with age.[214] No diurnal variations of GHB in CSF have been reported; daily variations in the same patient amount to 10% in both ventricular and lumbar CSF.[214]

γ-Hydroxybutyric acid has been measured in CSF but not in serum[214] by a GC/MS assay that has a sensitivity of 0.02 nmol/ml.[216]

3.1.3. Norepinephrine

Norepinephrine is the major adrenergic neurotransmitter within the central nervous system. Attempts to elevate endogenous brain norepinephrine by systemic administration have not been successful.

Increased oral intake of tyrosine, the dietary precursor of norepinephrine, appears to augment neuronal catecholamine release[217] or synthesis[218] when amino acids competing for transport are reduced or central decarboxylation is inhibited.[217] However, morphological and enzymatic barrier mechanisms normally impede such perturbations.[332] Intravenously injected norepinephrine is unable to cross the blood–brain barrier.[219,220] Although a significant relationship exists between plasma and CSF concentrations of norepinephrine, acute norepinephrine elevations in the plasma are not reflected in the CSF.[221] Intravenous infusions of labeled norepinephrine in monkeys evoke peak CSF levels of labeled norepinephrine that are equal to only 2% of the peak plasma concentrations.[221] Thus, the correlation between CSF and plasma norepinephrine cannot be explained by penetration of blood norepinephrine into CSF.[184,221] Peripheral sympathetic activity may account for this plasma–CSF correlation by making a minor contribution to the CSF norepinephrine content via innervation to the major subarachnoid blood vessels at the base of the brain (circle of Willis).[222] Overall, the majority of norepinephrine in the CSF is central in origin.[135,142]

The concentration of norepinephrine in the plasma in normotensive healthy volunteers is 304 ± 20 pg/ml,[223] and that in instantly frozen (-56 to $-70°C$) 4-ml aliquots of lumbar CSF taken after 12 ml of lumbar CSF drainage is 373 ± 38 pg/ml.[140] Thus, the plasma-to-lumbar CSF ratio for norepinephrine is approximately 0.8.

The central noradrenergic system in man includes the locus coeruleus and the lateral tegmental nucleus.[224] The locus coeruleus lies in the floor of the aqueduct and rostral fourth ventricle. This nucleus receives both β-endorphin and enkephalin peptidergic afferents,[225] which may suppress[226] its usually inhibitory output to the mesencephalic tegmentum, basal telencephalic areas representing the limbic system, cerebral cortex, cerebellum, and lateral column of the spinal cord.[224] The noradrenergic locus coeruleus participates in the regulation of levels of consciousness and arousal, behavior patterns such as reinforcement and learning, cerebral blood flow, and permeability of cerebral microvasculature. The lateral tegmental system innervates the medial hypothalamus and acts to augment thyrotropin-releasing hormone (TRH), gonadotropic hormones, LH[227] and possibly prolactin, and GH secretion[228] as well as probably inhibiting adrenocorticotropic (ACTH) and antidiuretic hormone (ADH, AVP) release.[227]

A concentration gradient for norepinephrine has been noted in man during continuous drainage of lumbar CSF such that the norepinephrine content in samples taken after 17 ml of CSF drainage is 30% higher than that in the first 5-ml sample[185] (Fig. 2). Additional lumbar CSF drainage was not associated with further increase of CSF norepinephrine levels.[185] The lower norepinephrine content in the initial CSF fractions was attributed to the minimal noradrenergic contribution of the cauda equina, whereas the higher concentrations noted in more rostral fractions represented the activity of noradrenergic tracts descending from brainstem nuclei such as the locus coeruleus.[184,185,229]

Norepinephrine synthesis[230] and its content in brain[231] are highest during the time of day when animals are least active, whereas the norepinephrine content in regions containing predominantly noradrenergic terminals is highest when the animals are most active. Accordingly, the midbrain,[233] hypothalamus,[281] and spinal cord[219] all exhibit significant circadian variations in their norepinephrine content. The concentrations of norepinephrine in ventricular CSF of monkeys[144,234,235] and in lumbar CSF in man[184,235] are maximal during light midafternoon hours and lowest in dark early morning hours. Norepinephrine infusion into the cerebral ventricles increases motor activity; thus, the diurnal rhythm of norepinephrine in CSF appears to parallel the functional effects of norepinephrine on activity.[184]

Although no definite correlation between sex and CSF norepinephrine levels has been observed, older patients have higher CSF norepinephrine concentrations.[184] Accordingly, plasma norepinephrine increases with age but does not depend on sex.[236]

3.1.3a. Investigational Considerations. Patients to be included in studies of noradrenergic metabolism involving CSF analysis should be carefully screened for concomitant illnesses that alter central noradrenergic activity.[142,184,193,194,237] Medications should be avoided, especially psychotropic, antihypertensive, and decongestant agents as well as bromocryptine, caffeine, probenecid, and drugs that alter brain tyrosine levels.[184,238] Obviously, patient populations should be age and sex matched.[140,142] Patient stress and physical activity should be minimized during patient preparation and CSF sampling periods.[140,142,184,239]

Cerebrospinal fluid samples should be obtained at the same time of day for all subjects in the study, and care should be taken to compare only similar CSF fractions,[142,143] thereby avoiding variations caused by circadian rhythms[234,235] and CSF concentration gradients,[135,142,184,185] respectively. Sampling techniques should be standardized, since lower CSF norepinephrine levels have been obtained when CSF is collected on ice (4°C)[139,141] than when it is frozen immediately on dry ice (-56°C).[140] Obviously, blood-contaminated CSF specimens should be discarded.

Sensitive radioenzymatic assays for norepinephrine have enabled its valid determination, and these have been verified by GC/MS.[240] These assay systems require the addition of preservatives such as reduced glutathione or dithiothreitol which are reducing agents. Although ascorbic acid would appear to be a better preservative by being both an acidifier and antioxidant, ascorbic acid is not suitable because it also serves as a weak substrate for the enzyme catechol-*O*-methyltransferase.[140]

3.1.3b. 3-Methoxy-4-hydroxyphenylethylene Glycol. The primary mechanism for termination of the action of norepinephrine is synaptic reuptake. Catabolism of central norepinephrine proceeds via the monoamine oxidase system to form the major metabolite, 3-methoxy-4-hydroxyphenylethylene glycol (MHPG), and lesser quantities of vanillylmandelic acid (VMA).[241] Accordingly, dietary manipulations that elevate the brain content of tyrosine, a precursor of norepinephrine, also raise brain MHPG levels.[217]

Since relatively little of intravenously injected labeled MHPG enters into lumbar CSF, the MHPG in CSF arises largely from central rather than peripheral noradrenergic metabolism.[242,243] Demonstrations of CSF MHPG elevations following noradrenergic neuronal stimulation[244] and a significant correlation between lumbar CSF norepinephrine and MHPG in man[185] suggest that CSF MHPG reflects central noradrenergic activity.[135,142]

In man, norepinephrine and MHPG are present in highest concentrations in structures adjacent to the third and fourth ventricles,[245] and, in monkeys, CSF MHPG levels are highest in the third and fourth ventricles.[243] A considerable portion of CSF MHPG is dependent on the activity of noradrenergic locus coeruleus neurons.[244] Similar MHPG levels have been noted in CSF obtained from the lateral ventricles and lumbar sac in man,[242,246] and early reports have denied the presence of a craniocaudal MHPG concentration gradient in draining lumbar CSF.[247] Patients with and without spinal canal blockage appear to have similar lumbar CSF MHPG levels[248]; thus, the spinal cord appears to contribute to lumbar CSF MHPG levels in man.[135,142] The lower-than-normal CSF MHPG concentration in patients with spinal cord transections[248] implies that the spinal source of MHPG in lumbar CSF may be the descending noradrenergic tracts. More recent investigations have demonstrated a 12% increase in MHPG up to the 17th milliliter of CSF removed during lumbar puncture[184,185] (Fig. 2), suggesting that such a small spinal MHPG gradient may have been obliterated in early studies by the slower MHPG turnover and lack of control of physical activity.[135] Recently, the concentration of MHPG in pools of the initial 12 ml of lumbar CSF in 31 normal individuals has been reported to be approximately 8.1 ± 0.2 ng/ml[141] or about 40 times as great as that of its parent compound, norepinephrine, in the 20th milliliter of lumbar CSF.[184,185] Somewhat higher values have also been suggested.[249]

The activity of catechol-*O*-methyltransferase and monoamine oxidase, the enzymes associated with the major degradative pathways of norepinephrine, fluctuate in a daily rhythm.[250] However, the levels of MHPG recovered in ventricular CSF do not appear to bear any temporal relationship to actual tissue production rates,[234] suggesting that fluctuations in MHPG formation may be buffered by the large exchangeable pool of MHPG already present in the intracellular and extracellular spaces.[251,252]

3.1.3c. Investigational Considerations. The MHPG content in CSF is usually measured by GC/MS methods[253–255]; however, recently, high-pressure liquid chromatographic procedures[256] have been advocated to simplify the methodology. Most methods require that preservative, usually ascorbic acid (2 mg/ml), be added to the CSF at the time of sampling prior to ultracold storage.

Although CSF MHPG concentrations correlate with CSF norepinephrine levels,[184,185] the blunted rostrocaudal concentration gradient and slower turnover[184,185] as well as the absence of expected circadian rhythms[234] and age correlations[249,257] of MHPG in CSF detract from its investigational value as an accurate reflection of steady-state central noradrenergic metabolism.[142,143] The combination of CSF norepinephrine and MHPG determinations may, however, yield some insight into CSF norepinephrine turnover. In contrast, urinary

MHPG is not considered a useful alternative for the study of centrally produced MHPG.[258,259]

3.1.3d. Vanillylmandelic Acid. The concentration of 4-hydroxy-3-methoxymandelic acid (vanillylmandelic acid, VMA) in lateral ventricular CSF in monkeys has been reported to be threefold higher than that in lumbar CSF, and the highest levels have been observed in third and fourth ventricular CSF.[243] Vanillylmandelic acid represents 8% of the total noradrenergic metabolites in lumbar CSF; thus, by extrapolation, VMA may represent as much as 20% of the major noradrenergic metabolites in lateral ventricular CSF.[260] However, the presence of this reported rostrocaudal gradient for CSF VMA is not universally accepted.[247] A reproducible diurnal pattern in VMA content has been observed in the ventricular CSF of monkeys.[234]

3.1.4. Dopamine

Dopamine-containing neurons in rodents connect the pars compacta of the substantia nigra to the striatum, and the nucleus interpeduncularis to the nucleus accumbens, olfactory tuberculum, and amygdala. Dopaminergic neurons also project from the arcuate and periventricular nuclei of the hypothalamus to the median eminence.[261] The highest levels of dopamine in man are located in the nucleus accumbens and caudate nucleus, but plentiful quantities reside in the olfactory area, hypothalamus, and substantia nigra.[245] In contrast, dopamine is almost absent in the spinal cord.[262]

The nigrostriatal pathway primarily modulates motor function, and the dopaminergic infundibular system inhibits the release of prolactin, melanocyte-stimulating hormone (MSH), and possibly TSH but stimulates GH release.[227] The exact function of the mesolimbic system is not known.

Dopamine is generally regarded as an inhibitory neurotransmitter and does not readily cross the blood–brain[263,264] or blood–CSF[265] barrier. Current methods of quantification of free or conjugated dopamine in CSF lack the sensitivity to measure this neurotransmitter without precursor stimulation in lumbar CSF[266,267] but have detected acid-hydrolazable conjugates of dopamine in ventricular CSF.[266]

3.1.4a. Tyrosine. Tyrosine is the dietary precursor that undergoes two hydroxylations in the synthesis of dopamine and norepinephrine. High degrees of correlation between brain tyrosine levels and brain catechol synthesis have been observed following the administration of decarboxylase inhibitors.[217,268] However, in the untreated organism, the feedback changes in tyrosine hydroxylase activity may minimize the effect of naturally occurring alterations in brain tyrosine concentration on central catecholamine synthesis.[269] Consequently, no overall relationship between the concentrations of tyrosine and dopamine metabolites in ventricular CSF exists.[270]

3.1.4b. L-3,4-Dihydroxyphenylalanine. Parentally administered L-3,4-dihydroxyphenylalanine (L-DOPA), an intermediary precursor of dopamine, crosses the blood–brain barrier.[263,271] Labeled L-DOPA and dopamine in brain increase to maximal levels within 30 min following intravenous injection of labeled L-DOPA.[271] The dopamine formed from the exogenously administered L-DOPA becomes part of the functional endogenous dopamine pool in the brain.

3.1.4c. Homovanillic Acid. The major metabolite of central dopamine is 3-methoxy-4-hydroxyphenylacetic acid (homovanillic acid, HVA). Rapid elevations in both labeled brain and CSF HVA are evoked by the intravenous administration of labeled L-DOPA.[271,272] Systemically injected dopamine or HVA usually does not elevate CSF HVA in dogs[273] or cats[274] unless large intravenous HVA doses are administered.[275] The induced release of HVA into lateral ventricular CSF during substantia nigral stimulation[276] and the correlation of HVA concentrations in the caudate nucleus with CSF HVA levels[273] provide additional evidence that the origin of CSF HVA is the brain parenchyma. However, brain capillary walls may also contribute to the HVA content of CSF.[277]

The ratio of HVA in the caudate nucleus to that in lateral ventricular CSF is 7.9 in normal dogs.[273] The amount of HVA entering the CSF represents only approximately 30% of the dopamine turnover of the caudate nucleus and may originate largely in the part adjacent to the ventricular surface.[278] Labeled HVA reaches a maximum level at 2 to 4 h in the cisternal CSF of patients given radioactive L-DOPA but does not attain its maximal concentration in lumbar CSF until 8 h.[272] The rostrocaudal CSF circulation time may contribute to part of this lag in the HVA accumulation in lumbar CSF.[279] The concentration of HVA in pools of the initial 12 ml of lumbar CSF in 42 normal volunteers is approximately 35.1 ± 2.3 ng/ml,[141] although higher values have been reported.[249]

The HVA concentrations in man are highest in the substantia nigra, caudate nucleus, olfactory region, and especially the nucleus accumbens.[245] Marked rostrocaudal HVA concentration gradients have been observed in dogs,[134,265] monkey,[243] and human[134,243,247,280,282] CSF (Fig. 4). The highest HVA concentrations are found in lateral ventricular CSF. Lumbar CSF HVA levels are greatly reduced or undetectable below the site of spinal canal blockage[248,282–284] and are low in patients whose lateral ventricles have become isolated by obliteration of the foramina of Monro.[283] Although most of the HVA in CSF originates from the brain surrounding the lateral ventricles, the HVA content in lumbar CSF reflects only a portion of the total central dopamine metabolism.[285]

Nocturnal increases in brain dopamine levels have been observed in night-active rodents.[286,287] Similarly, in day-active primates, the concentration of HVA in lateral ventricular CSF fluctuates in a circadian fashion, with maximum turnover occurring during the light hours and minimum turnover occurring during the dark hours.[144,288]

The activity of tyrosine hydroxylase and DOPA decarboxylase, enzymes that synthesize catecholamines, declines with advancing age in adults.[289] Significant inverse correlations have been noted between age and CSF levels of

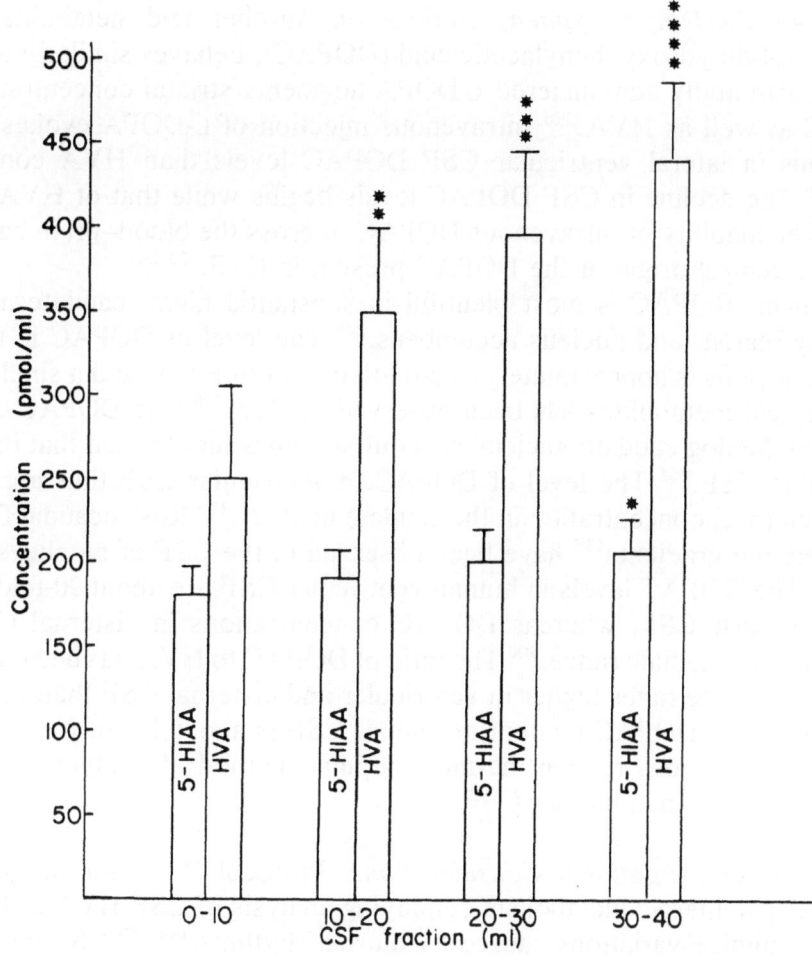

Fig. 4. 5-Hydroxyindoleacetic acid (5-HIAA) and homovanillic acid (HVA) concentrations (± SEM) in serial samples of lumbar CSF in man (data taken with permission from Sjöstrom *et al.*[247]). *Mean level of 5-HIAA in the 30th–40th ml CSF fraction is significantly higher ($P < 0.05$, Wilcoxon matched-pairs rank test) than that in first 10-ml CSF aliquot. Mean levels of HVA in **10th–20th ml, ***20th–30th ml, and ****30th–40th ml CSF fractions were significantly higher ($P < 0.05$, 0.01, and 0.001, respectively; two-tailed paired Student's *t*-test) than that in the first 10-ml CSF sample. Reprinted from Wood[135] with permission.

tetrahydrobiopterin, a cofactor involved in the hydroxylation of tyrosine.[232] In addition, the activity of the catabolic enzyme monoamine oxidase increases with age.[290] Accordingly, the brain content of dopamine[291,292] and HVA decreases with increasing age.[291] Although preliminary reports[293,294] suggested that CSF biogenic amine metabolite levels are increased in older patients, some recent studies have documented a strong negative correlation between HVA concentrations[257,295] or probenecid-induced HVA accumulations[206] in CSF and age, whereas other reports have observed no CSF HVA correlation with age.[249]

The presence of sex variations in CSF HVA has not been universally accepted[294,295] despite the demonstrations of higher probenecid-evoked accumulations of CSF HVA in women.[206]

3.1.4d. 3,4-Dihydroxyphenylacetic Acid. Another acid metabolite of dopamine, 3,4-dihydroxyphenylacetic acid (DOPAC), behaves similarly to HVA in CSF. Parentally administered L-DOPA augments striatal concentrations of DOPAC as well as HVA.[296] Intravenous injection of L-DOPA evokes earlier elevations in lateral ventricular CSF DOPAC levels than HVA concentrations.[265] The decline in CSF DOPAC levels begins while that of HVA is still rising. The inability of intravenous DOPAC to cross the blood–brain barrier[297] implies a central origin of the DOPAC present in CSF.[135,265]

In man, DOPAC is most plentiful in substantia nigra, caudate nucleus, olfactory region, and nucleus accumbens.[245] The level of DOPAC in the dog caudate nucleus is approximately one-tenth of that in HVA, and a similar ratio of these acid metabolites has been observed in CSF.[298] The DOPAC concentration in the dog caudate nucleus is about six times higher than that in lateral ventricular CSF.[265] The level of DOPAC in ventricular CSF thus appears to be related to its concentration in the caudate nucleus.[135] Rostrocaudal DOPAC concentration gradients[135] have been observed in the CSF of monkeys[299] and man.[300] The DOPAC levels in human ventricular CSF are about 20-fold higher than in lumbar CSF, whereas DOPAC concentrations in cisternal CSF are within an intermediate range.[300] The ratio of DOPAC to HVA has been reported to be about three times higher in ventricular and cisternal CSF than in lumbar CSF. Since the DOPAC content of lumbar CSF is about 1% of that of HVA, HVA determinations remain the most popular method of studying dopamine metabolism in human lumbar CSF.

3.1.4e. Investigational Considerations. Protocols[142,143] for the study of central dopaminergic metabolism employing analysis of CSF HVA or DOPAC should minimize variations caused by diurnal rhythms,[144,288] CSF concentration gradients,[135,142,243,247,280] age,[257,295] and possibly sex.[206] Cervical spondylosis or spinal canal blockage may lower lumbar CSF dopamine metabolite concentrations by disturbing the flow of CSF through the spinal subarachnoid space.[135,142,248,282–284] In contrast, exercise[239,301] and failure to enforce recumbency[302] before CSF sampling may result in higher HVA levels. Stress-[143,303] or drug-induced[271,272,304–308] alterations in CSF HVA levels should also be avoided.

Cerebrospinal fluid sample preparation usually requires the addition of preservatives such as ascorbic acid (2 mg/ml) prior to ultracold storage. Several assay techniques are currently available for the quantification of dopamine metabolites in CSF.[253,254,256,296,300,309–316]

3.1.5. Serotonin

Serotonergic neurons located in the dorsal and median raphe nuclei of the pons and mesencephalon send rostral projections to essentially all regions of the telencephalon and diencephalon. The serotonergic tracts that descend in the spinal cord originate in the medulla oblongata.[317] In man, serotonin (5-hydroxytryptamine) content is highest in the midbrain, with similarly high levels present in the substantia nigra and the pons. Serotonin concentrations in the hypothalamus, caudate nucleus, thalamus, and amygdala are approximately one-third as high as those in the pons.[245] Stimulation of serotonin-containing neurons may induce either excitation or depression depending on the particular region of origin.[317] Iontophoretically applied serotonin evokes excitatory responses more commonly in areas of the brain and spinal cord in which the density of serotonergic terminals is low or scattered.[318] Serotonergic activity has been associated with elevations in body temperature,[319] sleep induction and maintenance,[320] decreased pain sensitivity,[321–323] vasoconstriction,[324] and increased vascular permeability.[325] Serotonin has also been implicated in the suppression of gonodotropin[326] and possibly ADH output[227] as well as augmentation of prolactin[327,328] and possibly GH[328] and adrenocorticotropin-releasing factor (ACTH-RH) release.[227] Serotonergic neurons projecting from the raphe nuclei to intraparenchymal blood vessels may modulate the brain microcirculation.[329]

The blood–brain barrier is relatively impermeable to serotonin[264,330] however, at high intravenous dosages,[331] serotonin crosses both the structural and enzymatic barrier mechanisms,[332] thereby possibly accounting for the positive cerebral arteriovenous difference for this indoleamine.[333] Once in brain, exogenous serotonin has a half-life of 5 to 10 min, being more readily metabolized than eliminated.[331] Diurnal rhythms have been demonstrated for brain serotonin.[287]

Unfortunately, evaluation of serotonin pharmacokinetics in normal CSF has been impeded by the lack of suitable assays for CSF serotonin. Most studies of serotonin metabolism in CSF have relied on data obtained by CSF precursor or metabolite determinations,[142] although a few have equated total CSF 5-hydroxyindole content following monoamine oxidase inhibitors with serotonin.[334]

3.1.5a. L-Tryptophan. L-Tryptophan is the essential dietary precursor of serotonin. Daily rhythms in brain tryptophan content appear to be altered by both lighting and feeding schedules.[286] Tryptophan in plasma may compete with other neutral amino acids for transport into the brain.[269,286,335,336] Tryptophan availability during peripartum development determines subsequent brain tryptophan metabolism.[337] Tryptophan hydroxylase, the rate-limiting enzyme in the synthesis of serotonin, is not saturated at physiological levels of tryptophan.[338] Systemic administration of tryptophan increases the serotonin content of the brain[269,335,339,340] and its release into ventricular perfusates.[341]

Tryptophan crosses both the blood–brain and blood–CSF barriers.[342,343] As with brain, the tryptophan concentration in CSF varies with the dietary content of tryptophan.[344] A concentration gradient for CSF tryptophan has been reported to exist between the ventricles and spinal subarachnoid spaces[343]; however, tryptophan levels are similar in cisternal and lumbar CSF among patients with[284] or without[345] blocked spinal canals. In addition, CSF tryptophan levels are unaltered by the spinal CSF mixing that occurs with subarachnoid air injections during pneumoencephalography.[343,345] Thus, tryptophan in lumbar CSF may be derived almost totally from spinal sources,[135] whereas that in cisternal CSF reflects brain tryptophan content.[344,346]

Diurnal rhythms for tryptophan occur in blood[347,348] and brain.[335,349,350] Accordingly, both brain and CSF tryptophan are higher during the night in nocturnally active rodents.[346,350] Brain serotonin[351] and its metabolism[206] appear to be higher in females than males. The tryptophan content in CSF is higher in women than men,[345] thus suggesting greater precursor availability in women. No significant relationship occurs between CSF tryptophan levels and age.[345]

Tryptophan is also a precursor of tryptamine, a trace amine, which is catabolized to indoleacetic acid (IAA). The content of IAA in CSF reflects the central metabolism of tryptamine.[350,352] Concentrations of IAA are significantly higher in cisternal than lumbar CSF[345] and may be higher in lumbar CSF obtained from females than in that of males.[345]

3.1.5b. L-5-Hydroxytryptophan. L-5-Hydroxytryptophan (5-HTP), the precursor of serotonin, appears to cross the blood–brain barrier and influence brain serotonin metabolism.[353–356] Systemically injected 5-HTP augments the release of serotonin into ventricular perfusates.[341] Oral administration of 5-HTP[357] has been shown to elevate CSF serotonin metabolite concentrations in man.[354–356] Most of the brain aromatic L-amino acid decarboxylase, the enzyme that converts 5-HTP to serotonin, resides outside serotonergic neurons[358]; therefore, much of the serotonin produced from exogenously administered 5-HTP is synthesized extraneuronally, possibly in glial cells.[359,360]

3.1.5c. 5-Hydroxyindoleacetic Acid. The regional distribution of 5-hydroxyindoleacetic acid (5-HIAA) in the human brain parallels that of its parent, serotonin.[245] The crossing of the blood–brain[134,360,361] or the blood-CSF[362] barrier by systemically administered 5-HIAA cannot be demonstrated unless the free 5-HIAA in the blood is elevated and the active transport of 5-HIAA from the CSF to blood is inhibited by probenecid.[363]

Caudate nucleus,[273] brainstem,[364] and spinal cord[342] concentrations of 5-HIAA correlate with 5-HIAA levels in lateral ventricular, cisternal, and lumbar CSF, respectively. The ratio of 5-HIAA in the caudate nucleus to that in lateral ventricular CSF is 1.4 in normal dogs.[273] Approximately 10% of brain 5-HIAA is eliminated via the CSF pathway.[365] Although only about 22% of 5-HIAA formed in the spinal cord diffuses into lumbar CSF,[366] small changes in the spinal cord 5-HIAA content are more readily detected in lumbar CSF than in the spinal cord tissue itself.[366]

Tryptophan-evoked elevations in brain 5-HIAA levels[339,342,364] are reflected in corresponding alterations in CSF 5-HIAA concentrations.[342,347,364,367] The accumulation of tryptophan in lumbar CSF begins as early as 2 h after tryptophan administration; however, the accumulation of 5-HIAA occurs at 6 h.[367] Cisternal CSF elevations in tryptophan and 5-HIAA evoked in rats by tryptophan loading occur at 1 and 2 h, respectively.[350] No significant relationship between tryptophan and 5-HIAA has been demonstrated in the CSF of untreated patients[343]; however, significant correlations between these two substances have been noted in both the brains and CSF of tryptophan-treated rats.[344]

Rostrocaudal concentration gradients of CSF 5-HIAA exist in dogs[134,265] and man[134,247,280,343,368]; the ventricular/lumbar CSF 5-HIAA ratio in man has been reported to be approximately 5.[134] This 5-HIAA gradient in human lumbar CSF is less pronounced than that of lumbar HVA, suggesting possible 5-HIAA contributions from the spinal cord[247] (Fig. 4). Patients with apparent blockage of CSF flow in the spinal subarachnoid space have lumbar 5-HIAA levels similar to those in patients without evidence of obstructed spinal CSF flow.[248] In addition, spinal cord transection does not appear to alter lumbar CSF 5-HIAA concentrations.[248] The degree of contribution of the spinal cord to the 5-HIAA content of lumbar CSF is controversial.[135,142] Several reports suggest that approximately 70% of the lumbar CSF 5-HIAA has its origin rostral to the foramen magnum;[369,370] however, others maintain that almost all of the 5-HIAA in lumbar CSF arises from spinal serotonin metabolism.[366,371,372]

The concentrations of 5-HIAA in pools of the initial 12 ml of lumbar CSF in 42 normal volunteers has been reported to be 17.1 $\pm$ 1.0 ng/ml,[141] although somewhat higher levels have also been reported.[249] These 5-HIAA concentrations have been shown to exhibit circadian rhythms.[373]

The higher concentrations of 5-HIAA in rat CSF during the night[350] are consistent with the higher nocturnal levels of 5-HIAA found in rat brain.[349] The reported decrease in CSF 5-HIAA levels with advancing age[257,374] has not been universally confirmed.[206,249,293–295,374]

Women have been found to have significantly higher brain 5-HIAA levels than men,[375] a finding consistent with higher brain monoamine oxidase among women.[290] Accordingly, women have been reported to have significantly higher probenecid-induced accumulations of 5-HIAA, and postmenopausal amenorrheic women may have higher CSF 5-HIAA levels than males or menstruating females.[206] The presence of these sex variations in CSF 5-HIAA levels remains controversial.[294,295]

3.1.5d. Investigational Considerations. Patients undergoing preparation for CSF studies of serotonin metabolism should be screened for complicating disease[376] or dietary disorders[337,377] and should avoid abnormal precursor or neutral amino acid intake,[335,336,339,340] antipsychotic agents,[206] sedatives,[378] ethanol,[379] catabolic enzyme or reuptake inhibitors,[323] GABA transaminase inhibitors,[98,307,380] and, when appropriate, probenecid[305,363,366,381,382] and anticonvulsant drugs.[187,345,383,384] Exercise restrictions,[239,301] enforcement of recumbency,[302] and avoidance of sleep deprivation[385] or stress[270,303,385] help to minimize CSF variations.[143] Care should be taken to confine CSF sampling to the same time of day and to use similar CSF fractions as well as to use age- and sex-matched populations.[135,142,143]

Preservatives such as ascorbic acid (2 mg/ml) should be mixed with the CSF sample prior to ultracold storage.[143] Several assay systems are available for the determination of serotonin precursor and metabolite concentrations of CSF.[309,310,315,386–389a]

Since close correlations have been demonstrated between the 5-HIAA concentrations in ventricular[273] or cisternal CSF[364] and brain 5-HIAA content as well as between lumbar CSF 5-HIAA concentrations and spinal cord 5-

HIAA content,[366,372] but not vice versa,[366] lumbar CSF 5-HIAA may not adequately reflect brain serotonin metabolism, especially if the spinal cord has been compromised by disease or injury.[368]

3.1.6. Cyclic Nucleotides

The central nervous system contains large amounts of cyclic nucleotides and the enzymes for their synthesis and destruction.[391-395] The cyclic nucleotides, adenosine 3',5'-monophosphate (cyclic AMP) and guanosine 3',5'-monophosphate (cyclic GMP), are formed from adenosine triphosphate (ATP) and guanosine triphosphate (GTP) by the action of the membrane-bound enzymes adenylate and guanylate cyclase, respectively. Generally, adenylate cyclase is present in high amounts in the cortical gray matter of the cerebrum and cerebellum and the subcortical gray matter structures.[396] More specifically, dopamine-responsive adenylate cyclase activity is found primarily in the caudate nucleus and peripheral sympathetic ganglia,[391] whereas norepinephrine-responsive adenylate cyclase is widespread, yet highest in cerebellum.[397] Guanylate cyclase is present in high amounts in gray matter throughout the central nervous system.[398]

The tissue distribution of cyclic AMP is relatively uniform among major brain regions,[399-401] whereas that of cyclic GMP is uneven, with concentrations being much higher in the cerebellum than in the cerebrum, brainstem, or spinal cord.[393,401,402] Within the cerebrum and brainstem, the endogenous levels of cyclic AMP are 10 to 50 times that of cyclic GMP; however, the content of these cyclic nucleotides in cerebellar tissue is approximately equal.[395] Immunohistochemical investigations[403,404] have demonstrated that most of the brain cyclic AMP is localized in neuronal cell bodies of the gray matter. At present, the cellular or subcellular localization of cyclic GMP is controversial,[393] since early speculation that cyclic GMP is located in cerebellar Purkinje cells[405] has not been confirmed by recent histochemical studies.[406]

Intracellular cyclic AMP and cyclic GMP probably act as second messengers for the action of several neurotransmitters and hormones.[391,392,394,395,407] Cyclic nucleotides may also mediate metabolic functions[408] and postsynaptic electrochemical events.[409]

Cerebrospinal fluid bathes the brain and spinal cord and is in functional continuity with the brain extracellular fluid.[4,135] Since major mechanisms of termination of action of cyclic nucleotides include extrusion from the cells[410] and hydrolysis,[411] quantification of the cyclic nucleotides within the CSF may reflect alterations in intracellular cyclic nucleotide metabolism.[97]

3.1.6a. Cyclic Adenosine 3',5'-Monophosphate.

Intravenously injected cyclic AMP does not normally penetrate into the CSF of rabbits.[412] Early reports of a correlation between the baseline plasma and CSF cyclic AMP concentrations in man suggested that cyclic AMP might pass through the blood–brain barrier and be measurable in the CSF.[413] However, recent clinical studies demonstrate a blood–CSF barrier for cyclic AMP which prevents acute alterations in CSF cyclic AMP levels during systemic pharmacological manipulations that markedly raise plasma cyclic AMP concentrations.[97,414] Accordingly, pharmacologically induced perturbations of brain cyclic AMP[369] are appropriately reflected in alterations of cyclic AMP levels in rat ventricular[412,415] or cisternal[416] CSF. In addition, inhibitors of the central nucleotide 3',5'-phosphodiesterase located almost exclusively in postsynaptic dendritic process membranes[417] block the conversion of cyclic nucleotides to 5'-monophosphates and elevate CSF cyclic AMP in man.[97] Thus, the central nervous system appears to be the origin of the cyclic AMP in CSF.[97,135,142]

Recent studies employing standardized protocols, radioimmunoassay, and myopathy patients free of central nervous system disease, endocrinopathy, or exogenous medication have found the concentration of cyclic AMP in lumbar CSF to be 10.4 ± 0.5 pmol/ml or about one-half to two-thirds that in plasma.[97] Recently, cyclic AMP levels of 19.6 ± 4.6 pmol/ml have been determined by protein binding assay in the lumbar CSF of 41 normal individuals.[141]

Early studies suggesting differences between ventricular and lumbar CSF cyclic AMP concentrations[418,419] have not been confirmed by more recent investigations.[420,421] No rostrocaudal concentration gradient for cyclic AMP has been observed in continuously draining lumbar CSF[97,135,142] or following spinal subarachnoid air injections;[419,422] thus, the cyclic AMP content in lumbar CSF may not be totally dependent on the brain as the major source of this nucleotide but may also reflect spinal nucleotide metabolism.[135,419]

The cyclic AMP concentrations in ventricular CSF in monkeys exhibit diurnal rhythms, reaching a maximum during the light hours and a minimum during the dark hours.[144,423,424] This cyclic AMP rhythm is in phase with the daily pattern of turnover of norepinephrine,[234] dopamine,[288] and probably serotonin,[141,424] which are known to affect adenylate cyclase activity in neuronal tissues.[391,397]

Communications have indicated that old rats have higher base-line CSF cyclic AMP concentrations than younger rats[425] and that children have lower CSF cyclic AMP levels than adults.[426] Female rats have been reported to have higher cyclic AMP levels than male rats.[425] In man, the tendency for cyclic AMP levels in lumbar CSF to be higher in females than males has been reported to disappear after the age of 56, suggesting a hormonal basis for this sex difference.[427] Other communications have not noted significant relationships between CSF cyclic AMP concentrations and age, sex, pulse, or blood pressure.[97,428,429]

3.1.6b. Cyclic Guanosine 3',5'-Monophosphate. Large standardized evaluations of medication-free myopathy patients lacking central nervous system involvement or endocrinopathy have reported the concentration of cyclic GMP in lumbar CSF to be 2.4 ± 0.3 pmol/ml or about one-half to two-thirds that in plasma.[97] No rostrocaudal concentration gradient for cyclic GMP has been noted in serial sampling of lumbar CSF,[97] suggesting that lumbar CSF cyclic GMP levels may not accurately reflect intracranial cyclic nucleotide metabolism.[135,142] The concentration of cyclic GMP in lumbar CSF does not appear to correlate with age, sex, pulse rate, or blood pressure.[97]

3.1.6c. Investigational Considerations. Patients to be included in CSF studies of cyclic nucleotide metabolism should be free of both intracranial and spinal disorders as well as endocrinopathies or medications affecting catecholamine, adenylate or guanylate cyclase, or phosphodiesterase activity.[97,142,143,430,431] Consideration should be given to the reported correlations between cyclic AMP and cyclic GMP or GABA levels in lumbar CSF.[97] Concentrations of cyclic AMP in ventricular CSF correlate inversely with the grade of traumatic coma,[432] whereas that for cyclic GMP correlates positively with intracranial pressure.[421,432]

Collection of CSF for cyclic nucleotides should be performed at the same time of day to avoid variations secondary to circadian rhythms.[144,423,424] These CSF samples do not require preservatives but are best collected in unsiliconized borosilicate vials which can be frozen immediately and stored at ultracold temperatures.[97] Freezing and thawing the CSF sample twice significantly lowers its cyclic GMP but not its cyclic AMP content;[97] thus, the specimen should be aliquoted prior to the initial freezing if cyclic AMP and cyclic GMP are to be assayed at different times.[143]

Cyclic AMP in CSF has been measured by protein-binding assays,[433–435] radioimmunoassays,[97,436–440] or enzyme-coupling assays.[441] The radioimmunoassay technique is easiest and most reproducible and is capable of measuring cyclic GMP as well as cyclic AMP[97] without concern about interfering substances.[433,434]

3.1.7. Clearance of Acid Metabolites and Cyclic Nucleotides and its Blockade by Probenecid

The metabolic products of dopamine and serotonin in the central nervous system are weak acids, portions of which enter the CSF from brain extracellular fluid. Recirculatory ventricular perfusion data suggest that the egress of 5-HIAA from CSF occurs by diffusion, a saturable transport system, and bulk flow of CSF.[362] The ability of the isolated choroid plexus to actively accumulate 5-HIAA against a concentration gradient is blocked by organic acids[442] such as probenecid[442,443] and homovanillic acid.[442] Following probenecid administration, the gradient between lateral ventricular and cisternal CSF levels for 5-HIAA and HVA has been shown to be reduced by the relative elevation of cisternal metabolite concentrations, suggesting that the fourth ventricle is a major site of the transport system.[381] Accordingly, the tissue-to-medium ratio for 5-HIAA is significantly greater for the fourth ventricular than for the lateral ventricular choroid plexus.[442] The tenfold greater 5-HIAA clearance from cortical subarachnoid space perfusates than from ventriculocisternal perfusates indicates that the major site of acid metabolite clearance may be the rich capillary beds of the cerebral cortex rather than the choroid plexus.[444] Both the cerebral[444] and spinal[382] subarachnoid clearances of 5-HIAA are also sensitive to probenecid inhibition. The relative contributions of these clearance mechanisms remain controversial.[305,445]

Perfusion studies have demonstrated relatively high egress rates for CSF MHPG from both ventricles and subarachnoid spaces which were not diminished by probenecid.[252] Thus, MHPG appears to be cleared from the central nervous system by a rapid, unblockable diffusional mechanism.

The egress of cyclic AMP from CSF is inhibited by probenecid in animals[412] and man,[97,428,430,446] suggesting that cyclic nucleotide clearance from CSF occurs via a weakly anionic organic molecule carrier system.[97] Recent clinical investigations of cyclic nucleotide blockade by probenecid have demonstrated the ratio of the final cyclic AMP concentrations in CSF to their respective baseline cyclic AMP levels to be proportional to the final CSF probenecid concentrations.[97] Although linear relationships have been observed between either cyclic AMP or cyclic GMP accumulation ratios and the CSF probenecid levels, the slope of the least-squares regression line for cyclic AMP was greater than that of cyclic GMP, suggesting possible differences in their transport mechanisms or kinetics of inhibition.[97]

3.1.7a. Investigative Considerations. Determinations of the accumulation of 5-HIAA and HVA over a finite period of time following effective probenecid blockade of their CSF egress have been postulated to reflect primarily the rate of production of these metabolites from their parent amines, serotonin and dopamine, thereby providing an index of central amine transmitter activity. Accordingly, alterations in probenecid-induced metabolite accumulations are consistent with pathological or pharmacological manipulations of serotonin and dopamine metabolism.[206,447,448]

Unfortunately, observed metabolite levels after probenecid administration reflect variable blockade as well as variable central amine metabolism.[305] Contrary to early reports,[449] probenecid-induced transport blockade is incomplete, even at doses that result in high CSF probenecid levels.[305] In fact, the positive correlation of 5-HIAA and HVA concentrations with CSF probenecid levels has been shown to account for 14% and 24% of the total variance in the levels of these respective metabolites in CSF.[305]

Several methods to correct for the effect of CSF probenecid variations on CSF amine metabolite accumulations have been proposed. Assuming that increased metabolite levels in CSF are caused by higher probenecid concentrations, then the 14–24% of the total variation in CSF metabolite levels associated with variations in CSF probenecid concentrations could be removed statistically through an analysis of covariance, provided the metabolite–probenecid interaction is similar across significant subgroups of the population.[305] Commonly employed corrections accomplished by dividing metabolite levels[187,295,384] or their ratios[97] by the CSF probenecid level may overcorrect for the probenecid-associated variations in metabolite levels.[305] Regardless of method of correction, such accumulation studies require the determination of CSF probenecid levels[142,143] by electron capture–gas chromatography.[450]

The probenecid content in CSF may be reduced by concomitant lithium[305] or anticonvulsant[451] therapy. Variations in the slopes of the probenecid–metabolite regression lines suggest individual or group differences in probenecid pharmacokinetic or transport systems.[305] In addition, probenecid may actually alter the central turnover of the neurotransmitter under study.[238]

No apparent threshold probenecid level in CSF is required before inhibition of cyclic nucleotide clearance occurs, and the CSF probenecid-dependent blockade of this transport process does not become total despite high doses of probenecid.[97] Thus, the probenecid test may be used to evaluate the central turnover of cyclic nucleotides provided that their accumulations are corrected for variations in the degree of probenecid blockade.[142,143]

4. PEPTIDES, STEROIDS, AND OTHER HORMONES

Anterior pituitary function is modulated by hypothalamic releasing and inhibiting hormones which are secreted into the hypophyseal portal blood vessels in the median eminence to stimulate or inhibit the release of individual pituitary hormones.[227] These peptidergic neurons which synthesize and release the various releasing hormones within the hypothalamus are in close proximity to both putative transmitter synapses and the third ventricle. Evidence suggests that both neurotransmitters[452–454] and other brain peptides[455] may be involved in controlling the release of releasing and inhibiting hormones. Table VII lists the hormonal response of the pituitary to neurotransmitters and brain peptides instilled into ventricular CSF.[453,455,456]

Recently, the CSF has been implicated as a pathway in neuroendocrine integration.[4,458–460] Ventricular CSF may serve (1) to remove biologically active or metabolically inactivated hormones from the central nervous system, (2) to centrally distribute active hormones, and/or (3) to deliver biologically active hormones of central origin to peripheral target organs.[458] Potentially, endocrine tissue can be transplanted to the subarachnoid space and release trophic hormones directly into the CSF, thereby augmenting active hormone activity.[461]

Table VII
Effect of Intraventricular Administration of Neurotransmitters and Brain Peptides on Pituitary Hormone Release[a]

Neurotransmitter or peptide	Prolactin	GH	TSH	FSH	LH	ACTH
Acetylcholine	−	+	−	+	+	+
γ-Aminobutyric acid	+,−	+	−	0	+	+
Norepinephrine	?+	+	+		+	−
Dopamine	−	+	−		+,−	0
Serotonin	+	+	0		+,−	+
Melatonin	+				−	
Histamine	+				?+	
Somatostatin	0	+	−	−	−	
Cholecystokinin	+	+	−	0	−	
Gastrin	−	+	−	0	−	
Vasoactive intestinal peptide	+	+	0	0	+	
Substance P	?+	+	0	0	+	
Neurotensin	−	+	0	0	−	
Opioids	+	+	−	0	−,0	

[a] Abbreviations: +, stimulation; 0, no effect; −, inhibition; GH, growth hormone; TSH, thyrotropin; FSH, follicle-stimulating hormone; LH, luteinizing hormone; ACTH, adrenocorticotropin. Data derived from McCann et al.,[453] McCaan,[455] Lumpkin et al.,[456] and Taché et al.[456a]

Fig. 5. Schematic diagram of possible relationship between hypothalamic neurons containing releasing hormones and third ventricle (VENT). (A) Neuron releasing hormones directly onto hypothalamo–hypophyseal portal capillaries; (B) neuron delivering hormones into ventricular CSF which transports hormones to median eminence tanycyte (specialized ependyma) through which hormones pass to portal system; (C) hormone or neurotransmitter released by neuron and reaching third ventricular CSF by diffusion or bulk flow between lining ependymal cells. Reprinted from Pollay[465] with permission.

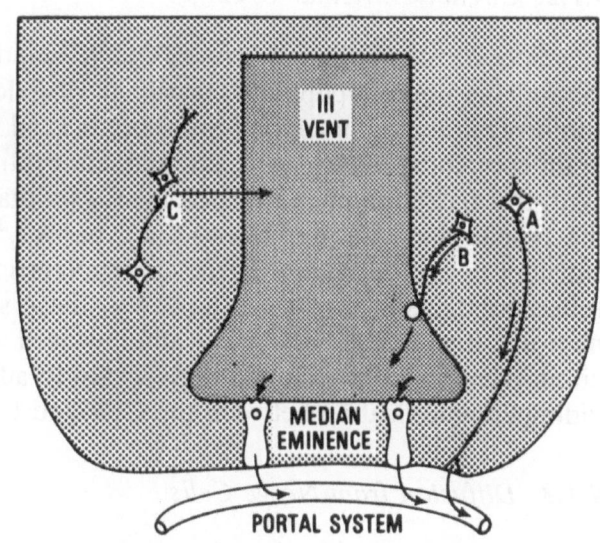

4.1. Entry into Cerebrospinal Fluid

Reviews have proposed that hormones and peptides reach the CSF by several physiological[457] and pathological[462] routes. Normal entry mechanisms include trans-median-eminence transport, circumventricular organ transfer, diffusion from nerve cells, and transport across the blood–brain barrier.[457]

4.1.1. Trans-Median-Eminence Transport

4.1.1a. Ependymal Tanycyte Theory. Neurohormone-releasing factors are synthesized in the hypothalamus and released into the extracellular fluid or into the CSF by neurons that make axonal contact with specialized "secretory" ependymal cells[463,464] (Fig. 5). These factors are carried via the CSF to the median eminence where stimulation of the dendritic processes of the "CSF-contacting neurons" occurs[463] or are transferred to the hypophyseal portal system by specialized ependymal cells called "tanycytes."[466,467]

The apices of the ependymal tanycytes form the floor of the third ventricle, and their opposite ends terminate near the primary capillaries of the hypophyseal portal vessels.[468] Thyroxine, ACTH, LH, prolactin, and steroid hormones are transferred from rat ventricular CSF to the hypophyseal portal system.[467,469] This ependymal transport is known to be preserved after disconnection of the hypothalamus from the median eminence and to be modified by peripheral hormone levels (feedback).

4.1.1b. Retrograde Portal Vessel Transport. The pituitary gland may secrete to the brain[470] by retrograde transport up the portal vessels.[471] Bidirectional transport of peptides occurs within the median eminence[468,472] which provides feedback from the pituitary to the CSF and on to the hypothalamus and distant parts of the brain.[457]

4.1.2. Circumventricular Organs

Specialized midline structures which include the median eminence, neural lobe of the neurohypophysis, organum vasculosum of the lamina terminalis, subfornical organ, subcommissural organ, and area postrema are areas of the brain containing tanycytes[473] and high concentrations of peptide hormones.[457] With the exception of the subcommissural organ, these structures, along with the pineal gland, lack a blood–brain barrier.[457] Intravenously administered substances are deposited in the circumventricular organs in concentrations reflecting those in the blood; however, these substances are not taken up by neurons. In contrast, substances administered via the CSF are taken up by the circumventricular organs and transported to adjacent capillaries, thereby providing communication between the CSF and the systemic circulation.[474]

4.1.3. Diffusion from Nerve Cells

Since the ependymal surfaces of the cerebral ventricles do not limit exchange between CSF and brain,[4,475] a true brain–CSF barrier does not exist. Thus, CSF concentrations of neuropeptides may reflect neuronal release and diffusion into CSF.

4.1.4. Transport across Blood–Brain Barrier

With the exception of the circumventricular organs, the brain has been reported to be relatively impermeable to intravenously injected peptides[476–478] except prolactin,[479] calcitonin,[480,481] and insulin.[482] Recent sensitive methods[483] have found that small but significant amounts of unbound peptides may enter the brain extracellular space from blood. This exposure of the brain to peripheral neuropeptides explains the central effects of systemically administered peptides reported in conscious animals and suggests feedback of peripheral neuropeptides on the brain which regulates their release into the circulation.[483]

In contrast to the most peptide hormones, steroid hormones that are not bound to plasma proteins freely cross the blood–brain barrier as functions of their plasma concentration.[484,485] In addition, albumin-bound but not globulin-bound steroids are transported to some extent into brain extracellular fluid.[486]

4.2. Endogenous and Exogenous Hormones and Neuropeptides in Cerebrospinal Fluid

The direct evidence that endogenous neuropeptides are present in CSF[462,487] and that their CSF concentrations respond appropriately to physiological stimuli[452–455] have recently been reviewed,[457] suggesting the role of CSF as a conduit for neuroendocrine regulation.[4,458–460] Evidence also implies that steroid hormones in CSF may participate in feedback control of the hypothalamus.[488]

References on the subject of the concentrations of these hormones in the plasma and lumbar CSF of control patients and normal volunteers are listed in Table VIII.

4.2.1. Thyrotropin-Releasing Hormone

Intraventricular administration of TRH in rabbits evokes hyperthermia and behavioral excitation as well as inhibition of the hypothermic effect of morphine.[506] Injection of TRH into rat cisternal CSF produces hyperthermia which can be antagonized by an intracisternal injection of the tetradecapeptide bombesin.[507] Stimulation of colonic activity[508] and suppression of feeding and drinking[509] occur in rats following the administration of TRH into ventricular CSF.

Lateral ventricular or intravenous administration of large doses of TRH evokes similar elevations in TSH release,[510] but lower TRH doses yield larger plasma TSH levels following intravenous than after similar intraventricular injections.[511] In contrast, more labeled TRH is retrievable from portal blood following lateral ventricular than after intravenous injections of TRH.[512] Tanycytes of the median eminence are able to concentrate intraventricularly administered TRH.[513] Thus, tanycytes function to transport TRH from the ventricle to the portal veins from which it stimulates the pituitary to release TSH; however, the physiological importance of TRH in CSF is unresolved.

The concentration of TRH, an excitatory neuroactive peptide,[514] is reported to be 40.2 ± 6.9 pg/ml as determined by radioimmunoassay in human lumbar CSF.[490] Other investigators employing radioimmunoassay have reported 4.9 ± 0.7 pg/ml of TRH-like activity in the lumbar CSF of 17 control individuals.[490a] No variations have been reported with respect to the sex of the individual or time of day of CSF sampling.[490] The immunoreactive TRH content in human cisternal CSF has been found to range from 60 to 490 pg/ml.[515]

4.2.1a. Thyroid-Stimulating Hormone (Thyrotropin).

The concentration of TSH in the lumbar CSF of 43 normal individuals has been reported to be 2.65 ± 0.2 μU/ml, whereas that in plasma was 5.95 ± 0.30 μU/ml; thus, the CSF/plasma ratio was 0.44 for TSH.[493] Cerebrospinal fluid levels of TSH apparently correlate with plasma TSH concentrations as determined by radioimmunoassay.

4.2.1b. Thyroid Hormones.

Central nervous system function is profoundly altered during the myxedematous coma or thyrotoxic agitation that may accompany deficient or elevated plasma levels of thyroid hormones. In addition, thyroid hormones regulate the rate of transport of both glucose and amino acids into the brain as well as protein synthesis.[516]

Intravenously infused labeled thyroxine (T_4) and triiodothyronine (T_3) penetrate both the blood–brain and blood–CSF barriers.[517] Although nearly all of the plasma T_3 and T_4 is bound to protein, the percentage of free T_3 is greater than that of free T_4 in both plasma and CSF,[494,517] probably secondary to the lesser protein affinity of T_3. Triiodothyronine enters the brain extra-

Table VIII

Concentrations of Hormones in Plasma and Lumbar CSF in Man[a]

Hormone	Units	Plasma concentration[a]	CSF concentration[a,e]	CSF/plasma ratio	Reference
Adrenocorticotropin	pg/ml	65 ± 15	73 ± 11 (15)[b]	1.12	487
		74 ± 10	98 ± 11 (22)[b]	1.32	489
Cortisol	fmol/ml		14.4 ± 0.3 (5)[c]		553
	μg/ml		14 ± 0.9 (29)[c]		141
	μg/100 ml		0.68 ± 0.08 (a.m.) (15)[b]		490
			0.38 ± 0.02 (p.m.) (12)[b]		
Stomatostatin	pg/ml		55 ± 6 (24)[c]		491
			42.9 ± 1.7 (16)[b]		457
			142 ± 8 (10)[b]		548a
			130.9 ± 10.8 (14)[b]		548b
			142 ± 8 (10)[c]		492
Growth hormone	ng/ml	1.95 ± 0.20	0.35 ± 0.03 (43)[b]	0.17	493
		6.7 ± 1.5	0.5 ± 0.1 (13)[b]	0.07	487
Thyrotropin-releasing hormone	pg/ml		40.2 ± 6.9 (9)[b]		490
			4.9 ± 0.7 (17)[b]		490a
Thyrotropin	μU/ml	5.95 ± 0.30	2.65 ± 0.20 (43)[b]	0.44	493
		4.0 ± 0.7	0.1 ± 0.0 (15)[b]	0.025	487
Thyroxine, total	ng/100 ml	7400	185 (pooled)[b]	0.025	494
Thyroxine, free		2.4	5.8	2.4	
Triiodothyronine, total	ng/100 ml	168	17.4 (pooled)[b]	0.1	494
Triiodothyronine, free		0.32	1.77	5.53	
Prolactin	ng/ml	25 ± 7	0.4 ± 0.1 (5)[b]	0.02	487
		7 ± 0.1	1.2 ± 0.1 (30)[b]	0.17	495
Luteinizing hormone	μg/liter		1.75 ± 0.23 (18)[b]		490a
	mIU/ml		5.7 ± 1.5 (6)[b]		496
	IU/liter		2.39 ± 0.13 (17)[b]		490a

	Units			Ref.
Follicle-stimulating hormone	mIU/ml		5.3 ± 1.2 (6)[b]	496
	IU/liter		1.67 ± 0.10 (17)[b]	490a
Testosterone, female	ng/100 ml		1.4 ± 0.3 (16)[b]	497
Testosterone, male			11.1 ± 1.1 (9)[b]	497
Estradiol, female	pg/ml		2.4 ± 0.4 (16)[b]	497
Estradiol, male			2.2 ± 0.2 (9)[b]	
Progesterone, female	ng/100 ml		4.5 ± 0.8 (16)[b]	497
Progesterone, male			3.9 ± 0.5 (9)[b]	
Melatonin	pg/ml	63 ± 8	59 ± 12 (8)[c]	498
Arginine vasopressin	pg/ml	2.8 ± 0.2	2.4 ± 0.2 (12)[b]	499
Insulin	μU/ml	38.0 ± 4.5	11.0 ± 1.3 (21)[b]	500
Cholecystokinin	pM		14.0 ± 3.2 (10)[b,d]	501
Gastrin	pM		3.4 ± 1.0 (10)[b,d]	501
Vasoactive intestinal polypeptide	pmol/ml	7.3	49.9 ± 4.9 (14)[c,d]	502
Calcitonin	pg/ml	89 ± 13	28 ± 3 (27)[c]	480
β-Melanocyte-stimulating hormone	ng/liter	16.1 ± 1.1	60.1 ± 8.9 (30)[b]	503
Substance P	fmol/ml		7.0 ± 0.6 (18)[c]	623, 624
	pg/ml		1.00 ± 0.13 (14)[b]	548b
β-Lipotropin	pg/ml	78.2 ± 13.7	89.3 ± 16.1 (4)[b]	637a
β-Endorphin	pg/ml	5.5 ± 2.3	17.9 ± 2.3 (4)[b]	637a
	ng/ml		6.1 ± 0.5 (9)[b]	625
	fmol/ml		22.3 ± 1.1 (5)[c]	553
Methionine-enkephalin	pmol/ml		3.12 ± 1.23 (10)[c]	505
	ng/ml		26.8 ± 0.8 (9)[b]	625
	pg/ml	69.3 (25)[b]	13.3 (7)[b]	643

[a] Mean ± standard error.
[b] Nonendocrine control patients.
[c] Normal volunteers.
[d] Personal communication from original source.
[e] Number of patients in parenthesis.

cellular fluid more readily than T_4 but is slower to equilibrate in CSF.[517] This difference in brain penetration may also be secondary to T_3 being less ionized at pH 7.4 than T_4 and thereby being more lipid soluble.

The mean total T_4 concentration in pooled lumbar CSF of euthyroid patients has been reported to be 185 ng/100 ml or 1/40 of the plasma level, and that for the total T_3 to be 17.4 ng/100 ml or one-tenth of the plasma level.[494] The absolute free T_4 in the CSF of these patients was 5.8 ng/100 ml, and the absolute free T_3 was 1.8 ng/100 ml, both exceeding the respective serum levels of 2.4 and 0.32 ng/100 ml.[494] The presence of higher free hormone levels in CSF than in serum means that the movement of free T_4 and T_3 from blood to CSF takes place against a concentration gradient, probably via a saturable, carrier-mediated active transport mechanism.[516,517]

4.2.1c. Investigational Considerations. Thyroxine-binding prealbumin and thyroxine-binding globulin are present in human CSF at concentrations of 1/12 and 1/75 that of plasma.[494] Although the total levels of these binding proteins are relatively limited, thyroid hormone transport in CSF qualitatively resembles that in serum. More specifically, phenytoin and salicylate inhibit the protein binding of thyroid hormones in both plasma and CSF[494] and thus tend to increase the concentrations of the free active T_4 and T_3 in both plasma and CSF.

4.2.2. Luteinizing Hormone-Releasing Hormone

Intraventricularly administered labeled LH-RH is taken up by the ependymal tanycytes lining the floor of the third ventricle[518] and then can be localized in the portal capillaries and the adenohypophysis.[469] Estrogen facilitates this LH-RH uptake by the median eminence.[519] Accordingly, LH-RH injected into either third ventricular or cisternal CSF evokes a rise in plasma LH concentrations.[520] Although intravenous injection of LH-RH induces higher elevations of plasma LH levels, intraventricular administration of this decapeptide appears to produce a more prolonged stimulation of LH release.[521]

The presence and variation of LH-RH in CSF remain controversial despite its localization in ependymal tanycytes of the median eminence.[522] The LH-RH concentration in third ventricular CSF has been reported to be 0.33 pg/ml in rats.[523] However, stimulation of the rat medial preoptic area, which causes an elevation of serum LH, fails to raise CSF LH-RH, suggesting that the CSF does not physiologically transport LH-RH to the median eminence.[523] Other communications[524,525] suggest that elevations in CSF LH-RH do occur after ovariectomy and deafferentation. Some investigators have not been able to detect LH-RH in ovine[526] and most normal human[527,528] CSF, whereas others have found 50–150 pg/ml in the ventricular CSF of hydrocephalic patients.[529]

4.2.2a. Luteinizing Hormone and Follicle-Stimulating Hormone. Gonadotropin hormones formed in the adenohypophysis are regulated by LH-RH and multiple feedback loops. Concentrations of LH in the lumbar CSF of nonendocrine control patients undergoing myelography have been reported to be 5.7 ± 1.5 mIU/ml and those of FSH to be 5.3 ± 1.2 mIU/ml.[496] Luteinizing hormone in monkey CSF appears to be higher at night than during daylight and correlates with plasma LH levels.[530]

4.2.2b. Gonadal Hormones.

Testosterone, estradiol, and progesterone profoundly affect central nervous system functions involving hypothalamic secretions, sex differentiation, and behavior.[531] Specific binding sites for these steroid hormones have been demonstrated in brain regions such as the hypothalamus and especially the adenohypophysis.[484,485,532]

The gonadal hormones are unchanged at physiological pH and are similar in size; thus, their degree of lipid solubility and protein binding would be expected to influence the rate and extent of their entry into CSF.[484] Almost 98% of these circulating steroid hormones are bound to protein,[533] being partitioned between low-affinity, high-capacity sites on albumin and high-affinity, low-capacity binding sites on the sex hormone-binding globulins. Recently steroid hormones bound to albumin but not those bound to the globulin protein have been shown to penetrate the blood–brain barrier.[486]

The degree of protein binding, rather than lipid solubility, appears to be the dominant factor in plasma-to-CSF transfer *in vivo*.[484] The completeness of steroid hormone binding inversely correlates with the ability of a steroid hormone to enter the CSF,[534] and steroid entry into CSF positively correlates with the free steroid hormone concentration in plasma.[484] Although clinical studies have demonstrated correlations between the plasma and CSF levels of testosterone, estradiol, and progesterone, individual concentrations of these gonadal hormones in CSF were approximately the same as the levels of each unbound steroid hormone in the plasma.[497]

The absolute concentrations of testosterone, estradiol, and progesterone reported to be present in the CSF of 16 female control patients are 1.4 ± 0.3 ng/100 ml, 2.4 ± 0.4 pg/ml, and 4.5 ± 0.8 ng/100 ml, respectively, and those in 9 male control patients are 11.1 ± 1.1 ng/100 ml, 2.2 ± 0.2 pg/ml, and 3.9 ± 0.5 ng/100 ml, respectively.[497] The difference in the mean CSF testosterone levels between females and males was significant. The percentages of the total plasma concentration of testosterone, estradiol, and progesterone in the CSF of 16 female control patients has been reported to be 2.9 ± 0.9%, 3.8 ± 0.5%, and 9.3 ± 0.8%, respectively, and those for nine male control patients to be 2.2 ± 0.2%, 4.3 ± 0.7%, and 13.2 ± 0.5%, respectively.[497] Significant sex differences in these percentages were noted for progesterone.

Both plasma and CSF testosterone concentrations vary in a daily rhythm, with higher levels occurring at night.[530]

4.2.3. Prolactin

Unlike most peptide hormones, intravenously injected prolactin passes across the blood–CSF barrier.[479] Prolactin concentrations in rat[535] and monkey[530] CSF parallel those in plasma. Accordingly, the CSF prolactin levels in nonendocrine patients have been found to be a function of the plasma concentrations,[536] necessitating the use of CSF/plasma ratios in clinical evaluations.[537] The site of prolactin entry into the CSF may be the choroid plexus,[536] although retrograde transport in portal vessels cannot be excluded.[538] No ventriculospinal concentration gradient exists for CSF prolactin.[539] The concentration of immunoreactive prolactin in the lumbar CSF of 5, 105, and 30 nonendocrine control patients has been reported to be 0.4 ± 0.1,[487] 0.78 ± 0.05,[539] and 1.2 ± 0.1 ng/ml,[495] respectively. Cerebrospinal fluid prolactin levels may be physiologically higher during pregnancy or stress[537] or evening hours[530] and are pharmacologically altered by dopaminergic agonists or antagonists.[540]

Although the exact function of CSF prolactin is unknown, recent studies have shown that elevation of CSF prolactin levels increases dopaminergic activity in the median eminence and temporarily suppresses endogenous prolactin secretion.[541] Thus, CSF prolactin may mediate prolactin autoregulation.

4.2.4. Somatostatin

Somatotropin release-inhibiting factor or somatostatin, a tetradecapeptide, has been found in high concentrations in the rat hypothalamus, especially within the median eminence and arcuate nucleus.[542] Somatostatin has also been localized in the medial basal hypothalamus and more specifically in the preoptic and anterior periventricular nuclei.[543] In addition, somatostatin has been found in the dorsal root ganglia and substantia gelatinosa.[544]

Microiontophoretic application of somatostatin in certain hypothalamic and extrahypothalamic structures depresses neuronal firing activity.[545] Somatostatin suppresses GH release from pituitary incubations or perfusions or after systemic administration[546] but paradoxically elevates GH release by ultrashort-loop feedback when injected into third ventricular CSF.[456] Intraventricular injections of somatostatin also decrease the plasma levels of LH, FSH, and TSH.[456] Injections of antisomatostatin antibodies into ventricular CSF in rats decreases the duration of strychnine-induced seizures and augments the central response to pentobarbital.[547] These studies suggest that the CSF may provide a conduit for the physiological action of somatostatin.[457]

Concentrations of immunoreactive somatostatin in the lumbar CSF of 24 normal individuals have been reported to be 55 ± 6 pg/ml,[491] which is in good agreement with published control groups,[548] although others have reported CSF somatostatin levels in 10 and 14 control patients to be 142 ± 8 pg/ml[548a] and 130.9 ± 10.8 pg/ml,[548b] respectively. Only 20% of the somatostatin content of normal CSF degrades over 24 h at 4°C; however, the presence of blood in CSF causes rapid degradation, probably because of the proteolytic activity of normal plasma which has entered the CSF along with the erythrocytes.[457] The daily rhythm of somatostatin in CSF appears to be variable, with higher hormone levels during darkness.[549]

4.2.4a. Growth Hormone. The GH concentrations in normal CSF are much lower than those in plasma.[487,550,551] The concentration of GH in the lumbar CSF of 43 control patients is 0.35 ± 0.03 ng/ml by radioimmunoassay, and the CSF/plasma ratio is 0.17.[493] Although significant correlations between CSF and plasma GH levels have been reported,[493] the acute changes in plasma GH concentrations occurring during pneumoencephalography are not reflected in CSF GH levels.[552] Thus, a relative blood–CSF barrier exists for GH in man.

4.2.5. Adrenocorticotropin

The precise concentration of ACTH in normal lumbar CSF is unknown. One group of investigators has reported the immunoreactive ACTH level in two groups of 13 and 15 nonendocrine control patients to be 27 ± 3[489] and 73 ± 11[487] pg/ml, respectively, thus indicating some interassay variability.[489] Another group has reported 14.4 ± 0.3 fmol/ml of ACTH-like activity in normal lumbar CSF.[553] The CSF concentrations of ACTH have been noted to be similar to simultaneous plasma levels,[487,489,554] but no correlation between CSF and plasma ACTH concentrations has been reported.[489,554] The acute rise in plasma ACTH levels observed during pneumoencephalography is not reflected in the CSF,[552] and the bolus intravenous administration of labeled ACTH in normal individuals fails to cause a significant rise in CSF ACTH.[489,554] These and similar studies in cats[489] suggest that the blood–CSF barrier is relatively impermeable to ACTH. Thus, the ACTH present in CSF is central in origin.[555]

As yet, diurnal variations in CSF ACTH have not been noted despite the presence of such rhythms for CSF cortisol.[490]

4.2.6. Cortisol

Approximately 95% of plasma cortisol is bound to protein in man.[556] Free cortisol, a steroid hormone, in plasma appears to cross the blood–CSF barrier with relative ease,[557] and the entry of cortisol into CSF correlates with the concentration of unbound cortisol in plasma.[484] The CSF/plasma ratio of cortisol is about 0.05,[556] and the level of free cortisol in CSF is similar to that in plasma.[484,556] The content of cortisol in CSF is thus less than that in plasma, and 80% is in the unbound form.[556] The cortisol concentration in the lumbar CSF of 29 normal subjects has been reported to be 14 ± 0.9 µg/100 ml.[141]

Cerebrospinal fluid cortisol concentrations in monkeys exhibit diurnal rhythms, with elevations occurring in the early portion of the light period[530,558] that are not altered by constant illumination or darkness.[558] Similarly, CSF cortisol levels in man have been found to be twofold higher in the morning than in the evening.[490]

The cortisol in ventricular CSF may play a role in the feedback regulation of ACTH. Administration of systemically ineffective doses of cortisol into ventricular CSF suppresses or prevents the typical stress-related serum elevations of ACTH.[488]

4.2.7. Melanocyte-Stimulating Hormone

Melanocyte-stimulating hormone (MSH) and related peptides affect avoidance and learning behavior,[559] increase the turnover of biogenic amines in brain,[560] and increase protein synthesis.[561] Melanocyte-stimulating hormone peptides have been localized in brain and persist after removal of the pituitary,[560] suggesting that the brain as well as the pituitary gland produces MSH peptides. Accordingly, hypophysectomy in rats has little effect on CSF α-MSH concentrations despite a fall in plasma α-MSH.[562] The lack of correlation between α-MSH levels in CSF and plasma suggests that the systemic circulation does not deliver α-MSH to the CSF[563]; however, later studies observed elevations in CSF α-MSH concentrations following the systemic administration of high levels of α-MSH.[563a] The β-MSH concentration of 60.1 ± 8.9 ng/liter in the CSF of 30 control subjects is significantly greater than the plasma level of 16.1 ± 1.1 ng/liter,[503] implying that MSH peptides may play a physiological role in central nervous system metabolism.

4.2.8. Vasopressin and Vasotocin

Antidiuretic hormone, arginine vasopressin (AVP), and the closely related nonapeptide hormone, arginine vasotocin (AVT), have been localized in high concentrations in the supraoptic nucleus, paraventricular nucleus, and median eminence, with significant levels residing in the arcuate, dorsomedial, and ventromedial nuclei.[564,565] These hormones promote the peripheral functions of water reabsorption by the distal renal tubules and uterine contraction, respectively. However, central AVP may facilitate memory processes[566] and regulate brain water permeability,[567] whereas brain AVT induces sleep.[568] The cross correlation between the AVP content and osmolality of plasma and CSF[569] suggests that osmolality may regulate AVP release into CSF. In addition, hemorrhage[570] and vagus nerve stimulation[571] increase the AVP content in CSF. Vasotocin release into CSF may be induced by sleep deprivation and rapid eye movement sleep[572] or the administration of LH-RH, TRH, somatostatin,[573] MSH-inhibiting factor,[574] and melatonin.[575] Many of these responses are mediated by serotonin.[576]

The AVP content in CSF appears central in origin.[569] Intravenous administration of AVP in man[499,569] and animals[570,577] raises plasma AVP concentrations without elevating CSF AVP levels. The majority of AVP in CSF may be secreted by neurons that are anatomically separate from those that secrete into blood, since CSF AVP levels are elevated following degeneration of the supraopticohypophyseal tract in patients with diabetes insipidus and low plasma AVP concentrations.[499,569] Although the exact location of the neurosecretory cells that release AVP into the CSF is unknown, extrahypothalamic AVP fibers have been immunohistochemically traced from the paraventricular nucleus to the lateral ventricle, the choroid plexus, and other areas of the rat brain.[578-580] Vasopressin is present in several hypothalamic nuclei other than those known to be responsible for the plasma AVP content.[564,565]

The concentration of immunoreactive AVP in lumbar CSF has been reported to be 2.4 ± 0.2 pg/ml in 12 control patients[499] and 1.8 ± 0.5 pg/ml in 34 normal volunteers.[141] Although continuous lumbar CSF drainage has revealed small rostrocaudal AVP concentration gradients, ventricular and lumbar CSF AVP levels are similar.[569] Vasopressin in CSF, but not in plasma, is relatively stable; thus, CSF–plasma comparisons can only be accomplished if the plasma is extracted shortly after collection.[569] In general, the CSF content of AVP is slightly but significantly lower than that in plasma.[499,569] Cat CSF AVP peaks in early morning hours.[580a] The concentration of AVT in CSF by bioassay is less than 5 μU antidiuretic activity/ml in conscious healthy adults[581] and is higher in newborns and infants.[582]

4.2.9. Angiotensin II

Angiotensin II (AII) acts on the central nervous system to increase water intake, vasopressin release, urinary sodium output, and blood pressure.[583] Renin substrate (angiotensinogen), iso-renin,[584] and AII are all present in brain,[586] especially in the anterior hypothalamus and median eminence.[585] Although these central actions AII may be induced by blood-borne AII entering the area postrema, the subfornical organ and the organum vasculosum which lack a blood–brain barrier,[585a] AII is probably not permeable to the blood–brain barrier, and thus this octapeptide must be locally synthesized in brain.[586,587]

Intraventricular administration of tonin, the serine protease that produces AII directly from its precursors, mimics the action of AII, suggesting that tonin may participate in the central regulation of water balance and blood pressure.[588] Central AII formation increases following the injection of renin but not renin substrate into ventricular CSF.[589] The augmentation of drinking, blood pressure, and vasopressin release evoked by injecting renin into ventricular CSF may be blocked by intraventricular saralasin acetate, an AII antagonist, suggesting that these physiological responses are mediated by the formation of AII.[590] Accordingly, infusion of AII into ventricular CSF promotes natriuresis, drinking behavior,[591] and plasma vasopressin elevation,[592] responses that may partially depend on local release of prostaglandins. The polydipsia induced by AII is threefold more active after intraventricular than following intravenous administration.[593] The level of AII in cisternal CSF of rats has been reported to be about 100 pg/ml[594] and 30 fmol/ml.[587] Angiotensin II has also been found in dog CSF[587,590] with a CSF/plasma ratio of 0.9.[587] Recent reports suggest that the AII content in human CSF may be an artifact of radioimmunoassay.[594a]

4.2.10. Calcitonin

Calcitonin, a peptide hormone whose sequence of 32 amino acids is species-specific, functions to lower plasma calcium and inorganic phosphorus concentrations by actions on bone and the kidney.[595] In addition, this peptide decreases gastrin and gastric acid secretion and increases small-bowel secretion of sodium, potassium, and water.[595] Calcitonin has been localized in the pituitary gland[596] and hypothalamus.[597] Although its central action requires clarification, calcitonin has been implicated in the modulation of serotoninergic and cholinergic activity,[598] arousal, sedation,[599,600] and analgesia.[601]

Calcitonin is synthesized and secreted by the C-cells of the thyroid gland,[596] and intravenously administered calcitonin enters the CSF of rabbits[481] and monkeys.[480] Although calcitonin appears to cross the blood–CSF barrier, its

levels in CSF and plasma do not correlate.[480,602] The concentration of immunoreactive calcitonin in the lumbar CSF of 27 normal individuals has been
reported to be 28 ± 3 pg/ml, and the CSF/plasma ratio to be 0.31.[480]

4.2.11. Cholecystokinin and Gastrin

Cholecystokinin (CCK), a tritriacontapeptide found both in the duodenum
and brain,[603,604] has been implicated in the promotion of satiety,[605–607] possibly
by acting on the hypothalamus or diencephalic noradrenergic feeding system.[608]
Injection of CCK into ventricular CSF decreases meal size and daily food
intake in sheep.[609] Levels of immunoreactive CCK in the lumbar CSF of ten
control patients have been reported to be 14.0 ± 3.2 pM.[501] The CCK content
in CSF appears to consist of several heterogeneous peptides.

Gastrin, a heptadecapeptide first isolated from gut, has also been localized
in the pituitary gland.[610] Concentrations of immunoreactive gastrin in the lumbar CSF of ten control patients undergoing myelography have been reported
to be 3.4 ± 1.0 pM.[501] The gastrin content in CSF consists of heterogeneous
forms.

4.2.12. Vasoactive Intestinal Polypeptide

Vasoactive intestinal polypeptide (VIP), consisting of a sequence of 28
amino acids, is present in brain arteries[502] and in high concentrations in the
hypothalamus.[611] Radioactivity cannot be detected in the CSF following the
intravenous administration of radiolabeled VIP, suggesting that VIP does not
cross the blood–CSF barrier.[612] Accordingly, the elevation of plasma VIP by
neostigmine in man may not be reflected in the CSF.[613] Intraventricular administration of VIP elicits increases in plasma prolactin levels in rats by inhibiting dopaminergic activity[614] and shivering and hyperthermic responses in
cats.[615] The concentration of VIP in the lumbar CSF of 14 control patients
undergoing myelography and pneumoencephalography have been reported to
be 49.9 ± 4.9 pmol/ml or approximately sevenfold greater than in normal
human plasma.[502]

4.2.13. Insulin

Insulin is permeable to the blood–CSF barrier in dogs.[482] Continuous
intravenous infusions of insulin induce a slow progressive rise in CSF insulin
levels; however, the CSF insulin content does not increase in proportion to
that of plasma insulin.[482] Thus, CSF insulin concentration appears not to be
maintained by passive diffusion but rather by a saturable transport system.
The insulin level has been found by radioimmunoassay[616] in the lumbar CSF
of 21 control patients to be 11 ± 1.3 μU/ml or approximately 29% of plasma
concentrations.[500] Insulin levels in the CSF were elevated in diabetic patients
with high plasma insulin concentrations but not in proportion to the insulin
content of the blood. The associated hypoglycorrachia probably results from
the peripheral hypoglycemia rather from than a central action of the insulin.[617]

4.2.14. Substance P

High concentrations of substance P (SP), an excitatory undecapeptide,[514] have been found in the mesencephalon, preoptic region, and substantia nigra of the rat brain.[618] In addition, SP has been localized to the dorsal roots of the spinal cord[619] and peripheral myelinated nerve fibers that traverse autonomic ganglia.[620] Intraventricular administration of SP may stimulate hypothalamic somatostatin release into the portal vessels, thereby decreasing GH secretion, and may potentiate the action of opiate-receptor stimulators on GH and prolactin release.[621] The concentrations of immunoreactive SP in the lumbar CSF of five pain-free control patients has been reported to be 25 to 45 pg/ml,[622] and that in 18 normal individuals to be 7.0 ± 0.6 fmol/ml.[623,624] Recently, levels of 1.00 ± 0.13 pg/ml were reported in the lumbar CSF of 14 control patients.[548b] The absence of a rostrocaudal SP concentration gradient[622–624] suggests that the spinal cord makes a major contribution to the SP content in lumbar CSF.

4.2.15. Endorphins and Enkephalins

The endogenous opioid peptides, endorphins and enkephalins, appear to be contained within the 91-amino-acid sequence of β-lipotropin. Reports concluding that these peptides do not cross the blood–brain barrier[476] have recently been disputed by others demonstrating that opioid peptides enter the central nervous system in rats at rates ranging from 1.4 to 3.9×10^{-6} cm/sec.[483] These significant permeabilities may allow these peptides to enter the brain extracellular fluid (and thus CSF) within a half-time of 3–11 min as a result of step increases in the plasma level of unbound endorphin or enkephalin.

These peptides do not appear to be degraded by proteolytic enzyme activity in human CSF.[625] Therefore, alterations in their CSF concentrations tend to reflect cellular metabolism and release rather than CSF enzyme activity. The opioid activity in CSF exhibits diurnal rhythms, with increased levels in the morning hours.[626]

4.2.15a. β-Endorphin. β-Endorphin (β-E) is the most potent and longest acting of the natural opioid agonists and is in highest immunoreactive concentrations in the adenohypophysis, anterior and posterior hypothalamus, septum, midbrain, pons, and medulla.[627,628] Little, if any, β-E activity is found in the neostriatum, hippocampus, and cerebral or cerebellar cortices.

Administration of β-E into ventricular CSF induces akinesia, analgesia,[629–632] hypothermia,[630] hyperglycemia,[631] respiratory depression,[633] electroconvulsive activity,[629] and elevations in blood pressure.[634] In addition, intraventricular injections of β-E augment the brain's synthesis of DOPA and 5-hydroxytryptophan,[635] the respective precursors of dopamine and serotonin, as well as stimulating prolactin and GH release with inhibition of TSH and LH secretion.[455] Interestingly, the hypothermia and cataplexy, but not the analgesia, induced by elevations in CSF β-E can be reversed by intraventricular injections of TRH, even after hypophysectomy. Thus, this TRH–β-E interaction may not be at the opiate receptor or mediated by the pituitary.[636] Naloxone will reverse most of these physiological responses evoked by β-E administration into the CSF.

Concentrations of immunoreactive β-E in the lumbar CSF of nine control patients has been reported to be 6.1 ± 0.5 ng/ml,[625] and that in five normal individuals to be 22.3 ± 1.1 fmol/ml.[553] The β-E content of CSF is higher than that of plasma, but CSF and plasma β-E levels have not been found to correlate.[504,637] In addition, endocrinopathic elevations in plasma β-E levels are not reflected in CSF, indicating the presence of a blood–CSF barrier for β-E.[504] These findings suggest that the central nervous system is the origin of β-E in CSF. Interestingly, positive correlations between the immunoreactive β-E and ACTH content in normal CSF have been reported.[553]

4.2.15b. Investigational Considerations. Recent studies have found that the actual β-E level in the CSF from four nonendocrine control patients was 17.9 ± 2.3 pg/ml, which corresponded to only 20% of β-E-like immunoreactivity.[637a] Gel chromatography has revealed that β-E-like immunoreactivity in human CSF consists of two components with elution positions compatible to those of β-E and β-lipotropin, respectively, and an additional larger molecule. These findings suggest that caution should be exercised in the interpretation of CSF β-E data obtained by current radioimmunoassay techniques. Current CSF assays employ both chromatography and radioimmunoassay.[637b]

4.2.15c. Methionine-Enkephalin and Leucine-Enkephalin. The amino acid sequences of pentapeptides methionine-enkephalin (M-e) and leucine-enkephalin (L-e) are contained within that of the β-E peptide; however, β-E degradation is not their physiological source. The regional distribution of M-e and L-e is different from that of β-E in both brain and the pituitary.[627,628] High concentrations of enkephalins have been found in the basal ganglia, especially the globus pallidus, neurohypophysis, and the hypothalamus, whereas minimal M-e resides in the adenohypophysis and cerebellum.[627,628,638]

Injection of M-e into ventricular CSF transiently induces responses similar to those evoked by β-E. Intraventricular M-e administration elicits analgesia,[639] stupor,[640] and electroconvulsive activity[641] by indirectly blocking inhibition[514] as well as promoting prolactin and GH release.[638,642]

Concentrations of immunoreactive M-e in the lumbar CSF of nine control patients have been reported to be 26.8 ± 0.8 ng/ml,[625] whereas that in ten normal individuals was 3.12 ± 1.23 pmol/ml,[505] although lower levels have also been reported.[643] Radioreceptor analysis of lumbar CSF from four normal individuals has yielded 2.6 ± 1.0 pmol/ml of M-e.[644] Unlike β-E, the M-e content in CSF is less than that in plasma.[643]

4.2.16. Melatonin

Lipophilic melatonin is derived from serotonin[645] and, among other functions, is involved in mediating pineal-dependent seasonal changes in the hypothalamic regulation of reproductive function.[646] Although the pineal gland may be the major source of melatonin, this hormone has been noted in the body fluids of pinealectomized rats,[647] suggesting the presence of extrapineal sources of melatonin. Its lipophilic nature may explain its ease of passing through the blood–CSF barrier.[648,649] Several routes of pineal melatonin entry into CSF have been postulated;[650] however, most pineal melatonin is secreted into the blood before entering the CSF, and direct secretion into the ventricles accounts for no more than 1% of the total melatonin entering the CSF.[649] Both the findings that the major route of secretion of melatonin in rats is into the blood draining into the confluens sinuum[651] and that no rostrocaudal CSF concentration gradient exists for melatonin in man[652,653] support the premise that most pineal melatonin is not released directly into third ventricular CSF.

Melatonin binding to plasma proteins may impede its entry into the CSF[654] and account for the relatively higher levels in primate plasma. Concentrations of melatonin in the lumbar CSF of eight normal male volunteers have been reported to be 59 ± 12 pg/ml.[498] The daytime CSF melatonin concentrations are negatively correlated with age in man, declining by approximately 50% between 15 and 50 years of age.[652] The melatonin content of CSF and plasma exhibit similar daily rhythms with higher levels at night.[146,648,655] This circadian rhythm appears to be regulated by the suprachiasmatic nucleus via the retinohypothalamic tract.[146]

Although the disappearance curve for labeled intravenous melatonin is biphasic in monkeys, that for intrathecal melatonin appears monophasic with a half-life of 30–40 min.[146] This rapid transfer of melatonin from CSF to plasma also occurs in sheep.[649]

5. ENZYMES

Normally, the activity of most enzymes is lower in CSF than in blood; thus, a major source of CSF enzymes may be central contamination from peripheral tissues. Additional sources include cells indigenous to the central nervous system as well as those, such as lymphocytes, that infiltrate the extracellular fluid compartment.[656]

5.1. Glycolytic and Mitochondrial Enzymes

The activity of enzymes that mediate the energy-producing reactions of anaerobic glycolysis and mitochondrial oxidative phosphorylation tend to reflect central nervous system activity.[657,658] Determinations of lactate dehydrogenase (LDH, E.C. 1.1.1.27) have been widely used for the assessment of central cellular damage, and its isozyme pattern in CSF tends to reflect its cell of origin.[390,659] Although CSF LDH alterations are not specific, LDH analysis of CSF may aid in the detection of structural central nervous system damage or recurrence of disease activity following therapy.[660] Cerebrospinal fluid LDH activity in control patients has been reported to range from 3–17 IU/ml (μmol/min per ml).[390] Creatine phosphokinase (CPK, E.C. 2.7.3.2) appears to be the most useful enzyme for assessing brain injury,[656,661] and its brain isozyme (BB) has been detected in CSF but not plasma.[662] Cerebrospinal fluid CPK activity in control patients is about 0–1 IU/ml.[663] Cerebrospinal fluid activity of glutamic–oxalacetic transaminase (GOT, E.C. 2.6.1.1),[664] aldolase (E.C. 4.1.2.7),[658,665] and adenylate kinase (E.C. 2.7.4.3)[666] may also reflect cellular damage in brain or the presence of leukocytes. Cerebrospinal fluid isocitric dehydrogenase (ICDH, E.C. 1.1.1.42) activity appears to reflect the degree of breakdown of the blood–brain barrier,[656,667] whereas the source of glutamic–pyruvic transaminase (GPT, E.C. 2.6.1.2) in CSF appears to be the plasma.[668]

5.2. Neurotransmitter Metabolism

Choline acetyltransferase or choline acetylase (ChAT, E.C. 2.3.1.6) has been localized in the cholinergic neurons of the cerebral cortex, caudate nucleus,[669] putamen, nucleus accumbens, and anterior perforated substance[292] and is responsible for the synthesis of acetylcholine. Since ChAT activity in CSF is higher than that in plasma but much lower than that of brain, the origin of ChAT in CSF may be central cholinergic neuronal activity or degeneration.[669–671] Choline acetyltransferase activity is higher in ventricular CSF than in lumbar CSF,[137] and CSF levels of 0.19 ± 0.04 and 0.15 ± 0.07 nmol/min per ml have been reported in five normal volunteers and four control patients, respectively.[671]

Acetylcholinesterase (AChE, E.C. 3.1.1.7), the enzyme that catabolizes acetylcholine, is present in brain in both membrane-bound and soluble forms,[672] and the exact source of its CSF activity is controversial. The levels of total cholinesterase (ChE) and AChE activity in CSF are 13 and 16 times greater, respectively, than those in plasma. The AChE-to-ChE activity ratio is higher in brain than in CSF or plasma.[673] Generally, neurons and erythrocytes contain most of the true ChE; however, the presence of leukocytes or erythrocytes in CSF has little effect on CSF ChE activity.[669] Cholinesterase activity appears to be higher in lumbar CSF than in ventricular CSF of patients with minimal pathology.[137] The AChE released into CSF by electrical or pharmacological brain stimulation is comprised of a single AChE isozyme,[670] suggesting that the augmentation of the AChE activity in CSF is secondary to central neuronal activity rather than to tissue damage or plasma enzyme transfer.[656,674]

Dopamine-β-hydroxylase (DBH, E.C. 1.14.2.1) catalyzes the formation of norepinephrine from dopamine.[675] Although a significant correlation between plasma and CSF DBH activity exists,[677] the plasma is unlikely to be the major source of CSF DBH[676] because the specific activity of DBH in CSF is 25 times higher than in plasma in man.[677] Concentrations of DBH in pools of the first 20 ml of lumbar CSF from 28 normal volunteers have been reported to be 0.60 ± 0.05 nmol/ml per hr.[141]

Both DBH and norepinephrine are present in noradrenergic neurons and are probably released into CSF by exocytosis.[678] Pharmacological alterations in central noradrenergic activity are appropriately reflected in changes in CSF DBH activity.[679] In addition, physiological augmentations of central noradrenergic metabolism evoke elevations in CSF DBH activity.[680] Thus, DBH activity in CSF appears to be an index of central noradrenergic activity.[676]

The methylating enzymes, catechol-O-methyltransferase (COMT. E.C.

2.1.1.6) and indolethylamine *N*-methyltransferase (INMT, E.C. 2.1.1.27a), involved in the catabolism of catecholamines and indoleamines have been detected in animal and human CSF.[681]

5.3. Lysosomal Enzymes

Polymorphonuclear leukocytes and macrophages as well as neurophils appear to be the origin of most of the lysosomal enzymes in CSF. Accordingly, increased lysosomal enzyme activity has been associated with the presence of inflammatory reactions or digestion of cell debris.[656,665]

Acid phosphatase (E.C. 3.1.3.1) has been localized to anterior horn cells[682] and CSF lymphocytes, whereas alkaline phosphatase is related to CSF granulocytes.[683]

β-Glucuronidase (E.C. 3.2.1.31) is found in both white and gray matter neurons and glia,[684] and its activity in normal lumbar CSF has been reported to be 28 ± 1 μg/hr per 100 ml.[685] Elevations in CSF β-glucuronidase result from autolysis, necrosis, and leukocytic phagocytosis[685] as well as the central alterations associated with demyelination or tumor growth.[686,687] Changes in neuronal activity do not appear to alter CSF β-glucuronidase activity.[656]

Lysozyme (E.C. 3.2.1.17), the enzyme that depolymerizes the bacterial walls of some bacteria, is not plentiful in brain, and its concentration in normal CSF is less than 1.5 μg/ml[688] or 69 μg/l.[689] The CSF activity of lysozyme is not altered by increases in plasma concentrations[689a] and is probably the result of leakage across the blood–brain barrier or CSF inflammatory cells.[656,688]

Other lysosomal enzymes, *N*-acetyl-β-glucosaminidase (E.C. 3.2.1.30), β-galactosidase (E.C. 3.2.1.23), α-mannosidase (E.C. 3.2.1.24), and arylsulfatase (E.C. 3.1.6.1) have been measured in CSF, but their wide variations in activity preclude their clinical application.

Enzymes mediating the synthesis and degradation of lipids have been demonstrated in CSF.[438] Cholesterol-esterifying enzyme (cholesterol acyltransferase, E.C. 2.3.1.26)[690] and cholesterol ester hydrolase (E.C. 3.1.1.13)[691] are present in CSF but not in plasma, suggesting that their activity in CSF may be derived from brain. The central origin of cholesterol ester hydrolase activity in CSF has recently been supported by observations[692] that its CSF activity of 23.8 ± 1.2 nmol/hr per 10 ml in 49 control patients was not different from that in patients with blood–brain barrier dysfunction and nonspecific increases in total CSF protein. The presence of the latter enzyme in CSF has not been universally confirmed.[693] Lecithin–cholesterol acyl transferase (E.C. 2.3.1.43) has been measured in human CSF in quantities less than those in plasma and appears not to be age dependent.[693] Phospholipases (E.C. 3.1.1.4) A_1, A_2, and traces of C have been demonstrated both in CSF and in the supernatant of brain homogenates,[694] implying that the phospholipase activity in CSF is derived from brain.[656]

5.4. Amino Acid and Protein Metabolism

Three peptidases, leucine aminopeptidase, glutamyl peptidase, and glutamyltranspeptidase, with distinct pH optima have been demonstrated in

CSF.[695] Leucine aminopeptidase (E.C. 3.4.11.1) activity in CSF appears to be proportional to CSF protein content and has been associated with myelin.[696] Acid proteinase (E.C. 3.4.4.23) has been isolated from supernatant fractions of CSF, whereas neutral proteinase (E.C. 3.4.24.4) appears to be associated with CSF macrophages but not lymphocytes.[697]

Glycosyl transferases which synthesize glycoproteins have been identified in CSF[698] and may be derived from brain.[699] The activity of these enzymes in CSF appears to be inhibited by metabolites of phenylalanine.[698]

5.5. Miscellaneous Enzymes

Adenosine 2',3'-cyclic nucleotide 3'-phosphohydrolase (CNPase, E.C. 3.1.4.37) is predominantly associated with myelin[700,701] and oligodendroglial plasma membranes[702] and is used as a marker enzyme for myelin breakdown.[656,703] Usually, CNPase activity in normal CSF is absent but rises with myelin degeneration or oligodendroglial disorders.[656,703]

Cyclic nucleotide phosphodiesterases (E.C. 3.1.4.17) in brain regulate the catabolism of cyclic AMP or cyclic GMP[97,392,395]; however, their function in CSF is unknown. These two individual phosphodiesterases have activities in CSF that are lower than in brain.[704] The hydrolysis of cyclic GMP in CSF occurs faster than that of cyclic AMP[97]; cyclic GMP inhibits the hydrolysis of cyclic AMP in CSF, but cyclic AMP does not affect the hydrolysis of CSF cyclic GMP.[704] Major biochemical characteristics of the cyclic GMP phosphodiesterase in brain and CSF are similar, suggesting that the origin of CSF phosphodiesterase activity is the brain.[704]

The formation of inosine and ammonia from adenosine is catalyzed by adenosine deaminase (E.C. 3.5.3.3). The activity of this enzyme is normally absent in CSF but is present in brain and plasma.[705] Its presence in CSF is usually associated with disorders that disrupt the blood–brain barrier.[705]

The activity of ribonuclease (E.C. 2.7.7.16), the enzyme that hydrolyzes ribonucleic acids, has been reported to be 269 ± 17 U/ml in the lumbar CSF of 33 normal subjects;[706] however, its source has not been defined.[707,708] The ribonuclease activity in CSF differs considerably from that in serum.[709]

5.6. Investigational Considerations

In review,[656] the major clinical applications of enzyme analysis include the detection of organic central nervous system disease, differential diagnosis of disorders, especially those involving the meninges, and evaluation of disease course or efficacy of therapy. Enzyme determinations in CSF may provide a noninvasive method of documenting these supportive, but usually not specific, alterations in central enzyme activity.

6. CONCLUSIONS

This chapter has listed and referenced the normal or control concentrations of the major constituents in CSF to provide a physiological base line on which pathological CSF alterations may be qualitatively and quantitatively evaluated. In addition, methodology for the reliable CSF collection, storage, preparation, and analysis has been discussed with respect to the individual, somatotopic, chronological, endocrinologic, pharmacological, and possible artifactual variations in CSF composition. These essential aspects, which may compromise the validity of CSF data, have been presented to aid the investigator in clinical and experimental protocol formulation and in elimination of possible sources of error. Detailed discussion and illustrations of the technical aspects of clinical and experimental CSF investigations[143] have been provided in the definitive reference texts, *Neurobiology of Cerebrospinal Fluid*.

ACKNOWLEDGMENTS. The author is grateful to Drs. Theodore A. Hare, Robert A. Fishman, and Michael Pollay for the preparation of illustrations and thanks Drs. S. J. Enna, Michael G. Ziegler, C. Raymond Lake, Benjamin Rix Brooks, Jonas Sode, Rolf Sjostrom, Jan Ekstedt, Erik Anggard, W. King Engel, Solomon H. Snyder, Floyd E. Bloom, Michael H. Ebert, Kalmon D. Post, Ivor M. D. Jackson, Lars Terenius, Lars von Knorring, Robert M. Post, James C. Ballenger, Pauline Lerner, Philip W. Gold, Gary L. Robertson, Richard P. White, Kenneth L. Davis, John G. Nutt, Jan Fahrenkrug, Jens F. Rehfeld, Josef E. Fischer, Wallace W. Tourtellotte, W. Richard Winn, Stanley I. Rapoport, Isidoor R. Leusen, Dorcas S. Fulton, Steven M. Reppert, Donald J. Jenden, and, especially, Naren L. Banik and Samuel M. McCann for their contributions to the data on which this review is based.

REFERENCES

1. Rapoport, S. I., 1975, *The Blood–Brain Barrier in Physiology and Medicine* Raven Press, New York, pp. 43–86.
2. Cserr, H. F., 1971, *Physiol. Rev.* **51**:273–311.
3. Lorenzo, A. V., 1977, *Exp. Eye Res. [Suppl.]* **25**:205–228.
4. Wood, J. H., 1980, *Neurobiology of Cerebrospinal Fluid 1* (J. H. Wood, ed.), Plenum Press, New York, pp. 1–16.
5. De Rougemont, J., Ames, A. III, Nesbett, F. B., and Hoffman, H. F., 1960, *J. Neurophysiol.* **23**:485–495.
6. Pollay, M., 1974, *Fed. Proc.* **33**:2065–2069.
7. Pappenheimer, J. F., Heisey, S. R., and Jordan, E. F., 1961, *Am. J. Physiol.* **200**:1–10.
8. Lorenzo, A. V., and Cutler, R. W. P., 1969, *J. Neurochem.* **16**:577–585.
9. Oldendorf, W. H., 1977, *Exp. Eye Res. [Suppl.]* **25**:177–190.
10. Cutler, R. W. P., 1980, *Neurobiology of Cerebrospinal Fluid 1*. (J. H. Wood, ed.), Plenum Press, New York, pp. 41–51.
11. Fishman, R. A., 1980, *Cerebrospinal Fluid in Diseases of the Nervous System*, W. B. Saunders, Philadelphia, pp. 168–252.
12. Pape, L., and Katzman, R., 1970, *Proc. Soc. Exp. Biol. Med.* **134**:430–433.
13. Bradbury, M. W. B., Stubbs, J., Hughes, I. E., and Parker, P., 1963, *Clin. Sci.* **23**:97–105.
14. Bekaert, J., and Demeester, G., 1954, *Exp. Med. Surg.* **12**:480–501.
15. Bito, L. A., and Davson, H., 1966, *Exp. Neurol.* **14**:264–280.
16. Naumann, H. N., 1958, *Proc. Soc. Exp. Biol. Med.* **98**:16–18.
17. Fuchs, C., 1976, *Ion and Enzyme Electrodes in Biology and Medicine* (M. Kessler, L. C. Clark, D. W. Lubbers, I. A. Silver, and W. Simon, eds.), University Park Press, Baltimore.
18. Goldstein, G. W., Romoff, M., Bogin, F., and Mossry, S. G., 1979, *J. Clin. Endocrinol. Metab.* **49**:58–62.

19. Jimerson, D. C., Wood, J. H., and Post, R. M., 1980, *Neurobiology of Cerebrospinal Fluid 1*. (J. H. Wood, ed.), Plenum Press, New York, pp. 743–749.

20. Hunter, G., and Smith, H. V., 1960, *Nature* **186**:161.

21. Chutkow, J. H., and Meyers, S., 1968, *Neurology (Minneap.)* **83**:963–974.

22. Kemeny, A., Boldizsar, H., and Pethes, G., 1961, *J. Neurochem.* **7**:218–227.

23. Heipertz, R., Eickhoff, K., and Karstens, K. H., 1979, *J. Neurol. Sci.* **40**:87–95.

24. Friedman, A., and Levinson, A., 1955, *Arch. Neurol. Psychiatry* **74**:424–440.

25. Bradbury, M. W. B., and Kleeman, C. R., 1969, *J. Physiol. (Lond.)* **204**:181–193.

26. Bourke, R. S., and Nelson, K. M., 1972, *J. Neurochem.* **19**:1225–1232.

27. Leusen, I. R., Weyne, J. J., and Demeester, G. M., 1982, *Neurobiology of Cerebrospinal Fluid 2* (J. H. Wood, ed.), Plenum Press, New York (in press).

28. Davies, D. G., 1976, *J. Appl. Physiol.* **40**:123–125.

29. Davies, D. G., 1977, *J. Appl. Physiol.* **43**:566–567.

30. Davies, D. G., 1978, *Regulation of Ventilation and Gas Exchange* (D. G. Davies and C. D. Barnes, eds.), Academic Press, New York, pp. 167–196.

31. Plum, F., and Posner, J. B., 1968, *Scand. J. Clin. Lab. Invest. [Suppl.]* **22**:1B.

32. Plum, F., and Price, R. W., 1973, *N. Engl. J. Med.* **289**:1346–1351.

33. Leusen, I., 1972, *Physiol. Rev.* **52**:1–56.

34. Loeschcke, H. H., 1980, *Neurobiology of Cerebrospinal Fluid 1* (J. H. Wood, ed.), Plenum Press, New York, pp. 29–40.

35. Levasseur, J. E., Wei, E. P., Kontos, H. A., and Patterson, J. L., 1979, *J. Appl. Physiol. Respir. Environ. Exercise Physiol.* **46**:89–95.

36. Simeone, F. A., Vinall, P. E., and Pickard, J. D., 1980, *Neurobiology of Cerebrospinal Fluid 1* (J. H. Wood, ed.), Plenum Press, New York, pp. 303–311.

36a. Mrowka, R., 1981, *J. Neurol. Sci.* **49**:181–191.

37. Merritt, H. H., and Fremont-Smith, F., 1938, *The Cerebrospinal Fluid*, W. B. Saunders, Philadelphia.

38. Sarff, L. D., Platt, L. H., and McCracken, G. H., 1976, *J. Pediatr.* **88**:473–477.

39. Fishman, R. A., 1963, *Trans. Am. Neurol. Assoc.* **88**:114–118.

40. Posner, J. B., and Plum, F., 1967, *Arch. Neurol.* **16**:492–496.

41. Siesjo, B. K., 1978, *Brain Energy Metabolism*, John Wiley & Sons, New York.

42. Humoller, F. L., Mahler, D. M., and Parker, M. M., 1966, *Int. J. Neuropsychiatry* **2**:293–297.

43. Perry, T. L., Hansen, S., and Kennedy, J., 1975, *J. Neurochem.* **24**:587–589.

44. McGale, E. H. F., Pye, I. F., Stonier, C., Hutchinson, E. C., and Aber, G. M., 1977, *J. Neurochem.* **29**:291–298.

45. Gjessing, L. R., Gjesdehl, P., Dietrichson, P., and Presthus, J., 1974, *Eur. Neurol.* **12**:33–37.

46. Goodnick, P. G., Evans, H. E., Dunner, D. L., and Fieve, R. R., 1980, *Biol. Psychiatry* **15**:557–563.

47. Poser, C. M., and Ho, B., 1972, *Arch. Neurol.* **26**:502–505.

48. Szilagyi, A. K., Lavinha, F., and Mardens, Y., 1974, *Acta Neurol. Belg.* **74**:329–336.

49. Perry, T. L., Hansen, S., Stedman, D., and Love, D., 1968, *J. Neurochem.* **15**:1203–1206.

50. Fishman, R. A., 1953, *Am. J. Physiol.* **179**:96–98.

51. Cutler, R. W. P., Watters, G. V., and Hammersted, J. P., 1970, *J. Neurol. Sci.* **10**:259–268.

52. Link, H., Zettervall, O., and Blemou, G., 1972, *J. Neurol.* **203**:119–132.

53. Felgenhauer, K., 1974, *Klin. Wochenschr.* **52**:1158–1164.

54. Rapoport, W. I., and Pettigrew, K. D., 1979, Microvasc. Res. **18**:105–119.

55. Rapoport, S. I., 1982, *Neurobiology of Cerebrospinal Fluid 2* (J. H. Wood, ed.), Plenum Press, New York (in press).

56. Lumsden, C., 1972, *Multiple Sclerosis: A Reappraisal* (D. McAlpine, C. Lumsden, and E. D. Acheson, eds.), Churchill Livingston, London, pp. 311–621.

57. Joseph, J. C., and Bermes, E. W., 1979, *Ann. Clin. Lab. Sci.* **9**:408–415.

58. Kleine, T. O., and Merten, B., 1980, *J. Clin. Chem. Clin. Biochem.* **18**:245–254.

59. Reiber, V. H., 1980, *J. Clin. Chem. Clin. Biochem.* **18**:123–127.

60. Scime, M. J., 1980, *Clin. Biochem.* **13**:144–145.

61. Lowry, O. H., Rosebrough, N. J., Farr, A. L., and Randall, R. J., 1951, *J. Biol. Chem.* **193**:265–275.

62. Kahn, S. N., Shortman, R. C., Khan, R. A., and Thompson, E. J., 1980, *J. Clin. Chem. Clin. Biochem.* **18**:23–26.
63. Tourtellotte, W. W., Tavolato, B., Parker, J. A., and Comiso, P., 1971, *Arch. Neurol.* **25**:345–350.
64. Mingioli, E. S., Strober, W., Tourtellotte, W. W., Whitaker, J. N., and McFarlin, D. E., 1978, *Neurology (Minneap.)* **28**:991–995.
65. Arseneault, J. J., 1980, *Clin. Chim. Acta.* **107**:73–84.
66. Allen, R. C., 1980, *Electrophoresis* **1**:32–37.
67. Laurenzi, M. A., and Link, H., 1979, *J. Neurol. Neurosurg. Psychiatry* **42**:368–372.
68. Nilsson, K., and Olsson, J. E., 1978, *Clin. Chem.* **24**:1134–1139.
69. Stibler, H., 1978, *J. Neurol. Sci.* **36**:273–288.
70. Stibler, H., 1979, *J. Neurol. Sci.* **42**:275–281.
71. Schultze, H. E., and Heremans, J. F., 1966, *Molecular Biology of the Human Proteins,* Volume I, Elsevier, Amsterdam, pp. 732–761.
72. Link, H., and Olsson, J. E., 1972, *Acta Neurol. Scand.* **48**:57–68.
73. Laterre, E. C., 1965, *Les Proteines du Liquide Cephalorachidien a L'Etat Normal et Pathologique,* Editions Arscia, Brussels, pp. 148–161.
74. Dencker, S. J., 1969, *J. Neurochem.* **16**:465–466.
75. Nerenberg, S. T., Prasad, R., and Rothman, M. E., 1978, *Neurology (Minneap.)* **28**:988–990.
76. Williams, A. C., Mingioli, E. S., McFarland, H. F., Tourtellotte, W. W., and McFarland, D. E., 1978, *Neurology (Minneap.)* **28**:996–998.
77. Walsh, M. J., Tourtellotte, W. W., and Potvin, A. R., 1982, *Neurobiology of Cerebrospinal Fluid 2* (J. H. Wood, ed.), Plenum Press, New York.
78. Reiber, H., 1980, *J. Neurol.* **224**:89–99.
79. Olsson, J. E., and Pettersson, B., 1976, *Acta Neurol. Scand.* **53**:308–322.
80. Tourtellotte, W. W., 1970, *J. Neurol. Sci.* **10**:279–304.
81. Hershey, L. A., and Trotter, J. L., 1980, *Ann. Neurol.* **8**:426–434.
82. Trotter, J. L., and Brooks, B. R., 1980, *Neurobiology of Cerebrospinal Fluid 1* (J. H. Wood, ed.), Plenum Press, New York, pp. 465–486.
83. Link, H., and Zellervall, O., 1970, *Clin. Exp. Immunol.* **6**:425–438.
84. Dziegielewska, K. M., Evans, C. A., Malinowska, D. H., Mollgard, K., Raynolds, M. L., and Saunders, N. R., 1980, *J.Physiol. (Lond.)* **300**:457–465.
85. Raker, R., Hegyi, T., and Koenigsberger, M. R., 1977, *Ann. Neurol.* **2**:259.
86. Lofberg, H., Grubb, A. O., Sreger, T., and Olsson, J. E., 1980, *J. Neurol.* **223**:159–170.
87. Kobatake, K., Shinohara, Y., and Yoshimura, S., 1980, *J. Neurol. Sci.* **47**:273–283.
87a. Eeg-Olofsson, O., Link, H., and Wigertz, A., 1981, *Acta Pediatr. Scand.* **70**:167–170.
88. Miller, O. H., Jaworski, A. A., Silverman, A. C., and Elwood, M. J., 1954, *Am. J. Med. Sci.* **228**:510–519.
89. Ahonen, A., Myllyla, V. V., and Hokkanen, E., 1978, *Acta Neurol. Scand.* **57**:358–365.
90. Weisner, B., and Bernhardt, W., 1978, *J. Neurol. Sci.* **37**:205–214.
91. Fishman, R. A., Ransohoff, A. J., and Osserman, E. F., 1958, *J. Clin. Invest.* **37**:1419–1428.
92. Fossan, G. O., and Larsen, J. L., 1979, *Eur. Neurol.* **18**:140–144.
93. Muting, D., Heinze, J., Reikowski, J., Betzien, G., Schwartz, M., and Schmidt, F., 1968, *Clin. Chim. Acta* **19**:391–395.
94. Davidson, J. S. D., and Jennings, D. B., 1980, *Can. J. Physiol. Pharmacol.* **58**:550–556.
95. Cockrill, J. R., 1931, *Arch. Neurol. Psychiatry* **25**:1297–1305.
96. Carlsson, C., and Dencker, S. J., 1973, *Acta Neurol. Scand.* **49**:39–56.
97. Brooks, B. R., Wood, J. H., Diaz, M., Czerwinski, C., Georges, L. P., Sode, J., Ebert, M. H., and Engel, W. K., 1980, *Neurobiology of Cerebrospinal Fluid 1* (J. H. Wood, ed.), Plenum Press, New York, pp. 113–139.
98. Enna, S. J., Ziegler, M. G., Lake, C. R., Wood, J. H., Brooks, B. R., and Butler, I. J., 1980, *Neurobiology of Cerebrospinal Fluid 1* (J. H. Wood, ed.), Plenum Press, New York, pp. 189–196.
99. Bachrach, U., 1978, *Advances in Polyamine Research,* Volume 2 (R. A. Campbell, D. R. Morris, D. Bartos, G. D. Davies, and F. Bartos, eds.), Raven Press, New York, pp. 5–11.
100. Seiler, N., 1977, *Clin. Chem.* **23**:1519–1526.

101. Fulton, D. S., Levin, V. A., Lubich, W. P., Wilson, C. B., and Marton, L. J., 1982, *Neurobiology of Cerebrospinal Fluid 2* (J. H. Wood, ed.), Plenum Press, New York (in press).
102. Marton, L. J., Heby, O., Levin, V. A., Lubich, W. P., and Wilson, C. B., 1976, *Cancer Res.* **36:**973–977.
103. Bartos, D., Campbell, R. A., Bartos, F., and Grettie, D. P., 1975, *Cancer Res.* **35:**2056–2060.
104. Harik, S. I., and Marton, L. J., 1981, *Arch. Neurol.* **38:**91–94.
105. Illingworth, D. R., and Glover, J., 1971, *J. Neurochem.* **18:**769–776.
106. Berry, J. F., Logothetis, J., and Boris, M., 1965, *Neurology (Minneap.)* **15:**1089–1094.
107. Pedersen, H. F., 1974, *Acta Neurol. Scand.* **50:**171–182.
108. Fumagalli, R., and Paoletti, P., 1971, *Neurology (Minneap.)* **21:**1149–1156.
109. Dencker, S. J., and Swahn, B., 1961, *Acta Psychiatry Scand.* **36:**325–336.
110. White, R. P., Hagen, A. A., and Robertson, J. T., 1982, *Neurobiology of Cerebrospinal Fluid 2* (J. H. Wood, ed.), Plenum Press, New York (in press).
111. Moncada, S., and Vane, J. R., 1979, *N. Engl. J. Med.* **300:**1142–1147.
112. White, R. P., Hagen, A. A., Morgan, H., Dawson, W. W., and Robertson, J. T., 1975, *Stroke* **6:**52–57.
113. Horton, E. W., 1972, *Monographs on Endocrinology: Prostaglandins,* Springer-Verlag, New York, pp. 117–147.
114. Wolfe, L. S., and Coceani, F., 1979, *Annu. Rev. Physiol.* **41:**669–684.
115. Holmes, S. W., 1970, *Br. J. Pharmacol.* **38:**653–658.
116. Bito, L. Z., and Baroody, R., 1974, *Prostaglandins,* **7:**131–140.
117. Levin, E., Sepulveda, F. V., and Yudilevich, D. L., 1974, *Nature* **249:**266–268.
118. Hagen, A. A., Gerber, J. N., Sweeley, C. C., White, R. P., and Robertson, J. T., 1977, *Stroke* **8:**672–675.
119. Hagen, A. A., Gerber, J. N., Sweeley, C. C., White, R. P., and Robertson, J. T., 1977, *Stroke* **8:**236–238.
120. Carasso, R. L., Vardi, J., Rabay, J. M., Zor, U., and Streifler, M., 1977, *J. Neurol. Neurosurg. Psychiatry* **40:**967–969.
121. LaTorre, E., Patrono, C., Fortuna, A., and Grossi-Belloni, D., 1974, *J. Neurosurg.* **41:**293–299.
122. Mathe, A. A., Sedvall, G., Wiesel, F. A., and Nyback, H., 1980, *Lancet* **1:**16–18.
123. Philipp-Dormston, W. K., and Siefert, R., 1975, *Klin. Wechenschr.* **53:**1167–1168.
124. Cory, H. T., Lascelles, P. T., Millard, B. J., Snedden, W., and Wilson, B., 1976, *Biomed. Mass Spectrom.* **3:**117–121.
125. Wolfe, L. S., and Mamer, O. A., 1979, *Prostaglandins* **9:**183–192.
126. Aizawa, Y., and Yamada, K., 1976, *Prostaglandins* **11:**43–50.
127. Harms, H. H., Wardeh, G., and Mulder, A. H., 1979, *Neuropharmacology* **18:**577–580.
128. Heistad, D. D., and Marcus, M. L., 1979, *Physiologist* **22:**54.
129. Sattin, A., and Rall, T. W., 1970, *Mol. Pharmacol.* **6:**13–23.
130. Berne, R. M., Rubio, R., and Curnish, R. P., 1974, *Circ. Res.* **35:**262–271.
131. Cornford, E. M., and Oldendorf, W. H., 1975, *Biochim. Biophys. Acta* **394:**211–219.
132. Winn, H. R., Rubio, R., and Berne, R. M., 1982, *Neurobiology of Cerebrospinal Fluid 2* (J. H. Wood, ed.), Plenum Press, New York (in press).
133. Meberg, A., and Sangstad, O. D., 1978, *Scand. J. Clin. Lab. Invest.* **38:**437–440.
134. Moir, A. T. B., Ashcroft, G. W., Crawford, T. B. B., Eccleston, D., and Guldberg, H. C., 1970, *Brain* **93:**357–368.
135. Wood, J. H., 1980, *Neurobiology of Cerebrospinal Fluid 1* (J. H. Wood, ed.), Plenum Press, New York, pp. 53–62.
136. Welch, M. J., Markham, C. H., and Jenden, D. J., 1976, *J. Neurol. Neurosurg. Psychiatry* **39:**367–374.
137. Haber, B., and Grossman, R. G., 1980, *Neurobiology of Cerebrospinal Fluid 1* (J. H. Wood, ed.), Plenum Press, New York, pp. 345–350.
138. Hare, T. A., Wood, J. H., Ballenger, J. C., and Post, R. M., 1979, *Lancet* **2:**534–535.
139. Wood, J. H., Ziegler, M. G., Lake, C. R., Shoulson, I., Brooks, B. R., and Van Buren, J. M., 1977, *Ann. Neurol.* **1:**94–99.

140. Lake, C. R., Ballenger, J. C., Ziegler, M. G., Post, R. M., Polinsky, R. J., Wood, J. H., Williams, A. C., and Ebert, M. H., 1982, *Psychiatr. Res.* (in press).
141. Ballenger, J. C., Post, R. M., and Goodwin, F. K., 1982, *Neurobiology of Cerebrospinal Fluid 2* (J. H. Wood, ed.), Plenum Press, New York (in press).
142. Wood, J. H., 1980, *Neurology (Minneap.)* 30:645–651.
143. Wood, J. H., 1980, *Neurobiology of Cerebrospinal Fluid 1* (J. H. Wood, ed.), Plenum Press, New York, pp. 71–96.
144. Perlow, M. J., and Lake, C. R., 1980, *Neurobiology of Cerebrospinal Fluid 1* (J. H. Wood, ed.), Plenum Press, New York, pp. 63–69.
145. Moore-Ede, M. C., Shemelzer, W. S., Kass, D. A., and Herd, J. A., 1976, *Fed. Proc.* 35:2333–2380.
146. Reppert, S. M., Perlow, M. J., and Klein, D. C., 1980, *Neurobiology of Cerebrospinal Fluid 1* (J. H. Wood, ed.), Plenum Press, New York, pp. 579–589.
147. Tower, D. B., and McEachern, D., 1949, *Can. J. Res.* 27:105–119.
148. Duvoisin, R. C., and Dettbarn, W. D., 1967, *Neurology (Minneap.)* 17:1077–1081.
149. Davis, K. L., Berger, P. A., Hollister, L. E., DoAmaral, J. R., and Barchas, J. D., 1977, *Cholinergic Mechanisms and Psychopharmacology* (D. J. Jenden, ed.), Plenum Press, New York, pp. 755–779.
150. Freeman, J. J., Choi, R. L., and Jenden, D. J., 1975, *J. Neurochem.* 24:729–734.
151. McCaman, R. E., and Stetzler, J., 1977, *J. Neurochem.* 28:669–671.
152. Jenden, D. J., 1979, *Brain Acetylcholine and Neuropsychiatric Disorders* (K. L. Davis, and P. A. Bergers, eds.), Plenum Press, New York, pp. 483–513.
153. Lanman, R. C., and Schanker, L. S., 1980, *J. Pharmacol. Exp. Ther.* 215:563–568.
154. Schuberth, J., and Jenden, D. J., 1975, *Brain Res.* 84:245–256.
155. Aquilonius, S. M., Nystrom, B., Schuberth, J., and Sundwall, A., 1972, *J. Neurol. Neurosurg. Psychiatry* 35:720–725.
156. Dreifuss, J. J., Kelly, J. S., and Krnjevic, K., 1969, *Exp. Brain Res.* 9:137–154.
157. Krnjevic, K., and Schwartz, S., 1967, *Exp. Brain Res.* 3:320–336.
158. Hare, T. A., Manyam, N. V. B., and Glaeser, B. S., 1980, *Neurobiology of Cerebrospinal Fluid 1* (J. H. Wood, ed.), Plenum Press, New York, pp. 171–187.
159. Levin, E., Garcia-Argiz, C. A., and Nogueira, G. J., 1966, *J. Neurochem.* 13:979–988.
160. Martin, D. L., 1976, *GABA in Nervous System Function* (E. Roberts, T. N. Chase and D. B. Tower, eds.), Raven Press, New York, pp. 347–386.
161. Storm-Methiesen, J., Fonnum, F., and Malthe-Sorenesen, O., 1976. *GABA in Nervous System Function* (E. Roberts, T. N. Chase and D. B. Tower, eds.), Raven Press, New York, pp. 387–394.
162. Snodgrass, S. R., and Lorenzo, A. V., 1973, *J. Neurochem.* 20:761–769.
163. Hare, T. A., and Manyam, N. V. B., 1980, *Anal. Biochem.* 101:349–355.
164. Hare, T. A., Wood, J. H., Manyam, N. V. B., Ballenger, J. C., Post, R. M., and Gerner, R. H., 1980, *Brain Res. Bull. [Suppl.]* 5:721–724.
165. Enna, S. J., Wood, J. H., and Snyder, S. H., 1977, *J. Neurochem.* 28:1121–1124.
166. Wood, J. H., Glaeser, B. S., Enna, S. J., and Hare, T. A., 1978, *J. Neurochem.* 30:291–293.
167. Berry, H. C., and Steiner, J. C., 1979, *Neurology (Minneap.)* 29:535.
168. Hespe, W., Roberts, E., and Prins, H., 1969, *Brain Res.* 14:663–671.
169. Perry, T. L., and Hansen, S., 1973, *J. Neurochem.* 21:1167–1175.
170. Purpura, D. P., Girado, M., Smith, T. G., and Gomez, J. A., 1958, *Proc. Soc. Exp. Biol.* 97:348–353.
171. Roberts, E., Lowe, I. P., Guth, L., and Jelinek, B., 1958, *J. Exp. Zool.* 138:313–328.
172. Biswas, B., and Barlsson, A., 1978, *Psychopharmacology* 59:91–94.
173. Tower, D. B., 1961, *Clinical Pathology of the Nervous System* (J. Folch-Pi, ed.), Pergamon Press, Oxford, London, pp. 307–344.
174. Böhlen, P., Huot, S., and Palfreyman, M. G., 1979, *Brain Res.* 167:297–350.
175. Loscher, W., 1979, *J. Neurochem.* 32:1587–1591.
176. Böhlen, P., Huot, S., Mellet, M., and Palfreyman, M. G., 1980, *Brain Res. Bull. [Suppl]* 5:905–908.

177. Wood, J. H., Hare, T. A., Enna, S. J., and Manyam, N. V. B., 1980, *Brain Res. Bull.* [*Suppl*] **5:**111–114.
178. Enna, S. J., Bennett, J. P., Bylund, D. B., Crese, I., Burt, D. R., Charness, M. E., Yamamura, H. I., Simantov, R., and Snyder, S. H., 1977, *J. Neurochem.* **28:**233–236.
179. Fahn, S., and Côté, L. J., 1968, *J. Neurochem.* **15:**209–213.
180. Storm-Mathisen, J., and Fonnum, F., 1971, *J. Neurochem.* **18:**1105–1111.
181. Duggan, A. W., and McLennan, H., 1971, *Brain Res.* **25:**188–191.
182. Fahn, S., 1976, *GABA in Nervous System Function* (E. Roberts, T. N. Chase, and D. B. Tower, eds.), Raven Press, New York, pp. 169–186.
183. Graham, L. T., Shank, R. P., Werman, R., and Aprison, M. H., 1967, *J. Neurochem.* **14:**465–472.
184. Ziegler, M. G., Lake, C. R., Wood, J. H., and Ebert, M. H., 1980, *Neurobiology of Cerebrospinal Fluid 1* (J. H. Wood, ed.), Plenum Press, New York, pp. 141–152.
185. Ziegler, M. G., Wood, J. H., Lake, C. R., and Kopin, I. J., 1977, *Am. J. Psychiatry* **134:**565–568.
186. Manyam, N. V. B., Katz, L., Hare, T. A., Gerber, J. C., and Grossman, M. H., 1980, *Arch. Neurol.* **37:**352–355.
187. Wood, J. H., and Brooks, B. R., 1980, *Neurobiology of Cerebrospinal Fluid 1* (J. H. Wood, ed.), Plenum Press, New York, pp. 259–278.
188. Wood, J. H., Hare, T. A., Glaeser, B. S., Ballenger, J. C., and Post, R. M., 1979, *Neurology (Minneap.)* **29:**1203–1208.
189. Wood, J. H., Hare, T. A., Glaeser, B. S., Brooks, B. R., Ballenger, J. C., and Post, R. M., 1980, *Brain Res. Bull.* [*431Suppl*] **5:**747–753.
190. Lott, I. T., Coulombe, T., DiPaolo, R. V., Richardson, E. P., and Levy, H. J., 1978, *Neurology (Minneap.)* **28:**47–54.
191. Perry, T. L., Hansen, S., and Kloster, M., 1973, *N. Engl. J. Med.* **288:**337–342.
192. Van Gelder, N. M., Sherwin, A. L., and Rasmussen, T., 1972, *Brain Res.* **40:**385–393.
193. Brooks, B. R., Ziegler, M. G., Lake, C. R., Wood, J. H., Enna, S. J., and Engel, W. K., 1980, *Brain Res. Bull.* [*Suppl*] **5:**765–768.
194. Ziegler, M. G., Brooks, B. R., Lake, C. R., Wood, J. H., and Enna, S. J., 1980, *Neurology (Minneap.)* **30:**98–101.
195. Perlow, M. J., Enna, S. J., O'Brien, P. J., Hoffman, H. J., and Wyatt, R. J., 1979, *J. Neurochem.* **32:**265–268.
196. Hare, T. A., Wood, J. H., Manyam, N. V. B., Gerner, R. H., Ballenger, J. C., and Post, R. M., 1982, *Arch. Neurol.* (in press).
197. Hare, T. A., Wood, J. H., and Manyam, B. V., 1981, *Arch. Neurol.* **38:**491–494.
198. Chase, T. N., and Tamminga, C. A., 1979, *GABA-Neurotransmitters: Pharmacological, Biochemical and Pharmacological Aspects* (P. Krogsgaard, J. Scheel-Kruger, and H. Kofod, eds.), Academic Press, New York, pp. 283–294.
199. Vijayan, E., and McCann, S. M., 1978, *Endocrinology* **103:**1888–1893.
200. Vijayan, E., and McCann, S. M., 1978, *Brain Res.* **155:**35–43.
201. Locatelli, V., Cocchi, D., Frigerio, C., Betti, R., Krogsgaard, P., Racagni, G., and Muller, E. E., 1979, *Endocrinology* **105:**778–785.
202. Negro-Vilar, A., Vijayan, E., and McCann, S. M., 1980, *Brain Res. Bull.* **5:**239–244.
203. Gillis, R. A., DiMicco, J. A., Williford, D. J., Hamilton, B. L., and Gale, K. N., 1980, *Brain Res. Bull.* [*Suppl.*] **5:**303–315.
204. Grossman, M. H., Hare, T. A., Manyam, N. V. B., Glaeser, B. S., and Wood, J. H., 1980, *Brain Res.* **182:**99–106.
205. Hare, T. A., Grossman, M. H., Wood, J. H., Glaeser, B. S., and Manyam, N. V. B., 1980, *Brain Res. Bull.* [*Suppl*] **5:**725–729.
206. Post, R. M., Ballenger, J. C., and Goodwin, F. K., 1980, *Neurobiology of Cerebrospinal Fluid 1* (J. H. Wood, ed.), Plenum Press, New York, pp. 685–717.
207. Patsalos, P. N., and Lascelles, P. T., 1981, *J. Neurochem.* **36:**688–695.
208. Perry, T. L., Hansen, S., Schier, G. M., and Halpern, B., 1977, *J. Neurochem.* **29:**791–795.
209. Kish, S. J., Perry, T. L., and Hansen, S., 1979, *J. Neurochem.* **32:**1629–1636.

210. Böhlen, P., Tell, G., Schechter, P. J., Koch-Weser, J., Agid, Y., Coquillat, G., Chazot, G., and Fischer, C., 1980, *Brain Res. Bull.* [*Suppl*] **5**:761–764.

211. Doherty, J. D., Hattox, S. E., Snead, O. C., and Roth, R. H., 1978, *J. Pharmacol. Exp. Ther.* **207**:130–139.

212. Snead, O. C., Yu, R. K., and Huttenlocher, P. R., 1976, *Neurology (Minneapo.)* **26**:51–56.

213. Shumate, J. S., and Snead, O. C., 1979, *Res. Commun. Chem. Pathol. Pharmacol.* **25**:241–256.

214. Snead, O. C., Brown, G. B., and Morawetz, R. B., 1981, *N. Engl. J. Med.* **304**:93–95.

215. Snead, O. C., and Bearden, L. J., 1980, *Neurology (Minneap.)* **30**:832–838.

216. Doherty, J. D., Snead, O. C., and Roth, R. H., *Anal. Biochem.* **69**:268–277.

217. Wurtman, R. J., Scally, M. C., Gibson, C. J., and Hefti, F., 1979, *Catecholamines: Basic and Clinical Frontiers* (E. Usdin, I. J. Kopin, and J. D., Barchas, eds.), Pergamon Press, New York, pp. 64–66.

218. Gibson, C. J., and Wurtman, R. J., 1978, *Life Sci.* **22**:1399–1406.

219. Reis, D. J., and Wurtman, R., 1968, *Life Sci.* **7**:91–98.

220. Weil-Hulherbe, H., Axelrod, J., and Tomchick, R., 1959, *Science* **129**:1226–1227.

221. Ziegler, M. G., Lake, C. R., Wood, J. H., Brooks, B. R., and Ebert, M. H., 1977, *J. Neurochem.* **28**:677–679.

222. Sato, S., and Suzuki, J., 1975, *J. Neurosurg.* **43**:559–568.

223. Lake, C. R., Ziegler, M. G., Coleman, M. D., and Kopin, I. J., 1977, *N. Engl. J. Med.* **296**:208–209.

224. Smith, B. H., and Sweet, W. H., 1978, *Neurosurgery* **3**:109–119.

225. Watson, S. J., Richard, C. S., Ciaranello, R. D., and Barchas, J. D., 1980, *Peptides* **1**:23–30.

226. Bird, S. J., Atweh, S. F., and Kuhar, M. J., 1976, *Opiates and Endogenous Opioid Peptides* (H. W. Kosterlitz, ed.), Elsevier/North Holland Biomedical Press, Amsterdam, pp. 109–204.

227. Brown, G. M., Friend, W. C., and Chambers, J. W., 1979, *Clinical Neuroendocrinology: A Pathophysiological Approach* (G. Tolis, F. Labrie, J. B. Martin, and F. Naftolin, eds.), Raven Press, New York, pp. 47–81.

228. Vijayan, E., and McCann, S. M., 1978, *Neuroendocrinology* **25**:150–165.

229. Levin, B. E., and Hubschmann, O. R., 1980, *Neurology (Minneap.)* **30**:65–71.

230. Zigmond, M. H., and Wurtman, R. J., 1970, *J. Pharmacol. Exp. Ther.* **172**:416–422.

231. Collu, R., Jequier, J. C., Letarte, J., Leboeuf, G., and Ducharne, J. R., 1973, *Can. J. Physiol. Pharmacol.* **51**:890–892.

232. Lovenberg, W., Levine, R. A., Robinson, D. S., Ebert, M., Williams, A. C., and Calne, D. B., 1979. *Science* **204**:624–626.

233. Reis, D. J., Weinbren, M., and Corvelli, A., 1968, *J. Pharmacol. Exp. Ther.* **164**:135–145.

234. Perlow, M., Ebert, M. H., Gordon, E. K., Ziegler, M. G., Lake, C. R., and Chase, T. M., 1978, *Brain Res.* **139**:101–113.

235. Ziegler, M. G., Lake, C. R., Wood, J. H., and Ebert, M. H., 1976, *Nature* **264**:656–658.

236. Ziegler, M. G., Lake, C. R., and Kopin, I. J., 1976, *Nature* **261**:333–335.

237. Teychenne, P. F., Lake, C. R., and Ziegler, M. G., 1980, *Neurobiology of Cerebrospinal Fluid I* (J. H. Wood, ed.), Plenum Press, New York, pp. 197–206.

238. Lake, C. R., Wood, J. H., Ziegler, M. G., Ebert, M. H., and Kopin, I. J., 1978, *Arch. Gen. Psychiatry* **35**:237–240.

239. Post, R. M., Allen, F. H., and Ommaya, A. K., 1974, *Life Sci.* **14**:1885–1894.

240. Ziegler, M. G., Lake, C. R., and Kopin, I. J., 1976, *Brain Res.* **108**:436–440.

241. Cooper, J. R., Bloom, F. E., and Roth, R. H., 1978, *The Biochemical Basis of Neuropharmacology*, Oxford University Press, New York.

242. Chase, T. N., Gordon, E. K., and Ng, L. K. Y., 1973, *J. Neurochem.* **21**:581–587.

243. Gordon, E., Perlow, M., Oliver, J., Ebert, M., and Kopin, I., 1975, *J. Neurochem.* **25**:347–349.

244. Adér, J.-P., Aizenstein, M. L., Postema, F., and Korf, J., 1979, *J. Neural. Transm.* **46**:279–290.

245. MacKay, A. V. P., Yates, C. M., Wright, A., Hamilton, P., and Davies, P., 1978, *J. Neurochem.* **30**:841–848.

246. Gordon, E. K., and Oliver, J., 1971, *Clin. Chim. Acta* **35**:145–150.
247. Sjöström, R., Ekstedt, J., and Änggard, E., 1975, *J. Neurol. Neurosurg. Psychiatry* **38**:666–668.
248. Post, R. M., Goodwin, F. K., Gordon, E., and Watkin, D. M., 1973, *Science* **179**:897–899.
249. Sedvall, G., Fyro, B., Gullberg, B., Nyback, H., Wiesel, R.-A., and Wode-Helgodt, B., 1980, *Br. J. Psychiatry* **136**:366–374.
250. Sharman, D. F., 1973, *Br. Med. Bull.* **29**:110–115.
251. Kessler, J. A., Fenstermacher, J. D., and Patlak, C. S., 1976, *Brain Res.* **102**:131–141.
252. Wolfson, L. E., and Escrira, A., 1976, *Neurology (Minneap.)* **26**:781–784.
253. Gordon, E. K., Oliver, J., Black, K., and Kopin, I. J., 1974, *Biochem. Med.* **11**:32–40.
254. Karoum, F., Gillin, J. C., Wyatt, R. J., and Costa, E., 1975, *Biochem. Mass Spectrom.* **2**:183–189.
255. Muskiet, F. A. J., Jeuring, H. J., Korf, J., Sedvall, G., Westerink, B. H. C., Teelken, A. W., and Wolthers, B. G., 1979, *J. Neurochem.* **32**:191–194.
256. Langlais, P. J., McEntee, W. J., and Bird, E. D., 1980, *Clin. Chem.* **26**:786–788.
257. Seifert, W. E., Foxx, J. L., and Butler, I. J., 1980, *Ann. Neurol.* **8**:38–42.
258. Blombery, P. A., Kopin, I. J., Gordon, E. K., Markey, S. P., and Ebert, M. H., 1980, *Arch. Gen. Psychiatry* **37**:1095–1098.
259. Karoum, F., Wyatt, R., and Costa, E., 1974, *J. Neurochem.* **13**:165–176.
260. Jimerson, D. C., Gordon, E. K., Post, R. M., and Goodwin, F. K., 1975, *Brain Res.* **99**:434–439.
261. Lindvall, O., and Bjorklund, A., 1978, *Adv. Biochem. Psychopharmacol.* **19**:1–23.
262. Magnusson, T., and Rosengren, E., 1963, *Experientia* **19**:229–230.
263. Friedman, A. H., and Everett, G. M., 1964, *Adv. Pharmacol.* **3**:83–127.
264. Oldendorf, W. H., 1971, *Am. J. Physiol.* **221**:1629–1639.
265. Guldberg, H. C., and Yates, C. M., 1968, *Br. J. Pharmacol.* **33**:457–471.
266. Sharpless, N. S., Tyce, G. M., Thal, L. J., Waltz, J. M., Tabbaddor, K., and Wolfson, L. T., 1980, *Soc. Neurosci. Abstr.*
267. Tyce, G. M., Sharpless, N. S., Kerr, F. W. L., and Muenter, M. D., 1980, *J. Neurochem.* **34**:210–212.
268. Wurtman, R. J., Larin, F., Mostafapour, S., and Fernstrom, J. D., 1974, *Science* **185**:183–184.
269. Wurtman, R. J., Cohen, E. L., and Fernstrom, J. D., 1977, *Neuroregulators and Psychiatric Disorders* (E. Usdin, D. A. Hamburg, and J. D. Barchas, eds.), Oxford University Press, New York, pp. 103–121.
270. Bridges, P. K., Bartlett, J. R., Sepping, P., Kentamaneni, B. D., and Curzon, G., 1976, *Psychol. Med.* **6**:399–405.
271. Extein, I., Roth, R. H., and Bowers, M. B., 1974, *Biol. Psychiatry* **9**:161–170.
272. Pletsches, A., Bartholini, G., and Tissot, R., 1967, *Brain Res.* **4**:106–109.
273. Guldberg, H. C., 1969, *Metabolism of Amines in the Brain* (G. Hooper, ed.), Macmillan, London, pp. 55–64.
274. Bartholini, G., Pletscher, A., and Tissot, R., 1966, *Experientia* **22**:609–610.
275. Jakupcevic, M., Bulat, M., and Lackovic, Z., 1978, *Jugosl. Physiol. Pharmacol. Acta* **14**:100–102.
276. Portig, P. J., and Vogt, M., 1969, *J. Physiol. (Lond.)* **204**:687–715.
277. Bartholini, G., Pletscher, A., and Tissot, R., 1971, *Brain Res.* **27**:163–168.
278. Sourkes, T. L., 1973, *J. Neural. Transm.* **34**:153–157.
279. DiChiro, G., Hammock, M. K., and Bleyer, W. A., 1976, *Neurology (Minneap.)* **26**:1–8.
280. Garelis, E., and Sourkes, T. L., 1974, *J. Neurol. Neurosurg. Psychiatry* **37**:704–710.
281. Manshardt, J., and Wurtman, R. J., 1968, *Nature* **217**:574–575.
282. Curzon, G., Gumpert, E. J. W., and Sharpe, D. M., 1971, *Nature [New Biol.)* **231**:189–191.
283. Garelis, E., and Sourkes, T. L., 1973, *J. Neurol. Neurosurg. Psychiatry* **36**:625–629.
284. Young, S. N., Lal, S., Martin, J. B., Ford, R. M., and Sourkes, T. L., 1973, *Psychiatr. Neurol. Neurchir.* **76**:439–444.
285. Kessler, J. A., Fenstermacher, J. D., and Patlak, C. S., 1976, *Neurology (Minneap.)* **26**:434–440.
286. Bobillier, P., and Mouret, J. R., 1971, *Int. J. Neurosci.* **2**:271–282.

287. Hillier, J. G., Martin, P. R., and Redfern, P. H., 1975, *J. Pharm. Pharmacol. [Suppl]* **27**:40P.
288. Perlow, M. J., Gordon, E. K., Ebert, M. H., Hoffman, H. J., and Chase, T. N., 1977, *J. Neurochem.* **28**:1381–1383.
289. McGeer, P. L., and McGeer, E. G., 1976, *J. Neurochem.* **26**:65–76.
290. Robinson, D. S., Sourkes, T. L., Nies, A., Harris, L. S., Spector, S., Barlett, D. L., and Kaye, I. S., 1977, *Arch. Gen. Psychiatry* **34**:89–92.
291. Carlsson, A., 1976, *The Basal Ganglia* (M. D. Yahr, ed.), Raven Press, New York, pp. 181–189.
292. Spokes, E. G. S., 1979, *Brain* **102**:333–346.
293. Bowers, M. B., and Gerbode, F. A., 1968, *Nature* **219**:1256–1257.
294. Gottfries, C. G., Gottfries, I., Johansson, B., Olsson, R., Persson, T., Roos, B. E., and Sjöström, R., 1971, *Neuropharmacology* **10**:665–672.
295. Cohen, D. J., Shaywitz, B. A., Young, J. G., and Bowers, M. B., 1980, *Neurobiology of Cerebrospinal Fluid 1* (J. H. Wood, ed.), Plenum Press, New York, pp. 665–683.
296. Anden, N.-E., Roos, B.-E., and Werdinius, B., 1963, *Life Sci.* **2**:448–458.
297. Carlsen, A., and Hillarp, N.-A., 1962, *Acta Physiol. Scand.* **55**:95–100.
298. Ashcroft, G. W., Crawford, T. B. B., Dow, R. C., and Guldberg, H. C., 1968, *Br. J. Pharmacol. Chemother.* **33**:441–456.
299. Gordon, E. K., Markey, S. P., Sherman, R. L., and Kopin, I. J., 1976, *Life Sci.* **18**:1285–1292.
300. Wiesel, F.-A., 1975, *Neurosci. Lett.* **1**:219–224.
301. Post, R. M., Kotin, J., Goodwin, F. K., and Gordon, E. K., 1973, *Am. J. Psychiatry* **130**:67–72.
302. Davidson, D. W., Pullar, I. A., Mawdsley, C., Kinloch, N., and Yates, C. M., 1977, *J. Neurol. Neurosurg. Psychiatry* **40**:741–745.
303. Griande, M., and Radulovacki, M., 1976, *J. Neurochem.* **26**:1301–1302.
304. Bowers, M. B., 1970, *Life Sci.* **9**:691–694.
305. Ebert, M. H., Kartzinel, R., Cowdry, R. W., and Goodwin, F. K., 1980, *Neurobiology of Cerebrospinal Fluid 1* (J. H. Wood, ed.), Plenum Press, New York, pp. 97–112.
306. Kirstein, L., Bowers, M. B., and Heninger, G., 1976, *Biol. Psychiatry* **11**:421–434.
307. Kukino, K., and Deguchi, T., 1977, *Chem. Pharm. Bull. (Tokyo)* **25**:2257–2262.
308. Papeschi, R., Molina-Negro, P., Sourkes, T. L., and Erba, G., 1972, *Neurology (Minneap.)* **22**:1151–1159.
309. Anderson, G. M., Young, J. G., and Cohen, D. J., 1979, *J. Chromatogr.* **164**:501–505.
310. Änggard, E., Sjöquist, B., and Sjöström, R., 1970, *J. Chromatogr.* **50**:251–259.
311. Gerbode, F. A., and Bowers, M. B., 1968, *J. Neurochem.* **15**:1053–1055.
312. Jimerson, D. C., Gordon, E. K., Post, R. M., and Goodwin, F. K., 1978, *Commun. Psychopharmacol.* **2**:343–350.
313. Korf, J., Ottema, S., and Van der Veen, I., 1971, *Anal. Biochem.* **40**:187–191.
314. Sjöquist, B., Lindstrom, B., and Änggard, E., 1973, *Life Sci.* **13**:1655–1664.
315. Swahn, C. G., Sandgarde, B., Wiesel, F.-A., and Sedvall, G., 1976, *Psychopharmacology* **48**:147–152.
316. Voght, W., Jacob, K., and Ohnesorge, A.-B., 1980, *J. Chromatogr.* **199**:191–197.
317. Smith, B. H., and Sweet, W. H., 1978, *Neurosurgery* **3**:257–272.
318. Aghajanian, G. K., Haigler, H. J., and Bennett, J. L., 1975, *Handbook of Psychopharmacology, Biogenic Amine Receptors*, Volume 6 (L. L. Iverson, S. D. Iverson, and S. H. Snyder, eds.), Plenum Press, New York, pp. 63–96.
319. Myers, R. D., 1973, *Serotonin and Behavior* (J. Barchas and E. Usdin, eds.), Academic Press, New York, pp. 293–302.
320. Bremer, F., 1977, *Ann. Neurol.* **2**:1–6.
321. Akil, H., and Liebeskind, J. C., 1975, *Brain Res.* **94**:279–296.
322. Hosobuchi, Y., and Bloom, F. E., 1982, *Neurobiology of Cerebrospinal Fluid 2* (J. H. Wood, ed.), Plenum Press, New York (in press).
323. Johansson, F., vonKnorring, L., Sedvall, G., and Terenius, L., 1980, *Psychiatr. Res.* **2**:167–172.
324. Von Essen, C., 1972, *J. Pharm. Pharmacol.* **24**:668.
325. Swank, R. L., and Hisson, W., 1964, *Arch. Neurol.* **10**:468–472.

326. Kamberi, I. A., Mical, R. S., and Porter, J. C., 1970, *Endocrinology* **87**:1–12.

327. Kamberi, I. A., Mical, R. S., and Porter, J. C., 1971, *Endocrinology* **88**:1288–1293.

328. Wehrenberg, W. B., McNicol, D., Frantz, A. G., and Ferin, M., 1980, *Endocrinology* **107**:1747–1750.

329. Reinhard, J. F., Liebmann, J. E., Schlosberg, A. J., and Moskowitz, M. A., 1979, *Science* **206**:85–87.

330. Axelrod, J., and Inscoe, J. K., 1963, *J. Pharmacol. Exp. Ther.* **141**:161–165.

331. Bulat, M., and Supek, Z., 1968, *J. Neurochem.* **15**:383–389.

332. Hardebo, J. E., and Owman, C., 1980, *Ann. Neurol.* **8**:1–11.

333. Welch, K. M. A., Meyer, J. S., and Kwant, S., 1972, *J. Neurochem.* **19**:1079–1087.

334. Brodner, R. A., Dohrmann, G. J., Roth, R. H., and Rubin, R. A., 1980, *Surg. Neurol.* **13**:337–343.

335. Fernstrom, J. D., and Wurtman, R. J., 1972, *Science* **178**:414–416.

336. Oldendorf, W. H., and Seabo, J., 1976, *Am. J. Physiol.* **230**:94–98.

337. Miller, M., and Resnick, O., 1980, *Exp. Neurol.* **67**:298–314.

338. Eccleston, D., Ashcroft, G. W., and Crawford, T. B. B., 1965, *J. Neurochem.* **12**:493–503.

339. Ashcroft, G. W., Eccleston, D., and Crawford, T. B. B., 1965, *J. Neurochem.* **12**:483–492.

340. Fernstrom, J. D., and Wurtman, R. J., 1971, *Science* **173**:149–152.

341. Ternaux, J. P., Boireau, A., Bourgoin, S., Hamon, M., Hery, F., and Glowinski, J., 1976, *Brain Res.* **101**:533–548.

342. Eccleston, D., Ashcroft, G. W., Moir, A. T. B., Parker-Rhodes, A., Lutz, W., and O'Mahoney, D. P., 1968, *J. Neurochem.* **15**:947–957.

343. Young, S. N., Garelis, E., Lal, S., Martin, J. B., Molina-Negro, P., and Sourkes, T. L., 1974, *J. Neurochem.* **22**:777–779.

344. Modigh, K., 1975, *J. Neurochem.* **25**:351–352.

345. Young, S. N., Gauthier, S., Anderson, G. M., and Purdy, W. C., 1980, *J. Neurol. Neurosurg. Psychiatry* **43**:438–445.

346. Young, S. N., Etienne, P., and Sourkes, T. L., 1976, *J. Neurol. Neurosurg. Psychiatry* **39**:239–243.

347. Rapoport, M. I., and Beisel, W. R., 1968, *J. Clin. Invest.* **47**:934–939.

348. Rapoport, M. I., Feigin, R. D., Bruton, J., and Beisel, W. R., 1966, *Science* **153**:1642–1644.

349. Hery, F., Chouvet, G., Kan, J. P., Pugol, J. F., and Glowinski, J., 1977, *Brain Res.* **123**:137–145.

350. Young, S. N., Anderson, G. M., and Purdy, W. C., 1980, *J. Neurochem.* **34**:309–315.

351. Giulian, D., Pohorecky, L. A., and McEwen, B. S., 1973, *Endocrinology* **93**:1329–1335.

352. Young, S. N., Anderson, G. M., Gauthier, S., and Purdy, W. C., 1981, *J. Neurochem.* **34**:1087–1092.

353. Everett, G. M., 1974, *Adv. Biochem. Psychopharmacol.* **10**:261–262.

354. Guilleminault, C., Tharp, B. R., and Cousin, D., 1973, *J. Neurol. Sci.* **18**:435–441.

355. Trimble, M., Chadwick, D., Reynolds, E. H., and Marsden, C. D., 1975, *Lancet* **1**:583.

356. Von Woert, M. H., and Sethy, V. H., 1975, *Neurology (Minneap.)* **25**:135–140.

357. Magnussen, I., Nielsen-Kudsk, F., 1980, *Acta Pharmacol. Toxicol. (Kbh.)* **46**:257–262.

358. Kuhar, M. J., Roth, R. H., and Aghajanian, G. K., 1971, *Brain Res.* **35**:167–176.

359. Corrodi, H., Fuxe, K., and Hökfelt, T., 1967, *J. Pharm. Pharmacol.* **19**:433–438.

360. Moir, A. T. B., and Eccleston, D., 1968, *J. Neurochem.* **15**:1093–1108.

361. Roos, B. E., 1962, *Life Sci.* **1**:25–27.

362. Ashcroft, G. W., Dow, R. C., and Moir, A. T. B., 1968, *J. Physiol. (Lond.)* **199**:397–425.

363. Bulat, M., and Zivkovic, B., 1973, *J. Pharm. Pharmacol.* **25**:178–179.

364. Bowers, M. B., 1970, *J. Neurochem.* **17**:827–828.

365. Meek, J. L., and Neff, N. J., 1973, *Neuropharmacology* **12**:497–499.

366. Bulat, M., and Zivkovic, B., 1978, *J. Physiol. (Lond.)* **275**:191–197.

367. Eccleston, D., Ashcroft, G. W., Crawford, T. B. B., Stanton, J. B., Wood, D., and McTurk, P. H., 1970, *J. Neurol. Neurosurg. Psychiatry* **33**:269–272.

368. Jakupcevic, M., Lackovic, Z., Stefoski, D., and Bulat, M., 1977, *J. Neurol. Sci.* **31**:165–177.

369. Garelis, E., and Neff, N. H., 1974, *Science* **183**:532–533.

370. Weir, R. L., Chase, R. N., Ng, L. K. Y., and Kopin, I. J., 1973, *Brain Res.* **52**:409–412.

371. Bulat, M., 1977, *Brain Res.* **122**:388–391.
372. Bulat, M., Lackovic, Z., Jakupcevic, M., and Damjanov. I., 1974, *Science* **185**:527–528.
373. Nicoletti, F., Raffaele, R., Falsaperle, A., and Paci, R., 1981, *Eur. Neurol.* **20**:8–12.
374. Anderson, H., and Roos, B.-E., 1969, *Acta Paediatr. Scand.* **58**:601–608.
375. Gottfries, C. G., Roos, B.-E., and Winblad, B., 1974, *Acta Psychiatr. Scand.* **50**:496–507.
376. Hare, T. A., and Wood, J. H., 1982, *Handbook of Neurochemistry*, 2nd ed. Volume 10 (A. Lajtha, ed.), Plenum Press, New York (in press).
377. Botez, M. I., Young, S. N., Bachevalier, J., and Gauthier, S., 1979, *Nature* **278**:182–183.
378. Chase, T. N., Katz, R. I., and Kopin, I. J., 1970, *Neuropharmacology* **9**:103–108.
379. Tabakoff, B., Bulat, M., and Anderson, R. A., 1975, *Nature* **254**:708–710.
380. Fahn, S., 1979, *Adv. Neurol.* **26**:117.
381. Guldberg, H. C., Ashcroft, G. W., and Crawford, T. B. B., 1966, *Life Sci.* **5**:1571–1575.
382. Zivkovic, B., and Bulat, M., 1971, *J. Pharm. Pharmacol.* **23**:539.
383. Green, A. R., and Grahame-Smith, D. G., 1975, *Neuropharmacology* **14**:107–113.
384. Shaywitz, B. A., Cohen, D. J., and Bowers, M. B., 1980, *Neurobiology of Cerebrospinal Fluid 1* (J. H. Wood, ed.), Plenum Press, New York, pp. 219–236.
385. Livrea, P., Di Reda, L., Puca, F. M., Genco, S., Specchio, L. M., and Papagno, G., 1977, *Eur. Neurol.* **16**:280–285.
385a. Hyyppä, M. T., and Kangasniemi, P., 1977, *Headache* **17**:25–27.
386. Anderson, G. M., and Purdy, W. C., 1979, *Anal. Chem.* **51**:283–286.
387. Ashcroft, G. W., and Sharman, D. F., 1962, *Br. J. Pharmacol.* **19**:153–160.
388. Beck, O., and Hesselgren, T., 1980, *J. Chromatogr.* **181**:100–102.
389. Dombro, R., and Hutson, D. G., 1980, *Clin. Chim. Acta* **100**:231–237.
389a. Korf, J., and Valkenburgh, T., 1969, *Clin. Chem. Acta* **26**:301–306.
390. Nelson, P. V., Carey, W. F., and Pollard, A. C., 1975, *J. Clin. Pathol.* **28**:828–833.
391. Daly, J. W., 1977, *Cyclic Nucleotides in the Nervous System*, Plenum Press, New York.
392. Daly, J. W., 1977, *Int. Rev. Neurobiol.* **20**:105–168.
393. Ferrendelli, J. A., 1978, *Adv. Cyclic Nucleotide Res.* **9**:453–464.
394. Ferrendelli, J. A., 1975, *Cyclic Nucleotides in Disease* (B. Weiss, ed.), University Park Press, Baltimore, pp. 377–390.
395. Nathanson, J. A., 1977, *Physiol. Rev.* **57**:157–256.
396. Weiss, B., and Costa, E., 1968, *Biochem. Pharmacol.* **17**:2107–2116.
397. Ferrendelli, J. A., Kinscherf, D. A., and Chang, M. M., 1975, *Brain Res.* **84**:63–73.
398. Nakayawa, K., and Sano, M., 1974, *J. Biol. Chem.* **249**:4207–4211.
399. Cramer, H. M., Paul, M. I., Silvergeld, S., and Forn, J., 1971, *J. Neurochem.* **18**:1605–1608.
400. Schmidt, M. J., Schmidt, D. E., and Robison, G. A., 1971, *Science* **173**:1142–1143.
401. Steiner, A. L., Ferrendelli, J. A., and Kipnis, D. M., 1972, *J. Biol. Chem.* **247**:1121–1124.
402. Goldberg, N. D., O'Dea, R. F., and Haddox, M. K., 1973, *Adv. Cyclic Nucleotide Res.* **3**:155–223.
403. Bloom, F. E., Hoffer, B. J., Battenberg, E. F., Siggins, G. R., Steiner, A. L., Parker, C. W., and Wedner, H. J., 1972, *Science* **177**:436–438.
404. Wedner, H. J., Hoffer, B. J., Battenberg, E. F., Steiner, A. L., Parker, C. W., and Bloom, F. E., 1972, *J. Histochem. Cytochem.* **20**:293–295.
405. Mao, C. C., Guidetti, A., and Landis, S., 1975, *Brain Res.* **90**:335–339.
406. Rubin, E. H., and Ferrendelli, J. A., 1977, *J. Neurochem.* **29**:43–51.
407. Peake, G. T., Wilson, M. C., and Ratner, A., 1975, *Cyclic Nucleotides in Disease* (B. Weiss, ed.), University Park Press, Baltimore, pp. 227–255.
408. Phillis, J. W., 1977, *Can. J. Neurol. Sci.* **4**:151–195.
409. Kebabian, J. W., 1977, *Adv. Cyclic Nucleotide Res.* **8**:421–508.
410. Rindler, M. J., Bashor, M. M., Spitzer, N., and Saier, H. H., 1978, *J. Biol. Chem.* **253**:5431–5436.
411. Gorin, E., and Brenner, T., 1976, *Biochim. Biophys. Acta* **451**:20–28.
412. Sebens, J. B., and Korf, J., 1975, *Exp. Neurol.* **46**:333–344.
413. Dascombe, M. J., and Milton, A. S., 1975, *Br. J. Pharmacol.* **54**:254P–255P.
414. Brooks, B. R., Engel, W. K., and Sode, J., 1977, *Arch. Neurol.* **34**:468–469.
415. Korf, J., Boer, P. H., and Fekkes, D., 1976, *Brain Res.* **113**:551–561.

416. Kiessling, M., Lindl, T., and Cramer, H., 1975, *Arch. Psychiatr. Nervenkr.* **220**:325–333.
417. Florendo, N. T., Barrnett, R. J., and Greengard, P., 1971, *Science* **173**:745–747.
418. Cramer, H., Renaud, B., and Ortega-Suhr Kamp, E., 1975, *J. Clin. Chem. Clin. Biochem.* **13**:245.
419. Tsang, D., Lal, S., Sourkes, T. L., Ford, R. M., and Aronoff, A., 1976, *J. Neurol. Neurosurg. Psychiatry* **39**:1186–1190.
420. Fleischer, A. S., Rudman, D. R., Fresh, C. B., and Tindall, G. T., 1977, *J. Neurosurg.* **47**:517–524.
421. Rudman, D., O'Brien, M. S., McKinney, A. S., Hoffman, J. C., and Patterson, J. H., 1976, *J. Clin. Endocrinol. Metab.* **42**:1088–1097.
422. Myllylä, V. V., Vapaatalo, H., Hokkanen, E., and Heikkinen, E. R., 1974, *Eur. Neurol.* **12**:28–32.
423. Katz, J. B., Valases, C., Catravas, G. N., and Wright, S. J., 1978, *Life Sci.* **22**:445–450.
424. Perlow, M. J., Festoff, B., Gordon, E. K., Ebert, M. H., Johnson, D. K., and Chase, T. N., 1977, *Brain Res.* **126**:391–396.
425. Clarenbach, P. A., Wenzel, D. C., and Cramer, H. L., 1978, *Eur. Neurol.* **17**:83–86.
426. Myllylä, V. V., Heikkinen, E. R., Similä, S., Hokkanen, E., and Vapaatalo, H., 1975, *Z. Kinderheilkd.* **118**:259–264.
427. Heikkinen, E. R., Myllylä, V. V., Vapaatalo, H., and Hokkanen, E., 1974, *Eur. Neurol.* **11**:270–280.
428. Cramer, H., Goodwin, F. K., Post, R. M., and Bunny, W. E., 1972, *Lancet* **1**:1346–1347.
429. Geisler, A., Bech, P., Johannesen, M., and Rafaelsen, O. J., 1976, *Neuropsychobiology* **2**:211–220.
430. Cramer, H., Ng, L. K. Y., and Chase, T. N., 1973, *Arch. Neurol.* **29**:197–199.
431. Post, R. M., Cramer, H., and Goodwin, F. K., 1977, *Neuroregulators and Psychiatric Disorders* (E. Usdin, D. A. Hamberg and J. D. Barchas, eds.), Oxford University Press, New York, pp. 464–469.
432. Fleischer, A. S., and Tindall, G. T., 1980, *Neurobiology of Cerebrospinal Fluid 1* (J. H. Wood; ed.), Plenum Press, New York, pp. 337–344.
433. Gilman, A. G., 1972, *Adv. Cyclic Nucleotide Res.* **2**:9–24.
434. Gilman, A. G., 1970, *Proc. Natl. Acad. Sci. U.S.A.* **67**:305–312.
435. Tovey, K. C., Oldham, K. G., and Whelan, J. A. M., 1974, *Clin. Chim. Acta* **56**:221–234.
436. Goldberg, M. L., 1977, *Clin. Chem.* **23**:576–580.
437. Harper, J. F., and Brooker, G., 1975, *J. Cyclic Nucleotide Res.* **1**:207–218.
438. Steiner, A. L., Wehmann, R. E., Parker, C. W., and Kipnis, D. M., 1972, *Adv. Cyclic Nucleotide Res.* **2**:51–61.
439. Tihon, C., Goren, M. B., Spitz, E., and Rickenberg, H. V., 1977, *Anal. Biochem.* **80**:652–653.
440. Zimmerman, T. P., Winston, M. S., and Chu, L. C., 1976, *Anal. Biochem.* **71**:79–95.
441. Goldberg, N. D., O'Toole, A. G., and Haddox, M. K., 1972, *Adv. Cyclic Nucleotide Res.* **2**:63–80.
442. Cserr, H., and Van Dyke, D. H., 1971, *Am. J. Physiol.* **220**:718–723.
443. Forn, J., 1972, *Biochem. Pharmacol.* **21**:619–624.
444. Wolfson, L. I., Katzman, R., and Escriva, A., 1974, *Neurology (Minneap.)* **24**:772–779.
445. Burns, D., London, J., Brunswick, D. J., Pring, M., Garkinkel, D., Rabinowitz, J. L., and Mendels, J., 1976, *Biol. Psychiatry* **11**:125–157.
446. Cramer, H., Ng, L. K. Y., and Chase, T. N., 1972, *J. Neurochem.* **19**:1601–1602.
447. Chase, T. N., 1980, *Neurobiology of Cerebrospinal Fluid 1* (J. H. Wood, ed.), Plenum Press, New York, pp. 207–218.
448. Goodwin, F. K., Post, R. M., Dunner, D. L., and Gordon, E. K., 1973, *Am. J. Psychiatry* **130**:73–79.
449. Neff, N. H., Tozer, T. N., and Brodie, B. B., 1967, *J. Pharmacol. Exp. Ther.* **158**:214.
450. Watson, E., and Wilk, S., 1973, *J. Neurochem.* **21**:1569–1571.
451. Sharpless, N. S., 1981, *Neuropharmacology*, **20**:211–216.
452. Krulich, L., 1979, *Annu. Rev. Physiol.* **41**:603–615.
453. McCann, S. M., Krulich, L., Ojeda, S. R., Negro-Vilar, A., and Vijayan, E., 1979, *Central Regulation of the Endocrine System* (K. Fine, T. Hökfelt, and R. Luft, eds.), Plenum Press, New York, pp. 329–347.

454. Weiner, R. I., and Ganong, W. F., 1978, *Physiol. Rev.* **58**:905–976.
455. McCann, S. M., 1980, *Neuroendocrinology* **31**:355–363.
456. Lumpkin, M. D., Negro-Vilar, A., and McCann, S. M., 1981, *Science* **211**:1072–1074.
456a. Tache, Y., Charpenet, G., Chretien, M., and Collu, R., 1979, *Central Nervous System Effects of Hypothalamic Hormones and Other Peptides*. (R. Collu, A. Barbeau, J. R. Ducharne, and J.-G. Rachefort, eds.), Raven Press, New York, pp. 301–313.
457. Jackson, I. M. D., 1980, *Neurobiology of Cerebrospinal Fluid 1* (J. H. Wood, ed.), Plenum Press, New York, pp. 625–650.
458. Knigge, K. M., Morris, M., Scott, D. E., Joseph, S. A., Notter, M., Schock, D., and Krobisch-Dudley, G., 1975, *Fluid Environment of the Brain* (H. F. Cserr, J. D. Fenstermacher, and V. Fencl, eds.), Academic Press, New York, pp. 237–253.
459. Knigge, K. M., Scott, D. E., Kobayashi, H., and Ishii, S. (eds.), 1975, *Brain–Endocrine Interaction, II, The Ventricular System in Neuroendocrine Mechanisms*, S. Karger, Basel.
460. Rodriguez, E. M., 1976, *J. Endocrinol.* **71**:407–443.
461. Perlow, M. J., 1981, *Brain Res. Bull.* **6**:171–176.
462. Post, K. D., Biller, B. J., and Jackson, I. M. D., 1980, *Neurobiology of Cerebrospinal Fluid 1* (J. H. Wood, ed.), Plenum Press, New York, pp. 591–604.
463. Vigh, B., and Vigh-Teichman, I., 1973, *Int. Rev. Cytol.* **35**:189–251.
464. Vigh-Teichman, I., and Vigh, B., 1970, *Aspects of Neuroendocrinology* (W. Bargmann and B. Scharrer, eds.), Springer-Verlag, Berlin, pp. 329–337.
465. Pollay, M., 1979, *Contemporary Neurosurgery 14* (G. T. Tindall and D. M. Long, eds.), Williams & Wilkins, Baltimore.
466. Joseph, S. A., and Knigge, K. M., 1978, *The Hypothalamus* (S. Reichlin, R. J. Baldissarini, and J. B. Martin, eds.), Raven Press, New York, pp. 15–47.
467. Uremura, H., Asai, T., Nozaki, M., and Kobayashi, H., 1975, *Cell Tissue Res.* **160**:443–452.
468. Willkowski, W., 1968, *Z. Zellforsch.* **86**:111–128.
469. Kobayashi, H., 1975, *Brain–Endocrine Interaction, II, The Ventricular System* (K. M. Knigge, D. E. Scott, H. Kobayashi, and S. Ishii, eds.), S. Karger, Basel, pp. 109–122.
470. Bergland, R. M., Davis, S. L., and Page, R. B., 1977, *Lancet* **2**:276–277.
471. Oliver, C., Mical, R. S., and Porter, J. C., 1977, *Endocrinology* **101**:598–604.
472. Nakai, Y., and Naito, N., 1974, *J. Electron Microsc. (Tokyo)* **23**:19–32.
473. Weindl, A., 1973, *Frontiers in Neuroendocrinology* (W. F. Ganong and L. Martini, eds.), Oxford University Press, London, pp. 1–32.
474. Kizer, J. S., Palkovits, M., and Brownstern, M. J., 1976, *Endocrinology* **98**:311–317.
475. Wald, A., Hochwald, G. M., and Gandhi, M., 1978, *Brain Res.* **151**:283–290.
476. Cornford, E. M., Braun, L. D., Crane, P. D., and Oldendorf, W. H., 1978, *Endocrinology* **103**:1297–1303.
477. Miyachi, Y., Mecklenburg, R. S., Hansen, J. W., and Lipsett, M. B., 1973, *J. Clin. Endocrinol.* **37**:63–67.
478. Redding, T. W., and Schally, A. V., 1972, *Neuroendocrinology* **9**:250–256.
479. Clemens, J. A., and Sawyer, B. D., 1974, *Exp. Brain Res.* **21**:399–402.
480. Becker, K. L., Silva, O. L., Post, R. M., Ballenger, J. C., Carman, J. S., Snider, R. H., and Moore, C. F., 1980, *Brain Res.* **194**:598–602.
481. Stekolinikov, L. I., and Abdukarimov, A., 1969, *Biofizika* **14**:921–925.
482. Margolis, R. H., and Altszuler, N., 1967, *Nature* **215**:1375–1376.
483. Rapoport, S. I., Klee, W. A., Pettigrew, K. D., and Ohno, K., 1980, *Science* **207**:84–86.
484. Marynick, S. P., Smith, G. B., Ebert, M. H., and Loriaux, L. D., 1977, *Endocrinology* **101**:562–567.
485. Marynick, S. P., Wood, J. H., and Loriaux, L. D., 1980, *Neurobiology of Cerebrospinal Fluid 1* (J. H. Wood, ed.), Plenum Press, New York, pp. 605–611.
486. Pardridge, W. M., and Mietus, L. J., 1979, *J. Clin. Invest.* **64**:145–154.
487. Kendall, J. W., Seaich, J. L., Allen, J. P., and Vander Laan, W. P., 1975, *Brain–Endocrine Interaction II: The Ventricular System* (K. M. Knigge, D. E. Scott, H. Kobayashi, and S. Ishii, eds.), S. Karger, Basel, pp. 313–325.
488. Kendall, J. W., Jacobs, J. J., and Kramer, R. M., 1971, *Brain–Endocrine Interaction: Median Eminence: Structure and Function* (K. M. Knigge, D. E. Scott, and A. Weidl, eds.), S. Karger, Basel, pp. 342–349.

489. Allen, J. P., Kendall, J. W. McGilvra, R., and Vancura, C., 1974, *J. Clin. Endocrinol. Metab.* **38:**586–593.

490. Shambaugh, G. E., III, Wilber, J. F., Montoya, E., Reider, H., and Blonsky, E. R., 1975, *J. Clin. Endocrinol. Metab.* **41:**131–134.

490a. Hyyppä, M. T., Liira, J., and Languik, V.-A., 1978, *Ann. Clin. Res.* **10:**133–138.

491. Kronheim, S., Berelowitz, M., and Pimstone, B. L., 1977, *Clin. Endocrinol.* **6:**411–415.

492. Sørensen, K. V., Christensen, S. E., Dupont, E., Hansen, A. P., Pedersen, E., and Orskov, H., 1980, *Acta Neurol. Scand.* **61:**186–191.

493. Schaub, C., Bluet-Pajot, M. T., Szikla, G., Lornet, C., and Talairach, J., 1977, *J. Neurol. Sci.* **31:**123–131.

494. Hagen, G. A., and Elliott, W. J., 1973, *J. Clin. Endocrinol. Metab.* **37:**415–422.

495. Schroeder, L. L., Johnson, J. C., and Malarkey, W. B., 1976, *J. Clin. Endocrinol. Metab.* **43:**1255–1260.

496. Luboshitzky, R., and Barzilai, D., 1978, *Acta Endocrinol. (Kbh.)* **87:**673–680.

497. Bäckström, T., Carstensen, H., and Södergard, R., 1976, *J. Steroid Biochem.* **7:**469–472.

498. Arendt, J., Wetterburg, L., Heyden, T., Sizenenko, P. C., and Paunier, L., *Horm. Res.* **8:**65–75.

499. Jenkins, J. S., Mather, H. M., and Ang, V., 1980, *J. Clin. Endocrinol. Metab.* **50:**364–367.

500. Greco, A. V., Ghirlanda, G., Fedeli, G., and Gambassi, G., *Eur. Neurol.* **3:**303–307.

501. Rehfeld, J. F., and Kruse-Larsen, C., 1978, *Brain Res.* **155:**19–26.

502. Fahrenkrug, J., Schaffalitzky de Muckadell, O. B., and Fahrenkrug, A., 1977, *Brain Res.* **124:**581–584.

503. Smith, A. G., and Shuster, S., 1976, *Lancet* **1:**1321–1322.

504. Nakao, K., Nakai, Y., Oki, S., Matsubara, S., Konishi, T., Nishitani, H., and Imura, H., 1980, *J. Clin. Endocrinol. Metab.* **50:**230–233.

505. Akil, H., Watson, S. J., Sullivan, S., and Barchas, J. D., 1978, *Life Sci.* **23:**121–126.

506. Horita, A., and Carino, M. A., 1975, *Psychopharmacol. Commun.* **1:**403–414.

507. Brown, M., River, J., Kobayashi, R., and Vale, W., 1978, *Gut Hormones* (S. R. Bloom, ed.), Churchill-Livingston, Edinburgh, pp. 550–558.

508. Smith, J. R., LaHann, T. R., Chesnut, R. M., Carino, M. A., and Horita, A., 1976, *Science* **196:**660–662.

509. Vijayan, E., and McCann, S. M., 1977, *Endocrinology* **100:**1727–1730.

510. Kendall, J. W., Rees, L. H., and Kramer, R., 1971, *Endocrinology* **88:**1503–1506.

511. Gordon, J., Bollinger, J., and Reichlin, S., 1972, *Endocrinology* **91:**696–701.

512. Oliver, C., Ben-Jonathan, N., Mical, R. S., and Porter, S. C., 1975, *Endocrinology* **97:**1138–1143.

513. Calas, A., 1975, *Brain–Endocrine Interaction II, The Ventricular System* (K. M. Knigge, D. E. Scott, H. Kobayashi, and S. Ishii, eds.), S. Karger, Basel, pp. 54–69.

514. Nicoll, R. A., Alger, B. E., and Jahr, C. E., 1980, *Proc. R. Soc. Lond. [Biol.]* **210:**133–149.

515. Oliver, C., Charvet, J. P., Codaccioni, J. L., Vagne, J., and Porter, J. C., 1974, *Lancet* **1:**873.

516. Pardridge, W. M., 1979, *Endocrinology* **105:**605–612.

517. Hagen, G. A., and Solberg, L. A., 1974, *Endocrinology* **95:**1398–1410.

518. Goldgefter, L., 1976, *Cell Tissue Res.* **168:**411–418.

519. Recabarren, S. E., and Wheaton, J. E., 1978, *Neuroendocrinology* **27:**1–8.

520. Ondo, J. G., Eskay, R. L., Mical, R. S., and Porter, J. C., 1973, *Endocrinology* **93:**231–237.

521. Ben-Jonathan, N., Mical, R. S., and Porter, J. C., 1974, *Endocrinology* **95:**18–25.

522. Zimmerman, E. A., Koylowski, G. P., and Scott, D. E., 1975, *Brain–Endocrine Interaction II, The Ventricular System* (K. M. Knigge, D. E. Scott, H. Kobayashi, and S. Ishii, eds.), S. Karger, Basel, pp. 123–134.

523. Cramer, O. M., and Barraclough, C. A., 1975, *Endocrinology* **96:**913–921.

524. Joseph, S. A., Sorrentino, S., and Sundberg, D. K., 1975, *Brain–Endocrine Interaction II, The Ventricular System* (K. M. Knigge, D. E. Scott, H. Kobayashi, and S. Ishii, eds.), S. Karger, Basel, pp. 306–321.

525. Morris, M., Tandy, B., Sundberg, D. K., and Knigge, K. M., 1975, *Neuroendocrinology* **18:**131–135.

526. Coppings, R. J., Malven, P. V., and Ramirez, V. D., 1977, *Proc. Soc. Exp. Biol. Med.* **154:**219–223.

527. Gunn, A., Fraser, H. M., Jeffcoate, S. L., Holland, D. T., and Jeffcoate, W. J., 1974, *Lancet* 1:1057.

528. Miyake, A., Kurachi, H., Kawamura, Y., Aono, T., Kurachi, K., and Yanogida, T., 1980, *Endocrinol. Jpn.* 27:117–119.

529. Rolandi, E., Barreca, T., Mastuizo, P., Gianrossi, R., Palleri, A., and Perria, C., 1976, *Lancet* 1:1080.

530. Puri, C. P., Puri, V., David, G. F. X., and Anand Kumar, T. C., 1980, *Brain Res.* 200:377–387.

531. Challis, J. R. G., Naftolin, F., Davies, I. J., Ryan, K. J., and Lanman, T., 1976, *Subcellular Mechanisms in Reproductive Neuroendocrinology* (F. Naftolin, K. J. Ryan, and J. Davies, eds.), Elsevier/North Holland Biomedical Press, Amsterdam, pp. 247–261.

532. McEwen, B. S., 1976, *Subcellular Mechanisms in Reproductive Neuroendocrinology* (F. Naftolin, K. J., Ryan, and J. Davies, eds.), Elsevier/North Holland Biomedical Press, Amsterdam, pp. 277–304.

533. Westphal, U. (ed.), 1971, *Steroid–Protein Interactions*, Springer-Verlag, New York.

534. Marynick, S. P., Havens, W. W., Ebert, M. H., and Loriaux, L. D., 1976, *Endocrinology* 99:400–405.

535. Logins, I. S., and MacLeod, R. M., 1977, *Brain Res.* 132:477–483.

536. Assies, J., Schellekens, P. M., and Touber, J. L., 1978, *J. Clin. Endocrinol. Metab.* 46:576–586.

537. Jordan, R. M., McDonald, S. D., Stevens, E. A., and Kendall, J. W., 1979, *Arch. Intern. Med.* 139:208–211.

538. Assies, J., Schellekens, P. M., and Touber, J. L., 1978, *Clin. Endocrinol. (Oxf.)* 8:487–491.

539. Latvala, M., Lüra, J., Langvik, V.-A., Jaykkä, H., Vapalahta, M., and Hyyppä, M. T., 1980, *Life Sci.* 26:1479–1484.

540. Jimerson, D. C., Post, R. M., Skyler, J., and Bunney, W. E., 1976, *J. Pharm. Pharmacol.* 28:845–847.

541. Nicholson, G., Greeley, G. H., Humm, J., Youngblood, W. W., and Kizer, J. S., 1980, *Brain Res.* 190:447–457.

542. Brownstein, M., Arimura, A., Sato, H., Schelly, A. V., and Kizer, J. S., 1975, *Endocrinology* 96:1456–1461.

543. Alpert, L. C., Brawer, J. R., Patel, Y. C., and Reichlin, S., 1976, *Endocrinology* 98:255–258.

544. Hökfelt, T., Elde, R., Johansson, O., Luft, R., and Arimura, A., 1975, *Neurosci. Lett.* 1:231–235.

545. Renaud, L. P., Martin, J. B., and Brazeau, P., 1975, *Lancet* 225:233–235.

546. Hall, R., Snow, M., Scanlon, M., Mora, B., and Gomez-Pan, A., 1978, *Metabolism* 27:1257–1262.

547. Chihara, K., Arimura, A., Chihara, M., and Schally, A. V., 1978, *Endocrinology* 103:912–916.

548. Patel, Y. C., Rao, K., and Reichlin, S., 1977, *N. Engl. J. Med.* 296:529–533.

548a. Sørensen, K. V., Christensen, S. E., Dupont, E., Hansen, A. P., Pedersen, E., and Orskov, H., 1980, *Acta Neurol. Scand.* 61:186–191.

548b. Cramer, H., Kohler, J., Oepen, G., Schomburg, G., and Schröter, E., 1981, *J. Neurol.* 225:183–187.

549. Arnold, M. A., Perlow, M. J., Reppert, S. M., Rorstad, O. P., and Martin, J. B., 1980, *Ann. Neurol.* 8:104.

550. Linfoot, J. A., Garcia, J. F., Wei, W., Fink, R., Sarin, R., Born, J. L., and Laurence, J. H., 1970, *J. Clin. Endocrinol. Metab.* 31:230–232.

551. Thomas, F. J., Lloyd, J. M., and Thomas, M. J., 1972, *J. Clin. Pathol.* 25:774–782.

552. Allen, J. P., Kendall, J. W., McGilvra, R., Lamorena, T. L., and Castro, A., 1974, *Arch. Neurol.* 31:325–328.

553. Nakao, K., Oki, S., Tanaka, I., Horii, K., Nakai, Y., Furui, T., Masonori, F., Kuwayama, A., Kageyama, N., and Imura, H., 1980, *J. Clin. Invest.* 66:1383–1390.

554. Jordan, R. M., Kendall, J. W., Seaich, J. L., Allen, J. P., Paulsen, C. A., Kerber, C. W., and VanderLaan, W. P., 1976, *Ann. Intern. Med.* 85:49–55.

555. Kendall, J., and Orwoll, E., 1980, *Frontiers of Neuroendocrinology*, Volume 6 (L. Martini and W. F. Ganong, eds.), Raven Press, New York, pp. 33–65.

556. Carroll, B. J., Curtis, G. C., and Mendels, J., 1976, *Psychol. Med.* 6:235–244.

557. Uete, T., Nishimura, S., Ohya, H., and Tatebayashi, Y., 1970, *J. Clin. Endocrinol. Metab.* **30**:208–214.

558. Perlow, M. J., Reppert, S. M., Boyart, R. M., and Klein, D. C., *Neuroendocrinology* **32**:193–196.

559. Kastin, A. J., Sandman, C. A., Stratton, L. O., Schally, A. V., and Miller, L. H., 1975, *Prog. Brain Res.* **42**:143–150.

560. Leonard, B. E., Kafoe, W. F., Thody, A. J., and Shuster, S., 1976, *J. Neurosci. Res.* **2**:39–45.

561. Rudman, D., Scott, J. W., DelRio, A. E., Houser, D. H., and Sheen, S., 1974, *Am. J. Physiol.* **226**:682–686.

562. Rudman, D., DelRio, A. E., Hollins, B. M., Houser, D. H., Keeling, M. E., Sutin, J., Scott, J. W., Sears, R. A., and Rosenberg, M. L., 1973, *Endocrinology* **92**:372–379.

563. Thody, A. J., DeRotte, A. A., and Van Wimersma Greidanus, T. J. B., 1979, *Brain Res. Bull.* **4**:213–216.

563a. De Rotte, A. A., Bouman, H. J., and Van Wimersma Greidanus, T. J. B., 1980, *Brain Res. Bull.* **4**:213–216.

564. George, J. M., 1978, *Science* **200**:342–343.

565. George, J. M., and Jacobowitz, D. M., 1975, *Brain Res.* **93**:363–366.

566. DeWied, D., 1976, *Life Sci.* **19**:685–690.

567. Raichle, M. E., and Grubb, R. L., 1978, *Brain Res.* **143**:191–194.

568. Pavel, S., Psatta, D., and Goldstein, R., 1977, *Brain Res. Bull.* **2**:251–254.

569. Luerssen, T. G., and Robertson, G. L., 1980, *Neurobiology of Cerebrospinal Fluid I* (J. H. Wood, ed.), Plenum Press, New York, pp. 613–623.

570. Vorherr, H., Bradbury, M. W. B., Hoghoughi, M., and Kleeman, C. R., 1968, *Endocrinology* **83**:246–250.

571. Heller, H., Hasan, S. H., and Saifi, A. W., 1968, *J. Endocrinol.* **41**:273–280.

572. Pavel, S., Goldstein, R., Papovicin, Carfariu, O., Foldes, A., and Farkas, E., 1979, *Waking Sleeping* **3**:347–352.

573. Goldstein, R., and Pavel, S., 1977, *J. Endocrinol.* **75**:175–176.

574. Pavel, S., Goldstein, R., Gheorghui, C., and Calb, M., 1977, *Science* **197**:179–180.

575. Pavel, S., and Goldstein, R., 1979, *J. Endocrinol.* **82**:1–6.

576. Pavel, S., 1979, *Prog. Brain Res.* **52**:445–458.

577. Zaidi, S. M., and Heller, H., 1974, *Endocrinology* **60**:195–196.

578. Brownfield, M. S., and Kozlowski, G. P., 1977, *Cell Tissue Res.* **178**:111–127.

579. Buigs, R. M., Swaab, D. F., Dogterom, J., and van Leeuwen, F. W., 1978, *Cell Tissue Res.* **186**:423–433.

580. Schultz, W. J., Brownfield, M. S., and Kozlowski, G. P., 1977, *Cell Tissue Res.* **178**:129–141.

580a. Reppert, S. M., Artman, H. G., Swaminathan, S., and Fisher, D. A., 1981, *Science* **213**:1256–1257.

581. Pavel, S., 1970, *J. Clin. Endocrinol. Metab.* **31**:369–371.

582. Pavel, S., *J. Clin. Endocrinol. Metab.* **50**:271–273.

583. Severs, N. B., and Daniels-Severs, A. E., 1973, *Pharmacol. Rev.* **25**:415–449.

584. Ganten, D., Marquey-Julio, A., Granger, P., Hayduk, K., Karsunky, K. P., Boucher, R., and Genest, J., 1971, *Am. J. Physiol.* **221**:1733–1751.

585. Quinlan, J. T., and Phillips, M. I., 1981, *Brain Res.* **205**:212–218.

585a. Simpson, J. B., 1981, *Neuroendocrinology.* **32**:248–256.

586. Phillips, M. I., 1978, *Neuroendocrinology* **25**:354–377.

587. Schelling, P., Ganten, U., Sponer, G., Unger, T., and Ganten, D., 1980, *Neuroendocrinology* **31**:297–308.

588. Kondo, K., Garcia, R., Boucher, R., and Genest, J., 1980, *Brain Res.* **200**:437–441.

589. Reid, I. A., and Moffat, B., 1978, *Endocrinology* **103**:1494–1498.

590. Reid, I. A., 1976, *Regulation of Blood Pressure by the Central Nervous System* (G. Onesti, M. Fernandes, and K. E. Kim, eds.), Grune & Stratton, New York, pp. 161–174.

591. Halperin, E. A., Summy-Long, J. Y., Keil, L. C., and Severs, W. B., 1981, *Brain Res.* **205**:219–221.

592. Yamamoto, M., Share, L., and Shade, R. E., 1978, *Neuroendocrinology* **25**:166–173.

593. Epstein, A. N., Fitzsimmons, J. T., and Rolls, B. J., 1970, *J. Physiol. (Lond.)* **210**:457–474.

594. Simpson, J. B., Saad, W. A., and Epstein, A. N., 1976, *Regulation of Blood Pressure by the*

Central Nervous System (G. Onesti, M. Fernandes, and K. E. Kim, eds.), Grune & Stratton, New York, pp. 191–202.

594a. Semple, P. F., MacRae, W. A., and Morton, J. J., 1980, *Clin. Sci.* **59**:61s–64s.

595. Austin, L. A., and Heath, H., 1981, *N. Engl. J. Med.* **304**:269–278.

596. Deftos, L. F., Burton, D., Bone, H. G., Catherwood, B. D., Parthemore, J. G., Moore, R. Y., Minick, S., and Guillemin, R., 1978, *Life Sci.* **23**:743–748.

597. Becker, K. L., Snider, R. H., Moore, C. F., Monagham, K. G., and Silva, O. L., 1979, *Acta Endocrinol. (Kbh.)* **92**:746–751.

598. Nakhla, A. M., and Nandi Majumdar, A. P., 1978, *Biochem. J.* **170**:445–448.

599. Carman, J. S., and Wyatt, R. J., 1979, *Arch. Gen. Psychiatry* **32**:72–75.

600. Pecile, A., Ferri, S., Braga, P. C., and Olgiati, V. R., 1975, *Experientia* **31**:332–333.

601. Braga, P., Ferri, S., Santagostino, A., Olgiati, V. R., and Pecile, A., 1978, *Life Sci.* **22**:971–977.

602. Pavlinac, D. M., Lenhard, L. W., Parthemore, J. G., and Deftos, L. J., 1980, *J. Clin. Endocrinol. Metab.* **50**:717–720.

603. Innis, R. B., Correa, F. M. A., Uhl, R., Schneider, B., and Snyder, S. H., 1979, *Proc. Natl. Acad. Sci. U.S.A.* **76**:521–525.

604. Straus, E., and Yalow, R. S., 1978, *Proc. Natl. Acad. Sci. U.S.A.* **75**:486–489.

605. Schanger, M. C., Jacobson, E. D., and Dafny, N., 1978, *Neuroendocrinology* **25**:329–342.

606. Smith, G. P., and Gibbs, J., 1975, *Pharmacol. Biochem. Behav.* **3**(Suppl.):135–138.

607. Straus, E., and Yalow, R. S., 1979, *Science* **203**:68–69.

608. McCaleb, M. L., and Myers, R. D., 1980, *Peptides* **1**:47–49.

609. Della-Fera, M. A., and Baile, C. A., 1980, *Peptides* **1**:51–54.

610. Rehfeld, J. F., 1978, *Nature* **271**:771–773.

611. Emson, P. C., Fahrenkrug, J., Schaffalizky de Muckadell, O. B., Jessell, T. M., and Iversen, L. L., 1978, *Brain Res.* **143**:174–178.

612. Ebeid, A. M., Smith, A. R., Escourrou, J., Murry, P., and Fischer, J. E., 1978, *J. Surg. Res.* **25**:538–541.

613. Ebeid, A. M., Attia, R. R., Sundavam, P., and Fischer, J. E., 1979, *Am. J. Surg.* **137**:123–127.

614. Kato, Y., Iwasaki, Y., Iwasaki, J., Abe, H., Yanaihara, N., and Imura, H., 1978, *Endocrinology* **103**:554–558.

615. Clark, W. G., Lipton, J. M., and Said, S. I., 1978, *Neuropharmacology* **17**:883–885.

616. Hales, C. N., and Randle, P. J., 1963, *Biochem. J.* **88**:137–146.

617. Margolis, R. U., and Altszuler, N., 1968, *Proc. Soc. Exp. Biol. Med.* **127**:1122–1125.

618. Brownstein, M. J., Mroz, E. A., Kizer, J. S., Palkovits, M., and Leeman, S. E., 1976, *Brain Res.* **116**:299–305.

619. Hökfelt, T., Johansson, O., Kellerth, J. O., Ljungdahl, A., Nilsson, G., Nygords, A., and Pernow, B., 1977, *Substance P* (U. S. von Euler and B. Pernow, eds.), Raven Press, New York, pp. 117–145.

620. Hökfelt, T., Elfvin, L. G., Schultzberg, M., Goldstein, M., and Nilsson, G., 1977, *Brain Res.* **132**:29–41.

621. Chihara, K., Arimura, A., Coy, D. H., and Schally, A. V., 1978, *Endocrinology* **102**:281–290.

622. Hosobuchi, Y., Emson, P., and Iversen, L. L., 1982, *Proceedings of World Congress on Pain* (J. Bonica, ed.), Raven Press, New York.

623. Nutt, J. G., Mroz, E. A., Leeman, S. E., Williams, A. C. Engel, W. K., and Chase, T. N., 1980, *Neurology (Minneap.)* **30**:1280–1285.

624. Nutt, J. G., 1982, *Neurobiology of Cerebrospinal Fluid 2* (J. H. Wood, ed.), Plenum Press, New York (in press).

625. Burbach, J. P. H., Loeber, J. G., Verhoef, J., deKloet, E. R., van Ree, J. M., and deWied, D., 1979, *Lancet* **2**:480–481.

626. Naber, D., Cohen, R. M., Pickar, D., Kalin, N. H., Davis, G., Pert, C. B., and Bunney, W. E., 1981, *Life Sci.* **28**:931–935.

627. Gramsch, C., Höllt, V., Mehraein, P., Pasi, A., and Hery, A., 1979, *Brain Res.* **171**:261–270.

628. Rosier, J., and Bloom, F., 1979, *Adv. Biochem. Psychopharmacol.* **20**:165–185.

629. Bloom, F. E., and Segal, D. S., 1980, *Neurobiology of Cerebrospinal Fluid 1* (J. H. Wood, ed.), Plenum Press, New York, pp. 651–664.

630. Bloom, F. E., Segal, D., Ling, N., and Guillemin, R., 1976, *Science* **194**:630–632.

631. Feldberg, W., and Smyth, D. G., 1976, *J. Physiol. (Lond.)* **260**:30P–31P.
632. Hosobuchi, Y., and Li, C. H., 1978, *Commun. Psychopharmacol.* **2**:33–37.
633. Moss, I. R., and Friedman, E., 1978, *Life Sci.* **23**:1271–1276.
634. Feldberg, W., and Wei, E., 1978, *J. Physiol. (Lond.)* **280**:18P.
635. Garcia-Sevilla, J. A., Ahtee, L., Mognusson, T., and Carlsson, A., 1978, *J. Pharm. Pharmacol.* **30**:613–621.
636. Holaday, J. W., Tseng, L.-F., Loh, H. H., and Li, C. H., 1978, *Life Sci.* **22**:1537–1544.
637. Jeffcoate, W. J. L., McLoughlin, L., Hope, J., Rees, L. H., Ratter, S. J., Lowry, P. J., and Besser, G. M., 1978, *Lancet* **2**:119–121.
637a. Nakao, K., Nakai, Y., Oki, S., Matsubara, S., Konishi, T., Nishitani, H., and Imura, H., 1980, *J. Clin. Endocrinol. Metab.* **50**:230–233.
637b. McLoughlin, L., Lowry, P. J., Ratter, S. J. Hope, J., Besser, G. M., and Rees, L. H., 1981, *Neuroendocrinology.* **32**:209–212.
638. Miller, R. J., and Cuatrecasas, P., 1979, *Adv. Biochem. Psychopharmacol.* **20**:187–225.
639. Belluzzi, J. D., Grant, N., Garsky, V., Sarantakis, D., Wise, C. D., and Stein, L., 1976, *Nature* **260**:625–626.
640. Tortella, F. C., Noreton, J. E., and Khazan, N., 1978, *J. Pharmacol. Exp. Ther.* **206**:636–642.
641. Urca, G., Frenk, H., Liebeskind, J. C., and Taylor, A. N., 1977, *Science* **197**:83–86.
642. Dupont, A., Cusan, L., Labrie, F., Coy, D. H., and Li, C. H., 1977, *Biochem. Biophys. Res. Commun.* **75**:76–82.
643. Clement-Jones, V., Lowry, P. J., Rees, L. H., and Besser, G. M., 1980, *J. Endocrinol.* **86**:231–243.
644. Furui, T., Kageyama, N., Haga, T., Ichiyama, A., and Fukushima, M., 1980, *Pain* **9**:63–72.
645. Minneman, K. P., and Wurtman, R. J., 1976, *Annu. Rev. Pharmacol. Toxicol.* **16**:33–51.
646. Tamarkin, L., Hollister, C., Lefebure, N. G., and Goldman, B. D., 1977, *Science* **198**:953–955.
647. Ozaki, Y., and Lynch, H. J., 1976, *Endocrinology* **99**:641–644.
648. Reppert, S. M., Perlow, M. J., Tamarkin, L., and Klein, D. C., 1979, *Endocrinology* **104**:295–301.
649. Rollag, M. D., Morgan, R. J., and Niswender, G. D., 1978, *Endocrinology* **102**:1–8.
650. Reiter, R. J., Voughan, M. K., and Blask, D. E., 1975, *Brain–Endocrine Interaction II, The Ventricular System* (K. M. Knigge, D. E. Scott, H. Kobayashi, and J. Ishii, eds.), S. Karger, Basel, pp. 337–354.
651. Withyachumnarnkul, B., and Knigge, K. M., 1980, *Neuroendocrinology* **30**:382–388.
652. Brown, G. M., Young, S. N., Gauthier, S., Tsui, H., and Grota, L. J., 1979, *Life Sci.* **25**:929–936.
653. Vaughan, G. M., McDonald, S. A., Jordon, R. M., Allen, J. P., Bohmfalk, G. L., Abou-Samra, M., and Story, J. L., 1978, *J. Clin. Endocrinol. Metab.* **47**:220–223.
654. Cardinali, A. P., Lynch, H. J., and Wurtman, R. J., 1972, *Endocrinology* **91**:1213–1218.
655. Hedlund, L., Lischko, M. M., Rollag, M. D., and Niswender, G. D., 1977, *Science* **195**:686–687.
656. Banik, N. L., and Hogan, E. L., 1982, *Neurobiology of Cerebrospinal Fluid 2* (J. H. Wood, ed.), Plenum Press, New York.
657. Herschkowitz, N., and Cumings, J. N., 1964, *J. Neurol. Neurosurg. Psychiatry* **27**:247–250.
658. Klun, B., 1974, *J. Neurosurg.* **41**:224–228.
659. Van der Helm, H. J., Zondag, H. A., and Klein, F., 1963, *Clin. Chim. Acta* **8**:193–196.
660. Viallard, J. L., Gaulne, J., Dalens, B., and Dastugue, B., 1978, *Clin. Chim. Acta* **89**:405–409.
661. Nordby, H. K., Tveit, B., and Ruud, I., 1975, *Acta Neurochir. (Wien)* **32**:209–217.
662. Sherwin, A. L., Norris, J. W., and Bulcke, J. A., 1969, *Neurology (Minneap.)* **19**:993–999.
663. Katz, R. M., and Leibman, W. Z., 1970, *Am. J. Dis. Child.* **120**:543–546.
664. Kaltiala, E. H., Heikkinen, E. S., Karki, N. T., and Larmi, T. K., 1968, *Acta Neurol. Scand.* **44**:124–129.
665. Sitzmann, F. C., 1969, *Z. Kinderheilkd.* **106**:76–88.
666. Ronquist, G., and Frithy, G., 1979, *Eur. Neurol.* **18**:101–110.
667. Van Rymenant, M., and Robert, J., 1960, *Cancer* **13**:878–881.
668. Wilcock, A. R., Sharpe, D. M., and Goldberg, D. M., 1973, *J. Neurol. Sci.* **20**:97–101.
669. Fonum, F., 1973, *Brain Res.* **62**:497–507.

670. Aquilonius, S. M., and Eckernas, S. A., 1976, *J. Neurochem.* **27**:317–318.
671. Johnson, S., and Domino, E. F., 1971, *Clin. Chim. Acta* **35**:421–428.
672. Hollunger, E. G., and Nikalasson, B. H., 1973, *J. Neurochem.* **20**:821–826.
673. Yaksh, T. L., Felele, L. A., and Yamamura, H. I., 1973, *Experientia* **30**:38–39.
674. Davis, K. L., 1982, *Neurobiology of Cerebrospinal Fluid 2* (J. H. Wood, ed.), Plenum Press, New York (in press).
675. Kaufman, S., and Friedman, S., 1965, *Pharmacol. Rev.* **17**:71–99.
676. Major, L. F., Lerner, P., Dendel, P. S., and Post, R. M., 1982, *Neurobiology of Cerebrospinal Fluid 2* (J. H. Wood, ed.), Plenum Press, New York (in press).
677. Lerner, P., Goodwin, F. K., van Kammen, D. P., Post, R. M., Major, L. F., Ballenger, J. C., and Lovenberg, W., 1978, *Biol. Psychiatry* **13**:685–694.
678. Fujita, K., Maruta, X., Teradaire, R., Beppu, H., Shinpo, K., Maeno, Y., Ito, T., Nagatsu, T., and Kato, T., 1977, *J. Neurochem.* **29**:1141–1142.
679. Lerner, P., Dendel, P. S., and Major, L. F., 1979, *Am. Soc. Neurochem.* **10**:192.
680. DePotter, W. P., 1979, *Neuroscience* **1**:523–529.
681. Narasimhachari, N., and Lin, R. L., 1975, *Brain Res.* **87**:126–129.
682. Colling, K. G., and Rossiter, R. J., 1950, *Can. J. Res.* **28**:56–68.
683. Shuttleworth, E. C., and Allen, N., 1968, *Neurology (Minneap.)* **18**:534–542.
684. Conchie, J., Hay, A. J., and Levy, G. A., 1961, *Biochem. J.* **79**:324–330.
685. Allen, N., and Reagan, E., 1964, *Arch. Neurol.* **11**:114–154.
686. Schold, S. C., and Bullard, D. E., 1980, *Neurobiology of Cerebrospinal Fluid 1* (J. H. Wood, ed.), Plenum Press, New York, pp. 549–559.
687. Shuttleworth, E., and Allen, N., 1980, *Arch. Neurol.* **37**:684–687.
688. Klockars, M., Reitamo, S., Webber, T., and Kerttula, Y., 1978, *Acta Med. Scand.* **203**:71–74.
689. Terent, A., Hallgren, R., Venge, P., and Bergstrom, K., 1981, *Stroke* **12**:40–46.
689a. Newman, J., Josephson, A. S., Cacatian, A., and Tsang, A., 1974, *Lancet* **2**:756–757.
690. Johnson, R., and Shah, S. N., 1979, *Brain Res.* **162**:353–357.
691. Shah, S. N., and Johnson, R., 1978, *Exp. Neurol.* **58**:68–73.
692. Reiber, H., and Voss, W., 1980, *J. Neurochem.* **34**:1324–1326.
693. Illingworth, D. R., and Glover, J., 1970, *Biochim. Biophys. Acta* **220**:610–613.
694. Illingworth, D. R., and Glover, J., 1979, *Biochemistry* **115**:16.
695. Swinnen, J., 1967, *Clin. Chim. Acta* **17**:255–263.
696. Green, J. P., and Perry, M., 1963, *Neurology (Minneap.)* **13**:924–926.
697. Cuzner, M. L., Davison, A. N., and Rudge, P., 1979, *Ann. Neurol.* **4**:337–344.
698. Ko, G. K. W., Raghupathy, E., and McKean, C. M., 1973, *Can. J. Biochem.* **51**:1460–1469.
699. Brunngraber, E., 1982, *Neurobiology of Cerebrospinal Fluid 2* (J. H. Wood, ed.), Plenum Press, New York (in press).
700. Banik, N. L., and Davison, A. N., 1969, *Biochem. J.* **115**:1051–1062.
701. Kurihara, T., and Tsukada, Y., 1967, *J. Neurochem.* **14**:1167–1174.
702. Poduslo, S. E., 1975, *J. Neurochem.* **24**:647–654.
703. Banik, N. L., Mauldin, L., and Hogan, E. L., 1979, *Ann. Neurol.* **5**:539–541.
704. Hidaka, H., Shibuya, A., Asano, T., and Hare, F., 1975, *J. Neurochem.* **25**:49–53.
705. Hankiewicz, J., and Lesniak, M., 1972, *Enzymologia Biol. Clin.* **43**:385–395.
706. Rabin, E. Z., Weinberg, V., and Tattrie, B., 1977, *Can. J. Neurol. Sci.* **4**:125–130.
707. Bien, A., 1968, *Neurol. Neurochir. Pol.* **18**:479–483.
708. Kovacs, E., 1953, *Can. J. Med. Sci.* **31**:437–446.
709. Blank, A., and Dekker, C. A., 1981, *Biochem.* **20**:2261–2267.

Index